Sociology in Our Times
THE ESSENTIALS

Diana Kendall — Fourth Edition

Baylor University

THOMSON
*
WADSWORTH

Australia • Canada • Mexico • Singapore
Spain • United Kingdom • United States

THOMSON
WADSWORTH

Executive Editor: Eve Howard
Development Editor: Natalie Cornelison
Acquisitions Editor: Sabra Horne
Assistant Editor: Stephanie Monzon
Editorial Assistant: Melissa Walter
Technology Project Manager: Dee Dee Zobian
Marketing Manager: Matthew Wright
Advertising Project Manager: Linda Yip
Project Manager, Editorial Production: Ritchie Durdin
Print/Media Buyer: Doreen Suruki

Permissions Editor: Joohee Lee
Production Service: Greg Hubit Bookworks
Text Designer: Jeanne Calabrese
Photo Researcher: Linda Rill
Copy Editor: Donald Pharr
Cover Designer: Jeanne Calabrese
Cover Image: Marina Jefferson/© Getty Images
Text and Cover Printer: Quebecor World/Versailles
Compositor: Thompson Type

For more information about our products, contact us at:
Thomson Learning Academic
Resource Center
1-800-423-0563
For permission to use material from this text, contact us by:
Phone: 1-800-730-2214
Fax: 1-800-730-2215
Web: http://www.thomsonrights.com

Library of Congress Control Number: 2002114549

Wadsworth/Thomson Learning
10 Davis Drive
Belmont, CA 94002-3098
USA

Asia
Thomson Learning
5 Shenton Way #01-01
UIC Building
Singapore 068808

Australia/New Zealand
Thomson Learning
102 Dodds Street
Southbank, Victoria 3006
Australia

Canada
Nelson
1120 Birchmount Road
Toronto, Ontario M1K 5G4
Canada

Europe/Middle East/Africa
Thomson Learning
High Holborn House
50/51 Bedford Row
London WC1R 4LR
United Kingdom

Student Edition with InfoTrac College Edition:
ISBN 0-534-60957-0

Instructor's Edition: ISBN 0-534-60958-9

To Terrence Kendall for his invaluable contributions

Brief Contents

Contents

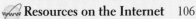
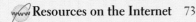

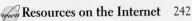
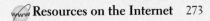

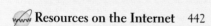

Preface

Welcome to the fourth edition of *Sociology in Our Times: The Essentials*! The twenty-first century offers unprecedented challenges and opportunities for each of us as individuals and for our larger society and world. In the United States, we can no longer take for granted the peace and economic prosperity that many—but far from all—people were able to enjoy in previous decades. However, even as some things change, others remain the same, and among the things that have not changed are the significance of education and the profound importance of understanding how and why people act the way they do, of how societies grapple with issues such as economic hardship and the threat of terrorist attacks and war, and why many of us seek stability in our social institutions—including family, religion, education, government, and media—even at times when we believe that these same institutions might benefit from certain changes.

Like previous editions of this widely read text, the fourth edition of *Sociology in Our Times: The Essentials* is a cutting-edge book that highlights the relevance of sociology. It does this in at least two ways: (1) by including a diversity of classical and contemporary theory, interesting and relevant research, and lived experiences that accurately mirror the diversity in society itself; and (2) by showing students that sociology involves important questions and issues that they confront both personally and vicariously (for example, through the media). This text speaks to a wide variety of students and captures their interest by taking into account their concerns and perspectives. The research used in this text includes the best work of classical and established contemporary sociologists—including many women and people of color—and it weaves an inclusive treatment of *all* people into the examination of sociology in *all* chapters. Through the use of the latest theory and research, *Sociology in Our Times: The Essentials* not only provides students with the most relevant information about sociological thinking but also helps students consider the significance of the interlocking nature of class, race, and gender in all aspects of social life.

I would encourage you to read a chapter in the book and judge for yourself the writing style, which I have sought to make both accessible and engaging for students and instructors. Concepts and theories are presented in a straightforward and understandable way, and the wealth of concrete examples and lived experiences woven throughout the chapters makes the relevance of sociological theory and research abundantly clear to students.

Organization of This Text

Sociology in Our Times: The Essentials, fourth edition, contains sixteen carefully written, well-organized chapters to introduce students to the best of sociological thinking. The length of the text makes full coverage of the book possible in the time typically allocated to the introductory course. As a result, students are not purchasing a book that contains numerous chapters which the instructor may not have the time or desire to cover.

Sociology in Our Times: The Essentials is divided into five parts.

Part 1 establishes the foundation for studying society and social life. **Chapter 1** introduces students to the *sociological perspective* and the *research* *process.* The chapter sets forth the major theoretical perspectives used by sociologists in analyzing compelling social issues and provides a thorough description of both quantitative and qualitative methods of sociological research. In **Chapter 2,** *culture* is spotlighted as either a stabilizing force or a force that can generate discord, conflict, and even violence in societies. Cultural diversity is discussed as a contemporary issue, and unique coverage is given to popular culture and leisure and to divergent perspectives on popular culture. **Chapter 3** looks at *socialization* and presents an innovative analysis of gender and racial–ethnic socialization.

Part 2 examines social groups and social control. **Chapter 4** examines *social structure and interaction in everyday life,* using homelessness as a sustained example of the dynamic interplay of social structure and human agency in people's daily lives. Unique to this chapter are discussions of the sociology of emotions and of personal space as viewed through the lenses of race, class, gender, and age. **Chapter 5** analyzes *groups and organizations,* including innovative forms of social organization and ways in which organizational structures may differentially affect people based on race, class, gender, and age. **Chapter 6** examines how *deviance and crime* emerge in societies, using diverse theoretical approaches to describe the nature of deviance, crime, and the criminal justice system.

Part 3 focuses on social differences and social inequality, looking at issues of class, race/ethnicity, and sex/gender. **Chapter 7** addresses the issue of *global stratification* and examines differences in wealth and poverty in rich and poor nations around the world. Explanations for these differences are discussed. **Chapter 8** extends the discussion to *class and stratification in the United States.* The chapter analyzes the causes and consequences of inequality and poverty, including a discussion of the ideology and accessibility of the American Dream. The focus of **Chapter 9** is *race and ethnicity,* including an illustration of the historical relationship (or lack of it) between sports and upward mobility by persons from diverse racial–ethnic groups. A thorough analysis of prejudice, discrimination, theoretical perspectives, and the experiences of diverse racial and ethnic groups is presented, along with global racial and ethnic issues. **Chapter 10** examines *sex and gender,* with special emphasis on gender stratification in historical perspective. Linkages between gender socialization and contemporary gender inequality are described and illustrated by lived experiences and perspectives on body image.

Part 4 offers a systematic discussion of social institutions, making students more aware of the importance of social institutions and showing how a problem in one often has a significant influence on others. *Families and intimate relationships* are explored in **Chapter 11,** which includes both U.S. and global perspectives on family relationships, a view of families throughout the life course, and a discussion of diversity in contemporary U.S. families. *Education and religion* are presented in **Chapter 12,** and important sociological theories pertaining to these pivotal social institutions are presented. **Chapter 13** discusses *politics and the economy in global perspective,* highlighting the global context in which contemporary political and economic systems operate and showing the intertwining nature of politics, the economy, and global media outlets. **Chapter 14** analyzes *health, health care, and disability* in both U.S. and global perspectives. Among the topics included are social epidemiology, lifestyle factors influencing health and illness, health care organization in the United States and other nations, social implications of advanced medical technology, and holistic and alternative medicine. This chapter is unique in that it contains a thorough discussion of the sociological perspectives on disability and of social inequalities based on disability.

Part 5 surveys social dynamics and social change. **Chapter 15** examines *population and urbanization,* looking at demography, global population change, and the process and consequences of urbanization. Special attention is given to race- and class-based segregation in urban areas and the crisis in health care in central cities. **Chapter 16** concludes the text with an innovative analysis of *collective behavior, social movements, and social change.* Environmental activism is used as a sustained example to help students grasp the importance of collective behavior and social movements in producing social change.

Distinctive Features

The following special features are specifically designed to reflect the themes of relevance and diversity in *Sociology in Our Times: The Essentials*, as well as to support students' learning.

Interesting and Engaging Lived Experiences Throughout Chapters

Authentic first-person accounts are used as opening vignettes and throughout each chapter to create interest and give concrete meaning to the topics being discussed. Lived experiences provide opportunities for students to examine social life beyond their own experiences and for instructors to systematically incorporate into lectures and discussions an array of interesting and relevant topics demonstrating to students the value of applying sociology to their everyday lives. Some examples of the lived experiences include the following:

- Kurt Cobain's suicide note, describing how a person about to commit suicide might feel, shows how such beliefs can be linked to sociological theories and research on pressing social issues such as suicide (Chapter 1, "The Sociological Perspective and Research Process").
- George Pataki, New York governor, discussing his first trip to the World Trade Center disaster site following the 2001 terrorist attacks (Chapter 5, "Groups and Organizations").
- Sociologist Felix M. Padilla, interviewing a gang member as part of Padilla's ethnographic research on why people join gangs (Chapter 6, "Deviance and Crime").
- John Cronin and Robert F. Kennedy, Jr., environmental activists, describing their work to save the Hudson River from environmental degradation (Chapter 16, "Collective Behavior, Social Movements, and Social Change").

Focus on the Relationship Between Sociology and Everyday Life

Each chapter has a brief quiz that relates the sociological perspective to the pressing social issues presented in the opening vignette. (Answers are provided on a subsequent page.) Topics such as these will pique students' interest:

- "How Much Do You Know About Homeless Persons?" (Chapter 4, "Society, Social Structure, and Interaction")
- "How Much Do You Know About Privacy in Groups and Organizations?" (Chapter 5, "Groups and Organizations")
- "How Much Do You Know About Body Image and Gender?" (Chapter 10, "Sex and Gender")
- "How Much Do You Know About Migration?" (Chapter 15, "Population and Urbanization")

Emphasis on the Importance of a Global Perspective

Sociology in Our Times: The Essentials analyzes our interconnected world and reveals how the sociological imagination extends beyond national borders. Global implications of all topics are examined throughout each chapter and in the Sociology in Global Perspective box found in many chapters. Here are a few examples:

- "Comparing Suicide Statistics from Different Nations" (Chapter 1, "The Sociological Perspective and Research Process")
- "Homelessness as an International Problem: The Plight of the 'Street Children'" (Chapter 4, "Social Structure and Interaction in Everyday Life")
- "Opposition, Resistance, and the Women of Afghanistan" (Chapter 10, "Sex and Gender")
- "Urban Migration and the 'Garbage Problem'" (Chapter 15, "Population and Urbanization")

Focusing on "Changing Times: Media and Technology" to Encourage Critical Thinking

A significant benefit of a sociology course is encouraging critical thinking about such things as how media and technological changes influence our daily lives. Here are the topics of a few Media/Technology boxes that will foster critical-thinking skills:

- "A Wired World?" (Chapter 7, "Global Stratification")
- "Media Stars and the American Dream" (Chapter 8, "Social Class in the United States")
- "The Reproductive Revolution: Who Are My Parents?" (Chapter 11, "Families and Intimate Relationships")
- "How the Internet Has Changed Environmental Activism" (Chapter 16, "Collective Behavior, Social Movements, and Social Change")

Applying the Sociological Imagination to Social Policy

The Sociology and Social Policy boxes in selected chapters help students understand the connection between sociology and social policy issues in society. Here are a few of the topics in these interesting and informative boxes:

- "The Debate Over Regulation of Entertainment Products with Violent Content" (Chapter 3, "Socialization")
- "The Continuing Debate Over Affirmative Action" (Chapter 9, "Race and Ethnicity")
- "The Ongoing Debate Over School Vouchers" (Chapter 12, "Education and Religion")
- "The Electoral College: Is School Over for This Body?" (Chapter 13, "Politics and the Economy in Global Perspective")

"You Can Make a Difference" Helps Get Students Involved in Each Chapter

The You Can Make a Difference boxes address the ways in which students can find out how the chapter theme affects their lives. For example,

- "Taking a Stand Against Overspending" (Chapter 2) makes students aware of the warning signs of overspending and amassing large credit card debts.
- "Creating Small Communities of Our Own Within Large Organizations" (Chapter 5) passes on ideas for building microcommunities of informal friendships and small groups within larger organizations.
- "Global Networking to Reduce World Hunger and Poverty" (Chapter 7) and "Feeding the Hungry" (Chapter 8) address the issue of hunger and how everyday people can con-

tribute in some small way to alleviating the suffering of others.

- "Creating Access to Information Technologies for Workers with Disabilities" (Chapter 13) describes how cost-efficient changes in the workplace can make it easier for some persons with a disability to be employed. Some students will eventually become owners or supervisors who are in a position to make decisions that provide greater access to meaningful employment for all people.
- "Recycling for Tomorrow" (Chapter 16) discusses how we can make a difference by reusing or recycling products and the packaging in which they come rather than creating more excess waste than already exists.

Unparalleled Study Aids

Study Aids in Each Chapter That Are New to This Edition

- **Reflections** boxes allow students to critically analyze central ideas and sociological concerns as they apply to issues raised in the chapter.
- **Checkpoints** question students on key factual material and definitions in the preceding section, thus serving as a tool to enhance student comprehension.
- A **running glossary** in the margin highlights key terms that students should know after reading the chapter.
- **Summary charts** categorize and contrast the major theories on the specific topics presented in a chapter.

Other In-Text Learning Features

Sociology in Our Times: The Essentials includes a number of other pedagogical aids to promote students' mastery of sociological perspectives:

- **Chapter Outlines.** A concise outline at the beginning of each chapter gives students an overview of major topics and a convenient aid for review.
- **Questions and Issues.** After the opening lived experience in each chapter, a series of introductory questions invites students to think about the major topics discussed in this

chapter. *Sociology in Our Times: The Essentials* also features a "Chapter Focus Question" linking the chapter topic to the compelling social theme analyzed in the chapter.

- **Key Terms.** Major concepts and key terms are concisely defined and highlighted in bold print within the text flow to avoid disrupting students' reading. These concepts and terms are also listed at the end of the chapters and in the glossary at the back of the text.
- **Questions for Critical Thinking.** Each chapter concludes with "Questions for Critical Thinking" to encourage students to reflect on important issues, to develop their own critical-thinking skills, and to highlight how ideas presented in one chapter often build on those developed previously.
- **End-of-Chapter Summaries in Question-and-Answer Format.** Chapter summaries provide a built-in review for students by re-examining material covered in the chapter in an easy-to-read question-and-answer format to review, highlight, and reinforce the most important concepts and issues discussed in each chapter.

Internet Learning Aids

- The **Resources on the Internet** section provides chapter-related web sites that encourage students to engage in research on various topics discussed in the chapter.

Innovations

The fourth edition of *Sociology in Our Times: The Essentials* builds on the best from previous editions while providing students with new insights for their times.

Census Profiles, a new feature in most chapters, provides recent data from the U.S. Census as applied to the chapter in which they are located. Topics include "Languages Spoken in U.S. Households" (Chapter 2), "Age of the U.S. Population" (Chapter 3), "Computer and Internet Access in U.S. Households" (Chapter 4), and "Disability and Employment Status" (Chapter 14).

Chapter 1 provides a new and interesting section on comparing sociology with other social sciences. The new Census Profiles feature shows "How

People in the United States Self Identify as to Race and Hispanic Origin." Chapter 1 features more recent theoretical insights, including postmodernist, not typically found in other introductory texts.

Chapter 2 has a new opening lived experience in which a college student describes her first experience with a credit card. This chapter looks at consumerism as a key part of culture in contemporary societies. It includes a new Sociology and Everyday Life box: "How Much Do You Know About Consumption and Credit Cards?" as well as a new You Can Make a Difference box: "Taking a Stand Against Overspending."

Chapter 3 has a new section on recent symbolic interactionist perspectives on human development. A new Social Policy box challenges students to think about the debate over regulation of entertainment products with violent content.

In **Chapter 4,** a new Sociology in Global Perspective box discusses homelessness of children as an international problem. A new Census Profiles looks at the extent of computer and Internet access in U.S. households.

Chapter 5 applies the sociological imagination to an examination of groups, organizations, and privacy issues in contemporary societies. All-new boxes in the chapter include Sociology and Everyday Life—"How Much Do You Know About Privacy in Groups and Organizations?"—and Social Policy—"Computer Privacy in the Workplace." An interesting new opening lived experience deals with the challenge that organizations in New York City faced as they sought to get back up and running after the 2001 terrorist attacks.

Chapter 6 includes the latest available crime statistics, and a new opening lived experience features an interview of a gang member by the sociologist Felix M. Padilla. There is a new discussion of how terrorist attacks are a form of violent deviant behavior, and a new section on terrorism and crime.

The content of Chapters 7 and 8 has been shifted so that **Chapter 7** now focuses on global stratification. This chapter examines wealth and poverty in global perspective, particularly through the eyes of women who have been carrying water for their families to drink or working on the global assembly line in low-income nations. It includes a discussion of Manuel Castells's (1998) analysis of the *Fourth World,* which he uses to describe the "multiple black holes of social exclusion" around the world. The chapter features a Media/Technology box ("A Wired World?") that highlights the difference between the "haves" and the "have nots" in regard to new technologies such as the Internet and the World Wide Web.

Chapter 8 now focuses on the U.S. system of social stratification. A new Changing Times: Media and Technology box, "Media Stars and the American Dream," looks at how the upward mobility of a few celebrity entertainers tends to reinforce the idea that anyone can get ahead in the U.S. class system.

In **Chapter 9,** new data on race and ethnicity are set forth in this cutting-edge chapter that examines the relationship between sports and racial–ethnic groups throughout U.S. history. A new Social Policy box invites students to consider the continuing debate over affirmative action in colleges and universities.

Chapter 10 features a new Global Perspective box on "Opposition, Resistance, and the Women of Afghanistan" as well as an expanded discussion of the gendered division of labor in postindustrial societies. All-new data and a new Census Profiles on single mothers with children under age eighteen raise some of the crucial gender-related issues discussed in the chapter.

In **Chapter 11,** the changing nature of the family is explored. All figures and tables have been updated with the latest available information on such topics as household composition. A new Census Profiles gives information on household composition in 1970 and 2000, and there are new sections on domestic violence and on children in foster care. A new You Can Make a Difference box describes how individuals can provide hope and help for children.

Chapter 12, Education and Religion, starts with a discussion of separation of church and state in the United States. It contains a new Social Policy box on "The Ongoing Debate Over School Vouchers" and a new Census Profiles showing the educational achievement of persons aged twenty-five and over.

Chapter 13 focuses on the effect that the intertwining of politics, the economy, and the media has on the United States and other nations. The opening lived experience is by a journalist who describes what it was like to go as a "media ride-along" with law enforcement officials as they made arrests. This chapter includes a new Census Profiles on political representation and shifts in the U.S. population. A new Social Policy box asks, "The Electoral

College: Is School Over for This Body?" The chapter discusses the 2000 presidential election and takes into account problems the United States faced in the aftermath of the 2001 terrorist attacks.

Chapter 14, on health, health care, and disability, features an opening lived experience in which a first-year college student describes her experience with chronic illness. A number of new features have been included in this chapter. The Census Profiles highlights disability and employment status, and an updated and expanded Social Policy box discusses the pros and cons of government-funded medical care such as Medicare and Medicaid.

Chapter 15 has been extensively rewritten to look at the issue of global migration as it affects population change. The chapter begins with a lived experience from an immigrant worker in the United States and features a new Sociology in Everyday Life box on migration. A new figure shows the projected growth of the world's population between 2000 and 2030. Other new sections include new theories of population change and international migration, problems in global cities, and the effects of terrorism and economic problems on global cities. New boxes look at urban migration and the garbage problem (Sociology in Global Perspective) and how students can work toward better communities (You Can Make a Difference).

The final chapter, **Chapter 16,** elaborates on why sociologists study collective behavior and social movements. New to this edition are a section on social constructionist theory (frame analysis) and a section on new social movement theory. A Media/Technology box describes how the Internet has changed environmental activism. A new You Can Make a Difference box discusses how recycling can reduce environmental degradation.

Supplements

Sociology in Our Times: The Essentials, Fourth Edition, is accompanied by a wide array of supplements prepared to create the best learning environment inside as well as outside the classroom for both the instructor and the student. All the continuing supplements for *Sociology in Our Times: The Essentials,* Fourth Edition, have been thoroughly revised and updated, and several are new to this edition. I invite you to take full advantage of the teaching and learning tools available to you.

Supplements for the Instructor

INSTRUCTOR'S EDITION. The *Instructor's Edition* contains a visual walk-through of the text, a Resource Integration Guide, and the Technology Demo and Virtual Explorations and Interactive Maps CD-ROMs (see below for CD-ROM descriptions). The visual walk-through provides an overview of the key features, themes, and supplements, while the Resource Integration Guide provides a chapter-by-chapter supplement correlation chart that enables instructors to easily coordinate and integrate the accompanying supplements into their course.

TECHNOLOGY DEMO CD-ROM (NEW 2004 VERSION). Do you want to use an online component for your class? Do you want your students to have access to InfoTrac College Edition? Do you wonder just exactly what many of Wadsworth's technology products do? If so, this CD-ROM is for you. The Technology Demo CD-ROM introduces and demonstrates the key sociology technology supplements that Wadsworth offers. The demo will give you both an overview of what will work well for your needs and a more detailed demonstration of exactly how to use each product. The Technology Demo CD-ROM comes in each *Instructor's Edition* of the text.

INSTRUCTOR'S MANUAL NOTEBOOK. This newly expanded instructor's manual, by D. R. Wilson of Houston Baptist University, provides instructors with the opportunity to integrate their personal notes into a three-hole punched and tabbed comprehensive notebook containing a wealth of comprehensive teaching resources. The enhanced manual offers chapter-specific summaries, extensive lecture outlines, and creative lecture and teaching suggestions. The manual has a special emphasis on active learning, with student learning objectives grouped according to Bloom's taxonomy, suggested discussion questions, class activities, student projects, and interactive Internet activities and InfoTrac College Edition exercises for each chapter. Also included is a list of additional print, video, and online resources, including a table of contents for the CNN Today Sociology Video Series and concise user guides for both InfoTrac College Edition and WebTutor.

TEST BANK. This test bank, by Stephen C. Light of State University of New York, Plattsburgh, consists of 75–100 multiple–choice questions and 20–25 true/false questions for each chapter of the text, all with answer explanations and page references to the text. Each multiple-choice item has the question type (fact, concept, or concept application) indicated. Also included are 5–10 short-answer and 3–5 essay questions for each chapter.

EXAMVIEW® COMPUTERIZED TESTING. Create, deliver, and customize tests and study guides (both print and online) in minutes with this easy-to-use assessment and tutorial system. ExamView offers both a Quick Test Wizard and an Online Test Wizard that guide you step by step through the process of creating tests. The test appears on screen exactly as it will print or display online. Using ExamView's complete word-processing capabilities, you can enter an unlimited number of new questions or edit existing questions.

MULTIMEDIA PRESENTATION MANAGER FOR SOCIOLOGY. The easy way to great multimedia lectures! This one-stop digital library and presentation tool helps you assemble, edit, and present custom lectures. The CD-ROM brings together art (figures, tables, photos) from the text itself, pre-assembled PowerPoint lecture slides, and CNN video. You can use the materials as they are or add your own materials for a truly customized lecture presentation.

DEMONSTRATING SOCIOLOGY CD-ROM SHOWCASE PRESENTATIONAL SOFTWARE. ShowCase is a powerful yet easy-to-use software program that enables instructors to do "live" demos of statistical analysis right in the classroom. Using mapping, scatterplots, cross-tabulations, and univariates, instructors can show students how sociologists ask and answer questions using sociological theory. A resource book with detailed "scripts" for using Show-Case in class accompanies the CD-ROM.

TRANSPARENCY ACETATES FOR INTRODUCTORY SOCIOLOGY 2004. A set of four-color acetates consisting of tables and figures from Wadsworth's introductory sociology texts is available to help prepare lecture presentations. Free to qualified adopters.

TEACHING TIPS FOR INTRODUCTORY SOCIOLOGY. This booklet contains tips on course goals and syllabi, lecture preparation, exams, class exercises, research projects, and course evaluations. It is an invaluable tool for first-time instructors of the introductory course and for veteran instructors in search of new ideas.

VIDEO: *SOCIAL ISSUES FOR SOCIOLOGY IN OUR TIMES.* This compelling video presentation of sociological concepts emphasizes the human drama, societal issues, and social problems in our lives in these times. It can be used to spark in-class discussions or facilitate lectures for instructors using a variety of texts. The rich array of 2–5-minute segments has been carefully selected from the library of Films for the Humanities and Sciences.

CNN TODAY VIDEOS: SOCIOLOGY (AVAILABLE IN NORTH AMERICA ONLY). Now you can integrate the up-to-the-minute programming power of CNN and its affiliate networks right into your course, with CNN Today Videos. Updated yearly, these course-specific videos can help you launch a lecture, spark a discussion, or demonstrate an application—using the top-notch business, science, consumer, and political reporting of the CNN networks. Produced by Turner Learning, Inc., these 45-minute videos show your students how the principles they learn in the classroom apply to the stories they see on television. Special adoption conditions apply.

VIDEO: *AMERICA'S NEW WAR: CNN LOOKS AT TERRORISM.* This great discussion starter includes sixteen 2–5-minute segments featuring CNN news footage, commentator remarks, and speeches dealing with terrorist attacks on U.S. targets throughout the world. Topics include anthrax and biological warfare, new security measures, Osama bin Laden, al Qaeda, asset freezing, homeland defense, renewed patriotism, new weapons of terrorism, the bombing of U.S. embassies in Kenya and Tanzania, the American psyche, and the Arab American response to recent events. Special adoption conditions apply.

WADSWORTH SOCIOLOGY VIDEO LIBRARY. Qualified adopters may select full-length videos from an extensive library of offerings drawn from excellent educational video sources. This large selection of thought-provoking films is available to adopters based on adoption size. Special adoption conditions may apply.

Supplements for the Student

VIRTUAL EXPLORATIONS AND INTERACTIVE MAPS CD-ROM (NEW 2004 VERSION). An excellent supplementary product for students who not only want to learn and remember the material covered in *Sociology in Our Times: The Essentials* but also to explore and research the principles in action today. Students will find three exciting features:

- The ASA-award-winning Virtual Explorations introduces students to many of the exciting resources for sociology available on the World Wide Web. Students are directed to important web sites where they will evaluate or analyze the information presented. Students will then be asked an in-depth series of questions about the information presented and will e-mail their responses to their instructor. The web sites represent a broad range: official U.S. and U.N. sites, academic sites, advocacy sites, and others. These explorations are dynamically illustrated with enriching photos, videos, charts, and graphs.
- New to this CD-ROM are interactive sociological data maps. For each main topic of the book, students will be able to color a map of the United States or the world and predict specific statistical findings based on a color key. Once they've colored the map, students will click "submit," and a data map with the true statistics will appear directly beside the map colored by their assumptions. Students will exercise their sociological imagination and learn that their assumptions about certain portions of the world or the United States are often incorrect.
- This CD-ROM also links students directly to the text's companion web site, where they can study the material from the text with a large

assortment of resources, including practice quizzes. Students are also linked directly to MicroCase Online, InfoTrac College Edition, and CNN videos for each chapter of the text.

STUDY GUIDE. This student study tool by Kathryn S. Mueller and Diana Kendall of Baylor University contains both a brief and detailed chapter outline, chapter summary, learning objectives, key terms and key people with page references to the text, diversity issue focus questions, student projects and activities, and Internet and InfoTrac exercises for each chapter. The guide also includes practice tests consisting of 20–25 multiple-choice and 10–15 true/false questions, all with page references to the text, as well as 5 short-answer/essay questions for each chapter.

PRACTICE TESTS. This booklet, by Stephen C. Light, State University of New York, Plattsburgh, contains 40–50 multiple-choice questions and 10–15 true/false questions for each chapter of the text, with page references and answer explanations to help students test their knowledge of the chapter concepts.

UNDERSTANDING SOCIETY: INTRODUCTORY READINGS. The emphasis of this collection is on articles that students will both understand and find intriguing. The collection is less rigid than the other readers now on the market and includes articles with a variety of styles and perspectives—a global perspective is apparent throughout. Like *Sociology in Our Times: The Essentials*, this reader has a strong focus on diversity. The book also features current research and presents a balance of classic and contemporary readings that professors teaching the course find important.

TEN QUESTIONS: A SOCIOLOGICAL PERSPECTIVE, FIFTH EDITION. This brief book employs a unique approach to introducing and examining sociological principles by posing and answering in each chapter such questions as "What does it mean to be human?" "Are human beings free?" and "Why is there misery in the world?" The book examines the philosophies of classical sociologists such as Marx, Weber, Durkheim, Mead, and Berger, and looks at how the field of sociology has approached these questions over the past 150 years.

WADSWORTH CLASSIC READINGS IN SOCIOLOGY, THIRD EDITION. This best-selling reader contains a series of classic articles written by key sociologists that will complement *Sociology in Our Times: The Essentials* and serve as a foundation by giving students the opportunity to read original works that teach the fundamental ideas of sociology. Free when bundled with *Sociology in Our Times: The Essentials*, this brief reader is a convenient and painless way to bring the "great thinkers" to students.

SOCIOLOGICAL FOOTPRINTS: INTRODUCTORY READINGS IN SOCIOLOGY, NINTH EDITION. The primary objective of this mainstream anthology is to provide a link between theoretical sociology and everyday life by presenting a comprehensive balance of classical, contemporary, popular, and multicultural articles and sociological studies.

SOCIOLOGICAL ODYSSEY: CONTEMPORARY READINGS. This reader speaks to the common issues of the introductory course. It includes articles that are contemporary, that are based on new research, and that demonstrate the new sociological issues in the world today, as well as articles that are near to students' experiences, highly readable, and based on the everyday concerns that influence their lives.

EXPERIENCING POVERTY: VOICES FROM THE BOTTOM. This reader is unique in its comprehensive understanding of poverty, incorporating the experiences of the impoverished in many situations, including people who are homeless, housed, working, nonworking, single, married, urban, rural, born in the United States, first-generation Americans, young, and old. Most significantly, crucial dimensions of inequality—race and gender—are interwoven throughout the selected readings.

APPLIED SOCIOLOGY: TERMS, TOPICS, TOOLS, AND TASKS. This concise, user-friendly book by award-winning sociology professor Stephen F. Steele of Anne Arundel Community College and Jammie Price of the University of North Carolina–Wilmington addresses a common question of many introductory sociology students: "What can I do with sociology?" The book introduces students to sociology as an active and relevant way to understand human social interaction by offering a clear, direct linkage between sociology and its practical use. It focuses on the core concepts in sociology (terms and topics), illustrates contemporary and practical skills used by sociologists to investigate these concepts (tools), and then contains concrete exercises for learning and applying these skills (tasks). The book also includes brief sections on using sociology to make a difference in the community and on developing a career in sociology. *Applied Sociology* is an ideal supplement to add an applied component to the introductory sociology course.

INFOTRAC® COLLEGE EDITION. Four months of *free* anywhere, anytime access to InfoTrac College Edition, the online library, is automatically packaged with *Sociology in Our Times: The Essentials.* The new and improved InfoTrac College Edition puts cutting-edge research and the latest headlines at students' fingertips, giving them access to an entire online library for the cost of one book! This fully searchable database offers more than 20 years' worth of full-text articles (more than 10 million) from almost 4,000 diverse sources, such as academic journals, newsletters, and up-to-the-minute publications including *Time, Newsweek, Science, Forbes,* and *USA Today.*

STUDENT GUIDE TO INFOTRAC COLLEGE EDITION FOR SOCIOLOGY. This valuable guide consists of exercises based on twenty-three core subjects vital to the study of sociology. As students use InfoTrac College Edition's huge database of articles to complete the exercises, they learn to refine their online research skills and strengthen their understanding of important sociological concepts.

CENSUS 2000: A STUDENT GUIDE FOR SOCIOLOGY. Wadsworth has created *Census 2000: A Student Guide for Sociology* to help instructors and students keep up with the steady stream of Census 2000 material. The 2000 Census presents a bounty of information and data that are key to understanding the changing nature of U.S. society. Keeping up with the flood of releases is a major challenge for both teachers and students alike. This guide contains a brief introduction to the Census, important trends that the Census reveals, critical-thinking exercises, test questions, and a number of demographic graphs.

TERRORISM: AN INTERDISCIPLINARY PERSPECTIVE, **SECOND EDITION.** The horrible terrorist attacks of September 11, 2001, and the world's subsequent response to those acts create a need for students to better understand the nature of terrorism and the developments that may lead to it. The expertise of several Wadsworth criminal justice, sociology, and anthropology textbook authors provides the most up-to-date essays related to terrorism. The free web site/booklet includes a background to terrorism, the history of Middle Eastern terrorism, the intersection of religion and terrorism, the role of globalization, and domestic responses and repercussions.

ONLINE CHAPTERS. Two online chapters are available to augment coverage of two important topics in sociology today—school violence and the environment. Visit the text's companion web site for additional information on accessing these two online chapters and bundling a printed version of the chapters with the text.

Suburban Youth and School Violence explores the concept of youth culture and the various sociological and criminological theories that postulate why youth become deviant.

Environmental Sociology examines such topics as the development of environmental sociology, social theory and the environment, environmental issues and problems, and the future of the environment.

RESEARCHING SOCIOLOGY ON THE INTERNET, **SECOND EDITION.** Written by D. R. Wilson, Houston Baptist University, and David L. Carlson, Texas A&M University, this guide is designed to assist sociology students with doing research on the Internet. Part One contains general information necessary to get started and answers questions about security, the type of sociology material available on the Internet, the information that is reliable and the sites that are not, the best ways to find research, and the best links to take students where they want to go. Part Two looks at each main discipline in the area of sociology and refers students to sites where the most enlightening research can be obtained.

SOCIOLOGY VIRTUAL EXPLORATIONS EXERCISE WORKBOOK. Virtual Explorations have been printed and packaged for easy-to-use assignments that students can turn in. Virtual Explorations are highly developed Internet exercises available on the Wadsworth web site. The explorations lead students through specific questions related to useful sites for a particular chapter in *Sociology in Our Times: The Essentials*. For example, the U.S. Census site is very useful for sociology. However, the site is enormous and often difficult to navigate for a first-time user. The Explorations guide students through this maze and enable, empower, and educate them to "do" sociology. These powerful activities are an excellent way to get involved in and excited about your sociology course. Great for online assignments!

DISCOVERING SOCIOLOGY: USING MICRO-CASE® EXPLORIT®, SECOND EDITION. This best-selling software-based workbook by Steven E. Barkan lets students explore dozens of sociological topics and issues, using data from the United States and around the world. With the workbook and accompanying ExplorIt software and data sets, students won't just read about what other sociologists have done; they will discover sociology for themselves. *Discovering Sociology* will add an exciting dimension to the introductory course. IBM-compatible only (Windows 95 or above).

DOING SOCIOLOGY: A GLOBAL PERSPECTIVE, USING MICROCASE® EXPLORIT® (WITH CD-ROM), FOURTH EDITION. This software/workbook package by sociologist Rodney Stark shows students what it takes to do real sociological research, using the same data and techniques employed by professional researchers. The workbook's step-by-step approach includes explanations of basic research concepts and methods, expanded exercises, and suggestions for independent research projects, effectively guiding students through the research process and offering them a real sense of what sociologists do. IBM-compatible only (Windows 95 or above).

Internet-Based Supplements

WEBTUTOR™ ADVANTAGE. WebTutor is a content-rich, Web-based teaching and learning tool that helps students succeed by taking the course beyond classroom boundaries to an anywhere, anytime environment. WebTutor is rich with study and mastery tools, communication tools, and course content. Professors can use Web-Tutor to provide virtual office hours, post syllabi, set up threaded discussions, track student progress with the quizzing material, and more.

WEBTUTOR™ ADVANTAGE PLUS. WebTutor Advantage Plus offers everything that WebTutor Advantage offers *plus* a complete e-book version of the text. Students can highlight key information, take notes within the page margins, and access hyperlinks.

MYCOURSE 2.0. Ask Wadsworth about this new *free* online course builder! Whether you want only the easy-to-use tools to build it or the content to furnish it, this simple solution for a custom course web site allows you to assign, track, and report on student progress; load your syllabus; and more. Contact your representative for details.

KENDALL'S COMPANION WEB SITE AT WADSWORTH'S VIRTUAL SOCIETY (HTTP:// WWW.WADSWORTH.COM/SOCIOLOGY). At Virtual Society: Wadsworth's Sociology Resource Center, click on the student companion web site for *Sociology in Our Times: The Essentials* to find useful learning resources for each chapter of the book. Some of those resources include

- Tutorial practice quizzes that can be scored and e-mailed to the instructor
- Internet links and exercises
- CNN Video exercises
- InfoTrac College Edition exercises
- Flashcards of the text's glossary
- Crossword puzzles
- Essay questions
- Learning objectives
- MicroCase exercises
- Virtual Explorations
- And much more!

Acknowledgments

Sociology in Our Times: The Essentials would not have been possible without the insightful critiques of these colleagues, who have reviewed some or all of this book. My profound thanks to each one for engaging in this time-consuming process:

Margret Choka, *Pellissippi State Community College*

Brooks Cowan, *University of Vermont*

Russell Curtis, *University of Houston*

Harold Dorton, *Southwest Texas State University*

Cindy Epperson, *St Louis Community College at Meramec*

Andrea Fontana, *UNLV*

Howard Gabennesch, *University of Southern Indiana*

Art Jipson, *University of Dayton*

Joachim Kibirige, *Missouri Western State College*

Amy Krull, *Western Kentucky University*

Barbara Richardson, *Eastern Michigan University*

Joseph Scimecca, *George Mason University*

James Unnever, *Radford University*

Richard Valencia, *Fresno City College*

Carl Wahlstrom, *Genesee Community College*

N. Ree Wells, *Missouri Southern State College*

Jerry Williams, *Stephen Austin State*

Previous Reviewers of *Sociology in Our Times: The Essentials*

Jan Abu-Shakrah, Portland Community College; Ted Alleman, Penn State University; Ginna Babcock, University of Idaho; Kay Barches, Skyline College; Sampson Lee Blair, University of Oklahoma; Barbara Bollmann, Community College of Denver; Joni Boye-Bearman, Wayne State College; John Brady, Texas Lutheran University; Valerie Brown, Case Western Reserve University; Deborah Carter, Johnson C. Smith University; Lillian Daughaday, Murray State University; Kevin Delaney, Temple University; Judith Dilorio, Indiana University–Purdue University; Kevin Early, Oakland University; Don Ecklund, Lincoln Land Community College; Julian Thomas Euell, Ithaca College; Grant Farr, Portland State University; Joe R. Feagin, University of Florida; Kathryn M. Feltey, University of Akron; William Finlay, University of Georgia; Mark Foster, Johnson County Community College; Richard Gale, University of Oregon; DeAnn K. Gauthier, University of Southwestern Louisiana; Michael Goslin, Tallahassee Community College; Anna Hall, Delgado Community College; Gary D. Hampe, University of Wyoming; Carole Hill,

Shelton State Community College; William Kelly, University of Texas, Austin; Bud Khleif, University of New Hampshire; Michael B. Kleiman, University of South Florida; Janet Koenigamen, College of St. Benedict; Monica Kumba, Buffalo State University; Yechiel Lahavy, Atlantic Community College; Abraham Levine, El Camino College; Diane Levy, University of North Carolina, Wilmington; Stephen Lilley, Sacred Heart University; Michael Lovaglia, University of Iowa; Akbar Mahdo, Ohio Wesleyan University; M. Cathey Maze, Franklin University; Jane McCandless, State University of West Georgia; John Molan, Alabama State University; Betty Morrow, Florida International University; Tina Mougouris, San Jacinto College; Dale Parent, Southeastern Louisiana University; Ellen Rosengarten, Sinclair Community College; Phil Rutledge, University of North Carolina, Charlotte; Marcia Texler Segal, Indiana University Southeast; Dorothy Smith, State University of New York–Plattsburg; Jackie Stanfield, Northern Colorado University; Mary Texeira, California State University–San Bernardino; Gary Tiedeman, Oregon State University; Steven Vasser, Mankato State University; Henry Walker, Cornell University; Susan Waller, University of Central Oklahoma; and John Zipp, University of Wisconsin–Milwaukee.

I deeply appreciate the energy, creativity, and dedication of the many people responsible for the development and production of *Sociology in Our Times: The Essentials*. I wish to thank Wadsworth Publishing Company's Susan Badger, Eve Howard, Sabra Horne, Natalie Cornelison, and Ritchie Durdin for their enthusiasm and insights throughout the development of this text. Many other people worked hard on the production of the fourth edition of *Sociology in Our Times: The Essentials*, especially Greg Hubit and Donald Pharr. I am extremely grateful to them. My most profound thanks go to my husband, Terrence Kendall, who made invaluable contributions to *Sociology in Our Times*.

I invite you to send your comments and suggestions about this book to me in care of:

Wadsworth Publishing Company
10 Davis Drive
Belmont, CA 94002

About the Author

Diana Kendall received a Ph.D. from the University of Texas at Austin, where she was invited to membership in Phi Kappa Phi Honor Society. Her areas of specialization and primary research interests are sociological theory, race/class/gender studies, and the sociology of medicine. In addition to *Sociology in Our Times*, she is the author of *The Power of Good Deeds: Privileged Women and the Social Reproduction of the Upper Class* (Rowman & Littlefield, 2002) and—along with Jane Lothian Murray and Rick Linden—of *Sociology in Our Times: Canadian Edition*, Second Edition (Nelson Canada, 2001). Her articles and presented papers primarily focus on the scholarship of teaching and on an examination of U.S. women of the upper classes across racial and ethnic groups.

Diana Kendall is currently a sociology professor at Baylor University, where she has taught a variety of courses, including Introduction to Sociology, Sociological Theory (undergraduate and graduate), Sociology of Medicine, and Race, Class, and Gender. Previously, she enjoyed many years of teaching sociology and serving as chair of the Social and Behavioral Science Division at Austin Community College.

Professor Kendall has been actively involved in national and regional sociological associations. She has served a term on the council of the ASA Section on Undergraduate Education and was recently a finalist for the Hans O. Mauksch Distinguished Contributions to Undergraduate Education Award. Professor Kendall is also a member of the Sociologists for Women in Society, the Society for the Study of Social Problems, and the Southwestern Sociological Association.

Jose Ortega/SIS

The Sociological Perspective and Research Process

1

Putting Social Life into Perspective
Why Study Sociology?
The Sociological Imagination
The Importance of a Global Sociological Imagination

Comparing Sociology with Other Social Sciences

The Development of Sociological Thinking
Early Thinkers: A Concern with Social Order and Stability
Differing Views on the Status Quo: Stability Versus Change
The Beginnings of Sociology in the United States

Contemporary Theoretical Perspectives
Functionalist Perspectives
Conflict Perspectives
Symbolic Interactionist Perspectives
Postmodern Perspectives

The Sociological Research Process
The "Conventional" Research Model
A Qualitative Research Model

Research Methods
Survey Research
Secondary Analysis of Existing Data
Field Research
Experiments

Ethical Issues in Sociological Research

This note should be pretty easy to understand. All the wording's from the Punk Rock 101. . . . I haven't felt the excitement of listening to as well as creating music, along with reading and writing for too many years now. . . . I've tried everything that's in my power to appreciate it, and I do. . . . I'm too sensitive. I need to be slightly numb in order to regain the enthusiasm I had as a child. . . . There's good in all of us and I simply love people too much. So much that it makes me feel too . . . sad. . . . I have it good, very good, and I'm grateful. But, since the age of seven, I've become hateful toward all humans in general. . . . I'm too much of an erratic, moody baby! I don't have passion anymore, and so remember, it's better to burn out than to fade away.

—Peace, love, empathy, Kurt Cobain (qtd. in Etkind, 1997: 38–39; originally appeared in *Rolling Stone,* June 2, 1994)

When the fire goes out you better learn to fake it. It's better to rise than fade away.

—After Cobain's suicide, Courtney Love responded to his death with these lyrics in *Celebrity Skin,* an album that she and her band recorded.

Paul Morse/CORBIS

Suicides of celebrities such as Kurt Cobain (pictured here with his wife, Courtney Love) call our attention to significant social facts related to suicide rates in contemporary societies. Sociologists ask questions such as why highly successful people, such as Cobain, commit suicide.

The first example set forth above is a suicide statement left by Kurt Cobain—founder of the band Nirvana, which is credited with inventing the grunge sound—who apparently found himself trapped in a downward spiral of depression and drug addiction. Why did Cobain commit suicide? It appeared that he had everything to live for. Millions of Nirvana albums had been sold, he was married to celebrity Courtney Love, and wealth and success were his. In the aftermath of his death, Love issued her reply to his feeling that "it's better to burn out than to fade away" by suggesting that people should go on with their lives even if "the fire goes out" for a period of time. According to Love's philosophy, "It's better to rise than fade away," and apparently she has done this as she has continued with her best-selling albums and successful career.

Although Cobain and Love are only two persons in a world comprising more than six billion people, the issues that are raised by Cobain's death are of importance in many social contexts. They bring us to a larger sociological question: Why does anyone commit suicide? Is suicide purely an individual phenomenon, or is it related to the social environments and societies in which people live?

In this chapter, we examine how sociological theories and research can help us understand the seemingly individualistic act of taking one's own life. We will see how sociological theory and research methods might be used to answer complex questions, and we will wrestle with some of the difficulties that sociologists experience as they study human behavior.

Questions and Issues

Chapter Focus Question: How do sociological theory and research add to our knowledge of human societies and social issues such as suicide?

What is the sociological imagination?

Why were early thinkers concerned with order and stability?

What are the assumptions behind each of the contemporary theoretical perspectives?

What are the main steps in the sociological research process?

Why is a code of ethics necessary for sociological research?

Putting Social Life into Perspective

Sociology is the systematic study of human society and social interaction. It is a *systematic* study because sociologists apply both theoretical perspectives and research methods (or orderly approaches) to examinations of social behavior. Sociologists study human societies and their social interactions in order to develop theories of how human behavior is shaped by group life and how, in turn, group life is affected by individuals.

Why Study Sociology?

Sociology helps us gain a better understanding of ourselves and our social world. It enables us to see how behavior is largely shaped by the groups to which we belong and the society in which we live. A *society* is a large social grouping that shares the same geographical territory and is subject to the same political authority and dominant cultural expectations, such as the United States, Mexico, or Nigeria. Examining the world order helps us understand that each of us is affected by *global interdependence*—a relationship in which the lives of all people are intertwined closely and any one nation's problems are part of a larger global problem. Environmental problems are an example: People throughout the world share the same biosphere—the zone of the earth's surface and atmosphere that sustains life. When environmental degradation, such as removing natural resources or polluting the air and water, takes place in one region, it may have an adverse effect on people around the globe.

Individuals can make use of sociology on a more personal level. Sociology enables us to move beyond established ways of thinking, thus allowing us to gain new insights into ourselves and to develop a greater awareness of the connection between our own "world" and that of other people. According to the sociologist Peter Berger (1963: 23), sociological inquiry helps us see that "things are not what they seem." Sociology provides new ways of approaching problems and making decisions in everyday life. For this reason, people with a knowledge of sociology are employed in a variety of fields that apply sociological insights to everyday life (see Figure 1.1).

Figure 1.1 Fields That Use Social Science Research

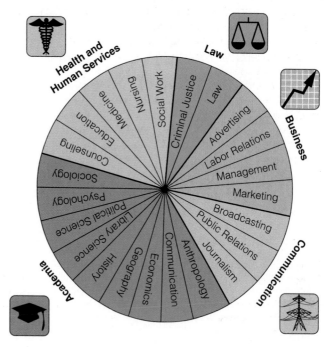

In many careers, including jobs in academia, business, communication, health and human services, and law, the ability to analyze social science research is an important asset.

Source: Based on Katzer, Cook, and Crouch, 1991.

Sociology promotes understanding and tolerance by enabling each of us to look beyond intuition, common sense, and our personal experiences. Many of us rely on intuition or common sense gained from personal experience to help us understand our daily lives and other people's behavior. *Commonsense knowledge* guides ordinary conduct in everyday life. However, many commonsense notions are actually myths. A *myth* is a popular but false notion that may be used, either intentionally or unintentionally, to perpetuate certain beliefs or "theories" even in the light of conclusive evidence to the contrary. Before reading on, take the quiz in

Sociology   the systematic study of human society and social interaction.

Society   a large social grouping that shares the same geographical territory and is subject to the same political authority and dominant cultural expectations.

Box 1.1 Sociology in Everyday Life

How Much Do You Know About Suicide?

True	False		
T	F	1.	For people thinking of suicide, it is difficult, if not impossible, to see the bright side of life.
T	F	2.	People who talk about suicide don't do it.
T	F	3.	Once people contemplate or attempt suicide, they must be considered suicidal for the rest of their lives.
T	F	4.	In the United States, suicide occurs on the average of one every seventeen minutes.
T	F	5.	Accidents and injuries sustained by teenagers and young adults may indicate suicidal inclinations.
T	F	6.	Over half of all suicides occur in adult women between the ages of 25 and 65.
T	F	7.	Older women have lower rates of both attempted and completed suicide than older men.
T	F	8.	Children don't know enough to be able to intentionally kill themselves.
T	F	9.	Suicide rates for African Americans are higher than for white Americans.
T	F	10.	More teenagers and young adults die from suicide than from cancer, heart disease, AIDS, birth defects, stroke, pneumonia, influenza, and chronic lung disease combined.

Answers on page 6.

Box 1.1, which lists a number of commonsense notions about suicide.

By contrast, sociologists strive to use scientific standards, not popular myths or hearsay, in studying society and social interaction. They use systematic research techniques and are accountable to the scientific community for their methods and the presentation of their findings. Whereas some sociologists argue that sociology must be completely value free—free from distorting subjective (personal or emotional) bias—others do not think that total objectivity is an attainable or desirable goal when studying human behavior. However, all sociologists attempt to discover patterns or commonalities in human behavior. When they study suicide, for example, they look for recurring patterns of behavior even though *individual* people usually commit the acts and *other individuals* suffer as a result of these actions.

Consequently, sociologists seek out the multiple causes and effects of suicide or other social is-

sues. They analyze the impact of the problem not only from the standpoint of the people directly involved but also from the standpoint of the effects of such behavior on all people.

The Sociological Imagination

Sociologist C. Wright Mills (1959b) described sociological reasoning as the *sociological imagination*— **the ability to see the relationship between individual experiences and the larger society.** This awareness enables us to understand the link between our personal experiences and the social contexts in which they occur. The sociological imagination helps us distinguish between personal troubles and social (or public) issues. *Personal troubles* are private problems that affect individuals and the networks of people with whom they associate regularly. As a result, those problems must be solved by individuals within their immediate social settings. For example, one person being unemployed may be a personal trouble. *Public issues* are problems that affect large numbers of people and often require solutions at the societal level. Widespread unemployment as a result of economic changes such as plant closings is an example of a public issue. The sociological imagination helps us place seemingly personal troubles, such as losing one's job or feeling like committing suicide, into a larger social context, where we can distinguish whether and how personal troubles may be related to public issues.

SUICIDE AS A PERSONAL TROUBLE Many of our individual experiences may be largely beyond our own control. They are determined by society as a whole—by its historical development and its organization. In everyday life, we do not define personal experiences in these terms. If a person commits suicide, many people consider it to be the result of his or her own personal problems. Historically, the commonsense view of suicide was that it was a sin, a crime, and a mental illness (Evans and Farberow, 1988).

Let's return to the case of Kurt Cobain, whose suicide note explained that he had lost his passion for life. In doing so, he expressed a typically individualistic view that his problem was purely personal and that there was nothing that he could do about it.

SUICIDE AS A PUBLIC ISSUE We can use the sociological imagination to look at the problem of suicide as a public issue—a societal problem. Early sociologist Emile Durkheim refused to accept commonsense explanations of suicide. In what is probably the first sociological study to use scientific research methods, he related suicide to the issue of cohesiveness (or lack of cohesiveness) in society instead of viewing suicide as an isolated act that could be understood only by studying individual personalities or inherited tendencies. In *Suicide* (1964b/1897), Durkheim documented his contention that a high suicide rate was symptomatic of large-scale societal problems.

The Importance of a Global Sociological Imagination

Although existing sociological theory and research provide the foundation for sociological thinking, we must reach beyond past studies that have focused primarily on the United States to develop a more comprehensive *global* approach for the future. In the twenty-first century, we face important challenges in a rapidly changing nation and world. The world's *high-income countries* **are nations with highly industrialized economies; technologically advanced industrial, administrative, and service occupations; and relatively high levels of national and personal income.** Some examples are the United States, Canada, Australia, New Zealand, Japan, and the countries of Western Europe.

As compared with other nations of the world, many high-income nations have a high standard of living and a lower death rate due to advances in nutrition and medical technology. However, everyone living in a so-called high-income country does not necessarily have a high income or an outstanding quality of life. In contrast, *middle-income countries* **are nations with industrializing economies, particularly in urban areas, and moderate levels of national and personal income.** Some examples of middle-income countries are the nations of Eastern Europe and many Latin American countries, where nations such as Brazil and Mexico are industrializing rapidly.

Low-income countries **are primarily agrarian nations with little industrialization and low levels of national and personal income.** Examples of low-income countries are many of the nations of Africa and Asia, particularly India and the People's Republic of China, where people typically work the land and are among the poorest in the world. However, generalizations are difficult to

---

Sociological imagination C. Wright Mills's term for the ability to see the relationship between individual experiences and the larger society.

High-income countries (sometimes referred to as **industrial countries**) nations with highly industrialized economies; technologically advanced industrial, administrative, and service occupations; and relatively high levels of national and personal income.

Middle-income countries (sometimes referred to as **developing countries**) nations with industrializing economies and moderate levels of national and personal income.

Low-income countries (sometimes referred to as **underdeveloped countries**) nations with little industrialization and low levels of national and personal income.

Box 1.1

Answers to the Sociology Quiz on Suicide

1. **True.** To people thinking of suicide, an acknowledgment that there is a bright side only confirms and conveys the message that they have failed; otherwise, they could see the bright side of life as well.

2. **False.** Some people who talk about suicide do kill themselves. Warning signals of possible suicide attempts include talk of suicide, the desire not to exist anymore, and despair.

3. **False.** Most people think of suicide for only a limited amount of time. When the crisis is over and the problems leading to suicidal thoughts are resolved, people usually cease to think of suicide as an option.

4. **True.** A suicide occurs on the average of every seventeen minutes in the United States; however, this differs with respect to the sex, race/ethnicity, and age of the individual. For example, men are more than three times more likely to kill themselves than are women.

5. **True.** Accidents, injuries, and other types of life-threatening behavior may be signs that a person is on a course of self-destruction. One study concluded that the incidence of suicide was twelve times higher among adolescents and young adults who had been previously hospitalized because of an injury.

6. **False.** Just the opposite is true: Over half of all suicides occur in adult men, aged 25 to 65.

7. **True.** In the United States, as in other countries, suicide rates are the highest among men over age 70. One theory of why this is true asserts that older women may have a more flexible and diverse coping style than do older men.

8. **False.** Children do know how to intentionally hurt or kill themselves. They may learn the means and methods from television, movies, or other people. However, the National Center for Health Statistics (the agency responsible for compiling suicide statistics) does not recognize suicides under the age of 10; they are classified as accidents, despite evidence that young children have taken their own lives.

9. **False.** Suicide rates are much higher among white Americans than African Americans. For example, in 1995 the overall U.S. suicide rate was 11.9 per 100,000 population. For white males, the rate was 21.4; for African American males, it was 11.9. For white females, it was 4.8; for African American females, the rate was 2.0.

10. **True.** Suicide is a leading cause of death among teenagers and young adults. It is the third leading cause of death among young people between 15 and 24 years of age, following unintentional injuries and homicide.

Sources: Based on Levy and Deykin, 1989; Patros and Shamoo, 1989; Leenaars, 1991; National Center for Health Statistics, 1997; and National Institute of Mental Health, 2002.

make because there are wide differences in income and standards of living within many nations (see Chapter 7, "Global Stratification").

In middle- and low-income countries, suicide patterns vary widely from those in high-income nations. For example, some analysts suggest that a re-

cent increase in suicide among the Kaiowá Indians in Brazil can be linked to the loss of their land (which was previously used to grow crops to feed their families) to large, transnational agribusinesses. With loss of land has come loss of culture. To the Kaiowá, land is more than a means of surviving; it is the support for a social life that is directly linked to their system of belief and knowledge (Schemo, 1996).

Throughout this text, we will continue to develop our sociological imaginations by examining social life in the United States and other nations. The future of our nation is deeply intertwined with the future of all other nations of the world on economic, political, environmental, and humanitarian levels. We buy many goods and services that were produced in other nations, and we sell much of what we produce to the people of other nations. Peace in other nations is important if we are to ensure peace within our borders. Famine, unrest, and brutality in other regions of the world must be of concern to people in the United States. Moreover, fires, earthquakes, famine, or environmental pollution in one nation typically has an adverse influence on other nations as well. Global problems contribute to the large influx of immigrants who arrive in the United States annually. These immigrants bring with them a rich diversity of language, customs, religions, and previous life experiences; they also contribute to dramatic population changes that will have a long-term effect on this country. Developing a better understanding of diversity and tolerance for people who are different from us is important for our personal, social, and economic well-being.

Whatever your race/ethnicity, class, sex, or age, are you able to include in your thinking the perspectives of people who are quite dissimilar in experiences and points of view? Before answering this question, a few definitions are in order. *Race* is a term used by many people to specify groups of people distinguished by physical characteristics such as skin color; in fact, there are no "pure" racial types, and the concept of race is considered by most sociologists to be a social construction that

people use to justify existing social inequalities. *Ethnicity* refers to the cultural heritage or identity of a group and is based on factors such as language or country of origin. *Class* is the relative location of a person or group within the larger society, based on wealth, power, prestige, or other valued resources. *Sex* refers to the biological and anatomical differences between females and males. By contrast, *gender* refers to the meanings, beliefs, and practices associated with sex differences, referred to as *femininity* and *masculinity* (Scott, 1986).

In forming your own global sociological imagination and in seeing the possibilities for sociology in the twenty-first century, it will be helpful to understand how the study of sociology compares with other social sciences.

Comparing Sociology with Other Social Sciences

In this chapter, we will discuss how sociologists examine social life and social problems such as suicide. Let's briefly examine how other social sciences investigate human relationships, and then compare this discipline to sociology.

Anthropologists and sociologists are interested in studying human behavior; however, there are differences between the two disciplines. *Anthropology* seeks to understand human existence over geographic space and evolutionary time (American Anthropological Association, 2001), whereas sociology seeks to understand contemporary social organization, relations, and change. Some anthropologists focus on the beginnings of human history, millions of years ago, whereas others primarily study contemporary societies. Anthropology is divided into four main subfields: sociocultural, linguistic, archaeological, and biological anthropology. Cultural anthropologists focus on culture and its many manifestations, including art, religion, and politics. Linguistic anthropologists primarily study language because culture itself depends on language. Archaeologists are interested in discovering and analyzing material artifacts—such as cave paintings, discarded stone tools, and abandoned baskets—from which they piece together a record of social life in earlier societies and assess what this information adds to our knowledge of contemporary cultures. Biological (or physical) anthropologists study the biological origins, evolutionary development, and genetic diversity

✓ **Checkpoints**

Checkpoint 1.1

1. What is sociology?
2. How can we use the sociological imagination?
3. What is the difference between a personal trouble and a public issue?
4. What are the major categories of countries by type of economy?

of primates, including Homo sapiens (human beings). Clearly, there are overlapping areas of research in which anthropologists and sociologists investigate similar topics such as the effect that culture has on the social life of people in a society.

Psychology is the systematic study of behavior and mental processes—what occurs in the mind. Psychologists focus not only on behavior that is directly observable, such as talking, laughing, and eating, but also on mental processes that cannot be directly observed, such as thinking and dreaming. For psychologists, behavior and mental processes are interwoven; therefore, to understand behavior, they examine the emotions that underlie people's actions.

Psychology is a diverse field. Some psychologists work in clinical settings, where they diagnose and treat psychological disorders; others practice in schools, where they are concerned with the intellectual, social, and emotional development of schoolchildren. Still other psychologists work in business, industry, and other work-related settings. Another branch of psychology—social psychology—is similar to sociology in that it emphasizes how social conditions affect individual behavior. Social psychological perspectives on human development are useful to sociologists who study the process of socialization (see Chapter 3, "Socialization"). However, a distinction between psychology and sociology is the extent to which most psychological studies focus on internal factors relating to the individual in their explanations of human behavior, whereas sociological research examines the effects of groups, organizations, and social institutions on social life.

Unlike the other social sciences we have discussed, *economics* concentrates primarily on a single institution in society—the economy (see Chapter 13, "Politics and the Economy in Global Perspective"). Economists attempt to explain how the limited resources of a society are allocated among competing demands. Economics is divided into two different branches. Macroeconomics looks at such things as the total amount of goods and services produced by a society; microeconomics studies such issues as decisions made by individual businesses. However, a distinction between economics and sociology is that economists focus on the complex workings of economic systems (such as monetary policy, inflation, and the national debt), whereas sociologists focus on a number of social institutions, one of which is the economy.

✓Checkpoints

Checkpoint 1.2

5.   How does sociology differ from (a) anthropology, (b) psychology, (c) economics, and (d) political science?

Political science is the academic discipline that studies political institutions such as the state, government, and political parties. Political scientists study power relations and seek to determine how power is distributed in various types of political systems. Some political scientists focus primarily on international relations and similarities and differences in the political institutions across nations. Other political scientists look at the political institutions of a particular country. Political scientists concentrate on political institutions, whereas sociologists study these institutions within the larger context of other social institutions, such as families, religion, education, and the media.

Clearly, the areas of interest and research in the social sciences overlap in that the goal of scholars, teachers, and students is to learn more about human behavior, including its causes and consequences. In applied sociology, there is increasing collaboration among researchers across disciplines to develop a more holistic, integrated view of how human behavior and social life take place in societies. However, as we next trace the development of sociological thinking, we examine specific ways in which sociology has its own realm of knowledge and theoretical perspectives that inform our analysis of social life, and how we can use that knowledge in our everyday lives.

The Development of Sociological Thinking

Throughout history, social philosophers and religious authorities have made countless observations about human behavior. However, the idea of observing how people lived, finding out what they thought, and doing so in a systematic manner that could be verified did not take hold until the nine-

As the Industrial Revolution swept through the United States in the nineteenth century, sights like this became increasingly common. The Remington Arms Works of Bridgeport, Connecticut, is emblematic of the factory system that shifted the base of the U.S. economy from agriculture to manufacturing. What new technologies are transforming the U.S. economy in the twenty-first century?

teenth century and the social upheaval brought about by industrialization and urbanization.

Industrialization **is the process by which societies are transformed from dependence on agriculture and handmade products to an emphasis on manufacturing and related industries.** This process occurred first during the Industrial Revolution in Britain between 1760 and 1850, and was soon repeated throughout Western Europe. By the mid-nineteenth century, industrialization was well under way in the United States. Massive economic, technological, and social changes occurred as machine technology and the factory system shifted the economic base of these nations from agriculture to manufacturing. A new social class of industrialists emerged in textiles, iron smelting, and related industries. Many people who had labored on the land were forced to leave their tightly knit rural communities and sacrifice well-defined social relationships to seek employment as factory workers in the emerging cities, which became the centers of industrial work.

Urbanization **is the process by which an increasing proportion of a population lives in cities rather than in rural areas.** Although cities existed long before the Industrial Revolution, the development of the factory system led to a rapid increase in both the number of cities and the size of their populations. People from very diverse backgrounds

Industrialization the process by which societies are transformed from dependence on agriculture and handmade products to an emphasis on manufacturing and related industries.

Urbanization the process by which an increasing proportion of a population lives in cities rather than in rural areas.

worked together in the same factory. At the same time, many people shifted from being *producers* to being *consumers*. For example, families living in the cities had to buy food with their wages because they could no longer grow their own crops to consume or to barter for other resources. Similarly, people had to pay rent for their lodging because they could no longer exchange their services for shelter. These living and working conditions led to the development of new social problems: inadequate housing, crowding, unsanitary conditions, poverty, pollution, and crime. Wages were so low that entire families—including very young children—were forced to work, often under hazardous conditions and with no job security. As these conditions became more visible, a new breed of social thinkers turned its attention to trying to understand why and how society was changing.

Early Thinkers: A Concern with Social Order and Stability

At the same time that urban problems were growing worse, natural scientists had been using reason, or rational thinking, to discover the laws of physics and the movement of the planets. Social thinkers started to believe that by applying the methods developed by the natural sciences, they might discover the laws of human behavior and apply these laws to solve social problems.

AUGUSTE COMTE The French philosopher Auguste Comte (1798–1857) coined the term *sociology* from the Latin *socius* ("social, being with others") and the Greek *logos* ("study of") to describe a new science that would engage in the study of society. Even though he never actually conducted sociological research, Comte is considered by some to be the "founder of sociology." Comte's theory that societies contain *social statics* (forces for social order and stability) and *social dynamics* (forces for conflict and change) continues to be used, although not in these exact terms, in contemporary sociology.

Drawing heavily on the ideas of his mentor, Count Henri de Saint-Simon, Comte stressed that the methods of the natural sciences should be applied to the objective study of society. Saint-Simon's primary interest in studying society was social reform, but Comte sought to unlock the secrets of society so that intellectuals like him could become the new secular (as contrasted with religious) "high priests" of society (Nisbet, 1979). For Comte, the best policies involved order and authority. He envisioned that a new consensus would emerge on social issues and that the new science of sociology would play a significant part in the reorganization of society (Lenzer, 1998).

Comte's philosophy became known as *positivism*—a belief that the world can best be understood through scientific inquiry. Comte believed that objective, bias-free knowledge was attainable only through the use of science rather than religion. However, scientific knowledge was "relative knowledge," not absolute and final. Comte's positivism had two dimensions: (1) methodological—the application of scientific knowledge to both physical and social phenomena—and (2) social and political—the use of such knowledge to predict the likely results of different policies so that the best one could be chosen.

Social analysts have praised Comte for his advocacy of sociology and his contributions to positivism. His insights regarding linkages between the social structural elements of society (such as family, religion, and government) and social thinking in specific historical epochs were useful to later sociologists. However, a number of contemporary sociologists argue that Comte, among others, brought about an overemphasis on the "natural science model," which has been detrimental to sociology (Vaughan, Sjoberg, and Reynolds, 1993). Still others state that sociology, while claiming to be "scientific" and "objective," has focused on the experiences of a privileged few, to the exclusion by class, gender, race, ethnicity, and age of all others (Harding, 1986; Collins, 1990).

HARRIET MARTINEAU Comte's works were made more accessible for a wide variety of scholars through the efforts of the British sociologist Harriet Martineau (1802–1876). Until recently, Martineau received no recognition in the field of sociology, partly because she was a woman in a male-dominated discipline and society. Not only did she translate and condense Comte's works, but she was also an active sociologist in her own right. Martineau studied the social customs of Britain and the United States, analyzing the consequences of industrialization and capitalism. In *Society in America*

Clockwise from top left: Auguste Comte, Harriet Martineau, Karl Marx, W .E. B. Du Bois, Emile Durkheim, Max Weber.

(1962/1837), she examined religion, politics, child rearing, slavery, and immigration in the United States, paying special attention to social distinctions based on class, race, and gender. Her works explore the status of women, children, and "sufferers" (persons who are considered to be criminal, mentally ill, handicapped, poor, or alcoholic).

Based on her reading of Mary Wollstonecraft's *A Vindication of the Rights of Women* (1974/1797),

Martineau advocated racial and gender equality. She was also committed to creating a science of society that would be grounded in empirical observations

Positivism a term describing Auguste Comte's belief that the world can best be understood through scientific inquiry.

and widely accessible to people. She argued that sociologists should be impartial in their assessment of society but that it is entirely appropriate to compare the existing state of society with the principles on which it was founded (Lengermann and Niebrugge-Brantley, 1998).

Recently, some scholars have argued that Martineau's place in the history of sociology should be as a founding member of this field of study, not just as the translator of Auguste Comte's work (Hoecker-Drysdale, 1992; Lengermann and Niebrugge-Brantley, 1998). Others have highlighted her influence in spreading the idea that societal progress could be brought about by the spread of democracy and the growth of industrial capitalism (Polanyi, 1944). Martineau believed that a better society would emerge if women and men were treated equally, enlightened reform occurred, and cooperation existed among people in all social classes (but led by the middle class).

HERBERT SPENCER Unlike Comte, who was strongly influenced by the upheavals of the French Revolution, the British social theorist Herbert Spencer (1820–1903) was born in a more peaceful and optimistic period in his country's history. Spencer's major contribution to sociology was an evolutionary perspective on social order and social change. Although the term *evolution* has various meanings, evolutionary theory should be taken to mean "a theory to explain the mechanisms of organic/social change" (Haines, 1997: 81). According to Spencer's Theory of General Evolution, society, like a biological organism, has various interdependent parts (such as the family, the economy, and the government) that work to ensure the stability and survival of the entire society.

Spencer believed that societies develop through a process of "struggle" (for existence) and "fitness" (for survival), which he referred to as the "survival of the fittest." Because this phrase is often attributed to Charles Darwin, Spencer's view of society is known as *social Darwinism*—**the belief that those species of animals, including human beings, best adapted to their environment survive and prosper, whereas those poorly adapted die out.** Spencer equated this process of *natural selection* with progress, because only the "fittest" members of society would survive the competition, and the "unfit" would be filtered out of society. Based on this belief, he strongly op-

posed any social reform that might interfere with the natural selection process and, thus, damage society by favoring its least worthy members.

Critics have suggested that many of his ideas have serious flaws. For one thing, societies are not the same as biological systems; people are able to create and transform the environment in which they live. Moreover, the notion of the survival of the fittest can easily be used to justify class, racial–ethnic, and gender inequalities and to rationalize the lack of action to eliminate harmful practices that contribute to such inequalities. Not surprisingly, Spencer's "hands-off" view was applauded by wealthy industrialists of his day. John D. Rockefeller, who gained monopolistic control of much of the U.S. oil industry early in the twentieth century, maintained that the growth of giant businesses was merely the "survival of the fittest" (Feagin and Feagin, 1997).

EMILE DURKHEIM French sociologist Emile Durkheim (1858–1917) was an avowed critic of some of Spencer's views while incorporating others into his own writing. Durkheim stressed that people are the product of their social environment and that behavior cannot be understood fully in terms of *individual* biological and psychological traits. He believed that the limits of human potential are *socially* based, not *biologically* based. As Durkheim saw religious traditions evaporating in his society, he searched for a scientific, rational way to provide for societal integration and stability (Hadden, 1997).

In his work *The Rules of Sociological Method* (1964a/1895), Durkheim set forth one of his most important contributions to sociology: the idea that societies are built on social facts. **Social facts are patterned ways of acting, thinking, and feeling that exist *outside* any one individual but that exert social control over each person.** Durkheim believed that social facts must be explained by other social facts—by reference to the social structure rather than to individual attributes.

Reflections

Reflections 1.1

Is Durkheim's concept of anomie still relevant to the study of contemporary suicides?

Durkheim observed that rapid social change and a more specialized division of labor produce *strains* in society. These strains lead to a breakdown in traditional organization, values, and authority and to a dramatic increase in **anomie—a condition in which social control becomes ineffective as a result of the loss of shared values and of a sense of purpose in society.** According to Durkheim, anomie is most likely to occur during a period of rapid social change. In *Suicide* (1964b/1897), he explored the relationship between anomic social conditions and suicide, as discussed in Chapter 2.

Durkheim's contributions to sociology are so significant that he has been referred to as *"the* crucial figure in the development of sociology as an academic discipline [and as] one of the deepest roots of the sociological imagination" (Tiryakian, 1978: 187). He has long been viewed as a proponent of the scientific approach to examining social facts that lie outside individuals. He is also described as the founding figure of the functionalist theoretical tradition.

Although they acknowledge Durkheim's important contributions, some critics note that his emphasis on societal stability, or the "problem of order"—how society can establish and maintain social stability and cohesiveness—obscured the *subjective meaning* that individuals give to social phenomena such as religion, work, and suicide. From this view, overemphasis on *structure* and the determining power of "society" resulted in a corresponding neglect of *agency* (the beliefs and actions of the actors involved) in much of Durkheim's theorizing (Zeitlin, 1997).

Differing Views on the Status Quo: Stability Versus Change

Together with Karl Marx, Max Weber, and Georg Simmel, Durkheim established the course of modern sociology. We will look first at Marx's and Weber's divergent thoughts about conflict and social change in societies and then at Georg Simmel's microlevel analysis of society.

KARL MARX In sharp contrast to Durkheim's focus on the stability of society, German economist and philosopher Karl Marx (1818–1883) stressed that history is a continuous clash between conflicting ideas and forces. He believed that conflict—especially class conflict—is necessary in order to produce social change and a better society. For Marx, the most important changes are economic. He concluded that the capitalist economic system was responsible for the overwhelming poverty that he observed in London at the beginning of the Industrial Revolution (Marx and Engels, 1967/1848).

In the Marxian framework, *class conflict* is the struggle between the capitalist class and the working class. The capitalist class, or *bourgeoisie*, comprises those who own and control the means of production—the tools, land, factories, and money for investment that form the economic basis of a society. The working class, or *proletariat*, is composed of those who must sell their labor because they have no other means to earn a livelihood. From Marx's viewpoint, the capitalist class controls and exploits the masses of struggling workers by paying less than the value of their labor. This exploitation results in workers' *alienation*—a feeling of powerlessness and estrangement from other people and from oneself. Marx predicted that the working class would become aware of its exploitation, overthrow the capitalists, and establish a free and classless society.

Marx is regarded as one of the most profound sociological thinkers, one who combined ideas derived from philosophy, history, and the social sciences into a new theoretical configuration. However, his social and economic analyses have also inspired heated debates among generations of social scientists. Central to his view was the belief that society should not just be studied but should also be changed, because the *status quo* (the existing state of society) involved the oppression of most of the population by a small group of wealthy people. Those who believe that sociology should be value free are uncomfortable with Marx's

Social Darwinism Herbert Spencer's belief that those species of animals, including human beings, best adapted to their environment survive and prosper, whereas those poorly adapted die out.

Social facts Emile Durkheim's term for patterned ways of acting, thinking, and feeling that exist *outside* any one individual but that exert social control over each person.

Anomie Emile Durkheim's designation for a condition in which social control becomes ineffective as a result of the loss of shared values and of a sense of purpose in society.

Richard Pasley/Stock Boston

According to the sociologist Georg Simmel, society is a web of patterned interactions among people. If we focus on the behavior of individuals in isolation, such as any one of the members of this women's rowing team, we may miss the underlying forms that make up the "geometry of social life."

advocacy of what some perceive to be radical social change. Scholars who examine society through the lens of race, gender, and class believe his analysis places too much emphasis on class relations, often to the exclusion of issues regarding race/ethnicity and gender.

MAX WEBER German social scientist Max Weber (pronounced VAY-ber) (1864–1920) was also concerned about the changes brought about by the Industrial Revolution. Although he disagreed with Marx's idea that economics is *the* central force in social change, Weber acknowledged that economic interests are important in shaping human action. Even so, he thought that economic systems were heavily influenced by other factors in a society.

Unlike many early analysts who believed that values could not be separated from the research process, Weber emphasized that sociology should be *value free* — research should be conducted in a scientific manner and should exclude the researcher's personal values and economic interests (Turner, Beeghley, and Powers, 1998). However,

Weber realized that social behavior cannot be analyzed by the objective criteria that we use to measure such things as temperature or weight. Although he recognized that sociologists cannot be totally value free, Weber stressed that they should employ *verstehen* (German for "understanding" or "insight") to gain the ability to see the world as others see it. In contemporary sociology, Weber's idea is incorporated into the concept of the sociological imagination (discussed earlier in this chapter). Weber was also concerned that large-scale organizations (bureaucracies) were becoming increasingly oriented toward routine administration and a specialized division of labor, which he believed were destructive to human vitality and freedom. According to Weber, rational bureaucracy, rather than class struggle, is the most significant factor in determining the social relations between people in industrial societies. From this view, bureaucratic domination can be used to maintain powerful (capitalist) interests in society. As discussed in Chapter 5 ("Groups and Organizations"), Weber's work on bureaucracy has had a far-reaching impact.

Weber made significant contributions to modern sociology by emphasizing the goal of value-free inquiry and the necessity of understanding how others see the world. He also provided important insights on the process of rationalization, bureaucracy, religion, and many other topics. In his writings, Weber was more aware of women's issues than many of the scholars of his day. Perhaps his awareness at least partially resulted from the fact that his wife, Marianne Weber, was an important figure in the women's movement in Germany in the early twentieth century (Roth, 1988).

GEORG SIMMEL At about the same time that Durkheim was developing the field of sociology in France, the German sociologist Georg Simmel (pronounced ZIM-mel) (1858–1918) was theorizing about society as a web of patterned interactions among people. The main purpose of sociology, according to Simmel, should be to examine these social interaction processes within groups. In *The Sociology of Georg Simmel* (1950/1902–1917), he analyzed how social interactions vary depending on the size of the social group. He concluded that interaction patterns differed between a *dyad,* a social group with two members, and a *triad,* a social group with three members. He developed *formal sociology*, an approach that focuses attention on the universal recurring social forms that underlie the varying content of social interaction. Simmel referred to these forms as the "geometry of social life." He also distinguished between *forms* of social interaction (such as cooperation or conflict) and the *content* of social interaction in different contexts (for example, between leaders and followers).

Like the other social thinkers of his day, Simmel analyzed the impact of industrialization and urbanization on people's lives. He concluded that class conflict was becoming more pronounced in modern industrial societies. He also linked the increase in individualism, as opposed to concern for the group, to the fact that people now had many cross-cutting "social spheres"—membership in a number of organizations and voluntary associations—rather than having the singular community ties of the past. Simmel also assessed the costs of "progress" on the upper-class city dweller, who, he believed, had to develop certain techniques to survive the overwhelming stimulation of the city. Simmel's ultimate concern was to protect the autonomy of the individual in society.

Simmel's contributions to sociology are significant. He wrote more than thirty books and numerous essays on diverse topics, leading some critics to state that his work is fragmentary and piecemeal. However, his thinking has influenced a wide array of sociologists, including the members of the "Chicago School" in the United States.

The Beginnings of Sociology in the United States

From Western Europe, sociology spread in the 1890s to the United States, where it thrived as a result of the intellectual climate and the rapid rate of social change. The first departments of sociology in the United States were located at the University of Chicago and at Atlanta University, then an African American school.

THE CHICAGO SCHOOL The first department of sociology in the United States was established at the University of Chicago, where the faculty was instrumental in starting the American Sociological Society (now known as the American Sociological Association). Robert E. Park (1864–1944), a member of the Chicago faculty, asserted that urbanization has a disintegrating influence on social life by producing an increase in the crime rate and in racial and class antagonisms that contribute to the segregation and isolation of neighborhoods (Ross, 1991). George Herbert Mead (1863–1931), another member of the faculty at Chicago, founded the symbolic interaction perspective, which is discussed later in this chapter.

JANE ADDAMS Jane Addams (1860–1935) is one of the best-known early women sociologists in the United States because she founded Hull House, one of the most famous settlement houses, in an impoverished area of Chicago. Throughout her career, she was actively engaged in sociological endeavors: She lectured at numerous colleges, was a charter member of the American Sociological Society, and published a number of articles and books. Addams was one of the authors of *Hull-House Maps and Papers*, a ground-breaking book that used a methodological technique employed by sociologists for the next forty years (Deegan, 1988). She was also awarded a Nobel Prize for her assistance to the underprivileged.

W. E. B. DU BOIS AND ATLANTA UNIVERSITY The second department of sociology in the United States was founded by W. E. B. Du Bois (1868–1963) at Atlanta University. He created a laboratory of sociology, instituted a program of systematic research, founded and conducted regular sociological conferences on research, founded two journals, and established a record of valuable publications. His classic work, *The Philadelphia Negro: A Social Study* (1967/1899), was based on his research into Philadelphia's African American community and stressed the strengths and weaknesses of a community wrestling with overwhelming social problems. Du Bois was one of the first scholars to note that a dual heritage creates conflict for people of color. He called this duality *double-consciousness*—the identity conflict of being both a black and an American. Du Bois pointed out that although people in this country espouse such values as democracy, freedom, and equality, they also accept racism and group discrimination. African Americans are the victims of these conflicting values and the actions that result from them (Benjamin, 1991).

✓ **Checkpoints**

Checkpoint 1.3

6. Why were industrialization and urbanization important to the development of sociological thinking?
7. Who is referred to as the founder of sociology, and what were his central ideas about how we can understand social life?
8. Why is Harriet Martineau an important figure in early sociological thinking?
9. What was Herbert Spencer's major contribution to sociology?
10. What are social facts, as identified by Emile Durkheim?
11. What were the key ideas about class of (a) Karl Marx, (b) Max Weber, and (c) Georg Simmel?
12. Who was the first well-known African American sociologist, and what were his contributions?

Contemporary Theoretical Perspectives

Given the many and varied ideas and trends that influenced the development of sociology, how do contemporary sociologists view society? Some see it as basically a stable and ongoing entity; others view it in terms of many groups competing for scarce resources; still others describe it based on the everyday, routine interactions among individuals. Each of these views represents a method of examining the same phenomena. Each is based on general ideas about how social life is organized and represents an effort to link specific observations in a meaningful way. Each uses a *theory*—a set of logically interrelated statements that attempts to describe, explain, and (occasionally) predict social events. Each theory helps interpret reality in a distinct way by providing a framework in which observations may be logically ordered. Sociologists refer to this theoretical framework as a *perspective*—an overall approach to or viewpoint on some subject. Three major theoretical perspectives have been predominant in U.S. sociology: the functionalist, conflict, and symbolic interactionist perspectives. Other perspectives, such as postmodernism, have emerged and gained acceptance among some social thinkers more recently.

Before turning to the specifics of these perspectives, we should note that some theorists and theories do not neatly fit into any of these perspectives.

Functionalist Perspectives

Also known as *functionalism* and *structural functionalism*, **functionalist perspectives are based on the assumption that society is a stable, orderly system.** This stable system is characterized by *societal consensus*, whereby the majority of members share a common set of values, beliefs, and behavioral expectations. According to this perspective, a society is composed of interrelated parts, each of which serves a function and (ideally) contributes to the overall stability of the society. Societies develop social structures, or institutions, that persist because they play a part in helping society survive. These institutions include the family, education, government, religion, and the economy. If anything adverse happens to one of these institutions or parts, all other parts are affected and the system no longer functions properly. As Durkheim noted, rapid social change and a more specialized division of labor produce *strains* in society that lead to a breakdown in these traditional institutions and may result in social problems such as an increase in crime and suicide rates.

TALCOTT PARSONS AND ROBERT MERTON
Talcott Parsons (1902–1979), perhaps the most influential contemporary advocate of the functionalist perspective, stressed that all societies must provide for meeting social needs in order to survive. Parsons (1955) suggested, for example, that a division of labor (distinct, specialized functions) between husband and wife is essential for family stability and social order. The husband/father performs the *instrumental tasks*, which involve leadership and decision-making responsibilities in the home and employment outside the home to support the family. The wife/mother is responsible for the *expressive tasks*, including housework, caring for the children, and providing emotional support for the entire family. Parsons believed that other institutions, including school, church, and government, must function to assist the family and that all institutions must work together to preserve the system over time (Parsons, 1955).

Functionalism was refined further by Robert K. Merton (b. 1910), who distinguished between mani-

Bob Daemmrich/The Image Works

Shopping malls are a reflection of a consumer society. A manifest function of a shopping mall is to sell goods and services to shoppers; however, a latent function may be to provide a communal area in which people can visit with friends and eat. For this reason, food courts have proven to be a boon in shopping malls around the globe.

fest and latent functions of social institutions. *Manifest functions* **are intended and/or overtly recognized by the participants in a social unit.** In contrast, *latent functions* **are unintended functions that are hidden and remain unacknowledged by participants.** For example, a manifest function of education is the transmission of knowledge and skills from one generation to the next; a latent function is the establishment of social relations and networks. Merton noted that all features of a social system may not be functional at all times; *dysfunctions* are the undesirable consequences of any element of a society. A dysfunction of education in the United States is the perpetuation of gender, racial, and class inequalities. Such dysfunctions may threaten the capacity of a society to adapt and survive (Merton, 1968).

APPLYING A FUNCTIONAL PERSPECTIVE TO SUICIDE How might functionalists analyze the problem of suicide among young people? Most functionalists emphasize the importance of shared

moral values and strong social bonds to a society. When rapid social change or other disruptive conditions occur, moral values may erode, and people may become more uncertain about how to act and about whether or not their life has meaning.

In sociologist Donna Gaines's (1991) study of the suicide pact of four teenagers in Bergenfield, New Jersey, she concluded that Durkheim's

Theory a set of logically interrelated statements that attempts to describe, explain, and (occasionally) predict social events.

Functionalist perspectives the sociological approach that views society as a stable, orderly system.

Manifest functions functions that are intended and/or overtly recognized by the participants in a social unit.

Latent functions unintended functions that are hidden and remain unacknowledged by participants.

description of both fatalistic and anomic suicide could be applied to the suicides of some teenagers. In regard to fatalistic suicide, people sometimes commit suicide "because they have lost the ability to dream" (Gaines, 1991: 253). According to Gaines, the four teenagers who took their lives in Bergenfield were bound closely together in a shared predicament: They wanted to get out, but they couldn't. As one young man noted, "They were beaten down as far as they could go" because they lacked confidence in themselves and were fearful of the world they faced. In other words, they felt trapped, and to them—without having a sense of meaningful choices—the only way out was to commit suicide (Gaines, 1991: 253). But, according to Gaines, fatalistic suicide does not completely explain the experiences of suicidal young people. Young people may also engage in anomic suicide, where the individual does not feel connected to the society. In fact, as Gaines (1991: 253) notes, "the glue that holds the person to the group isn't strong enough; social bonds are loose, weak, or absent. To be anomic is to feel disengaged, adrift, alienated. Like you don't fit in anywhere, there is no place for you: in your family, your school, your town—in the social order."

Barry King/Liaison Agency

As one of the wealthiest and most-beloved entertainers in the world, Oprah Winfrey is an example of Max Weber's concept of *prestige*—a positive estimation of honor.

Conflict Perspectives

According to *conflict perspectives*, **groups in society are engaged in a continuous power struggle for control of scarce resources.** Conflict may take the form of politics, litigation, negotiations, or family discussions about financial matters. Simmel, Marx, and Weber contributed significantly to this perspective by focusing on the inevitability of clashes between social groups. Today, advocates of the conflict perspective view social life as a continuous power struggle among competing social groups.

MAX WEBER AND C. WRIGHT MILLS As previously discussed, Karl Marx focused on the exploitation and oppression of the proletariat (the workers) by the bourgeoisie (the owners or capitalist class). Max Weber recognized the importance of economic conditions in producing inequality and conflict in society, but he added *power* and *prestige* as other sources of inequality. Weber (1968/1922) defined *power* as the ability of a person within a social relationship to carry out his or her

own will despite resistance from others, and *prestige* as a positive or negative social estimation of honor (Weber, 1968/1922).

C. Wright Mills (1916–1962), a key figure in the development of contemporary conflict theory, encouraged sociologists to get involved in social reform. He contended that value-free sociology was impossible because social scientists must make value-related choices—including the topics they investigate and the theoretical approaches they adopt. Mills encouraged everyone to look beneath everyday events in order to observe the major resource and power inequalities that exist in society. He believed that the most important decisions in the United States are made largely behind the scenes by the *power elite*—a small clique composed of top corporate, political, and military officials. Mills's power elite theory is discussed in Chapter 13 ("Politics and the Economy in Global Perspective").

The conflict perspective is not one unified theory but rather encompasses several branches. One branch is the neo-Marxist approach, which views struggle between the classes as inevitable and as a prime source of social change. A second branch focuses on racial–ethnic inequalities and the continued exploitation of members of some racial–ethnic groups. A third branch is the feminist perspective, which focuses on gender issues (Feagin and Feagin, 1997).

THE FEMINIST APPROACH A feminist theoretical approach (or "feminism") directs attention to women's experiences and the importance of gender as an element of social structure. This approach is based on the belief that "women and men are equal and should be equally valued as well as have equal rights" (Basow, 1992). According to feminists (including many men as well as women), we live in a *patriarchy*, a system in which men dominate women and in which things that are considered to be "male" or "masculine" are more highly valued than those considered to be "female" or "feminine." The feminist perspective assumes that gender is socially created, rather than determined by one's biological inheritance, and that change is essential in order for people to achieve their human potential without limits based on gender. It also assumes that society reinforces social expectations through social learning, which is acquired through social institutions such as education, religion, and the political and economic structure of society. Some feminists argue that women's subordination can end only after the patriarchal system becomes obsolete. However, note that feminism is not one single, unified approach. Rather, there are several feminist perspectives, which are discussed in Chapter 10 ("Sex and Gender").

APPLYING CONFLICT PERSPECTIVES TO SUICIDE How might advocates of a conflict approach explain suicide among teenagers and young people?

Social Class Although many other factors may be present, social class pressures may affect rates of suicide among young people. According to some social analysts using a conflict framework, certain U.S. teenagers may perceive that they have no future because they see few educational or employment opportunities in our technologically oriented society. As a result, young people from low-income or working-class family backgrounds may perceive that they are among the

Reflections

Reflections 1.2

Why are class, gender, and race important factors in analyzing differences in suicide rates?

most powerless people in our society. However, class-based inequality alone cannot explain suicide pacts among teenagers in suburbs where there has been a rash of teenage suicides among affluent young people (Colt, 1991).

Gender In North America, females are more likely to attempt suicide, whereas males are more likely to actually take their own life. Analysts using a feminist perspective might be interested in determining why women, more than men, are likely to attempt suicide when they are young (under thirty years of age) and socioeconomically disadvantaged. Despite the fact that women's suicidal behavior has traditionally been attributed to problems in their interpersonal relationships, such as loss of a boyfriend, lover, or husband, some analysts believe that we must examine social structural pressures that are brought to bear on young women and how these may contribute to their behavior—for example, cultural assumptions about women and what their multiple roles should be in the family, education, and the workplace. Women also experience unequal educational and employment opportunities that may contribute to feelings of powerlessness and alienation. Recent research shows that there are persistent gender gaps in U.S. employment, politics, education, and other areas of social life that tend to adversely affect women more than men (Benokraitis and Feagin, 1995).

Race Racial subordination may be a factor in many suicides. (Figure 1.2 displays U.S. suicides in terms of race and sex.) This fact is most glaringly

Conflict perspectives the sociological approach that views groups in society as engaged in a continuous power struggle for control of scarce resources.

Figure 1.2 Suicide Rates by Race and Sex

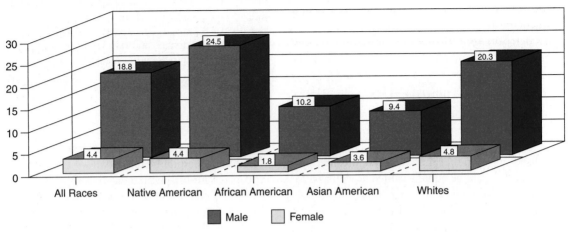

Rates are for U.S. suicides and indicate the number of deaths by suicide for every 100,000 people in each category for 1998 except for Native Americans and Asian Americans, where rates are for 1997.

Source: Computed by author based on National Center for Health Statistics, 1999: National Center for Injury Prevention and Control, 1999.

reflected in the extremely high rate of suicide among Native Americans, who constitute less than one-half of one percent of the U.S. population. The rate of suicide among young Native American males on government-owned reservations is especially high and usually involves particularly lethal methods such as shotguns and hanging (U.S. Congress, 1990). Most research has focused on individualistic reasons why young Native Americans commit suicide; however, analysts using a race-and-ethnic framework focus on the effect of social inequalities and racial discrimination on suicidal behavior. For example, researchers studying suicide among the Inuit (formerly known as Eskimos) of Alaska, Canada, and Greenland concluded that suicide rates are linked to the cultural conflict between traditional Inuit culture and the dominant culture in which Inuit must live in order to have educational or economic opportunities (Colt, 1991). However, some Native Americans remain optimistic by holding on to their most cherished traditions and ways (Allen, 1997).

Symbolic Interactionist Perspectives

The conflict and functionalist perspectives have been criticized for focusing primarily on macrolevel analysis. A *macrolevel analysis* examines whole societies, large-scale social structures, and social systems instead of looking at important social dynamics in individuals' lives. Our third perspective, symbolic interactionism, fills this void by examining people's day-to-day interactions and their behavior in groups. Thus, symbolic interactionist approaches are based on a *microlevel analysis*, which focuses on small groups rather than large-scale social structures.

We can trace the origins of this perspective to the Chicago School, especially George Herbert Mead and the sociologist Herbert Blumer (1900–1986), who is credited with coining the term *symbolic interactionism*. According to *symbolic interactionist perspectives*, society is the sum of the interactions of individuals and groups. Theorists using this perspective focus on the process of *interaction*—defined as immediate reciprocally oriented communication between two or more people—and the part that symbols play in giving meaning to human communication. A *symbol* is anything that meaningfully represents something else. Examples of symbols include signs, gestures, written language, and shared values. Symbolic interaction occurs when people communicate through the use of symbols; for example, a gift of food—a cake or a casserole—to a newcomer in a neighborhood is a symbol of welcome and friendship. But symbolic communication occurs in a variety of forms, including facial gestures, posture, tone of voice, and other symbolic gestures (such as a handshake or a clinched fist).

Symbols are instrumental in helping people derive meanings from social interactions. In social encounters, each person's interpretation or definition of a given situation becomes a *subjective reality* from that person's viewpoint. We often assume that what we consider to be "reality" is shared by others; however, this assumption is often incorrect. Subjective reality is acquired and shared through agreed-upon symbols, especially language. If a person shouts "Fire!" in a crowded movie theater, for example, that language produces the same response (attempting to escape) in all of those who hear and understand it. When people in a group do not share the same meaning for a given symbol, however, confusion results: People who do not know the meaning of the word *fire* will not know what the commotion is about. How people *interpret* the messages they receive and the situations they encounter becomes their subjective reality and may strongly influence their behavior.

APPLYING SYMBOLIC INTERACTIONIST PERSPECTIVES TO SUICIDE Social analysts applying a symbolic interactionist framework to the study of suicide would focus on a microlevel analysis of the people's face-to-face interactions and the roles they played in society. In our efforts to interact with others, we define the situation according to our own subjective reality. This applies to suicide just as it does to other types of conduct.

From this point of view, a suicide attempt may be a way of moving toward other people—in the form of a cry for help and personal acceptance—rather than a move toward death (Colt, 1991). People may attempt to communicate in such desperate ways because other forms of communication have failed. For example, a thirteen-year-old Illinois girl slashed her wrists shortly before she knew her mother was expected to come into her room; she wanted to show her parents how upset and unhappy she was about their pending divorce. As the girl later said, "I didn't really want to die. I just hoped and prayed that if Mom and Dad knew how upset and unhappy I was, Dad would move back in" (Giffin and Felsenthal, 1983: 19).

Postmodern Perspectives

According to *postmodern perspectives*, existing theories have been unsuccessful in explaining social life in contemporary societies that are characterized by postindustrialization, consumerism, and global communications. Postmodern social theorists reject the theoretical perspectives we have previously discussed, as well as how those thinkers created the theories (Ritzer, 2000b). These theorists oppose the grand narratives that characterize modern thinking and believe that boundaries should not be placed on academic disciplines—such as philosophy, literature, art, and the social sciences—because much could be learned by sharing ideas.

Just as functionalist, conflict, and symbolic interactionist perspectives emerged in the aftermath of the Industrial Revolution, postmodern theories emerged after World War II (in the late 1940s) and reflected the belief that some nations were entering a period of postindustrialization. Postmodern (or "postindustrial") societies are characterized by an *information explosion* and an economy in which large numbers of people either provide or apply information, or are employed in service jobs (such as fast-food server or health care worker). There is a corresponding *rise of a consumer society* and the emergence of a *global village*, in which people around the world communicate with one another by electronic technologies such as television, telephone, fax, e-mail, and the Internet.

Today, postmodern theory remains an emerging perspective in the social sciences. How influential will this approach be? It remains to be seen what influence postmodern thinkers will have on

---

Macrolevel analysis an approach that examines whole societies, large-scale social structures, and social systems.

Microlevel analysis sociological theory and research that focus on small groups rather than on large-scale social structures.

Symbolic interactionist perspectives the sociological approach that views society as the sum of the interactions of individuals and groups.

Postmodern perspectives the sociological approach that attempts to explain social life in modern societies that are characterized by postindustrialization, consumerism, and global communications.

Concept Table 1.A The Major Theoretical Perspectives

Perspective	Analysis Level	View of Society
Functionalist	Macrolevel	Society is composed of interrelated parts that work together to maintain stability within society. This stability is threatened by dysfunctional acts and institutions.
Conflict	Macrolevel	Society is characterized by social inequality; social life is a struggle for scarce resources. Social arrangements benefit some groups at the expense of others.
Symbolic Interactionist	Microlevel	Society is the sum of the interactions of people and groups. Behavior is learned in interaction with other people; how people define a situation becomes the foundation for how they behave.
Postmodernist	Macrolevel/ Microlevel	Societies characterized by postindustrialization, consumerism, and global communications bring into question existing assumptions about social life and the nature of reality.

the social sciences. Although this approach opens up broad new avenues of inquiry by challenging existing perspectives and questioning current belief systems, it also tends to ignore many of the central social problems of our time—such as inequalities based on race, class, and gender, and global political and economic oppression (Ritzer, 2000b).

APPLYING POSTMODERN PERSPECTIVES TO SUICIDE   Do information technologies such as computers and the Internet bind people together, or do they create a world where people feel only tenuously involved in collective life and interpersonal relations? Such questions might be asked by postmodern analysts in a study of people's suicidal behavior.

According to some postmodern thinkers, we face a high-tech world in which images, ideas, and identity-forming material pass before our eyes at an accelerated pace. As a result, we are required to develop our own coherent narrative by which we understand ourselves and the social events that take place around us (Gergen, 1991; Harvey, 1989). In the aftermath of the 1997 Heaven's Gate mass suicide in California, for example, some social analysts highlighted what they perceived to be the role of the media, the Internet, and the World Wide Web in

that event. Media representations of the mass suicide often depicted the Internet and cyberspace as having *agency* (as possessing motives and being able to "act" as if they were human beings). Referring to the Heaven's Gate group as a "computer cult," *CNN Interactive* titled its story "The Internet as a God and Propaganda Tool for Cults." From a postmodern approach, representations such as this reflect a world in which individuals and the media may establish "fake" realities and pseudo-explanations in the absence of *real* knowledge about events or their causes. In future chapters, we use a postmodernist framework to examine topics such as cultural ideas, language use, issues of race/class/gender, the political economy, and individual self-conceptions.

Each of the four sociological perspectives we have examined involves different assumptions. Consequently, each leads us to ask different questions and to view the world somewhat differently. Different aspects of reality are the focus of each approach. Whereas functionalism emphasizes social cohesion and order, conflict approaches focus primarily on social tension and change. In contrast, symbolic interactionism primarily examines people's interactions and shared meanings in everyday life. Postmodernism challenges all of the other perspectives and questions current belief systems. Concept Table 1.A

✓Checkpoints

Checkpoint 1.4

13. What is the functionalist perspective?
14. What are manifest and latent functions?
15. How do conflict perspectives differ from functionalist approaches?
16. How do symbolic interactionist perspectives differ from other theoretical approaches?
17. What are postmodernist perspectives?

reviews all four of these perspectives. Throughout this book, we will be using these perspectives as lenses through which to view our social world.

The Sociological Research Process

Research is the process of systematically collecting information for the purpose of testing an existing theory or generating a new one. What is the relationship between sociological theory and research? The relationship between theory and research has been referred to as a continuous cycle, as shown in Figure 1.3 (Wallace, 1971).

Not all sociologists conduct research in the same manner. Some researchers primarily engage in quantitative research whereas others engage in qualitative research. With *quantitative research,* **the goal is scientific objectivity, and the focus is on data that can be measured numerically.** Quantitative research typically emphasizes complex statistical techniques. Most sociological studies on suicide have used quantitative research. They have compared rates of suicide with almost every conceivable variable, including age, sex, race/ethnicity, education, and even sports participation (see Lester, 1992). For example, researchers in one study examined the effects of church membership, divorce, and migration on suicide rates in the United States at various times between 1933 and 1980. The study concluded that suicide rates were higher where divorce rates were higher, migration higher, and church membership lower (see Breault, 1986). (The box on pages 24–25, "Understanding Statistical Data Presentations," explains how to read nu-

Figure 1.3 The Theory and Research Cycle

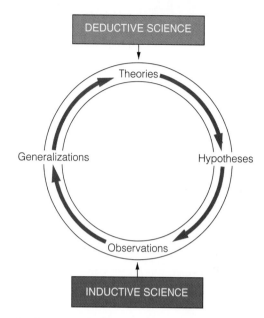

The theory and research cycle can be compared to a relay race; although all participants do not necessarily start or stop at the same point, they share a common goal—to examine all levels of social life.

Source: Adapted from Walter Wallace, *The Logic of Science in Sociology*, New York: Aldine de Gruyter, 1971.

merical tables, how to interpret the data and draw conclusions, and how to calculate ratios and rates.)

With *qualitative research,* **interpretive descriptions (words) rather than statistics (numbers) are used to analyze underlying meanings and patterns of social relationships.** An example of qualitative research is a study in which the researchers systematically analyzed the contents of suicide notes to look for recurring themes (such as feelings of despair or failure) in such notes to determine if any patterns

Quantitative research sociological research methods that are based on the goal of scientific objectivity and that focus on data that can be measured numerically.

Qualitative research sociological research methods that use interpretive description (words) rather than statistics (numbers) to analyze underlying meanings and patterns of social relationships.

Understanding Statistical Data Presentations

Are men or women more likely to commit suicide? Are suicide rates increasing or decreasing? Such questions may be answered in numerical terms. Sociologists often use statistical tables as a concise way to present data because such tables convey a large amount of information in a relatively small space; Table 1 gives an example. To understand a table, follow these steps:

1. *Read the title.* The title indicates the topic. From the title, "U.S. Suicides, by Sex and Method Used, 1970 and 1999," we learn that the table shows relationships between two variables: sex and method of suicide used. It also indicates that the table contains data for two different time periods: 1970 and 1999.
2. *Check the source and other explanatory notes.* In this case, the sources are the *U.S. Census Bureau, 2000c,* and *American Association of Suicidology, 2001.* Checking the source helps determine its reliability and timeliness. The first footnote indicates that the table includes only people who reside in the United States. The next two footnotes provide more information about exactly what is included in each category.
3. *Read the headings for each column and each row.* The main column headings in Table 1 are "Method," "Males," and "Females." These latter two column headings are divided into two groups: 1970 and 1999. The columns present information (usually

numbers) arranged vertically. The rows present information horizontally. Here, the row headings indicate suicide methods.
4. *Examine and compare the data.* To examine the data, determine what units of measurement have been used. In Table 1, the figures are numerical counts (for example, the total number of reported female suicides by poisoning in 1999 was 2,066) and percentages (for example, in 1999, poisoning accounted for 36.0 percent of all female suicides reported). A *percentage,* or proportion, shows how many of a given item there are in every one hundred. Percentages allow us to compare groups of different sizes. For example, percentages show the proportion of people who used each method, thus giving a more meaningful comparison.
5. *Draw conclusions.* By looking for patterns, some conclusions can be drawn from Table 1.
   a. *Determining the increase or decrease.* Between 1970 and 1999, reported male suicides increased from 16,629 to 23,458—an increase of 6,829 (but keep in mind that the overall male population also increased!)—while female suicides decreased by 1,110 (even as the female population increased). This represents a *total* increase (for males) and decrease (for females) in suicides for the two years being compared. The *amount* of increase or decrease can be stated as a percentage: Total male suicides were about 41.1 percent

could be found that would help in understanding why people kill themselves (Leenaars, 1988).

The "Conventional" Research Model

Research models are tailored to the specific problem being investigated and the focus of the researcher. Both quantitative research and qualitative research contribute to our knowledge of society and human social interaction, and involve a series of steps as shown in Figure 1.4. We will now trace the steps in the "conventional" research model, which focuses on quantitative research. Then we will describe an alternative model that emphasizes qualitative research.

1. *Select and define the research problem.* Sometimes, a specific experience such as knowing someone who committed suicide can trigger your interest in a topic. Other times, you might select topics to fill gaps or challenge misconceptions in existing research or to test a specific theory (Babbie, 2001). Emile Durkheim selected suicide because he wanted to demonstrate the importance of *society* in situations that might appear to be arbitrary acts by individuals. Suicide was a suitable topic because it was widely believed that suicide was a uniquely individualistic act. However, Durkheim emphasized that *suicide rates* provide better explanations for suicide than do *individual acts* of

Table 1 U.S. Suicides, by Sex and Method Used, 1970 and 1999^a

Method	Males		Females	
	1970	1999	1970	1999
Total	16,629	23,458	6,851	5,741
Firearms	9,704	14,479	2,068	2,120
(% of total)	(58.4)	(61.7)	(30.2)	(36.9)
Poisoning^b	3,299	2,827	3,285	2,066
(% of total)	(19.8)	(12.1)	(48.0)	(36.0)
Hanging/strangulation^c	2,422	4,490	831	937
(% of total)	(14.6)	(19.1)	(12.1)	(16.3)
Other	1,204	1,662	667	618
(% of total)	(7.2)	(7.1)	(9.7)	(10.8)

^{a}Excludes deaths of nonresidents of the United States.
^{b}Includes solids, liquids, and gases.
^{c}Includes suffocation.

Sources: U.S. Census Bureau, 2000c; American Association of Suicidology, 2001.

higher in 1999, calculated by dividing the total increase (6,829) by the earlier (lower) number. Total female suicides were about 16.2 percent lower in 1999, calculated by dividing the total decrease (1,110) by the earlier (higher) number.
b. *Drawing appropriate conclusions.* The number of female suicides by firearms increased about 2.5 percent between 1970 and 1999; the number for poisoning dropped by 37 percent. We might conclude that more women preferred firearms over poisoning as a means of killing themselves in 1999 than in 1970. Does that mean more women gained access to guns? That taking one's life with a firearm became more acceptable? Such generalizations do not take into account that we are looking at *fatal* suicides. If we had accurate data for *nonfatal* suicidal behavior (attempted suicide), we would likely find a higher rate for poisoning than for firearms.

suicide. He reasoned that if suicide were purely an individual act, then the rate of suicide (the relative number of people who kill themselves each year) should be the same for every group regardless of culture and social structure (see Box 2.2 for a current example). Moreover, Durkheim wanted to know why there were different rates of suicide—whether factors such as religion, marital status, sex, and age had an effect on social cohesion.

2. *Review previous research.* Before beginning the research, it is important to analyze what others have written about the topic. You should determine where gaps exist and note mistakes to avoid. When Durkheim began his

study, very little sociological literature existed to review; however, he studied the works of several moral philosophers, including Henry Morselli (1975/1881).

3. *Formulate the hypothesis* (if applicable). You may formulate a *hypothesis*—a statement of the expected relationship between two or more variables. A *variable* is any concept with

Variable in sociological research, any concept with measurable traits or characteristics that can change or vary from one person, time, situation, or society to another.

Figure 1.4 Steps in Sociological Research

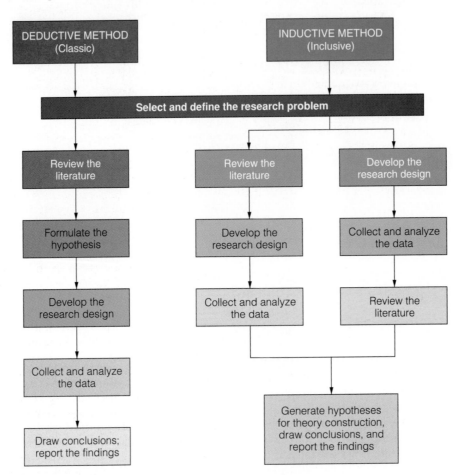

measurable traits or characteristics that can change or vary from one person, time, situation, or society to another. The most fundamental relationship in a hypothesis is between a dependent variable and one or more independent variables (see Figure 1.5). The *independent variable* is presumed to be the cause of the relationship; the *dependent variable* is assumed to be caused by the independent variable(s) (Babbie, 2001). Durkheim's hypothesis stated that the rate of suicide varies *inversely* with the degree of social integration. In other words, a low degree of social integration (the independent variable) may "cause" or "be related to" a high rate of suicide (the dependent variable).

Not all social research uses hypotheses. If you plan to conduct an explanatory study (showing a cause-and-effect relationship), you

likely will want to formulate one or more hypotheses to test theories. If you plan to conduct a descriptive study, however, you will be less likely to do so, since you may desire only to describe social reality or provide facts.

4. *Develop the research design.* You must determine the unit of analysis to be used in the study. A *unit of analysis* is *what* or *whom* is being studied (Babbie, 2001). In social science research, individuals, social groups (such as families, cities, or geographic regions), organizations (such as clubs, labor unions, or political parties), and social artifacts (such as books, paintings, or weddings) may be units of analysis. Durkheim's unit of analysis was social groups, not individuals, because he believed that the study of individual cases of suicide would not explain the rates of suicide in various European countries.

Box 1.2 Sociology in Global Perspective

Comparing Suicide Statistics from Different Nations

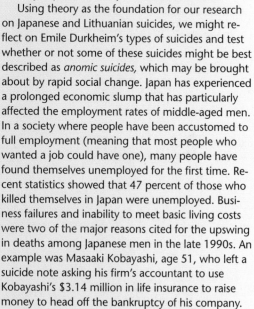

We're often told that this is the age of convergence. As global optimists see it, free-market capitalism, democratic norms, and maybe even a better appreciation for the sanctity of life are gaining ascendancy across the borders. In the rich world at least, you get the sense that a country's unique way of looking at the world will eventually be submerged by these big global trends.

There's some truth to that. But every once in a while here in Japan, I'm abruptly reminded that some things about this remarkable culture I'll never begin to fathom. To my mind the most fundamental one is the prevalence of suicide in a nation that boasts some of the highest living standards and longest life expectancies in the world.
—Brian Bremner (2000), Tokyo bureau chief for *Business Week*, stating his concern about the "suicide epidemic" in Japan

When sociologists select and define a research problem, they may look at a current social phenomenon such as statistical trends in suicides or how the rates of suicide compare across nations. If we look at the rates of suicide per 100,000 people in various countries, will these rates differ? In fact, the answer to this question is "yes." There is a wide disparity among suicide rates in various nations. For example, Lithuania has a suicide rate of 48 per 100,000 people, as compared with the United States, which has a suicide rate of 12 per 100,000 people. Can any patterns be identified in regard to Lithuanian suicides? One that has been identified is that the suicide rate has risen in Lithuania over the past three decades as that nation has gone through a lengthy period of economic, political, and social upheaval.

We might ask this: Is Lithuania's suicide rate increase due to that social upheaval? Japan has also had an upswing in the number of reported suicides, culminating in the late 1990s, when there was an unprecedented 35-percent surge in suicides. The suicide rate in Japan (24 per 100,000 people) is half of that of Lithuania but twice that of the United States. To put the suicide rate in Japan in perspective, that country has three times as many reported suicide victims as it does traffic fatalities. In 1999, men ac-

counted for 71 percent of all suicides in Japan, with 40 percent of the men being in their forties and fifties, and many of them having recently experienced major financial difficulties.

Using theory as the foundation for our research on Japanese and Lithuanian suicides, we might reflect on Emile Durkheim's types of suicides and test whether or not some of these suicides might be best described as *anomic suicides*, which may be brought about by rapid social change. Japan has experienced a prolonged economic slump that has particularly affected the employment rates of middle-aged men. In a society where people have been accustomed to full employment (meaning that most people who wanted a job could have one), many people have found themselves unemployed for the first time. Recent statistics showed that 47 percent of those who killed themselves in Japan were unemployed. Business failures and inability to meet basic living costs were two of the major reasons cited for the upswing in deaths among Japanese men in the late 1990s. An example was Masaaki Kobayashi, age 51, who left a suicide note asking his firm's accountant to use Kobayashi's $3.14 million in life insurance to raise money to head off the bankruptcy of his company.

Some social analysts have suggested that additional factors come into play in explaining the high rates of suicide in any nation. Although there is no consensus on the factors associated with suicide in Japan, some analysts attribute the problem to "cultural factors" such as a belief that the group is more important than the individual and that, under certain conditions, suicide is an honorable deed. Examples include the samurai warriors who committed hara-kiri and the kamikaze pilots of World War II. Other analysts believe that the two main Japanese religions—Shintoism and Buddhism—are related to the high rate of suicides because these religions have no moral prohibition on self killing. However, it is important for us to realize that these views about cultural differences as factors remain nothing more than assumptions until more social scientists conduct systematic research on these issues and are able to more conclusively demonstrate that cause-and-effect relationships exist. Which of Durkheim's types of suicide do you think might be applicable for a study of suicide in Japan?

Comparative Suicide Rates

Country	Rate
Lithuania	48
Hungary	38
Japan	24
U.S.	12

Suicides per 100,000 People

Source: *BBC News*, 1999.

Sources: Based on *BBC News*, 1999; Bremner, 2000; Lamar, 2000; and Lev, 1998.

Figure 1.5 Hypothesized Relationships Between Variables

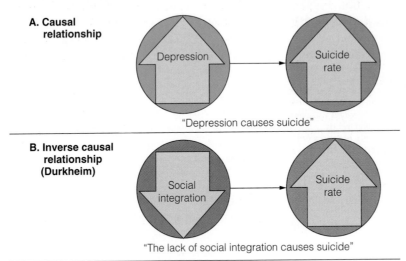

A. Causal relationship

Depression → Suicide rate

"Depression causes suicide"

B. Inverse causal relationship (Durkheim)

Social integration → Suicide rate

"The lack of social integration causes suicide"

A causal hypothesis connects one or more independent (causal) variables with a dependent (affected) variable. The diagram illustrates three hypotheses about the causes of suicide. To test these hypotheses, social scientists would need to operationalize the variables (define them in measurable terms) and then investigate whether the data support the proposed explanation.

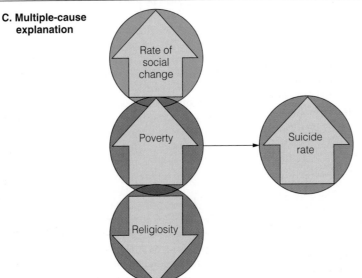

C. Multiple-cause explanation

Rate of social change / Poverty / Religiosity → Suicide rate

"Many factors interact to cause suicide"

5. *Collect and analyze the data.* You must decide what population will be observed or questioned and then carefully select a sample. A *sample* is the people who are selected from the population to be studied; the sample should accurately represent the larger population. A *representative sample* is a selection from a larger population that has the essential characteristics of the total population. For example, if you interviewed five students selected haphazardly from your sociology class, they would not be representative of your

school's total student body. By contrast, if you selected five students from the total student body by a random sample, they would be closer to being representative (although a random sample of five students would be too small to yield much useful data).

Validity and reliability may be problems in research. *Validity* **is the extent to which a study or research instrument accurately measures what it is supposed to measure.** A recurring issue in studies that analyze the relationship between religious beliefs and suicide is whether "church membership" is an accurate indicator of a person's religious beliefs. In fact, one person may be very religious yet not belong to a specific church, whereas another person may be a member of a church yet not hold very deep religious beliefs. *Reliability* **is the extent to which a study or research instrument yields consistent results** when applied to different individuals at one time or to the same individuals over time. Sociologists have found that different interviewers get different answers from the people being interviewed. For example, how might interviews with college students who have contemplated suicide be influenced by the interviewers themselves?

Once you have collected your data, it must be analyzed. *Analysis* is the process through which data are organized so that comparisons can be made and conclusions drawn. Sociologists use many techniques to analyze data. After collecting data from vital statistics for approximately 26,000 suicides, Durkheim analyzed his data according to four distinctive categories of suicide. *Egoistic suicide* occurs among people who are isolated from any social group. By contrast, *altruistic suicide* occurs among individuals who are excessively integrated into society (for example, military leaders who kill themselves after defeat in battle). *Anomic suicide* results from a lack of social regulation, whereas *fatalistic suicide* results from excessive regulation and oppressive discipline (for example, slaves).

6. *Draw conclusions and report the findings.* After analyzing the data, your first step in drawing conclusions is to return to your hypothesis or research objective to clarify how the data relate both to the hypothesis and to the larger issues being addressed. At this stage, you note the limitations of the study, such as problems with the sample, the influence of variables over which you had no control, or variables that your study was unable to measure.

Reporting the findings is the final stage. The report generally includes a review of each step taken in the research process in order to make the study available for *replication*—the repetition of the investigation in substantially the same way that it was originally conducted. Social scientists generally present their findings

in papers at professional meetings and publish them in technical journals and books. In reporting his findings in *Suicide* (1964b/1897), Durkheim concluded that the suicide rate of a group is a social fact that cannot be explained in terms of the personality traits of individuals.

We have traced the steps in the "conventional" research process (based on deduction and quantitative research). But what steps might be taken in an alternative approach based on induction and qualitative research?

A Qualitative Research Model

Although the same underlying logic is involved in both quantitative and qualitative sociological research, the *styles* of these two models are very different (King, Keohane, and Verba, 1994). As previously stated, qualitative research is more likely to be used when the research question does not easily lend itself to numbers and statistical methods. As compared to a quantitative model, a qualitative approach often involves a different type of research question and a smaller number of cases.

How might qualitative research be used to study suicidal behavior? In studying different rates of suicide among women and men, for example, the social psychologist Silvia Canetto (1992) questioned whether existing theories and quantitative research provided an adequate explanation for gender differences in suicidal behavior and decided that she would explore alternate explanations. Analyzing previous research, Canetto learned that most studies linked suicidal behavior in women to problems in their personal relationships, particularly with men. By contrast, most studies of men's suicides focused on their performance and found that men are more likely to be suicidal when their self-esteem and independence are threatened. According to Canetto's analysis, gender differences in

Validity in sociological research, the extent to which a study or research instrument accurately measures what it is supposed to measure.

Reliability in sociological research, the extent to which a study or research instrument yields consistent results when applied to different individuals at one time or to the same individuals over time.

Spencer Grant/Stock Boston

Qualitative researchers often generate their own data by using research methods such as the focus group. In this photo, the focus group is being observed through a one-way mirror. What are the strengths of such an approach? What are the limitations?

suicidal behavior are more closely associated with beliefs about and cultural expectations for men and women rather than purely interpersonal crises.

As in Canetto's study, researchers using a qualitative approach may engage in *problem formulation* to clarify the research question and to formulate questions of concern and interest to people participating in the research (Reinharz, 1992). To create a research design for Canetto's study, we might start with the proposition that studies have attributed women's and men's suicidal behavior to the wrong causes. Next, we might decide to interview individuals who have attempted suicide. Our research design might develop a collaborative approach in which the participants are brought into the research design process, not just treated as passive objects to be studied (Reinharz, 1992).

Although Canetto did not gather data in her study, she reevaluated existing research, concluding that alternate explanations of women's and men's suicidal behavior are justified from existing data.

In a qualitative approach, the next step is collecting and analyzing data to assess the validity of the starting proposition. Data gathering is the foundation of the research. Researchers pursuing a qualitative approach tend to gather data in natural settings, such as where the person lives or works, rather than in a laboratory or other research setting. Data collection and analysis frequently occur concurrently, and the analysis draws heavily on the language of the persons studied, not the researcher.

Research Methods

How do sociologists know which research method to use? Are some approaches best for a particular problem? *Research methods* are specific strategies

✓ Checkpoints

Checkpoint 1.5

18. What are the steps in the conventional research model?
19. How does qualitative research differ from quantitative research?

Bill Aron/PhotoEdit

Conducting surveys and polls is an important means of gathering data from respondents. Some surveys take place on street corners; however, increasingly, such surveys are done by telephone, Internet, or other means.

or techniques for systematically conducting research. We will look at four of these methods: survey research, analysis of existing statistical data, field research, and experiments.

Survey Research

A *survey* is a poll in which the researcher gathers facts or attempts to determine the relationships among facts. Surveys are the most widely used research method in the social sciences because they make it possible to study things that are not directly observable—such as people's attitudes and beliefs—and to describe a population too large to observe directly (Babbie, 2001). Researchers frequently select a representative sample (a small group of respondents) from a larger population (the total group of people) to answer questions about their attitudes, opinions, or behavior. *Respondents* are people who provide data for analysis through interviews or questionnaires. The Gallup and Harris polls are among the most

widely known large-scale surveys; however, government agencies such as the U.S. Census Bureau conduct a variety of surveys as well.

Unlike many polls that use various methods of gaining a representative sample of the larger population, the Census Bureau attempts to gain information from all persons in the United States. The decennial census occurs every ten years, in the years ending in "0." The purpose of this census is to count the population and housing units of the entire United States. The population count determines how seats in the U.S. House of Representatives are apportioned; however, census figures are also used in formulating public policy and in planning and decision making in the private sector. The Census Bureau attempts to survey the

Survey a poll in which the researcher gathers facts or attempts to determine the relationships among facts.

CENSUS PROFILES

How People in the United States Self-Identify as to Race and Hispanic Origin

→ NOTE: Please answer BOTH Questions 5 and 6.

5. **Is this person Spanish/Hispanic/Latino?** *Mark* ⊠ *the "No" box if **not** Spanish/Hispanic/Latino.*

☐ No, not Spanish/Hispanic/Latino   ☐ Yes, Puerto Rican

☐ Yes, Mexican, Mexican Am., Chicano   ☐ Yes, Cuban

☐ Yes, other Spanish/Hispanic/Latino — *Print group.* ⌐

| | | | | | | | | | | | | | | | | | | | | |

6. **What is this person's race?** *Mark* ⊠ *one or more races* to indicate what this person considers himself/herself to be.

☐ White

☐ Black, African Am., or Negro

☐ American Indian or Alaska Native — *Print name of enrolled or principal tribe.* ⌐

| | | | | | | | | | | | | | | | | | | | | |

☐ Asian Indian   ☐ Japanese   ☐ Native Hawaiian

☐ Chinese   ☐ Korean   ☐ Guamanian or Chamorro

☐ Filipino   ☐ Vietnamese   ☐ Samoan

☐ Other Asian — *Print race.* ⌐   ☐ Other Pacific Islander — *Print race.* ⌐

| | | | | | | | | | | | | | | | | | | | | |

☐ Some other race — *Print race.* ⌐

| | | | | | | | | | | | | | | | | | | |

Shown above are two questions from the 2000 census. As you can see from Question 6, a person could mark one or more races; however (see below), nearly 98 percent of all respondents reported only one race. Most (75.1 percent) reported "white" alone, whereas 12.3 percent reported "black" or African American alone. It is more difficult to compare data pertaining to "race" with the data from Question 5 regarding Hispanic or Latino origin. As you can see, Latinos/as (now constituting about 12.5 percent of the population) can be of any race or of multiple races. Thus, it would be incorrect to assume that 24.8 percent of the population is either African American or Latino/a: There is some overlap between the two categories.

Race and Hispanic or Latino	Percent of Total Population
Race	
One race	97.6
White	75.1
Black or African American	12.3
American Indian or Alaska Native	0.9
Asian	3.6
Native Hawaiian and Other Pacific Islander	0.1
Some other race	5.5
Two or more races	2.4
Hispanic or Latino	
Hispanic or Latino	12.5
Not Hispanic or Latino	87.5

Source: U.S. Census Bureau, 2001d.

entire U.S. population by using two forms—a "short form" of questions asked of *all* respondents and a "long form" that contains additional questions asked of a *representative sample* of about one in six respondents. Statistics from the Census Bureau provide information that sociologists use in their research. An example is shown in the Census Profiles feature: "How People in the United States Self-Identify as to Race and Hispanic Origin." Note that because of recent changes in the methods used to collect data by the Census Bureau, information on race from the 2000 census is not directly comparable with data from earlier censuses.

Survey data are collected by using questionnaires and interviews. A *questionnaire* is a printed research instrument containing a series of items to which subjects respond. Items are often in the form of statements with which the respondent is asked to "agree" or "disagree." Questionnaires may be administered by interviewers in face-to-face encounters or by telephone, but the most commonly used technique is the *self-administered questionnaire*, which is either mailed to the respondent's home or administered to groups of respondents gathered at the same place at the same time. For example, the sociologist Kevin E. Early (1992) used survey data collected through questionnaires to test his hypothesis that suicide rates are lower among African Americans than among white Americans due to the influence of the black church. Data from questionnaires filled out by members of six African American churches in Florida supported Early's hypothesis that the church buffers some African Americans against harsh social forces—such as racism—that might otherwise lead to suicide.

Survey data may also be collected by interviews. An *interview* is a data collection encounter in which an interviewer asks the respondent questions and records the answers. Survey research often uses *structured interviews*, in which the interviewer asks questions from a standardized questionnaire. Structured interviews tend to produce uniform or replicable data that can be elicited time after time by different interviews. For example, in addition to surveying congregation members, Early (1992) conducted interviews with pastors of African American churches to determine the pastors' opinions about the extent to which the

Bob Daemmrich/The Image Works

Today, many researchers conduct secondary analysis of existing data that have been generated by sources such as the U.S. Census Bureau. Regional data banks, such as this one in Austin, Texas, collect and process data for the Census Bureau.

African American church reinforces values and beliefs that discourage suicide.

Survey research is useful in describing the characteristics of a large population without having to interview each person in that population. In recent years, computer technology has enhanced researchers' ability to do *multivariate analysis*—research involving more than two independent variables. For example, to assess the influence of religion on suicidal behavior among African Americans, a researcher might look at the effects of age, sex, income level, and other variables all at once to determine which of these independent variables influences suicide the most or least and how influential each variable is relative to the others. However, a weakness of survey research is the use of standardized questions; this approach tends to force respondents into categories in which they may or may not belong. Moreover, survey research relies on self-reported information, and some people may be less than truthful, particularly on emotionally charged issues such as suicide.

Secondary Analysis of Existing Data

In *secondary analysis*, **researchers use existing material and analyze data that were originally collected by others.** Existing data sources include public records, official reports of organizations or government agencies, and *raw data* collected by other researchers. For example, Durkheim used vital statistics (death records) that were originally collected for other purposes to examine the relationship among variables such as age, marital status, and the circumstances surrounding the person's suicide.

Interview a research method using a data collection encounter in which an interviewer asks the respondent questions and records the answers.

Secondary analysis a research method in which researchers use existing material and analyze data that were originally collected by others.

Addison Geary

Sociologist Elijah Anderson's fourteen-year study of two Philadelphia neighborhoods—one populated by low-income African Americans, the other racially mixed but becoming increasingly middle- to upper-income and white—is an example of ethnographic research.

Secondary analysis also includes ***content analysis***—**the systematic examination of cultural artifacts or various forms of communication to extract thematic data and draw conclusions about social life.** Among the materials studied are written records (such as books, diaries, poems, and graffiti), narratives and visual texts (such as movies, television shows, advertisements, and greeting cards), and material culture (such as music, art, and even garbage). In content analysis, researchers look for regular patterns, such as the frequency of suicide as a topic on television talk shows.

One strength of secondary analysis is that data are readily available and inexpensive. Another is that because the researcher often does not collect the data personally, the chances of bias may be reduced. In addition, the use of existing sources makes it possible to analyze longitudinal data (things that take place over a period of time or at several different points in time) to provide a historical context within which to locate original research. However, secondary analysis has inherent problems. For one thing, the researcher does not always know if the data are incomplete, unauthentic, or inaccurate.

Field Research

Field research is the study of social life in its natural setting: observing and interviewing people where they live, work, and play. Some kinds of behavior can be best studied by "being there"; a fuller understanding can be developed through observations, face-to-face discussions, and participation in events. Researchers use these methods to generate *qualitative* data: observations that are best described verbally rather than numerically.

Sociologists who are interested in observing social interaction as it occurs may use ***participant observation***—**the process of collecting systematic observations while being part of the activities of the group that the researcher is studying.** Participant observation generates more "inside" information than simply asking questions or observing from the outside. For example, to learn more about how coroners make a ruling of "suicide" in connection with a death and to analyze what (if any) effect such a ruling has on the accuracy of "official" suicide statistics, the sociologist Steve Taylor (1982) engaged in participant observation at a coroner's office over a six-month period. As he followed a number of cases from the initial report of death through the various stages of investigation, Taylor learned that it was important to "be around" so that he could listen to discussions and ask the coroners questions because intuition and guesswork play a large part in some decisions to rule a death as a suicide.

Another approach to field research is the *ethnography*—a detailed study of the life and activities of a group of people by researchers who may live with that group over a period of years (Feagin, Orum, and Sjoberg, 1991). Unlike participant observation, ethnographic studies usually take place over a longer period of time. For example, the sociologist Elijah Anderson (1990) conducted a study in two areas of a major city—one African American and low-income, the other racially mixed but becoming increasingly middle- to upper-income and white. As Anderson spent numerous hours on the streets, talking and listening to the people, he was able to document changes in residents' everyday lives brought about by increased drug abuse, loss of jobs, decreases in city services despite increases in taxes, and the eventual exodus of middle-income people from the central city.

Experiments

An *experiment* is a carefully designed situation in which the researcher studies the impact of certain variables on subjects' attitudes or behavior. Experiments are designed to create "real-life" situations, ideally under controlled circumstances, in which the influence of different variables can be modified and measured. Conventional experiments require that subjects be divided into two groups: an experimental group and a control group. The *experimental group* contains the subjects who are exposed to an independent variable (the experimental condition) to study its effect on

them. The *control group* contains the subjects who are not exposed to the independent variable. For example, the sociologist Arturo Biblarz and colleagues (1991) examined the effects of media violence and depictions of suicide on attitudes toward suicide by showing one group of subjects (an experimental group) a film about suicide, while a second (another experimental group) saw a film about violence, and a third (the control group) saw a film containing neither suicide nor violence. The research found some evidence that people exposed to suicidal acts or violence in the media may be more likely to demonstrate an emotional state favorable to suicidal behavior, particularly if they are already "at risk" for suicide.

Researchers may use experiments when they want to demonstrate that a cause-and-effect relationship exists between variables. In order to show that a change in one variable causes a change in another, three conditions must be satisfied: (1) a correlation between the two variables must be shown to exist (*correlation* exists when two variables are associated more frequently than could be expected by chance), (2) the independent variable must have occurred prior to the dependent variable, and (3) any change in the dependent variable must not have been due to an extraneous variable—one outside the stated hypothesis.

Content analysis the systematic examination of cultural artifacts or various forms of communication to extract thematic data and draw conclusions about social life.

Participant observation a research method in which researchers collect data while being part of the activities of the group being studied.

Ethnography a detailed study of the life and activities of a group of people by researchers who may live with that group over a period of years.

Experiment a research method involving a carefully designed situation in which the researcher studies the impact of certain variables on subjects' attitudes or behavior.

Correlation a relationship that exists when two variables are associated more frequently than could be expected by chance.

Reflections

Reflections 1.3

What ethical issues should sociologists consider when they are conducting a study of suicidal behavior?

The major advantage of an experiment is the researcher's control over the environment and the ability to isolate the experimental variable. Since many experiments require relatively little time and money and can be conducted with limited numbers of subjects, it is possible for researchers to replicate an experiment several times by using different groups of subjects (Babbie, 2001). Perhaps the greatest limitation of experiments is that they are artificial: Social processes that are set up by researchers or that take place in a laboratory setting are often not the same as real-life occurrences.

Ethical Issues in Sociological Research

The study of people ("human subjects") raises vital questions about ethical concerns in sociological research. Researchers are required to obtain written "informed consent" statements from the persons they study—but what constitutes "informed consent"? How do researchers protect the identity and confidentiality of their sources?

The American Sociological Association (ASA) *Code of Ethics* (1997) sets forth certain basic standards that sociologists must follow in conducting research. Among these standards are the following:

1. Researchers must endeavor to maintain objectivity and integrity in their research by disclosing their research findings in full and including all possible interpretations of the data (even when these interpretations do not support their own viewpoints).
2. Researchers must safeguard the participants' right to privacy and dignity while protecting them from harm.

3. Researchers must protect confidential information provided by participants, even when this information is not considered to be "privileged" (legally protected, as is the case between doctor and patient and between attorney and client) and legal pressure is applied to reveal this information.
4. Researchers must acknowledge research collaboration and assistance they receive from others and disclose all sources of financial support.

Sociologists are obligated to adhere to this code and to protect research participants; however, many ethical issues arise in conducting research. For example, the sociologist William Zellner (1978) wanted to look at fatal single-occupant automobile accidents to determine if some drivers were actually committing suicide. To examine this issue further, he sought to interview the family, friends, and acquaintances of persons killed in single-car crashes to determine if the deaths were possibly intentional. To recruit respondents, Zellner told them that he hoped the research would reduce the number of automobile accidents in the future. He did not mention that he suspected "autocide" might have occurred in the case of their friend or loved one. From his data, Zellner concluded that at least 12 percent of the fatal single-occupant crashes were suicides—and that these crashes sometimes also killed or critically injured other people as well. However, Zellner's research raised important research questions: Was his research unethical? Did he misrepresent the reasons for his study?

In this chapter, we have looked at how theory and research work together to provide us with in-

✓ Checkpoints

Checkpoint 1.6

20. What is survey research, and when is it most useful?
21. What are the central approaches used in (a) secondary analysis of data, (b) field research, and (c) experiments?
22. What is the ASA *Code of Ethics*?

Box 1.3 You Can Make a Difference
Responding to a Cry for Help

Chad felt that he knew Frank quite well. After all, they had been roommates for two years at State U. As a result, Chad was taken aback when Frank became very withdrawn, sleeping most of the day and mumbling about how unhappy he was. One evening, Chad began to wonder whether he needed to do something since Frank had begun to talk about "ending it all" and saying things like "the world will be better off without me." If you were in Chad's place, would you know the warning signs that you should look for? Do you know what you might do to help someone like Frank?

The American Suicide Foundation, a national nonprofit organization dedicated to funding research, education, and treatment programs for depression and suicide prevention, suggests that each of us should be aware of these warning signs of suicide:

- *Talking about it.* Be alert to such statements as "Everyone would be better off without me." Sometimes, individuals who are thinking about suicide speak as if they are saying good-bye.
- *Making plans.* The person may do such things as giving away valuable items, paying off debts, and otherwise "putting things in order."
- *Showing signs of depression.* Although most depressed people are not suicidal, most suicidal people are depressed. Serious depression tends to be expressed as a loss of pleasure or withdrawal from activities that a person has previously enjoyed. It is especially important to note whether five of the following symptoms are present almost every day for several weeks: change in appetite or weight, change in sleeping patterns, speaking or moving with unusual speed or slowness, loss of interest in usual activities, decrease in sexual drive, fatigue, feelings of guilt or worthlessness, and indecisiveness or inability to concentrate.

The possibility of suicide must be taken seriously: Most people who commit suicide give some warning to family members or friends. Instead of attempting to argue the person out of suicide or saying "You have so much to live for," let the person know that you care and understand, and that his or her problems can be solved. Urge the person to see a school counselor, a physician, or a mental health professional immediately. If you think the person is in imminent danger of committing suicide, you should take the person to an emergency room or a walk-in clinic at a psychiatric hospital. It is best to remain with the person until help is available.

For more information about suicide prevention, contact any of the following agencies:

- American Suicide Foundation, 1045 Park Avenue, New York, NY 10028. (800) 531-4477 or (212) 410-1111.
- SOS (Survivors of Suicide), The Link Counseling Center, 348 Mt. Vernon Highway, N.E., Sandy Springs, GA 30328. (404) 256-9797.

On the Internet:

- Suicide Awareness Voices of Education (**http://www.save.org**) is a resource index with links to other valuable resources, such as "Questions Most Frequently Asked on Suicide," "Symptoms of Depression and Danger Signs of Suicide," and "What to Do If Someone You Love Is Suicidal."
- American Association of Suicidology (**http://www.suicidology.org**) is another resource with links to topics such as "If You Are Considering Suicide," "Crisis Centers in Your Area," and "Understanding and Helping the Suicidal Person."

sights on human behavior. Theory and research are the "lifeblood" of sociology. Theory provides the framework for analysis; research provides opportunities for us to use our sociological imagination to generate new knowledge. Our challenge today is to find new ways of integrating knowledge and action and to include all people in the theory and research process in order to help fill the gaps in our existing knowledge about social life and how it is shaped by gender, race, class, age, and the broader social and cultural context in which everyday life occurs (Cancian, 1992). Each of us can and should find new ways to integrate knowledge and action into our daily lives (see Box 1.3).

Chapter Review

What is sociology, and how can it help us understand ourselves and others?
Sociology is the systematic study of human society and social interaction. We study sociology to understand how human behavior is shaped by group life and, in turn, how group life is affected by individuals. Our culture tends to emphasize individualism, and sociology pushes us to consider more complex connections between our personal lives and the larger world.

What is the sociological imagination?
According to C. Wright Mills, the sociological imagination helps us understand how seemingly personal troubles, such as suicide, are actually related to larger social forces. It is the ability to see the relationship between individual experiences and the larger society.

What factors contributed to the emergence of sociology as a discipline?
Industrialization and urbanization increased rapidly in the late eighteenth century, and social thinkers began to examine the consequences of these powerful forces. Auguste Comte coined the term *sociology* to describe a new science that would engage in the study of society.

What are the major contributions of early sociologists such as Durkheim, Marx, and Weber?
The ideas of Emile Durkheim, Karl Marx, and Max Weber helped lead the way to contemporary sociology. Durkheim argued that societies are built on social facts, that rapid social change produces strains in society, and that the loss of shared values and purpose can lead to a condition of anomie. Marx stressed that within society there is a continuous clash between the owners of the means of production and the workers, who have no choice but to sell their labor to others. According to Weber, sociology should be value free and people should become more aware of the role that bureaucracies play in daily life.

How did Simmel's perspective differ from that of other early sociologists?
Whereas other sociologists primarily focused on society as a whole, Simmel explored small social groups and argued that society was best seen as a web of patterned interactions among people.

What are the major contemporary sociological perspectives?
Functionalist perspectives assume that society is a stable, orderly system characterized by societal consensus. Conflict perspectives argue that society is a continuous power struggle among competing groups, often based on class, race, ethnicity, or gender. Symbolic interactionist perspectives focus on how people make sense of their everyday social interactions, which are made possible by the use of mutually understood symbols. By contrast, postmodern theorists believe that entirely new ways of examining social life are needed and that it is time to move beyond functionalist, conflict, and symbolic interactionist approaches.

How does quantitative research differ from qualitative research?
Quantitative research focuses on data that can be measured numerically (comparing rates of suicide, for example). Qualitative research focuses on interpretive description (words) rather than statistics to analyze underlying meanings and patterns of social relationships.

What are the key steps in the conventional research process?
A conventional research process based on deduction and the quantitative approach has these key steps: (1) selecting and defining the research problem; (2) reviewing previous research; (3) formulating the hypothesis, which involves constructing variables; (4) developing the research design; (5) collecting and analyzing the data; and (6) drawing conclusions and reporting the findings.

What steps are often taken by researchers using the qualitative approach?
A researcher taking the qualitative approach might (1) formulate the problem to be studied instead of creating a hypothesis, (2) collect and analyze the data, and (3) report the results.

What are the major types of research methods?
The main types of research methods are surveys, secondary analysis, field research, and experiments. Surveys are polls used to gather facts about people's attitudes, opinions, or behaviors; a representative sample of respondents provides data through questionnaires or interviews. In secondary analysis, researchers analyze existing data, such as a government census, or cultural artifacts, such as a diary. In field research, sociologists study social life in its natural setting through participant observation, interviews, and ethnography. Through experiments, researchers study the impact of certain variables on their subjects.

Key Terms

anomie 13
conflict perspectives 18
content analysis 33
correlation 35
ethnography 35
experiment 35
functionalist perspectives 16
high-income countries 5
industrialization 9
interview 32
latent functions 17
low-income countries 5
macrolevel analysis 20
manifest functions 17
microlevel analysis 20
middle-income countries 5
participant observation 34
positivism 10
postmodern perspectives 21
qualitative research 23
quantitative research 23
reliability 29
secondary analysis 33
social Darwinism 12
social facts 12
society 3
sociological imagination 4
sociology 3
survey 31
symbolic interactionist perspectives 20
theory 16
urbanization 9
validity 28
variable 25

Questions for Critical Thinking

1. What does C. Wright Mills mean when he says that the sociological imagination helps us "to grasp history and biography and the relations between the two within society"? (Mills, 1959b: 6).

2. As a sociologist, how would you remain objective yet still see the world as others see it? Would you make subjective decisions when trying to understand the perspectives of others?

3. Early social thinkers were concerned about stability in times of rapid change. In our more global world, is stability still a primary goal? Or is constant conflict important for the well-being of all humans? Use the conflict and functionalist perspectives to bolster your analysis.

Resources on the Internet

Chapter-Related Web Sites
The following web sites have been selected for their relevance to the topics in this chapter. These sites are among the more stable, but please note that web site addresses change frequently.

"Census Sampling Confusion"
http://www.sciencenews.org/sn_arc99/3_6_99/bob1.htm
The U.S. Census is different from almost all other kinds of social research in that a sample is not selected. The goal of the census is to count every single person living in the United States. Officials at the U.S. Census Bureau proposed using sampling procedures for the 2000 census in order to obtain a more accurate count. Their efforts were rejected by the U.S. Supreme Court. Read more about the story in Ivars Peterson's article at this site.

Sampling Theory
http://library.thinkquest.org/10030/8stsp.htm
For a more in-depth explanation of sampling procedures, visit this ThinkQuest site, which was designed by students.

Dead Sociologists' Index
http://www2.pfeiffer.edu/~Iridener/DSS/INDEX.HTML
The Dead Sociologists' Index contains biographical information as well as useful summaries of the ideas of the social theorists covered in this chapter.

Marxists.org Internet Archive
http://www.marxists.org
The Marxists Internet Archive contains original work from Marx and his philosophical/political soul mates.

"Qualitative Methods: Part Three"
http://www.ship.edu/~cgboeree/qualmeththree.html
This site contains a very interesting exploration and elaboration of qualitative research methodology along with some great activities and examples.

A Sociological Tour Through Cyberspace
http://www.trinity edu/mkearl/
Take a sociological tour through cyberspace and explore the numerous web sites related to a variety of topics, such as demography, gender, and mass media.

"Watch *Friends* and Improve Your Sociological Imagination"
http://www.polity.co.uk/giddens/pdfs/friends.pdf
This site is a fascinating combination of sociological methods and popular culture.

Everything Postmodern

http://www.ebbflux.com/postmodern/

This site is the most comprehensive list of links to post-modern web sites on the Internet.

American Sociological Association Code of Ethics

http://www.asanet.org/members/ecoderev.html

Explore the American Sociological Association's *Code of Ethics* at this site.

ASA Public Area

http://www.asanet.org/public/public.html

The American Sociological Association has a public area on its web site. From here, you can access helpful articles on such subjects as what sociology is, careers in sociology, and research data on the discipline. Check out the careers in the sociology link as well.

Companion Web Site for This Book

Virtual Society: The Wadsworth Sociology Resource Center

Visit **http://sociology.wadsworth.com** and click on the page for Kendall, *Sociology in Our Times: The Essentials*, Fourth Edition, to access a wide range of enrichment material to aid in your study of sociology. Click on the Student Resources section of the web site. Next, select from the pull-down menu the chapter that you are presently studying. Among the useful options for self-study are chapter objectives, flashcards, practical study tips, and practice tests for each chapter.

MicroCase Online

From the Virtual Society home page, click on MicroCase Online to access book-specific MicroCase exercises that allow you to further explore sociological issues and principles presented in the text.

InfoTrac College Edition

Another unique option available to you at the Student Resources section of the companion web site is InfoTrac College Edition, an online library with access to hundreds of scholarly and popular periodicals. Below are suggested search terms for this chapter. Results from these and other searches are found at the site.

- Search keywords: *class, race, and gender*. Look for articles that deal with these key sociological factors in combination. How many articles did your search produce? Note the range of articles that your search delivers.
- Search keyword: *August Comte*. The chapter discusses a number of influential thinkers in the development of sociology. Conduct a subject-guide search for one of these founders, Auguste Comte. Read the encyclopedia excerpt. Next, select another historical figure discussed in the text, and do the same.
- Search keyword: *postmodernism*. Try to locate at least two articles in which postmodernism is discussed from a sociological perspective as opposed to other uses of the term.

Virtual Explorations CD-ROM

Go to the Virtual Explorations CD-ROM to begin the interactive exercise for this chapter. This Virtual Exploration will introduce you to some of the exciting resources for sociology on the World Wide Web. You will be guided through an exercise that employs related web sites on the sociological perspective and the research process. Answer the questions, and e-mail your responses to your instructor.

Robert Neubecker/SIS

Culture

2

Culture and Society
Material Culture and Nonmaterial Culture
Cultural Universals

Components of Culture
Symbols
Language
Values
Norms

Technology, Cultural Change, and Diversity
Cultural Change
Cultural Diversity
Culture Shock
Ethnocentrism and Cultural Relativism

A Global Popular Culture?
High Culture and Popular Culture
Forms of Popular Culture

Sociological Analysis of Culture
Functionalist Perspectives
Conflict Perspectives
Symbolic Interactionist Perspectives
Postmodernist Perspectives

Culture in the Future

41

I got my first credit card, an American Express, a few weeks before the mall trip. . . . And let me confess up front that I have always loved shopping and buying things for myself, even when I couldn't afford them. *Especially* when I couldn't afford them. So my story of the mall trip is a story of pleasure, but also a story about denial—for I spent money I did not have, as usual, and I could only enjoy that while denying I was doing it. . . . I looked long and lovingly at a red and white plaid flannel dress—it would have been perfect with a white t-shirt and black docmartin boots. But it was $75.00, two weeks worth of groceries or several much-needed university press books. So I knew I could not buy any new clothes. [Instead,] I bought a video. . . .

Using my credit card was really easy. I just handed it over to the young man at the cash register, who had me sign a slip of paper. Then, the video was mine. . . . I think my urge to call this "losing my virginity" is just one more way to deny what actually happened to me. . . . Denial is a fairly predictable process—whatever it is that you repress to get your pleasure always comes back to haunt you. Credit card denial works like that: first you buy something for "free," then your monthly statement comes with an itemized list which tells you exactly what you'll have to pay for your pleasure. . . .

—Annalee Newitz (1993) describing her first experience using a credit card, which occurred when she was a University of California–Berkeley graduate student

Like millions of college students in the United States and other high-income nations, Annalee quickly learned both the liberating and constraining aspects of living in a "consumer society" where credit cards play a crucial role in everyday life. Sociologists are interested in studying the *consumer society,* which refers to a society, such as ours, in which discretionary consumption is a mass phenomenon among people across diverse income categories. In the consumer society, purchasing goods and services is not in the exclusive province of the rich or even the middle classes; people in all but the lowest income categories may spend extensive amounts of time, energy, and money shopping, while amassing larger credit card debts in the process (see Baudrillard, 1998/1970; Ritzer, 1995; Schor, 1999). Did you know that recent surveys show that people in this country go to shopping centers more often than they go to church? Are you aware that U.S. teenagers spend more time at malls than anywhere besides school or home? (Twitchell, 1999).

What all of this means to sociologists is that shopping and consumption are integral components of culture in the United States and other high-income countries. What is culture? *Culture* **is the knowledge, language, values, customs, and material objects that are passed from person to person and from one generation to the next in a human group or society.** As previously defined, a *society* is a large social grouping that occupies the same geographic territory and is subject to the same political authority and dominant cultural expectations. Whereas a society is composed of people, a culture is composed of ideas, behavior, and material possessions. Society and culture are interdependent; neither could exist without the other.

Although people throughout history have had to consume necessities such as food and water, in contemporary societies such as ours, consumption of both essential and nonessential goods and services has become

a way of life among those who are able to afford such activities—and even among some people who cannot afford them. According to sociologists, shopping and consumption are processes that extend beyond individual choices and are rooted in larger structural conditions in the social, political, and economic order. Our ability to engage in 24-hour-a-day consumption has been aided by the growth of "cathedrals of consumption," which include mega-shopping malls, Disney World-like theme parks, fast-food restaurants, home shopping television networks, cybermalls on the Internet, and the omnipresent credit card industry (see Baudrillard, 1998/1970; Gottdiener, 1997; Ritzer, 1999).

How are social relations and social meanings shaped by what people in a given society produce and how they consume? What national and worldwide social processes shape the production and consumption of goods, services, and information? In this chapter, we examine society and culture, with special attention to how material culture is related to our beliefs, values, and actions. We also analyze culture from functionalist, conflict, symbolic interactionist, and postmodern perspectives. Before reading on, test your knowledge of the relationship between culture and the consumption of goods and services by answering the questions in Box 2.1.

Questions and Issues

Chapter Focus Question: What part does culture play in shaping people and the social relations in which they participate?

What are the essential components of culture?

To what degree are we shaped by popular culture?

How do subcultures and countercultures reflect diversity within a society?

How do the various sociological perspectives view culture?

Box 2.1 Sociology and Everyday Life

How Much Do You Know About Consumption and Credit Cards?

TRUE FALSE

T   F   1. Less than half of all undergraduate students at four-year colleges have at least one credit card.

T   F   2. The average debt owed on undergraduate college students' credit cards is more than $2,000.

T   F   3. In the United States, it is illegal to offer incentives such as free T-shirts and Frisbees to encourage students to apply for credit cards.

T   F   4. Some consumer activist groups want Congress to adopt a measure requiring people under age 21 to get parental approval or show that they have sufficient income prior to obtaining a credit card.

T   F   5. More than one million people in this country file for bankruptcy each year.

T   F   6. If we added up everyone's credit card balances in the United States, we would find that the total amount owed is more than $555 billion.

T   F   7. More than 100,000 people under the age of 25 file for personal bankruptcy each year.

T   F   8. The sociology of consumption focuses primarily on how individuals determine what they are going to purchase and how much they will spend.

Answers on page 44.

Culture and Society

How important is culture in determining how people think and act on a daily basis? Simply stated, culture is essential for our individual survival and our communication with other people. We rely on culture because we are not born with the information we need to survive. We do not know how to take care of ourselves, how to behave, how to dress, what to eat, which gods to worship, or how to make or spend money. We must learn about culture through interaction, observation, and imitation in order to participate as members of the group (Samovar and Porter, 1991). Sharing a common culture with others simplifies day-to-day interactions. However, we must also understand other cultures and the world views therein.

Just as culture is essential for individuals, it is also fundamental for the survival of societies. Culture has been described as "the common denominator that makes the actions of individuals intelligible to the group" (Haviland, 1993: 30). Some system of rule making and enforcing necessarily exists in all societies. What would happen, for example, if *all* rules and laws in the United States suddenly disappeared? At a basic level, we need rules in order to navigate our bicycles and cars through traffic. At a more abstract level, we need laws to establish and protect our rights.

In order to survive, societies need rules about civility and tolerance toward others. We are not born knowing how to express certain kinds of feelings toward others. When a person shows kindness or hatred toward another individual, some people may say "Well, that's just human nature" when explaining this behavior. Such a statement is built on the

Culture   the knowledge, language, values, customs, and material objects that are passed from person to person and from one generation to the next in a human group or society.

Box 2.1

Answers to the Sociology Quiz on Consumption and Credit Cards

1. **False.** About 78 percent of college students have at least one credit card, and 32 percent have four or more cards.
2. **True.** The average debt on undergraduate college students' credit cards in 2001 was about $2,748.
3. **False.** Aggressive marketing of credit cards to college students and others is not illegal; the credit card industry routinely pays colleges and universities fees to rent tables for campus solicitations, and alumni groups offer "affinity" cards linked to the schools.
4. **True.** Organizations such as the Consumer Federation of America, the Consumers Union, and the U.S. Public Interest Research Group have pointed out the problem of credit card debt among college students and encouraged Congress to adopt a measure requiring age or income requirements on the issuance of credit cards.
5. **True.** About 1.3 million people in the United States file for bankruptcy each year.
6. **True.** The total would be about $555 billion, with $450 billion of that amount being owed on general-use cards such as Visa, Master-Card, Discovery, and American Express.
7. **True.** One study estimated that 120,000 people under the age of 25 filed for personal bankruptcy in the year 2000.
8. **False.** The sociology of consumption does not focus on individual decisions about purchasing products as much as it offers us a way to learn more about society as a whole, including how consumption affects people's social identities and the society in which we live.

Sources: Based on Associated Press, 1999; Hoover, 2001; Schor, 1999; and Sullivan, Warren and Westbrook, 2000.

---

assumption that what we do as human beings is determined by *nature* (our biological and genetic makeup) rather than *nurture* (our social environment)—in other words, that our behavior is instinctive. An *instinct* is an unlearned, biologically determined behavior pattern common to all members of a species that predictably occurs whenever certain environmental conditions exist. For example, spiders do not learn to build webs. They build webs because of instincts that are triggered by basic biological needs such as protection and reproduction.

Humans do not have instincts. What we most often think of as instinctive behavior can actually be attributed to reflexes and drives. A *reflex* is an unlearned, biologically determined involuntary response to some physical stimuli (such as a sneeze after breathing some pepper in through the nose or the blinking of an eye when a speck of dust gets in

it). *Drives* are unlearned, biologically determined impulses common to all members of a species that satisfy needs such as those for sleep, food, water, or sexual gratification. Reflexes and drives do not determine how people will behave in human societies; even the expression of these biological characteristics is channeled by culture. For example, we may be taught that the "appropriate" way to sneeze (an involuntary response) is to use a tissue or turn our head away from others (a learned response). Similarly, we may learn to sleep on mats or in beds. Most contemporary sociologists agree that culture and social learning, not nature, account for virtually all of our behavior patterns.

Since humans cannot rely on instincts in order to survive, culture is a "tool kit" for survival. According to the sociologist Ann Swidler (1986: 273), culture is a "tool kit of symbols, stories, rituals, and

Figure 2.1 Hand Gestures with Different Meanings in Other Societies

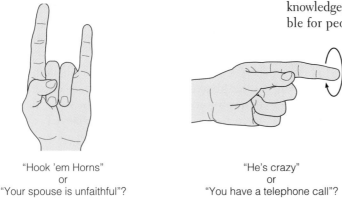

"Hook 'em Horns"
or
"Your spouse is unfaithful"?

"He's crazy"
or
"You have a telephone call"?

"Okay"
or
"I'll kill you"?

As international travelers and businesspeople have learned, hand gestures may have very different meanings in different cultures.

world views, which people may use in varying configurations to solve different kinds of problems." The tools we choose will vary according to our own personality and the situations we face. We are not puppets on a string; we make choices from among the items in our own "tool box."

Material Culture and Nonmaterial Culture

Our cultural tool box is divided into two major parts: material culture and nonmaterial culture (Ogburn, 1966/1922). *Material culture* **consists of the physical or tangible creations that members of a society make, use, and share.** Initially, items of material culture begin as raw materials or resources such as ore, trees, and oil. Through technology,

these raw materials are transformed into usable items (ranging from books and computers to guns and tanks). Sociologists define *technology* as the knowledge, techniques, and tools that make it possible for people to transform resources into usable forms, and the knowledge and skills required to use them after they are developed. From this standpoint, technology is both concrete and abstract. For example, technology includes a pair of scissors and the knowledge and skill necessary to make them from iron, carbon, and chromium (Westrum, 1991). At the most basic level, material culture is important because it is our buffer against the environment. For example, we create shelter to protect ourselves from the weather and to provide ourselves with privacy. Beyond the survival level, we make, use, and share objects that are interesting and important to us. Why are you wearing the particular clothes you have on today? Perhaps you're communicating something about yourself, such as where you attend school, what kind of music you like, or where you went on vacation.

Nonmaterial culture **consists of the abstract or intangible human creations of society that influence people's behavior.** Language, beliefs, values, rules of behavior, family patterns, and political systems are examples of nonmaterial culture. Even the gestures that we use in daily conversation are part of the nonmaterial culture in a society. As many international travelers and businesspeople have learned, it is important to know what gestures mean in various nations (see Figure 2.1). Although the "hook 'em Horns" sign—the pinky and index finger raised up and the middle two fingers folded down—is used by fans to express their support for University of Texas at Austin sports teams, for millions of

Material culture   a component of culture that consists of the physical or tangible creations (such as clothing, shelter, and art) that members of a society make, use, and share.

Nonmaterial culture   a component of culture that consists of the abstract or intangible human creations of society (such as attitudes, beliefs, and values) that influence people's behavior.

Reflections

Reflections 2.1

How does an emphasis on consumerism influence our attitudes about the types of material culture that we value the highest?

Italians the same gesture means "Your spouse is being unfaithful." In Argentina, rotating one's index finger around the front of the ear means "you have a telephone call," but in the United States it usually suggests that a person is "crazy" (Axtell, 1991). Similarly, making a circle with your thumb and index finger indicates "OK" in the United States, but in Tunisia it means "I'll kill you!" (Samovar and Porter, 1991).

As the example of hand gestures shows, a central component of nonmaterial culture is *beliefs*—the mental acceptance or conviction that certain things are true or real. Beliefs may be based on tradition, faith, experience, scientific research, or some combination of these. Faith in a supreme being and trust in another person are examples of beliefs. We may also have a belief in items of material culture. When we travel by airplane, for instance, we believe that it is possible to fly at 33,000 feet and to arrive at our destination even though we know that we could not do this without the airplane itself.

Cultural Universals

Because all humans face the same basic needs (such as for food, clothing, and shelter), we engage in similar activities that contribute to our survival. Anthropologist George Murdock (1945: 124) compiled a list of over seventy *cultural universals*—**customs and practices that occur across all societies.** His categories included appearance (such as bodily adornment and hairstyles), activities (such as sports, dancing, games, joking, and visiting), social institutions (such as family, law, and religion), and customary practices (such as cooking, folklore, gift giving, and hospitality). Whereas these general customs and practices may be present in all cultures, their specific forms vary from one group to another and from one time to another within the same group. For example, although telling jokes may be a universal practice, what is considered to be a joke in one society may be an insult in another.

How do sociologists view cultural universals? In terms of their functions, cultural universals are useful because they ensure the smooth and continual operation of society (Radcliffe-Brown, 1952). A society must meet basic human needs by providing food, shelter, and some degree of safety for its members so that they will survive. Children and other new members (such as immigrants) must be taught the ways of the group. A society must also settle disputes and deal with people's emotions. All the while, the self-interest of individuals must be balanced with the needs of society as a whole. Cultural universals help fulfill these important functions of society.

From another perspective, however, cultural universals are not the result of functional necessity; these practices may have been *imposed* by members of one society on members of another. Similar customs and practices do not necessarily constitute cultural universals. They may be an indication that a conquering nation used its power to enforce certain types of behavior on those who were defeated (Sargent, 1987). Sociologists might ask questions such as "Who determines the dominant cultural patterns?" For example, although religion is a cultural universal, traditional religious practices of indigenous peoples (those who first live in an area) have often been repressed and even stamped out by subsequent settlers or conquerors who hold political and economic power over them. However, many people believe there is cause for optimism in the United States because the democratic ideas of this nation provide more guarantees of religious freedom than do some other nations. For example, Josif Shrayman, aged sixty-two, who recently became a U.S. citizen, believes that he has found freedom here that he never previously had as a Jew in Ukraine: "For Jews it's very good here. God has helped me. America is a free country. I celebrate

✓ Checkpoints

Checkpoint 2.1

1. Why is culture important in the daily lives of people?
2. Do sociologists believe that humans have instincts?
3. How does material culture differ from non-material culture?
4. How do sociologists view cultural universals?

Rob Nelson/Zuma

James Nelson/Ston—Getty Images

Jean-Marie Truchet/Stone—Getty Images

Shelter is a universal type of material culture, but it comes in a wide variety of shapes and forms. What might be some of the reasons for the similarities and differences you see in these cross-cultural examples?

the Jewish holidays. I go to synagogue every day. I eat kosher" (qtd. in Dugger, 1997: 27).

Components of Culture

Even though the specifics of individual cultures vary widely, all cultures have four common non-material cultural components: symbols, language,

values, and norms. These components contribute to both harmony and strife in a society.

Cultural universals customs and practices that occur across all societies.

Spencer Grant/PhotoEdit

PhotoEdit

Michael Greenlar/The Image Works

The customs and rituals associated with weddings are one example of nonmaterial culture. What can you infer about beliefs and attitudes concerning marriage in the societies represented by these photographs?

Symbols

A *symbol* **is anything that meaningfully represents something else.** Culture could not exist without symbols because there would be no shared meanings among people. Symbols can simultaneously produce loyalty and animosity, and love and hate. They help us communicate ideas such as love or patriotism because they express abstract concepts with visible objects.

For example, flags can stand for patriotism, nationalism, school spirit, or religious beliefs held by members of a group or society. They also can be a source of discord and strife among people, as evidenced by recent controversies over the Confederate flag, the banner of the southern states during the Civil War. To some people, this flag symbolizes the "Old South" or "American history."

To others, such as members of the Ku Klux Klan, it symbolizes "white supremacy." To still others, it symbolizes slavery. Beginning in the 1990s and continuing into the twenty-first century, the U.S. Senate and various state governments have been confronted with the issue of what to do about this emotion-ridden symbol. In 1993, when the United Daughters of the Confederacy asked Congress to renew the design patent on their logo (which is a laurel wreath encircling the Confederate flag), they asserted that the flag was part of their cultural heritage, not a racist symbol. Former U.S. Senator Carol Moseley-Braun, an African American, delivered the following argument against renewing the patent:

On this issue there can be no consensus. It is an outrage. It is an insult.

AP/Wide World Photos

Controversy continues over the use of symbols such as the Confederate flag. Members of the South Carolina Council of Conservative Citizens, shown here, were very displeased when the flag was removed from atop the dome of the state capitol in Columbia, South Carolina. Other groups applauded the removal of the flag, which they believed was a racist symbol. How would sociologists explain the divergent meanings of such a symbol?

It is absolutely unacceptable to me and to millions of Americans, black or white, that we would put the imprimatur of the United States Senate on a symbol of this kind of idea. . . .

This is no small matter. This is not a matter of little old ladies walking around doing good deeds. There is no reason why these little old ladies cannot do good deeds anyway. If they choose to wave the Confederate flag, that certainly is their right. . . .

[But] a flag that symbolized slavery should not be underwritten, underscored, adopted, approved by the United States Senate. (qtd. in Clymer, 1993: A10)

Other conflicts over use of the Confederate flag have occurred in Columbia, South Carolina, where it has recently been removed from the dome of the state capitol, where it flew for four decades (Tanner, 1994), and at the University of Mississippi, where one supporter described the symbolic nature of the flag as follows:

[L]et us freely admit that all symbols can be (and usually are) many things to many people. To adopt the notion that any and all displays of the Confederate flag are symbols of racism and slavery is not only ridiculous, it also displays an abysmal ignorance of U.S. history and a lack of understanding of human behavior. . . .

What then is the Confederate flag? It is many things to many people. It has been used (and shall continue to be used) by those who espouse evil endeavors. It has been used by racists, South and North, but it is also recognized by thousands of others as a symbol of the American South, a short-lived government, an era now "Gone With the Wind," and a cause for which thousands of their ancestors died. With our 20th-century eye-glasses, we may look back upon this time, with its scourge of human slavery, rhetoric about state sovereignty, federalism, etc., and easily judge a few of the issues. But we dishonor the memories and sacrifices (on both sides) of this time if we allow simplistic flapdoodle to obscure the real issues of that great conflict that divided but also forged us together as a great nation. (Ezell, 1993: B2)

To both of these people, the Confederate flag is a meaningful symbol for some value, belief, or institution. In an effort to compromise on this emotional issue, political leaders in several states have decided to move the Confederate battle flag from highly visible locations such as atop the state capitol dome in South Carolina to less prominent locations on or around public buildings.

Symbols can stand for love (a heart on a valentine), peace (a dove), or hate (a Nazi swastika), just as words can be used to convey these meanings. Symbols can also transmit other types of ideas. A siren is a symbol that denotes an emergency situation and sends the message to clear the way immediately. Gestures are also a symbolic form of communication—a movement of the head, body, or hands can express our ideas or feelings to others. For example, in the United States, pointing toward your chest with your thumb or finger is a symbol for "me."

Symbols affect our thoughts about class. For example, how a person is dressed or the kind of car that he or she drives is often at least subconsciously used as a measure of that individual's economic

Symbol anything that meaningfully represents something else.

Joe Pololio/Stone—Getty Images

Does the language people use influence their perception of reality? According to social scientists, the words we use strongly influence our relationships with other people. What kind of interaction is taking place in this photo?

standing or position. With regard to clothing, although many people wear casual clothes on a daily basis, where the clothing was purchased is sometimes used as a symbol of social status. Were the items purchased at K-Mart, Old Navy, Abercrombie & Fitch, or Saks Fifth Avenue? What indicators are there on the items of clothing—such as the Nike *swoosh*, some other logo, or a brand name—that indicate something about the status of the product? Automobiles and their logos are also symbols that have cultural meaning beyond the shopping environment in which they originate.

Symbols may also affect our beliefs about race and ethnicity. Although black and white are not truly colors at all, the symbolic meanings associated with these labels permeate society and affect everyone. English-language scholar Alison Lurie (1981: 184) suggests that it is incorrect to speak of "whites" and "blacks." She notes that "pinkish-tan persons . . . have designated themselves the 'White race' while affixing the term 'Black' [to people] whose skin is some shade of brown or gold." The result of this "semantic sleight of hand" has been the association of pinkish-tan skin with virtue and cleanliness, and "brown or golden skin with evil, dirt and danger" (Lurie, 1981: 184).

Language

Language is a set of symbols that expresses ideas and enables people to think and communicate with one another. Verbal (spoken) language and nonverbal (written or gestured) language help us describe reality. One of our most important human attributes is the ability to use language to share our experiences, feelings, and knowledge with others. Language can create visual images in our head, such as "the kittens look like little cotton balls" (Samovar and Porter, 1991). Language also allows people to distinguish themselves from outsiders and maintain group boundaries and solidarity (Farb, 1973).

Language is not solely a human characteristic. Other animals use sounds, gestures, touch, and smell to communicate with one another, but they use signals with fixed meanings that are limited to the immediate situation (the present) and that cannot encompass past or future situations. For example, chimpanzees can use elements of Standard American Sign Language and manipulate physical objects to make "sentences," but they are not physically endowed with the vocal apparatus needed to form the consonants required for verbal language. As a result, nonhuman animals cannot transmit the more complex aspects of culture to their offspring. Humans have a unique ability to manipulate symbols to express abstract concepts and rules and thus to create and transmit culture from one generation to the next.

Reflections

Reflections 2.2

Do the items we wear and the vehicles we drive become symbols of who we are or what we aspire to become?

LANGUAGE AND SOCIAL REALITY   Does language *create* or simply *communicate* reality? Anthropological linguists Edward Sapir and Benjamin Whorf have suggested that language not only expresses our thoughts and perceptions but also influences our perception of reality. According to the *Sapir–Whorf hypothesis*, **language shapes the view of reality of its speakers** (Whorf, 1956; Sapir, 1961). If people are able to think only through language, then language must precede thought.

If language actually shapes the reality we perceive and experience, then some aspects of the world are viewed as important and others are virtually neglected because people know the world only in terms of the vocabulary and grammar of their own language. For example, the Eskimo language has over twenty words associated with snow, making it possible for people to make subtle distinctions regarding different types of snowfalls. A language such as English, which does not make these distinctions, considers them less important.

If language does create reality, are we trapped by our language? Many social scientists agree that the Sapir–Whorf hypothesis overstates the relationship between language and our thoughts and behavior patterns. Although they acknowledge that language has many subtle meanings and that words used by people reflect their central concerns, most sociologists contend that language may *influence* our behavior and interpretation of social reality but does not *determine* it.

LANGUAGE AND GENDER   What is the relationship between language and gender? What cultural assumptions about women and men does language reflect? Scholars have suggested several ways in which language and gender are intertwined:

- The English language ignores women by using the masculine form to refer to human beings in general (Basow, 1992). For example, the word *man* is used generically in words like *chairman* and *mankind*, which allegedly include both men and women.
- Use of the pronouns *he* and *she* affects our thinking about gender. Pronouns show the gender of the person we *expect* to be in a particular occupation. For instance, nurses, secretaries, and schoolteachers are usually referred to as *she*, but doctors, engineers, electricians, and presidents are usually referred to as *he* (Baron, 1986).
- Words have positive connotations when relating to male power, prestige, and leadership; when related to women, they carry negative

overtones of weakness, inferiority, and immaturity (Epstein, 1988: 224). Table 2.1 shows how gender-based language reflects the traditional acceptance of men and women in certain positions, implying that the jobs are different when filled by women rather than men.

- A language-based predisposition to think about women in sexual terms reinforces the notion that women are sexual objects. Women are often described by terms such as *fox*, *broad*, *bitch*, *babe*, or *doll*, which ascribe childlike or even petlike characteristics to them. By contrast, men have performance pressures placed on them by being defined in terms of their sexual prowess, such as *dude*, *stud*, and *hunk* (Baker, 1993).

Gender in language has been debated and studied extensively in recent years, and some changes have occurred. The preference of many women to be called *Ms.* (rather than *Miss* or *Mrs.* in reference to their marital status) has received a degree of acceptance in public life and the media (Maggio, 1988). Many organizations and publications have established guidelines for the use of nonsexist language and have changed titles such as *chairman* to *chair* or *chairperson*. "Men Working" signs in many areas have been replaced with "People Working" (Epstein, 1988). Some occupations have been given "genderless" titles, such as *firefighter* or *flight attendant* (Maggio, 1988). Yet many people resist change, arguing that the English language is being ruined (Epstein, 1988). To develop a more inclusive and equitable society, many scholars suggest that a more inclusive language is needed (see Basow, 1992).

LANGUAGE, RACE, AND ETHNICITY   Language may create and reinforce our perceptions about race and ethnicity by transmitting preconceived ideas about the superiority of one category of people over another. Let's look at a few images

Language   a set of symbols that expresses ideas and enables people to think and communicate with one another.

Sapir–Whorf hypothesis   the proposition that language shapes the view of reality of its speakers.

Table 2.1 Language and Gender

Male Term	Female Term	Neutral Term
Teacher	Teacher	Teacher
Chairman	Chairwoman	Chair, chairperson
Congressman	Congresswoman	Representative
Policeman	Policewoman	Police officer
Fireman	Lady fireman	Firefighter
Airline steward	Airline stewardess	Flight attendant
Race car driver	Woman race car driver	Race car driver
Wrestler	Lady/woman wrestler	Wrestler
Professor	Female/woman professor	Professor
Doctor	Lady/woman doctor	Doctor
Bachelor	Spinster/old maid	Single person
Male prostitute	Prostitute	Prostitute
Welfare recipient	Welfare mother	Welfare recipient
Worker/employee	Working woman	Worker/employee
Janitor/maintenance man	Maid/cleaning lady	Custodial attendant

Sources: Adapted from Korsmeyer, 1981: 122; and Miller and Swift, 1991.

conveyed by words in the English language in regard to race/ethnicity:

- Words may have more than one meaning and create and reinforce negative images. Terms such as *blackhearted* (malevolent) and expressions such as *a black mark* (a detrimental fact) and *Chinaman's chance of success* (unlikely to succeed) associate the words *black* or *Chinaman* with negative associations and derogatory imagery. By contrast, expressions such as *that's white of you* and *the good guys wear white hats* reinforce positive associations with the color white.
- Overtly derogatory terms such as *nigger, kike, gook, honkey, chink, spic,* and other racial–ethnic slurs have been "popularized" in movies, music, comic routines, and so on. Such derogatory terms are often used in conjunction with physical threats against persons.
- Words are frequently used to create or reinforce perceptions about a group. For example,

Native Americans have been referred to as "savage" and "primitive," and African Americans have been described as "uncivilized," "cannibalistic," and "pagan."
- The "voice" of verbs may minimize or incorrectly identify the activities or achievements of people of color. For example, the use of the passive voice in the statement "African Americans *were given* the right to vote" ignores how African Americans *fought* for that right. Active-voice verbs may also inaccurately attribute achievements to people or groups. Some historians argue that cultural bias is shown by the very notion that "Columbus discovered America" — given that America was already inhabited by people who later became known as Native Americans (see Stannard, 1992; Takaki, 1993).

In addition to these concerns about the English language, problems also arise when more than one language is involved. Consider an incident in

Florida described by journalist Carlos Alberto Montaner, a native of Cuba:

> I was walking quietly with my wife on a sidewalk in Miami Beach. We were speaking Spanish, of course, because that is our language. Suddenly, we were accosted by a spry little old lady, wearing a baseball cap and sneakers, who told us: "Talk English. You are in the United States." She continued on her way at once, without stopping to see our reaction. The expression on her face, curiously, was not that of somebody performing a rude action, but of someone performing a sacred patriotic duty. (Montaner, 1992: 163)

In recent decades, the United States has experienced rapid changes in language and culture. Data gathered by the U.S. Census Bureau (see "Census Profiles: Languages Spoken in U.S. Households") indicate that although more than 82 percent of the people in this country speak only English at home, more than 17 percent speak a language other than English, with the largest portion (estimated to be slightly over 10 percent of the U.S. population) speaking Spanish at home. Previous studies have shown that French, German, Italian, Polish, Chinese, Tagalog, Korean, and Vietnamese are among the top languages other than English and Spanish spoken at home in U.S. households.

From the functionalist perspective, a shared language is essential to a common culture; language is a stabilizing force in society. This view was expressed by Mauro E. Mujica (1993: 101), chairman of the organization called U.S. English:

> I am proud of my heritage. Yet when I emigrated to the United States from Chile in 1965 to study architecture at Columbia University, I knew that to succeed I would have to adopt the language of my new home.
>
> As in the past, it is critical today for new immigrants to learn English as quickly as possible. And that's so they can benefit from the many economic opportunities that this land has to offer. I believe so much in this concept that when elected to head an organization that promotes the use of English, I eagerly accepted.

Language is an important means of cultural transmission. Through language, children learn about their cultural heritage and develop a sense of personal identity in relationship to their group. For example, Latinos/as in New Mexico and south Texas use *dichos*—proverbs or sayings that are unique to the Spanish language—as a means of

CENSUS PROFILES Languages Spoken in U.S. Households

a. Does this person speak a language other than English at home?
☐ Yes
☐ No → *Skip to 12*

b. What is this language?
| | | | | | | | | | | | | | | | | | | | | |

(For example: Korean, Italian, Spanish, Vietnamese)

c. How well does this person speak English?
☐ Very well
☐ Well
☐ Not well
☐ Not at all

Shown above is another question from the 2000 census. For a number of years, the U.S. Census Bureau has gathered data on the languages spoken in U.S. households. People who speak a language other than English at home are asked not only to indicate which other language they speak but also how well they speak English. Based on a sample of respondents, the Census Bureau released the following information regarding language. Do you think that changes in the languages spoken in this country will bring about other significant changes in U.S. culture? Why or why not?

	Total estimated percentage	Percentage of whom speak English "less than well"
English only	82.36	
Language other than English	17.64	43.44
Spanish	10.51	46.61
Other Indo-European Languages[a]	3.73	32.75
Asian & Pacific Islander languages	2.70	49.50
Other languages	0.70	29.47

[a]Most other languages spoken in Europe and many of those spoken in southwestern Asia and India.

Source: Calculated by the author based on U.S. Census Bureau, 2001b.

Stephen Ferry/Liaison—Getty Images

Rapid changes in language and culture in the United States are reflected in this street scene. How do functionalist and conflict theorists' views regarding language differ?

expressing themselves and as a reflection of their cultural heritage. Examples of *dichos* include *Anda tu camino sin ayuda de vecino* ("Walk your own road without the help of a neighbor") and *Amor de lejos es para pendejos* ("A long-distance romance is for fools"). *Dichos* are passed from generation to generation as a priceless verbal tradition whereby people can give advice or teach a lesson (Gandara, 1995).

Conflict theorists view language as a source of power and social control; language perpetuates inequalities between people and between groups because words are used (whether or not intentionally) to "keep people in their place." As the linguist Deborah Tannen (1993: B5) has suggested, "The devastating group hatreds that result in so much suffering in our own country and around the world are related in origin to the small intolerances in our everyday conversations—our readiness to attribute good intentions to ourselves and bad intentions to others." Language, then, is a reflection of our feelings and values.

Values

Values **are collective ideas about what is right or wrong, good or bad, and desirable or undesirable in a particular culture** (Williams, 1970). Values do not dictate which behaviors are appropriate and which ones are not, but they provide us with the criteria by which we evaluate people, objects, and events. Values typically come in pairs of positive and negative values, such as being brave or cowardly, hardworking or lazy. Since we use values to justify our behavior, we tend to defend them staunchly (Kluckhohn, 1961).

CORE AMERICAN VALUES   Do we have shared values in the United States? Sociologists disagree about the extent to which all people in this country share a core set of values. Functionalists tend to believe that shared values are essential for societies and have conducted most of the research on this issue. Sociologist Robin M. Williams, Jr. (1970), has identified ten core values as being important to people in the United States:

1. *Individualism.* People are responsible for their own success or failure. Individual ability and hard work are the keys to success.
2. *Achievement and success.* Personal achievement results from successful competition with others.
3. *Activity and work.* People who are industrious are praised for their achievement; those perceived as lazy are ridiculed.
4. *Science and technology.* People in the United States have a great deal of faith in science and technology. They expect scientific and technological advances ultimately to control nature, the aging process, and even death.
5. *Progress and material comfort.* The material comforts of life include not only basic necessities (such as adequate shelter, nutrition, and medical care) but also the goods and services that make life easier and more pleasant.
6. *Efficiency and practicality.* People want things to be bigger, better, and faster.
7. *Equality.* Overt class distinctions are rejected in the United States. However, "equality" has been defined as "equality of *opportunity*"—an assumed equal chance to achieve success— not as "equality of *outcome.*"
8. *Morality and humanitarianism.* Aiding others, especially following natural disasters (such as floods or hurricanes), is seen as a value.
9. *Freedom and liberty.* Individual freedom is highly valued in the United States. The idea of freedom includes the right to private ownership of property, the ability to engage in private enterprise, freedom of the press, and other freedoms that are considered to be "basic" rights.

Lou Dematteis/Reuters/CORBIS

Baseball star Barry Bonds continues to break home run records. Which core American values are reflected in sports such as baseball?

10. *Racism and group superiority.* People value their own racial or ethnic group above all others. Such feelings of superiority may lead to discrimination; slavery and segregation laws are classic examples. Many people also believe in the superiority of their country and that "the American way of life" is best.

As you can see from this list, some core values may contradict others.

VALUE CONTRADICTIONS All societies, including the United States, have value contradictions. *Value contradictions* are values that conflict with one another or are mutually exclusive (achieving one makes it difficult, if not impossible, to achieve another). Core values of morality and humanitarianism may conflict with values of individual achievement and success. For example, humanitarian values reflected in welfare and other government aid programs for people in need come into conflict with values emphasizing hard work and personal achievement.

IDEAL CULTURE VERSUS REAL CULTURE
What is the relationship between values and human behavior? Sociologists stress that a gap always exists between ideal culture and real culture in a society. *Ideal culture* refers to the values and standards of behavior that people in a society profess to hold. *Real culture* refers to the values and standards of behavior that people actually follow. For example, we may claim to be law-abiding (ideal cultural value) but smoke marijuana (real cultural behavior), or we may regularly drive over the speed limit but think of ourselves as "good citizens."

Most of us are not completely honest about how well we adhere to societal values. In a University of Arizona study known as the "Garbage Project," household waste was analyzed to determine

Values collective ideas about what is right or wrong, good or bad, and desirable or undesirable in a particular culture.

David Falconer/SuperStock

Crowded conditions exist around the world, yet certain norms prevail in everyday life. Is the behavior of the people in this Osaka, Japan, train station a reflection of formal or informal norms?

the rate of alcohol consumption in Tucson. People were asked about their level of alcohol consumption, and in some areas of the city, they reported very low levels of alcohol use. However, when these people's garbage was analyzed, researchers found that over 80 percent of those households consumed some beer, and more than half discarded eight or more empty beer cans a week (Haviland, 1993). Obviously, this study shows a discrepancy between ideal cultural values and people's actual behavior.

Norms

Values provide ideals or beliefs about behavior but do not state explicitly how we should behave. Norms, on the other hand, do have specific behavioral expectations. *Norms* are established rules of behavior or standards of conduct. *Prescriptive norms* state what behavior is appropriate or acceptable. For example, persons making a certain amount of money are expected to file a tax return and pay any taxes they owe. Norms based on custom direct us to open a door for a person carrying a heavy load. By contrast, *proscriptive norms* state what behavior is inappropriate or unacceptable. Laws that prohibit us from driving over the speed limit and "good manners" that preclude you from reading a newspaper

during class are examples. Prescriptive and proscriptive norms operate at all levels of society, from our everyday actions to the formulation of laws.

FORMAL AND INFORMAL NORMS Not all norms are of equal importance; those that are most crucial are formalized. *Formal norms* are written down and involve specific punishments for violators. Laws are the most common type of formal norms; they have been codified and may be enforced by sanctions. *Sanctions* are rewards for appropriate behavior or penalties for inappropriate behavior. Examples of *positive sanctions* include praise, honors, or medals for conformity to specific norms. *Negative sanctions* range from mild disapproval to the death penalty.

Norms considered to be less important are referred to as *informal norms*—unwritten standards of behavior understood by people who share a common identity. When individuals violate informal norms, other people may apply informal sanctions. *Informal sanctions* are not clearly defined and can be applied by any member of a group (such as frowning at someone or making a negative comment or gesture).

FOLKWAYS Norms are also classified according to their relative social importance. *Folkways* are informal norms or everyday customs that may be violated without serious consequences within a particular culture (Sumner, 1959/1906). They provide rules for conduct but are not considered to be essential to society's survival. In the United States, folkways include using underarm deodorant, brushing our teeth, and wearing appropriate clothing for a specific occasion. Often, folkways are not enforced; when they are enforced, the resulting sanctions tend to be informal and relatively mild.

MORES Other norms are considered to be highly essential to the stability of society. *Mores* are strongly held norms with moral and ethical connotations that may not be violated without serious consequences in a particular culture. Since mores (pronounced MOR-ays) are based on cultural values and are considered to be crucial to the well-being of the group, violators are subject to more severe negative sanctions (such as ridicule, loss of employment, or imprisonment) than are those who fail to adhere to folkways. The strongest mores are referred to as taboos. *Taboos* are mores

✓ Checkpoints

Checkpoint 2.2

5. What are symbols, and how do they affect our thoughts and beliefs?
6. How does language affect our perceptions of gender, race, and ethnicity?
7. What are the core American values?
8. What are the main types of norms?

so strong that their violation is considered to be extremely offensive and even unmentionable. Violation of taboos is punishable by the group or even, according to certain belief systems, by a supernatural force. The incest taboo, which prohibits sexual or marital relations between certain categories of kin, is an example of a nearly universal taboo.

LAWS *Laws* are formal, standardized norms that have been enacted by legislatures and are enforced by formal sanctions. Laws may be either civil or criminal. *Civil law* deals with disputes among persons or groups. Persons who lose civil suits may encounter negative sanctions such as having to pay compensation to the other party or being ordered to stop certain conduct. *Criminal law*, on the other hand, deals with public safety and well-being. When criminal laws are violated, fines and prison sentences are the most likely negative sanctions, although in some states the death penalty is handed down for certain major offenses.

Technology, Cultural Change, and Diversity

Cultures do not generally remain static. There are many forces working toward change and diversity. Some societies and individuals adapt to this change, whereas others suffer culture shock and succumb to ethnocentrism.

Cultural Change

Societies continually experience cultural change at both material and nonmaterial levels. Changes in technology continue to shape the material culture of society. *Technology* refers to the knowledge, techniques, and tools that allow people to transform resources into usable forms and the knowledge and skills required to use what is developed. Although most technological changes are primarily modifications of existing technology, *new technologies* refers to changes that make a significant difference in many people's lives. Examples of new technologies include the introduction of the printing press more than 500 years ago and the advent of computers and electronic communications in the twentieth century. The pace of technological change has increased rapidly in the past 150 years, as contrasted with the 4,000 years prior to that, during which humans advanced from digging sticks and hoes to the plow.

All parts of culture do not change at the same pace. When a change occurs in the material culture of a society, nonmaterial culture must adapt to that change. Frequently, this rate of change is uneven, resulting in a gap between the two. Sociologist William F. Ogburn (1966/1922) referred to this disparity as *cultural lag*—a gap between the

Norms   established rules of behavior or standards of conduct.

Sanctions   rewards for appropriate behavior or penalties for inappropriate behavior.

Folkways   informal norms or everyday customs that may be violated without serious consequences within a particular culture.

Mores   strongly held norms with moral and ethical connotations that may not be violated without serious consequences in a particular culture.

Taboos   mores so strong that their violation is considered to be extremely offensive and even unmentionable.

Laws   formal, standardized norms that have been enacted by legislatures and are enforced by formal sanctions.

Technology   the knowledge, techniques, and tools that allow people to transform resources into a usable form and the knowledge and skills required to use what is developed.

Cultural lag   William Ogburn's term for a gap between the technical development of a society (material culture) and its moral and legal institutions (nonmaterial culture).

technical development of a society and its moral and legal institutions. In other words, cultural lag occurs when material culture changes faster than nonmaterial culture, thus creating a lag between the two cultural components. For example, at the material cultural level, the personal computer and electronic coding have made it possible to create a unique health identifier for each person in the United States. Based on available technology (material culture), it would be possible to create a national data bank that included everyone's individual medical records from birth to death. Using this identifier, health providers and insurance companies could rapidly transfer medical records around the globe, and researchers could access unlimited data on people's diseases, test results, and treatments. However, the availability of this technology does not mean that it will be accepted by people who believe (nonmaterial culture) that such a national data bank constitutes an invasion of privacy and could easily be abused by others. The failure of nonmaterial culture to keep pace with material culture is linked to social conflict and societal problems. As in the above example, such changes are often set in motion by discovery, invention, and diffusion.

Discovery is the process of learning about something previously unknown or unrecognized. Historically, discovery involved unearthing natural elements or existing realities, such as "discovering" fire or the true shape of the earth. Today, discovery most often results from scientific research. For example, discovery of a polio vaccine virtually eliminated one of the major childhood diseases. A future discovery of a cure for cancer or the common cold could result in longer and more productive lives for many people.

As more discoveries have occurred, people have been able to reconfigure existing material and nonmaterial cultural items through invention. *Invention* is the process of reshaping existing cultural items into a new form. Guns, video games, airplanes, and First Amendment rights are examples of inventions that positively or negatively affect our lives today.

When diverse groups of people come into contact, they begin to adapt one another's discoveries, inventions, and ideas for their own use. *Diffusion* is the transmission of cultural items or social practices from one group or society to another through such

means as exploration, military endeavors, the media, tourism, and immigration. To illustrate, piñatas can be traced back to the twelfth century, when Marco Polo brought them back from China, where they were used to celebrate the springtime harvest, to Italy, where they were filled with costly gifts in a game played by the nobility. When the piñata traveled to Spain, it became part of Lenten traditions. In Mexico, it was used to celebrate the birth of the Aztec god Huitzilopochtli (Burciaga, 1993). Today, children in many countries squeal with excitement at parties as they swing a stick at a piñata. In today's "shrinking globe," cultural diffusion moves at a very rapid pace as countries continually seek new markets for their products.

Cultural Diversity

Cultural diversity refers to the wide range of cultural differences found between and within nations. Cultural diversity between countries may be the result of natural circumstances (such as climate and geography) or social circumstances (such as level of technology and composition of the population). Some nations—such as Sweden—are referred to as *homogeneous societies*, meaning that they include people who share a common culture and are typically from similar social, religious, political, and economic backgrounds. By contrast, other nations—including the United States—are referred to as *heterogeneous societies*, meaning that they include people who are dissimilar in regard to social characteristics such as religion, income, or race/ethnicity (see Figure 2.2).

Immigration contributes to cultural diversity in a society. Throughout its history, the United States has been a nation of immigrants. Over the past 175 years, more than 55 million "documented" (legal) immigrants have arrived here; innumerable people have also entered the country as

Reflections

Reflections 2.3

How are cultural change and cultural diversity reflected in consumer patterns in the United States and other nations?

Figure 2.2 Heterogeneity of U.S. Society

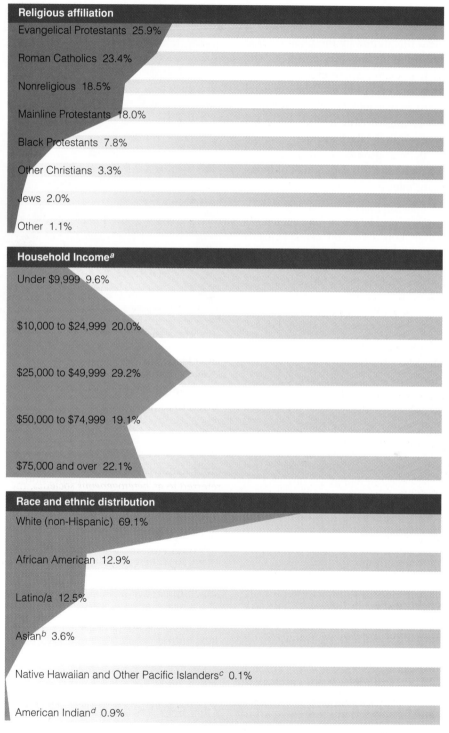

Religious affiliation
- Evangelical Protestants 25.9%
- Roman Catholics 23.4%
- Nonreligious 18.5%
- Mainline Protestants 18.0%
- Black Protestants 7.8%
- Other Christians 3.3%
- Jews 2.0%
- Other 1.1%

Household Income^a
- Under $9,999 9.6%
- $10,000 to $24,999 20.0%
- $25,000 to $49,999 29.2%
- $50,000 to $74,999 19.1%
- $75,000 and over 22.1%

Race and ethnic distribution
- White (non-Hispanic) 69.1%
- African American 12.9%
- Latino/a 12.5%
- Asian^b 3.6%
- Native Hawaiian and Other Pacific Islanders^c 0.1%
- American Indian^d 0.9%

Throughout history, the United States has been heterogenous. Today, we represent a wide diversity of social categories, including our religious affiliations, income levels, and racial–ethnic categories.

^{a}In Census Bureau terminology, a household consists of people who occupy a housing unit.
^{b}Includes Chinese, Filipino, Japanese, Asian Indian, Korean, Vietnamese, and other Asians.
^{c}Includes Native Hawaiian, Guamanian or Chamorro, Samoan, and other Pacific Islanders.
^{d}Includes American Indians, Eskimos, and Aleuts.

Source: U.S. Census Bureau, 2001g.

undocumented immigrants. Immigration can cause feelings of frustration and hostility, especially in people who feel threatened by the changes that large numbers of immigrants may produce (Mydans, 1993). Often, people are intolerant of those who are different from themselves. When societal tensions rise, people may look for others on whom they can place blame—or single out persons because they are the "other," the "outsider," the one who does not "belong." Ronald Takaki, an ethnic studies scholar, described his experience of being singled out as an "other":

> I had flown from San Francisco to Norfolk and was riding in a taxi to my hotel to attend a conference on multiculturalism. . . . My driver and I chatted about the weather and the tourists. . . . The rearview mirror reflected a white man in his forties. "How long have you been in this country?" he asked. "All my life," I replied, wincing. "I was born in the United States." With a strong southern drawl, he remarked: "I was wondering because your English is excellent!" Then, as I had many times before, I explained: "My grandfather came here from Japan in the 1880s. My family has been here, in America, for over a hundred years." He glanced at me in the mirror. Somehow I did not look "American" to him; my eyes and complexion looked foreign. (Takaki, 1993: 1)

Have you ever been made to feel like an "outsider"? Each of us receives cultural messages that may make us feel good or bad about ourselves or may give us the perception that we "belong" or "do not belong." Can people overcome such feelings in a culturally diverse society such as the United States? Some analysts believe it is possible to communicate with others despite differences in race, ethnicity, national origin, age, sexual orientation, religion, social class, occupation, leisure pursuits, regionalism, and so on. People who differ from the dominant group may also find reassurance and social support in a subculture or a counterculture.

SUBCULTURES A *subculture* is a category of **people who share distinguishing attributes, beliefs, values, and/or norms that set them apart in some significant manner from the dominant culture.** Emerging from the functionalist tradition, this concept has been applied to distinctions ranging from ethnic, religious, regional, and age-based categories to those categories presumed to be "de-

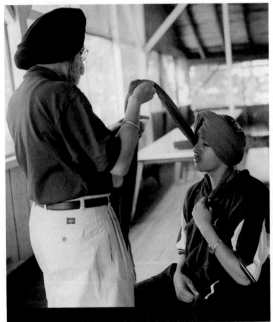

In heterogeneous societies such as the United States, people from very diverse cultures encourage their children to learn about their heritage. At Camp Robin Hood, a Sikh camp in Pennsylvania, children learn turban-tying and other practices of their religious tradition.

viant" or marginalized from the larger society. In the broadest use of the concept, thousands of categories of people residing in the United States might be classified as participants in one or more subcultures, including Native Americans, Muslims, Generation Xers, and motorcycle enthusiasts. However, many sociological studies of subcultures have limited the scope of inquiry to more visible, distinct subcultures such as the Old Order Amish and ethnic enclaves in large urban areas—to see how subcultural participants interact with the dominant U.S. culture.

The Old Order Amish Having arrived in the United States in the early 1700s, members of the Old Order Amish have fought to maintain their distinct identity. Today, over 75 percent of the more than 100,000 Amish live in Pennsylvania, Ohio, and Indiana, where they practice their religious beliefs and remain a relatively closed social network. According to sociologists, this religious

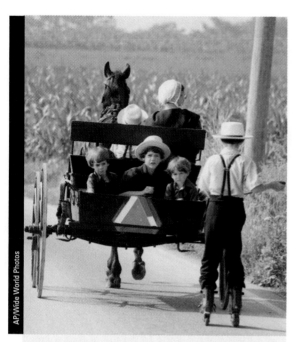

AP/Wide World Photos

Take a close look at the boy's feet as he follows the horse-drawn carriage of his Old Order Amish family. Are modernization and consumerism a threat to the way of life of subcultures such as the Amish? Why or why not?

community is a subculture because its members share values and norms that differ significantly from those of people who primarily identify with the dominant culture. The Amish have a strong faith in God and reject worldly concerns. Their core values include the joy of work, the primacy of the home, faithfulness, thriftiness, tradition, and humility. The Amish hold a conservative view of the family, believing that women are subordinate to men, birth control is unacceptable, and wives should remain at home. Children (about seven per family) are cherished and seen as an economic asset: They help with the farming and other work. Many of the Old Order Amish speak Pennsylvania Dutch (a dialect of German) as well as English. They dress in traditional clothing, live on farms, and rely on the horse and buggy for transportation.

The Amish are aware that they share distinctive values and look different from other people; these differences provide them with a collective identity and make them feel close to one another

(Kephart and Zellner, 1994). The belief system and group cohesiveness of the Amish remain strong despite the intrusion of corporations and tourists, the vanishing farmlands, and increasing levels of government regulation in their daily lives (Kephart and Zellner, 1994).

Ethnic Subcultures Some people who have unique shared behaviors linked to a common racial, language, or nationality background identify themselves as members of a specific subculture whereas others do not. Examples of ethnic subcultures include African Americans, Latinos/Latinas (Hispanic Americans), Asian Americans, and Native Americans. Some analysts include "white ethnics" such as Irish Americans, Italian Americans, and Polish Americans. Others also include Anglo Americans (Caucasians).

Although people in ethnic subcultures are dispersed throughout the United States, a concentration of members of some ethnic subcultures is visible in many larger communities and cities. For example, Chinatowns, located in cities such as San Francisco, Los Angeles, and New York, are one of the more visible ethnic subcultures in the United States. In San Francisco, over 100,000 Chinese Americans live in a twenty-four-block "city" within a city, which is the largest Chinese community outside of Asia. Traditionally, the core values of this subculture have included loyalty to others and respect for one's family. Obedience to parental authority, especially the father's, is expected, and sexual restraint and control over one's emotions in public are also highly valued.

By living close to one another and clinging to their original customs and language, first-generation immigrants can survive the abrupt changes they experience in material and nonmaterial cultural patterns. In New York City, for example, Korean Americans and Puerto Rican Americans constitute distinctive subcultures, each with its own food, music, and personal style. In San Antonio, Mexican Americans enjoy different food and music than do Puerto Rican Americans or other groups.

Subculture a group of people who share a distinctive set of cultural beliefs and behaviors that differs in some significant way from that of the larger society.

Even as global travel and the media make us more aware of people around the world, the distinctiveness of the Yanomamö in South America remains apparent. Are people today more or less likely than those in the past to experience culture shock upon encountering diverse groups of people such as these Yanomamö?

Gerhard Hümer/Contrast/Liaison—Getty Images

Subcultures provide opportunities for expression of distinctive lifestyles, as well as sometimes helping people adapt to abrupt cultural change. Subcultures can also serve as a buffer against the discrimination experienced by many ethnic or religious groups in the United States. However, some people may be forced by economic or social disadvantage to remain in such ethnic enclaves.

COUNTERCULTURES Some subcultures actively oppose the larger society. *A counterculture is a group that strongly rejects dominant societal values and norms and seeks alternative lifestyles* (Yinger, 1960, 1982). Young people are most likely to join countercultural groups, perhaps because younger persons generally have less invested in the existing culture. Examples of countercultures include the beatniks of the 1950s, the flower children of the 1960s, the drug enthusiasts of the 1970s, and members of nonmainstream religious sects, or cults. Some countercultures (such as the Ku Klux Klan, militias, neo-Nazi skinheads, and the Nation of Islam) engage in revolutionary political activities.

Culture Shock

Culture shock is the disorientation that people feel when they encounter cultures radically different from their own and believe they cannot depend on their own taken-for-granted assumptions about life. When people travel to another society, they may not know how to respond to that setting. For example, Napoleon Chagnon (1992) described his initial shock at seeing the Yanomamö (pronounced yah-noh-MAH-mah) tribe of South America on his first trip in 1964.

The Yanomamö (also referred to as the "Yanomami") are a tribe of about 20,000 South American Indians who live in the rain forest. Although Chagnon traveled in a small aluminum motorboat for three days to reach these people, he was not prepared for the sight that met his eyes when he arrived:

> I looked up and gasped to see a dozen burly, naked, sweaty, hideous men staring at us down the shafts of their drawn arrows. Immense wads of green tobacco were stuck between their lower teeth and lips, making them look even more hideous, and strands of dark-green slime dripped from their nostrils—strands so long that they reached down to their pectoral muscles or drizzled down their chins and stuck to their chests and bellies. We arrived as the men were blowing ebene, a hallucinogenic drug, up their noses. . . . I was horrified. What kind of welcome was this for someone who had come to live with these people and learn their way of life—to become friends with them? But when they recognized Barker [a guide], they put their weapons down and returned to their chanting, while keeping a nervous eye on the village entrances. (Chagnon, 1992: 12–14)

The Yanomamö have no written language, system of numbers, or calendar. They lead a nomadic lifestyle, carrying everything they own on their backs. They wear no clothes and paint their bodies; the women insert slender sticks through holes in the lower lip and through the pierced nasal septum. In other words, the Yanomamö—like the members of thousands of other cultures around the world—live in a culture very different from that of the United States.

Ethnocentrism and Cultural Relativism

When observing people from other cultures, many of us use our own culture as the yardstick by which we judge their behavior. Sociologists refer to this approach as *ethnocentrism*—**the practice of judging all other cultures by one's own culture** (Sumner, 1959/1906). Ethnocentrism is based on the assumption that one's own way of life is superior to all others. For example, most schoolchildren are taught that their own school and country are the best. The school song, the pledge to the flag, and the national anthem are forms of *positive ethnocentrism.* However, *negative ethnocentrism* can also result from constant emphasis on the superiority of one's own group or nation. Negative ethnocentrism is manifested in derogatory stereotypes that ridicule recent immigrants whose customs, dress, eating habits, or religious beliefs are markedly different from those of dominant-group members. Long-term U.S. residents who are members of racial and ethnic minority groups, such as Native Americans,

African Americans, and Latinas/os, have also been the target of ethnocentric practices by other groups.

An alternative to ethnocentrism is *cultural relativism*—**the belief that the behaviors and customs of any culture must be viewed and analyzed by the culture's own standards.** For example, the anthropologist Marvin Harris (1974, 1985) uses cultural relativism to explain why cattle, which are viewed as sacred, are not killed and eaten in India, where widespread hunger and malnutrition exist. From an ethnocentric viewpoint, we might conclude that cow worship is the cause of the hunger and poverty in India. However, according to Harris, the Hindu taboo against killing cattle is very important to their economic system. Live cows are more valuable than dead ones because they have more important uses than as a direct source of food. As part of the ecological system, cows consume grasses of little value to humans. Then they produce two valuable resources—oxen (the neutered offspring of cows) to power the plows and manure (for fuel and fertilizer)—as well as milk, floor covering, and leather. As Harris's study reveals, culture must be viewed from the standpoint of those who live in a particular society.

Cultural relativism also has a downside. It may be used to excuse customs and behavior (such as cannibalism) that may violate basic human rights. Cultural relativism is a part of the sociological imagination; researchers must be aware of the customs and norms of the society they are studying and then spell out their background assumptions so that others can spot possible biases in their studies. However, according to some social scientists, issues

✓ Checkpoints

Checkpoint 2.3

9. How does technology bring about cultural change?
10. What is cultural lag?
11. How do population changes contribute to cultural diversity?
12. What is a subculture, and what are several examples of subcultures?
13. When are people most likely to experience culture shock?
14. What is the difference between ethnocentrism and cultural relativism?

Counterculture   a group that strongly rejects dominant societal values and norms and seeks alternative lifestyles.

Culture shock   the disorientation that people feel when they encounter cultures radically different from their own and believe they cannot depend on their own taken-for-granted assumptions about life.

Ethnocentrism   the assumption that one's own culture and way of life are superior to all others.

Cultural relativism   the belief that the behaviors and customs of any culture must be viewed and analyzed by the culture's own standards.

surrounding ethnocentrism and cultural relativism may become less distinct in the future as people around the globe increasingly share a common popular culture. Others, of course, disagree with this perspective. Let's see what you think.

A Global Popular Culture?

Before taking this course, what was the first thing you thought about when you heard the term *culture*? In everyday life, culture is often used to describe the fine arts, literature, and classical music. When people say that a person is "cultured," they may mean that the individual has a highly developed sense of style or aesthetic appreciation of the "finer" things.

High Culture and Popular Culture

Some sociologists use the concepts of high culture and popular culture to distinguish between different cultural forms. *High culture* consists of classical music, opera, ballet, live theater, and other activities usually patronized by elite audiences, composed primarily of members of the upper-middle and upper classes, who have the time, money, and knowledge assumed to be necessary for its appreciation. In the United States, high culture is often viewed as being international in scope, arriving in this country through the process of diffusion, because many art forms originated in European nations or other countries of the world. By contrast, much of U.S. popular culture is often thought of as "homegrown" in this country. **Popular culture consists of activities, products, and services that are assumed to appeal primarily to members of the middle and working classes.** These include rock concerts, spectator sports, movies, and television soap operas and situation comedies. Although we will distinguish between "high" and "popular" culture in our discussion, it is important to note that some social analysts believe that the rise of a consumer society in which luxury items have become more widely accessible to the masses has greatly reduced the great divide between activities and possessions associated with wealthy people or a social elite (see Huyssen, 1984; Lash and Urry, 1994).

However, most sociological examinations of high culture and popular culture focus primarily on the link between culture and social class. French sociologist Pierre Bourdieu's (1984) *cultural capital theory* views high culture as a device used by the dominant class to exclude the subordinate classes. According to Bourdieu, people must be trained to appreciate and understand high culture. Individuals learn about high culture in upper-middle- and upper-class families and in elite education systems, especially higher education. Once they acquire this trained capacity, they possess a form of cultural capital. Persons from poor and working-class backgrounds typically do not acquire this cultural capital. Since knowledge and appreciation of high culture are considered a prerequisite for access to the dominant class, its members can use their cultural capital to deny access to subordinate-group members and thus preserve and reproduce the existing class structure (but see Halle, 1993).

Forms of Popular Culture

Three prevalent forms of popular culture are fads, fashions, and leisure activities. A *fad* is a temporary but widely copied activity followed enthusiastically by large numbers of people. Most fads are short-lived novelties (Garreau, 1993). According to the sociologist John Lofland (1993), fads can be divided into four major categories. First, *object fads* are items that people purchase despite the fact that they have little use or intrinsic value. Recent examples include *Star Wars Episode II: Attack of the Clones* action figures such as Jedi Master Yoda, other toys, trading cards, clothing, cartoons, and snack foods. Second, *activity fads* include pursuits such as body piercing, "surfing" the Internet, and fad diets where a person eats large quantities of one specific food such as grapefruit or popcorn. Third are *idea fads*, such as New Age ideologies. Fourth are *personality* fads, for example those surrounding celebrities such as Jennifer Lopez, Tiger Woods, Michael Jordan, and Brad Pitt.

A *fashion* is a currently valued style of behavior, thinking, or appearance that is longer lasting and more widespread than a fad. Examples of fashion are found in many areas, including child rearing, education, arts, clothing, music, and sports. Soccer is an example of a fashion in sports. Until recently, only schoolchildren played soccer in the United States. Now it has become a popular sport, perhaps in part because of immigration from Latin America and other areas of the world where soccer is widely played.

Sung Park

Activity fads such as crowd surfing are particularly popular with young people. Why are such fads often short-lived?

Like soccer, other forms of popular culture move across nations. In fact, popular culture is the United States' second largest export (after aircraft) to other nations (Rockwell, 1994). Of the world's 100 most-attended films in the 1990s, for example, 88 were produced by U.S.-based film companies. Likewise, music, television shows, novels, and street fashions from the United States have become a part of many other cultures. In turn, people in this country continue to be strongly influenced by popular culture from other nations. For example, contemporary music and clothing in the United States reflect African, Caribbean, and Asian cultural influences, among others.

Will the spread of popular culture produce a homogeneous global culture? Critics argue that the world is not developing a global culture; rather, other

✓ Checkpoints

Checkpoint 2.4

15. What are the differences between high culture and popular culture?
16. What are the three major forms of popular culture?

cultures are becoming westernized. Political and religious leaders in some nations oppose this process, which they view as *cultural imperialism* — the extensive infusion of one nation's culture into other nations (see Box 2.2). For example, some view the widespread infusion of the English language into countries that speak other languages as a form of cultural imperialism. On the other hand, the concept of cultural imperialism may fail to take into account various cross-cultural influences. For example, cultural diffusion of literature, music, clothing, and food has occurred on a global scale. A global culture, if it comes into existence, will most likely include components from many societies and cultures.

Sociological Analysis of Culture

Sociologists regard culture as a central ingredient in human behavior. Although all sociologists share a similar purpose, they typically see culture through somewhat different lenses as they are guided by different theoretical perspectives in their research. What do these perspectives tell us about culture?

Functionalist Perspectives

As previously discussed, functionalist perspectives are based on the assumption that society is a stable, orderly system with interrelated parts that

Popular culture the component of culture that consists of activities, products, and services that are assumed to appeal primarily to members of the middle and working classes.

Cultural imperialism the extensive infusion of one nation's culture into other nations.

Box 2.2 Sociology in Global Perspective

Popular Culture, the Internet, and Cultural Imperialism

- In Hanoi, a photographer in Lenin Park charges the equivalent of 50 cents (U.S. currency) for taking a person's picture standing beside a human-size Mickey Mouse doll. ("Mouse Makes the Man," 1995)
- Euro Disney, outside Paris, is the single largest tourist attraction in France, far surpassing the Eiffel Tower and the Louvre. (Kraft, 1994)
- During Moscow's December holiday season, parents can buy their children nonalcoholic champagne in bottles decorated with Disney's Pocahontas and the Lion King. (Hilsenrath, 1996)
- Ninety percent of worldwide traffic on the Internet is in English; the Internet is anchored in the United States, and the vast majority of web sites are based in this country. (Harvard Law School Seminar, 1996)

Are these examples of cultural imperialism and the "Americanization" of the world's cultures? According to Richard Kuisel (1993), *Americanization* is not the central issue when people around the globe go to U.S.-inspired amusement parks or surf the Internet. As Kuisel (1993: 4) states, "Although the phenomenon is still described as Americanization, it has become increasingly disconnected from America. Perhaps it would be better described as the coming of the consumer society." However, other global analysts disagree with Kuisel's assertion and suggest that we should think about how we would feel if we were in this situation:

What if one day you woke up, turned on the radio and could not find an American song? What if you went to the movies and the only films were foreign? What if you wanted to buy a book and you found that the only

American works were located in a small "Americana" section of the store? Furthermore, what if you came home to find your kids glued to the television for back-to-back reruns of a French soap opera and two German police dramas? It sounds foreign, even silly to an American. Ask a Canadian, a German, or a Greek the same question, however, and you will get a different reaction. (Harvard Law School Seminar, 1996: 1)

In nations such as France, which is extremely proud of its cultural identity, many people believe that the effects of cultural imperialism are evident in the types of entertainment and other cultural products available to them in everyday life. According to this view, a nation's *cultural products,* including its books, films, television programs, and other modes of communication, define its *identity.* As a result, some believe that "the Internet will have a huge impetus in the Anglification of the world and that English, currently the unofficial language of world commerce, will become that of world culture" (Harvard Law School Seminar, 1996: 5). Leaders in some nations have become concerned that the Internet constitutes a new tool of U.S. cultural imperialism and have formed groups such as France's "La Francophonie," an organization created to preserve the use of French in cyberspace.

Although the ideas and products of many nations have also permeated U.S. culture and influenced our consumption patterns, many social analysts believe that U.S. culture has had a greater influence on other nations' cultures than other nations' cultures have had on ours. Some social analysts believe that American culture is likely to become the "second culture" of people around the globe (Kuisel, 1993). Is the commercialized diffusion of popular culture beneficial to people in other nations? Or is it a form of cultural imperialism? What do you think?

---

serve specific functions. Anthropologist Bronislaw Malinowski (1922) suggested that culture helps people meet their *biological needs* (including food and procreation), *instrumental needs* (including law and education), and *integrative needs* (including religion and art). Societies in which people share a common language and core values are more likely to have consensus and harmony.

How might functionalist analysts view popular culture? According to many functionalist theo-

rists, popular culture serves a significant function in society in that it may be the "glue" that holds society together. Regardless of race, class, sex, age, or other characteristics, many people are brought together (at least in spirit) to cheer teams competing in major sporting events such as the Super Bowl or the Olympic Games. Television and the Internet help integrate recent immigrants into the mainstream culture, whereas longer-term residents may become more homogenized as a result of seeing the same images and being exposed to the same beliefs and values (Gerbner et al., 1987).

However, functionalists acknowledge that all societies have dysfunctions that produce a variety of societal problems. When a society contains numerous subcultures, discord results from a lack of consensus about core values. In fact, popular culture may undermine core cultural values rather than reinforce them (Christians, Rotzoll, and Fackler, 1987). For example, movies may glorify crime, rather than hard work, as the quickest way to get ahead. According to some analysts, excessive violence in music videos, movies, and television programs may be harmful to children and young people (Medved, 1992). From this perspective, popular culture may be a factor in antisocial behavior as seemingly diverse as hate crimes and fatal shootings in public schools.

A strength of the functionalist perspective on culture is its focus on the needs of society and the fact that stability is essential for society's continued survival. A shortcoming is its overemphasis on harmony and cooperation. This approach also fails to fully account for factors embedded in the structure of society—such as class-based inequalities, racism, and sexism—that may contribute to conflict among people in the United States or to global strife.

Conflict Perspectives

Conflict perspectives are based on the assumption that social life is a continuous struggle in which members of powerful groups seek to control scarce resources. According to this approach, values and norms help create and sustain the privileged position of the powerful in society while excluding others. As early conflict theorist Karl Marx stressed, ideas are *cultural creations* of a society's most powerful members. Thus, it is possible for political, economic, and social leaders to use *ideology*—an integrated system of ideas that is external to, and coercive of, people—to maintain their positions of dominance in a society. As Marx stated,

> The ideas of the ruling class are in every epoch the ruling ideas, i.e., the class which is the ruling material force in society, is at the same time, its ruling intellectual force. The class, which has the means of material production at its disposal, has control at the same time over the means of mental production. . . . The ruling ideas are nothing more than the ideal expression of the dominant material relationships, the

Reflections

Reflections 2.4

How does popular culture promote consumption of commodities?

> dominant material relationships grasped as ideas. (Marx and Engels, 1970/1845–46: 64)

Many contemporary conflict theorists agree with Marx's assertion that ideas, a nonmaterial component of culture, are used by agents of the ruling class to affect the thoughts and actions of members of other classes.

How might conflict theorists view popular culture? Some conflict theorists believe that popular culture, which originated with everyday people, has been largely removed from their domain and has become nothing more than a part of the capitalist economy in the United States (Gans, 1974; Cantor, 1980, 1987). From this approach, media conglomerates such as AOL Time Warner, Disney, and Viacom create popular culture, such as films, television shows, and amusement parks, in the same way that they would produce any other product or service. Creating new popular culture also promotes consumption of *commodities*—objects outside ourselves that we purchase to satisfy our human needs or wants (Fjellman, 1992). According to contemporary social analysts, consumption—even of things that we do not necessarily need—has become prevalent at all social levels, and some middle- and lower-income individuals and families now use as their frame of reference the lifestyles of the more affluent in their communities. As a result, many families live on credit in order to purchase the goods and services that they would like to have or that keep them on the competitive edge with their friends, neighbors, and co-workers (Schor, 1999). However, others may decide not to overspend, instead seeking to make changes in their lives and encouraging others to do likewise (see Box 2.3).

Other conflict theorists examine the intertwining relationship among race, gender, and popular culture. According to the sociologist K. Sue Jewell (1993), popular cultural images are often linked to negative stereotypes of people of color, particularly African American women. Jewell believes

Box 2.3 You Can Make a Difference
Taking a Stand Against Overspending

I never like what's already in my closet. I only like what I'm about to buy. . . . I'm thinking of converting my guest bedroom into one big closet. Who needs friends when you have clothes?
—actor Jennifer Tilly (quoted in *People,* 1999: 106–107)

My little girl had a friend visit her who was really into Guess jeans. My nine-year-old didn't even know what Guess jeans were. Well, after that kid left, that was all she talked about. She had to have a pair of Guess jeans.
—a mother (quoted in Schor, 1999: 68)

People's consumer desires and the urge to overspend are often fueled by others. Shopping and spending are routinely glorified in the media. A recent *People* magazine cover story touted "Hollywood's Super Shoppers," including celebrities such as Jennifer Tilly (quoted above), Arnold Schwarzenegger, and Charles Oakley, who like to "shop till they drop." Although celebrities are frequently used to "pitch" products in commercials, our family members and friends (such as the nine-year-old whose friend "sold" her on the Guess jeans) can be even stronger advocates for certain consumer goods and services.

Is it possible to take a stand against overconsumption and overspending? Recently, some people have been choosing *voluntary simplicity* or *downshifting* in an effort to simplify their lives, including reconsidering where they work, what they purchase, and how much they spend. Downshifting has been defined as "the practice of replacing busyness with business. It is the deliberate act of opting out of conspicuous consumption and possession of possessions which end up possessing you, and rejection of non-job sacrifices such as missing reading the children a story at night because of the pursuit of job promotions in the career rat race" (Buckingham, 1999). Many reassessments of life go into such decisions, including seeking a less stressful lifestyle, wanting more time to spend with family and friends, and having a concern about the environment. However, most "voluntary downshifters" also decide that they want to reduce the "clutter" (an excess of material possessions) in their lives and seek ways to decide to forgo excessive consumerism. According to the economist Juliet Schor (1999), there are several things that downshifters and others can do to avoid overspending, including (1) *controlling desire* by gaining knowledge of the process of consumption and its effect on people; (2) *helping make exclusivity uncool* by demystifying the belief that people are "better" for having purchased extraordinarily expensive items; (3) *exercising voluntary restraints on competitive consumption* by encouraging friends and acquaintances

to decelerate spending on presents and other purchases; (4) *learning to share,* particularly expensive items such as a lawn mower or a boat; (5) *becoming an educated consumer;* and (6) *avoiding the use of shopping as a form of therapy.* Since overspending is often linked with credit card use, other social analysts suggest that a person should have no more than one credit card and that the entire balance on the card should be paid off each month.

Can an individual make a difference by using these suggestions? Certainly, these ideas may change an individual's shopping and overspending habits. However, if we apply C. Wright Mills's (1959b) *sociological imagination,* including the distinction between personal troubles and public issues (as discussed in Chapter 1), we see that these suggestions focus exclusively on what *individuals* can do to change their own behavior. As a result, this approach may be somewhat useful but may still overlook the larger structural factors that contribute to people's overspending. According to the sociologist George Ritzer (1999), individual actions in this regard are likely to fail as long as there are no changes in the larger society, particularly in the cathedrals of consumption, advertisers, credit card companies, and other businesses that have a vested interest in promoting hyperconsumption. Some organizations suggest that the only way to reduce overconsumption and credit card debt is through activism, such as getting the age limit raised at which people can be issued their first credit card. In this view, those who want to make a difference could also become involved in advocating social change. If you would like to know more about the simplicity or downshifting movements, here are several organizations to contact:

- Center for a New American Dream, 6930 Carroll Avenue, Suite 900, Takoma Park, MD 20912.
- The Media Foundation, 1243 W. 7th Avenue, Vancouver, BC V6H 1B7, Canada. Online: **http://www.adbusters.org**
- Co-op America, 1612 K Street, Washington, DC 20006. Online: **http://www.coopamerica.org**

The following are web sites with links to other resources. Note that some may sell books or other products as well as providing information:

- Frugal Living Resources: **http://www.igc.apc.org/frugal**
- Living Lightly on the Earth: **http://www.scn.org/earth/lightly**

Sources: Based on Buckingham, 1999; *People,* 1999; Ritzer, 1999; and Schor, 1999.

that cultural images depicting African American women as mammies or domestics—such as those previously used in Aunt Jemima Pancake ads and recent resurrections of films like *Gone with the Wind*—affect contemporary black women's economic prospects in profound ways (Jewell, 1993).

A strength of the conflict perspective is that it stresses how cultural values and norms may perpetuate social inequalities. It also highlights the inevitability of change and the constant tension between those who want to maintain the status quo and those who desire change. A limitation is its focus on societal discord and the divisiveness of culture.

Symbolic Interactionist Perspectives

Unlike functionalists and conflict theorists, who focus primarily on macrolevel concerns, symbolic interactionists engage in a microlevel analysis that views society as the sum of all people's interactions. From this perspective, people create, maintain, and modify culture as they go about their everyday activities. Symbols make communication with others possible because they provide us with shared meanings.

According to some symbolic interactionists, people continually negotiate their social realities. Values and norms are not independent realities that automatically determine our behavior. Instead, we reinterpret them in each social situation we encounter. However, the classical sociologist Georg Simmel warned that the larger cultural world—including both material culture and nonmaterial culture—eventually takes on a life of its own apart from the actors who daily re-create social life. As a result, individuals may be more controlled by culture than they realize. Simmel (1990/1907) suggested that money is an example of how people may be controlled by their culture. According to Simmel, people initially create money as a means of exchange, but then money acquires a social meaning that extends beyond its purely economic function. Money becomes an end in itself, rather than a means to an end. Today, we are aware of the relative "worth" not only of objects but also of individuals. Many people revere wealthy entrepreneurs and highly paid celebrities, entertainers, and sports figures for the amount of money they make, not for their intrinsic qualities. According to Simmel (1990/1907), money makes it possible for us to *relativize* everything, including our relationships with other people. When social life can be reduced to money, people become cynical, believing that anything—including people,

Jim McHugh/CORBIS—Outline

During a visit to the Prada boutique in Beverly Hills, film star Jennifer Tilly admits that she views shopping as a mission. How might Georg Simmel's ideas about money be used to explain contemporary consumerism?

objects, beauty, and truth—can be bought if we can pay the price. Although Simmel acknowledged the positive functions of money, he believed that the social interpretations people give to money often produce individual feelings of cynicism and isolation.

A symbolic interactionist approach highlights how people maintain and change culture through their interactions with others. However, interactionism does not provide a systematic framework for analyzing how we shape culture and how it, in turn, shapes us. It also does not provide insight into how shared meanings are developed among people, and it does not take into account the many situations in which there is disagreement on meanings. Whereas the functional and conflict approaches tend to overemphasize the macrolevel workings of society, the interactionist viewpoint often fails to take these larger social structures into account.

Postmodernist Perspectives

Postmodernist theorists believe that much of what has been written about culture in the Western world is Eurocentric—that it is based on the uncritical

In recent years, there has been a significant increase in the number of immigrants who have become U.S. citizens. However, an upsurge in anti-immigrant sentiment has put pressure on the Border Patrol and the Immigration and Naturalization Service, which are charged with enforcing immigration laws.

assumption that European culture (including its dispersed versions in countries such as the United States, Australia, and South Africa) is the true, universal culture in which all the world's people ought to believe (Lemert, 1997). By contrast, postmodernists believe that we should speak of *cultures*, rather than *culture*.

However, Jean Baudrillard, one of the best-known French social theorists, believes that the world of culture today is based on *simulation*, not reality. According to Baudrillard, social life is much more a spectacle that simulates reality than reality itself. Many people gain "reality" from the media or cyberspace. For example, consider the many U.S. children who, upon entering school for the first time, have already watched more hours of television than the total number of hours of classroom instruction they will encounter in their entire school careers (Lemert, 1997). Add to this the number of hours that some will have spent playing computer games or surfing the Internet. Baudrillard refers to this social creation as *hyperreality*—a situation in which the *simulation* of reality is more real than the thing itself. For Baudrillard, everyday life has been captured by the signs and symbols generated to represent it, and we ultimately relate to simulations and models as if they were reality. Baudrillard (1983) uses Disneyland as an example

of a simulation that conceals the reality that exists outside rather than inside the boundaries of the artificial perimeter. According to Baudrillard, Disney-like theme parks constitute a form of seduction that substitutes symbolic (seductive) power for real power, particularly the ability to bring about social change. From this perspective, amusement park "guests" may feel like "survivors" after enduring the rapid speed and gravity-defying movements of the roller coaster rides or see themselves as "winners" after surviving fights with hideous cartoon villains on the "dark rides" when they have actually experienced the substitution of an *appearance* of power over their lives for the *absence* of real power. Similarly, the anthropologist Stephen M. Fjellman (1992) studied Disney World in Orlando, Florida, and noted that people may forget, at least briefly, that the outside world can be threatening while they stroll Disney World's streets without fear of crime or automobiles. Although this freedom may be temporarily empowering, it also may lull people into accepting a "worldview that presents an idealized United States as heaven. . . . How nice if they could all be like us—with kids, a dog, and General Electric appliances—in a world whose only problems are avoiding Captain Hook, the witch's apple, and Toad Hall weasels" (Fjellman, 1992: 317).

Concept Table 2.A Analysis of Culture

Components of Culture	Symbol	Anything that meaningfully represents something else.
	Language	A set of symbols that expresses ideas and enables people to think and communicate with one another.
	Values	Collective ideas about what is right or wrong, good or bad, and desirable or undesirable in a particular culture.
	Norms	Established rules of behavior or standards of conduct.
Sociological Analysis of Culture	Functionalist Perspectives	Culture helps people meet their biological, instrumental, and expressive needs.
	Conflict Perspectives	Ideas are a cultural creation of society's most powerful members and can be used by the ruling class to affect the thoughts and actions of members of other classes.
	Symbolic Interactionist Perspectives	People create, maintain, and modify culture during their everyday activities; however, cultural creations can take on a life of their own and end up controlling people.
	Postmodern Perspectives	Much of culture today is based on simulation of reality (e.g., what we see on television) rather than reality itself.

In their examination of culture, postmodernist social theorists make us aware of the fact that no single perspective can grasp the complexity and diversity of the social world. They also make us aware that reality may not be what it seems. According to the postmodernist view, no one authority can claim to know social reality, and we should deconstruct— take apart and subject to intense critical scrutiny— existing beliefs and theories about culture in hopes of gaining new insights (Ritzer, 1997).

Although postmodern theories of culture have been criticized on a number of grounds, we will mention only three. One criticism is postmodern-

ism's lack of a clear conceptualization of ideas. Another is the tendency to critique other perspectives as being "grand narratives," whereas postmodernists offer their own varieties of such narratives. Finally, some analysts believe that postmodern analyses of culture lead to profound pessimism about the future.

Concept Table 2.A reviews the components of culture as well as how the four major perspectives view culture.

Culture in the Future

As we have discussed in this chapter, many changes are occurring in the United States. Increasing cultural diversity can either cause long-simmering racial and ethnic antagonisms to come closer to a boiling point or result in the creation of a truly "rainbow culture" in which diversity is respected and encouraged.

In the future, the issue of cultural diversity will increase in importance, especially in schools. Multicultural education that focuses on the contributions of a wide variety of people from different backgrounds will continue to be an issue of controversy from kindergarten through college. In the Los Angeles school district, for example, students speak

✓**Checkpoints**

Checkpoint 2.5

17. How do each of the following perspectives view culture: (a) functionalist, (b) conflict, (c) symbolic interactionist, and (d) postmodernist?
18. What are the strengths and weaknesses of each of these viewpoints?

more than 114 different languages and dialects. Schools will face the challenge of embracing widespread cultural diversity while conveying a sense of community and national identity to students.

Technology will continue to have a profound effect on culture. Television and radio, films and videos, and electronic communications will continue to accelerate the flow of information and expand cultural diffusion throughout the world. Global communication devices will move images of people's lives, behavior, and fashions instantaneously among almost all nations. Increasingly, computers and cyberspace will become people's window on the world and, in the process, promote greater integration or fragmentation among nations. Integration occurs when there is a widespread acceptance of ideas and items—such as democracy, rock music, blue jeans, and McDonald's hamburgers—among cultures. By contrast, fragmentation occurs when people in one culture disdain the beliefs and actions of other cultures. As a force for both cultural integration and fragmentation, technology will continue to revolutionize communications, but most of the world's population will not participate in this revolution.

From a sociological perspective, the study of culture helps us not only understand our own "tool kit" of symbols, stories, rituals, and world views but also expand our insights to include those of other people of the world, who also seek strategies for enhancing their own lives. If we understand how culture is used by people, how cultural elements constrain or further certain patterns of action, what aspects of our cultural heritage have enduring effects on our action, and what specific historical changes undermine the validity of some cultural patterns and give rise to others, we can apply our sociological imagination not only to our own society but to the entire world as well (see Swidler, 1986).

Chapter Review

What is culture?
Culture is the knowledge, language, values, and customs passed from one generation to the next in a human group or society. Culture may be either material or nonmaterial. Material culture consists of the physical creations of society. Nonmaterial culture is more abstract and reflects the ideas, values, and beliefs of a society.

What are cultural universals?
Cultural universals are customs and practices that exist in all societies and include activities and institutions such as storytelling, families, and laws. However, specific forms of these universals vary from one cultural group to another.

What are the four nonmaterial components of culture that are common to all societies?
These components are symbols, language, values, and norms. Symbols express shared meanings; through them, groups communicate cultural ideas and abstract concepts. Language is a set of symbols through which groups communicate. Values are a culture's collective ideas about what is acceptable or not acceptable. Norms are the specific behavioral expectations within a culture.

What are the main types of norms?
Folkways are norms that express the everyday customs of a group, whereas mores are norms with strong moral and ethical connotations and are essential to the stability of a culture. Laws are formal, standardized norms that are enforced by formal sanctions.

What are high culture and popular culture?
High culture consists of classical music, opera, ballet, and other activities usually patronized by elite audiences. Popular culture consists of the activities, products, and services of a culture that appeal primarily to members of the middle and working classes.

What causes cultural change in societies?
Cultural change takes place in all societies. Change occurs through discovery and invention, and through diffusion, which is the transmission of culture from one society or group to another.

How is cultural diversity reflected in society?
Cultural diversity is reflected through race, ethnicity, age, sexual orientation, religion, occupation, and so forth. A diverse culture also includes subcultures and countercultures. A subculture has distinctive ideas and behaviors that differ from the larger society to which it belongs. A counterculture rejects the dominant societal values and norms.

What are culture shock, ethnocentrism, and cultural relativism?
Culture shock refers to the anxiety people experience when they encounter cultures radically different from their own. Ethnocentrism is the assumption that one's

own culture is superior to others. Cultural relativism views and analyzes another culture in terms of that culture's own values and standards.

How do the major sociological perspectives view culture?

A functionalist analysis of culture assumes that a common language and shared values help produce consensus and harmony. According to some conflict theorists, culture may be used by certain groups to maintain their privilege and exclude others from society's benefits. Symbolic interactionists suggest that people create, maintain, and modify culture as they go about their everyday activities. Postmodern thinkers believe that there are many cultures within the United States alone. In order to grasp a better understanding of how popular culture may simulate reality rather than be reality, postmodernists believe that we need a new way of conceptualizing culture and society.

Key Terms

counterculture 62
cultural imperialism 65
cultural lag 57
cultural relativism 63
cultural universals 46
culture 42
culture shock 62
ethnocentrism 63
folkways 56
language 50
laws 57
material culture 45
mores 56
nonmaterial culture 45
norms 56
popular culture 64
sanctions 56
Sapir–Whorf hypothesis 51
subculture 60
symbol 48
taboos 56
technology 57
values 54

Questions for Critical Thinking

1. Would it be possible today to live in a totally separate culture in the United States? Could you avoid all influences from the mainstream popular culture or from the values and norms of other cultures?

How would you be able to avoid any change in your culture?

2. Do fads and fashions reflect and reinforce or challenge and change the values and norms of a society? Consider a wide variety of fads and fashions: musical styles; computer and video games and other technologies; literature; and political, social, and religious ideas.

3. You are doing a survey analysis of recent immigrants to the United States to determine the effects of popular culture on their views and behavior. What are some of the questions you would use in your survey?

 ## Resources on the Internet

Chapter-Related Web Sites
The following web sites have been selected for their relevance to the topics in this chapter. These sites are among the more stable, but please note that web site addresses change frequently.

The Library of Congress: American Memory
http://memory.loc.gov/ammem/amhome.html
American Memory is a gateway to rich primary source materials relating to the history and culture of the United States. The site offers more than 7,000,000 digital items from more than 100 historical collections.

World Civilizations
http://www.wsu.edu/~dee/WORLD.HTM
Sponsored by Washington State University, this online education site has an excellent resources section that you can use to broaden your understanding of other cultures.

Native American Culture and American Society
http://www.djmcadam.com/ojibwe.html
A brief description of the historical context of Native American populations and a number of links to specific groups and historical events.

Center for the Study of White Culture: A Multiracial Organization
http://www.euroamerican.org/
The Center for the Study of White American Culture supports cultural exploration and self-discovery among white Americans. It encourages a dialogue among all racial and cultural groups concerning the role of white American culture in the larger American society.

The African-American Mosaic
http://www.loc.gov/exhibits/african/intro.html
A Library of Congress resource guide for the study of black history and culture.

Sarah Zupko's Cultural Studies Center
http://www.popcultures.com/
This is an academic-friendly site created and maintained by a media producer and syndicated columnist. It is the starting place for any study of pop culture.

Popular Culture Association and American Culture Association
http://www2.h-net.msu.edu/~pcaaca/pop.html
This site advertises an academic journal to which you can subscribe. The site is useful to students because it provides a number of categories and topics within the popular culture movement.

The Literature & Culture of the American 1950s
http://www.english.upenn.edu/~afilreis/50s/home.html
This site contains an extensive reading list.

American Cultural History: The Twentieth Century
http://www.nhmccd.edu/contracts/lrc/kc/decades.html
Web guides are provided by Kingwood College Library for each decade of the last century.

The Institute for American Values
http://fatherfamilylink.gse.upenn.edu/org/iav/mission.htm
This site is sponsored by the National Center on Fathers and Families at the University of Pennsylvania Graduate School of Education.

Companion Web Site for This Book

Virtual Society: The Wadsworth Sociology Resource Center
Visit **http://sociology.wadsworth.com** and click on the page for Kendall, *Sociology in Our Times: The Essentials*, Fourth Edition, to access a wide range of enrichment material to aid in your study of sociology. Click on the Student Resources section of the web site. Next, select from the pull-down menu the chapter that you are presently studying. Among the useful options for self-study are chapter objectives, flashcards, practical study tips, and practice tests for each chapter.

MicroCase Online
From the Virtual Society home page, click on MicroCase Online to access book-specific MicroCase exercises that allow you to further explore sociological issues and principles presented in the text.

InfoTrac College Edition
Another unique option available to you at the Student Resources section of the companion web site is InfoTrac College Edition, an online library with access to hundreds of scholarly and popular periodicals. Below are suggested search terms for this chapter. Results from these and other searches are found at the site.

• Search keywords: *hate crime*. There have been a number of highly publicized incidents of hate crime in the United States. From the articles your search produces, find several that discuss legislative efforts to deal with the effects of hate crime.
• Search keyword: *bilingualism*. Bilingualism in the classroom is a highly controversial issue in public education. Look for articles that discuss issues of bilingualism in the classroom and in the workplace.
• Search keyword: *subculture*. Use subculture as a subject-guide search. Are there any articles that deal with a subculture along with a form of popular culture such as music? What did you learn about the particular subculture discussed in the article?

Virtual Explorations CD-ROM
Go to the Virtual Explorations CD-ROM to begin the interactive exercise for this chapter. This Virtual Exploration will introduce you to some of the exciting resources for sociology on the World Wide Web. You will be guided through an exercise that employs related web sites on culture. Answer the questions, and e-mail your responses to your instructor.

Socialization

Why Is Socialization Important Around the Globe?
Human Development: Biology and Society
Problems Associated with Social Isolation and
    Maltreatment

**Social Psychological Theories of Human
    Development**
Freud and the Psychoanalytic Perspective
Piaget and Cognitive Development
Kohlberg and the Stages of Moral Development
Gilligan's View on Gender and Moral Development

Sociological Theories of Human Development
Cooley and the Looking-Glass Self
Mead and Role-Taking
Recent Symbolic Interactionist Perspectives

Agents of Socialization
The Family
The School
Peer Groups
Mass Media

Gender and Racial–Ethnic Socialization

Socialization Through the Life Course
Infancy and Childhood
Adolescence
Adulthood
Late Adulthood and Ageism

Resocialization
Voluntary Resocialization
Involuntary Resocialization

Socialization in the Future

I think my mom and dad were boyfriend and girlfriend for a couple of years, but they were apart by the time I was born. . . . The earliest memory I have of my father isn't pleasant. I was three years old. It was afternoon, and I was wearing a pair of jeans and these cute Mickey Mouse suspenders, a favorite play outfit. My mom and I were standing in the kitchen, doing the laundry. . . . Suddenly the door swung open and there was this man standing there. I yelled, "Daddy!" Even though I didn't know what he looked like, I just automatically knew it was him.

He paused in the doorway, like he was making a dramatic entrance, and I think he said something, but he was so drunk, it was unintelligible. It sounded more like a growl. We stood there, staring at him. I was so excited to see him. I was just coming to the age where I noticed that I didn't have a father like everyone else, and I wanted one. I didn't really know what my dad was like, but I learned real fast.

In a blur of anger he roared into the room and threw my mom down on the ground. Then he turned on me. I didn't know what was happening. I was still excited to see him, still hearing the echo of my gleeful yell, "Daddy!" when he picked me up and threw me into the wall. Luckily, half of my body landed on a big sack of laundry, and I wasn't hurt. But my dad didn't even look back at me. He turned and grabbed a bottle of tequila, shattered a bunch of glasses all over the floor, and then stormed out of the house. . . . And that was it. That was the first time I remember seeing my dad.

—actress Drew Barrymore, describing her first encounter with her father (Barrymore, 1994: 185–187)

AP/Wide World Photos

Actress Drew Barrymore's description of childhood maltreatment in her family makes us aware of the importance of early socialization in all our lives. As an adult, she has experienced many happier times.

The process of socialization is of major significance to sociologists. Although most children are nurtured, trusted, and loved by their parents, Barrymore's experience is not an isolated incident: Large numbers of children experience maltreatment at the hands of family members or other caregivers such as baby-sitters or day-care workers. Child maltreatment includes physical abuse, sexual abuse, physical neglect, and emotional mistreatment of children and young adolescents (Zuravin, 1991). Such maltreatment is of interest to sociologists because it has a serious impact on a child's social growth, behavior, and self-image—all of which develop within the process of socialization. By contrast, children who are treated with respect by their parents are more likely to develop a positive self-image and learn healthy conduct because their parents provide appropriate models of behavior.

In this chapter, we examine why socialization is so crucial, and we discuss both sociological and social psychological theories of human development. We look at the dynamics of socialization—how it occurs and what shapes it. Throughout the chapter, we focus on positive and negative aspects of the socialization process. Before reading on, test your knowledge of socialization and child care by taking the quiz in Box 3.1.

Questions and Issues

Chapter Focus Question: What happens when children do not have an environment that supports positive socialization?

What purpose does socialization serve?

How do individuals develop a sense of self?

How does socialization occur?

Who experiences resocialization?

Box 3.1 Sociology and Everyday Life

How Much Do You Know About Early Socialization and Child Care?

True	False	
T	F	1. In the United States, full-day child care often costs as much per year as college tuition at a public college or university.
T	F	2. The cost of child care is a major problem for many U.S. families.
T	F	3. After-school programs have greatly reduced the number of children who are home alone after school.
T	F	4. The average annual salary of a child-care worker is less than the average yearly salaries for funeral attendants, bellhops, and garbage collectors.
T	F	5. All states require teachers in child-care centers to have training in their field and to pass a licensing examination.
T	F	6. In a family in which child abuse occurs, all the children are likely to be victims.
T	F	7. It is against the law to fail to report child abuse.
T	F	8. Some people are "born" child abusers whereas others learn abusive behavior from their family and friends.

Answers on page 78.

Why Is Socialization Important Around the Globe?

Socialization **is the lifelong process of social interaction through which individuals acquire a self-identity and the physical, mental, and social skills needed for survival in society.** It is the essential link between the individual and society. Socialization enables each of us to develop our human potential and to learn the ways of thinking, talking, and acting that are necessary for social living.

Socialization is essential for the individual's survival and for human development. The many people who met the early material and social needs of each of us were central to our establishing our own identity. During the first three years of our life, we begin to develop both a unique identity and the ability to manipulate things and to walk. We acquire sophisticated cognitive tools for thinking and for analyzing a wide variety of situations, and we learn effective communication skills. In the process, we begin a relatively long socialization process that culminates in our inte-

gration into a complex social and cultural system (Garcia Coll, 1990).

Socialization is also essential for the survival and stability of society. Members of a society must be socialized to support and maintain the existing social structure. From a functionalist perspective, individual conformity to existing norms is not taken for granted; rather, basic individual needs and desires must be balanced against the needs of the social structure. The socialization process is most effective when people conform to the norms of society because they believe this is the best course of action. Socialization enables a society to "reproduce" itself by passing on its culture from one generation to the next.

Although the techniques used to teach newcomers the beliefs, values, and rules of behavior are

Socialization the lifelong process of social interaction through which individuals acquire a self-identity and the physical, mental, and social skills needed for survival in society.

Box 3.1

Answers to the Sociology Quiz on Early Socialization and Child Care

1. **True.** Full-day child care typically costs between $4,000 and $10,000 per child per year, which is as much or more than tuition at many public colleges and universities.
2. **True.** Child care outside the home is a major financial burden, particularly for the one out of every three families with young children but with income of less than $25,000 a year.
3. **False.** Although after-school programs have slightly reduced the number of children at home alone after school, about seven million school-age children are alone because many communities do not have such programs.
4. **True.** The average salary for a child-care worker is only $15,430 per year, which is less than the yearly salaries for people in many other employment categories.
5. **False.** Although all states require haircutters and manicurists to have about 1,500 hours of training at an accredited school, 32 states do not require teachers in child-care centers to have any training in their field prior to teaching children.
6. **False.** In some families, one child may be the victim of repeated abuse whereas others are not.
7. **True.** In the United States, all states have reporting requirements for child maltreatment; however, compliance with these legal mandates has been inconsistent. Some states have requirements that everyone who suspects abuse or neglect must report it. Other states mandate reporting only by certain persons, such as medical personnel and child-care providers.
8. **False.** No one is "born" to be an abuser. People learn abusive behavior from their family and friends.

Source: Based on Children's Defense Fund, 2001.

somewhat similar in many nations, the *content* of socialization differs greatly from society to society. How people walk, talk, eat, make love, and wage war are all functions of the culture in which they are raised. At the same time, we are also influenced by our exposure to subcultures of class, race, ethnicity, religion, and gender. In addition, each of us has unique experiences in our family and friendship groupings. The kind of human being that we become depends greatly on the particular society and social groups that surround us at birth and during early childhood. What we believe about ourselves, our society, and the world does not spring full-blown from inside ourselves; rather, we learn these things from our interactions with others.

Human Development: Biology and Society

What does it mean to be "human"? To be human includes being conscious of ourselves as individuals with unique identities, personalities, and relationships with others. As humans, we have ideas, emotions, and values. We have the capacity to think and to make rational decisions. But what is the source of "humanness"? Are we born with these human characteristics, or do we develop them through our interactions with others?

When we are born, we are totally dependent on others for our survival. We cannot turn ourselves over, speak, reason, plan, or do many of the things that are associated with being human. Although we can nurse, wet, and cry, most small mammals can also do those things. As discussed in Chapter 2, we humans differ from nonhuman animals because we lack instincts and must rely on learning for our survival. Human infants have the potential to develop human characteristics if they are exposed to an adequate socialization process.

Every human being is a product of biology, society, and personal experiences—that is, of heredity and environment or, in even more basic terms, "nature" and "nurture." How much of our development can be explained by socialization? How much by our genetic heritage? Sociologists focus on how humans design their own culture and transmit it from generation to generation through socialization. By contrast, sociobiologists assert that nature, in the form of our genetic makeup, is a major factor in shaping human behavior. *Sociobiology* is the systematic study of how biology affects social behavior (Wilson, 1975). According to the zoologist Edward O. Wilson, who pioneered sociobiology, genetic inheritance underlies many forms of social behavior, such as war and peace, envy and concern for others, and competition and cooperation. Most sociologists disagree with the notion that biological principles can be used to explain all human behavior. Obviously, however, some aspects of our physical makeup—such as eye color, hair color, height, and weight—are largely determined by our heredity.

How important is social influence ("nurture") in human development? There is hardly a single behavior that is not influenced socially. Except for simple reflexes, most human actions are social, either in their causes or in their consequences. Even solitary actions such as crying or brushing our teeth are ultimately social. We cry because someone has hurt us. We brush our teeth because our parents (or dentist) told us it was important. Social environment probably has a greater effect than heredity on the way we develop and the way we act. However, heredity does provide the basic material from which other people help to mold an individual's human characteristics.

Our biological and emotional needs are related in a complex equation. Children whose needs are met in settings characterized by affection, warmth, and closeness see the world as a safe and comfortable place and see other people as trustworthy and helpful. By contrast, infants and children who receive less-than-adequate care or who are emotionally rejected or abused often view the world as hostile and have feelings of suspicion and fear.

Problems Associated with Social Isolation and Maltreatment

Social environment, then, is a crucial part of an individual's socialization. Even nonhuman primates such as monkeys and chimpanzees need social contact with others of their species in order to develop properly. As we will see, appropriate social contact is even more important for humans.

ISOLATION AND NONHUMAN PRIMATES

Researchers have attempted to demonstrate the effects of social isolation on nonhuman primates raised without contact with others of their own species. In a series of laboratory experiments, the psychologists Harry and Margaret Harlow (1962, 1977) took infant rhesus monkeys from their mothers and isolated them in separate cages. Each cage contained two nonliving "mother substitutes" made of wire, one with a feeding bottle attached and the other covered with soft terry cloth but without a bottle. The infant monkeys instinctively clung to the cloth "mother" and would not abandon it until hunger drove them to the bottle attached to the wire "mother." As soon as they were full, they went back to the cloth "mother" seeking warmth, affection, and physical comfort.

The Harlows' experiments show the detrimental effects of isolation on nonhuman primates. When the young monkeys were later introduced to other members of their species, they cringed in the corner. Having been deprived of social contact with other monkeys during their first six months of life, they never learned how to relate to other monkeys or to become well-adjusted adults—they were fearful of or hostile toward other monkeys (Harlow and Harlow, 1962, 1977).

Because humans rely more heavily on social learning than do monkeys, the process of socialization is even more important for us.

Sociobiology the systematic study of how biology affects social behavior.

Bill Aron/PhotoEdit

Bob Daemmrich/The Image Works

What are the consequences to children of isolation and physical abuse, as contrasted with social interaction and parental affection? Sociologists emphasize that social environment is a crucial part of an individual's socialization.

ISOLATED CHILDREN Of course, sociologists would never place children in isolated circumstances so that they could observe what happened to them. However, some cases have arisen in which parents or other caregivers failed to fulfill

Reflections

Reflections 3.1

Recent cases of isolated children have shown that the issues in this chapter remain relevant. What part should social institutions such as religion, education, and/or government play in assisting childhood victims of abuse or neglect?

their responsibilities, leaving children alone or placing them in isolated circumstances. From analysis of these situations, social scientists have documented cases in which children were deliberately raised in isolation. A look at the lives of two children who suffered such emotional abuse provides important insights into the importance of a positive socialization process and the negative effects of social isolation.

Anna Born in 1932 to an unmarried, mentally impaired woman, Anna was an unwanted child. She was kept in an attic-like room in her grandfather's house. Her mother, who worked on the farm all day and often went out at night, gave Anna just enough care to keep her alive; she received no other care. Sociologist Kingsley Davis (1940) described Anna's condition when she was found in 1938:

> [Anna] had no glimmering of speech, absolutely no ability to walk, no sense of gesture, not the least capacity to feed herself even when the food was put in front of her, and no comprehension of cleanliness. She was so apathetic that it was hard to tell whether or not she could hear. And all of this at the age of nearly six years.

When she was placed in a special school and given the necessary care, Anna slowly learned to walk, talk, and care for herself. Just before her death at the age of ten, Anna reportedly could follow directions, talk in phrases, wash her hands, brush her teeth, and try to help other children (Davis, 1940).

Genie Almost four decades later, Genie was found in 1970 at the age of thirteen. She had been locked in a bedroom alone, alternately strapped down to a child's potty chair or straitjacketed into a sleeping bag, since she was twenty months old. She had been fed baby food and beaten with a wooden paddle when she whimpered. She had not heard the sounds of human speech because no one talked to her and there was no television or radio in her room (Curtiss, 1977; Pines, 1981). Genie was placed in a pediatric hospital, where one of the psychologists described her condition:

> At the time of her admission she was virtually unsocialized. She could not stand erect, salivated continuously, had never been toilet-trained and had no control over her urinary or bowel functions. She was unable to chew solid food and had the weight, height and appearance of a child half her age. (Rigler, 1993: 35)

In addition to her physical condition, Genie showed psychological traits associated with neglect, as described by one of her psychiatrists:

> If you gave [Genie] a toy, she would reach out and touch it, hold it, caress it with her fingertips, as though she didn't trust her eyes. She would rub it against her cheek to feel it. So when I met her and she began to notice me standing beside her bed, I held my hand out and she reached out and took my hand and carefully felt my thumb and fingers individually, and then put my hand against her cheek. She was exactly like a blind child. (Rymer, 1993: 45)

Extensive therapy was used in an attempt to socialize Genie and develop her language abilities (Cur-

Alexandra Boulat/SIPA Press

What can we learn from this street child in Bucharest about the effects of social isolation and neglect on young people?

tiss, 1977; Pines, 1981). These efforts met with limited success: In the 1990s, Genie was living in a board-and-care home for retarded adults (see Angier, 1993; Rigler, 1993; Rymer, 1993).

Why do we discuss children who have been the victims of maltreatment in a chapter that looks at the socialization process? The answer lies in the fact that such cases are important to our understanding of the socialization process because they show the importance of this process and reflect how detrimental social isolation and neglect can be to the well-being of people.

CHILD MALTREATMENT What do the terms *child maltreatment* and *child abuse* mean to you? When asked what constitutes child maltreatment, many people first think of cases that involve severe physical injuries or sexual abuse. However, neglect is the most frequent form of child maltreatment (Dubowitz et al., 1993). Child neglect occurs

✓**Checkpoints**

Checkpoint 3.1

1. What is socialization, and why is it essential for both individuals and societies?
2. How is human development affected by both biology and society?
3. What problems are associated with social isolation and maltreatment?

when children's basic needs—including emotional warmth and security, adequate shelter, food, health care, education, clothing, and protection—are not met, regardless of cause (Dubowitz et al., 1993: 12). Neglect often involves acts of omission (where parents or caregivers fail to provide adequate physical or emotional care for children) rather than acts of commission (such as physical or sexual abuse). Of course, what constitutes child maltreatment differs from society to society.

Social Psychological Theories of Human Development

Over the past hundred years, a variety of psychological and sociological theories have been developed to not only explain child abuse but also to describe how a positive process of socialization occurs. Let's look first at several social psychological theories that focus primarily on how the individual personality develops.

Freud and the Psychoanalytic Perspective

The basic assumption in Sigmund Freud's (1924) psychoanalytic approach is that human behavior and personality originate from unconscious forces within individuals. Freud (1856–1939), who is known as the founder of psychoanalytic theory, developed his major theories in the Victorian era, when biological explanations of human behavior were prevalent. It was also an era of extreme sexual repression and male dominance when compared to contemporary U.S. standards. Freud's

theory was greatly influenced by these cultural factors, as reflected in the importance he assigned to sexual motives in explaining behavior. For example, Freud based his ideas on the belief that people have two basic tendencies: the urge to survive and the urge to procreate.

According to Freud (1924), human development occurs in three states that reflect different levels of the personality, which he referred to as the *id, ego,* and *superego.* **The *id* is the component of personality that includes all of the individual's basic biological drives and needs that demand immediate gratification.** For Freud, the newborn child's personality is all id, and from birth the child finds that urges for self-gratification—such as wanting to be held, fed, or changed—are not going to be satisfied immediately. However, id remains with people throughout their life in the form of *psychic energy,* the urges and desires that account for behavior. By contrast, the second level of personality—the *ego*—develops as infants discover that their most basic desires are not always going to be met by others. **The *ego* is the rational, reality-oriented component of personality that imposes restrictions on the innate pleasure-seeking drives of the id.** The ego channels the desire of the id for immediate gratification into the most advantageous direction for the individual. The third level of personality—the superego—is in opposition to both the id and the ego. **The *superego,* or conscience, consists of the moral and ethical aspects of personality.** It is first expressed as the recognition of parental control and eventually matures as the child learns that parental control is a reflection of the values and moral demands of the larger society. When a person is well adjusted, the ego successfully manages the opposing forces of the id and the superego. Figure 3.1 illustrates Freud's theory of personality.

Although subject to harsh criticism, Freud's theory made people aware of the importance of early childhood experiences, including abuse and neglect. His theories have also had a profound influence on contemporary mental health practitioners and on other human development theories.

Piaget and Cognitive Development

Jean Piaget (1896–1980), a Swiss psychologist, was a pioneer in the field of cognitive (intellectual) development. Cognitive theorists are interested in

Figure 3.1 Freud's Theory of Personality

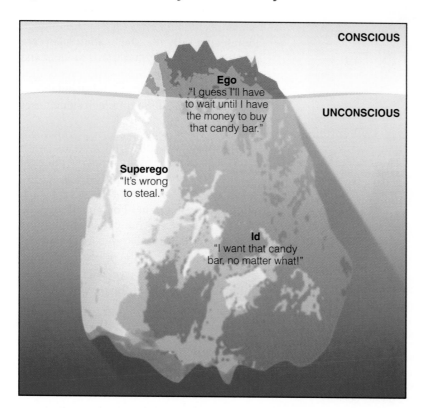

CONSCIOUS

Ego
"I guess I'll have to wait until I have the money to buy that candy bar."

UNCONSCIOUS

Superego
"It's wrong to steal."

Id
"I want that candy bar, no matter what!"

This illustration shows how Freud might picture a person's internal conflict over whether to commit an antisocial act such as stealing a candy bar. In addition to dividing personality into three components, Freud theorized that our personalities are largely unconscious—hidden from our normal awareness. To dramatize his point, Freud compared conscious awareness (portions of the ego and superego) to the visible tip of an iceberg. Most of personality—including the id, with its raw desires and impulses—lies submerged in our subconscious.

how people obtain, process, and use information—that is, in how we think. Cognitive development relates to changes over time in how we think.

Piaget (1954) believed that in each stage of development (from birth through adolescence), children's activities are governed by their perception of the world around them. His four stages of cognitive development are organized around specific tasks that, when mastered, lead to the acquisition of new mental capacities, which then serve as the basis for the next level of development. Piaget emphasized that all children must go through each stage in sequence before moving on to the next one, although some children move through them faster than others.

1. *Sensorimotor stage* (birth to age two). During this period, children understand the world only through sensory contact and immediate action because they cannot engage in symbolic thought or use language. Toward the end of the second year, children comprehend

object permanence; in other words, they start to realize that objects continue to exist even when the items are out of sight.

2. *Preoperational stage* (age two to seven). In this stage, children begin to use words as mental symbols and to form mental images. However, they still are limited in their ability to use logic

Id Sigmund Freud's term for the component of personality that includes all of the individual's basic biological drives and needs that demand immediate gratification.

Ego according to Sigmund Freud, the rational, reality-oriented component of personality that imposes restrictions on the innate pleasure-seeking drives of the id.

Superego Sigmund Freud's term for the conscience, consisting of the moral and ethical aspects of personality.

Tony Freeman/PhotoEdit

Psychologist Jean Piaget identified four stages of cognitive development, including the preoperational stage, in which children have limited ability to realize that physical objects may change in shape or appearance. To prove his point, Piaget showed children two different-sized beakers, each containing the same amount of water, and asked the children to determine which one contained the greater quantity.

to solve problems or to realize that physical objects may change in shape or appearance while still retaining their physical properties. For example, Piaget showed children two identical beakers filled with the same amount of water. After the children agreed that both beakers held the same amount of water, Piaget poured the water from one beaker into a taller, narrower beaker and then asked them about the amounts of water in each beaker. Those still in the preoperational stage believed that the taller beaker held more water because the water line was higher than in the shorter, wider beaker.

3. *Concrete operational stage* (age seven to eleven). During this stage, children think in terms of tangible objects and actual events. They can draw conclusions about the likely physical consequences of an action without always having to try the action out. Children

begin to take the role of others and start to empathize with the viewpoints of others.

4. *Formal operational stage* (age twelve through adolescence). By this stage, adolescents are able to engage in highly abstract thought and understand places, things, and events they have never seen. They can think about the future and evaluate different options or courses of action.

Piaget provided useful insights on the emergence of logical thinking as the result of biological maturation and socialization. However, critics have noted several weaknesses in Piaget's approach to cognitive development. For one thing, the theory says little about individual differences among children, nor does it provide for cultural differences. For another, as the psychologist Carol Gilligan (1982) has noted, Piaget did not take into account how gender affects the process of social development.

Kohlberg and the Stages of Moral Development

Lawrence Kohlberg (b. 1927) elaborated on Piaget's theories of cognitive reasoning by conducting a series of studies in which children, adolescents, and adults were presented with moral dilemmas that took the form of stories. Based on his findings, Kohlberg (1969, 1981) classified moral reasoning into three sequential levels:

1. *Preconventional level* (age seven to ten). Children's perceptions are based on punishment and obedience. Evil behavior is that which is likely to be punished; good conduct is based on obedience and avoidance of unwanted consequences.

2. *Conventional level* (age ten through adulthood). People are most concerned with how they are perceived by their peers and with how one conforms to rules.

3. *Postconventional level* (few adults reach this stage). People view morality in terms of individual rights; "moral conduct" is judged by principles based on human rights that transcend government and laws.

Although Kohlberg presents interesting ideas about the moral judgments of children, some crit-

SuperStock

According to Lawrence Kohlberg, peers play a significant role in the development of moral reasoning. How might study groups influence students' behavior with regard to honesty or dishonesty in the classroom?

ics have challenged the universality of his stages of moral development. They have also suggested that the elaborate "moral dilemmas" he used are too abstract for children. In one story, for example, a husband contemplates stealing for his critically ill wife medicine that he cannot afford. When questions are made simpler, or when children and adolescents are observed in natural (as opposed to laboratory) settings, they often demonstrate sophisticated levels of moral reasoning (Darley and Shultz, 1990; Lapsley, 1990).

Gilligan's View on Gender and Moral Development

Psychologist Carol Gilligan (b. 1936) is one of the major critics of Kohlberg's theory of moral development. According to Gilligan (1982), Kohlberg's model was developed solely on the basis of research with male respondents, and women and men often have divergent views on morality based on differences in socialization and life experiences. Gilligan believes that men become more concerned with law and order but that women tend to analyze social re-

lationships and the social consequences of behavior. For example, in Kohlberg's story about the man who is thinking about stealing medicine for his wife, Gilligan argues that male respondents are more likely to use *abstract standards* of right and wrong, whereas female respondents are more likely to be concerned about what *consequences* his stealing the drug might have on the man and his family. Does this constitute a "moral deficiency" on the part of either women or men? Not according to Gilligan.

To correct what she perceived to be a male bias in Kohlberg's research, Gilligan (1982) examined morality in women by interviewing twenty-nine pregnant women who were contemplating having an abortion. Based on her research, Gilligan concluded that Kohlberg's stages do not reflect the ways that many women think about moral problems. As a result, Gilligan identified three stages in female moral development. In stage 1, the woman is motivated primarily by selfish concerns ("This is what I want . . . this is what I need"). In stage 2, she increasingly recognizes her responsibility to others. In stage 3, she makes a decision based on her desire to do the greatest good for both herself and for others. Gilligan argued that men are socialized to make moral decisions based on a justice perspective ("What is the fairest thing to do?") whereas women are socialized to make such decisions on a care and responsibility perspective ("Who will be hurt least?").

Subsequent research that directly compared women's and men's reasoning about moral dilemmas has supported some of Gilligan's assertions but not others. For example, some other researchers have not found that women are more compassionate than men (Tavris, 1993). Overall, however, Gilligan's argument that people make moral decisions according to both abstract principles of justice and principles of compassion and care is an important contribution to our knowledge about moral reasoning. Her book *In a Different Voice* (1982) also made social scientists more aware that the same situation may be viewed quite differently by men and women.

✓Checkpoints

Checkpoint 3.2

4. What are the three stages in human development identified by Sigmund Freud?
5. What are the key factors in Jean Piaget's perspective on cognitive development?
6. What do the theories of Lawrence Kohlberg and Carol Gilligan tell us about moral development?

Sociological Theories of Human Development

Although social scientists acknowledge the contributions of psychoanalytic and psychologically based explanations of human development, sociologists believe that it is important to bring a sociological perspective to bear on how people develop an awareness of self and learn about the culture in which they live. According to a sociological perspective, we cannot form a sense of self or personal identity without intense social contact with others. The self represents the sum total of perceptions and feelings that an individual has of being a distinct, unique person—a sense of who and what one is. When we speak of the "self," we typically use words such as *I, me, my, mine,* and *myself* (Cooley, 1998/1902). This sense of self (also referred to as *self-concept*) is not present at birth; it arises in the process of social experience. *Self-concept* **is the totality of our beliefs and feelings about ourselves** (Gecas, 1982). Four components make up our self-concept: (1) the physical self ("I am tall"), (2) the active self ("I am good at soccer"), (3) the social self ("I am nice to others"), and (4) the psychological self ("I believe in world peace"). Between early and late childhood, a child's focus tends to shift from the physical and active dimensions of self toward the social and psychological aspects (Lippa, 1994). Self-concept is the foundation for communication with others; it continues to develop and change throughout our lives (Zurcher, 1983).

Our *self-identity* is our perception about what kind of person we are. As we have seen, socially isolated children do not have typical self-identities because they have had no experience of "humanness." According to symbolic interactionists, we do not know who we are until we see ourselves as we believe that others see us. We gain information about the self largely through language, symbols, and interaction with others. Our interpretation and evaluation of these messages are central to the social construction of our identity. However, we are not just passive reactors to situations, programmed by society to respond in fixed ways. Instead, we are active agents who develop plans out of the pieces supplied by culture and attempt to execute these plans in social encounters (McCall and Simmons, 1978).

The perspectives of symbolic interactionists Charles Horton Cooley and George Herbert Mead help us understand how our self-identity is developed through our interactions with others.

Cooley and the Looking-Glass Self

According to the sociologist Charles Horton Cooley (1864–1929), the *looking-glass self* **refers to the way in which a person's sense of self is derived from the perceptions of others.** Our looking-glass self is not who we actually are or what people actually think about us; rather, it is based on our perception of *how* other people think of us (Cooley, 1998/1902). Cooley asserted that we base our perception of who we are on how we think other people see us and on whether this opinion seems good or bad to us.

As Figure 3.2 shows, the looking-glass self is a self-concept derived from a three-step process:

1. We imagine how our personality and appearance will look to other people. We may imagine that we are attractive or unattractive, heavy or slim, friendly or unfriendly, and so on.
2. We imagine how other people judge the appearance and personality that we think we present. This step involves our perception of how we think they are judging us. We may be correct or incorrect!
3. We develop a self-concept. If we think the evaluation of others is favorable, our self-concept is enhanced. If we think the evaluation is unfavorable, our self-concept is diminished. (Cooley, 1998/1902)

According to Cooley, we use our interactions with others as a mirror for our own thoughts and

Figure 3.2 How the Looking-Glass Self Works

We imagine how we appear to other people.

We imagine how other people judge the appearance that we think we present.

If we think the evaluation is favorable, our self-concept is enhanced.

If we think the evaluation is unfavorable, our self-concept is diminished.

in order to understand the world from that person's or group's point of view. Role-taking often occurs through play and games, as children try out different roles (such as being mommy, daddy, doctor, or teacher) and gain an appreciation of them. First, people come to take the role of the other (role-taking). By taking the roles of others, the individual hopes to ascertain the intention or direction of the acts of others. Then the person begins to construct his or her own roles (role-making) and to anticipate other individuals' responses. Finally, the person plays at her or his particular role (role-playing) (Marshall, 1998).

According to Mead (1934), in the early months of life, children do not realize that they are separate from others. However, they do begin early on to see a mirrored image of themselves in others. Shortly after birth, infants start to notice the faces of those around them, especially the significant others, whose faces start to have meaning because they are associated with experiences like feeding and cuddling. *Significant others* **are those persons whose care, affection, and approval are especially desired and who are most important in the development of the self.** Gradually, we distinguish ourselves from our caregivers and begin to perceive ourselves in contrast to them. As we develop language skills and learn to understand symbols, we begin to develop a

actions; our sense of self depends on how we interpret what they do and say. Consequently, our sense of self is not permanently fixed; it is always developing as we interact with others in the larger society. For Cooley, self and society are merely two sides of the same coin: "Self and society go together, as phases of a common whole. I am aware of the social groups in which I live as immediately and authentically as I am aware of myself" (Cooley, 1963/1909: 8–9). Accordingly, the self develops only through contact with others, just as social institutions and societies do not exist independently of the interaction of individuals (Schubert, 1998). By developing the idea of the looking-glass self, Cooley made us aware of the mutual interrelationship between the individual and society—namely, that society shapes people and people shape the society.

Mead and Role-Taking

George Herbert Mead (1863–1931) extended Cooley's insights by linking the idea of self-concept to *role-taking*—the process by which a person mentally assumes the role of another person or group

Self-concept the totality of our beliefs and feelings about ourselves.

Looking-glass self Charles Horton Cooley's term for the way in which a person's sense of self is derived from the perceptions of others.

Role-taking the process by which a person mentally assumes the role of another person in order to understand the world from that person's point of view.

Significant others those persons whose care, affection, and approval are especially desired and who are most important in the development of the self.

According to sociologist George Herbert Mead, the self develops through three stages. In the preparatory stage, children imitate others; in the play stage, children pretend to take the roles of specific people; and in the game stage, children become aware of the "rules of the game" and the expectations of others.

self-concept. When we can represent ourselves in our minds as objects distinct from everything else, our self has been formed.

Mead (1934) divided the self into the "I" and the "me." The "I" is the subjective element of the self and represents the spontaneous and unique traits of each person. The "me" is the objective element of the self, which is composed of the internalized attitudes and demands of other members of society and the individual's awareness of those demands. Both the "I" and the "me" are needed to

form the social self. The unity of the two constitutes the full development of the individual. According to Mead, the "I" develops first, and the "me" takes form during the three stages of self development:

1. During the *preparatory stage*, up to about age three, interactions lack meaning, and children largely imitate the people around them. At this stage, children are preparing for role-taking.

2. In the *play stage*, from about age three to five, children learn to use language and other sym-

bols, thus enabling them to pretend to take the roles of specific people. At this stage, they begin to see themselves in relation to others, but they do not see role-taking as something they have to do.

3. During the *game stage*, which begins in the early school years, children understand not only their own social position but also the positions of others around them. In contrast to play, games are structured by rules, are often competitive, and involve a number of other "players." At this time, children become concerned about the demands and expectations of others and of the larger society. Mead used the example of a baseball game to describe this stage because children, like baseball players, must take into account the roles of all the other players at the same time. Mead's concept of the ***generalized other* refers to the child's awareness of the demands and expectations of the society as a whole or of the child's subculture.**

Is socialization a one-way process? No, according to Mead. Socialization is a two-way process between society and the individual. Just as the society in which we live helps determine what kind of individuals we will become, we have the ability to shape certain aspects of our social environment and perhaps even the larger society.

How useful are symbolic interactionist perspectives such as Cooley's and Mead's in enhancing our understanding of the socialization process? Certainly, this approach contributes to our understanding of how the self develops. Cooley's idea of the looking-glass self makes us aware that our perception of how we think others see us is not always correct. Mead extended Cooley's ideas by emphasizing the cognitive skills acquired through role-taking. His concept of the generalized other helps us see that the self is a social creation. According to Mead (1934: 196), "Selves can only exist in defi-

nite relations to other selves. No hard-and-fast line can be drawn between our own selves and the selves of others." However, the viewpoints of symbolic interactionists such as Cooley and Mead have certain limitations. Sociologist Anne Kaspar (1986) suggests that Mead's ideas about the social self may be more applicable to men than to women because women are more likely to experience inherent conflicts between the meanings they derive from their personal experiences and those they take from the culture, particularly in regard to balancing the responsibilities of family life and paid employment. (Concept Table 3.A summarizes the major theories of human development.)

Recent Symbolic Interactionist Perspectives

The symbolic interactionist approach emphasizes that socialization is a collective process in which children are active and creative agents, not passive recipients of the socialization process. From this view, childhood is a *socially constructed* category (Adler and Adler, 1998). Children are capable of actively constructing their own shared meanings as they acquire language skills and accumulate interactive experiences (Qvortrup, 1990). According to the "orb web model" of the sociologist William A. Corsaro (1985, 1997), children's cultural knowledge reflects not only the beliefs of the adult world but also the unique interpretations and aspects of the children's own peer culture. Corsaro (1992: 162) states that *peer culture* is "a stable set of activities or routines, artifacts, values, and concerns that children produce and share." This peer culture emerges through interactions as children "borrow" from the adult culture but transform it so that it fits their own situation. Using ethnographic studies of U.S. and Italian preschoolers, Corsaro found that very young children engage in predictable patterns of interaction. For example, when playing together, children often permit some children to gain access to their group and play area while preventing others

Reflections

Reflections 3.2

How is socialization related to the development of a positive self-concept?

Generalized other George Herbert Mead's term for the child's awareness of the demands and expectations of the society as a whole or of the child's subculture.

Concept Table 3.A Psychological and Sociological Theories of Human Development

Social Psychological Theories of Human Development	Freud's psychoanalytic perspective	Children first develop the id (drives and needs), then the ego (restrictions on the id), and then the superego (moral and ethical aspects of personality).
	Piaget's cognitive development	Children go through four stages of cognitive (intellectual) development, going from understanding only through sensory contact to engaging in highly abstract thought.
	Kohlberg's stages of moral development	People go through three stages of moral development, from avoidance of unwanted consequences to viewing morality based on human rights.
	Gilligan: gender and moral development	Women go through stages of moral development, from personal wants to the greatest good for themselves and others.
Sociological Theories of Human Development	Cooley's looking-glass self	A person's sense of self is derived from his or her perception of how others view him or her.
	Mead's three stages of self-development	In the preparatory stage, children imitate the people around them; in the play stage, children pretend to take the roles of specific people; and in the game stage, children learn the demands and expectations of roles.

from becoming a part of their group. Children also play "approach–avoidance" games in which they alternate between approaching a threatening person or group and then running away. In fact, Corsaro (1992) believes that the peer group is the most significant public realm for children. (Peer groups as agents of socialization are discussed later in the chapter.) This approach contributes to our knowledge about human development because it focuses on group life rather than individuals. Researchers using this approach "look at social relations, the organization and meanings of social situations, and the collective practices through which children create and recreate key constructs in their daily interactions" (Adler and Adler, 1998: 10; see also Thorne, 1993; Eder, 1995).

✓ Checkpoints

Checkpoint 3.3

7. What is the self-concept, and why is it central to our identity?
8. What are the three steps in Charles Horton Cooley's looking-glass self?
9. What are the three stages in role-taking as identified by George Herbert Mead?

Agents of Socialization

Agents of socialization **are the persons, groups, or institutions that teach us what we need to know in order to participate in society.** We are exposed to many agents of socialization throughout our lifetime; in turn, we have an influence on those socializing agents and organizations. Here, we look at the most pervasive ones in childhood—the family, the school, peer groups, and the mass media.

Michael Newman/PhotoEdit

As this birthday celebration attended by three generations of family members illustrates, socialization enables society to "reproduce" itself.

The Family

The family is the most important agent of socialization in all societies. From infancy, our families transmit cultural and social values to us. As discussed later in this book, families in the United States vary in size and structure. Some families consist of two parents and their biological children, whereas others consist of a single parent and one or more children. Still other families reflect changing patterns of divorce and remarriage, and an increasing number are made up of same-sex partners and their children.

Theorists using a functionalist perspective emphasize that families serve important functions in society because they are the primary locus for the procreation and socialization of children. Most of us form an emerging sense of self and acquire most of our beliefs and values within the family context. We also learn about the larger dominant culture (including language, attitudes, beliefs, values, and norms) and the primary subcultures to which our parents and other relatives belong.

Families are also the primary source of emotional support. Ideally, people receive love, understanding, security, acceptance, intimacy, and companionship within families (Benokraitis, 1999). The role of the family is especially significant because young children have little social experience beyond the family's boundaries; they have no basis for comparison or for evaluating how they are treated by their own family.

To a large extent, the family is where we acquire our specific social position in society. From

Agents of socialization   the persons, groups, or institutions that teach us what we need to know in order to participate in society.

birth, we are a part of the specific racial, ethnic, class, religious, and regional subcultural grouping of our family. Studies show that families socialize their children somewhat differently based on race, ethnicity, and class (Kohn, 1977; Kohn et al., 1990; Harrison et al., 1990). For example, sociologist Melvin Kohn (1977; Kohn et al., 1990) has suggested that social class (as measured by parental occupation) is one of the strongest influences on what and how parents teach their children. On the one hand, working-class parents, who are closely supervised and expected to follow orders at work, typically emphasize to their children the importance of obedience and conformity. On the other hand, parents from the middle and professional classes, who have more freedom and flexibility at work, tend to give their children more freedom to make their own decisions and to be creative. Kohn concluded that differences in the parents' occupations were a better predictor of child-rearing practices than was social class itself.

Whether or not Kohn's findings are valid today, the issues he examined make us aware that not everyone has the same family experiences. Many factors—including our cultural background, nation of origin, religion, and gender—are important in determining how we are socialized by family members and others who are a part of our daily life.

Conflict theorists stress that socialization contributes to false consciousness—a lack of awareness and a distorted perception of the reality of class as it affects all aspects of social life. As a result, socialization reaffirms and reproduces the class structure in the next generation rather than challenging the conditions that presently exist. For example, children in low-income families may be unintentionally socialized to believe that acquiring an education and aspiring to lofty ambitions are pointless because of existing economic conditions in the family (Ballantine, 2000). By contrast, middle- and upper-income families typically instill ideas of monetary and social success in children while encouraging them to think and behave in "socially acceptable" ways.

The School

As the amount of specialized technical and scientific knowledge has expanded rapidly and as the amount of time that children are in educational settings has increased, schools continue to play an enormous role in the socialization of young people. For many people, the formal education process is an undertaking that lasts up to twenty years.

As the number of one-parent families and families in which both parents work outside the home has increased dramatically, the number of children in day-care and preschool programs has also grown rapidly. Currently, more than 50 percent of all U.S. preschool children are in day care, either in private homes or institutional settings, and this percentage continues to climb (Children's Defense Fund, 2001). Generally, studies have found that quality day-care and preschool programs have a positive effect on the overall socialization of children. These programs provide children with the opportunity to have frequent interactions with teachers and to learn how to build their language and literacy skills. High-quality programs have a positive effect on the academic performance of children, particularly those from low-income families. For example, several states with pre-kindergarten programs reported an increase in children's math and reading scores, school attendance records, and parents' involvement in their children's education (Children's Defense Fund, 2001).

Although schools teach specific knowledge and skills, they also have a profound effect on children's self-image, beliefs, and values. As children enter school for the first time, they are evaluated and systematically compared with one another by the teacher. A permanent, official record is kept of each child's personal behavior and academic activities. From a functionalist perspective, schools are responsible for (1) socialization, or teaching students to be productive members of society; (2) transmission of culture; (3) social control and personal development; and (4) the selection, training, and placement of individuals on different rungs in the society (Ballantine, 2000).

In contrast, conflict theorists assert that students have different experiences in the school system depending on their social class, their racial–ethnic background, the neighborhood in which they live, their gender, and other factors. According to the sociologists Samuel Bowles and Herbert Gintis (1976), much of what happens in school amounts to teaching a hidden curriculum in which children learn to be neat, to be on time, to be quiet, to wait their turn, and to remain attentive to their work.

Joseph Schuyler/Stock Boston

Day-care centers have become important agents of socialization for increasing numbers of children. Today, more than 50 percent of all U.S. preschool children are in day care of one kind or another.

Thus, schools do not socialize children for their own well-being but rather for their later roles in the work force, where it is important to be punctual and to show deference to supervisors. Students who are destined for leadership or elite positions acquire different skills and knowledge than those who will enter working-class and middle-class occupations (see Cookson and Persell, 1985).

Peer Groups

As soon as we are old enough to have acquaintances outside the home, most of us begin to rely heavily on peer groups as a source of information and approval about social behavior (Lips, 1989). A *peer group* **is a group of people who are linked by common interests, equal social position, and (usually) similar age.** In early childhood, peer groups are often composed of classmates in day care, preschool, and elementary school. Recent studies have found that preadolescence—the latter part of the elementary school years—is an age pe-

riod in which children's peer culture has an important effect on how children perceive themselves and how they internalize society's expectations (Adler and Adler, 1998). In adolescence, peer groups are typically made up of people with similar interests and social activities. As adults, we continue to participate in peer groups of people with whom we share common interests and comparable occupations, income, and/or social position.

Peer groups function as agents of socialization by contributing to our sense of "belonging" and our feelings of self-worth. As early as the preschool years, peer groups provide children with an opportunity for successful adaptation to situations such as gaining access to ongoing play, protecting shared activities from intruders, and building solidarity

Peer group a group of people who are linked by common interests, equal social position, and (usually) similar age.

The pleasure of participating in activities with friends is one of the many attractions of adolescent peer groups. What groups have contributed the most to your own sense of belonging and self-worth?

Michael Newman/PhotoEdit

and mutual trust during ongoing activities (Corsaro, 1985; Rizzo and Corsaro, 1995). Unlike families and schools, peer groups provide children and adolescents with some degree of freedom from parents and other authority figures (Corsaro, 1992). Although peer groups afford children some degree of freedom, they also teach cultural norms such as what constitutes "acceptable" behavior in a specific situation. Peer groups simultaneously reflect the larger culture and serve as a conduit for passing on culture to young people. As a result, the peer group is both a product of culture and one of its major transmitters (Elkin and Handel, 1989).

Is there such a thing as "peer pressure"? Individuals must earn their acceptance by their peers by conforming to a given group's norms, attitudes, speech patterns, and dress codes. When we conform to our peer group's expectations, we are rewarded; if we do not conform, we may be ridiculed or even expelled from the group. Conforming to the demands of peers frequently places children and adolescents at cross-purposes with their parents. Sociologist William A. Corsaro (1992) notes that children experience strong peer pressure even during their preschool years. For example, children are frequently under pressure to obtain certain valued material possessions (such as toys, videotapes, clothing, or athletic shoes); they then pass this pressure on to their parents through emotional pleas to purchase the desired items. In this way, adult caregivers learn about the latest fads and

fashions from children, and they may contribute to the peer culture by purchasing the items desired by the children (Corsaro, 1992).

Mass Media

An agent of socialization that has a profound impact on both children and adults is the *mass media*, composed of large-scale organizations that use print or electronic means (such as radio, television, film, and the Internet) to communicate with large numbers of people. The media function as socializing agents in several ways: (1) they inform us about events; (2) they introduce us to a wide variety of people; (3) they provide an array of viewpoints on current issues; (4) they make us aware of products and services that, if we purchase them, will supposedly help us to be accepted by others; and (5) they entertain us by providing the opportunity to live vicariously (through other people's experiences). Although most of us take for granted that the media play an important part in contemporary socialization, we frequently underestimate the enormous influence this agent of socialization may have on children's attitudes and behavior.

Recent studies have shown that, on average, U.S. children are spending more time each year in front of TV sets, computers, and video games. According to the Annenberg Public Policy Center (University of Pennsylvania) study on media in the

Reflections

Reflections 3.3

How do the mass media influence our thinking about problems associated with childhood?

home, "The introduction of new media continues to transform the environment in American homes with children. . . . Rather than displacing television as the dominant medium, new technologies have supplemented it, resulting in an aggregate increase in electronic media penetration and use by America's youth" (qtd. in Dart, 1999: A5). It is estimated that U.S. children spend 2.5 hours per day watching television programs and about 2 hours with computers, video games, or a VCR, which adds up to about 1,642 hours per year (Dart, 1999). By contrast, U.S. children spend about 1,000 hours per year in school. Considering television watching time alone, by the time that students graduate from high school, they will have spent more time in front of the television set than sitting in the classroom (American Academy of Child and Adolescent Psychiatry, 1997; Dart, 1999). Perhaps it is no surprise that the Annenberg researchers found that 93 percent of children between the ages of 10 and 17 knew that Homer, Bart, and Maggie are characters on the animated Fox series *The Simpsons* whereas only 63 percent could name the current vice president of the United States (Dart, 1999).

Parents, educators, social scientists, and public officials have widely debated the consequences of young people watching that much television. Television has been praised for offering numerous positive experiences to children. Some scholars suggest that television (when used wisely) can enhance children's development by improving their language abilities, concept-formation skills, and reading skills and by encouraging prosocial development (Winn, 1985). However, other studies have shown that children and adolescents who spend a lot of time watching television often have lower grades in school, read fewer books, exercise less, and are overweight (American Academy of Child and Adolescent Psychiatry, 1997).

Of special concern to many people is the issue of television violence. It is estimated that the typical young person who watches 28 hours of televi-

sion a week will have seen 16,000 simulated murders and 200,000 acts of violence by the time he or she reaches age 18. A report by the American Psychological Association states that about 80 percent of all television programs contain acts of violence, and commercial television for children is 50 to 60 times more violent than prime time television for adults. For example, some cartoons average more than 80 violent acts per hour (APA Online, 2000). The violent content of media programming and the marketing and advertising practices of mass media industries that routinely target children under age 17 have come under the scrutiny of government agencies such as the Federal Trade Commission due to concerns raised by parents and social analysts (see Box 3.2 for further discussion).

In addition to concerns about violence in television programming, motion pictures, and electronic games, television shows have been criticized for projecting negative images of women and people of color. Although the mass media have changed some of the roles that they depict women as playing (such as showing Xena, Warrior Princess, who is able to vanquish everything that stands in her way), even these newer images tend to reinforce existing stereotypes of women as sex symbols because of the clothing they wear in their action adventures. Throughout this text, we will look at additional examples of how the media—ranging from advertising and television programs to video games and the Internet—socialize all of us, particularly when we are young, in ways that we may or may not realize.

Gender and Racial–Ethnic Socialization

Gender socialization is the aspect of socialization that contains specific messages and practices concerning the nature of being female or male in a specific group or society. Gender socialization is

✓ Checkpoints

Checkpoint 3.4

10. Why are the following considered to be major agents of socialization: (a) families, (b) schools, (c) peer groups, and (d) mass media?

Box 3.2 Sociology and Social Policy

The Debate Over Regulation of Entertainment Products with Violent Content

Do the motion picture, music recording, and electronic game industries promote products that they themselves acknowledge warrant parental caution in situations where children make up a substantial percentage of the audience? If so, are the advertisements for those products intended to attract children and teenagers?

In a report issued in September 2000, the Federal Trade Commission ("FTC") answered "yes" to both those questions. The report found that, despite criticism, these media industries still routinely target children under age seventeen in marketing products that their own ratings systems deem inappropriate for such audiences or that warrant parental caution due to violent content: Advertisements for violent video games, movies, and music were aired during television shows for younger audiences and printed in teen magazines. The FTC report also noted that children under age seventeen are frequently able to buy tickets to R-rated movies without being accompanied by an adult or to purchase music recordings and electronic games that contain parental advisory labels or are supposed to be sold only to an older audience.

Many parents are quite concerned about the amount of violence on television, in films, and in other forms of entertainment. They are especially concerned about how the media target young audiences when the media are advertising these products. Parents and social analysts assert that ratings systems—such as those for movies and television programs—should be strengthened and enforced in order to protect young audiences from gratuitous depictions of violence.

Although advocates of change argue that "the government" should do something about this, the FTC report did not suggest this. The report recommends only that the motion picture, music recording, and electronic games industry—*voluntarily*— try to do something about this problem. To many critics, this is like having "a fox guard the chicken house." Why do media marketers target young audiences? The reason is clear: Young people are the ones who spend money on these media products, and those who create media products hope to give these audiences what they think the young consumers want.

Why did the FTC report stop short of recommending that "the government" do something to reduce or eliminate inappropriate content targeted at younger audiences? The First Amendment to the United States Constitution prohibits the government from making any law or regulation that would ban or indirectly tend to suppress—to "chill"—free speech or expression. This provision in the nation's Bill of Rights has been interpreted by the courts over the history of this nation in a manner intended to protect individuals from government attempts to suppress political, ideological, or scientific ideas or information. However, there obviously have to be limits on anyone's right to free speech: A person should not be allowed to yell "fire" in a crowded theater, for example. Generally speaking, the courts allow the government to place limits on free speech only to the extent that the government proves that the restriction promotes a compelling government interest and is narrowly tailored to promote that interest. If a less-restrictive alternative would serve the government's purpose, the government must use that alternative.

Is the violent content of media programming and the marketing and advertising practices of mass media industries that routinely target children under age seventeen a compelling government interest? What are the best alternatives to government regulation of the media? What do you think?

Source: Based on Federal Trade Commission, 2000.

important in determining what we *think* the "preferred" sex of a child should be and in influencing our beliefs about acceptable behaviors for males and females.

In some families, gender socialization starts before birth. Parents who learn the sex of the fetus through ultrasound or amniocentesis often purchase color-coded and gender-typed clothes, toys, and nursery decorations in anticipation of their daughter's or son's arrival. After birth, parents may respond differently toward male and female infants; they often play more roughly with boys and talk more lovingly to girls (Eccles, Jacobs, and Harold, 1990). Throughout childhood and adolescence, boys and girls are typically assigned different household chores and given different privileges (such as how late they may stay out at night).

When we look at the relationship between gender socialization and social class, the picture becomes more complex. Although some studies have found less-rigid gender stereotyping in higher-income families (Seegmiller, Suter, and Duviant, 1980; Brooks-Gunn, 1986), others have found more (Bardwell, Cochran, and Walker, 1986). One study found that higher-income families are more likely than low-income families to give "male-oriented" toys (which develop visual spatial and problem-solving skills) to children of both sexes (Serbin et al., 1990). Working-class families tend to adhere to more-rigid gender expectations than do middle-class families (Canter and Ageton, 1984; Brooks-Gunn, 1986).

We are limited in our knowledge about gender socialization practices among racial–ethnic groups because most studies have focused on white, middle-class families. In a study of African American families, the sociologist Janice Hale-Benson (1986) found that children typically are not taught to think of gender strictly in "male–female" terms. Both daughters and sons are socialized toward autonomy, independence, self-confidence, and nurturance of children (Bardwell, Cochran, and Walker, 1986). Sociologist Patricia Hill Collins (1990) has suggested that "othermothers" (women other than a child's biological mother) play an important part in the gender socialization and motivation of African American children, especially girls. Othermothers often serve as gender role models and encourage women to become activists on behalf of their children and community (Collins, 1990). By contrast, studies of Korean American and Latino/a families have found more traditional gender socialization (Min, 1988), although some evidence indicates that this pattern may be changing (Jaramillo and Zapata, 1987).

Schools, peer groups, and the media also contribute to our gender socialization. From kindergarten through college, teachers and peers reward gender-appropriate attitudes and behavior. Sports reinforce traditional gender roles through a rigid division of events into male and female categories. The media are also a powerful source of gender socialization; starting very early in childhood, children's books, television programs, movies, and music provide subtle and not-so-subtle messages about "masculine" and "feminine" behavior. Gender socialization is discussed in more depth in Chapter 10 ("Sex and Gender").

In addition to gender-role socialization, we receive racial socialization throughout our lives. *Racial socialization* **is the aspect of socialization that contains specific messages and practices concerning the nature of one's racial or ethnic status** as it relates to (1) personal and group identity, (2) intergroup and interindividual relationships, and (3) position in the social hierarchy. Racial socialization includes direct statements regarding race, modeling behavior (wherein a child imitates the behavior of a parent or other caregiver), and indirect activities such as exposure to specific objects, contexts, and environments that represent one's racial–ethnic group (Thornton et al., 1990).

The most important aspects of our racial identity and attitudes toward other racial–ethnic groups are passed down in our families from generation to generation. As discussed in Chapter 2, some of the core values of U.S. society may support racist beliefs. As the sociologist Martin Marger (1994: 97) notes, "Fear of, dislike for, and antipathy toward one group or another is learned in much the same way that people learn to eat with a knife or fork rather than with their bare hands or to respect others' privacy in personal matters." These beliefs can be transmitted in subtle and largely unconscious ways; they do not have to be taught directly or intentionally. Scholars have found that ethnic values and attitudes begin to crystallize among children as young as age four (Goodman, 1964; Porter, 1971). By this age, the society's ethnic hierarchy has become apparent to the child (Marger, 2000). Some minority parents feel that racial socialization is essential because it provides children with the skills and abilities they will need to survive in the larger society (Hale-Benson, 1986).

Gender socialization the aspect of socialization that contains specific messages and practices concerning the nature of being female or male in a specific group or society.

Racial socialization the aspect of socialization that contains specific messages and practices concerning the nature of one's racial or ethnic status.

✓Checkpoints

Checkpoint 3.5

11. What is gender socialization?
12. What is racial socialization, and how is it passed down from one generation to the next?

Socialization Through the Life Course

Why is socialization a lifelong process? Throughout our lives, we continue to learn. Each time we experience a change in status (such as becoming a college student or getting married), we learn a new set of rules, roles, and relationships. Even before we achieve a new status, we often participate in *anticipatory socialization*—the process by which knowledge and skills are learned for future roles. Many societies organize social activities according to age and gather data regarding the age composition of the people who live in that society. For example (see "Census Profile: Age of the U.S. Population"), the U.S. Census Bureau gathers and maintains those data in the United States. Some societies have distinct *rites of passage*, based on age or other factors, that publicly dramatize and validate changes in a person's status. In the United States and other industrialized societies, the most common categories of age are infancy, childhood, adolescence, and adulthood (often subdivided into young adulthood, middle adulthood, and older adulthood).

Infancy and Childhood

Some social scientists believe that a child's sense of self is formed at a very early age and that it is difficult to change this self-perception later in life. Symbolic interactionists emphasize that during infancy and early childhood, family support and guidance are crucial to a child's developing self-concept. In some families, children are provided with emotional warmth, feelings of mutual trust, and a sense of security. These families come closer to our ideal cultural belief that childhood should

CENSUS PROFILES Age of the U.S. Population

❹ What is this person's age and what is this person's date of birth?

Age on April 1, 2000

☐ ☐

Print numbers in boxes.

Month Day Year of Birth

☐ ☐ ☐ ☐ ☐ ☐ ☐ ☐

Just as age is a crucial variable in the socialization process, the U.S. Census Bureau gathers data about people's age so that the government and other interested parties will know how many individuals residing in this country are in different age categories. This chapter examines how a person's age is related to socialization and one's life experiences. Shown below is a depiction of the nation's population in the year 2000, separated into three broad age categories.

Below age 25   Age 25 to 54   Age 55 and above

Can age be a source of social cohesion among people? Why might age differences produce conflict among individuals in different age groups? What do you think?

Source: U.S. Census Bureau, 2001f.

be a time of carefree play, safety, and freedom from economic, political, and sexual responsibilities. However, other families reflect the discrepancy between cultural ideals and reality—children grow up in a setting characterized by fear, danger, and risks that are created by parental neglect, emotional maltreatment, or premature economic and sexual demands (Knudsen, 1992).

Social psychological theories of human development typically suggest that individuals go through stages of cognitive and moral development. According to Lawrence Kohlberg, during the conventional level of moral development, people are most concerned with how they are perceived by their peers.

Abused children often experience low self-esteem, an inability to trust others, feelings of isolationism and powerlessness, and denial of their feelings. However, the manner in which parental abuse affects children's ongoing development is subject to much debate and uncertainty. For example, some scholars and therapists assert that the intergenerational hypothesis—the idea that abused children will become abusive parents—is valid, but others have found little support for this hypothesis (Knudsen, 1992).

Adolescence

In industrialized societies, the adolescent (or teenage) years represent a buffer between childhood and adulthood. In the United States, no specific rites of passage exist to mark children's move into adulthood; therefore, young people have to pursue their own routes to self-identity and adulthood (Gilmore, 1990). Anticipatory socialization is often associated with adolescence,

during which many young people spend much of their time planning or being educated for future roles they hope to occupy. However, other adolescents (such as eleven- and twelve-year-old mothers) may have to plunge into adult responsibilities at this time.

Adolescence is often characterized by emotional and social unrest. In the process of developing their own identities, some young people come into conflict with parents, teachers, and other authority figures who attempt to restrict their freedom. Adolescents may also find themselves caught between the demands of adulthood and their own lack of financial independence and experience in the job market. The experiences of individuals during adolescence vary according to race, class, and gender. Based on their

Anticipatory socialization the process by which knowledge and skills are learned for future roles.

family's economic situation, some young people move directly into the adult world of work. However, those from upper-middle-class and upper-class families may extend adolescence into their late twenties or early thirties by attending graduate or professional school and then receiving additional advice and financial support from their parents as they start their own families, careers, or businesses.

Adulthood

One of the major differences between child socialization and adult socialization is the degree of freedom of choice. If young adults are able to support themselves financially, they gain the ability to make more choices about their own lives. In early adulthood (usually until about age forty), people work toward their own goals of creating meaningful relationships with others, finding employment, and seeking personal fulfillment. Of course, young adults continue to be socialized by their parents, teachers, peers, and the media, but they also learn new attitudes and behaviors. For example, when we marry or have children, we learn new roles as partners or parents. Adults often learn about fads and fashions in clothing, music, and language from their children. Parents in one study indicated that they had learned new attitudes and behaviors about drug use, sexuality, sports, leisure, and racial–ethnic issues from their college-age children (Peters, 1985).

Workplace (occupational) socialization is one of the most important types of adult socialization. Sociologist Wilbert Moore (1968) divided occupational socialization into four phases: (1) career choice, (2) anticipatory socialization (learning different aspects of the occupation before entering it), (3) conditioning and commitment (learning the "ups" and "downs" of the occupation and remaining committed to it), and (4) continuous commitment (remaining committed to the work even when problems or other alternatives may arise). This type of socialization tends to be most intense immediately after a person makes the transition from school to the workplace; however, this process continues throughout our years of employment. In the future, many people will experience continuous workplace socialization as a result

of having more than one career in their lifetime (Lefrançois, 1999).

Between the ages of forty and sixty-five, people enter middle adulthood, and many begin to compare their accomplishments with their earlier expectations. This is the point at which people either decide that they have reached their goals or recognize that they have attained as much as they are likely to achieve.

Some sociologists subdivide late adulthood into three categories: (1) the "young-old" (ages 65 to 74), (2) the "old-old" (ages 75 to 85), and (3) the "oldest-old" (over age 85) (see Moody, 1998). Although these are somewhat arbitrary divisions, the "young-old" are less likely to suffer from disabling illnesses, whereas some of the "old-old" are more likely to suffer such illnesses (Belsky, 1990). However, a recent study found that the prevalence of disability among those 85 and over decreased during the 1980s due to better health care.

In older adulthood, some people are quite happy and content; others are not. Erik Erikson noted that difficult changes in adult attitudes and behavior occur in the last years of life, when people experience decreased physical ability, lower prestige, and the prospect of death. Older adults in industrialized societies have experienced *social devaluation*—wherein a person or group is considered to have less social value than other persons or groups. Social devaluation is especially acute when people are leaving roles that have defined their sense of social identity and provided them with meaningful activity (Achenbaum, 1978).

Late Adulthood and Ageism

Negative images regarding older persons reinforce *ageism*—prejudice and discrimination against people on the basis of age, particularly against older persons (Butler, 1975). Ageism against older persons is rooted in the assumption that people become unattractive, unintelligent, asexual, unemployable, and mentally incompetent as they grow older (Comfort, 1976).

Ageism is reinforced by stereotypes, whereby people have narrow, fixed images of certain groups. One-sided and exaggerated images of older people are used repeatedly in everyday life. Older persons are often stereotyped as thinking and moving

Roberto Soncin Gerometta/Photo 20-20

Various societies treat older people differently as a result of how people are socialized to think about those in late adulthood. Do you believe that ageism is a problem in the United States? Why or why not?

slowly; as being bound to themselves and their past, unable to change and grow; as being unable to move forward and often moving backward (Belsky, 1990). They are viewed as cranky, sickly, and lacking in social value (Atchley, 2000); as egocentric and demanding; as shallow, enfeebled, aimless, and absentminded (Belsky, 1990).

Negative images also contribute to the view that women are "old" ten or fifteen years sooner than men (Bell, 1989). The multi-billion-dollar cosmetics industry helps perpetuate the myth that age reduces the "sexual value" of women but increases it for men. Men's sexual value is defined more in terms of personality, intelligence, and earning power than by physical appearance. For women, however, sexual attractiveness is based on

youthful appearance. By idealizing this "youthful" image of women and playing up the fear of growing older, sponsors sell thousands of products that claim to prevent the "ravages" of aging.

Although not all people act on appearances alone, Patricia Moore, an industrial designer, found that many do. At age 27, Moore disguised herself as an 85-year-old woman by donning age-appropriate clothing and placing baby oil in her eyes to create the appearance of cataracts. With the help of a makeup artist, Moore supplemented the "aging process" with latex wrinkles, stained teeth, and a gray wig. For three years, "Old Pat Moore" went to various locations, including a grocery store, to see how people responded to her:

> When I did my grocery shopping while in character, I learned quickly that the Old Pat Moore behaved—and was treated—differently from the Young Pat Moore. When I was 85, people were more likely to jockey ahead of me in the checkout line. And even more interesting, I found that when it happened, I didn't say anything to the offender, as I certainly would at age 27. It seemed somehow, even to me, that it was okay for them to do this to the Old Pat Moore, since they were undoubtedly busier than I was anyway. And further, they apparently thought it was okay, too! After all, little old ladies have plenty of time, don't they? And then when I did get to the checkout counter, the clerk might start yelling, assuming I was deaf, or becoming immediately testy, assuming I would take a long time to get my money out, or would ask to have the price repeated, or somehow become confused about the transaction. What it all added up to was that people feared I would be trouble, so they tried to have as little to do with me as possible. And the amazing thing is that I began almost to believe it myself. . . . I think perhaps the worst thing about aging may be the overwhelming sense that everything around you is letting you know that you are not terribly important any more. (Moore with Conn, 1985: 75–76)

Social devaluation a situation in which a person or group is considered to have less social value than other individuals or groups.

Ageism prejudice and discrimination against people on the basis of age, particularly against older people.

Larry Kolvoord/The Image Works

People in military training are resocialized through extensive, grueling military drills and maneuvers. What new values and behaviors are learned in marching drills such as this?

If we apply our sociological imagination to Moore's study, we find that "Old Pat Moore's" experiences reflect what many older persons already know—it is other people's *reactions* to their age, not their age itself, that place them at a disadvantage.

Many older people buffer themselves against ageism by continuing to view themselves as being in middle adulthood long after their actual chronological age would suggest otherwise. Other people begin a process of resocialization to redefine their own identity as mature adults.

✓Checkpoints

Checkpoint 3.6

13. What is anticipatory socialization, and when is it most likely to occur?
14. What is a major difference between childhood socialization and adult socialization?
15. Why is workplace (occupational) socialization one of the most important types of adult socialization?
16. What is ageism, and how does it affect people?

Resocialization

Resocialization is the process of learning a new and different set of attitudes, values, and behaviors from those in one's background and previous experience. Resocialization may be voluntary or involuntary. In either case, people undergo changes that are much more rapid and pervasive than the gradual adaptations that socialization usually involves.

Voluntary Resocialization

Resocialization is voluntary when we assume a new status (such as becoming a student, an employee, or a retiree) of our own free will. Sometimes, voluntary resocialization involves medical or psychological treatment or religious conversion, in which case the person's existing attitudes, beliefs, and behaviors must undergo strenuous modification to a new regime and a new way of life. For example, resocialization for adult survivors of emotional/physical child abuse includes extensive therapy in order to form new patterns of thinking and action, somewhat like Alcoholics Anonymous and its twelve-step program, which has become the basis for many other programs dealing with addictive behavior (Parrish, 1990).

Involuntary Resocialization

Involuntary resocialization occurs against a person's wishes and generally takes place within a *total institution*—a place where people are isolated from the rest of society for a set period of time and come under the control of the officials who run the institution (Goffman, 1961a). Military boot camps, jails

Reflections

Reflections 3.4

Would it be possible to somehow resocialize parents who do not provide social support for their children?

and prisons, concentration camps, and some mental hospitals are total institutions. In these settings, people are totally stripped of their former selves—or depersonalized—through a degradation ceremony (Goffman, 1961a). For example, inmates entering prison are required to strip, shower, and wear assigned institutional clothing. In the process, they are searched, weighed, fingerprinted, photographed, and given no privacy even in showers and restrooms. Their official identification becomes not a name but a number. In this abrupt break from their former existence, they must leave behind their personal possessions and their family and friends. The depersonalization process continues as they are required to obey rigid rules and to conform to their new environment.

After stripping people of their former identities, the institution attempts to build a more compliant person. A system of rewards and punishments (such as providing or withholding television or exercise privileges) encourages conformity to institutional norms. Some individuals may be rehabilitated; others become angry and hostile toward the system that has taken away their freedom. Although the assumed purpose of involuntary resocialization is to reform persons so that they will conform to societal standards of conduct after their release, the ability of total institutions to modify offenders' behavior in a meaningful manner has been widely questioned. In many prisons, for example, inmates may conform to the norms of the prison or of other inmates, but little relationship exists between those norms and the laws of society.

Socialization in the Future

In the future, the family is likely to remain the institution that most fundamentally shapes and nurtures personal values and self-identity. However, parents may increasingly feel overburdened by this responsibility, especially without societal support—such as high-quality, affordable child care—and more education in parenting skills. Some analysts have suggested that there will be an increase in known cases of child maltreatment and in the number of children who experience delayed psychosocial development, learning difficulties, and emotional and behavioral problems (see Box 3.3 for suggestions on how to prevent child maltreatment). Such analysts attribute these increases to the dramatic changes occurring in the size, structure, and economic stability of families.

A central value-oriented issue facing parents and teachers as they attempt to socialize children is the growing dominance of the mass media and other forms of technology. For example, interactive television and computer networking systems will enable children to experience many things outside their own homes and schools and to communicate regularly with people around the world. If futurists are correct in predicting that ideas and information and access to them will be the basis for personal, business, and political advancement in the twenty-first century, persons without access to computers and other information technology will become even more disadvantaged. This prediction raises important issues about the effects of social class and race on the socialization process. Socialization—a lifelong learning process—can no longer be viewed as a "glance in the rearview mirror" or as a reaction to some previous experience. With the rapid pace of technological change, we must not only learn about the past but also learn how to anticipate the future—and consider its consequences (Westrum, 1991).

Resocialization the process of learning a new and different set of attitudes, values, and behaviors from those in one's background and previous experience.

Total institution Erving Goffman's term for a place where people are isolated from the rest of society for a set period of time and come under the control of the officials who run the institution.

Box 3.3 You Can Make a Difference
Helping a Child Reach Adulthood

After Tina—one of your best friends—moves into a large apartment complex near her university, she keeps hearing a baby cry at all hours of the day and night. Although the crying is coming from the apartment next to Tina's, she never sees anyone come or go from it. On several occasions, she knocks on the door, but no one answers. At first Tina tries to ignore the situation, but eventually she can't sleep or study because the baby keeps crying. Tina decides she must take action and asks you, "What do you think I ought to do?" What advice could you give Tina?

Like Tina, many of us do not know if we should get involved in other people's lives. We also do not know how to report child maltreatment. However, social workers and researchers suggest that bystanders must be willing to get involved in cases of possible abuse or neglect to save a child from harm by others. They also note the importance of people knowing how to report incidents of maltreatment:

• *Report child maltreatment.* Cases of child maltreatment can be reported to any social service or law enforcement agency.

It has been said that it takes a village to raise a child. In contemporary societies, it takes many people pulling together to help a child have a safe and happy childhood and a productive adulthood. Are there ways in which you, like the man in this photo, can help a young person in your community?

• *Identify yourself to authorities.* Although most agencies are willing to accept anonymous reports, many staff members prefer to know your name, address, telephone number, and other basic information so that they can determine that you are not a self-interested person such as a hostile relative, ex-spouse, or vindictive neighbor.

• *Follow up with authorities.* Once an agency has validated a report of child maltreatment, the agency's first goal is to stop the neglect or abuse of that child, whose health and safety are paramount concerns. However, intervention also has long-term goals. Sometimes, the situation can be improved simply by teaching the parents different values about child rearing or by pointing them to other agencies and organizations that can provide needed help. Other times, it may be necessary to remove the child from the parents' custody and place the child in a foster home, at least temporarily. Either way, the situation for the child will be better than if he or she had been left in an abusive or neglectful home environment.

So the best advice for Tina—or for anyone else who has reason to believe that child maltreatment is occurring—is to report it to the appropriate authorities. In most telephone directories, the number can be located in the government listings section. Here are some other resources for help:

• Child Help USA offers a 24-hour crisis hot line, national information, and referral network for support groups and therapists and for reporting suspected abuse: 15757 North 78th Street, Scottsdale, AZ 85260. (800) 422-4453.

• Child Welfare League of America, a Washington, D.C., association of nearly 800 public and private nonprofit agencies, serves as an advocacy group for children who have experienced maltreatment: 440 First Street, NW, Suite 310, Washington, DC 20001-2085. (202) 638-2952.

• The National Committee to Prevent Child Abuse (NCPCA) is a nonprofit organization that seeks the prevention of child abuse and the promotion of public education about the causes and consequences of child maltreatment. For the location of the chapter nearest you, contact P.O. Box 2866, Chicago, IL 60690. (800) 556-2722.

• On the Internet, the National Center for Missing and Exploited Children provides brochures about child safety and child protection upon request:

http://www.missingkids.org/index.html

Chapter Review

What is socialization, and why is it important for human beings?

Socialization is the lifelong process through which individuals acquire their self-identity and learn the physical, mental, and social skills needed for survival in society. The kind of person we become depends greatly on what we learn during our formative years from our surrounding social groups and social environment.

How much of our unique human characteristics comes from heredity and how much from our social environment?

As individual human beings, we have unique identities, personalities, and relationships with others. Individuals are born with some of their unique physical characteristics; other characteristics and traits are gained during the socialization process. Each of us is a product of two forces: (1) heredity, referred to as "nature," and (2) the social environment, referred to as "nurture." Whereas biology dictates our physical makeup, the social environment largely determines how we develop and behave.

Why is social contact essential for human beings?

Social contact is essential in developing a self, or self-concept, which represents an individual's perceptions and feelings of being a distinct or separate person. Much of what we think about ourselves is gained from our interactions with others and from what we perceive that others think of us.

What are the main social psychological theories on human development?

According to Sigmund Freud, the self emerges from three interrelated forces: the id, the ego, and the superego. When a person is well adjusted, the three forces act in balance. Jean Piaget identified four cognitive stages of development: (1) sensorimotor, in which children understand the world only through sensory contact and immediate action; (2) preoperational, the ability to use words as mental symbols; (3) concrete operational, in which children think in terms of tangible objects and actual events; and (4) formal operational, the ability to engage in abstract thought.

How do sociologists believe that we develop a self-concept?

According to Charles Horton Cooley's concept of the looking-glass self, we develop a self-concept as we see ourselves through the perceptions of others. Our initial sense of self is typically based on how families perceive and treat us. George Herbert Mead suggested that we develop a self-concept through role-taking and learning the rules of social interaction. According to Mead, the self is divided into the "I" and the "me." The "I" represents the spontaneous and unique traits of each person. The "me" represents the internalized attitudes and demands of other members of society.

What is Lawrence Kohlberg's perspective on moral development?

Lawrence Kohlberg classified moral development into stages; certain levels of cognitive development are essential before corresponding levels of moral reasoning may occur. By contrast, Carol Gilligan suggested that there are male–female differences regarding morality and identified three stages of female moral development.

What are the primary agents of socialization?

The agents of socialization include the family, schools, peer groups, and the media. Our families, which transmit cultural and social values to us, are the most important agents of socialization in all societies, serving these functions: (1) procreating and socializing children, (2) providing emotional support, and (3) assigning social position. Schools primarily teach knowledge and skills but also have a profound influence on the self-image, beliefs, and values of children. Peer groups contribute to our sense of belonging and self-worth, and are a key source of information about acceptable behavior. The media function as socializing agents by (1) informing us about world events, (2) introducing us to a wide variety of people, and (3) providing an opportunity to live vicariously through other people's experiences.

Are socialization practices universal?

No, social class, gender, and race are all determining factors in socialization practices. Social class is one of the strongest influences on what and how parents teach their children. Gender socialization strongly influences what we believe to be acceptable behavior for females and males. Racial socialization concerns messages and practices about the nature of racial status as it relates to personal and group identity, relationships within groups and between individuals, and position in the social hierarchy.

When does socialization end?

Socialization is ongoing throughout the life course. We learn knowledge and skills for future roles through anticipatory socialization. Parents are socialized by their own children, and adults learn through workplace socialization. Resocialization is the process of learning new attitudes, values, and behaviors, either voluntarily or involuntarily.

Key Terms

ageism 100
agents of socialization 90
anticipatory socialization 98
ego 82
gender socialization 95
generalized other 89
id 82
looking-glass self 86
peer group 93
racial socialization 97
resocialization 102
role-taking 87
self-concept 86
significant others 87
social devaluation 100
socialization 77
sociobiology 79
superego 82
total institution 102

Questions for Critical Thinking

1. Consider the concept of the looking-glass self. How do you think others perceive you? Do you think most people perceive you correctly?
2. What are your "I" traits? What are your "me" traits? Which ones are stronger?
3. What are some different ways that you might study the effect of toys on the socialization of children? How could you isolate the toy variable from other variables that influence children's socialization?
4. Is the attempted rehabilitation of criminal offenders—through boot camp programs, for example—a form of socialization or resocialization?

 # Resources on the Internet

Chapter-Related Web Sites
The following web sites have been selected for their relevance to the topics in this chapter. These sites are among the more stable, but please note that web site addresses change frequently.

Gilligan's *In a Different Voice*
http://www.stolaf.edu/people/huff/classes/
handbook/Gilligan.html
An extensive examination of Carol Gilligan's theories of moral development in women.

Secret of the Wild Child
http://www.pbs.org/wgbh/nova/transcripts/2112gchild.html
This is a transcript of a PBS *Nova* documentary on Genie, a little girl who was socially isolated during her childhood, and the attempts to resocialize her. (See page 81 in this chapter.)

Mass Media—Television and News Coverage: Selected Bibliography
http://www.wfu.edu/~louden/
Political%20Communication/Bibs/MEDIA.html
An extensive bibliography from the Communication Department at Wake Forest University.

Mid-Life Crisis: Recent Research
http://www.hope.edu/academic/psychology/335/
webrep2/crisis.html
A literature review on research related to mid-life. The site includes a link to another report on mid-life transition based on the Meyers–Briggs research.

"Surviving Teenage Moods"
http://www.parentingteens.com/commun14.shtml
This article, by Susan Knox, has helpful advice for parents who are trying to improve communication with their teens.

Education and Gender Differences
http://serendip.brynmawr.edu/sci_edu/education/
genderdiff.html
Links to a number of studies that have investigated gender socialization in school.

Grandparents Day 2002: Sept. 8
http://www.census.gov/Press-Release/www/2002/
cb02ff14.html
Census 2000 data on grandparents in America.

Companion Web Site for This Book

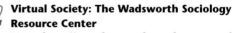 ### Virtual Society: The Wadsworth Sociology Resource Center
Visit **http://sociology.wadsworth.com** and click on the page for Kendall, *Sociology in Our Times: The Essentials*, Fourth Edition, to access a wide range of enrichment material to aid in your study of sociology. Click on the Student Resources section of the web site. Next, select from the pull-down menu the chapter that you are presently studying. Among the useful options for self-study are chapter objectives, flashcards, practical study tips, and practice tests for each chapter.

MicroCase Online

From the Virtual Society home page, click on MicroCase Online to access book-specific MicroCase exercises that allow you to further explore sociological issues and principles presented in the text.

InfoTrac College Edition

Another unique option available to you at the Student Resources section of the companion web site is InfoTrac College Edition, an online library with access to hundreds of scholarly and popular periodicals. Below are suggested search terms for this chapter. Results from these and other searches are found at the site.

- Search keywords: *child abuse*. Search for articles that address the causes of child abuse and neglect, and write a review of these articles in which you summarize the most important causes.
- Search keywords: *gender socialization*. Locate articles that deal with gender socialization among young people. What agents of socialization are discussed in the articles you found?

Virtual Explorations CD-ROM

Go to the Virtual Explorations CD-ROM to begin the interactive exercise for this chapter. This Virtual Exploration will introduce you to some of the exciting resources for sociology on the World Wide Web. You will be guided through an exercise that employs related web sites on socialization. Answer the questions, and e-mail your responses to your instructor.

Social Structure and Interaction in Everyday Life

Christopher Bladwin/SIS

Social Structure: The Macrolevel Perspective

Components of Social Structure
Status
Roles
Groups
Social Institutions

Societies: Changes in Social Structure
Durkheim: Mechanical and Organic Solidarity
Tönnies: *Gemeinschaft* and *Gesellschaft*
Industrial and Postindustrial Societies

Social Interaction: The Microlevel Perspective
Social Interaction and Meaning
The Social Construction of Reality
Ethnomethodology
Dramaturgical Analysis
The Sociology of Emotions
Nonverbal Communication

Changing Social Structure and Interaction in the Future

4

I began Dumpster diving [scavenging in a large garbage bin] about a year before I became homeless. . . . The area I frequent is inhabited by many affluent college students. I am not here by chance; the Dumpsters in this area are very rich. Students throw out many good things, including food. In particular they tend to throw everything out when they move at the end of a semester, before and after breaks, and around midterm, when many of them despair of college. So I find it advantageous to keep an eye on the academic calendar. I learned to scavenge gradually, on my own. Since then I have initiated several companions into the trade. I have learned that there is a predictable series of stages a person goes through in learning to scavenge.

At first the new scavenger is filled with disgust and self-loathing. He is ashamed of being seen and may lurk around, trying to duck behind things, or he may dive at night. (In fact, most people instinctively look away from a scavenger. By skulking around, the novice calls attention to himself and arouses suspicion. Diving at night is ineffective and needlessly messy.) . . . That stage passes with experience. The scavenger finds a pair of running shoes that fit and look and smell brand-new. . . . He begins to understand: People throw away perfectly good stuff, a lot of perfectly good stuff.

At this stage, Dumpster shyness begins to dissipate. The diver, after all, has the last laugh. He is finding all manner of good things that are his for the taking. Those who disparage his profession are the fools, not he.

Bob Collins/The Image Works

All activities in life—including scavenging in garbage bins and living "on the streets"—are social in nature.

—Author Lars Eighner recalls his experiences as a Dumpster diver while living under a shower curtain in a stand of bamboo in a public park. Eighner became homeless when he was evicted from his "shack" after being unemployed for about a year. (Eighner, 1993: 111–119)

ighner's "diving" activities reflect a specific pattern of social behavior. All activities in life—including scavenging in garbage bins and living "on the streets"—are social in nature. Homeless persons and domiciled persons (those with homes) live in social worlds that have predictable patterns of social interaction. *Social interaction* is the **process by which people act toward or respond to other people and is the foundation for all relationships and groups in society.** In this chapter, we look at the relationship between social structure and social interaction. In the process, homelessness is used as an example of how social problems occur and how they may be perpetuated within social structures and patterns of interaction.

Social structure is the complex framework of **societal institutions (such as the economy, politics, and religion) and the social practices (such as rules and social roles) that make up a society and that**

organize and establish limits on people's behavior. This structure is essential for the survival of society and for the well-being of individuals because it provides a social web of familial support and social relationships that connects each of us to the larger society. Many homeless people have lost this vital linkage. As a result, they often experience a loss of personal dignity and a sense of moral worth because of their "homeless" condition (Snow and Anderson, 1993).

Who are the homeless? Before reading on, take the quiz on homelessness in Box 4.1. The characteristics of the homeless population in the United States vary widely. Among the homeless are single men, single women, and families. In recent years, families have accounted for almost half of the homeless population (U.S. Conference of Mayors, 2000). Further, people of color are overrepresented among the homeless. In 2000, African Americans made up 50 percent of the homeless population, whites (Caucasians)

Box 4.1 Sociology and Everyday Life

How Much Do You Know About Homeless Persons?

True False

True	False		
T	F	1.	Most homeless people choose to be homeless.
T	F	2.	Homelessness is largely a self-inflicted condition.
T	F	3.	Homeless people do not work.
T	F	4.	Most homeless people are mentally ill.
T	F	5.	Homeless people typically panhandle (beg for money) so that they can buy alcohol or drugs.
T	F	6.	Most homeless people are heavy drug users.
T	F	7.	A large number of homeless persons are dangerous.
T	F	8.	Homeless persons have existed throughout the history of the United States.
T	F	9.	One out of every four homeless persons is a child.
T	F	10.	Some homeless people have attended college and graduate school.

Answers on page 112.

30 percent, Latinas/os (Hispanics) 10 percent, Native Americans 2 percent, and Asian Americans 1 percent (U.S. Conference of Mayors, 2000). These percentages obviously vary across communities and in different areas of the country.

Homeless persons come from all walks of life. They include undocumented workers, parolees, runaway youths and children, Vietnam veterans, and the elderly. They live in cities, suburbs, and rural areas (U.S. Department of Agriculture, 1996). Contrary to popular myths, most of the homeless are not on the streets by choice or because they were deinstitutionalized by mental hospitals. Not all of the homeless are unemployed. About 26 percent of homeless people hold full- or part-time jobs but earn too little to find an affordable place to live (U.S. Conference of Mayors, 2000).

Questions and Issues

Chapter Focus Question: How is homelessness related to the social structure of a society?

What are the components of social structure?

How do societies maintain social solidarity and continue to function in times of rapid social change?

Why do societies have shared patterns of social interaction?

How are daily interactions similar to being onstage?

Do positive changes in society occur through individual efforts or institutional efforts?

Social Structure: The Macrolevel Perspective

Social structure provides the framework within which we interact with others. This framework is an orderly, fixed arrangement of parts that together make up the whole group or society (see Figure 4.1). As defined in Chapter 1, a *society* is a large social grouping that shares the same geographical territory and is subject to the same political author-

ity and dominant cultural expectations. At the macrolevel, the social structure of a society has several essential elements: social institutions, groups, statuses, roles, and norms.

Functional theorists emphasize that social structure is essential because it creates order and predictability in a society (Parsons, 1951). Social structure is also important for our human development. As we saw in Chapter 3, we develop a self-concept as we learn the attitudes, values, and behaviors of the people around us. When these at-

Figure 4.1 Social Structure Framework

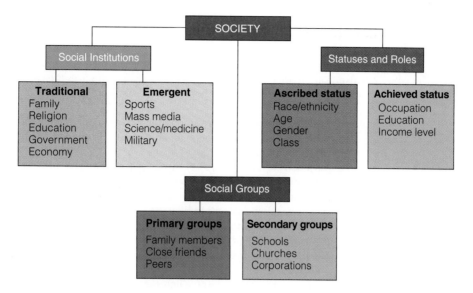

titudes and values are part of a predictable structure, it is easier to develop that self-concept.

Social structure gives us the ability to interpret the social situations we encounter. For example, we expect our families to care for us, our schools to educate us, and our police to protect us. When our circumstances change dramatically, most of us feel an acute sense of anxiety because we do not know what to expect or what is expected of us. For example, newly homeless individuals may feel disoriented because they do not know how to function in their new setting. The person is likely to ask questions: "How will I survive on the streets?" "Where do I go to get help?" "Should I stay at a shelter?" "Where can I get a job?" Social structure helps people make sense out of their environment even when they find themselves on the streets. As sociologists David Snow and Leon Anderson (1993) suggest in their study of unattached, homeless men, survival strategies are the product of the interplay between the resourcefulness and ingenuity of the homeless and local political and ecological constraints.

In addition to providing a map for our encounters with others, social structure may limit our options and place us in arbitrary categories not of our own choosing. Conflict theorists maintain that there is more to social structure than is readily visible and that we must explore the deeper, underlying structures that determine social rela-

tions in a society. For example, Karl Marx suggested that the way economic production is organized is the most important structural aspect of any society. In capitalistic societies, where a few people control the labor of many, the social structure reflects a system of relationships of domination among categories of people (for example, owner–worker and employer–employee).

Social structure creates boundaries that define which persons or groups will be the "insiders" and which will be the "outsiders." *Social marginality* is the state of being part insider and part outsider in the social structure. Sociologist Robert Park (1928) coined this term to refer to persons (such as immigrants) who simultaneously share the life and traditions of two distinct groups. Social marginality results in stigmatization. A *stigma* is any physical or social attribute or sign that so devalues a person's social identity that it disqualifies that person

Social interaction the process by which people act toward or respond to other people; the foundation for all relationships and groups in society.

Social structure the complex framework of societal institutions and the social practices that make up a society and that organize and establish limits on people's behavior.

Box 4.1

Answers to the Sociology Quiz on Homeless Persons

1. **False.** Less than 6 percent of all homeless people are that way by choice.
2. **False.** Most homeless persons did not inflict upon themselves the conditions that produced their homelessness. Some are the victims of child abuse or violence.
3. **False.** Many homeless people are among the working poor. Minimum-wage jobs do not pay enough for an individual to support a family or pay inner-city rent.
4. **False.** Most homeless people are not mentally ill; estimates suggest that about one-fourth of the homeless are emotionally disturbed.
5. **False.** Many homeless persons panhandle to pay for food, a bed at a shelter, or other survival needs.
6. **False.** Most homeless people are not heavy drug users. Estimates suggest that about one-third of the homeless are substance abusers. Many of these are part of the same one-fourth of the homeless who are mentally ill.
7. **False.** Although an encounter with a homeless person occasionally ends in tragedy, most homeless persons are among the least threatening members of society. They are more often the victims of crime, not the perpetrators.
8. **True.** Scholars have found that homelessness has always existed in the United States. However, the number of homeless persons has increased or decreased with fluctuations in the national economy.
9. **True.** Families with children are the fastest growing category of homeless persons in the United States. The number of such families nearly doubled between 1984 and 1989, and continues to do so. Many homeless children are alone. They may be runaways or "throwaways" whose parents do not want them to return home.
10. **True.** Some homeless persons have attended college and graduate school, and many have completed high school.

Sources: Based on Kroloff, 1993; Liebow, 1993; Snow and Anderson, 1993; U.S. Conference of Mayors, 2000; Vissing, 1996; and Waxman and Hinderliter, 1996.

from full social acceptance (Goffman, 1963b). A convicted criminal, wearing a prison uniform, is an example of a person who has been stigmatized; the uniform says that the person has done something wrong and should not be allowed unsupervised outside the prison walls.

✓ Checkpoints

Checkpoint 4.1

1. What do sociologists mean by the term *social structure*?

Components of Social Structure

The social structure of a society includes its social positions, the relationships among those positions, and the kinds of resources attached to each of the positions. Social structure also includes all the groups that make up society and the relationships among those groups (Smelser, 1988). We begin by

Reflections

Reflections 4.1

What structural factors in society may increase the rate of homelessness?

examining the social positions that are closest to the individual.

Status

A *status* **is a socially defined position in a group or society characterized by certain expectations, rights, and duties.** Statuses exist independently of the specific people occupying them (Linton, 1936); the statuses of professional athlete, rock musician, professor, college student, and homeless person all exist exclusive of the specific individuals who occupy these social positions. For example, although thousands of new students arrive on college campuses each year to occupy the status of first-year student, the status of college student and the expectations attached to that position have remained relatively unchanged for the past century.

Does the term *status* refer only to high-level positions in society? No, not in a sociological sense. Although many people equate the term *status* with high levels of prestige, sociologists use it to refer to all socially defined positions—high rank and low rank. For example, both the position of director of the Department of Health and Human Services in Washington, D.C., and that of a homeless person who is paid about five dollars a week (plus bed and board) to clean up the dining room at a homeless shelter are social statuses (see Snow and Anderson, 1993).

Take a moment to answer the question "Who am I?" To determine who you are, you must think about your social identity, which is derived from the statuses you occupy and is based on your status set. A status set comprises all the statuses that a person occupies at a given time. For example, Maria may be a psychologist, a professor, a wife, a mother, a Catholic, a school volunteer, a Texas resident, and a Mexican American. All of these socially defined positions constitute her status set.

ASCRIBED STATUS AND ACHIEVED STATUS
Statuses are distinguished by the manner in which we acquire them. An *ascribed status* **is a social position conferred at birth or received involuntarily later in life, based on attributes over which the individual has little or no control, such as race/ethnicity, age, and gender.** For example, Maria is a female born to Mexican American parents; she was assigned these statuses at birth. She is an adult and—if she lives long enough—will someday become an "older adult," which is an ascribed status received involuntarily later in life. An *achieved status* **is a social position a person assumes voluntarily as a result of personal choice, merit, or direct effort.** Achieved statuses (such as occupation, education, and income) are thought to be gained as a result of personal ability or successful competition. Most occupational positions in modern societies are achieved statuses. For instance, Maria voluntarily assumed the statuses of psychologist, professor, wife, mother, and school volunteer. However, not all achieved statuses are positions most people would want to attain; for example, being a criminal, a drug addict, or a homeless person is a negative achieved status.

Ascribed statuses have a significant influence on the achieved statuses that we occupy. Race/ethnicity, gender, and age affect each person's opportunity to acquire certain achieved statuses. Those who are privileged by their positive ascribed statuses are more likely to achieve the more prestigious positions in a society. Those who are disadvantaged by their ascribed statuses may more easily acquire negative achieved statuses.

Status a socially defined position in a group or society characterized by certain expectations, rights, and duties.

Ascribed status a social position conferred at birth or received involuntarily later in life based on attributes over which the individual has little or no control, such as race/ethnicity, age, and gender.

Achieved status a social position that a person assumes voluntarily as a result of personal choice, merit, or direct effort.

Mary Kate Denny/PhotoEdit

Robert Brenner/PhotoEdit

Sociologists believe that being rich or poor may be a master status in the United States. How do the lifestyles of these two women differ based on their master statuses?

MASTER STATUS If we occupy many different statuses, how can we determine which is the most important? Sociologist Everett Hughes has stated that societies resolve this ambiguity by determining master statuses. **A *master status* is the most important status a person occupies;** it dominates all of the individual's other statuses and is the overriding ingredient in determining a person's general social position (Hughes, 1945). Being poor or rich is a master status that influences many other areas of life, including health, education, and life opportunities. Historically, the most common master statuses for women have related to positions in the family, such as daughter, wife, and mother. For men, occupation has usually been the most important status, although occupation is increasingly a master status for many women as well. "What do you do?" is one of the first questions most people ask when meeting another. Occupation provides important clues to a person's educational level, income, and family background. An individual's race/ethnicity may also constitute a master status in a society in which dominant-group members single out members of other groups as "inferior" on the basis of real or alleged physical, cultural, or nationality characteristics (see Feagin and Feagin, 2003).

Master statuses are vital to how we view ourselves, how we are seen by others, and how we interact with others. Justice Ruth Bader Ginsburg is both a U.S. Supreme Court justice and a mother.

Which is her master status? Can you imagine how she would react if attorneys arguing a case before the Supreme Court treated her as if she were a mother rather than a justice? Lawyers wisely use "justice" as her master status and act accordingly.

Master statuses confer high or low levels of personal worth and dignity on people. These are not characteristics that we inherently possess; they are derived from the statuses we occupy. For those who have no residence, being a homeless person readily becomes a master status regardless of the person's other attributes. Homelessness is a stigmatized master status that confers disrepute on its occupant because domiciled people often believe a homeless person has a "character flaw." The circumstances under which someone becomes homeless determine the extent to which that person is stigmatized. For example, individuals who become homeless as a result of natural disasters (such as a hurricane or a brush fire) are not seen as causing their homelessness or as being a threat to the community. Thus, they are less likely to be stigmatized. However, in cases in which homeless persons are viewed as the cause of their own problems, they are more likely to be stigmatized and marginalized by others. Snow and Anderson (1993: 199) observed the effects of homelessness as a master status:

It was late afternoon, and the homeless were congregated in front of [the Salvation Army shelter] for dinner. A school bus approached that was packed with Anglo junior high school students being bused from an eastside barrio school to their upper-middle and upper-class homes in the city's northwest neighborhoods. As the bus rolled by, a fusillade of coins came flying out the windows, as the students made obscene gestures and shouted, "Get a job." Some of the homeless gestured back, some scrambled for the scattered coins—mostly pennies—others angrily threw the coins at the bus, and a few seemed oblivious to the encounter. For the passing junior high schoolers, the exchange was harmless fun, a way to work off the restless energy built up in school; but for the homeless it was a stark reminder of their stigmatized status and of the extent to which they are the objects of negative attention.

STATUS SYMBOLS When people are proud of a particular social status that they occupy, they often choose to use visible means to let others know about their position. *Status symbols* **are material signs that inform others of a person's specific status.** For example, just as wearing a wedding ring proclaims that a person is married, owning a Rolls-Royce announces that one has "made it." As we saw in Chapter 2, achievement and success are core U.S. values. For this reason, people who have "made it" tend to want to display symbols to inform others of their accomplishments.

In our daily lives, status symbols both announce our statuses and aid our interactions with others. In a hospital affiliated with a medical school, the length and color of a person's uniform coat (worn over street clothing) indicates the individual's status within the medical center. Physicians wear longer white coats, medical students wear shorter white coats, laboratory technicians wear short blue coats, and so forth.

Status symbols for the domiciled and for the homeless may have different meanings. Among affluent persons, a full shopping cart in the grocery store and bags of merchandise from expensive department stores indicate a lofty financial position. By contrast, among the homeless, bulging shopping bags and overloaded grocery carts suggest a completely different status. Carts and bags are essential to street life; there is no other place to keep things, as shown by this description of Darian, a homeless woman in New York City:

The possessions in her postal cart consist of a whole house full of things, from pots and pans to books, shoes, magazines, toilet articles, personal papers and clothing, most of which she made herself. . . .

Because of its weight and size, Darian cannot get the cart up over the curb. She keeps it in the street near the cars. This means that as she pushes it slowly up and down the street all day long, she is living almost her entire life directly in traffic. She stops off along her route to sit or sleep for awhile and to be both stared at as a spectacle and to stare back. Every aspect of her life including sleeping, eating, and going to the bathroom is constantly in public view. . . . [S]he has no space to call her own and she never has a moment's privacy. Her privacy, her home, is her cart with all its possessions. (Rousseau, 1981: 141)

For homeless women and men, possessions are not status symbols as much as they are a link with the past, a hope for the future, and a potential source of immediate cash. As Snow and Anderson (1993: 147) note, selling personal possessions is not uncommon among most social classes; members of the working and middle classes hold garage sales, and those in the upper classes have estate sales. However, when homeless persons sell their personal possessions, they do so to meet their immediate needs, not because they want to "clean house."

Roles

A role is the dynamic aspect of a status. Whereas we occupy a status, we play a role (Linton, 1936). **A *role* is a set of behavioral expectations associated with a given status.** For example, a carpenter (employee) hired to remodel a kitchen is not expected to sit down uninvited and join the family (employer) for dinner.

Role expectation **is a group's or society's definition of the way a specific role ought to be played.**

Master status the most important status that a person occupies.

Status symbol a material sign that informs others of a person's specific status.

Role a set of behavioral expectations associated with a given status.

Role expectation a group's or society's definition of the way that a specific role *ought* to be played.

By contrast, *role performance* **is how a person actually plays the role.** Role performance does not always match role expectation. Some statuses have role expectations that are highly specific, such as that of surgeon or college professor. Other statuses, such as friend or significant other, have less-structured expectations. The role expectations tied to the status of student are more specific than those of being a friend. Role expectations are typically based on a range of acceptable behavior rather than on strictly defined standards.

Our roles are relational (or complementary); that is, they are defined in the context of roles performed by others. We can play the role of student because someone else fulfills the role of professor. Conversely, to perform the role of professor, the teacher must have one or more students.

Role ambiguity occurs when the expectations associated with a role are unclear. For example, it is not always clear when the provider–dependent aspect of the parent–child relationship ends. Should it end at age eighteen or twenty-one? When a person is no longer in school? Different people will answer these questions differently depending on their experiences and socialization, as well as on the parents' financial capability and psychological willingness to continue contributing to the welfare of their adult children.

Ellen Senisi/The Image Works

Parents sometimes experience role conflict when they are faced with societal expectations that they will earn a living for their family and that they will also be good parents to their children. This father seeks to take care of his sick daughter at the same time that he knows that he has responsibilities at work that he should be fulfilling.

ROLE CONFLICT AND ROLE STRAIN   Most people occupy a number of statuses, each of which has numerous role expectations attached. For example, Charles is a student who attends morning classes at the university, and he is an employee at a fast-food restaurant, where he works from 3:00 to 10:00 P.M. He is also Stephanie's boyfriend, and she would like to see him more often. On December 7, Charles has a final exam at 7:00 P.M., when he is supposed to be working. Meanwhile, Stephanie is pressuring him to take her to a movie. To top it off, his mother calls, asking him to fly home because his father is going to have emergency surgery. How can Charles be in all these places at once? Such experiences of role conflict can be overwhelming.

Role conflict **occurs when incompatible role demands are placed on a person by two or more statuses held at the same time.** When role conflict occurs, we may feel pulled in different directions. To deal with this problem, we may prioritize

our roles and first complete the one we consider to be most important. Or we may compartmentalize our lives and "insulate" our various roles (Merton, 1968). That is, we may perform the activities linked to one role for part of the day and then engage in the activities associated with another role in some other time period or elsewhere. For example, under routine circumstances, Charles would fulfill his student role for part of the day and his employee role for another part of the day. In his current situation, however, he is unable to compartmentalize his roles.

Role conflict may occur as a result of changing statuses and roles in society. Research has found that women who engage in behavior that is gender-typed as "masculine" tend to have higher rates of role conflict than those who engage in traditional "feminine" behavior (Basow, 1992). According to the sociologist Tracey Watson (1987), role conflict can sometimes be attributed not to the roles themselves but to the pressures that people feel when they do not fit into culturally prescribed roles. In her study of women athletes in college sports programs, Watson found role conflict in the traditionally incongruent identities of being a woman and being an athlete. Even though the women athletes in her study wore makeup and presented a conventional image when they were not on the basketball court, their peers in school still saw them as "female jocks," thus leading to role conflict.

Whereas role conflict occurs between two or more statuses (such as being homeless and being a temporary employee of a social service agency), role strain takes place within one status. *Role strain* **occurs when incompatible demands are built into a single status that a person occupies** (Goode, 1960). For example, many women experience role strain in the labor force because they hold jobs that are "less satisfying and more stressful than men's jobs since they involve less money, less prestige, fewer job openings, more career roadblocks, and so forth" (Basow, 1992: 192). Similarly, married women may experience more role strain than married men because of work overload, marital inequality with their spouse, exclusive parenting responsibilities, unclear expectations, and lack of emotional support.

Recent social changes may have increased role strain for men. In the family, men's traditional position of dominance has eroded as more women have entered the paid labor force and demanded more assistance in child-rearing and homemaking responsibilities. Role strain may occur among African American men who have internalized North American cultural norms regarding masculinity yet find it very difficult (if not impossible) to attain cultural norms of achievement, success, and power because of racism and economic exploitation (Basow, 1992).

Sexual orientation, age, and occupation are frequently associated with role strain. Lesbians and gay men often experience role strain because of the pressures associated with having an identity heavily stigmatized by the dominant cultural group (Basow, 1992). Women in their thirties may experience the

highest levels of role strain; they face a large amount of stress in terms of role demands and conflicting work and family expectations (Basow, 1992). Dentists, psychiatrists, and police officers have been found to experience high levels of occupation-related role strain, which may result in suicide. (The concepts of role expectation, role performance, role conflict, and role strain are illustrated in Figure 4.2.)

Individuals frequently distance themselves from a role they find extremely stressful or otherwise problematic. *Role distancing* occurs when people consciously foster the impression of a lack of commitment or attachment to a particular role and merely go through the motions of role performance (Goffman, 1961b). People use distancing techniques when they do not want others to take them as the "self" implied in a particular role, especially if they think the role is "beneath them." While Charles is working in the fast-food restaurant, for example, he does not want people to think of him as a "loser in a dead-end job." He wants them to view him as a college student who is working there just to "pick up a few bucks" until he graduates. When customers from the university come in, Charles talks to them about what courses they are taking, what they are majoring in, and what professors they have. He does not discuss whether the bacon cheeseburger is better than the chili burger. When Charles is really involved in role distancing, he tells his friends that he "works there but wouldn't eat there."

ROLE EXIT *Role exit* **occurs when people disengage from social roles that have been central to their self-identity** (Ebaugh, 1988). Sociologist

Role performance how a person *actually* plays a role.

Role conflict a situation in which incompatible role demands are placed on a person by two or more statuses held at the same time.

Role strain a condition that occurs when incompatible demands are built into a single status that a person occupies.

Role exit a situation in which people disengage from social roles that have been central to their self-identity.

Figure 4.2 Role Expectation, Performance, Conflict, and Strain

Role Expectation: a group's or society's definition of the way a specific role *ought* to be played.

Role Performance: how a person *actually* plays a role.

Role Conflict: occurs when incompatible demands are placed on a person by two or more statuses held at the same time.

Role Strain: occurs when incompatible demands are built into a single status that the person holds.

When playing the role of "student," do you sometimes personally encounter these concepts?

Helen Rose Fuchs Ebaugh studied this process by interviewing ex-convicts, ex-nuns, retirees, divorced men and women, and others who had exited voluntarily from significant social roles. According to Ebaugh, role exit occurs in four stages. The first stage is doubt, in which people experience frustration or burnout when they reflect on their existing roles. The second stage involves a search for alternatives; here, people may take a leave of absence from their work or temporarily separate from their marriage partner. The third stage is the turning point at which people realize that they must take some final action, such as quitting their job or getting a divorce. The fourth and final stage involves the creation of a new identity.

Exiting the "homeless" role is often very difficult. The longer a person remains on the streets, the more difficult it becomes to exit this role. Personal resources diminish over time. Possessions are often stolen, lost, sold, or pawned. Work experience and skills become outdated, and physical disabilities that prevent individuals from working are likely to develop. However, a number of homeless people are able to exit this role. For example, Christopher, a former crack addict who had lived in New York subway stations, was able to become a domiciled person after completing a drug rehabilitation program and receiving assistance from a community service agency:

> I felt like I had reached the end of my life. I felt awful talking to people with dirty tennis shoes and ripped pants, smelling bad. I couldn't look anybody in the eye. Once you're taking drugs, nobody respects you, especially you. (qtd. in R. Kennedy, 1993: A15)

Of course, many of the homeless do not beat the odds and exit this role. Instead, they shift their focus from role exiting to survival on the streets.

For many years, capitalism has been dominated by powerful "old-boy" social networks. Professional women have increasingly turned to sporting activities such as golf or tennis to create social networks that enhance their business opportunities.

Groups

Groups are another important component of social structure. To sociologists, a *social group* consists of two or more people who interact frequently and share a common identity and a feeling of interdependence. Throughout our lives, most of us participate in groups, from our families and childhood friends, to our college classes, to our work and community organizations, and even to society.

Primary and secondary groups are the two basic types of social groups. A *primary group* is a small, less specialized group in which members engage in face-to-face, emotion-based interactions over an extended period of time. Primary groups include our family, close friends, and school- or work-related peer groups. By contrast, a *secondary group* is a larger, more specialized group in which members engage in more impersonal, goal-oriented relationships for a limited period of time. Schools, churches, and corporations are examples of secondary groups. In secondary groups, people have few, if any, emotional ties to one another. Instead, they come together for some specific, practical purpose, such as getting a degree or a paycheck. Secondary groups are more specialized than primary ones; individuals relate to one another in terms of specific roles (such as professor and student) and more limited activities (such as course-related endeavors). Primary and secondary groups are further discussed in Chapter 5 ("Groups and Organizations").

Social solidarity, or cohesion, relates to a group's ability to maintain itself in the face of obstacles. Social solidarity exists when social bonds, attractions, or other forces hold members of a group in interaction over a period of time (Jary and Jary, 1991). For example, if a local church is destroyed by fire and congregation members still

Social group a group that consists of two or more people who interact frequently and share a common identity and a feeling of interdependence.

Primary group Charles Horton Cooley's term for a small, less specialized group in which members engage in face-to-face, emotion-based interactions over an extended period of time.

Secondary group a larger, more specialized group in which members engage in more impersonal, goal-oriented relationships for a limited period of time.

worship together in a makeshift setting, then they have a high degree of social solidarity.

Many of us build social networks that involve our personal friends in primary groups and our acquaintances in secondary groups. A *social network* is a series of social relationships that links an individual to others. Social networks work differently for men and women, for different races/ethnicities, and for members of different social classes. Traditionally, people of color and white women have been excluded from powerful "old-boy" social networks (Kanter, 1993/1977; McPherson and Smith-Lovin, 1982, 1986). At the middle- and upper-class levels, individuals tap social networks to find employment, make business deals, and win political elections. However, social networks typically do not work effectively for poor and homeless individuals. Snow and Anderson (1993) found that homeless men have fragile social networks that are plagued with instability. Homeless men often do not even know one another's "real" names.

Sociological research on the homeless has noted the social isolation experienced by people on the streets. Sociologist Peter H. Rossi (1989) found that a high degree of social isolation exists because the homeless are separated from their extended family and former friends. Rossi noted that among the homeless who did have families, most either did not wish to return or believed that they would not be welcome. Most of the avenues for exiting the homeless role and acquiring housing are intertwined with the large-scale, secondary groups that sociologists refer to as formal organizations.

A *formal organization* **is a highly structured group formed for the purpose of completing certain tasks or achieving specific goals.** Many of us spend most of our time in formal organizations such as colleges, corporations, or the government. In Chapter 5 ("Groups and Organizations"), we analyze the characteristics of bureaucratic organizations; however, at this point we should note that these organizations are a very important component of social structure in all industrialized societies. We expect such organizations to educate us, solve our social problems (such as crime and homelessness), and provide us work opportunities.

Many formal organizations today have been referred to as "people-processing" organizations. For example, the Salvation Army and other caregiver groups provide services for the homeless and others in need. However, these organizations must work with limited monetary resources and at the same time maintain some control over their clientele. This control is necessary in order to provide their services in an orderly and timely fashion, according to a major at the Salvation Army:

> I'll sleep and feed almost anybody, but such help requires that they be deserving. Some people would say I'm cold-hearted, but I rule with an iron hand. I have to because these guys need to respect authority. . . . The experience of working with these guys has taught us the necessity of rules in order to avoid problems. (qtd. in Snow and Anderson, 1993: 81)

Because of rules and policies, the "Sally" (as the Salvation Army is sometimes called) tends to close its doors to those who are currently inebriated, are chronic drunks, or are viewed as "troublemakers." Likewise, a number of women's shelters have restrictions and regulations that some of the women feel deprive them of their personhood. One shelter used to require a compulsory gynecological examination of its residents (Golden, 1992). Another required that the women be out of the building by 7:00 A.M. and not return before 7:00 P.M. Fearful of violence among shelter residents or between residents and staff, many shelters use elaborate questionnaires and interviews to screen out potential disruptive clients. Those who are supposed to benefit from the services of such shelters may find the experience demeaning and alienating. Nevertheless, organizations such as the Salvation Army and women's shelters do help people within the limited means they have available.

Social Institutions

At the macrolevel of all societies, certain basic activities routinely occur—children are born and socialized, goods and services are produced and distributed, order is preserved, and a sense of purpose is maintained (Aberle et al., 1950; Mack and Bradford, 1979). Social institutions are the means by which these basic needs are met. A *social institution* **is a set of organized beliefs and rules that establishes how a society will attempt to meet its basic social needs.** In the past, these needs have centered around five basic social institutions: the family, religion, education, the economy, and the government or politics. Today, mass media, sports, science and medicine, and the military are also considered to be social institutions.

What is the difference between a group and a social institution? A group is composed of specific, identifiable people; an institution is a standardized way of doing something. The concept of "family" helps to distinguish between the two. When we talk about "your family" or "my family," we are referring to a specific family. When we refer to the family as a social institution, we are talking about ideologies and standardized patterns of behavior that organize family life. For example, the family as a social institution contains certain statuses organized into well-defined relationships, such as husband–wife, parent–child, and brother–sister. Specific families do not always conform to these ideologies and behavior patterns.

Functional theorists emphasize that social institutions exist because they perform five essential tasks:

1. *Replacing members.* Societies and groups must have socially approved ways of replacing members who move away or die. The family provides the structure for legitimated sexual activity—and thus procreation—between adults.
2. *Teaching new members.* People who are born into a society or move into it must learn the group's values and customs. The family is essential in teaching new members, but other social institutions educate new members as well.
3. *Producing, distributing, and consuming goods and services.* All societies must provide and distribute goods and services for their members. The economy is the primary social institution fulfilling this need; the government is often involved in the regulation of economic activity.
4. *Preserving order.* Every group or society must preserve order within its boundaries and protect itself from attack by outsiders. The government legitimates the creation of law enforcement agencies to preserve internal order and the existence of some form of military for external defense.
5. *Providing and maintaining a sense of purpose.* In order to motivate people to cooperate with one another, a sense of purpose is needed.

Although this list of functional prerequisites is shared by all societies, the institutions in each society per-

form these tasks in somewhat different ways depending on their specific cultural values and norms.

Conflict theorists agree with functionalists that social institutions are originally organized to meet basic social needs. However, they do not believe that social institutions work for the common good of everyone in society. For example, the homeless lack the power and resources to promote their own interests when they are opposed by dominant social groups. This is a problem not only in the United States but also for homeless people, especially children and youth, throughout the world (see Box 4.2). From the conflict perspective, social institutions such as the government maintain the privileges of the wealthy and powerful while contributing to the powerlessness of others (see Domhoff, 1983, 1990). For example, U.S. government policies in urban areas have benefited some people but exacerbated the problems of others. Urban renewal and transportation projects caused the destruction of low-cost housing and put large numbers of people "on the street" (Katz, 1989).

✓ Checkpoints

Checkpoint 4.2

2. What are the differences between ascribed and achieved statuses?
3. What is a master status, and what are several examples of such a status?
4. What is a status symbol, and what part do status symbols play in our everyday lives?
5. How does role expectation differ from role performance?
6. How does role conflict differ from role strain?
7. What is an example of role exit?
8. What is the central difference between a primary group and a secondary group?
9. What are the essential tasks of social institutions?

Formal organization a highly structured group formed for the purpose of completing certain tasks or achieving specific goals.

Social institution a set of organized beliefs and rules that establishes how a society will attempt to meet its basic social needs.

Box 4.2 Sociology in Global Perspective

Homelessness as an International Problem: The Plight of the "Street Children"

The eyes of the homeless, those who really need help, are the same eyes in any country—eyes that can see every door is closed to them.

—Valery Sokolov, a journalist and president of the Nochlyazhka Charitable Foundation in St. Petersburg, Russia, at the time he made this statement, knew what it means to be homeless: During six years of homelessness, he slept in a railroad station and sometimes in the forest (Spence 1997).

Homelessness is a widespread problem around the globe. In cities such as St. Petersburg, Vancouver, Tokyo, Paris, Sydney, and New York, some homeless individuals and families are quite visible to others as they sleep in train stations, on the sidewalks, in the doorways, and near warm air vents. But not all homeless people are so highly visible: Some reside in temporary shelters, sleep on the spare beds or sofas of friends and relatives, or are the "hidden homeless" who, in nations like Australia, camp in caravans or live in cars.

Who are the homeless in other nations? A partial answer is that homeless people share many commonalities regardless of where they reside. According to recent reports from a variety of nations, the faces of homelessness have changed in recent decades. As a study of homelessness in Australia stated,

> The old, derelict wino on the park bench has been joined by younger men, unemployed, and hopeless; by the confused and mentally ill, frightened by the pace of activity surrounding them; by women and children, desperate to escape violent and destructive domestic situations; by young people, cast off by families who can't cope or don't care. (Harrison 2001)

As this statement suggests, people from all walks of life may be affected by homelessness. A particularly pressing issue in many countries is the growing number of families with children and "chronically homeless children," typically between the ages of twelve and eighteen, who reside on the streets or move between shelters, street life, and squatting (Harrison 2001).

Let's specifically focus our attention on the plight of the so-called "street children." The World Health Organization's Street Children Project identified four categories of young people who fit the definition of "street children":

1. Children living on the streets, whose immediate concerns are survival and shelter.
2. Children who are detached from their families and living in temporary shelter, such as abandoned houses and other buildings (hostels/refuges/shelters), or moving about between friends.
3. Children who remain in contact with their families but because of poverty, overcrowding, or sexual or physical abuse within the family will spend some nights, or most days, on the streets.
4. Children who are in institutional care, who come from a situation of homelessness and are at risk of returning to a homeless existence.

If our children are the future of our nations, what can be said for those children who do not have a home to call their own? Some analysts believe that the plight of "street children" is especially dire in that they experience alienation and exclusion from society's mainstream structures and systems, including schools, churches, and other organizations that may provide stability and hope to youth. In addition, "street children" may participate in significant antisocial and self-destructive behavior, and have a profound distrust for social services and welfare systems that might, in some cases, make it possible for them to leave a life of homelessness. Whether in England, Australia, Russia, Norway, Canada, the United States, or other nations of the world, homelessness, particularly of children and youth, marks the failure of existing social institutions to adequately provide for the next generation and puts people in a devalued master status in which they lack the power and resources to promote their own interests and in which survival becomes their primary objective.

Similarly, the shift in governmental policies toward the mentally ill and welfare recipients has resulted in more people struggling—and often failing—to find affordable housing. Meanwhile, many wealthy and privileged bankers, investors, developers, and builders have benefited at the expense of the low-income casualties of those policies.

Societies: Changes in Social Structure

Changes in social structure have a dramatic impact on individuals, groups, and societies. Social arrangements in contemporary societies have grown more complex with the introduction of new technology, changes in values and norms, and the rapidly shrinking "global village." How do societies maintain some degree of social solidarity in the face of such changes? As you may recall from Chapter 1, theorists using a functionalist perspective focus on the stability of societies and the importance of equilibrium even in times of rapid social change. By contrast, conflict perspectives highlight how societies go through continuous struggles for scarce resources and how innovation, rebellion, and conquest may bring about social change. Sociologists Emile Durkheim and Ferdinand Tönnies developed typologies to explain the processes of stability and change in the social structure of societies. A *typology* is a classification scheme containing two or more mutually exclusive categories that are used to compare different kinds of behavior or types of societies.

Durkheim: Mechanical and Organic Solidarity

Emile Durkheim (1933/1893) was concerned with the question "How do societies manage to hold together?" He asserted that preindustrial societies are held together by strong traditions and by the members' shared moral beliefs and values. As societies industrialized and developed more specialized economic activities, social solidarity came to be rooted in the members' shared dependence on one another. From Durkheim's perspective, social solidarity derives from a society's social structure, which, in turn, is based on the society's division of labor. *Division of labor* refers to how the various tasks of a society are divided up and performed. People in diverse societies (or in the same society at different points in time) divide their tasks somewhat differently, based on their own history, physical environment, and level of technological development.

To explain social change, Durkheim categorized societies as having either mechanical or organic solidarity. *Mechanical solidarity* **refers to the social cohesion of preindustrial societies, in which there is minimal division of labor and people feel united by shared values and common social bonds.** Durkheim used the term *mechanical solidarity* because he believed that people in such preindustrial societies feel a more or less automatic sense of belonging. Social interaction is characterized by face-to-face, intimate, primary-group rela-

tionships. Everyone is engaged in similar work, and little specialization is found in the division of labor.

Organic solidarity **refers to the social cohesion found in industrial (and perhaps postindustrial) societies, in which people perform very specialized tasks and feel united by their mutual dependence.** Durkheim chose the term *organic solidarity* because he believed that individuals in industrial societies come to rely on one another in much the same way that the organs of the human body function interdependently. Social interaction is less personal, more status oriented, and more focused on specific goals and objectives. People no longer rely on morality or shared values for social solidarity; instead, they are bound together by practical considerations.

Tönnies: *Gemeinschaft* and *Gesellschaft*

Sociologist Ferdinand Tönnies (1855–1936) used the terms *Gemeinschaft* and *Gesellschaft* to characterize the degree of social solidarity and social control found in societies. He was especially concerned about what happens to social solidarity in a society when a "loss of community" occurs.

The *Gemeinschaft* **(guh-MINE-shoft) is a traditional society in which social relationships are based on personal bonds of friendship and kinship and on intergenerational stability.** These relationships are based on ascribed rather than achieved status. In such societies, people have a commitment to the entire group and feel a sense of togetherness. Tönnies (1963/1887) used the German term *Gemeinschaft* because it means "commune" or "community"; social solidarity and social control are maintained by

---

Mechanical solidarity Emile Durkheim's term for the social cohesion in preindustrial societies, in which there is minimal division of labor and people feel united by shared values and common social bonds.

Organic solidarity Emile Durkheim's term for the social cohesion found in industrial societies, in which people perform very specialized tasks and feel united by their mutual dependence.

Gemeinschaft (guh-MINE-shoft) a traditional society in which social relationships are based on personal bonds of friendship and kinship and on intergenerational stability.

Figure 4.3 *Gemeinschaft* and *Gesellschaft* Societies

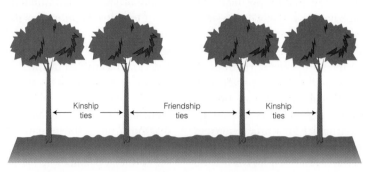

If we compare societies to trees, a *Gemeinschaft* society would be made up of the various family trees (clans or kinship groups) and how they are related to one another.

By contrast, a *Gesellschaft* society would be made up of clumps of trees (kinship/friendship ties) that are important, but each of them has a more specialized relationship and may not be committed to the others, just as an individual tree and the pot it is growing in could be moved somewhere else.

the community. Members have a strong sense of belonging, but they also have very limited privacy.

By contrast, the ***Gesellschaft* (guh-ZELL-shoft) is a large, urban society in which social bonds are based on impersonal and specialized relationships, with little long-term commitment to the group or consensus on values.** In such societies, most people are "strangers" who perceive that they have very little in common with most other people. Consequently, self-interest dominates, and little consensus exists regarding values. Tönnies (1963/1887) selected the German term *Gesellschaft* because it means "association"; relationships are based on achieved statuses, and interactions among people are both rational and calculated. See Figure 4.3.

In *Gesellschaft* societies such as the United States, a prevailing core value is that people should be able to take care of themselves. Thus, many people view the homeless as "throwaways"—as beyond help or as having already had enough done for them by society. Some argue that the homeless

made their own bad decisions, which led them into alcoholism or drug addiction, and should be held responsible for the consequences of their actions. In this sense, homeless people serve as a visible example to others to "follow the rules" lest they experience a similar fate (see White, 1992).

Alternative explanations for homelessness in *Gesellschaft* societies have been suggested. Elliot Liebow (1993) notes that homelessness is rooted in poverty; overwhelmingly, homeless people are poor people who come from poor families. Home-

Reflections

Reflections 4.2

Is homelessness more likely to be a major problem in *Gesellschaft* societies than in earlier societies?

lessness is a "social class phenomenon, the direct result of a steady, across-the-board lowering of the standard of living of the working class and lower class" (Liebow, 1993: 224). As the standard of living falls, those at the bottom rungs of society are plunged into homelessness. The problem is exacerbated by a lack of jobs. Of those who find work, a growing number work full time, year-round, but remain poor because of substandard wages. Structural changes in society—such as a shift from an industrial to a postindustrial economy—also affect people's job opportunities and social well-being.

Industrial and Postindustrial Societies

Industrial societies are based on technology that mechanizes production. Take a look around you: Most of what you see would not exist if it were not for industrialization. Cars, computers, electric lights, stereos, cell phones, and virtually every other possession we own is a product of an industrial society.

In industrial societies, a large proportion of the population lives in or near cities. Large corporations and government bureaucracies grow in size and complexity. The nature of social life changes as people come to know one another more as statuses than as individuals. In fact, a person's occupation becomes a key defining characteristic in industrial societies, meaning that those people who are unemployed do not share the same status markers as those who have jobs. Although industrial societies have brought about some of the greatest innovations in all of human history, they have also maintained and perpetuated some of the greatest problems, including violence; race-, class-, and gender-based inequalities; and environmental degradation.

The shift that has taken place toward a postindustrial society in the United States and some other high-income nations has produced new opportunities and problems as well. A *postindustrial society* is one in which technology supports a service- and information-based economy. Postindustrial societies are characterized by an *information explosion* and an economy in which large numbers of people either provide or apply information or are employed in service jobs (such as fast-food server or health care worker). For example, banking, law, and the travel industry are characteristic forms of employment in postindustrial

societies, whereas producing steel or automobiles is representative of employment in industrial societies.

Postindustrial societies produce knowledge that becomes a commodity. This knowledge can be leased or sold to others, or it can be used to generate goods, services, or more knowledge. In the previous types of societies we have examined, machinery or raw materials are crucial to how the economy operates. In postindustrial societies, the economy is based on involvement with people and communications technologies such as the mass media, computers, and the World Wide Web. For example, recent information from the U.S. Census Bureau indicates that more than half of all U.S. households have at least one computer (see "Census Profile: Computer and Internet Access in U.S. Households"). Some analysts refer to postindustrial societies as "service economies," based on the assumption that many workers provide services for others. Examples include home health care workers and airline flight attendants. However, most of the new service occupations pay relatively low wages and offer limited opportunities for advancement.

Some sociological theories and research focus on the systematic examination of societies at the macrolevel, including the effects of the postindustrial economy on other social institutions; however,

✓Checkpoints

Checkpoint 4.3

10. How does Durkheim distinguish between mechanical solidarity and organic solidarity?
11. How does the *Gemeinschaft* society differ from the *Gesellschaft* society?

Gesellschaft (guh-ZELL-shoft) a large, urban society in which social bonds are based on impersonal and specialized relationships, with little long-term commitment to the group or consensus on values.

Industrial societies societies based on technology that mechanizes production.

Postindustrial societies societies in which technology supports a service- and information-based economy.

CENSUS PROFILES
Computer and Internet Access in U.S. Households

The U.S. Census Bureau collects extensive data on U.S. households in addition to the questions it used for Census 2000. For example, Current Population Survey data, collected from about 50,000 U.S. households during 2000, show an increase in the percentage of homes with computers and access to the Internet, as the following figure illustrates:

Computers and Internet Access in the Home: 1984 to 2000
(Civilian noninstitutional population)

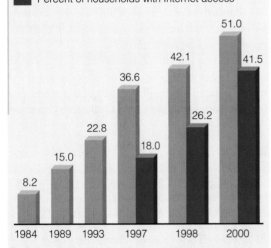

Note: Data on Internet access were not collected before 1997.

Since 1984, the first year in which the Census Bureau collected data on computer ownership and use, there has been more than a fivefold increase in the percentage of households with computers. However, in Chapter 8 ("Social Class in the United States"), we will see that computer ownership varies widely by income and educational level.

Source: Newburger, 2001.

other sociologists primarily use a microlevel perspective in their analysis of social interaction in contemporary societies.

Social Interaction: The Microlevel Perspective

So far in this chapter, we have focused on society and social structure from a macrolevel perspective, seeing how the structure of society affects the statuses we occupy, the roles we play, and the groups and organizations to which we belong. Functionalist and conflict perspectives provide a macrosociological overview because they concentrate on large-scale events and broad social features. For example, sociologists using the macrosociological approach to study the homeless might analyze how social institutions have operated to produce current conditions. By contrast, the symbolic interactionist perspective takes a microsociological approach, asking how social institutions affect our daily lives. We will now look at society from the microlevel perspective, which focuses on social interactions among individuals, especially face-to-face encounters.

Social Interaction and Meaning

When you are with other people, do you often wonder what they think of you? If so, you are not alone! Because most of us are concerned about the meanings that others ascribe to our behavior, we try to interpret their words and actions so that we can plan how we will react toward them (Blumer, 1969). We know that others have expectations of us. We also have certain expectations about them. For example, if we enter an elevator that has only one other person in it, we do not expect that individual to confront us and stare into our eyes. As a matter of fact, we would be quite upset if the person did so.

Social interaction within a given society has certain shared meanings across situations. For instance, our reaction would be the same regardless of *which* elevator we rode in *which* building. Sociologist Erving Goffman (1961b) described these shared meanings in his observation about two pedestrians approaching each other on a public sidewalk. He noted that each will tend to look at the other just long enough to acknowledge the other's presence. By the time they are about eight feet away from each other, both individuals will tend to look downward.

Sharply contrasting perceptions of the same reality are evident in these people's views about the death penalty in the United States.

Goffman referred to this behavior as *civil inattention*—the ways in which an individual shows an awareness that another is present without making this person the object of particular attention. The fact that people engage in civil inattention demonstrates that interaction does have a pattern, or *interaction order*, which regulates the form and processes (but not the content) of social interaction.

Does everyone interpret social interaction rituals in the same way? No. Race/ethnicity, gender, and social class play a part in the meanings we give to our interactions with others, including chance encounters on elevators or the street. Our perceptions about the meaning of a situation vary widely based on the statuses we occupy and our unique personal experiences. For example, sociologist Carol Brooks Gardner (1989) found that women frequently do not perceive street encounters to be "routine" rituals. They fear for their personal safety and try to avoid comments and propositions that are sexual in nature. African Americans may also feel uncomfortable in street encounters. A middle-class African American college student described his experiences walking home at night from a campus job:

> So, even if you wanted to, it's difficult just to live a life where you don't come into conflict with others. . . . Every day that you live as a black person you're reminded how you're perceived in society. You walk the streets at night; white people cross the streets. I've seen white couples and individuals dart in front of cars to not be on the same side of the street. Just the other day, I was walking down the street, and this white female with a child, I saw her pass a young white male about 20 yards ahead. When she saw me, she quickly dragged the child and herself across the busy street. . . . [When I pass,] white men tighten their grip on their women. I've seen people turn around and seem like they're going to take blows from me. . . . So, every day you realize [you're black]. Even though you're not doing anything wrong; you're just existing. You're just a person. But you're a black person perceived in an unblack world. (qtd. in Feagin, 1991: 111–112)

As this passage indicates, social encounters have different meanings for men and women, whites and people of color, and individuals from different social classes. Members of the dominant classes regard the poor, unemployed, and working class as less worthy of attention, frequently subjecting them to subtle yet systematic "attention deprivation" (Derber, 1983). The same can certainly be said about how members of the dominant classes "interact" with the homeless.

The Social Construction of Reality

If we interpret other people's actions so subjectively, can we have a shared social reality? Some symbolic interaction theorists believe that there is very little shared reality beyond that which is socially created. Symbolic interactionists refer to this as the *social construction of reality*—the process by which our

Social construction of reality the process by which our perception of reality is largely shaped by the subjective meaning that we give to an experience.

perception of reality is largely shaped by the subjective meaning that we give to an experience (Berger and Luckmann, 1967). This meaning strongly influences what we "see" and how we respond to situations.

As discussed previously, our perceptions and behavior are influenced by how we initially define situations: We act on reality as we see it. Sociologists describe this process as the *definition of the situation*, meaning that we analyze a social context in which we find ourselves, determine what is in our best interest, and adjust our attitudes and actions accordingly. This can result in a ***self-fulfilling prophecy***—a **false belief or prediction that produces behavior that makes the originally false belief come true** (Merton, 1968). An example would be a person who has been told repeatedly that she or he is not a good student; eventually, this person might come to believe it to be true, stop studying, and receive failing grades.

People may define a given situation in very different ways, a tendency demonstrated by the sociologist Jacqueline Wiseman (1970) in her study of "Pacific City's" skid row. She wanted to know how people who live or work on skid row (a run-down area found in all cities) felt about it. Wiseman found that homeless persons living on skid row evaluated it very differently from the social workers who dealt with them there. On the one hand, many of the social workers "saw" skid row as a smelly, depressing area filled with men who were "down-and-out," alcoholic, and often physically and mentally ill. On the other hand, the men who lived on skid row did not see it in such a negative light. They experienced some degree of satisfaction with their "bottle clubs [and a] remarkably indomitable and creative spirit"—at least initially (Wiseman, 1970: 18). As this study shows, we define situations from our own frame of reference, based on the statuses that we occupy and the roles that we play.

Dominant-group members with prestigious statuses may have the ability to establish how other people define "reality" (Berger and Luckmann, 1967: 109). Some sociologists have suggested that dominant groups, particularly higher-income white males in powerful economic and political statuses, perpetuate their own world view through ideologies that frequently are seen as "social reality." For example, the sociologist Dorothy E. Smith (1999) points out that the term "Standard North American Family" (meaning a heterosexual two-parent family) is an ideological code promulgated by the dominant group to identify how people's family life *should* be arranged. According to Smith (1999), this code plays a powerful role in determining how people in organizations such as the government and schools believe that a family should be. Likewise, the sociologist Patricia Hill Collins (1998) argues that "reality" may be viewed differently by African American women and other historically oppressed groups when compared to the perspectives of dominant-group members. Collins (1998) gives the example of how groups view economic and social justice. If the views of marginalized "outsiders" were taken into account by members of the dominant group, new angles of vision on problems such as race- and class-based injustice might bring about new social realities. However, according to Collins (1998), mainstream, dominant-group members sometimes fail to realize how much they could learn about "reality" from "outsiders." As these theorists state, social reality and social structure are often hotly debated issues in contemporary societies.

Ethnomethodology

How do we know how to interact in a given situation? What rules do we follow? Ethnomethodologists are interested in the answers to these questions. ***Ethnomethodology*** **is the study of the commonsense knowledge that people use to understand the situations in which they find themselves** (Heritage, 1984: 4). Sociologist Harold Garfinkel (1967) initiated this approach and coined the term: *ethno* for "people" or "folk" and *methodology* for "a system of methods." Garfinkel was critical of mainstream sociology for not recognizing the ongoing ways in which people create reality and produce their own world. Consequently, ethnomethodologists examine existing patterns of conventional behavior in order to uncover people's background expectancies—that is, their shared interpretation of objects and events, as well as their resulting actions (Zimmerman, 1992). According to ethnomethodologists, interaction is based on assumptions of shared expectancies. For example, when you are talking with someone, what expectations do you have that you will take turns? Based on your background expectancies, would you be surprised if the other person talked for an hour and never gave you a chance to speak?

To uncover people's background expectancies, ethnomethodologists frequently break "rules" or act as though they do not understand some basic rule of social life so that they can observe other people's responses. In a series of *breaching experiments*, Garfinkel assigned different activities to his students to see how breaking the unspoken rules of behavior created confusion. In one experiment, when students participating in the study were asked "How are you?" by persons not in the study, they were instructed to respond with very detailed accounts of their health and personal problems, as in this example:

ACQUAINTANCE: How are you?

STUDENT: How am I in regard to what? My health, my finances, my school work, my peace of mind, my . . .

ACQUAINTANCE (red in the face and suddenly out of control): Look! I was just trying to be polite. Frankly, I don't give a damn how you are. (Garfinkel, 1967: 44)

In this encounter, the acquaintance expected the student to use conventional behavior in answering the question. By acting unconventionally, the student violated background expectancies and effectively "sabotaged" the interaction.

The ethnomethodological approach contributes to our knowledge of social interaction by making us aware of subconscious social realities in our daily lives. However, a number of sociologists regard ethnomethodology as a frivolous approach to studying human behavior because it does not examine the impact of macrolevel social institutions—such as the economy and education—on people's expectancies. Women's studies scholars suggest that ethnomethodologists fail to do what they claim to do: look at how social realities are created. Rather, they take ascribed statuses (such as race, class, gender, and age) as "givens," not as *socially created* realities. For example, in the experiments that Garfinkel assigned to his students, he did not account for how gender affected their experiences. When Garfinkel asked students to reduce the distance between themselves and a nonrelative to the point that "their noses were almost touching," he ignored the fact that gender was as important to the encounter as was the proximity of the two persons. Scholars have recently emphasized that our expectations about reality are strongly influenced by our assumptions relating to gender, race, and social class (see Bologh, 1992).

Dramaturgical Analysis

Erving Goffman suggested that day-to-day interactions have much in common with being on stage or in a dramatic production. *Dramaturgical analysis is the study of social interaction that compares everyday life to a theatrical presentation.* Members of our "audience" judge our performance and are aware that we may slip and reveal our true character (Goffman, 1959, 1963a). Consequently, most of us attempt to play our role as well as possible and to control the impressions we give to others. *Impression management (presentation of self) refers to people's efforts to present themselves to others in ways that are most favorable to their own interests or image.*

For example, suppose that a professor has returned graded exams to your class. Will you discuss the exam and your grade with others in the class? If you are like most people, you probably play your student role differently depending on whom you are talking to and what grade you received on the exam. Your "presentation" may vary depending on the grade earned by the other person (your "audience"). In one study, students who all received high grades ("Ace–Ace encounters") willingly talked with one another about their grades and sometimes engaged in a little bragging about how they had "aced" the test. However, encounters between students who had received high grades and those who had received low or

Self-fulfilling prophecy the situation in which a false belief or prediction produces behavior that makes the originally false belief come true.

Ethnomethodology the study of the common-sense knowledge that people use to understand the situations in which they find themselves.

Dramaturgical analysis the study of social interaction that compares everyday life to a theatrical presentation.

Impression management (presentation of self) Erving Goffman's term for people's efforts to present themselves to others in ways that are most favorable to their own interests or image.

According to the sociologist Erving Goffman, our day-to-day interactions have much in common with a dramatic production. How does this scene from the Broadway play *The Producers* show impression management?

Jeff Christiansen/Reuters NewMedia, Inc./Corbis

failing grades ("Ace–Bomber encounters") were uncomfortable. The Aces felt as if they had to minimize their own grade. Consequently, they tended to attribute their success to "luck" and were quick to offer the Bombers words of encouragement. On the other hand, the Bombers believed that they had to praise the Aces and hide their own feelings of frustration and disappointment. Students who received low or failing grades ("Bomber–Bomber encounters") were more comfortable when they talked with one another because they could share their negative emotions. They often indulged in self-pity and relied on face-saving excuses (such as an illness or an unfair exam) for their poor performances (Albas and Albas, 1988).

In Goffman's terminology, *face-saving behavior* refers to the strategies we use to rescue our performance when we experience a potential or actual loss of face. When the Bombers made excuses for their low scores, they were engaged in face-saving; the Aces attempted to help them save face by asserting that the test was unfair or that it was only a small part of the final grade. Why would the Aces and Bombers both participate in face-saving behavior? In most social interactions, all role players have an interest in keeping the "play" going so that they can maintain their overall definition of the situation in which they perform their roles.

Goffman noted that people consciously participate in *studied nonobservance*, a face-saving technique in which one role player ignores the flaws in another's performance to avoid embarrass-

ment for everyone involved. Most of us remember times when we have failed in our role and know that it is likely to happen again; thus, we may be more forgiving of the role failures of others.

Social interaction, like a theater, has a front stage and a back stage. The *front stage* is the area where a player performs a specific role before an audience. The *back stage* is the area where a player is not required to perform a specific role because it is out of view of a given audience. For example, when the Aces and Bombers were talking with each other at school, they were on the "front stage." When they were in the privacy of their own residences, they were in "back stage" settings—they no longer had to perform the Ace and Bomber roles and could be themselves.

The need for impression management is most intense when role players have widely divergent or devalued statuses. As we have seen with the Aces and Bombers, the participants often play different roles under different circumstances and keep their various audiences separated from one another. If one audience becomes aware of other roles that a person plays, the impression being given at that time may be ruined. For example, homeless people may lose jobs or the opportunity to get them when their homelessness becomes known. One woman had worked as a receptionist in a doctor's office for several weeks but was fired when the doctor learned that she was living in a shelter (Liebow, 1993). However, the homeless do not passively

Ed Peters/Impact Visuals

Antonio Ribiero/Liaison—Getty Images

Are there different gender-based expectations about the kinds of emotions that men, as compared with women, are supposed to show in the United States? What feeling rules shape the emotions of the men in these two roles?

accept the roles into which they are cast. For the most part, they attempt—as we all do—to engage in impression management in their everyday life.

The dramaturgical approach helps us think about the roles we play and the audiences who judge our presentation of self. Like all other approaches, it has its critics. Sociologist Alvin Gouldner (1970) criticized this approach for focusing on appearances and not the underlying substance. Others have argued that Goffman's work reduces the self to "a peg on which the clothes of the role are hung" (see Burns, 1992) or have suggested that this approach does not place enough emphasis on the ways in which our everyday interactions with other people are influenced by occurrences within the larger so-

ciety. For example, if some members of Congress belittle the homeless as being lazy and unwilling to work, it may become easier for people walking down a street to do likewise. Goffman's defenders counter that he captured the essence of society because social interaction "turns out to be not only where most of the world's work gets done, but where the solid buildings of the social world are in fact constructed" (Burns, 1992: 380). Goffman's work was influential in the development of the sociology of emotions, a relatively new area of theory and research.

The Sociology of Emotions

Why do we laugh, cry, or become angry? Are these emotional expressions biological or social in nature? To some extent, emotions are a biologically given sense (like hearing, smell, and touch), but they are also social in origin. We are socialized to feel certain emotions, and we learn how and when to express (or not express) those emotions (Hochschild, 1983).

How do we know which emotions are appropriate for a given role? Sociologist Arlie Hochschild (1983) suggests that we acquire a set of *feeling rules*

Reflections

Reflections 4.3

Why is impression management important to many homeless people?

that shapes the appropriate emotions for a given role or specific situation. These rules include how, where, when, and with whom an emotion should be expressed. For example, for the role of a mourner at a funeral, feeling rules tell us which emotions are required (sadness and grief, for example), which are acceptable (a sense of relief that the deceased no longer has to suffer), and which are unacceptable (enjoyment of the occasion expressed by laughing out loud) (see Hochschild, 1983: 63–68).

Feeling rules also apply to our occupational roles. For example, the truck driver who handles explosive cargos must be able to suppress fear. Although all jobs place some burden on our feelings, *emotional labor* occurs only in jobs that require personal contact with the public or the production of a state of mind (such as hope, desire, or fear) in others (Hochschild, 1983). With emotional labor, employees must display only certain carefully selected emotions. For example, flight attendants are required to act friendly toward passengers, to be helpful and open to requests, and to maintain an "omnipresent smile" in order to enhance the customers' status. By contrast, bill collectors are encouraged to show anger and make threats to customers, thereby supposedly deflating the customers' status and wearing down their presumed resistance to paying past-due bills. In both jobs, the employees are expected to show feelings that are often not their true ones (Hochschild, 1983).

Emotional labor may produce feelings of estrangement from one's "true" self. C. Wright Mills (1956) suggested that when we "sell our personality" in the course of selling goods or services, we engage in a seriously self-alienating process. In other words, the "commercialization" of our feelings may dehumanize our work role performance and create alienation and contempt that spill over into other aspects of our life (Hochschild, 1983; Smith and Kleinman, 1989).

Those who are unemployed and homeless are also required to engage in emotional labor. Governmental agencies and nonprofit organizations that function as caregivers to the homeless sometimes require emotional labor (such as feelings of gratitude or penitence) from their recipients. Homeless people have been denied social services even when they were eligible and have been asked to leave shelters when they did not show the appropriate deference and gratitude toward staff members (Liebow, 1993).

Do all people experience and express emotions the same way? It is widely believed that women express emotions more readily than men; as a result, very little research has been conducted to determine the accuracy of this belief. In fact, women and men may differ more in the way they express their emotions than in their actual feelings (Fabes and Martin, 1991). Differences in emotional expression may also be attributed to socialization, for the extent to which men and women have been taught that a given emotion is appropriate (or inappropriate) to their gender no doubt plays an important part in their perceptions (Lombardo et al., 1983).

Social class is also a determinant in managed expression and emotion management. Emotional labor is emphasized in middle- and upper-class families. Since middle- and upper-class parents often work with people, they are more likely to teach their children the importance of emotional labor in their own careers than are working-class parents, who tend to work with things, not people (Hochschild, 1983). Race is also an important factor in emotional labor. People of color spend much of their life engaged in emotional labor because racist attitudes and discrimination make it continually necessary to manage one's feelings.

Clearly, Hochschild's contribution to the sociology of emotions helps us understand the social context of our feelings and the relationship between the roles we play and the emotions we experience. However, her thesis has been criticized for overemphasizing the cost of emotional labor and the emotional controls that exist outside the individual (Wouters, 1989). The context in which emotions are studied and the specific emotions examined are important factors in determining the costs and benefits of emotional labor.

Nonverbal Communication

In a typical stage drama, the players not only speak their lines but also convey information by nonverbal communication. In Chapter 2, we discussed the importance of language; now we will look at the messages we communicate without speaking. **Nonverbal communication is the transfer of information between persons without the use of words.** It includes not only visual cues (gestures, appearances) but also vocal features (inflection, volume, pitch) and environmental factors (use of

Lindsay Hebberd/Woodfin Camp & Associates

Momatiuk/Eastcott/Woodfin Camp & Associates

Misha Erwitt/Magnum Photos

Nonverbal communication may be thought of as an international language. What message do you receive from the facial expression, body position, and gestures of each of these people? Is it possible to misinterpret their messages?

space, position) that affect meanings (Wood, 1999). Facial expressions, head movements, body positions, and other gestures carry as much of the total meaning of our communication with others as our spoken words do (Wood, 1999).

Nonverbal communication may be intentional or unintentional. Actors, politicians, and salespersons may make deliberate use of nonverbal communication to convey an idea or "make a sale." We may also send nonverbal messages through gestures or facial expressions or even our appearance without intending to let other people know what we are thinking.

FUNCTIONS OF NONVERBAL COMMUNICATION Nonverbal communication often supplements verbal communication (Wood, 1999). Head and facial movements may provide us with information about other people's emotional states, and others receive similar information from us (Samovar and Porter, 1991). We obtain first impressions of others from various kinds of nonverbal

Nonverbal communication the transfer of information between persons without the use of speech.

communication, such as the clothing they wear and their body positions.

Our social interaction is regulated by nonverbal communication. Through our body posture and eye contact, we signal that we do or do not wish to speak to someone. For example, we may look down at the sidewalk or off into the distance when we pass homeless persons who look as if they are going to ask for money.

Nonverbal communication establishes the relationship among people in terms of their responsiveness to and power over one another (Wood, 1999). For example, we show that we are responsive toward or like another person by maintaining eye contact and attentive body posture and perhaps by touching and standing close. By contrast, we signal to others that we do not wish to be near them or that we dislike them by refusing to look them in the eye or stand near them. We can even express power or control over others through nonverbal communication. Goffman (1956) suggested that *demeanor* (how we behave or conduct ourselves) is relative to social power. People in positions of dominance are allowed a wider range of permissible actions than are their subordinates, who are expected to show deference. *Deference* is the symbolic means by which subordinates give a required permissive response to those in power; it confirms the existence of inequality and reaffirms each person's relationship to the other (Rollins, 1985).

FACIAL EXPRESSION, EYE CONTACT, AND TOUCHING Deference behavior is important in regard to facial expression, eye contact, and touching. This type of nonverbal communication is symbolic of our relationships with others. Who smiles? Who stares? Who makes and sustains eye contact? Who touches whom? All these questions relate to demeanor and deference; the key issue is the status of the person who is doing the smiling, staring, or touching relative to the status of the recipient (Goffman, 1967).

Facial expressions, especially smiles, also reflect gender-based patterns of dominance and subordination in society. Typically, white women have been socialized to smile and frequently do so even when they are not actually happy (Halberstadt and Saitta, 1987). Jobs held predominantly by women (including flight attendant, secretary, elementary schoolteacher, and nurse) are more closely associated with being pleasant and smiling than are "men's jobs." In addition to smiling more fre-

quently, many women tend to tilt their heads in deferential positions when they are talking or listening to others. By contrast, men tend to display less emotion through smiles or other facial expressions and instead seek to show that they are reserved and in control (Wood, 1999).

Women are more likely to sustain eye contact during conversations (but not otherwise) as a means of showing their interest in and involvement with others. By contrast, men are less likely to maintain prolonged eye contact during conversations but are more likely to stare at other people (especially men) in order to challenge them and assert their own status (Pearson, 1985).

Eye contact can be a sign of domination or deference. For example, in a participant observation study of domestic (household) workers and their employers, the sociologist Judith Rollins (1985) found that the domestics were supposed to show deference by averting their eyes when they talked to their employers. Deference also required that they present an "exaggeratedly subservient demeanor" by standing less erect and walking tentatively.

Touching is another form of nonverbal behavior that has many different shades of meaning. Gender and power differences are evident in tactile communication from birth. Studies have shown that touching has variable meanings to parents: Boys are touched more roughly and playfully, whereas girls are handled more gently and protectively (Condry, Condry, and Pogatshnik, 1983). This pattern continues into adulthood, with women touched more frequently than men. Sociologist Nancy Henley (1977) attributed this pattern to power differentials between men and women and to the nature of women's roles as mothers, nurses, teachers, and secretaries. Clearly, touching has a different meaning to women than to men (Stier and Hall, 1984). Women may hug and touch others to indicate affection and emotional support, but men are more likely to touch others to give directions, assert power, and express sexual interest (Wood, 1999). The "meaning" we give to touching is related to its "duration, intensity, frequency, and the body parts touching and being touched" (Wood, 1994: 162).

PERSONAL SPACE Physical space is an important component of nonverbal communication. Anthropologist Edward Hall (1966) analyzed the physical distance between people speaking to each other and found that the amount of personal space

Mark Antman/The Image Works

Michael Newman/PhotoEdit

Nonverbal communication is influenced by many factors, including a person's race, class, and gender. What messages are conveyed by the manner in which the people shown here greet each other?

people prefer varies from one culture to another. **_Personal space_ is the immediate area surrounding a person that the person claims as private.** Our personal space is contained within an invisible boundary surrounding our body, much like a snail's shell. When others invade our space, we may retreat, stand our ground, or even lash out, depending on our cultural background (Samovar and Porter, 1991).

Age, gender, kind of relationship, and social class are important factors in the allocation of personal space. Power differentials between people (including adults and children, men and women, and dominant-group members and people of color) are reflected in personal space and privacy issues. With regard to age, adults generally do not hesitate to enter the personal space of a child (Thorne, Kramarae, and Henley, 1983). Similarly, young children who invade the personal space of an adult tend to elicit a more favorable response than do older uninvited visitors (Dean, Willis, and la Rocco, 1976). The need for personal space appears to increase with age (Baxter, 1970; Aiello and Jones, 1971) although it may begin to decrease at about age forty (Heshka and Nelson, 1972).

For some people, the idea of privacy or personal space is an unheard-of luxury afforded only to those in the middle and upper classes. As we have seen in this chapter, homeless bag ladies may have as their only personal space the bags they carry or the shopping carts they push down the streets. Some of the homeless may try to "stake a claim" on a heat grate or the same bed in a shelter for more than one night, but such claims have dubious authenticity in a society in which the homeless are

Reflections

Reflections 4.4

Why is personal space a particularly difficult problem for homeless individuals?

Personal space   the immediate area surrounding a person that the person claims as private.

Concept Table 4.A Social Interaction: The Microlevel Perspective

Social interaction and meaning	In a given society, forms of social interaction have shared meanings, although these may vary to some extent based on race/ethnicity, gender, and social class.
Social construction of reality	The process by which our perception of reality is largely shaped by the subjective meaning that we give to an experience.
Ethnomethodology	Studying the commonsense knowledge that people use to understand the situations in which they find themselves makes us aware of subconscious social realities in daily life.
Dramaturgical analysis	The study of social interaction that compares everyday life to a theatrical presentation—it includes impression management (people's efforts to present themselves favorably to others).
Sociology of emotions	We are socialized to feel certain emotions, and we learn how and when to express (or not express) them.
Nonverbal communication	The transfer of information between persons without the use of speech, such as by facial expressions, head movements, and gestures.

assumed to own nothing and to have no right to lay claim to anything in the public domain.

In sum, all forms of nonverbal communication are influenced by gender, race, social class, and the personal contexts in which they occur. Although it is difficult to generalize about people's nonverbal behavior, we still need to think about our own nonverbal communication patterns. Recognizing that differences in social interaction exist is important. We should be wary of making value judgments—the differences are simply differences. Learning to understand and respect alternative styles of social interaction enhances our personal effectiveness by increasing the range of options we have for communicating with different people in diverse contexts and for varied reasons (Wood, 1999). (Concept Table 4.A summarizes the microlevel approach to social interaction.)

Changing Social Structure and Interaction in the Future

The social structure in the United States has been changing rapidly in recent decades. Currently, there are more possible statuses for persons to occupy and roles to play than at any other time in history. Although achieved statuses are considered very impor-

tant, ascribed statuses still have a significant effect on the options and opportunities that people have.

Ironically, at a time when we have more technological capability, more leisure activities and types of entertainment, and more quantities of ma-

✓ Checkpoints

Checkpoint 4.4

12. How does social interaction provide shared meaning across situations within a given society?
13. What is the social construction of reality?
14. What is an example of a self-fulfilling prophecy?
15. What does the ethnomethodological approach contribute to our understanding of social interaction?
16. What are the key ideas in Erving Goffman's dramaturgical analysis?
17. What does the sociology of emotions contribute to our understanding of human behavior?
18. What is the importance of nonverbal communication in social life?
19. What is personal space, and how do we use it in our interactions with others?

Box 4.3 You Can Make a Difference
Overcoming Compassion Fatigue

Homelessness is such an overwhelming problem that many of us believe there is nothing we can do to help homeless people. Rabbi Charles A. Kroloff (1993), senior rabbi of Temple Emanu-El in Westfield, New Jersey, disagrees. As an advocate for the homeless and founder of the Interfaith Council for the Homeless, he has written *54 Ways You Can Help the Homeless*, listing ways that everyday people can do something for homeless persons. Some of his most novel suggestions include how we can help children who are homeless. For example, Rabbi Kroloff believes that people can use their hobbies and personal interests—such as cooking, repairing, gardening, or photography—to help these children:

Jim Hubbard, a Washington, D.C., professional photographer whose specialty is photographing the homeless, developed "Shooting Back," a program that teaches homeless children photography. Each week, Hubbard teaches eight children how to use a camera and pro-

vides them with free film. According to Hubbard, the children rarely shoot images of decay. Instead, they prefer to take pictures of other children, particularly those leaping into swimming pools or playing in the water spray from fire hydrants. Hubbard explains that the children have an urgent need for housing, but they also need an opportunity to develop self-esteem. He has found that mastering the camera and seeing their own images in print have greatly increased their self-confidence and feelings of self-worth.

Like Jim Hubbard, each of us can make a difference for homeless children. We can contribute hands-on knowledge and experience by tutoring children at homeless shelters, or we can donate funds so that volunteers can provide the children with clothing, toiletries, school supplies, and toys. We can also look into groups such as Children First, which was founded by the Interfaith Council for the Homeless of Union County, New Jersey, and help support enrichment projects—workshops and field trips to the zoo, museums, and other settings where homeless children and their parents can participate together in the simple pleasures that many other parents and children take for granted. If you would like to know more about helping homeless persons, contact your local Salvation Army, Coalition for the Homeless, Interfaith group, or one of the following agencies:

AP/Wide World Photos

Volunteers across the nation seek to assist homeless people by providing them with food, clothing, and other goods and services. An example is volunteer Anne Corry, who is shown helping a homeless man select a new coat for the cold winter months in New York City. What outreach services for the homeless are available in your city?

- The National Alliance to End Homelessness, 1518 K Street, N.W., Suite 206, Washington, DC 20005. (202) 638-1526.
- National Coalition for the Homeless, 1012 Fourteenth Street, N.W. #600, Washington, DC 20005-3410. (202) 731-6444.

You can purchase *54 Ways You Can Help the Homeless* by writing to P.O. Box 712, Westfield, NJ 07091. (Profits are donated to organizations that serve homeless people.) On the Internet:

http://www.earthsystems.org/ways

---

terial goods available for consumption than ever before, many people experience high levels of stress, fear for their lives because of crime, and face problems such as homelessness. In a society that can send astronauts into space to perform complex scientific experiments, is it impossible to solve some of the problems that plague us here on Earth?

Individuals and groups often show initiative in trying to solve some of our pressing problems (see Box 4.3). For example, Ellen Baxter has single-handedly tried to create housing for hundreds of New York City's homeless by reinventing well-maintained, single-room-occupancy residential hotels to provide cheap lodging and social services

(Anderson, 1993). However, individual initiative alone will not solve all our social problems in the future. Large-scale, formal organizations must become more responsive to society's needs.

At the microlevel, we need to regard social problems as everyone's problem; if we do not, they have a way of becoming everyone's problem anyway. When we think about "the homeless," for example, we are thinking in a somewhat misleading manner. "The homeless" suggests a uniform set of problems and a single category of poor people. Jonathan Kozol (1988: 92) emphasizes that "their miseries are somewhat uniform; the squalor is uniform; the density of living space is uniform. [However, the] uniformity is in their mode of suffering, not in themselves."

What can be done about homelessness in the future? Martha R. Burt, director of the Urban Institute's 1987 national study of urban homeless shelter and soup kitchen users, notes that we must first become dissatisfied with explanations that see personal problems as the cause of homelessness. Many people in the past have suffered from poverty, mental illness, alcoholism, physical handicaps, and drug addiction, but they have not become homeless. Only structural changes can explain why we have a larger homeless population today than in earlier times. As Burt suggests,

> To undo the effects of changing structural factors we will have to address the factors themselves, not the vulnerabilities of the people caught by changing times. We can take a short-term approach, raising benefit levels and expanding eligibility to cover those most vulnerable to homelessness. Such actions would prevent homelessness rather than ameliorate it, and are therefore preferable to building emergency shelters. They would not, however, change the underlying conditions, and the need for public support would be likely to continue indefinitely. A far better approach is to fulfill our commitments to support people, such as those with severe mental illness, who cannot be expected to support themselves, and also address simultaneously the employer and the employee requirements for increasing productivity, by reshaping the work environment and improving education and training. If we succeed at this much larger agenda, we will solve not only the problem of homelessness, but also the problem of declining living standards for a much broader spectrum of American workers. (Burt, 1992: 225–226)

In sum, the future of this country rests on our collective ability to deal with major social problems at both the macrolevel and the microlevel of society.

Chapter Review

How does social structure shape our social interactions?
The stable patterns of social relationships within a particular society make up its social structure. Social structure is a macrolevel influence because it shapes and determines the overall patterns in which social interaction occurs. Social structure provides an ordered framework for society and for our interactions with others.

What are the main components of social structure?
Social structure comprises statuses, roles, groups, and social institutions. A status is a specific position in a group or society and is characterized by certain expectations, rights, and duties. Ascribed statuses, such as gender, class, and race/ethnicity, are acquired at birth or involuntarily later in life. Achieved statuses, such as education and occupation, are assumed voluntarily as a result of personal choice, merit, or direct effort. We occupy a status, but a role is the set of behavioral expectations associated with a given status. A social group consists of two or more people who interact frequently and share a common identity and sense of interdependence. A formal organization is a highly structured group formed to complete certain tasks or achieve specific goals. A social institution is a set of organized beliefs and rules that establishes how a society attempts to meet its basic needs.

What are the functionalist and conflict perspectives on social institutions?
According to functionalist theorists, social institutions perform several prerequisites of all societies: replace members; teach new members; produce, distribute, and consume goods and services; preserve order; and provide and maintain a sense of purpose. Conflict theorists suggest that social institutions do not work for the common good of all individuals: Institutions may enhance and uphold the power of some groups but exclude others, such as the homeless.

How do societies maintain stability in times of social change?
According to Emile Durkheim, although changes in social structure may dramatically affect individuals and groups, societies manage to maintain some degree of stability. People in preindustrial societies are united by mechanical solidarity because they have shared values and common social bonds. Industrial societies are characterized by organic solidarity, which refers to the cohesion that results when people perform specialized tasks and are united by mutual dependence.

How do *Gemeinschaft* and *Gesellschaft* societies differ in social solidarity?
According to Ferdinand Tönnies, the *Gemeinschaft* is a traditional society in which relationships are based on personal bonds of friendship and kinship and on intergenerational stability. The *Gesellschaft* is an urban society in which social bonds are based on impersonal and specialized relationships, with little group commitment or consensus on values.

Is all social interaction based on shared meanings?
Social interaction within a society, particularly face-to-face encounters, is guided by certain shared meanings of how we should behave. All meanings may not be shared—race/ethnicity, gender, and social class often influence people's perceptions of meaning.

What is the dramaturgical perspective?
According to Erving Goffman's dramaturgical analysis, our daily interactions are similar to dramatic productions. Presentation of self refers to efforts to present our own self to others in ways that are most favorable to our interests or self-image.

Why are feeling rules important?
Feeling rules shape the appropriate emotions for a given role or specific situation. Our emotions are not always private, and specific emotions may be demanded of us on certain occasions.

Key Terms

achieved status 113
ascribed status 113
dramaturgical analysis 129
ethnomethodology 128
formal organization 120
Gemeinschaft 123
Gesellschaft 124
impression management (presentation of self) 129
industrial society 125
master status 114
mechanical solidarity 123
nonverbal communication 132
organic solidarity 123
personal space 135
postindustrial society 125
primary group 119
role 115
role conflict 116
role exit 117
role expectation 115
role performance 116

role strain 117
secondary group 119
self-fulfilling prophecy 128
social construction of reality 127
social group 119
social institution 120
social interaction 109
social structure 109
status 113
status symbol 115

Questions for Critical Thinking

1. Think of a person you know well who often irritates you or whose behavior grates on your nerves (it could be a parent, friend, relative, teacher). First, list that person's statuses and roles. Then, analyze the person's possible role expectations, role performance, role conflicts, and role strains. Does anything you find in your analysis help to explain the irritating behavior? How helpful are the concepts of social structure in analyzing individual behavior?

2. Are structural problems responsible for homelessness, or are homeless individuals responsible for their own situation?

3. You are conducting field research on gender differences in nonverbal communication styles. How are you going to account for variations among age, race, and social class?

4. When communicating with other genders, races, and ages, is it better to express and acknowledge different styles or to develop a common, uniform style?

 ## Resources on the Internet

Chapter-Related Web Sites
The following web sites have been selected for their relevance to the topics in this chapter. These sites are among the more stable, but please note that web site addresses change frequently.

Homelessness
http://aspe.os.dhhs.gov/progsys/homeless/
The U.S. Department of Health and Human Services provides information and links to government research on homelessness in America.

Hunger and Homelessness: Waste Not, Want Not
http://www.project.org/guide/hunger.html
In this Project America site, learn about social programs and ideas to organize resources to help those who are homeless and hungry in America.

The Durkheim Pages
http://www.relst.uiuc.edu/durkheim/
Find in-depth information here about Emile Durkheim, the French sociologist who wrote about social solidarity.

Ferdinand Tönnies on *Gemeinschaft* and *Gesellschaft*
http://www2.pfeiffer.edu/~lridener/courses/
GEMEIN.HTML
Read for yourself Tönnies's theory about the changes in modern social structure.

Readings in Critical Dimensions in Dramaturgical Analysis
http://www.tryoung.com/CRITDRAM/
CRITDRAMATURGYREADINGS.htm
Here you will find a number of essays about the application of dramaturgical analysis to various aspects of social life, such as "Backstage at the White House."

"Ethnomethodology and the Study of Online Communities"
http://informationr.net/ir/4-1/paper50.html
This essay, by Steven R. Thomsen, explores the basic dimensions of online communities and the resulting need for scholars to rethink the assumptions behind long-accepted paradigms regarding the nature of social interaction, social bonding, and empirical experience.

Population in Emergency and Transitional Housing
http://www.census.gov/population/www/cen2000/
phc-t12.html
This U.S. Census Bureau site contains detailed reports on and analysis of the population living in shelters.

Companion Web Site for This Book

Virtual Society: The Wadsworth Sociology Resource Center
Visit **http://sociology.wadsworth.com** and click on the page for Kendall, *Sociology in Our Times: The Essentials*, Fourth Edition, to access a wide range of en-

richment material to aid in your study of sociology. Click on the Student Resources section of the web site. Next, select from the pull-down menu the chapter that you are presently studying. Among the useful options for self-study are chapter objectives, flashcards, practical study tips, and practice tests for each chapter.

MicroCase Online
From the Virtual Society home page, click on MicroCase Online to access book-specific MicroCase exercises that allow you to further explore sociological issues and principles presented in the text.

InfoTrac College Edition
Another unique option available to you at the Student Resources section of the companion web site is InfoTrac College Edition, an online library with access to hundreds of scholarly and popular periodicals. Below are suggested search terms for this chapter. Results from these and other searches are found at the site.

- Search keyword: *homelessness*. Conduct a search for articles on homelessness, focusing on research that addresses the causes of homelessness.
- Search keywords: *nonverbal communication*. Look for articles that address research on the differences between male and female nonverbal communication. What types of nonverbal communication are discussed in the articles that you found?

Virtual Explorations CD-ROM
Go to the Virtual Explorations CD-ROM to begin the interactive exercise for this chapter. This Virtual Exploration will introduce you to some of the exciting resources for sociology on the World Wide Web. You will be guided through an exercise that employs related web sites on social structure and interaction in everyday life. Answer the questions, and e-mail your responses to your instructor.

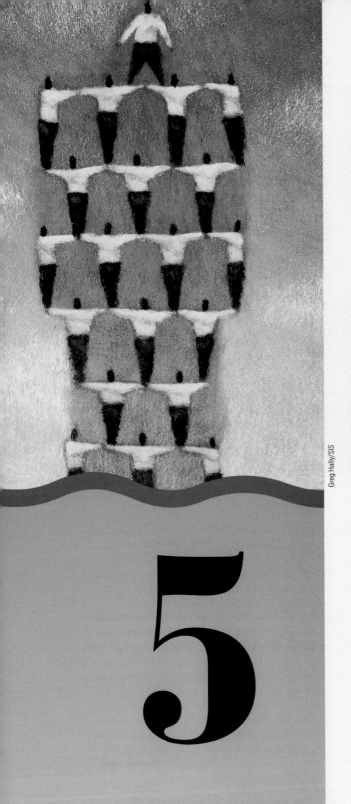

Greg Hally/SIS

Groups and Organizations

Social Groups
Groups, Aggregates, and Categories
Types of Groups

Group Characteristics and Dynamics
Group Size
Group Leadership
Group Conformity
Groupthink

Formal Organizations in Global Perspective
Types of Formal Organizations
Bureaucracies
Shortcomings of Bureaucracies
Bureaucracy and Oligarchy

Alternative Forms of Organization
Organizational Structure in Japan

Organizations in the Future

5

It's just incomprehensible to see what it was like down there. You know, I remember seeing one of these Cold War movies [that showed what things might be like] after the nuclear attacks with the Hollywood portrayal of a nuclear winter. It looked worse than that in downtown Manhattan, and it wasn't some grade "B" movie. It was life. It was real.

—New York Governor George Pataki describing his first trip to the World Trade Center disaster site following September 11, 2001 (qtd. in Nacos, 2002: 13)

AP/Wide World Photos

Following the terrorist attacks of 2001, the U.S. Senate Judiciary Committee conducted hearings in an effort to learn what information about terrorist activities and possible targets in this country had been available to federal agencies such as the Federal Bureau of Investigation and the Central Intelligence Agency in advance of September 11, and why the government had not acted on any available information in a manner that might have prevented the attacks from occurring on that fateful day. On June 6, 2002, the Judiciary Committee interviewed FBI agent Coleen Rowley, who had written a letter setting forth information that had been known by the FBI; she discussed her belief that the "culture" of that agency had prevented it from acting on what it knew (*New York Times*, 2002b):

AGENT ROWLEY: We have a culture in the FBI that there's a certain pecking order, and it's pretty strong. And it's very rare that someone picks up the phone and calls a rank or two above themselves. It would have to be only on the strongest reasons. Typically, you would have to . . . pick up the phone and talk to somebody who is at your rank. So when you have an item that requires review by a higher level, it's incumbent for you to go to a higher-level person in your office and then for that person to make a call. . . .

SENATOR GRASSLEY: In your letter [to the FBI director], you mention a culture of fear, especially a fear of taking action, and the problem of careerism. Could you talk about how this hurts investigations in the field, what the causes are, and what you think might fix these problems?

AGENT ROWLEY: [W]hen I looked up the definition [of careerism], I really said [it's] unbelievable how appropriate that is. I think the FBI does have a problem with that. And if I remember right, it means, "promoting one's career over integrity." So, when people make decisions, and it's basically so that [they] can get to the next level and not rock—either it's not rock the boat or do what a boss says without question. And either way that works, if you're making a decision to try to get to the next level, but you're not making that decision for the real right reason, that's a problem. . . .

According to sociologists, we need groups and organizations—just like we need culture and socialization—in order to live and participate in societies as we know them; however, there are both positive and negative aspects of groups, regardless of their size. For example, a small group may be made up only of people who are good friends and enjoy being with one another, yet the group may disband if one or more of its members

Box 5.1 Sociology and Everyday Life
How Much Do You Know About Privacy in Groups and Organizations?

True	False	
T	F	1. A fast-food restaurant can legally require all employees under the age of twenty-one to submit to periodic, unannounced drug testing.
T	F	2. Members of a high school football team can be required to submit to periodic, unannounced drug testing.
T	F	3. Parents of students at all U.S. colleges and universities are entitled to obtain a transcript of their children's college grades, regardless of the student's age.
T	F	4. A company has the right to keep its employees under video surveillance at all times while they are at the company's place of business—even in the company's restrooms.
T	F	5. A private club has the right to require an applicant for membership to provide his or her Social Security number as a condition of membership.
T	F	6. If a person applies for a job in a workplace that has more than twenty-five employees, the employer can require that person to provide medical information or take a physical examination prior to offering him or her a job.
T	F	7. Students at a church youth group meeting who hear one member of the group confess to an illegal act can be required to divulge what that member said.
T	F	8. A student's privacy is protected when using a computer, even if it is owned by the college or university, because deleting an e-mail or other document from a computer prevents anyone else from examining that document.

Answers on page 144.

feel left out and cease to participate. A large group may be able to accomplish tasks that a smaller number of people could never achieve, yet the larger group may become unwieldy due to its size, with the result that some members come to feel that it is no longer focused on its original goals. In each instance, part of the problem may be inadequate or ineffective communication.

When we encounter a major problem with an existing organization, we may wonder whether the group should be restructured or possibly replaced by some new structure that we believe might meet our needs more efficiently. Such has been the case in the aftermath of the terrorist attacks on the United States: The Federal Bureau of Investigation, the Central Intelligence Agency, and other governmental organizations have faced criticism about how they used information before these attacks—not only how crucial information flowed or failed to flow within the particular organization but also why such information was not better conveyed among supposedly cooperating agencies. This is an area in which sociologists have applied the sociological imagination by studying how bureaucracies and technology contribute to (or impede) information flow within and between large-scale organizations.

As we shall see in this chapter, bureaucratic structure is the most common organizational form in gov-

ernments, businesses, schools, and other institutions around the globe. Although there are many benefits from this organizational model, there are also significant limitations, including loss of personal privacy. Before we take a closer look at social groups and bureaucratic organizations, take the quiz in Box 5.1 on privacy in groups and organizations.

Questions and Issues

Chapter Focus Question: Why is it important for groups and organizations to enhance communication among participants and aid the flow of information while protecting the privacy of individuals?

What constitutes a social group?

How are groups and their members shaped by group size, leadership style, and pressures to conform?

What is the relationship between information and social organizations in societies such as ours?

What purposes does bureaucracy serve?

What alternative forms of organization exist as compared with the most widespread forms today?

Box 5.1

Answers to the Sociology Quiz on Privacy

1. **True.** In all but a few states, an employee in the private sector of the economy can be required to submit to a drug test even where nothing about the employee's job performance or history suggests illegal drug use. An employee who refuses can be terminated without legal recourse.

2. **True.** The U.S. Supreme Court has ruled that schools may require students to submit to random drug testing as a condition to participating in extracurricular activities such as sports teams, the school band, the future homemakers' club, the cheerleading squad, and the choir.

3. **False.** The Family Educational Right to Privacy Act, which allows parents of a student under age eighteen to obtain their child's grades, requires the student's consent once he or she has attained age eighteen; however, that law applies only to institutions that receive federal educational funds.

4. **False.** An employer may not engage in video surveillance of its employees in situations where they have a reasonable right of privacy. At least in the absence of a sign warning of such surveillance, employees have such a right in company restrooms.

5. **True.** Although the Privacy Act of 1974 makes it illegal for federal, state, and local governmental agencies to deny rights, privileges, or benefits to individuals who refuse to provide their Social Security number unless disclosure is required by law, no federal law extends this prohibition to private groups and organizations.

6. **False.** The Americans with Disabilities Act prohibits employers in workplaces with more than twenty-five employees from asking job applicants about medical information or requiring a physical examination prior to employment.

7. **True.** Although confidential communication made privately to a minister, priest, rabbi, or other religious leader (or to an individual the person reasonably believes to hold such a position) generally cannot be divulged without the consent of the person making the communication, this does not apply when other people are present who are likely to hear the statement.

8. **False.** Deleting an e-mail or other document from a computer does not actually remove it from the computer's memory. Until other files are entered that actually write over the space where the document was located, experts can retrieve the document that was deleted.

Social Groups

Three strangers are standing at a street corner waiting for a traffic light to change. Do they constitute a group? Five hundred women and men are first-year graduate students at a university. Do they constitute a group? In everyday usage, we use the word *group* to mean any collection of people. According to sociologists, however, the answer to these questions is no; individuals who happen to share a common feature or to be in the same place at the same time do not constitute social groups.

Groups, Aggregates, and Categories

As we saw in Chapter 4, a *social group* is a collection of two or more people who interact frequently with one another, share a sense of belonging, and have a feeling of interdependence. Several people waiting for a traffic light to change constitute an *aggregate*—**a collection of people who happen to be in the same place at the same time but share little else in common.** Shoppers in a department store and passengers on an airplane flight are also examples of aggregates. People in aggregates share a common purpose (such as purchasing items or arriving at their destination) but generally do not interact with one another, except perhaps briefly. The first-year graduate students, at least initially, constitute a *category*—**a number of people who may never have met one another but share a similar characteristic** (such as education level, age, race, or gender). Men and women make up categories, as do Native Americans and Latinos/as, and victims of sexual or racial harassment. Categories are not social groups because the people in them do not usually create a social structure or have anything in common other than a particular trait.

Occasionally, people in aggregates and categories form social groups. For instance, people within the category known as "graduate students" may become an aggregate when they get together for an orientation to graduate school. Some of them may form social groups as they interact with one another in classes and seminars, find that they have mutual interests and concerns, and develop a sense of belonging to the group.

Types of Groups

As you will recall from Chapter 4, groups have varying degrees of social solidarity and structure. This structure is flexible in some groups and more rigid in others. Some groups are small and personal; others are large and impersonal. We more closely identify with the members of some groups than we do with others.

COOLEY'S PRIMARY AND SECONDARY GROUPS Sociologist Charles H. Cooley (1963/1909) used the term *primary group* to describe a small, less specialized group in which members engage in face-to-face, emotion-based interactions over an extended period of time. We have primary relationships with other individuals in our primary groups—that is, with our *significant others*, who frequently serve as role models.

In contrast, you will recall, a *secondary group* is a larger, more specialized group in which the members engage in more impersonal, goal-oriented relationships for a limited period of time. The size of a secondary group may vary. Twelve students in a graduate seminar may start out as a secondary group but eventually become a primary group as they get to know one another and communicate on a more personal basis. Formal organizations are secondary groups, but they also contain many primary groups within them. For example, how many primary groups do you think there are within the secondary-group setting of your college?

SUMNER'S INGROUPS AND OUTGROUPS All groups set boundaries by distinguishing between insiders who are members and outsiders who are not. Sociologist William Graham Sumner (1959/1906) coined the terms *ingroup* and *outgroup* to describe people's feelings toward members of their own and other groups. An *ingroup* **is a group to which a person belongs and with which the person feels a sense of identity.** An *outgroup* **is a group to which a person does not belong and toward which the person may feel a sense of competitiveness or hostility.** Distinguishing between our ingroups and our outgroups helps us establish our individual identity and self-worth. Likewise, groups are solidified by ingroup and outgroup distinctions; the presence of an enemy or a hostile group binds members more closely together (Coser, 1956).

Group boundaries may be formal, with clearly defined criteria for membership. For example, a

---

Aggregate a collection of people who happen to be in the same place at the same time but share little else in common.

Category a number of people who may never have met one another but share a similar characteristic, such as education level, age, race, or gender.

Ingroup a group to which a person belongs and with which the person feels a sense of identity.

Outgroup a group to which a person does not belong and toward which the person may feel a sense of competitiveness or hostility.

Ted Grant/Canadian Sports Images

These Olympic teams graphically illustrate the concept of ingroups and outgroups. Each Olympic team can be seen as an ingroup that helps give its members a sense of belonging and identity—feelings that are strengthened by the presence of clearly defined outgroups (competing teams).

country club that requires an applicant for membership to be recommended by four current members and to pay a $25,000 initiation fee has clearly set requirements for its members. However, group boundaries are not always that formal. For example, friendship groups usually do not have clear guidelines for membership; rather, the boundaries tend to be very informal and vaguely defined.

Ingroup and outgroup distinctions may encourage social cohesion among members, but they also may promote classism, racism, sexism, and ageism. Ingroup members typically view themselves positively and members of outgroups negatively. These feelings of group superiority, or *ethnocentrism*, are somewhat inevitable. However, members of some groups feel more free than others to act on their beliefs. If groups are embedded in larger groups and organizations, the large organization may discourage such beliefs and their consequences (Merton, 1968). Conversely, organizations may covertly foster these ingroup/outgroup distinctions by denying their existence or by failing to take action when misconduct occurs.

REFERENCE GROUPS   Ingroups provide us not only with a source of identity but also with a point of reference. A *reference group* is a group that strongly influences a person's behavior and social attitudes, regardless of whether that individual is an actual member. When we attempt to evaluate our appearance, ideas, or goals, we automatically refer to the standards of some group. Sometimes, we will refer to our membership groups, such as family or friends. Other times, we will rely on groups to which we do not currently belong but that we might wish to join in the future, such as a social club or a profession. We may also have negative reference groups. For many people, the Ku Klux Klan and neo-Nazi skinheads are examples of negative reference groups because most people's racial attitudes compare favorably with such groups' blatantly racist behavior.

Reference groups help explain why our behavior and attitudes sometimes differ from those of our membership groups. We may accept the values and norms of a group with which we identify rather than one to which we belong. We may also act more like members of a group we want to join than members of groups to which we already be-

✓Checkpoints

Checkpoint 5.1

1. What is an aggregate?
2. What is an example of a category?
3. How do ingroups differ from outgroups?
4. What is a reference group?

long. In this case, reference groups are a source of anticipatory socialization. Many people have more than one reference group and often receive conflicting messages from these groups about how they should view themselves. For most of us, our reference-group attachments change many times during our life course, especially when we acquire a new status in a formal organization.

Group Characteristics and Dynamics

What purpose do groups serve? Why are individuals willing to relinquish some of their freedom to participate in groups? According to functionalists, people form groups to meet instrumental and expressive needs. *Instrumental*, or task-oriented, needs cannot always be met by one person, so the group works cooperatively to fulfill a specific goal. For example, think of how hard it would be to function as a one-person football team or to single-handedly build a skyscraper. Groups help members do jobs that are impossible to do alone or that would be very difficult and time-consuming at best. In addition to instrumental needs, groups also help people meet their *expressive*, or emotional, needs, especially those involving self-expression and support from family, friends, and peers.

Although not disputing that groups ideally perform such functions, conflict theorists suggest that groups also involve a series of power relationships whereby the needs of individual members may not be equally served. Symbolic interactionists focus on how the size of a group influences the kind of interaction that takes place among members. To many postmodernists, groups and organizations—like other aspects of postmodern societies—are generally characterized by superficiality and depthlessness in social relationships (Jameson, 1984). One postmodern

thinker who focuses on this issue is the literary theorist Fredric Jameson, whose works continue to have a significant influence on contemporary sociological theorizing. According to Jameson's (1984) analysis, not only are postmodern organizations (and societies as a whole) characterized by superficial relations and lack of depth; people also experience a waning of emotion because the world, and the people in it, have become more fragmented (Ritzer, 1997). For example, the sociologist George Ritzer (1997) examined fast-food restaurants using a postmodern approach and concluded that both restaurant employees and customers interact in extremely superficial ways that are largely scripted by large-scale organizations: The employees learn to follow scripts in taking and filling customers' orders (Leidner, 1993), whereas customers also respond with their own "recipied" action (Schutz, 1967/1932). According to Ritzer (1997: 226), "[C]ustomers are mindlessly following what they consider tried-and-true social recipes, either learned or created by them previously, on how to deal with restaurant employees and, more generally, how to work their way through the system associated with the fast-food restaurant."

We will now look at certain characteristics of groups, such as how size affects group dynamics.

Group Size

The size of a group is one of its most important features. Interactions are more personal and intense in a *small group*, a collectivity small enough for all

Reflections

Reflections 5.1

Why are intensified levels of surveillance of individuals and groups often characteristic of postmodern societies?

Reference group a group that strongly influences a person's behavior and social attitudes, regardless of whether that individual is an actual member.

Small group a collectivity small enough for all members to be acquainted with one another and to interact simultaneously.

Chip Henderson/Index Stock Imagery

Chip Henderson/Index Stock Imagery

According to the sociologist Georg Simmel, interaction patterns change when a third person joins a dyad—a group composed of two members. How might the conversation between these two people change when another person arrives to talk with them?

members to be acquainted with one another and to interact simultaneously. Sociologist Georg Simmel (1950/1902–1917) suggested that small groups have distinctive interaction patterns that do not exist in larger groups. According to Simmel, in a *dyad*—**a group composed of two members**—the active participation of both members is crucial to the group's survival. If one member withdraws from interaction or "quits," the group ceases to exist. Examples of dyads include two people who are best friends, married couples, and domestic partnerships. Dyads provide members with an intense bond and a sense of unity not found in most larger groups.

When a third person is added to a dyad, a *triad*, **a group composed of three members,** is formed. The nature of the relationship and interaction patterns changes with the addition of the third person. In a triad, even if one member ignores another or declines to participate, the group can still function. In addition, two members may unite to create a coalition that can subject the third member to group pressure to conform. A *coalition* is an alliance created in an attempt to reach a shared objective or goal. If two members form a coalition, the other member may be seen as an outsider or intruder.

As the size of a group increases beyond three people, members tend to specialize in different tasks, and everyday communication patterns change. For instance, in groups of more than six or seven people, it becomes increasingly difficult for everyone to take part in the same conversation; there-

fore, several conversations will likely take place simultaneously. Members are also likely to take sides on issues and form a number of coalitions. In groups of more than ten or twelve people, it becomes virtually impossible for all members to participate in a single conversation unless one person serves as moderator and guides the discussion. As shown in Figure 5.1, when the size of the group increases, the number of possible social interactions also increases.

Although large groups typically have less social solidarity than small ones, they may have more power. However, the relationship between size and power is more complicated than it might initially seem. The power relationship depends on both a group's *absolute* size and its *relative* size (Simmel, 1950/1902–1917; Merton, 1968). The absolute size is the number of members the group actually has; the relative size is the number of potential members. For example, suppose that three hundred people band together to "march on Washington" and demand enactment of a law on some issue that they feel is important. Although three hundred people is a large number in some contexts, opponents of this group would argue that the low turnout (compared with the number of people in this country) demonstrates that most people don't believe the issue is important. At the same time, the power of a small group to demand change may be based on a "strength in numbers" factor if the group is seen as speaking on behalf of a large number of other people (who are also voters).

Figure 5.1 Growth of Possible Social Interaction Based on Group Size

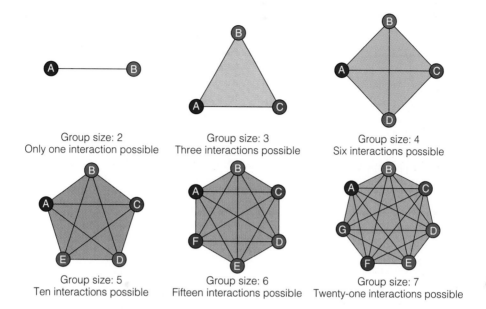

Group size: 2
Only one interaction possible

Group size: 3
Three interactions possible

Group size: 4
Six interactions possible

Group size: 5
Ten interactions possible

Group size: 6
Fifteen interactions possible

Group size: 7
Twenty-one interactions possible

Larger groups typically have more formalized leadership structures. Their leaders are expected to perform a variety of roles, some related to the internal workings of the group and others related to external relationships with other groups.

Group Leadership

What role do leaders play in groups? Leaders are responsible for directing plans and activities so that the group completes its task or fulfills its goals. Primary groups generally have informal leadership. For example, most of us do not elect or appoint leaders in our own families. Various family members may assume a leadership role at various times or act as leaders for specific tasks. In traditional families, the father or eldest male is usually the leader. However, in today's more diverse families, leadership and power are frequently in question, and power relationships may be quite different, as discussed later in this text. By comparison, larger groups typically have more formalized leadership structures. Their leaders are expected to perform a variety of roles, some related to the internal workings of the group and others related to external relationships with other groups. For example, leadership in secondary groups (such as colleges, governmental agencies, and corporations) involves

a clearly defined chain of command, with written responsibilities assigned to each position in the organizational structure.

LEADERSHIP FUNCTIONS Both primary and secondary groups have some type of leadership or positions that enable certain people to be leaders, or at least to wield power over others. From a functionalist perspective, if groups exist to meet the instrumental and expressive needs of their members, then leaders are responsible for helping the group meet those needs. *Instrumental leadership* **is goal or task oriented;** this type of leadership is most appropriate when the group's purpose is to complete a task or reach a particular goal. *Expressive leadership* **provides emotional support for members;** this type of leadership is most appropriate when the group is

---

Dyad a group composed of two members.

Triad a group composed of three members.

Instrumental leadership goal- or task-oriented leadership.

Expressive leadership an approach to leadership that provides emotional support for members.

dealing with emotional issues, and harmony, solidarity, and high morale are needed. Both kinds of leadership are needed for groups to work effectively.

LEADERSHIP STYLES Three major styles of leadership exist in groups: authoritarian, democratic, and laissez-faire. *Authoritarian leaders* **make all major group decisions and assign tasks to members.** These leaders focus on the instrumental tasks of the group and demand compliance from others. In times of crisis, such as a war or natural disaster, authoritarian leaders may be commended for their decisive actions. In other situations, however, they may be criticized for being dictatorial and for fostering intergroup hostility. By contrast, *democratic leaders* **encourage group discussion and decision making through consensus building.** These leaders may be praised for their expressive, supportive behavior toward group members, but they also may be blamed for being indecisive in times of crisis.

Laissez-faire literally means "to leave alone." *Laissez-faire leaders* **are only minimally involved in decision making and encourage group members to make their own decisions.** On the one hand, laissez-faire leaders may be viewed positively by group members because they do not flaunt their power or position. On the other hand, a group that needs active leadership is not likely to find it with this style of leadership, which does not work vigorously to promote group goals (White and Lippitt, 1953).

Studies of kinds of leadership and decision-making styles have certain inherent limitations. They tend to focus on leadership that is imposed externally on a group (such as bosses or political leaders) rather than leadership that arises within a group. Different decision-making styles may be more effective in one setting than another. For example, imagine attending a college class in which the professor asked the students to determine what should be covered in the course, what the course requirements should be, and how students should be graded. It would be a difficult and cumbersome way to start the semester; students might spend the entire semester negotiating these matters and never actually learn anything.

Group Conformity

To what extent do groups exert a powerful influence in our lives? Groups have a significant amount of influence over our values, attitudes, and behavior. In order to gain and then retain our membership in groups, most of us are willing to exhibit a high level of conformity to the wishes of other group members. *Conformity* **is the process of maintaining or changing behavior to comply with the norms established by a society, subculture, or other group.** We often experience powerful pressure from other group members to conform. In some situations, this pressure may be almost overwhelming.

In several studies (which would be impossible to conduct today for ethical reasons), researchers found that the pressure to conform may cause group members to say they see something that is contradictory to what they are actually seeing or to do something that they would otherwise be unwilling to do. As we look at two of these studies, ask yourself what you might have done if you had been involved in this research.

ASCH'S RESEARCH Pressure to conform is especially strong in small groups in which members want to fit in with the group. In a series of experiments conducted by Solomon Asch (1955, 1956), the pressure toward group conformity was so great that participants were willing to contradict their own best judgment if the rest of the group disagreed with them.

One of Asch's experiments involved groups of undergraduate men (seven in each group) who were allegedly recruited for a study of visual perception. All the men were seated in chairs. However, the person in the sixth chair did not know that he was the only actual subject; all the others were assisting the researcher. The participants were first shown a large card with a vertical line on it and then a second card with three vertical lines (see Figure 5.2). Each of the seven participants was asked to indicate which of the three lines on the second card was identical in length to the "standard line" on the first card.

In the first test with each group, all seven men selected the correct matching line. In the second trial, all seven still answered correctly. In the third trial, however, the actual subject became very uncomfortable when all the others selected the incorrect line. The subject could not understand what was happening and became even more confused as the others continued to give incorrect responses on eleven out of the next fifteen trials.

If you had been in the position of the subject, how would you have responded? Would you have

Figure 5.2 Asch's Cards

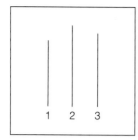

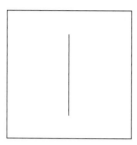

Although Line 2 is clearly the same length as the line in the lower card, Solomon Asch's research assistants tried to influence "actual" participants by deliberately picking Line 1 or Line 3 as the correct match. Many of the participants went along rather than risking the opposition of the "group."

Source: Asch, 1955.

continued to give the correct answer, or would you have been swayed by the others? When Asch (1955) averaged the responses of all fifty actual subjects who participated in the study, he found that about 33 percent routinely chose to conform to the group by giving the same (incorrect) responses as Asch's assistants. Another 40 percent gave incorrect responses in about half of the trials. Although 25 percent always gave correct responses, even they felt very uneasy and "knew that something was wrong." In discussing the experiment afterward, most of the subjects who gave incorrect responses indicated that they had known the answers were wrong but decided to go along with the group in order to avoid ridicule or ostracism.

After conducting additional research, Asch concluded that the size of the group and the degree of social cohesion felt by participants were important influences on the extent to which individuals respond to group pressure. In dyads, for example, the subject was much less likely to conform to an incorrect response from one assistant than in four-member groups. This effect peaked in groups of approximately seven members and then leveled off.

One contribution of Asch's research is the dramatic way in which it calls our attention to the power that groups have to produce a certain type of conformity. *Compliance* is the extent to which people say (or do) things so that they may gain the approval of other people. Certainly, Asch demonstrated that people will bow to social pressure in small-group settings. From a sociological perspective, however, the study was flawed because it involved deception about the purpose of the study and about the role of individual group members.

MILGRAM'S RESEARCH How willing are we to do something because someone in a position of authority has told us to do it? How far are we willing to go to follow the demands of that individual? Stanley Milgram (1963, 1974) conducted a series of controversial experiments to find answers to these questions about people's obedience to authority. *Obedience* is a form of compliance in which people follow direct orders from someone in a position of authority.

Milgram's subjects were men who had responded to an advertisement for participants in an experiment. When the first (actual) subject arrived, he was told that the study concerned the effects of punishment on learning. After the second subject (an assistant of Milgram's) arrived, the two men were instructed to draw slips of paper from a hat to get their assignments as either the "teacher" or the "learner." Because the drawing was rigged, the actual subject always became the teacher, and the assistant the learner. Next, the learner was strapped into a chair with protruding electrodes that looked something like an electric chair. The teacher was placed in an adjoining room and given a realistic-looking but nonoperative shock generator. The "generator's" control panel showed levels that went from "Slight Shock" (15 volts) on the left, to "Intense Shock" (255 volts) in the middle,

Authoritarian leaders people who make all major group decisions and assign tasks to members.

Democratic leaders leaders who encourage group discussion and decision making through consensus building.

Laissez-faire leaders leaders who are only minimally involved in decision making and who encourage group members to make their own decisions.

Conformity the process of maintaining or changing behavior to comply with the norms established by a society, subculture, or other group.

Figure 5.3 Results of Milgram's Obedience Experiment

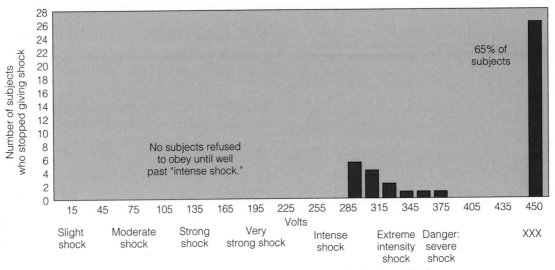

Even Milgram was surprised by subjects' willingness to administer what they thought were severely painful and even dangerous shocks to a helpless "learner."

Source: Milgram, 1963.

to "DANGER: SEVERE SHOCK" (375 volts), and finally "XXX" (450 volts) on the right.

The teacher was instructed to read aloud a pair of words and then repeat the first of the two words. At that time, the learner was supposed to respond with the second of the two words. If the learner could not provide the second word, the teacher was instructed to press the lever on the shock generator so that the learner would be punished for forgetting the word. Each time the learner gave an incorrect response, the teacher was supposed to increase the shock level by 15 volts. The alleged purpose of the shock was to determine if punishment improves a person's memory.

What was the maximum level of shock that a "teacher" was willing to inflict on a "learner"? The learner had been instructed (in advance) to beat on the wall between him and the teacher as the experiment continued, pretending that he was in intense pain. The teacher was told that the shocks might be "extremely painful" but that they would cause no permanent damage. At about 300 volts, when the learner quit responding at all to questions, the teacher often turned to the experimenter to see what he should do next. When the experimenter indicated that the teacher should give increasingly painful shocks, 65 percent of the teachers administered

shocks all the way up to the "XXX" (450-volt) level (see Figure 5.3). By this point in the process, the teachers were frequently sweating, stuttering, or biting on their lip. According to Milgram, the teachers (who were free to leave whenever they wanted to) continued in the experiment because they were being given directions by a person in a position of authority (a university scientist wearing a white coat).

What can we learn from Milgram's study? The study provides evidence that obedience to authority may be more common than most of us would like to believe. None of the "teachers" challenged the process before they had applied 300 volts. Almost two-thirds went all the way to what could have been a deadly jolt of electricity if the shock generator had been real. For many years, Milgram's findings were found to be consistent in a number of different settings and with variations in the research design (Miller, 1986).

This research once again raises some questions concerning research ethics. As was true of Asch's research, Milgram's subjects were deceived about the nature of the study in which they were asked to participate. Many of them found the experiment extremely stressful. Such conditions cannot be ignored by social scientists because subjects may receive lasting emotional scars from this kind of research. Today,

Figure 5.4 Janis's Description of Groupthink

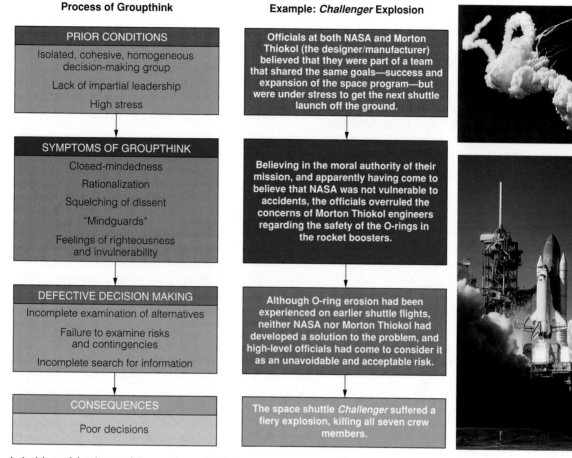

In Janis's model, prior conditions such as a highly homogenous group with committed leadership can lead to potentially disastrous "groupthink," which short-circuits careful and impartial deliberation. Events leading up to the tragic 1986 launch of the space shuttle *Challenger* have been cited as an example of this process.

Source: Lippa, 1994.

it would be virtually impossible to obtain permission to replicate this experiment in a university setting.

Groupthink

As we have seen, individuals often respond differently in a group context than they might if they were alone. Social psychologist Irving Janis (1972, 1989) examined group decision making among political experts and found that major blunders in U.S. history may be attributed to pressure toward group conformity. To describe this phenomenon, he coined the term *groupthink*—the process by which members of a cohesive group arrive at a decision that many individual members privately believe is un-

wise. Why not speak up at the time? Members usually want to be "team players." They may not want to be the ones who undermine the group's consensus or who challenge the group's leaders. Consequently, members often limit or withhold their opinions and focus on consensus rather than on exploring all of the options and determining the best course of action. Figure 5.4 summarizes the dynamics and results of groupthink.

Groupthink the process by which members of a cohesive group arrive at a decision that many individual members privately believe is unwise.

✓ Checkpoints

Checkpoint 5.2

5. How does group size affect social interaction?
6. How does instrumental leadership differ from expressive leadership?
7. What are the key leadership styles?
8. What did Solomon Asch find in his studies of group conformity?
9. What can we learn from Stanley Milgram's study of group conformity?
10. What is groupthink?

The tragic 1986 launch of the space shuttle *Challenger*—which exploded 73 seconds into its flight, killing all seven crew members—has been cited as an example of this process. On the day preceding the launch, engineers at the company responsible for designing and manufacturing the shuttle's solid rocket boosters became concerned that freezing temperatures at the launch site would interfere with the proper functioning of the O-ring seals in the boosters, but were overruled by high-level officials at the company and within NASA, where executives were impatient as the result of earlier delays. A presidential commission investigating the tragedy concluded that neither the manufacturer nor NASA responded adequately to warnings about the seals (Lippa, 1994).

Formal Organizations in Global Perspective

Over the past century, the number of formal organizations has increased dramatically in the United States and other industrialized nations. Previously, everyday life was centered in small, informal, primary groups, such as the family and the village. With the advent of industrialization and urbanization (as discussed in Chapter 1), people's lives became increasingly dominated by large, formal, secondary organizations. A *formal organization*, you will recall, is a highly structured secondary group formed for the purpose of achieving specific goals in the most efficient manner. Formal organizations (such as corporations, schools, and govern-ment agencies) usually keep their basic structure for many years in order to meet their specific goals.

Types of Formal Organizations

We join some organizations voluntarily and others out of necessity. Sociologist Amitai Etzioni (1975) classified formal organizations into three categories—normative, coercive, and utilitarian—based on the nature of membership in each.

NORMATIVE ORGANIZATIONS We voluntarily join *normative organizations* when we want to pursue some common interest or gain personal satisfaction or prestige from being a member. Political parties, ecological activist groups, religious organizations, parent–teacher associations, and college sororities and fraternities are examples of normative, or voluntary, associations.

Class, gender, and race are important determinants of a person's participation in a normative association. Class (socioeconomic status based on a person's education, occupation, and income) is the most significant predictor of whether a person will participate in mainstream normative organizations; membership costs may exclude some from joining. Those with higher socioeconomic status are more likely to be not only members but also active participants in these groups. Gender is also an important determinant. Half of the voluntary associations in the United States have all-female memberships; one-fifth are all male. However, all-male organizations usually have higher levels of prestige than do all-female ones (Odendahl, 1990).

Throughout history, people of all racial–ethnic categories have participated in voluntary organizations. However, the involvement of women in these groups has largely gone unrecognized. For example, African American women were actively involved in antislavery societies in the nineteenth century and in the civil rights movement in the twentieth century (see Scott, 1990). Other normative organizations focusing on civil rights, self-help, and philanthropic activities in which African American women and men have been involved include the National Association for the Advancement of Colored People (NAACP) and the Urban League. Similarly, Native American women have participated in the American Indian Movement, a group organized to fight problems ranging from police brutality to housing and employment discrimination (Feagin and Feagin, 2003). Mexican American women (as well as men) have held a wide range of leadership positions in La Raza Unida Party and the League of United Latin American Citizens, organizations oriented toward civic activities and protest against injustices (Amott and Matthaei, 1996). The activities of some of these organizations are discussed in more detail in Chapter 9 ("Race and Ethnicity").

One of the central characteristics of normative associations is that membership is voluntary—members function as unpaid workers. However, even when unpaid work has required education and experience similar to that required by wage labor, this work has been devalued (Daniels, 1988). Conflict theorists argue that the devalued nature of unpaid work derives from the fact that women have historically done most of this kind of work.

COERCIVE ORGANIZATIONS People do not voluntarily become members of *coercive organizations*—associations that people are forced to join. Total institutions, such as boot camps, prisons, and some mental hospitals, are examples of coercive organizations. As discussed in Chapter 3, the assumed goal of total institutions is to resocialize people through incarceration. These environments are characterized by restrictive barriers (such as locks, bars, and security guards) that make it impossible for people to leave freely. When people leave without being officially dismissed, their exit is referred to as an "escape."

UTILITARIAN ORGANIZATIONS We voluntarily join *utilitarian organizations* when they can provide us with a material reward that we seek. To make a living or earn a college degree, we must participate in organizations that can provide us these opportunities. Although we have some choice regarding where we work or attend school, utilitarian organizations are not always completely voluntary. For example, most people must continue to work even if the conditions of their employment are less than ideal. (Concept Table 5.A reviews types of groups, sizes of groups, and types of formal organizations.)

Bureaucracies

The bureaucratic model of organization remains the most universal organizational form in govern-ment, business, education, and religion. A *bureaucracy* **is an organizational model characterized by a hierarchy of authority, a clear division of labor, explicit rules and procedures, and impersonality in personnel matters.**

When we think of a bureaucracy, we may think of "buck-passing," such as occurs when we are directed from one office to the next without receiving an answer to our question or a solution to our problem. We also may view a bureaucracy in terms of "red tape" because of situations in which there is so much paperwork and so many incomprehensible rules that no one really understands what to do. However, bureaucracy was not originally intended to be this way; it was seen as a way to make organizations *more* productive and efficient.

Sociologist Max Weber (1968/1922) was interested in the historical trend toward bureaucratization that accelerated during the Industrial Revolution. To Weber, the bureaucracy was the most "rational" and efficient means of attaining organizational goals because it contributed to coordination and control. According to Weber, *rationality* **is the process by which traditional methods of social organization, characterized by informality and spontaneity, are gradually replaced by efficiently administered formal rules and procedures.** Bureaucracy can be seen in all aspects of our lives, from small colleges with perhaps a thousand students to multinational corporations employing many thousands of workers worldwide.

In his study of bureaucracies, Weber relied on an ideal-type analysis, which he adapted from the field of economics. An *ideal type* **is an abstract**

Bureaucracy an organizational model characterized by a hierarchy of authority, a clear division of labor, explicit rules and procedures, and impersonality in personnel matters.

Rationality the process by which traditional methods of social organization, characterized by informality and spontaneity, are gradually replaced by efficiently administered formal rules and procedures.

Ideal type an abstract model that describes the recurring characteristics of some phenomenon (such as bureaucracy).

Concept Table 5.A Characteristics of Groups and Organizations

Types of social groups	primary group	small, less specialized group in which members engage in face-to-face, emotion-based interaction over an extended period of time
	secondary group	larger, more specialized group in which members engage in more impersonal, goal-oriented relationships for a limited period of time
	ingroup	a group to which a person belongs and with which the person feels a sense of identity
	outgroup	a group to which a person does not belong and toward which the person may feel a sense of competitiveness or hostility
	reference group	a group that strongly influences a person's behavior and social attitudes, regardless of whether the person is actually a member
Group size	dyad	a group composed of two members
	triad	a group composed of three members
	formal organization	a highly structured secondary group formed for the purpose of achieving specific goals
Types of formal organizations	normative	organizations we join voluntarily to pursue some common interest or gain personal satisfaction or prestige by joining
	coercive	associations that people are forced to join (total institutions such as boot camps and prisons are examples)
	utilitarian	organizations we join voluntarily when they can provide us with a material reward that we seek

model that describes the recurring characteristics of some phenomenon (such as bureaucracy). To develop this ideal type, Weber abstracted the most characteristic bureaucratic aspects of religious, educational, political, and business organizations. For example, to develop an ideal type for bureaucracy in higher education, you would need to include the relationships among governing bodies (such as boards of regents or trustees), administrators, faculty, staff, and students. You would also have to include the rules and policies that govern the school's activities (such as admissions criteria, grading policies, and graduation requirements). Although no two schools would have exactly the same criteria, the ideal-type constructs would be quite similar. Weber acknowledged that no existing organization would exactly fit his ideal type of bureaucracy (Blau and Meyer, 1987).

IDEAL CHARACTERISTICS OF BUREAUCRACY Weber set forth several ideal-type characteristics of bureaucratic organizations. Although bureaucratic realities often differ from these ideal characteristics, Weber's model highlights the organizational efficiency and productivity that bureaucracies strive for. The model has been criticized for not taking into account informal networks (Roethlisberger and Dickson, 1939; Blau and Meyer, 1987). However, it is not surprising that these patterns are largely ignored in Weber's theory of bureaucracy; he constructed an idealized model that deliberately overlooked imperfections and unintended outcomes (Blau and Meyer, 1987).

Division of Labor Bureaucratic organizations are characterized by specialization, and each member has a specific status with certain assigned tasks to

Frank Pedrick/The Image Works

A ritual of college life is standing in line and waiting one's turn. Students are acutely aware of how academic bureaucracies operate. Although some students are critical of the "red tape" they experience, most realize that bureaucracy is necessary to enable the complex system of the university to operate smoothly.

fulfill. This division of labor requires the employment of specialized experts who are responsible for the effective performance of their duties.

Hierarchy of Authority In the sense that Weber described hierarchy of authority, or chain of command, it includes each lower office being under the control and supervision of a higher one. Sociologist Charles Perrow (1986) has noted that all groups with a division of labor are hierarchically structured. Although the chain of command is not always followed, "in a crunch, the chain is there for those higher up to use it." Authority that is distributed hierarchically takes the form of a pyramid; those few individuals at the top have more power and exercise more control than do the many at the lower levels. Hierarchy inevitably influences social interaction. Those who are lower in the hierarchy report to (and often take orders from) those above them in the organizational pyramid. Persons at the upper levels are responsible not only for their own actions but also for those of the individuals they supervise. Hierarchy has been described as

a graded system of interpersonal relationships, a society of unequals in which scarce rewards become even more scarce further down the hierarchy (Presthus, 1978).

Rules and Regulations Weber asserted that rules and regulations establish authority within an organization. These rules are typically standardized and provided to members in a written format. In theory, written rules and regulations offer clear-cut standards for determining satisfactory performance. They also provide continuity so that each new member does not have to reinvent the necessary rules and regulations.

Qualification-Based Employment Bureaucracies hire staff members and professional employees based on specific qualifications. Favoritism, family connections, and other subjective factors not relevant to organizational efficiency are not acceptable criteria for employment. Individual performance is evaluated against specific standards, and promotions are based on merit as spelled out in personnel policies.

Impersonality A detached approach should prevail toward clients so that personal feelings do not interfere with organizational decisions. Officials must interact with subordinates based on their official status, not on the officials' personal feelings.

CONTEMPORARY APPLICATIONS OF WEBER'S THEORY How well do Weber's theory of rationality and his ideal-type characteristics of bureaucracy withstand the test of time? Over 100 years later, many organizational theorists still modify and extend Weber's perspective. For example, the sociologist George Ritzer has applied Weber's theories to an examination of fast-food restaurants such as McDonald's. According to Ritzer, the process of "McDonaldization" has become a global phenomenon as four elements of rationality can be found in fast-food restaurants and other "speedy" or "jiffy" businesses (such as Sir Speedy Printing and Jiffy Lube). Ritzer (2000a: 433) identifies four dimensions of formal rationality—efficiency, predictability, emphasis on quantity rather than quality, and control through nonhuman technologies—that are found in today's fast-food restaurants:

> Efficiency means the search for the best means to the end; in the fast-food restaurant, the drive-through window is a good example of heightening the efficiency of obtaining a meal. Predictability means a world of no surprises; the Big Mac in Los Angeles is indistinguishable from the one in New York; similarly, the one we consume tomorrow or next year will be just like the one we eat today. Rational systems tend to emphasize quantity, usually large quantities, rather than quality. The Big Mac is a good example of this emphasis on quantity rather than quality. Instead of the human qualities of a chef, fast-food restaurants rely on nonhuman technologies like unskilled cooks following detailed directions and assembly-line methods applied to the cooking and serving of food. Finally, such a formally rational system brings with it various irrationalities, most notably the demystification and dehumanization of the dining experience.

INFORMAL STRUCTURE IN BUREAUCRACIES When we look at an organizational chart, the official, formal structure of a bureaucracy is readily apparent. In practice, however, a bureaucracy has patterns of activities and interactions that cannot be accounted for by its organizational chart. These have been referred to as *bureaucracy's other face* (Page, 1946).

How do people use the informal "grapevine" to spread information? Is this faster than the organization's official communication channels? Is it more or less accurate than official channels?

An organization's *informal structure* is composed of those aspects of participants' day-to-day activities and interactions that ignore, bypass, or do not correspond with the official rules and procedures of the bureaucracy. An example is an informal "grapevine" that spreads information (with varying degrees of accuracy) much faster than do official channels of communication, which tend to be slow and unresponsive. The informal structure has also been referred to as *work culture* because it includes the ideology and practices of workers on the job. It is the "informal, customary values and rules [that] mediate the formal authority structure of the workplace and distance workers from its impact" (Benson, 1983: 185). Workers create this work culture in order to confront, resist, or adapt to the constraints of their jobs, as well as to guide and interpret social relations on the job (Zavella, 1987). Today, computer networks and e-mail offer additional opportunities for workers to enhance or degrade their work culture. Some organizations have sought to control offensive communications so that workers will not be exposed to a hostile work environment brought about by colleagues, but such control has raised significant privacy issues (see Box 5.2).

Box 5.2 Sociology and Social Policy
Computer Privacy in the Workplace

I think I'm getting paranoid. When I'm at work, I think somebody is watching me. When I send an e-mail or search the Web, I wonder who knows about it besides me. Don't get me wrong. I get all my work done first, but when I have some spare time, I may play a computer game or check out some web site I'm interested in. I know that when I'm at work, my time belongs to the company, but somehow I still feel like it's an invasion of my privacy for some computer to monitor every single thing I do. I mean, I work for a company that makes cardboard boxes, not the CIA!

—a student in one of the author's classes, expressing her irritation over computer surveillance at work

Do employers really have the right to monitor everything that their employees do on company-owned computers? Generally speaking, the answer is yes, and the practice is widespread. A recent survey found that about one-half of all U.S. companies monitor their employees' e-mail and more than 60 percent of employers monitor Internet connections (American Management Association, 2001).

Employers assert not only that they have the right to engage in such surveillance but also that it may be necessary for them to do so for their own protection. As for their right to do so, they note that they own the computer, pay for the Internet service, and are paying the employee to spend his or her time on company business. As for it possibly being necessary for them to monitor their employees' computer use, employers argue that they may be held legally responsible for harassing or discriminatory e-mail sent on company computers and that surveillance is the only way to protect against such liability. As a result, many employers take the position, according to the Privacy Foundation's Stephen Keating (qtd. in Agonafir, 2002), that "You leave your First Amendment [privacy] rights at the door when you work for a private employer. That's the way it has always been."

In most instances, courts have upheld monitoring even when the employees were not aware of the surveillance, and according to the American Management Association (2001), about one-third of all U.S. employers who engage in computer surveillance do not advise their employees that they are doing so. (At the time of this writing, Connecticut is the only state that requires that employees be notified of monitoring before it can legally occur.) In several high-profile cases, the firing of an employee based on his or her "inappropriate" e-mail has been upheld even when the employer's stated policy was to *not* monitor employee e-mail.

Yet there are valid arguments against computer surveillance as well, and invasion of a worker's privacy is certainly one of them. When an employee makes a personal phone call while at work, and it is a local or toll-free call, the employee usually has a reasonable expectation of privacy—a reasonable belief that neither fellow workers nor his or her employer is eavesdropping on that call. How about "snail mail"? An employee has a reasonable expectation of privacy that the employer will not steam open a personal letter addressed to the employee, read it, and reseal the envelope. Why should an e-mail exchange with friends or relatives not be equally private and protected? If the employer is going to read an employee's e-mail or track the person's Internet activities, shouldn't the employer at least have to make sure that its workers are aware of that policy?

Regarding the argument that personal e-mail wastes the employer's time, privacy advocates state that most employees who exchange personal e-mails or surf the Internet while at work are either doing so on their own time during breaks or at least at times when they have a "free" moment or two. If an employee's productivity falls below expected levels, shouldn't that be obvious to his or her boss without snooping on the employee's e-mail?

Finally, with regard to the employer possibly being held responsible for its employees' actions, Chief Judge Edith H. Jones (qtd. in Gordon, 2001) of the U.S. Fifth Circuit Court has observed that "It seems highly disproportionate to inflict a monitoring program that may invade thousands of people's privacy for the sake of exposing a handful of miscreants." The need to prevent a crime or to protect a company against potential liability must be balanced against each individual's privacy rights.

Ultimately, this balancing must be done by the legislature or the courts. Since the terrorist attacks of September 11, 2001, numerous analysts have called for greater Internet security, whereas other analysts have argued that U.S. citizens are being asked to relinquish too many of their rights in the interest of security. There are no easy answers to this pressing social policy issue, but it should remain a concern for all who live in a democratic society. Where do you stand on this topic? Why?

Thomas K. Wanstall/The Image Works

Women in firefighting are less likely than men to be included in informal networks and more likely to be harassed on the job. What steps could be taken to reduce women's lack of networks in traditionally male-dominated occupations?

POSITIVE AND NEGATIVE ASPECTS OF IN-FORMAL STRUCTURE Is informal structure good or bad? Should it be controlled or encouraged? Two schools of thought have emerged with regard to these questions. One approach emphasizes control (or eradication) of informal groups; the other suggests that they should be nurtured. Traditional management theories are based on the assumption that people are basically lazy and motivated by greed. Consequently, informal groups must be controlled (or eliminated) in order to ensure greater worker productivity.

By contrast, the other school of thought asserts that people are capable of cooperation. Thus, organizations should foster informal groups that permit people to work more efficiently toward organizational goals. Chester Barnard (1938), an early organizational theorist, focused on the functional aspects of informal groups. He suggested that organizations are cooperative systems in which informal groups "oil the wheels" by providing understanding and motivation for participants. In other words, informal networks serve as a means of communication and cohesion among individuals, as well as protect the integrity of the individual (Barnard, 1938; Perrow, 1986).

The *human relations approach*, which is strongly influenced by Barnard's model, views informal networks as a type of adaptive behavior that workers engage in because they experience a lack of congruence between their own needs and the demands of the organization (Argyris, 1960). Organizations typically demand dependent, childlike behavior from their members and strive to thwart the members' ability to grow and achieve "maturity" (Argyris, 1962). At the same time, members have their own needs to grow and mature. Informal networks help workers fill this void.

Large organizations would be unable to function without strong informal norms and relations among participants (Blau and Meyer, 1987).

More recent studies have confirmed the importance of informal networks in bureaucracies. Whereas some scholars have argued that women and people of color receive fairer treatment in larger bureaucracies than they do in smaller organizations, others have stressed that they may be categorically excluded from networks that are important for survival and advancement in the organization (Kanter, 1993/1977; South et al., 1982; Benokraitis and Feagin, 1995; Feagin, 1991). A woman firefighter in New York City describes how detrimental, and even hazardous, it is for workers to be excluded from such informal networks because of race/ethnicity, gender, or other attributes:

I had sort of a "Pollyanna" view of how long it would take before women were really accepted in these nontraditional, very male-dominated jobs [such as being a firefighter]. One always thinks that once I and the other women prove that we can do the job well, people will just accept us and we'll all fit in. . . . [However,] I went to a firehouse where the men refused to eat with me. They would not talk to me. On one occasion my protective gear had been tampered with. It was always a big question as to whether in fact you have anyone there to back you up when you needed them. (PBS, 1992b)

White women and people of color who are employed in positions traditionally held by white men (such as firefighters, police officers, and factory workers) sometimes experience categorical exclusion from the informal structure. Not only do they

Figure 5.5 Characteristics and Effects of Bureaucracy

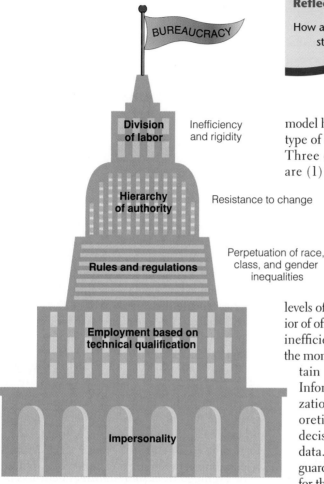

The very characteristics that define Weber's idealized bureaucracy can create or exacerbate the problems that many people associate with this type of organization. Can you apply this model to an organization with which you are familiar?

lack an informal network to "grease the wheels"; they may also be harassed and endangered by their co-workers. In sum, the informal structure is critical for employees—whether or not they are allowed to participate in it.

Shortcomings of Bureaucracies

As noted previously, Weber's description of bureaucracy was intentionally an abstract, idealized model of a rationally organized institution. However, the very characteristics that make up this "rational"

Reflections

Reflections 5.2

How are the rights of individuals affected by the structure of bureaucratic organizations?

model have a dark side that has frequently given this type of organization a bad name (see Figure 5.5). Three of the major problems of bureaucracies are (1) inefficiency and rigidity, (2) resistance to change, and (3) perpetuation of race, class, and gender inequalities (see Blau and Meyer, 1987).

INEFFICIENCY AND RIGIDITY Bureaucracies experience inefficiency and rigidity at both the upper and lower levels of the organization. The self-protective behavior of officials at the top may render the organization inefficient. One type of self-protective behavior is the monopolization of information in order to maintain control over subordinates and outsiders. Information is a valuable commodity in organizations. Budgets and long-range plans are theoretically based on relevant information, and decisions are made based on the best available data. However, those in positions of authority guard information because it is a source of power for them—others cannot "second-guess" their decisions without access to relevant (and often "confidential") information (Blau and Meyer, 1987).

This information blockage is intensified by the hierarchical arrangement of officials and workers. When those at the top tend to use their power and authority to monopolize information, they also fail to communicate with workers at the lower levels. As a result, they are often unaware of potential problems facing the organization and of high levels of worker frustration. Meanwhile, those at

Informal structure those aspects of participants' day-to-day activities and interactions that ignore, bypass, or do not correspond with the official rules and procedures of the bureaucracy.

the bottom of the structure hide their mistakes from supervisors, a practice that may ultimately result in disaster for the organization.

Policies and procedures also contribute to inefficiency and rigidity. Sociologists Peter M. Blau and Marshall W. Meyer (1987) have suggested that bureaucratic regulations are similar to bridges and buildings in that they are designed to withstand far greater stresses than they will ever experience. Accordingly, bureaucratic regulations are written in far greater detail than is necessary in order to ensure that almost all conceivable situations are covered. *Goal displacement* **occurs when the rules become an end in themselves rather than a means to an end, and organizational survival becomes more important than achievement of goals** (Merton, 1968). Administrators tend to overconform to the rules because their expertise is knowledge of the regulations, and they are paid to enforce them. Officials are most likely to emphasize rules and procedures when they fear that they may lose their jobs or a "spoils system" that benefits them. They also fear that if they bend the rules for one person, they may be accused of violating the norm of impersonality and engaging in favoritism (Blau and Meyer, 1987).

Inefficiency and rigidity occur at the lower levels of the organization as well. Workers often engage in *ritualism*; that is, they become most concerned with "going through the motions" and "following the rules." According to Robert Merton (1968), the term *bureaucratic personality* **describes those workers who are more concerned with following correct procedures than they are with getting the job done correctly.** Such workers are usually able to handle routine situations effectively but are frequently incapable of handling a unique problem or an emergency. Thorstein Veblen (1967/1899) used the term *trained incapacity* to characterize situations in which workers have become so highly specialized, or have been given such fragmented jobs to do, that they are unable to come up with creative solutions to problems. Workers who have reached this point also tend to experience bureaucratic alienation—they really do not care what is happening around them.

Sociologists have extensively analyzed the effects of bureaucracy on workers. Whereas some may become alienated, others may lose any identity apart from the organization. Sociologist William H.

Whyte, Jr. (1957) coined the term *organization man* to identify an individual whose life is controlled by the corporation. C. Wright Mills (1959a) suggested that employees become "cheerful robots" when they are controlled by an organization. Other scholars have argued that most workers do not reach these extremes.

RESISTANCE TO CHANGE   Once bureaucratic organizations are created, they tend to resist change. Although the formal structure may help an organization survive during periods of crisis, it may also undermine creativity and profitability. In a study of 152 New York architectural firms, for example, the sociologist Judith R. Blau (1984) found that the same characteristics that made these firms the most likely to survive an economic recession also made them the least likely to become highly profitable. Although the large, well-established firms had the corporate clients and the resources to withstand difficult economic times, their greater overhead made them less inclined than smaller firms to take risks. Some of the risky ventures taken by the smaller firms led to greater profits, yet others led to bankruptcy.

Resistance to change occurs in all bureaucratic organizations, including schools, trade unions, businesses, and government agencies. This resistance not only makes bureaucracies virtually impossible to eliminate but also contributes to bureaucratic enlargement. Because of the assumed relationship between size and importance, officials tend to press for larger budgets and more staff and office space. To justify growth, administrators and managers must come up with more tasks for workers to perform.

Resistance to change may also lead to incompetence. Based on organizational policy, bureaucracies tend to promote people from within the organization. As a consequence, a person who performs satisfactorily in one position is promoted to a higher level in the organization. Eventually, people reach a level that is beyond their own knowledge, experience, and capabilities.

PERPETUATION OF RACE, CLASS, AND GENDER INEQUALITIES   Some bureaucracies perpetuate inequalities of race, class, and gender because this form of organizational structure cre-

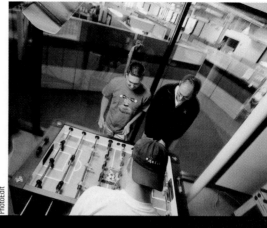

Tim Brown/Stone—Getty Images

PhotoEdit

Corporate employees differ widely in their manner and appearance, as shown here. How does the environment in which we work affect how we dress and act?

ates a specific type of work or learning environment. This structure was typically created for middle- and upper-middle-class white men, who for many years were the predominant organizational participants.

Racial and Ethnic Inequalities In a study of 209 middle-class African Americans from more than a dozen cities, the sociologist Joe R. Feagin (1991) found that *entry* into dominant white bureaucratic organizations should not be equated with thorough *integration*. Instead, many have experienced an internal conflict between the bureaucratic ideals of equal opportunity and fairness and the prevailing norms of discrimination and hostility that exist in many organizations. In another study of 125 African American bankers, interviewees repeatedly indicated that they were excluded from informal communications networks and were not "in on things" in their organization. Very few had mentors or anyone else highly placed within their organizations to take an interest in furthering their careers (Irons and Moore, 1985; Cose, 1993). Other research has found that people of color are more adversely affected than dominant-group members by hierarchical bureaucratic structures. These studies have been conducted in a number of organizational settings, ranging from medical schools to canning factories to corporations (see Kendall and Feagin, 1983; Zavella, 1987; and Collins, 1989).

Social Class Inequalities Like racial inequalities, social class divisions may be perpetuated in bureaucracies (Blau and Meyer, 1987). Sociologists have explored the impact of labor market conditions on the kinds of jobs and wages available to workers. The theory of a "dual labor market" has been developed to explain how social class distinctions are perpetuated through different types of employment. Middle- and upper-middle-class employees are more likely to work in industries characterized by higher wages, more job security, and opportunities for advancement. By contrast, poor and working-class employees work in industries characterized by low wages, lack of job security, and few opportunities for promotion. Even though the "dual economy" is not a perfect model for explaining class-based organizational inequalities, it does illuminate how individuals' employment not only re-

Goal displacement a process that occurs in organizations when the rules become an end in themselves rather than a means to an end, and organizational survival becomes more important than achievement of goals.

Bureaucratic personality a psychological construct that describes those workers who are more concerned with following correct procedures than they are with getting the job done correctly.

Karl Kummels/SuperStock

According to conflict theorists, members of the capitalist class benefit from the work of laborers such as these people harvesting cauliflower on a farm in California's Imperial Valley. How do low wages and lack of job security contribute to social class inequalities in the United States?

flects their position in the social class but also perpetuates it. Peter Blau and Marshall Meyer (1987: 160–161) conclude that "over time, then, organizational conditions reinforce social stratification. . . . Bureaucracies create profound differences in the life chances of the people working in them."

Gender Inequalities Gender inequalities are also perpetuated in bureaucracies. Sociologist Rosabeth Moss Kanter (1993/1977) analyzed how the power structure of bureaucratic hierarchies can negatively affect white women and people of color when they are underrepresented within an organization. In such cases, they tend to be more visible ("on display") and feel greater pressure not to make mistakes or stand out too much. They may also find it harder to gain credibility, particularly in management positions. As a result, they are more likely to feel isolated, to be excluded from informal networks, and to have less access to mentors and to power through alliances. By contrast, affluent white men are generally seen as being "one of the group." They find it easier to gain cred-

ibility, to join informal networks, and to find sponsorships (Kanter, 1993/1977).

Gender inequality in organizations has additional consequences. People who lack opportunities for integration and advancement tend to be pessimistic and to have lower self-esteem. They seek satisfaction away from work and are less likely to promote change at work. Believing that they have few opportunities, they resign themselves to staying put and surviving at that level. By contrast, those who enjoy full access to organizational opportunities tend to have high aspirations and high self-esteem. They feel loyalty to the organization and typically see their job as a means for mobility and growth.

In addition, women working in occupations and professions traditionally dominated by men have a greater likelihood of becoming the victims of sexual harassment. Reports of overt harassment of women have come from virtually all occupational areas, including the armed forces, coal mines, corporate offices, universities, and factories (Lott, 1994). For example, the U.S. Army concluded in one report that "sexual harassment exists throughout the Army, crossing gender, rank and racial lines" and that the military leadership was to blame (Shenon, 1997: A1).

Many women do not report incidents of harassment because they fear that they may lose their job or suffer retaliation from their boss or co-workers. When harassment occurs in the workplace, women may simply quit their jobs. Those who experience harassment in college may drop a class, change majors, or transfer to another institution in order to escape the harasser. Women of color face double jeopardy in that they may experience both racial discrimination and sexual harassment (Benokraitis and Feagin, 1995).

The bottom line is this: Sexual or racial harassment undermines the goal of equality in the workplace and in education. Consequently, organizations must be proactive in establishing guidelines for what is considered acceptable and unacceptable behavior. The elimination of sexual harassment must be viewed as desirable not only for moral, legal, and financial reasons but also for creating and maintaining a positive organizational atmosphere for all participants (see Riggs, Murrell, and Cutting, 1993). Ultimately, gender equality and racial equality in the workplace reflect the "ideal" characteristics of bureaucracy, as well as being guaranteed by law in the United States.

Bureaucracy and Oligarchy

Max Weber believed that bureaucracy was a necessary evil because it achieved coordination and control and thus efficiency in administration (Blau and Meyer, 1987). Sociologist Charles Perrow (1986) has suggested that bureaucracy produces a high standard of living for persons in industrialized countries because of its superiority as a "social tool over other forms of organization." Bureaucratic characteristics (such as a hierarchy of authority, a clear division of labor, explicit rules and procedures, and impersonality in personnel matters) may contribute to organizational efficiency, or they may produce gridlock.

However, Weber was not completely favorable toward bureaucracies. He believed that such organizations stifle human initiative and creativity, thus producing an "iron cage." Bureaucracy also places an enormous amount of unregulated and often unperceived social power in the hands of a very few leaders. Such a situation is referred to as an *oligarchy*—the rule of the many by the few.

Why do a small number of leaders at the top make all the important organizational decisions? According to the German political sociologist Robert Michels (1949/1911), all organizations encounter the ***iron law of oligarchy*—the tendency to become a bureaucracy ruled by the few.** His central idea was that those who control bureaucracies not only wield power but also have an interest in retaining their power. In his research, Michels studied socialist parties and labor unions in Europe before World War I and concluded that even some of the most radical leaders of these organizations had a vested interest in clinging to their power. In this case, if the leaders lost their power positions, they would become manual laborers once again.

According to Michels, the hierarchical structures of bureaucracies and oligarchies go hand in hand. On the one hand, power may be concentrated in the hands of a few people because rank-and-file members must inevitably delegate a certain amount of decision-making authority to their leaders. Leaders then have access to information that other members do not have. They also have "clout," which they may use to protect their

✓ Checkpoints

Checkpoint 5.3

11. How do normative organizations differ from coercive and utilitarian organizations?
12. What is bureaucracy?
13. Why was Max Weber interested in rationality?
14. What are the ideal characteristics of bureaucracy?
15. Why is it important to study informal structure in bureaucracies?
16. What are the major shortcomings of bureaucracies?
17. What is the relationship between bureaucracy and oligarchy?

own interests, sometimes at the expense of the interests of others. On the other hand, oligarchy may result when individuals have certain outstanding qualities that make it possible for them to manage, if not control, others. The members choose to look to their leaders for direction; the leaders are strongly motivated to maintain the power and privileges that go with their leadership positions.

Is the iron law of oligarchy correct? Many scholars believe that Michels overstated his case. The leaders in most organizations do not have unlimited power. Divergent groups within a large-scale organization often compete for power, and informal networks can be used to "go behind the backs" of leaders. In addition, members routinely challenge, and sometimes remove, their leaders when they are not pleased with their actions.

Alternative Forms of Organization

Many organizations have sought new and innovative ways to organize work more efficiently than the traditional hierarchical model. In the early 1980s, there was a movement in the United States to *humanize bureaucracy*—to establish an organizational

Iron law of oligarchy according to Robert Michels, the tendency of bureaucracies to be ruled by a few people.

The Japanese model of organization has viewed the workplace as an extension of the family, and workers have been encouraged to join in shared activities such as the daily exercise program in this office.

Mike Yamashita/Woodfin Camp & Associates

environment that develops rather than impedes human resources. More-humane bureaucracies are characterized by (1) less-rigid hierarchical structures and greater sharing of power and responsibility by all participants, (2) encouragement of participants to share their ideas and try new approaches to problem solving, and (3) efforts to reduce the number of people in dead-end jobs, train people in needed skills and competencies, and help people meet outside family responsibilities while still receiving equal treatment inside the organization (Kanter, 1983, 1985, 1993/1977). However, this movement may have been overshadowed by the perceived strengths of the Japanese system of macroplanning, which has included economic planning, teamwork, and job security (Hodson and Sullivan, 2002).

Organizational Structure in Japan

For several decades, the Japanese model of organization has been widely praised for its innovative structure. A number of social scientists and management specialists concluded that guaranteed lifetime employment and a teamwork approach to management were the major reasons that Japanese workers had been so productive since the end of World War II, when Japan's economy was in shambles. When the U.S. manufacturing sector weakened in the 1980s, the Japanese system was widely discussed as an alternative to the prevalent U.S. hierarchical organizational structure. Let's briefly compare the characteristics of large Japanese corporations with their U.S.-based counterparts.

LIFETIME EMPLOYMENT Until recently, many large Japanese corporations guaranteed their workers permanent employment after an initial probationary period. Thus, Japanese employees often remained with the same company for their entire career, whereas their American counterparts often changed employers every few years. Likewise, Japanese employers in the past had an obligation not to "downsize" by laying off workers or cutting their wages. Unlike top managers in the United States who gave themselves pay raises, bonuses, and so on even when their companies were financially strapped and laying off workers, Japanese managers took pay cuts. When financial problems occurred, Japanese workers were frequently reassigned or retrained by their company.

According to advocates, the Japanese system encourages worker loyalty and a high level of productivity. Managers move through various parts of the organization and acquire technical knowledge about the workings of many aspects of the corporation, unlike their U.S. counterparts, who tend to become highly specialized (Sengoku, 1985). Clearly, permanent employment has its advantages for workers. They do not have to worry about losing their jobs or having their wages cut. Employers benefit by not having to compete with one another to keep workers.

QUALITY CIRCLES Small work groups made up of about five to fifteen workers who meet regularly with one or two managers to discuss the group's performance and working conditions are known as *quality circles*. The purpose of this team approach to management is both to improve product quality and to lower product costs. Workers are motivated to save the corporation money because they, in turn, receive bonuses or higher wages for their efforts. Quality circles have been praised for creating worker satisfaction, helping employees develop their potential, and improving productivity (Ishikawa, 1984). Because quality circles focus on both productivity and worker satisfaction, they (at least ideally) meet the needs of both the corporation and the workers. Although many Japanese corporations have used quality circles to achieve better productivity, many U.S.-based corporations have instead implemented automation in hopes of controlling the quality of their products.

LIMITATIONS Will the Japanese organizational structure continue to work over a period of time? Recently, the notion of lifetime employment has become problematic as an economic recession forced some Japanese factories to close and other companies ran out of subsidiaries willing to take on workers for "reassignment" (Sanger, 1994: E5). Also, the sociologist Robert Coles (1979) has suggested that the Japanese model does not actually provide workers more control over the corporation. This model may give them more control over their own work, but production goals set by managers must still be met.

Although the possibility of implementing the Japanese model in U.S.-based corporations has been widely discussed, its large-scale acceptance is doubtful. Cultural traditions in Japan have focused on the importance of the group rather than the individual. Workers in the United States are not likely to embrace this idea because it directly conflicts

✓ Checkpoints

Checkpoint 5.4

18. How does organizational structure in Japan compare with that in the United States?
19. What are quality circles?
20. What are the major limitations of alternative organizational structures?

Reflections

Reflections 5.3

What steps might be taken to ensure an appropriate balance between the needs of organizations and the protection of the privacy of individuals?

with the value of individualism so strongly held by many in this country (Ouchi, 1981). U.S. workers are also unwilling to commit themselves to one corporation for their entire work life. Moreover, men typically fare much better than women in Japanese corporations in which, due to patriarchy, many women have found themselves excluded from career-track positions (Brinton, 1989).

In spite of these limitations, many organizations in the United States are turning to a more participatory style of management. The incentives for such changes exist because many of the corporations experience greater worker satisfaction and higher productivity and profits (see Florida and Kenney, 1991).

Organizations in the Future

What kind of "help wanted" ad might a company run in the future? One journalist has suggested that a want ad in the new century might read something like this: "WANTED: Bureaucracy basher, willing to challenge convention, assume big risks, and rewrite the accepted rules of industrial order" (Byrne, 1993: 76). Organizational theorists have suggested a *horizontal* model for corporations in which both hierarchy and functional or departmental boundaries would largely be eliminated. Seven key elements of the horizontal corporation have been suggested: (1) work would be organized around "core" processes, not tasks; (2) the hierarchy would be flattened; (3) teams would manage everything and be held accountable for measurable performance goals; (4) performance would be measured by customer satisfaction, not profits; (5) team performance would be rewarded; (6) employees would have regular contact with suppliers and customers; and

(7) all employees would be trained in how to use available information effectively to make their own decisions (Byrne, 1993).

In the horizontal structure, a limited number of senior executives would still fill support roles (such as finance and human resources) while everyone else would work in multidisciplinary teams and perform core processes (such as product development or sales generation). Organizations would have fewer layers between company heads and the staffers responsible for any given process. Performance objectives would be related to the needs of customers; people would be rewarded not just for individual performance but for skills development and team performance. If such organizations become a reality, organizational charts of the twenty-first century will more closely resemble a pepperoni pizza, a shamrock, or an inverted pyramid than the traditional pyramid-shaped stack of boxes connected by lines.

Regardless of the organizational structure, in times of crisis we learn how very important *people* are to the functioning of the collectivity. We also become acutely aware of the significance of effective communication with other people (see Box 5.3).

What is the best organizational structure for the future? Of course, this question is difficult to answer because it requires the ability to predict economic, political, and social conditions. Nevertheless, we can make several observations. Ultimately, everyone has a stake in seeing that organizations operate in as humane a fashion as possible and that channels for opportunity are widely available to all people regardless of race, gender, or class. Workers and students alike can benefit from organizational environments that make it possible for people to explore their joint interests without fear of losing their privacy or being pitted against one another in a competitive struggle for advantage.

Chapter Review

How do sociologists distinguish among social groups, aggregates, and categories?

Sociologists define a social group as a collection of two or more people who interact frequently, share a sense of belonging, and depend on one another. People who happen to be in the same place at the same time are considered an aggregate. Those who share a similar characteristic are considered a category. Neither aggregates nor categories are considered social groups.

How do sociologists classify groups?

Sociologists distinguish between primary groups and secondary groups. Primary groups are small and personal, and members engage in emotion-based interactions over an extended period. Secondary groups are larger and more specialized, and members have less personal and more formal, goal-oriented relationships. Sociologists also divide groups into ingroups, outgroups, and reference groups. Ingroups are groups to which we belong and with which we identify. Outgroups are groups we do not belong to or perhaps feel hostile toward. Reference groups are groups that strongly influence people's behavior whether or not they are actually members.

What is the significance of group size?

In small groups, all members know one another and interact simultaneously. In groups with more than three members, communication dynamics change and members tend to assume specialized tasks.

What are the major styles of leadership?

Leadership may be authoritarian, democratic, or laissez-faire. Authoritarian leaders make major decisions and assign tasks to individual members. Democratic leaders encourage discussion and collaborative decision making. Laissez-faire leaders are minimally involved and encourage members to make their own decisions.

What do experiments on conformity show us about the importance of groups?

Groups may have significant influence on members' values, attitudes, and behaviors. In order to maintain ties with a group, many members are willing to conform to norms established and reinforced by group members.

What are the strengths and weaknesses of bureaucracies?

A bureaucracy is a formal organization characterized by hierarchical authority, division of labor, explicit procedures, and impersonality. According to Max Weber, bureaucracy supplies a rational means of attaining organizational goals because it contributes to coordination and control. A bureaucracy also has an informal structure, which includes the daily activities and interactions that bypass the official rules and procedures. The informal structure may enhance productivity or may be counterproductive to the organization. Bureaucracies may be inefficient, resistant to change, and a vehicle for perpetuating class, race, and gender inequalities.

Box 5.3 You Can Make a Difference

Creating Small Communities of Our Own Within Large Organizations

We had 47 hours to get [ready for] September 13th, when the bond markets reopened and there was one situation that our technology department had that they spent more time on than anything else. . . . It was getting into the systems, [figuring out] the IDs of the systems because so many people had died and the people that knew how to get into those systems and who knew the backup . . . and the second emergency guys were all gone. The way they got into those systems? They sat around [in a] group, they talked about where they went on vacation, what their kids' names were, what their wives' names were, what their dogs' names were, you know, every imaginable thing about their personal life. And the fact that we knew things about their personal life to break into those IDs and into the systems to be able to get the technology up and running before the bond market opened, I think [that] is probably the number one connection between technology, communication, and sociology.

> —an executive describing how, following the September 11, 2001, terrorist attack on the World Trade Center, employees in companies that were devastated by the collapse of the towers and the loss of hundreds of co-workers found themselves needing to quickly re-establish communications networks and to get their operations up and running, not only for the profitability of their own corporations but also for the viability of the nation's economy (qtd. in Kelly and Stark, 2002: 2)

Prior to the attack, sociologist David Stark had been conducting an ethnographic study of the trading room of a major financial firm in the World Financial Center (which was adjacent to the World Trade Center). As a result, he was in a prime position to learn how organizations cope with unimaginable devastation. Like other sociologists who have studied disasters, Stark concluded that it was people, not technology, who saved the situation and made it possible to move forward. Although technology and risk-management plans are important, the microcommunities that people had established through their conversations ultimately saved the day. As the researchers noted, "No one said, 'Our technology saved us,' or, 'Our plan really worked.' To a person, they said, 'It was people'" (Schwartz, 2002: WK5).

Perhaps one lesson that we can learn from the experiences of others and from Stark's research is that each of us, through our everyday communications with other people, may establish significant microcommunities within the larger organizations of which we are a part. These microcommunities of informal friendships and smaller organized groups such as campus clubs not only give us a sense of belonging but may also provide vital support systems in difficult times. Each of us may be able to make a difference at our colleges and in the lives of others by trying to create a sense of community among our friends and colleagues by learning more about them—getting to know not only those people who are similar to us in outlook and world view but also those who may hold very different views on social issues and global concerns.

As we face the future, many analysts believe that lack of meaningful conversation among people from diverse backgrounds will produce a growing sense of "us" versus "them," a feeling of alienation and possibly even antagonism. But how can we improve communications with people we don't really know or understand? One analyst offers these suggestions (Drake, 2002):

- Strive for open and direct communication with the other person, yet make sure that you show respect for that person.
- Focus on the positive aspects of the other person, enhancing that person's self-esteem.
- Focus on the situation that you are discussing—don't let the conversation get personal.
- Share appropriate information with the other person, keeping both of you better informed.
- Look for "win–win" topics, where both of you benefit from the conversation.

Hopefully, none of us will ever face the same situation that the people in the World Financial Center did after the 2001 terrorist attacks—desperately needing to remember key details from our conversations with other people. However, each of us does face the smaller—but still significant—challenge of engaging in effective communication in the groups and organizations to which we belong in order to produce a better school, work, or community environment, and perhaps to enhance the quality of life for ourselves and for other people.

Are all formal organizations based on oligarchy?
An oligarchy is the rule of the many by the few. In bureaucracies with an oligarchical structure, those in control have not only power but also a great interest in maintaining that power. However, recent trends such as the Japanese model of organization emphasize a teamwork approach to management rather than an oligarchical structure.

Key Terms

aggregate 145
authoritarian leaders 150
bureaucracy 155
bureaucratic personality 162
category 145
conformity 150
democratic leaders 150
dyad 148
expressive leadership 149
goal displacement 162
groupthink 153
ideal type 155
informal structure 158
ingroup 145
instrumental leadership 149
iron law of oligarchy 165
laissez-faire leaders 150
outgroup 145
rationality 155
reference group 146
small group 147
triad 148

Questions for Critical Thinking

1. Who might be more likely to conform in a bureaucracy, those with power or those wanting more power?
2. Although there has been much discussion recently concerning what is and what is not sexual harassment, it has been difficult to reach a clear consensus on what behaviors and actions are acceptable. What are some specific ways that both women and men can avoid contributing to an atmosphere of sexual harassment in organizations? Consider team relationships, management and mentor relationships, promotion policies, attitudes, behavior, dress and presentation, and after-work socializing.
3. Do the insights gained from Milgram's research on obedience outweigh the elements of deception and stress that were forced on its subjects?

4. If you were forming a company based on humane organizational principles, would you base the promotional policies on merit and performance or on affirmative action goals?

 **Resources on the Internet**

Chapter-Related Web Sites
The following web sites have been selected for their relevance to the topics in this chapter. These sites are among the more stable, but please note that web site addresses change frequently.

U.S. Census Bureau Profile of Selected Economic Characteristics
http://factfinder.census.gov/bf/_lang=en_vt_name=DEC_2000_SF3_U_DP3_geo_id=01000US.html
Investigate the various occupational categories and how many people are working in each one.

Max Weber: On Bureaucracy
http://www.humanities.mq.edu.au/Ockham/y64l09.html
An online lecture, by R. J. Kilcullen, on Weber's ideas about bureaucracy.

Robert Michels, *Political Parties: A Sociological Study of the Oligarchical Tendencies of Modern Democracy*
http://religionanddemocracy.lib.virginia.edu/library/tocs/MicPoli.html
This site is an online edition of Robert Michels's famous book. You can search by chapter and read for yourself what the author has to say about the iron law of oligarchy.

Groupthink
http://www.abacon.com/commstudies/groups/groupthink.html
The site, by Tim Borchers of Moorhead State University, also contains an interactive activity and a quiz.

Center for Creative Leadership
http://www.ccl.org/index.shtml
This is a business site that offers leadership education and training services. Examine the different kinds of services available.

Social Groups
http://www.usi.edu/libarts/socio/chapter/groups/test.html
You can take a quick tour through the essential topics related to social groups. Each topic contains links to more-detailed information.

Companion Web Site for This Book

Virtual Society: The Wadsworth Sociology Resource Center

Visit **http://sociology.wadsworth.com** and click on the page for Kendall, *Sociology in Our Times: The Essentials*, Fourth Edition, to access a wide range of enrichment material to aid in your study of sociology. Click on the Student Resources section of the web site. Next, select from the pull-down menu the chapter that you are presently studying. Among the useful options for self-study are chapter objectives, flashcards, practical study tips, and practice tests for each chapter.

MicroCase Online

From the Virtual Society home page, click on MicroCase Online to access book-specific MicroCase exercises that allow you to further explore sociological issues and principles presented in the text.

InfoTrac College Edition

Another unique option available to you at the Student Resources section of the companion web site is InfoTrac College Edition, an online library with access to hundreds of scholarly and popular periodicals. Below are suggested search terms for this chapter. Results from these and other searches are found at the site.

- Search keywords: *sexual harassment*. Conduct a limit search to locate articles that address the issue of sexual harassment in formal organizations (for example, universities and corporations). In the articles that you found, did the organization play a role in resolving or contributing to the sexual harassment?
- Search keyword: *bureaucracy*. Conduct a subject guide search using bureaucracy. Select "subdivisions." How many subdivisions are there for bureaucracy? From the subdivision list, select "social aspects." How many articles did you find by doing this research?

Virtual Explorations CD-ROM

Go to the Virtual Explorations CD-ROM to begin the interactive exercise for this chapter. This Virtual Exploration will introduce you to some of the exciting resources for sociology on the World Wide Web. You will be guided through an exercise that employs related web sites on groups and organizations. Answer the questions, and e-mail your responses to your instructor.

Deviance and Crime

Theo Rudnack/SIS

What Is Deviance?
Who Defines Deviance?
What Is Social Control?

Functionalist Perspectives on Deviance
What Causes Deviance, and Why Is It Functional for Society?
Strain Theory: Goals and Means to Achieve Them
Opportunity Theory: Access to Illegitimate Opportunities

Symbolic Interactionist Perspectives on Deviance
Differential Association Theory and Differential
   Reinforcement Theory
Control Theory: Social Bonding
Labeling Theory

Conflict Perspectives on Deviance
Deviance and Power Relations
Deviance and Capitalism
Feminist Approaches
Approaches Focusing on Race, Class, and Gender

Postmodernist Perspectives on Deviance

Crime Classifications and Statistics
How the Law Classifies Crime
How Sociologists Classify Crime
Crime Statistics
Terrorism and Crime
Street Crimes and Criminals
Crime Victims

The Criminal Justice System
The Police
The Courts
Punishment

Deviance and Crime in the United States in the Future

The Global Criminal Economy

172

Felix (a sociologist): When did you join the Diamonds?

Flaco (a gang member): Four and a half years ago. I was a freshman in high school.

Felix: And what were the reasons for joining?

Flaco: Well, let me see. More or less the reason was because I was already hanging with them since I was small, so I was in the neighborhood. And I had a couple of real good buddies. And I had protection by them.

Felix: What exactly did your friends say to you?

Flaco: Not much. It wasn't like they were forcing me. They just told me to turn. It was no big deal.

Felix: And what did you do after this?

Flaco: I turned. I believed them. . . .

Felix: You mentioned earlier that your friends gave you protection. What's protection?

Flaco: Protection? I had backup. You have to have Folks on your side, 'cause, if you're not in one gang, you're not in another gang, and they always be asking if you got a sign, and if you're on this side, and you get rolled on anyway. And you have nobody on your side, so you're still gonna pick a favorite of your Folks or People. I don't know, kids could stay neutron, but it's very hard, and it's very rarely you see kids that are neutrons. And the ones that are neutrons— they still got a favorite; they like Folks better, or People. And the Folks or the People can't let you see them too much on the other side. Like, if you're a neutron, and you're in favor of Folks, if the Folks see you in People's neighborhoods, they're gonna think something about it. And then that's when you got problems.

Felix: So one of the reasons why you joined was for protection, for backup?

Mark Richards/PhotoEdit

Members of the group known as the "Grape Street Gang" typify how gang members use items of clothing such as the bandanas shown here and gang signs made with their hands to assert their identity with the group and solidarity with one another. Some people might view this conduct as deviant behavior, whereas many gang members view it as an act of conformity.

Flaco: Yeah. I liked chilling out with them, too.

Felix: What did you like about hanging out with them?

Flaco: Everything's fun. Sometimes we would ditch school or take a day off. We just go, we get a couple of cases of beer, we drink those, and then we'll go up to the top of the roofs and look over. Police come, and they chase us down. We just had a lot of fun, do a lot of kinds of things.

—sociologist Felix M. Padilla (1993: 80–81) interviewing a gang member as part of Padilla's ethnographic research on why people join gangs

Sociologists and criminologists typically define a *gang* as a group of people, usually young, who band together for purposes generally considered to be deviant or criminal by the larger society. Throughout the past century, gang behavior has been of special interest to sociologists (see Puffer, 1912), who generally agree that youth gangs can be found in many settings and among all racial and ethnic categories. The National Institute of Justice estimates that there are over 8,600 gangs with more than 378,000 members in the United States (Associated Press, 1997).

As unusual as it initially may sound, some important similarities exist between youth gangs and peer cliques, which are typically viewed as conforming to most social norms. At the most basic level, *cliques* are friendship circles, whose members identify one another as mutually connected (Adler and Adler, 1998). However, cliques are much more complex than this definition suggests. According to the sociologists Patricia A. Adler and Peter Adler (1998: 56), cliques "have a hierarchical structure, being dominated by leaders, and are exclusive in nature, so that not all individuals who desire membership are accepted." Moreover, sociologists have found that cliques function as "bodies of power" in schools by "incorporating the most popular individuals, offering the most exciting

Box 6.1 Sociology and Everyday Life

How Much Do You Know About Peer Cliques, Youth Gangs, and Deviance?

True	False	
T	F	1. According to some sociologists, deviance may serve a useful purpose in society.
T	F	2. Peer cliques on high school campuses have few similarities to youth gangs.
T	F	3. Most people join gangs to escape from broken homes caused by divorce or the death of a parent.
T	F	4. Juvenile gangs are an urban problem; few rural areas have problems with gangs.
T	F	5. Street crime has a much higher economic cost to society than crimes committed in executive suites or by government officials.
T	F	6. Persons aged fifteen to twenty-four account for almost half of all arrests for property crimes such as burglary, larceny, arson, and vandalism.
T	F	7. Recent studies have shown that peer cliques have become increasingly important to adolescents over the past two decades.
T	F	8. Gangs are an international problem.

Answers on page 176.

social lives, and commanding the most interest and attention from classmates" (Adler and Adler, 1998: 56).

In this chapter, we look at the relationship among conformity, deviance, and crime; even in times of national crisis and war, "everyday" deviance and crime occur as usual. People do not stop activities that might be viewed by others—or by law enforcement officials—as violating social norms. An example is gang behavior, which is used in this chapter as an example of deviant behavior. For individuals who find a source of identity, self-worth, and a feeling of protection by virtue of gang membership, no radical change occurs in daily life even as events around them may change. Youth gangs have been present in the United States for many years because they meet perceived needs of members. Some gangs may be thought of as being very similar to youth cliques

whereas other gangs engage in activities that constitute crime. Before reading on, take the quiz on peer cliques, youth gangs, and deviance in Box 6.1.

Questions and Issues

Chapter Focus Question: What do studies of peer cliques and youth gangs tell us about deviance?

What is deviant behavior?

When is deviance considered a crime?

What are the major theoretical perspectives on deviance?

How are crimes classified?

How does the criminal justice system deal with crime?

What Is Deviance?

Deviance is any behavior, belief, or condition that violates significant social norms in the society or group in which it occurs. We are most fa-

miliar with *behavioral* deviance, based on a person's intentional or inadvertent actions. For example, a person may engage in intentional deviance by drinking too much or robbing a bank, or in inadvertent deviance by losing money in a Las Vegas casino or laughing at a funeral.

Although most people think of a high-school clique as being far different from a gang, patterns of inclusion and exclusion operate similarly in both groups. The three young women talking to one another here are obviously good friends, but their clique does not include the other young woman standing nearby.

Spencer Grant/PhotoEdit

Although we usually think of deviance as a type of behavior, people may be regarded as deviant if they express a radical or unusual *belief system.* Members of cults (such as Moonies and satanists) or of far-right-wing or far-left-wing political groups may be considered deviant when their religious or political beliefs become known to people with more conventional cultural beliefs. However, individuals who are considered to be "deviant" by one category of people may be seen as conformists by another group. For example, adolescents in some peer cliques and youth gangs may shun mainstream cultural beliefs and values but routinely conform to subcultural codes of dress, attitude (such as defiant individualism), and behavior (Jankowski, 1991). Those who think of themselves as "Goths" may wear black trench coats, paint their fingernails black, and listen to countercultural musicians.

We often interpret other people's belief systems based on how they look. In Littleton, Colorado, many Columbine High School students and teachers had previously identified the two youthful killers and some members of their clique as Goths because of their appearance. However, after widespread media coverage of the "Goth link" to the Littleton tragedy, other people in the "Gothic community" stated that they were appalled by the killings, as well as the inference that the murderers had belonged to their culture (Taylor, 1999).

In addition to their behavior and beliefs, individuals may also be regarded as deviant because they possess a specific *condition or characteristic.* A wide range of conditions have been identified as "deviant," including being obese (Degher and Hughes, 1991; Goode, 1996) and having AIDS (Weitz, 2001). For example, research by the sociologist Rose Weitz (2001) has shown that persons with AIDS live with a stigma that affects their relationships with other people, including family members, friends, lovers, colleagues, and health care workers. Chapter 4 defines a *stigma* as any physical or social attribute or sign that so devalues a person's social identity that it disqualifies the person from full social acceptance (Goffman, 1963b). Based on this definition, the stigmatized person has a "spoiled identity" as a result of being negatively evaluated by others (Goffman, 1963b). To avoid or reduce stigma, many people seek to conceal the characteristic or condition that might lead to stigmatization.

Who Defines Deviance?

Are some behaviors, beliefs, and conditions inherently deviant? In commonsense thinking, deviance is often viewed as inherent in certain kinds of behavior or people. For sociologists, however,

Deviance any behavior, belief, or condition that violates significant social norms in the society or group in which it occurs.

Box 6.1

Answers to the Sociology Quiz on Peer Cliques, Youth Gangs, and Deviance

1. **True.** From Durkheim to contemporary functionalists, theorists have regarded some degree of deviance as functional for societies.
2. **False.** Many social scientists believe that there are striking similarities between adolescent cliques and youth gangs, including the demands that are placed on members in each category to conform to group norms pertaining to behavior, appearance, and people with whom one is allowed to associate.
3. **False.** Recent studies have found that people join gangs for a variety of reasons, including the desire to gain access to money, recreation, and protection.
4. **False.** Gangs are frequently thought of as an urban problem because central-city gangs organized around drug dealing have become prominent in recent years; however, gangs are found in rural areas throughout the country as well.
5. **False.** Although street crime—such as assault and robbery—often has a greater psychological cost, crimes committed by persons in top positions in business (such as accounting and tax fraud) or government (including the Pentagon) have a far greater economic cost, especially for U.S. taxpayers.
6. **True.** This age group accounts for about 46 percent of all arrests for property crimes, the most common crimes committed in the United States.
7. **True.** As more youths grow up in single-parent households or in households where both parents are employed, many adolescents have turned to members of their peer cliques to satisfy their emotional needs and to gain information.
8. **True.** Gangs are found in nations around the world. In countries such as Japan, youth gangs are often points of entry into adult crime organizations.

Sources: Based on Adler and Adler, 1997, 1998; Jankowski, 1991; and Inciardi, Horowitz, and Pottieger, 1993.

deviance is a formal property of social situations and social structure. As the sociologist Kai T. Erikson (1964: 11) explains,

> Deviance is not a property inherent in certain forms of behavior; it is a property conferred upon these forms by the audiences which directly or indirectly witness them. The critical variable in the study of deviance, then, is the social audience rather than the individual actor, since it is the audience which eventually determines whether or not any episode of behavior or any class of episodes is labeled deviant.

Based on this statement, we can conclude that deviance is *relative*—that is, an act becomes deviant when it is socially defined as such. Definitions of deviance vary widely from place to place, from time to time, and from group to group. Today, for

example, some women wear blue jeans and very short hair to college classes; some men wear an earring and long hair. In the past, such looks violated established dress codes in many schools, and administrators probably would have asked these students to change their appearance or leave school.

Deviant behavior also varies in its *degree of seriousness*, ranging from mild transgressions of folkways, to more serious infringements of mores, to quite serious violations of the law. Have you kept a library book past its due date or cut classes? If so, you have violated folkways. Others probably view your infraction as relatively minor; at most, you might have to pay a fine or receive a lower grade. Violations of mores—such as falsifying a college application or cheating on an examination—are viewed as more serious infractions and are punishable by stronger sanctions, such as academic probation or expulsion. Some forms of deviant behavior violate the criminal law, which defines the behaviors that society labels as criminal. A *crime* **is a behavior that violates criminal law and is punishable with fines, jail terms, and/or other negative sanctions.** Crimes range from minor offenses (such as traffic violations) to major offenses (such as murder). A subcategory, *juvenile delinquency,* **refers to a violation of law or the commission of a status offense by young people.** Note that the legal concept of juvenile delinquency includes not only crimes but also status offenses, which are illegal only when committed by younger people (such as cutting school or running away from home).

What Is Social Control?

Societies not only have norms and laws that govern acceptable behavior; they also have various mechanisms to control people's behavior. *Social control* **refers to the systematic practices that social groups develop in order to encourage conformity to norms, rules, and laws and to discourage deviance.** Social control mechanisms may be either internal or external. Internal social control takes place through the socialization process: Individuals *internalize* societal norms and values that prescribe how people should behave and then follow those norms and values in their everyday lives. By contrast, external social control involves the use of negative sanctions that proscribe certain behaviors and set forth the punishments for rule breakers and nonconformists. In contemporary societies, the criminal justice system, which includes the police, the courts, and the prisons, is the primary mechanism of external social control.

If most actions deemed deviant do little or no direct harm to society or its members, why is social control so important to groups and societies? Why are some people who engage in deviance punished

and others not? Why is the same belief or action punished in one group or society and not in another? These questions pose interesting theoretical concerns and research topics for sociologists and criminologists who examine issues pertaining to law, social control, and the criminal justice system. *Criminology* **is the systematic study of crime and the criminal justice system, including the police, courts, and prisons.**

The primary interest of sociologists and criminologists is not on questions of how crime and criminals can best be controlled. Instead, as the questions set forth above suggest, sociologists focus on social control as a social product. Likewise, when sociologists study deviance, they do not judge certain kinds of behavior or people as being "good" or "bad." Instead, they attempt to determine what types of behavior are defined as deviant, who does the defining, how and why people become deviants, and how society deals with deviants (Schur, 1983). Although sociologists have developed a number of theories to explain deviance and crime, no one perspective is a comprehensive explanation of all deviance. Each

Crime behavior that violates criminal law and is punishable with fines, jail terms, and other sanctions.

Juvenile delinquency a violation of law or the commission of a status offense by young people.

Social control systematic practices developed by social groups to encourage conformity to norms, rules, and laws and to discourage deviance.

Criminology the systematic study of crime and the criminal justice system, including the police, courts, and prisons.

theory provides a different lens through which we can examine aspects of deviant behavior.

Functionalist Perspectives on Deviance

As we have seen in previous chapters, functionalists focus on societal stability and the ways in which various parts of society contribute to the whole. According to functionalists, a certain amount of deviance contributes to the smooth functioning of society.

What Causes Deviance, and Why Is It Functional for Society?

Sociologist Emile Durkheim believed that deviance is rooted in societal factors such as rapid social change and lack of social integration among people. As you will recall, Durkheim attributed the social upheaval he saw at the end of the nineteenth century to the shift from mechanical to organic solidarity, which was brought about by rapid industrialization and urbanization. Although many people continued to follow the dominant morals (norms, values, and laws) as best they could, rapid social change contributed to *anomie*—a social condition in which people experience a sense of futility because social norms are weak, absent, or conflicting. According to Durkheim, as social integration (bonding and community involvement) decreased, deviance and crime increased. However, from his perspective, this was not altogether bad because he believed that deviance has positive social functions in terms of its consequences. For Durkheim (1964a/1895), deviance is a natural and inevitable part of all societies. Likewise, contemporary functionalist theorists suggest that deviance is universal because it serves three important functions:

1. *Deviance clarifies rules.* By punishing deviant behavior, society reaffirms its commitment to the rules and clarifies their meaning.
2. *Deviance unites a group.* When deviant behavior is seen as a threat to group solidarity and people unite in opposition to that behavior, their loyalties to society are reinforced.
3. *Deviance promotes social change.* Deviants may violate norms in order to get them changed. For example, acts of *civil disobedience*—includ-

ing lunch counter sit-ins and bus boycotts—were used to protest and eventually correct injustices such as segregated buses and lunch counters in the South. Students periodically stage campus demonstrations to call attention to perceived injustices, such as a tuition increase or the firing of a popular professor.

Functionalists acknowledge that deviance may also be dysfunctional for society. If too many people violate the norms, everyday existence may become unpredictable, chaotic, and even violent. If even a few people commit acts that are so violent that they threaten the survival of a society, then deviant acts move into the realm of the criminal and even the unthinkable. Of course, the examples that stand out in everyone's mind are the terrorist attacks on the United States and the fear that ensued as a result.

Although there are a wide array of contemporary functionalist theories regarding deviance and crime, many of these theories focus on social structure. For this reason, the first theory we will discuss is referred to as a structural functionalist approach. It describes the relationship between the society's economic structure and why people might engage in various forms of deviant behavior.

Strain Theory: Goals and Means to Achieve Them

Modifying Durkheim's (1964a/1895) concept of *anomie*, the sociologist Robert Merton (1938, 1968) developed strain theory. According to **strain theory, people feel strain when they are exposed to cultural goals that they are unable to obtain because they do not have access to culturally approved means of achieving those goals.** The goals may be material possessions and money; the approved means may include an education and jobs. When denied legitimate access to these goals, some people seek access through deviant means.

Merton identified five ways in which people adapt to cultural goals and approved ways of achieving them: conformity, innovation, ritualism, retreatism, and rebellion (see Table 6.1). According to Merton, *conformity* occurs when people accept culturally approved goals and pursue them through approved means. Persons who want to achieve success through conformity work hard,

Table 6.1 Merton's Strain Theory of Deviance

Mode of Adaptation	Method of Adaptation	Seeks Culture's Goals	Follows Culture's Approved Ways
Conformity	Accepts culturally approved goals; pursues them through culturally approved means	Yes	Yes
Innovation	Accepts culturally approved goals; adopts disapproved means of achieving them	Yes	No
Ritualism	Abandons society's goals but continues to conform to approved means	No	Yes
Retreatism	Abandons both approved goals and the approved means to achieve them	No	No
Rebellion	Challenges both the approved goals and the approved means to achieve them	No—seeks to replace	No—seeks to replace

save their money, and so on. Even people who find that they are blocked from achieving a high level of education or a lucrative career may take a lower-paying job and attend school part time, join the military, or seek alternative (but legal) avenues, such as playing the lottery, to "strike it rich."

Conformity is also crucial for members of middle- and upper-class teen cliques, who often gather in small groups to share activities and confidences. Some youths are members of a variety of cliques, and peer approval is of crucial significance to them—being one of the "in" crowd, not a "loner," is a significant goal for many teenagers. In the aftermath of the recent school shootings, for example, journalists trekked to school campuses to report that athletes ("jocks"), cheerleaders, and other "popular" students enforce the social code at high schools (Adler, 1999; Cohen, 1999).

Merton classified the remaining four types of adaptation as deviance:

- *Innovation* occurs when people accept society's goals but adopt disapproved means for achiev-

ing them. Innovations for acquiring material possessions or money cover a wide variety of illegal activities, including theft and drug dealing.
- *Ritualism* occurs when people give up on societal goals but still adhere to the socially approved means for achieving them. Ritualism is the opposite of innovation; persons who cannot obtain expensive material possessions or wealth may nevertheless seek to maintain the respect of others by being a "hard worker" or "good citizen."
- *Retreatism* occurs when people abandon both the approved goals and the approved means of achieving them. Merton included persons such as skid-row alcoholics and drug addicts

Strain theory the proposition that people feel strain when they are exposed to cultural goals that they are unable to obtain because they do not have access to culturally approved means of achieving those goals.

AP/Wide World Photos

Sociologist Robert Merton identified five ways in which people adapt to cultural goals and approved ways of achieving them. Consider The Scary Guy (now his legal name), who is covered head to foot with tattoos. Which of Merton's modes of adaptation might best explain The Scary Guy's views on social life?

in this category; however, not all retreatists are destitute. Some may be middle- or upper-income individuals who see themselves as rejecting the conventional trappings of success or the means necessary to acquire them.

- *Rebellion* occurs when people challenge both the approved goals and the approved means for achieving them and advocate an alternative set of goals or means. To achieve their alternative goals, rebels may use violence (such as vandalism or rioting) or nonviolent tactics (such as civil disobedience).

Which type of adaptation is demonstrated in the journalist Nathan McCall's (1994: 86) description of activities he engaged in when he was a gang member?

Hustling seemed like the thing to do. With Shell Shock [a fellow gang member] as my main partner, I tried every nickel-and-dime hustle I came across, focusing mainly on stealing. We stole everything that wasn't nailed down, from schoolbooks, which we sold at half price, to wallets, which we lifted from guys' rear pockets. We even stole gifts from under the Christmas tree of a girl we visited. . . .

Opportunity Theory: Access to Illegitimate Opportunities

Expanding on Merton's strain theory, sociologists Richard Cloward and Lloyd Ohlin (1960) suggested that for deviance to occur, people must have access to *illegitimate opportunity structures*—circumstances that provide an opportunity for people to acquire through illegitimate activities what they cannot achieve through legitimate channels. For example, gang members may have insufficient legitimate means to achieve conventional goals of status and wealth but have illegitimate opportunity structures—such as theft, drug dealing, or robbery—through which they can achieve these goals. In his study of the "Diamonds," a Chicago street gang whose members are second-generation Puerto Rican youths, sociologist Felix M. Padilla (1993) found that gang membership was linked to the members' belief that they might reach their aspirations by transforming the gang into a business enterprise. Coco, one of the Diamonds, explains the importance of sticking together in the gang's income-generating business organization:

We are a group, a community, a family—we have to learn to live together. If we separate, we will never have a chance. We need each other even to make sure that we have a spot for selling our supply [of drugs]. You know, there is people around here, like some opposition, that want to take over your *negocio* [business]. And they think that they can do this very easy. So we stick together, and that makes other people think twice about trying to take over what is yours. In our case, the opposition has never tried messing with our hood, and that's because they know it's protected real good by us fellas. (qtd. in Padilla, 1993: 104)

Reflections

Reflections 6.2

What do studies about gangs tell us about the need for young people to have access to legitimate opportunities?

Based on their research, Cloward and Ohlin (1960) identified three basic gang types—criminal, conflict, and retreatist—which emerge on the basis of what type of illegitimate opportunity structure is available in a specific area. The *criminal gang* is devoted to theft, extortion, and other illegal means of securing an income. For young men who grow up in a criminal gang, running drug houses and selling drugs on street corners make it possible for them to support themselves and their families as well as purchase material possessions to impress others. By contrast, *conflict gangs* emerge in communities that do not provide either legitimate or illegitimate opportunities. Members of conflict gangs seek to acquire a "rep" (reputation) by fighting over "turf" (territory) and adopting a value system of toughness, courage, and similar qualities. Unlike criminal and conflict gangs, members of *retreatist gangs* are unable to gain success through legitimate means and are unwilling to do so through illegal ones. As a result, the consumption of drugs is stressed and addiction is prevalent.

Sociologist Lewis Yablonsky (1997) has updated Cloward and Ohlin's findings on delinquent gangs. According to Yablonsky, today's gangs are more likely to use and sell drugs, and carry more lethal weapons than gang members did in the past. Today's gangs have become more varied in their activities and are more likely to engage in intraracial conflicts, with "black on black and Chicano on Chicano violence," whereas minority gangs in the past tended to band together to defend their turf from gangs of different racial and ethnic backgrounds (Yablonsky, 1997: 3).

How useful are social structural approaches such as opportunity theory and strain theory in explaining deviant behavior? Although there are weaknesses to these approaches, they focus our attention on one crucial issue: the close association between certain forms of deviance and social class position. According to criminologist Anne Campbell (1984: 267), gangs are a "microcosm of American society, a mirror image in which power, possession, rank, and role . . . are found within a subcultural life of poverty and crime."

Symbolic Interactionist Perspectives on Deviance

As we discussed in Chapter 3, symbolic interactionists focus on *social processes*, such as how people develop a self-concept and learn conforming behavior through socialization. According to this approach, deviance is learned in the same way as conformity—through interaction with others. Although there are a number of symbolic interactionist perspectives on deviance, we will examine three major approaches—differential association and differential reinforcement theories, control theory, and labeling theory.

Differential Association Theory and Differential Reinforcement Theory

How do people learn deviant behavior through their interactions with others? According to the sociologist Edwin Sutherland (1939), people learn the necessary techniques and the motives, drives, rationalizations, and attitudes of deviant behavior from people with whom they associate. *Differential association theory* states that people have a greater tendency to deviate from societal norms when they frequently associate with individuals

✓ Checkpoints

Checkpoint 6.2

5. What do functionalists believe causes deviance?
6. Why is deviance sometimes functional for society?
7. What are the modes of adaptation as identified in strain theory?
8. What is the key idea in opportunity theory?

Illegitimate opportunity structures circumstances that provide an opportunity for people to acquire through illegitimate activities what they cannot achieve through legitimate channels.

Differential association theory the proposition that individuals have a greater tendency to deviate from societal norms when they frequently associate with persons who are more favorable toward deviance than conformity.

who are more favorable toward deviance than conformity. From this approach, criminal behavior is learned within intimate personal groups such as one's family and peer groups. Learning criminal behavior also includes learning the techniques of committing crimes, as former gang member Nathan McCall explains:

> Sometimes I picked up hustling ideas at the 7-Eleven, which was like a criminal union hall: Crapshooters, shoplifters, stickup men, burglars, everybody stopped off at the store from time to time. While hanging up there one day, I ran into Holt. . . . He had a pocketful of cash, even though he had quit school and was unemployed. I asked him, "Yo, man, what you been into?"
>
> "Me and my partner kick in cribs and make a killin'. You oughta come go with us sometimes." . . . I hooked school one day, went with them, and pulled my first B&E [breaking and entering]. Before we went to the house, Hilliard . . . explained his system: "Look, man, we gonna split up and go to each house on the street. Knock on the door. If somebody answers, make up a name and act like you at the wrong crib. If nobody answers, we mark it for a hit." . . .
>
> After I learned the ropes, Shell Shock [another gang member] and I branched out, doing B&Es on our own. We learned to get in and out of houses in no time flat. (McCall, 1994: 93–94)

As McCall's orientation to breaking and entering shows, learning deviance may involve the acquisition of certain attitudes and the mastery of specialized techniques.

Differential association theory contributes to our knowledge of how deviant behavior reflects the individual's learned techniques, values, attitudes, motives, and rationalizations. It calls attention to the fact that criminal activity is more likely to occur when a person has frequent, intense, and long-lasting interactions with others who violate the law. However, it does not explain why many individuals who have been heavily exposed to people who violate the law still engage in conventional behavior most of the time.

Criminologist Ronald Akers (1990) has combined differential association theory with elements of psychological learning theory to create *differential reinforcement theory*, which suggests that both deviant behavior and conventional behavior are learned through the same social processes. Akers

starts with the fact that people learn to evaluate their own behavior through interactions with significant others. If the persons and groups that a particular individual considers most significant in his or her life define deviant behavior as being "right," the individual is more likely to engage in deviant behavior; likewise, if the person's most significant friends and groups define deviant behavior as "wrong," the person is less likely to engage in that behavior. This approach helps explain not only juvenile gang behavior but also how peer cliques on high school campuses have such a powerful influence on people's behavior. For example, when clique members at Glenbrook, a suburban Chicago high school, jealously "guarded" their favorite locations at the school, one student described her response to the powerful pressures to conform as follows:

> As an experiment . . . Lauren Barry, a pink-haired trophy-case kid at Glenbrook, switched identities with a well-dressed girl from "the wall." Barry walked around all day in the girl's expensive jeans and Doc Martens, carrying a shopping bag from Abercrombie & Fitch. "People kept saying, 'Oh, you look so pretty'," she recalls. "I felt really uncomfortable." It was interesting, but the next day, and ever since, she's been back in her regular clothes. (qtd. in Adler, 1999: 58)

Another such approach to studying deviance is control theory, which suggests that conformity is often associated with a person's bonds to other people.

Control Theory: Social Bonding

According to the sociologist Walter Reckless (1967), society produces pushes and pulls that move people toward criminal behavior; however, some people "insulate" themselves from such pressures by having positive self-esteem and good group cohesion. Reckless suggests that many people do not resort to deviance because of *inner containments*—such as self-control, a sense of responsibility, and resistance to diversions—and *outer containments*—such as supportive family and friends, reasonable social expectations, and supervision by others. Those with the strongest containment mechanisms are able to withstand external pressures that might cause them to participate in deviant behavior.

Bill Bachmann/PhotoEdit

Michael Newman/PhotoEdit

According to control theory, strong bonds—including close family ties—are a factor in explaining why many people do not engage in deviant behavior. Across class lines, how are early childhood experiences linked to juvenile and adult behavior?

Extending Reckless's containment theory, sociologist Travis Hirschi's (1969) social control theory is based on the assumption that deviant behavior is minimized when people have strong bonds that bind them to families, schools, peers, churches, and other social institutions. ***Social bond theory* holds that the probability of deviant behavior increases when a person's ties to society are weakened or broken.** According to Hirschi, social bonding consists of (1) *attachment* to other people, (2) *commitment* to conformity, (3) *involvement* in conventional activities, and (4) *belief* in the legitimacy of conventional values and norms. Although Hirschi did not include females in his study, others who have replicated that study with both females and males have found that the theory appears to apply to each (see Naffine, 1987).

What does control theory have to say about delinquency and crime? Control theories suggest that the probability of delinquency increases when a person's social bonds are weak and when peers promote antisocial values and violent behavior (Massey and Krohn, 1986). However, some critics assert that Hirschi was mistaken in his assumption that a weakened social bond leads to deviant be-

havior. The chain of events may be just the opposite: People who routinely engage in deviant behavior may find that their bonds to people who would be positive influences are weakened over time (Agnew, 1985; Siegel 1998). Or, as labeling theory suggests, people may engage in deviant and criminal behavior because of destructive social interactions and encounters (Siegel, 1998).

Labeling Theory

Labeling theory **states that deviance is a socially constructed process in which social control agencies designate certain people as deviants, and they, in turn, come to accept the label placed upon them and begin to act accordingly.** Based on the

Social bond theory the proposition that the probability of deviant behavior increases when a person's ties to society are weakened or broken.

Labeling theory the proposition that deviants are those people who have been successfully labeled as such by others.

symbolic interaction theory of Charles H. Cooley and George H. Mead (see Chapter 3), labeling theory focuses on the variety of symbolic labels that people are given in their interactions with others.

How does the process of labeling occur? The act of fixing a person with a negative identity, such as "criminal" or "mentally ill," is directly related to the power and status of those persons who *do the labeling* and those who are *being labeled*. Behavior, then, is not deviant in and of itself; it is defined as such by a social audience (Erikson, 1962). According to the sociologist Howard Becker (1963), *moral entrepreneurs* are often the ones who create the rules about what constitutes deviant or conventional behavior. Becker believes that moral entrepreneurs use their own perspectives on "right" and "wrong" to establish the rules by which they expect other people to live. They also label others as deviant. Often these rules are enforced on persons with less power than the moral entrepreneurs. Becker (1963: 9) concludes that the deviant is "one to whom the label has successfully been applied; deviant behavior is behavior that people so label."

As the definition of labeling theory suggests, several stages may occur in the labeling process. **Primary deviance refers to the initial act of rule breaking** (Lemert, 1951). However, if individuals accept the negative label that has been applied to them as a result of the primary deviance, they are more likely to continue to participate in the type of behavior that the label was initially meant to control. **Secondary deviance occurs when a person who has been labeled a deviant accepts that new identity and continues the deviant behavior.** For example, a person may shoplift an item of clothing from a department store but not be apprehended or labeled as a deviant. The person may subsequently decide to forgo such behavior in the future. However, if the person shoplifts the item, is apprehended, is labeled as a "thief," and subsequently accepts that label, then the person may shoplift items from stores on numerous occasions. A few people engage in *tertiary deviance,* **which occurs when a person who has been labeled a deviant seeks to normalize the behavior by relabeling it as nondeviant** (Kitsuse, 1980). An example would be drug users who believe that using marijuana or other illegal drugs is no more deviant than drinking alcoholic beverages and therefore should not be stigmatized.

Can labeling theory be applied to high school peer groups and gangs? In a classic study, the sociologist William Chambliss (1973) documented how the labeling process works in some high schools when he studied two groups of adolescent boys: the "Saints" and the "Roughnecks." Members of both groups were constantly involved in acts of truancy, drinking, wild parties, petty theft, and vandalism. Although the Saints committed more offenses than the Roughnecks, the Roughnecks were the ones who were labeled as "troublemakers" and arrested by law enforcement officials. By contrast, the Saints were described as being the "most likely to succeed," and none of the Saints was ever arrested. According to Chambliss (1973), the Roughnecks were more likely to be labeled as deviants because they came from lower-income families, did poorly in school, and were generally viewed negatively whereas the Saints came from "good families," did well in school, and were generally viewed positively. Although both groups engaged in similar behavior, only the Roughnecks were stigmatized by a deviant label.

How successful is labeling theory in explaining deviance and social control? One contribution of labeling theory is that it calls attention to the way in which social control and personal identity are intertwined: Labeling may contribute to the acceptance of deviant roles and self-images. Critics argue that this does not explain what caused the original acts that constituted primary deviance, nor does it provide insight into why some people accept deviant labels and others do not (Cavender, 1995).

Whereas symbolic interactionist perspectives are concerned with how people learn deviant

✓Checkpoints

Checkpoint 6.3

9. What is differential association theory?
10. What is the central idea in differential reinforcement theory?
11. What are the major components of social bonding?
12. What does labeling theory tell us about deviant behavior?
13. How does primary deviance differ from secondary and tertiary deviance?

Conflict theorists suggest that criminal law is unequally enforced along class lines. Consider two courtroom settings: one involving low-income defendants who are arraigned by a judge they see only on a television monitor and the other involving an affluent defendant who is facing the judge with his well-paid lawyer at his side.

behavior, identities, and social roles through interaction with others, conflict theorists emphasize the connections among social class, deviance, and social control.

Conflict Perspectives on Deviance

Who determines what kinds of behavior are deviant or criminal? Different branches of conflict theory offer somewhat divergent answers to this question. One branch emphasizes power as the central factor in defining deviance and crime: People in positions of power maintain their advantage by using the law to protect their interests. Another branch emphasizes the relationship between deviance and capitalism, whereas a third focuses on feminist perspectives and the confluence of race, class, and gender issues in regard to deviance and crime.

Deviance and Power Relations

Conflict theorists who focus on power relations in society suggest that the lifestyles considered deviant by political and economic elites are often defined as illegal. According to this approach, norms and laws are established for the benefit of those in power and do not reflect any absolute standard of right and wrong (Turk, 1969, 1977). As a result, the activities of poor and lower-income individuals are more likely to be defined as criminal than those of persons from middle- and upper-income backgrounds. Moreover, the criminal justice system is more focused on, and is less forgiving of, deviant and criminal behavior engaged in by people in specific categories. For example, research shows that young, single, urban males are more likely to be perceived as members of the *dangerous classes* and receive stricter sentences in criminal courts (Miethe and Moore, 1987). Power differentials are also evident in how victims of crime are treated. When the victims are wealthy, white, and male, law enforcement officials are more likely to put forth more extensive efforts to apprehend the perpetrator as contrasted with cases in which the victims are poor, black, and female (Smith, Visher,

Primary deviance the initial act of rule-breaking.

Secondary deviance the process that occurs when a person who has been labeled a deviant accepts that new identity and continues the deviant behavior.

Tertiary deviance deviance that occurs when a person who has been labeled a deviant seeks to normalize the behavior by relabeling it as nondeviant.

and Davidson, 1984). Recent research generally supports this assertion (Wonders, 1996).

Deviance and Capitalism

A second branch of conflict theory—Marxist/critical theory—views deviance and crime as a function of the capitalist economic system. Although the early economist and social thinker Karl Marx wrote very little about deviance and crime, many of his ideas are found in a critical approach that has emerged from earlier Marxist and radical perspectives on criminology. The critical approach is based on the assumption that the laws and the criminal justice system protect the power and privilege of the capitalist class. As you may recall from Chapter 1, Marx based his critique of capitalism on the inherent conflict that he believed existed between the capitalists (bourgeoisie) and the working class (proletariat). In a capitalist society, social institutions (such as law, politics, and education, which make up the superstructure) legitimize existing class inequalities and maintain the capitalists' superior position in the class structure. According to Marx, capitalism produces haves and have-nots, who engage in different forms of deviance and crime.

According to the sociologist Richard Quinney (1974, 1979, 1980), people with economic and political power define as criminal any behavior that threatens their own interests. The powerful use law to control those who are without power. For example, drug laws enacted early in the twentieth century were actively enforced in an effort to control immigrant workers, especially the Chinese, who were being exploited by the railroads and other industries (Tracy, 1980). By contrast, antitrust legislation passed at about the same time was seldom enforced against large corporations owned by prominent families such as the Rockefellers, Carnegies, and Mellons. Having antitrust laws on the books merely shored up the government's legitimacy by making it appear responsive to public concerns about big business (Barnett, 1979).

In sum, the Marxist/critical approach argues that criminal law protects the interests of the affluent and powerful. The way that laws are written and enforced benefits the capitalist class by ensuring that individuals at the bottom of the social class structure do not infringe on the property or threaten the safety of those at the top (Reiman, 1998). However, others assert that critical theorists have not shown that powerful economic and political elites actually manipulate lawmaking and law enforcement for their own benefit. Rather, people of all classes share a consensus about the criminality of certain acts. For example, laws that prohibit murder, rape, and armed robbery protect not only middle- and upper-income people but also low-income people, who are frequently the victims of such violent crimes (Klockars, 1979).

Feminist Approaches

Can theories developed to explain male behavior be used to understand female deviance and crime? According to feminist scholars, the answer is no. A new interest in women and deviance developed in 1975 when two books—Freda Adler's *Sisters in Crime* and Rita James Simons's *Women and Crime*—declared that women's crime rates were going to increase significantly as a result of the women's liberation movement. Although this so-called *emancipation theory* of female crime has been refuted by subsequent analysts, Adler's and Simons's works encouraged feminist scholars (both women and men) to examine more closely the relationship among gender, deviance, and crime. More recently, feminist scholars such as Kathleen Daly and Meda Chesney-Lind (1988) have developed theories and conducted research to fill the void in our knowledge about gender and crime. For example, in a recent study of the female offender, Chesney-Lind (1997) examined the cultural factors in women's lives that may contribute to their involvement in criminal behavior. Although there is no single feminist perspective on deviance and crime, three schools of thought have emerged.

Why do women engage in deviant behavior and commit crimes? According to the *liberal feminist approach*, women's deviance and crime are a rational response to the gender discrimination that women experience in families and the workplace. From this view, lower-income and minority women typically have fewer opportunities not only for education and good jobs but also for "high-end" criminal endeavors. As some feminist theorists have noted, a woman is no more likely to be a big-time drug dealer or an organized crime boss than she is to be a corporate director (Daly and Chesney-Lind, 1988; Simpson, 1989).

By contrast, the *radical feminist approach* views the cause of women's crime as originating in patriarchy (male domination over females). This approach focuses on social forces that shape women's lives and experiences and shows how exploitation may trigger deviant behavior and criminal activities. From this view, arrests and prosecution for crimes such as prostitution reflect our society's sexual double standard whereby it is acceptable for a man to pay for sex but unacceptable for a woman to accept money for such services. Although state laws usually view both the female prostitute and the male customer as violating the law, in most states the woman is far more likely than the man to be arrested, brought to trial, convicted, and sentenced.

The third school of feminist thought, the *Marxist (socialist) feminist approach*, is based on the assumption that women are exploited by both capitalism and patriarchy. Because most females have relatively low-wage jobs (if any) and few economic resources, crimes such as prostitution and shoplifting become a means to earn money or acquire consumer goods. However, instead of freeing women from their problems, prostitution institutionalizes women's dependence on men and results in a form of female sexual slavery (Vito and Holmes, 1994). Lower-income women are further victimized by the fact that they are often the targets of violent acts by lower-class males, who perceive themselves as being powerless in the capitalist economic system.

Some feminist scholars have noted that these approaches to explaining deviance and crime neglect the centrality of race and ethnicity and focus on the problems and perspectives of women who are white, middle and upper income, and heterosexual without taking into account the views of women of color, lesbians, and women with disabilities (Martin and Jurik, 1996).

Approaches Focusing on Race, Class, and Gender

Some recent studies have focused on the simultaneous effects of race, class, and gender on deviant behavior. In one study, the sociologist Regina Arnold (1990) examined the relationship between women's earlier victimization in their family and their subsequent involvement in the criminal justice system. Arnold interviewed African American

✓ Checkpoints

Checkpoint 6.4

14. What are the central ideas in conflict perspectives on deviance?
15. What is the unique focus of feminist approaches to understanding deviance and crime?
16. Why are race, class, and gender important issues in some studies of deviant behavior?

women serving criminal sentences and found that adolescent females are often "labeled and processed as deviants—and subsequently as criminals—for refusing to accept or participate in their own victimization." Arnold attributes many of the women's offenses to living in families in which sexual abuse, incest, and other violence left them few choices except to engage in deviance. Economic marginality and racism also contributed to their victimization: "To be young, Black, poor, and female is to be in a high-risk category for victimization and stigmatization on many levels" (Arnold, 1990: 156). According to Arnold, the criminal behavior of the women in her study was linked to class, gender, and racial oppression, which they experienced daily in their families and at school and work.

Feminist sociologists and criminologists believe that research on women as both victims and perpetrators of crime is long overdue. For example, few studies of violent crime—such as robbery and aggravated assault—have included women as subjects or respondents. Some scholars have argued that women are less motivated to commit such crimes, are not as readily exposed to attractive targets, and are more protected from being the victims of such crimes than men are (see Sommers and Baskin, 1993). Other scholars have stressed that research should integrate women into the larger picture of criminology (Simpson, 1989). For example, the sociologists Susan Ehrlich Martin and Nancy C. Jurik (1996) examined the role of women in the criminal justice system and found that women continue to experience significant barriers in justice occupations ranging from law enforcement to the legal profession.

Postmodernist Perspectives on Deviance

How might postmodernists view deviance and social control? Although the works of social theorist Michel Foucault defy simple categorization, *Discipline and Punish* (1979) might be considered somewhat postmodernist in its approach to explaining the intertwining nature of power, knowledge, and social control. In his study of prisons from the mid-1800s to the early 1900s, Foucault found that many penal institutions ceased torturing prisoners who disobeyed the rules and began using new surveillance techniques to maintain social control. Although the prisons appeared to be more humane in the post-torture era, Foucault contends that the new means of surveillance impinged more on prisoners and brought greater power to prison officials. To explain, he described the *Panoptican*—a structure that gives prison officials the possibility of complete observation of criminals at all times. For example, the Panoptican might be a tower located in the center of a circular prison from which guards can see all the cells. Although the prisoners know they can be observed at any time, they do not actually know when their behavior is being scrutinized. As a result, prison officials are able to use their knowledge as a form of power over the inmates. Eventually, the guards would not even have to be present all the time because prisoners would believe that they were under constant scrutiny by officials in the observation post. In this case, then, social control and discipline are based on the use of knowledge, power, and technology.

How does Foucault's perspective explain social control in the larger society? According to Foucault, technologies such as the Panoptican make widespread surveillance and disciplinary power possible in many settings, including the state-police network, factories, schools, and hospitals. However, he did not believe that discipline would sweep uniformly through society, due to opposing forces that would use their power to oppose such surveillance.

Foucault's view on deviance and social control has influenced other social analysts, including Shoshana Zuboff (1988), who views the computer as a modern Panoptican that gives workplace supervisors virtually unlimited capabilities for surveillance over subordinates.

We have examined functionalist, interactionist, conflict, and postmodernist perspectives on social control, deviance, and crime (see Concept Table 6.A). All of these explanations contribute to our understanding of the causes and consequences of deviant behavior; however, we now turn to the subject of crime itself.

Crime Classifications and Statistics

Crime in the United States can be divided into different categories. We will look first at the legal classifications of crime and then at categories typically used by sociologists and criminologists.

How the Law Classifies Crime

Crimes are divided into felonies and misdemeanors. The distinction between the two is based on the seriousness of the crime. A *felony* is a serious crime such as rape, homicide, or aggravated assault, for which punishment typically ranges from more than a year's imprisonment to death. A *misdemeanor* is a minor crime that is typically punished by less than one year in jail. In either event, a fine may be part of the sanction as well. Actions that constitute felonies and misdemeanors are determined by the legislatures in the various states; thus, their definitions vary from jurisdiction to jurisdiction.

The *Uniform Crime Report* (UCR) is the major source of information on crimes reported in the United States. The UCR has been compiled since 1930 by the Federal Bureau of Investigation based on information filed by law enforcement agencies throughout the country. When we read that the homicide rate in California is higher than the national average, for example, this information

✓Checkpoints

Checkpoint 6.5

17. What is the postmodernist view on deviance?
18. What is the Panoptican, and how is it related to the issue of deviance?

Concept Table 6.A Theoretical Perspectives on Deviance

	Theory	Key Elements
Functionalist Perspectives		
Robert Merton	Strain theory	Deviance occurs when access to the approved means of reaching culturally approved goals is blocked. Innovation, ritualism, retreatism, or rebellion may result.
Richard Cloward/Lloyd Ohlin	Opportunity theory	Lower-class delinquents subscribe to middle-class values but cannot attain them. As a result, they form gangs to gain social status and may achieve their goals through illegitimate means.
Symbolic Interactionist Perspectives		
Edwin Sutherland	Differential association	Deviant behavior is learned in interaction with others. A person becomes delinquent when exposure to law-breaking attitudes is more extensive than exposure to law-abiding attitudes.
Travis Hirschi	Social control/ social bonding	Social bonds keep people from becoming criminals. When ties to family, friends, and others become weak, an individual is most likely to engage in criminal behavior.
Howard Becker	Labeling theory	Acts are deviant or criminal because they have been labeled as such. Powerful groups often label less-powerful individuals.
Edwin Lemert	Primary/secondary deviance	Primary deviance is the initial act. Secondary deviance occurs when a person accepts the label of "deviant" and continues to engage in the behavior that initially produced the label.
Conflict Perspectives		
Karl Marx Richard Quinney	Critical approach	The powerful use law and the criminal justice system to protect their own class interests.
Kathleen Daly Meda Chesney-Lind	Feminist approach	Historically, women have been ignored in research on crime. Liberal feminism views women's deviance as arising from gender discrimination, radical feminism focuses on patriarchy, and socialist feminism emphasizes the effects of capitalism and patriarchy on women's deviance.
Postmodernist Perspective		
Michel Foucault	Knowledge is power	Power, knowledge, and social control are intertwined. In prisons, for example, new means of surveillance that make prisoners think they are being watched all the time give officials knowledge that inmates do not have. Thus, the officials have a form of power over the inmates.

Figure 6.1 Distribution of Arrests by Type of Offenses, 2000

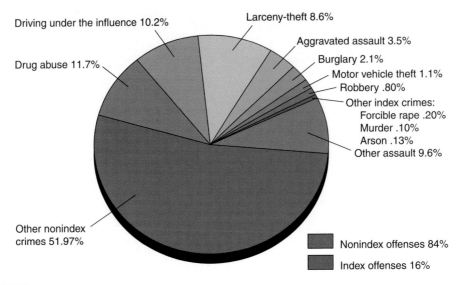

Source: FBI, 2002.

is usually based on UCR data. The UCR focuses on eight major crimes, called *index crimes*: murder, rape, robbery, assault, burglary (breaking into private property to commit a serious crime), motor vehicle theft, arson, and larceny (theft of property worth $50 or more). It also contains data on other types of crime. In 2001, about 9.5 million arrests were made in the United States for all criminal infractions (excluding traffic violations). Of those arrests, 16 percent were for index crimes, as shown in Figure 6.1. Although the UCR gives some indication of crime in the United States, the figures do not reflect the actual number and kinds of crimes, as will be discussed later.

How Sociologists Classify Crime

Sociologists categorize crimes based on how they are committed and how society views the offenses. We will examine four types: (1) conventional (street) crime, (2) occupational (white-collar) and corporate crime, (3) organized crime, and (4) political crime. As you read about these types of crime, ask yourself how you feel about them. Should each be a crime? How stiff should the sanctions be against each type? What personal characteristics (such as your race/ethnicity, class, and gender) influence your opinion?

CONVENTIONAL CRIME Most of the UCR's eight index crimes are conventional crimes. *Conventional (street) crime is all violent crime, certain property crimes, and certain morals crimes.* Obviously, all street crime does not occur on the street; it frequently occurs in the home, workplace, and other locations.

Violent crime consists of actions involving force or the threat of force against others, including murder, rape, robbery, and aggravated assault. Violent crimes are probably the most anxiety-provoking of all criminal behavior—most of us know someone who has been a victim of violent crime, or we have been so ourselves. Victims are often physically injured or even lose their lives; the psychological trauma may last for years after the event (Parker, 1995). Violent crime receives the most sustained attention from law enforcement officials and the media (see Warr, 1995). Media coverage particularly emphasizes the violence associated with gangs and the violent nature of gang members. However, although many people fear the violent stranger, the vast majority of murder victims are actually killed by someone whom they know: family members, friends, neighbors, or co-workers (Parker, 1995).

Nationwide, there is growing concern over juvenile violence. Beginning in 1988, juvenile violent crime arrest rates started to rise, a trend that

Figure 6.2 The FBI Crime Clock

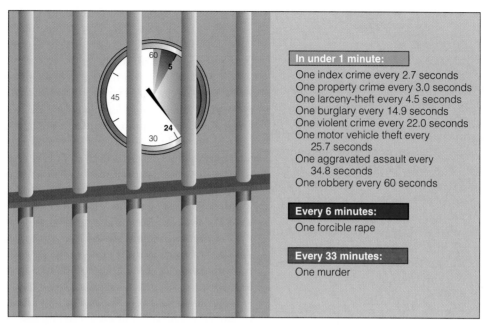

In under 1 minute:
One index crime every 2.7 seconds
One property crime every 3.0 seconds
One larceny-theft every 4.5 seconds
One burglary every 14.9 seconds
One violent crime every 22.0 seconds
One motor vehicle theft every
    25.7 seconds
One aggravated assault every
    34.8 seconds
One robbery every 60 seconds

Every 6 minutes:
One forcible rape

Every 33 minutes:
One murder

Source: FBI, 2002.

has been linked by some scholars to gang membership (see Inciardi, Horowitz, and Pottieger, 1993; Thornberry et al., 1993). Fear of violence is felt not only by the general public but by gang members themselves, as Charles Campbell commented:

> The generation I'm in is going to be lost. Of the circle of friends I grew up in, three are dead, four are in jail, and another is out of school and just does nothing. When he runs out of money he'll sell a couple bags of weed. . . .
>
> I would carry a gun because I am worried about that brother on the fringe. There are some people, there is nothing out there for them. They will blow you away because they have nothing to lose. There are no jobs out there. It's hard to get money to go to school. (qtd. in F. Lee, 1993: 21)

Property crimes include robbery, burglary, larceny, motor vehicle theft, and arson. Some offenses, such as robbery, are both violent crimes and property crimes. In the United States, a property crime occurs, on average, once every three seconds; a violent crime occurs, on average, once every twenty-two seconds (see Figure 6.2). In most property crimes, the primary motive is to obtain money or some other desired valuable (Cook, 1990).

"Morals" crimes involve an illegal action voluntarily engaged in by the participants, such as prostitution, illegal gambling, the private use of illegal drugs, and illegal pornography. Many people assert that such conduct should not be labeled as a crime; these offenses are often referred to as *victimless crimes* because they involve a willing exchange of illegal goods or services among adults (Schur, 1965). However, morals crimes can include children and adolescents as well as adults. Young children and adolescents may unwillingly become child pornography "stars" or prostitutes. Members of juvenile gangs often find selling drugs to be a lucrative business in which getting arrested is merely an occupational hazard. Fernando Morales, age sixteen, is an example of a young man who feels he has nothing to lose by selling drugs: "Sometimes [selling drugs] bothers me. But see, I'm a hustler. I got to look out for myself. I got to be making money. Forget [the customers]. If

Conventional (street) crime all violent crime, certain property crimes, and certain morals crimes.

you put that in your head, you're going to be caught out. You going to be a sucker. You going to be like them" (qtd. in Tierney, 1993: A1, A11).

OCCUPATIONAL AND CORPORATE CRIME

Although the sociologist Edwin Sutherland (1949) developed the theory of white-collar crime more than fifty years ago, it was not until the 1980s that the public became aware of its nature. *Occupational (white-collar) crime comprises illegal activities committed by people in the course of their employment or financial affairs.*

In addition to acting for their own financial benefit, some white-collar offenders become involved in criminal conspiracies designed to improve the market share or profitability of their companies. This is known as *corporate crime—illegal acts committed by corporate employees on behalf of the corporation and with its support.* Examples are antitrust violations; tax evasion; misrepresentations in advertising; infringements on patents, copyrights, and trademarks; price fixing; and financial fraud (Friedrichs, 1996). These crimes are a result of deliberate decisions made by corporate personnel to enhance resources or profits at the expense of competitors, consumers, and the general public.

Although people who commit occupational and corporate crimes can be arrested, fined, and sent to prison, many people often have not regarded such behavior as "criminal." People who tend to condemn street crime are less sure of how their own (or their friends') financial and corporate behavior should be judged. At most, punishment for such offenses has usually been a fine or a relatively brief prison sentence. The case of Michael Milken illuminates the typical sanctions for white-collar crimes. After his 1990s conviction for illegal insider trading, Milken was ordered to

pay $600 million in fines and restitution and was sentenced to ten years in prison, but he was released from prison after less than two years to do community service (Clines, 1993).

Until recently, public concern and media attention focused primarily on the street crimes disproportionately committed by persons who are poor, powerless, and nonwhite. In 2002, however, our attention at least temporarily shifted to crimes committed in corporate suites, such as fraud, tax evasion, and insider trading by executives at some large and well-known corporations. High-ranking executives at communications giant WorldCom admitted to overstating its financial situation by almost $4 billion and were indicted for securities and accounting fraud, among other charges. WorldCom, which owned the nation's second-largest long-distance telephone company and once had been a favorite of the stock market because its per-share price doubled annually for a number of years, filed bankruptcy. Accounting fraud charges forced other corporations, including energy trader Enron, into bankruptcy as well; Enron's outside auditor, big-five accounting firm Arthur Andersen, was convicted of obstruction of justice in connection with its handling of the account. Dennis Kozlowski, chief executive of Tyco International, was indicted for evading $1 million in sales tax on art purchases for his own collection (Berenson, 2002). Biopharmaceutical company ImClone's CEO, Sam Waskal, and some of his relatives and friends—including lifestyle expert Martha Stewart—were accused of insider trading in the sale of shares of that company prior to a Food and Drug Administration announcement driving down the price of the company's stock (Peyser, 2002).

Corporate crimes are often more costly in terms of money and lives lost than street crimes. Thousand of jobs and billions of dollars were lost as a result of corporate crime in the year 2002 alone. Deaths resulting from corporate crimes such as polluting the air and water, manufacturing defective products, and selling unsafe foods and drugs far exceed the number of deaths due to homicides each year. Other costs include the effect on the moral climate of society (Clinard and Yeager, 1980; Simon, 1996). Throughout the United States, the confidence of everyday people in the nation's economy has been shaken badly by the greedy and illegal behavior of corporate insiders.

Reflections

Reflections 6.3

Why do some juvenile gang members view getting arrested as nothing more than an "occupational hazard"?

Reuters/Teddy Blackburn/Archive Photos—Getty Images

Like his father before him, John Gotti, Jr. has been linked to organized crime and sentenced to prison for racketeering activities.

ORGANIZED CRIME *Organized crime* is a business operation that supplies illegal goods and services for profit. Premeditated, continuous illegal activities of organized crime include drug trafficking, prostitution, loan-sharking, money laundering, and large-scale theft such as truck hijackings (Simon, 1996). No single organization controls all organized crime; rather, many groups operate at all levels of society. Organized crime thrives because there is great demand for illegal goods and services. Criminal organizations initially gain control of illegal activities by combining threats and promises. For example, small-time operators running drug or prostitution rings may be threatened with violence if they compete with organized crime or fail to make required payoffs (Cressey, 1969).

Apart from their illegal enterprises, organized crime groups have infiltrated the world of legitimate business. Known linkages between legitimate businesses and organized crime exist in banking, hotels and motels, real estate, garbage collection, vending machines, construction, delivery and long-distance hauling, garment manufacture, in-

surance, stocks and bonds, vacation resorts, and funeral parlors (National Council on Crime and Delinquency, 1969). In addition, some law enforcement and government officials are corrupted through bribery, campaign contributions, and favors intended to buy them off.

POLITICAL CRIME The term *political crime* refers to illegal or unethical acts involving the usurpation of power by government officials, or illegal/unethical acts perpetrated against the government by outsiders seeking to make a political statement, undermine the government, or overthrow it. Government officials may use their authority unethically or illegally for the purpose of material gain or political power (Simon, 1996). They may engage in graft (taking advantage of political position to gain money or property) through bribery, kickbacks, or "insider" deals that financially benefit them. In the late 1980s, for example, several top Pentagon officials were found guilty of receiving bribes for passing classified information on to major defense contractors that had garnered many lucrative contracts from the government (Simon, 1996).

Other types of corruption have been costly for taxpayers, including dubious use of public funds and public property, corruption in the regulation of commercial activities (such as food inspection), graft in zoning and land use decisions, and campaign contributions and other favors to legislators that corrupt the legislative process. Whereas some political crimes are for personal material gain, others (such as illegal wiretapping and political "dirty

Occupational (white-collar) crime illegal activities committed by people in the course of their employment or financial affairs.

Corporate crime illegal acts committed by corporate employees on behalf of the corporation and with its support.

Organized crime a business operation that supplies illegal goods and services for profit.

Political crime illegal or unethical acts involving the usurpation of power by government officials, or illegal/unethical acts perpetrated against the government by outsiders seeking to make a political statement, undermine the government, or overthrow it.

tricks") are aimed at gaining or maintaining political office or influence.

Some acts committed by agents of the government against persons and groups believed to be threats to national security are also classified as political crimes. Four types of political deviance have been attributed to some officials: (1) secrecy and deception designed to manipulate public opinion, (2) abuse of power, (3) prosecution of individuals due to their political activities, and (4) official violence, such as police brutality against people of color or the use of citizens as unwilling guinea pigs in scientific research (Simon, 1996).

Political crimes also include illegal or unethical acts perpetrated against the government by outsiders seeking to make a political statement or to undermine or overthrow the government. Examples are treason, acts of political sabotage, and terrorist attacks on public buildings.

Crime Statistics

How useful are crime statistics as a source of information about crime? As mentioned previously, official crime statistics provide important information on crime; however, the data reflect only those crimes that have been reported to the police. Although the rates have been decreasing slightly during the past few years—except in 2001, when there was a slight rise—the UCR reflects that overall levels of crime have increased by nearly two-thirds over the past twenty-five years. However, this increase may reflect (at least partially) an increase in the number of crimes *reported*, not necessarily a change in the number of crimes *committed*. Why are some crimes not reported? People are more likely to report crime when they believe that something can be done about it (apprehension of the perpetrator or retrieval of their property, for example). About half of all assault and robbery victims do not report the crime because they may be embarrassed or fear reprisal by the perpetrator. Thus, the number of crimes reported to police represents only the proverbial "tip of the iceberg" when compared with all offenses actually committed. Official statistics are problematic in social science research because of these limitations.

The *National Crime Victimization Survey* was developed by the Bureau of Justice Statistics as an alternative means of collecting crime statistics. In this annual survey, the members of 100,000 randomly selected households are interviewed to determine whether they have been the victims of crime, even if the crime was not reported to the police. The most recent victimization survey indicates that 62 percent of all crimes are not reported to the police and are thus not reflected in the UCR (Lightblau, 1999).

Studies based on anonymous self-reports of criminal behavior also reveal much higher rates of crime than those found in official statistics. For example, self-reports tend to indicate that adolescents of all social classes violate criminal laws. However, official statistics show that those who are arrested and placed in juvenile facilities typically have limited financial resources, have repeatedly committed serious offenses, or both (Hindelang, Hirschi, and Weis, 1981; Steffensmeier and Streifel, 1991). Data collected for the Juvenile Court Statistics program also reflect class and racial bias in criminal justice enforcement. Not all children who commit juvenile offenses are apprehended and referred to court. Children from white, affluent families are more likely to have their cases handled outside the juvenile justice system (for example, a youth may be sent to a private school or hospital rather than to a juvenile correctional facility).

Many crimes committed by persons of higher socioeconomic status in the course of business are handled by administrative or quasi-judicial bodies, such as the Securities and Exchange Commission or the Federal Trade Commission, or by civil courts. As a result, many elite crimes are never classified as "crimes," nor are the businesspeople who commit them labeled as "criminals."

Terrorism and Crime

In 2001 the United States and other nations were confronted with a difficult prospect: how to deal with terrorism. As compared with crimes that are committed by the citizens of one country, how are sociologists and criminologists to explain world terrorism, which may have its origins in more than one nation and include diverse "cells" of terrorists who operate in a somewhat gang-like manner but are believed to be following directives from leaders elsewhere? In order to deal with the aftermath of the terrorist attacks on New York City and Wash-

Figure 6.3 Arrest Rates by Sex, 2001 (Selected Offenses)

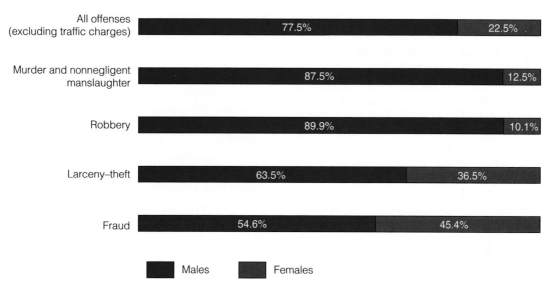

Source: FBI, 2002.

ington, D.C., government officials focused on "known enemies" of the United States, including Osama bin Laden and Saddam Hussein, the president of Iraq. The anthrax attacks that followed the September 11 terrorism were even more difficult to pinpoint; at one point, the FBI indicated that—despite following up on thousands of leads—it was no closer to knowing the source of these acts of deadly bioterrorism. The nebulous nature of the "enemy" and the limitations on any one government to identify and apprehend the perpetrators of such acts of terrorism have resulted in a "war on terrorism." Social scientists who use a rational choice approach suggest that terrorists are rational actors who constantly calculate the gains and losses of participation in violent—and sometimes suicidal—acts against others. Chapter 13 ("Politics and the Economy in Global Perspective") further discusses the issue of terrorism.

Street Crimes and Criminals

Given the limitations of official statistics, is it possible to determine who commits crimes? We have much more information available about conventional (street) crime than elite crime; therefore, statistics concerning street crime do not show who commits all types of crime. Gender, age, class, and race are important factors in official statistics pertaining to street crime.

GENDER AND CRIME Before considering differences in crime rates by males and females, three similarities should be noted. First, the three most common arrest categories for both men and women are driving under the influence of alcohol or drugs (DUI), larceny, and minor or criminal mischief types of offenses. These three categories account for about 40 percent of all male arrests and about 46 percent of all female arrests. Second, liquor law violations (such as underage drinking), simple assault, and disorderly conduct are middle-range offenses for both men and women. Third, the rate of arrests for murder, arson, and embezzlement is relatively low for both men and women (Steffensmeier and Allan, 1995).

The most important gender differences in arrest rates are reflected in the proportionately greater involvement of men in major property crimes (such as robbery and larceny–theft) and violent crime, as shown in Figure 6.3. In 2001, men accounted for almost 90 percent of robberies and murders and almost 64 percent of all larceny–theft arrests in the United States. Of those types of offenses, males under age 18 accounted for approximately 20 percent of the 2001 arrests. The property crimes for which women most frequently are arrested are nonviolent in nature, including shoplifting, theft of services, passing bad checks, credit card fraud, and employee pilferage. When women are arrested for serious violent and property crimes, they are typically seen as accomplices of the men who planned the crime and instigated its commission (Sommers and Baskin, 1993; Steffensmeier and Allan, 1995). However, one study found that some women play an active role in planning and carrying out robberies (Sommers and Baskin, 1993).

AGE AND CRIME Of all factors associated with crime, the age of the offender is one of the most significant. Arrest rates for index crimes are highest for people between the ages of 13 and 25, with the peak being between ages 16 and 17. In 2001, persons under age 25 accounted for more than 54 percent of all arrests for index crimes (FBI, 2002). Individuals under age 18 accounted for over 23 percent of all arrests for robbery and 29 percent of all arrests for larceny–theft.

Scholars do not agree on the reasons for this age distribution. In one study, the sociologist Mark Warr (1993) found that peer influences (defined as exposure to delinquent peers, time spent with peers, and loyalty to peers) tend to be more significant in explaining delinquent behavior than age itself.

The median age of those arrested for aggravated assault and homicide is somewhat older, generally in the late twenties. Typically, white-collar criminals are even older because it takes time to acquire both a high-ranking position and the skills needed to commit this type of nonindex crime.

Rates of arrest remain higher for males than females at every age and for nearly all offenses. This female-to-male ratio remains fairly constant across all age categories. The most significant gender difference in the age curve is for prostitution (a nonindex crime). In 2001, over 59 percent of all women arrested for prostitution were under age 35. For individuals over age 45, many more men than women are arrested for sex-related offenses (including procuring the services of a prostitute). This difference has been attributed to a more stringent enforcement of prostitution statutes when young females are involved (Chesney-Lind, 1986). It has also been suggested that opportunities for prostitution are greater for younger women. This

age difference may not have the same impact on males, who continue to purchase sexual services from young females or males (see Steffensmeier and Allan, 1995).

SOCIAL CLASS AND CRIME Individuals from all social classes commit crimes; they simply commit different kinds of crimes. Persons from lower socioeconomic backgrounds are more likely to be arrested for violent and property crimes. By contrast, persons from the upper part of the class structure generally commit white-collar or elite crimes, although only a very small proportion of these individuals will ever be arrested or convicted of a crime.

What about social class and recent violence by youths? Between 1994 and 1998, there were 173 violent deaths in U.S. schools (*Time*, 1999b). Most of these deaths were not attributed to lower-income, inner-city youths, as popular stereotypes might suggest. Instead, some of these acts of violence were perpetrated by young people who lived in houses that cost anywhere from $75,000 to $5 million or more (Gibbs, 1999).

Similarly, membership in today's youth gangs cannot be identified with just one social class. Increases in gang membership among middle-class suburban youths have been reported (Henneberger, 1993). However, the lower classes are heavily represented in central-city gangs. Today, females are more visible in some previously all-male gangs as well as in female gangs. Criminologist Carl Taylor (1993) found social class differences in the four types of female gangs in Detroit.

In any case, official statistics are not an accurate reflection of the relationship between social class and crime. Self-report data from offenders themselves may be used to gain information on family income, years of education, and occupational status; however, such reports rely on respondents to report information accurately and truthfully.

RACE AND CRIME In 2001, whites (including Latinos/as) accounted for about 70 percent of all arrests for index crimes, as shown in Figure 6.4. Compared with African Americans, arrest rates for whites were higher for nonviolent property crimes such as fraud (a nonindex crime) and larceny–theft but were lower for violent crimes such as robbery and murder. In 2001, whites accounted for about

Figure 6.4 Arrests by Race, 2001 (Selected Offenses)

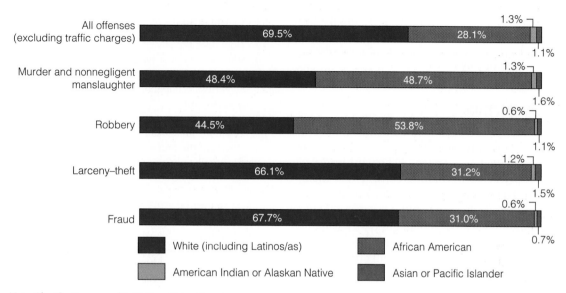

Note: Classifications as used in Uniform Crime Report.

Source: FBI, 2002.

66 percent of all arrests for property crimes and about 60 percent of arrests for violent crimes. African Americans accounted for over 37 percent of arrests for violent crimes and 31 percent of arrests for property crimes (FBI, 2002).

Although official arrest records reveal certain trends, these data tell us very little about the actual dynamics of crime by racial–ethnic category. According to official statistics, African Americans are overrepresented in arrest data (Harris, 1991). In 2001, African Americans made up about 12 percent of the U.S. population but accounted for almost 30 percent of all arrests. Latinos/as made up about 12 percent of the U.S. population and accounted for about 13 percent of all arrests. Over two-thirds of their offenses were for nonindex crimes such as alcohol- and drug-related offenses, and disorderly conduct. In 2001, about 1 percent of all arrests were of Asian Americans or Pacific Islanders, and about 1 percent were of Native Americans (designated in the UCR as "American Indian" or "Alaskan Native"). For the general population, the majority of arrests were for larceny–theft, assaults, vandalism, and alcohol- and drug-related violations (FBI, 2002).

Criminologist Coramae Richey Mann (1993) has argued that arrest statistics are not an accurate reflection of the crimes actually committed in our society. Reporting practices differ in accordance with race and social class. Arrest statistics reflect the UCR's focus on index crimes, and especially property crimes, which are committed primarily by low-income people. This emphasis on index crimes draws attention away from the white-collar and elite crimes committed by middle- and upper-income people (Harris, 1991). Police may also demonstrate bias and racism in their decisions regarding whom to question, detain, or arrest under certain circumstances (Mann, 1993).

Another reason that statistics may show a disproportionate number of people of color being arrested is because of the focus of law enforcement on certain types of crime and certain neighborhoods in which crime is considered more prevalent. As discussed previously, many poor, young, central-city males turn to forms of criminal activity due to their belief that no opportunities exist for them to earn a living wage through legitimate employment. Because of the trend of law enforcement efforts to focus on drug-related offenses, arrest rates for young people of color have risen rapidly. These young people are also more likely to live in central-city areas, where there are more police patrols to make arrests.

Finally, we should remember that being arrested does not mean that a person is guilty of the crime with which he or she has been charged. In the United States, individuals accused of crimes are, at least theoretically, "innocent until proven guilty" (Mann, 1993).

Crime Victims

Based on the National Crime Victimization Survey (NCVS), men are more likely to be victimized by crime, although women tend to be more fearful of crime, particularly crimes directed toward them, such as forcible rape (Warr, 1985). Victimization surveys indicate that men are the most frequent victims of most crimes of violence and theft. Among males who are now 12 years old, an estimated 89 percent will be the victims of a violent crime at least once during their lifetime, as compared with 73 percent of females. The elderly also tend to be more fearful of crime but are the least likely to be victimized. Young men of color between the ages of 12 and 24 have the highest criminal victimization rates.

A study by the Justice Department found that Native Americans are more likely to be victims of violent crimes than are members of any other racial category and that the rate of violent crimes against Native American women was nearly 50 percent higher than that for African American men (Butterfield, 1999). During the period covered in the study (from 1992 to 1996), Native Americans were the victims of violent crimes at a rate more than twice the national average. They were also more likely to be the victims of violent crimes committed by members of a race other than their own (Butterfield, 1999). There has been a shift over the past twenty years in which more Native Americans have moved from reservations to urban areas. In the cities they do not tend to live in segregated areas, so they come into contact more often with people of other racial and ethnic groups, whereas African Americans and whites are more likely to live in segregated areas of cities and commit violent crimes against people in their same racial or ethnic category. According to the survey, the average annual rate at which Native Americans were victims of crime—124 crimes per 1,000 people, ages 12 or older—is about two-and-a-half times the national average of 50 crimes per 1,000 people who are above the age of 12. By comparison, the average annual rate for whites was 49 crimes per 1,000 people,

for African Americans 61 per 1,000, and for Asian Americans 29 per 1,000 (Butterfield, 1999).

The burden of robbery victimization falls more heavily on some categories of people than others. NCVS data indicate that males are robbed at more than twice the rate of females. African Americans are more than three times as likely to be robbed as whites. Young people have a much greater likelihood of being robbed than do middle-aged and older persons. Persons from lower-income families are more likely to be robbed than people from higher-income families (Karmen, 1995).

The Criminal Justice System

The criminal justice system includes the police, the courts, and prisons. However, the term *criminal justice system* is misleading because it implies that law enforcement agencies and courts constitute one large, integrated system. Actually, the criminal justice system is a collection of "somewhat interrelated, semi-autonomous bureaucracies," each of which possesses considerable discretion to do as it wishes (Sheley, 1991: 334). *Discretion* refers to the use of personal judgment by police officers, prosecutors, judges, and other criminal justice system officials regarding whether and how to proceed in a given situation.

The Police

The police are responsible for crime control and maintenance of order. Not all "police officers" are employed by local police departments; they are employed in over 25,000 governmental agencies

Figure 6.5 Discretionary Powers in Law Enforcement

Police
Enforce specific laws
Investigate specific crimes
Search people, vicinities, buildings
Arrest or detain people

Prosecutors
File charges or petitions for judicial decision
Seek indictments
Drop cases
Reduce charges
Recommend sentences

Judges or Magistrates
Set bail or conditions for release
Accept pleas
Determine delinquency
Dismiss charges
Impose sentences
Revoke probation

ranging from local jurisdictions to federal levels. A central issue in policing is the use of discretion in the enforcement of laws (see Figure 6.5). Sociologist Jerome Skolnick (1975) has argued that police officers develop a sense of suspicion, social isolation, and solidarity because they must make these decisions in a dangerous environment. A New York City police officer describes this feeling:

> Guys . . . will eat you alive, even with the uniform on. They can sense fear, smell it like a dog smells it. Some of the mopes will come right out and tell you you're nothing, and you don't want that, oh no. If you're going to do this job, wear this uniform, you definitely don't want that. If it gets around that you're soft, that without your nightstick and gun you can't fight, that's bad. If you allow someone to smoke a joint in front of you or curse you out, word will spread throughout the neighborhood like a disease. You're a beat cop, out here every day, alone, so you set standards right away. . . . (Norman, 1993: 64)

Most officers feel that they must demand respect on the streets, but they also know that they must answer to their superiors, who expect them to handle situations "by the book."

The problem of discretion is most acute in the decision to use "deadly force" against an alleged offender. Generally, deadly force is allowed only in situations in which a suspect is engaged in a felony, is fleeing the scene of a felony, or is resisting arrest and has endangered someone's life

(Barlow, 1987). Rates of shootings by police vary widely from one jurisdiction to another; however, the percentage of shooting incidents involving African American suspects is disproportionately high. African Americans are from five to thirteen times more likely than whites to be killed by police officers (Takagi, 1979; Mann, 1993).

Differential treatment in terms of race contributes to a negative image of police among people of color (Anderson, 1990; Cose, 1993; Mann, 1993). Historically, criminal justice personnel at all levels have been white and male. Although the composition may be slowly changing, more than 83 percent of all police officers are white. No city has a proportion of African American or Latino/a police officers higher than the percentage of African Americans or Latinos/as who live in that city (Mann, 1993). Of the few Asian American police officers, most are employed in cities with large Asian American populations, such as Honolulu, San Francisco, Los Angeles, Chicago, and New York City (Mann, 1993). This lack of diversity has been attributed in part to officer recruitment practices that discriminate against people of color when police recruiters and personnel officers are white (non-Latino/a) (Carter, 1986). In states such as California, Japanese Americans and Chinese Americans have often been categorically excluded from police work by height and weight requirements based on white (non-Latino) male standards.

Women account for less than 10 percent of all police officers. In the past, women were excluded from these careers largely because of stereotypical beliefs that they were not physically and psychologically strong enough to enforce the law. Recent studies have indicated that as more females have entered police work, they still may receive lower evaluations from male administrators on their effectiveness even when objective measures show them to be equally effective on patrol (Reid, 1987).

The Courts

Criminal courts determine the guilt or innocence of those accused of committing a crime. In theory, justice is determined in an adversarial process in which the prosecutor (an attorney who represents the state) argues that the accused is guilty and the defense attorney asserts that the accused is innocent. In reality, judges wield a great deal of discretion. Working with prosecutors, they decide whom to release and whom to hold for further hearings, and what sentences to impose on those persons who are convicted.

Prosecuting attorneys also have considerable leeway in deciding which cases to prosecute and when to negotiate a plea bargain with a defense attorney. As cases are sorted through the legal machinery, a steady attrition occurs. At each stage, various officials determine what alternatives will be available for those cases still remaining in the system (Hills, 1971). These discretionary decisions often have a disproportionate impact on youthful offenders who are poor (see Box 6.2).

About 90 percent of criminal cases are never tried in court; instead, they are resolved by plea bargaining, a process in which the prosecution negotiates a reduced sentence for the accused in exchange for a guilty plea. Defendants (especially those who are poor and cannot afford to pay an attorney) are urged to plead guilty to a lesser crime in return for not being tried for the more serious crime for which they were arrested. Prison sentences given in plea bargains vary widely from one region to another and even from judge to judge within one state (see Barlow, 1987). For example, although women typically commit less serious crimes than men, they do not fare as well as men in negotiating or bargaining for sentence reductions (Chesney-Lind, 1986).

Punishment

Punishment **is any action designed to deprive a person of things of value (including liberty) because of some offense the person is thought to have committed** (Barlow, 1987). Punishment is seen as serving four functions:

1. *Retribution* imposes a penalty on the offender. Retribution is based on the premise that the punishment should fit the crime: The greater the degree of social harm, the more the offender should be punished. For example, an individual who murders should be punished more severely than one who shoplifts.
2. *Social protection* results from restricting offenders so that they cannot commit further crimes.
3. *Rehabilitation* seeks to return offenders to the community as law-abiding citizens; however, many prisons are seriously understaffed and underfunded in the few rehabilitation programs that exist. The job skills (such as agricultural work) that may be learned in prison typically do not transfer to the outside world, nor are offenders given any assistance in finding work that fits their skills once they are released.
4. *Deterrence* seeks to reduce criminal activity by instilling a fear of punishment. Recently, criminologists have debated whether imprisonment has a deterrent effect, given the fact that 30 to 50 percent of those released from prison become recidivists (previous offenders who commit new crimes).

Today, because of plea bargains, credits for "good time" served, and overcrowded conditions in many prisons, most convicted criminals are out on either probation (close supervision of their everyday lives in lieu of serving a prison term) or parole (early release from prison). If offenders violate the conditions of their probation or parole, they may be required to serve their full sentence in prison.

Once again, the disparate treatment of the poor and people of color is evident in the prison system. Whereas incarceration rates have risen steadily in recent years, the rate for minorities (with the exception of Asian Americans, who are the least likely to be incarcerated) has been disproportionately higher than that for whites. According to a recent report by the Justice Department, the

Box 6.2 Sociology and Social Policy
Juvenile Offenders and "Equal Justice Under the Law"

When you walk into the U.S. Supreme Court building in Washington, D.C., it is impossible to miss the engraved statement overhead: "Equal Justice Under the Law." Do young people, regardless of race, class, or gender, receive the same treatment under the law?

In courtrooms throughout the nation, judges have a wide range of discretion in their decisions regarding juveniles alleged to have committed some criminal or status offense. Whereas judges in television courtroom dramas are often African Americans, women, or members of other subordinate groups, "real-life" judges typically come from capitalist or managerial and professional backgrounds. Because more than 90 percent are white and most are male, their decisions may reflect a built-in class, racial, and gender bias.

Juvenile courts were established under a different premise than courts for adults. Under the doctrine of *parens patriae* (the state as parent), the stated purpose of juvenile courts has been to care for, rather than punish, youthful offenders. In theory, less weight is given to offenses and more weight to the youth's physical, mental, or social condition. The juvenile court seeks to change or resocialize offenders through treatment or therapy, not to punish them. Consequently, judges in juvenile courts are given relatively wide latitude, or discretion, in the decisions they mete out regarding young offenders.

Unlike adult offenders, juveniles are not always represented by legal counsel. A juvenile hearing is not a trial but rather an informal private hearing before a judge or probation officer with only the young person and a parent or guardian present. No jury is convened, and the juvenile offender does not cross-examine her or his accusers. In addition, the offender is not "sentenced"; rather, the case is "adjudicated" or "disposed of." Finally, the offender is not "punished" but instead may be placed in the custody of a youth authority in order to receive training, treatment, or care.

Because of judicial discretion, courts may treat juveniles differently based on gender. Considerable disparity exists in the disposition of juvenile cases, with much of the variation thought to result from judges' beliefs rather than objective facts in the case. Female offenders are more likely than males to be institutionalized for committing status offenses such as truancy, running away from home, and other offenses that serve as "buffer charges" for suspected sexual misconduct (Chesney-Lind, 1989).

Disparity also exists on the basis of race and class. Judges tend to see youths from white, middle- or upper-class families as being very much like their own children and to believe that the families will take care of the problem on their own. They may view juveniles from lower-income families or other racial–ethnic groups as delinquents in need of attention from authorities. Furthermore, some judges view gang members from impoverished central cities as "guilty by association" because of their companions.

The political climate may have an effect on how judges dispose of juvenile cases. In the process of dealing with the public perception that the juvenile justice system is too lenient, some judges may have inadvertently contributed to other problems. Many more youths have been remanded to overcrowded juvenile detention facilities that are unable to provide necessary educational, health, and social services. Based on a judge's discretion, many juvenile offenders are incarcerated under indeterminate sentences and placed in a detention facility that may serve merely as a school for adult criminality.

Sources: Based on Barlow, 1987; Chesney-Lind, 1989; and Inciardi, Horowitz, and Pottieger, 1993.

incarceration rate for African American men in 1996 was 3,096 per 100,000, which was eight times the rate for white men (370 per 100,000) and more than double the rate for Latinos (1,276 per 100,000) (Butterfield, 1998).

Disparate treatment of women is also evident. Women's prisons offer far fewer educational and training programs than do men's prisons (Reid, 1987). Medical facilities in many prisons lack essential elements of women's health care, such as

Punishment   any action designed to deprive a person of things of value (including liberty) because of some offense the person is thought to have committed.

Porter Gifford/Liaison—Getty Images

In recent years, military-style boot camps such as this one have been used as an alternative to prison and long jail terms for nonviolent offenders under age 30. Critics argue that structural solutions—not stop-gap measures—are needed to reduce crime.

The ex-slave states are more likely to execute criminals than are other states (see Figure 6.6). African Americans are eight to ten times more likely to be sentenced to death for homicidal rape than are whites (non-Latinos/as) who have committed the same crime (Marquart, Ekland-Olson, and Sorensen, 1994).

People who have lost relatives and friends as a result of criminal activity often see the death penalty as justified. However, capital punishment raises many doubts for those who fear that innocent individuals may be executed for crimes they did not commit. For still others, the problem of racial discrimination in the sentencing process poses troubling questions.

Deviance and Crime in the United States in the Future

Two pressing questions pertaining to deviance and crime will face us in the future: Is the solution to our "crime problem" more law and order? Is equal justice under the law possible?

Although many people in the United States agree that crime is one of the most important problems in this country, they are divided over what to do about it. Some of the frustration about crime might be based on unfounded fears; studies show that the overall crime rate has been decreasing slightly in recent years. However, it is difficult to be complacent with a murder rate twice what it was thirty years ago.

One thing is clear: The existing criminal justice system cannot solve the "crime problem." If roughly 20 percent of all crimes result in arrest, only half of those lead to a conviction in serious cases, and less than 5 percent of those result in a jail term, the "lock 'em up and throw the key away"

mammography to detect breast cancer. Psychotropic drugs are overused on women (who are likely to be perceived as being sick or mentally disturbed) to keep them "calm" (Chesney-Lind, 1986).

Historically, removal from the group has been considered one of the ultimate forms of punishment. For many years, capital punishment, or the death penalty, has been used in the United States as an appropriate and justifiable response to very serious crimes. About four thousand persons have been executed in the United States since 1930, when the federal government began collecting data on executions.

In 1972 the U.S. Supreme Court ruled (in *Furman v. Georgia*) that *arbitrary* application of the death penalty violates the Eighth Amendment to the Constitution but that the death penalty itself is not unconstitutional. In other words, determining who receives the death penalty and who receives a prison term for similar offenses should not be done on a "lotterylike" basis (Bowers, 1984). To be constitutional, the death penalty must be imposed for reasons other than the race/ethnicity, gender, and social class of the offender.

Figure 6.6 Death Row Census, April 2001

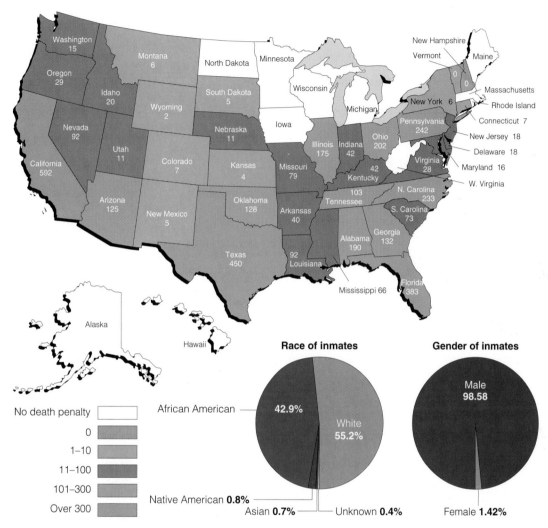

Death row inmates are heavily concentrated in certain states. African Americans, who make up about 12 percent of the U.S. population, account for approximately 43 percent of inmates on death row.

Sources: U.S. Department of Justice, 2000a; Death Penalty Institute of Oklahoma, 2001.

approach has little chance of succeeding. Nor does the high rate of recidivism among those who have been incarcerated speak well for the rehabilitative efforts of our existing correctional facilities. Reducing street crime may hinge on finding ways to short-circuit criminal behavior.

One of the greatest challenges is juvenile offenders, who may become the adult criminals of tomorrow. However, instead of military-style boot camps or other stopgap measures, *structural solutions*—such as more and better education and

jobs, affordable housing, more equality and less discrimination, and socially productive activities—are needed to reduce street crime. In the past, structural solutions such as these have made it possible for immigrants who initially committed street crimes to leave the streets, get jobs, and lead productive lives. Ultimately, the best approach for reducing delinquency and crime would be prevention: to work with young people *before* they become juvenile offenders to help them establish family relationships, build self-esteem, choose a

Box 6.3 You Can Make a Difference
Combating Delinquency and Crime at an Early Age

Linda Warsaw of San Bernardino, California, explains that she started Kids Against Crime after being inspired by an eight-year-old girl who insisted that the neighbor who had assaulted her be brought to trial:

> This little girl decided on her own that she wanted this person put away so no other children would have to go through the pain she'd gone through. It was hard for her, but she went through with the trial. After seeing her courage and conviction, I realized kids can make a difference. That's when I went home and drew up the proposal that started Kids Against Crime. (qtd. in Lappé and Du Bois, 1994: 147–148)

When you think about your community, what organization might you start—or volunteer to assist—that could enhance children's lives and help prevent gang violence and delinquency? Consider, for example, these programs related to Kids Against Crime:

- *A peer support hotline.* Peer support hotlines address issues and questions about gangs, drugs, crime, and personal problems.
- *Preventive education programs.* Skits and workshops on topics such as suicide, child abuse, teen pregnancy, and AIDS are presented at shopping malls, schools, and community centers.
- *Improvement projects for neighborhoods.* Children and young people are encouraged to participate in projects to clean up graffiti and improve neighborhoods.
- *Learning public life skills.* Programs include public speaking, planning, active listening, and dealing with the media (see Lappé and Du Bois, 1994).
- *Organizing young people for social change.* Volunteers work with children and young people to organize so that their voices can be heard. For

example, here is how an organizer for the Youth Action Program in New York, which seeks to involve poor teenagers in self-governing training programs in the construction trades to prevent them from becoming involved with gangs and crime, describes this process:

> When we were organizing to appear at the city council, calling for jobs for young people, we practiced every Wednesday night for three months. We practiced walking in an organized way, in single file, filling up every successive seat in a row rather than flowing in to the council in an undisciplined way. We practiced standing in unison and clapping together at the close of every speech of one of our supporters. Nobody wore hats or chewed gum. We knew that this degree of self-discipline, implying no threat but demonstrating internal unity, would have an impact. It did. (qtd. in Lappé and Du Bois, 1994: 62)

For additional information on programs for children and young people, contact one of these sources:

- Kids Against Crime, P.O. Box 22004, San Bernardino, CA 92406. (714) 882-1344.
- KIDS Consortium (Kids Involved Doing Service), P.O. Box 27, East Boothbay, ME 14544. (207) 633-3152.
- Youth Action Program, 1280 Fifth Avenue, New York, NY 10029. (212) 860-8170.
- YouthBuild, 55 Day Street, West Somerville, MA 02144. (617) 623-9900.
- On the Internet, the Center for the Future of Children provides sociological data and other useful information about the well-being of children:

http://futureofchildren.org

- For tips on making crime prevention a family matter, try this book: Susan Hull, *50 Simple Ways to Make Life Safer from Crime* (New York: Pocket Books, 1996).

career, and get an education that will help them pursue that career (see Box 6.3). Sociologist Elliott Currie (1998) has proposed that an initial goal in working to prevent delinquency and crime is to pinpoint specifically what kinds of preventive programs work and to establish priorities that make prevention possible. Among these priorities are preventing child abuse and neglect, enhancing children's intellectual and social development,

providing support and guidance to vulnerable adolescents, and working intensively with juvenile offenders (Currie, 1998).

Is equal justice under the law possible? As long as racism, sexism, classism, and ageism exist in our society, people will see deviant and criminal behavior through a selective lens. Researchers have found a close association between white racial prejudice and support for the death penalty

Reflections

Reflections 6.4

Why are issues pertaining to juvenile offenders among the greatest challenges facing the U.S. criminal justice system?

✓Checkpoints

Checkpoint 6.8

27. What are some structural solutions for criminal behavior?

(Aguirre and Baker, 1994). To solve the problems addressed in this chapter, we must ask ourselves what we can do to ensure the rights of the poor, of African Americans, and of other people of color. Many of us can counter classism, racism, sexism, and ageism where they occur. Perhaps the only way that the United States can have equal justice under the law (and, perhaps, less crime as a result) in the future is to promote social justice for individuals regardless of their race, class, gender, or age. The current system offers neither justice for many of the accused nor safety for society as a whole (J. Johnson, 1994).

The Global Criminal Economy

Consider this scenario:

> Con men operating out of Amsterdam sell bogus U.S. securities by telephone to Germans; the operation is controlled by an Englishman residing in Monaco, with his profits in Panama. Which police force should investigate? In which jurisdiction should a prosecution be mounted? There may even be a question about whether a crime has been committed, although if all the actions had taken place in a single country there would be little doubt. (United Nations Development Programme, 1999: 104)

As this example shows, international criminal activity poses new and interesting questions not only for those who are the victims of such actions but also for governmental agencies mandated to control crime.

Global crime—the networking of powerful criminal organizations and their associates in shared activities around the world—is a relatively new phenomenon (Castells, 1998). However, it is an extremely lucrative endeavor as criminal organizations have increasingly set up their operations on a transnational basis, using the latest communication and transportation technologies.

How much money and other resources change hands in the global criminal economy? Although the exact amount of profits and financial flows originating in the global criminal economy is impossible to determine, the 1994 United Nations Conference on Global Organized Crime estimated that about $500 billion (in U.S. currency) a year is accrued in the global trade in drugs alone. Today, profits from all kinds of global criminal activities are estimated to range from $750 billion to over $1.5 trillion a year (United Nations Development Programme, 1999). Some analysts believe that even these figures underestimate the true nature and extent of the global criminal economy (Castells, 1998). The highest income-producing activities of global criminal organizations include trafficking in drugs, weapons, and nuclear material; smuggling of things and people (including many migrants); trafficking in women and children for the sex industry; and trafficking in body parts such as corneas and major organs for the medical industry. Undergirding the entire criminal system is money laundering and various complex financial schemes and international trade networks that make it possible for people to use the resources they obtain through illegal activity for the purposes of consumption and investment in the ("legitimate") formal economy.

Who engages in global criminal activities? Although groups such as the *Cosa Nostra* have existed for many years in their countries of origin and perhaps surrounding territories, they have expanded rapidly in the era of global communications and rapid transportation networks. In some countries, criminal organizations maintain a quasi-legal existence and are visible in "legitimate" business and political activities. For example, the *Yakuza* in Japan have operated for many years

✓Checkpoints

Checkpoint 6.9

28. What is meant by the term *global crime*?

without much scrutiny by the Japanese government. Today, the *Yakuza* have exported their practice of blackmail and extortion of corporations to the United States and other nations, where they send in violent *provocateurs*, known as the *Sokaiya*, to intimidate Japanese executives living abroad (Castells, 1998). Among other global criminal organizations, these Japanese gang members have been able to operate in the United States by investing heavily in real estate and engaging in agreements with the U.S. Mafia and various Russian criminal groups for control over numerous illegal enterprises. According to the sociologist Manuel Castells (1998), networking and strategic alliances between criminal networks have been key factors in the success of the criminal organizations that have sought to expand their criminal activities over the past two decades. It is through such networks that the criminal economy thrives on a global basis, allowing participants to escape police control and live beyond the laws of any one nation.

Can anything be done about global crime? Recent studies have concluded that reducing global crime will require a global response, including the cooperation of law enforcement agencies, prosecutors, and intelligence services across geopolitical boundaries. However, this approach is problematic because countries such as the United States often have difficulty getting the various law enforcement agencies to cooperate within their own nation. Similarly, law enforcement agencies in high-income nations such as the United States and Canada are often suspicious of law enforcement agencies in low-income countries, believing that these law enforcement officers are corrupt. Regulation by the international community (for example, through the United Nations) would also be necessary to control global criminal activities such as international money laundering and trafficking in people and controlled substances such as drugs and weapons. However, development and enforcement of international agreements on activities such as the smuggling of migrants or the trafficking of women and children for the sex industry have been extremely limited thus far. Many analysts acknowledge that economic globalization has provided great opportunities for wealth through global organized crime (Castells, 1998; United Nations Development Programme, 1999).

Chapter Review

How do sociologists view deviance?
Sociologists are interested in what types of behavior are defined by societies as "deviant," who does that defining, how individuals become deviant, and how those individuals are dealt with by society.

What are the main functionalist theories for explaining deviance?
Functionalist perspectives on deviance include strain theory and opportunity theory. Strain theory focuses on the idea that when people are denied legitimate access to cultural goals, such as a good job or a nice home, they may engage in illegal behavior to obtain them. Opportunity theory suggests that for deviance to occur, people must have access to illegitimate means to acquire what they want but cannot obtain through legitimate means.

How do symbolic interactionists view deviance?
According to symbolic interactionists, deviance is learned through interaction with others. Differential association theory states that individuals have a greater tendency to deviate from societal norms when they frequently associate with persons who tend toward deviance instead of conformity. According to social control theories, everyone is capable of committing crimes, but social bonding (attachments to family and to other social institutions) keeps many from doing so. According to labeling theory, deviant behavior is that which is labeled deviant. The process of labeling is related to the power and status of those persons who do the labeling and those who are labeled. Generally, those in power label the behavior of others as deviant.

How do conflict and feminist perspectives explain deviance?
Conflict perspectives on deviance focus on inequalities in society. According to some conflict theorists, lifestyles considered deviant by those with political and economic power are often defined as illegal. Marxist conflict theorists link deviance and crime to the capitalist society, which divides people into haves and have-nots, leaving

crime as the only source of support for those at the bottom of the economic ladder. Feminist approaches to deviance focus on the relationship between gender and deviance. Liberal feminism explains female deviance as a rational response to gender discrimination experienced in work, marriage, and interpersonal relationships. Marxist/radical feminism suggests that patriarchy (male domination of females) contributes to female deviance, especially prostitution. Socialist feminism states that exploitation of women by patriarchy and capitalism is related to women's involvement in criminal acts such as prostitution and shoplifting.

What is the postmodernist view on deviance?
Postmodernist views on deviance focus on how control may be maintained through largely invisible forces such as the Panoptican, described by Michel Foucault. According to this approach, everyone—not just persons engaged in deviant or criminal behavior—is likely to be under surveillance by unseen persons. Some postmodern approaches point out the intertwining nature of knowledge, power, and technology while rejecting the grand narratives found in functionalist and conflict approaches.

How do sociologists classify crime?
Sociologists identify four main categories of crime: conventional, occupational, organized, and political. Conventional (street) crime includes violent crimes, property crimes, and morals crimes. Occupational (white-collar) crimes are illegal activities committed by people in the course of their employment or financial dealings. Organized crime is a business operation that supplies illegal goods and services for profit. Political crime refers to illegal or unethical acts involving the usurpation of power by government officials, or illegal or unethical acts perpetrated against the government by outsiders seeking to make a political statement, to undermine the government, or to overthrow it.

What are the main sources of crimes statistics?
Official crime statistics are taken from the Uniform Crime Report, which lists crimes reported to the police, and the National Crime Victimization Survey, which interviews households to determine the incidence of crimes, including those not reported to police. Studies show that many more crimes are committed than are officially reported.

How are age and class related to crime statistics?
Age is the key factor in crime statistics. In 2001, persons under age 25 accounted for more than 54 percent of all arrests for index crimes, whereas persons arrested for aggravated assault, homicide, and occupational crimes tend to be older. Persons from lower socioeconomic backgrounds are more likely to be arrested for violent and property crimes; white-collar crime is more likely to occur among upper socioeconomic classes.

Who are the most frequent victims of crime?
Young males of color between ages 12 and 24 have the highest criminal victimization rates. The elderly tend to be fearful of crime but are the least likely to be victimized.

How is discretion used in the criminal justice system?
The criminal justice system, including the police, the courts, and prisons, often has considerable discretion in dealing with offenders. The police often use discretion in deciding whether to act on a situation. Prosecutors and judges use discretion in deciding which cases to pursue and how to handle them.

Key Terms

conventional (street) crime 190
corporate crime 192
crime 177
criminology 177
deviance 174
differential association theory 181
illegitimate opportunity structures 180
juvenile delinquency 177
labeling theory 183
occupational (white-collar) crime 192
organized crime 193
political crime 193
primary deviance 184
punishment 200
secondary deviance 184
social bond theory 183
social control 177
strain theory 178
tertiary deviance 184

Questions for Critical Thinking

1. Does public toleration of deviance lead to increased crime rates? If people were forced to conform to stricter standards of behavior, would there be less crime in the United States?
2. Should so-called victimless crimes, such as prostitution and recreational drug use, be decriminalized? Do these crimes harm society?
3. As a sociologist armed with a sociological imagination, how would you propose to deal with the problem of crime in the United States? What programs

would you suggest enhancing? What programs would you reduce?

Resources on the Internet

Chapter-Related Web Sites

The following web sites have been selected for their relevance to the topics in this chapter. These sites are among the more stable, but please note that web site addresses change frequently.

These are government-sponsored sites that contain data related to a variety of crimes in the United States:

National Archive of Criminal Justice Data
http://www.icpsr.umich.edu/NACJD/

Bureau of Justice Statistics
http://www.ojp.usdoj.gov/bjs/welcome.html

Bureau of Justice Statistics: Crime and Victims Statistics
http://www.ojp.usdoj.gov/bjs/cvict.htm

Bureau of Justice Statistics: Statistics on Drugs and Crime
http://www.ojp.usdoj.gov/bjs/drugs.htm

U.S. Drug Enforcement Administration
http://www.usdoj.gov/dea/

Federal Bureau of Investigation
http://www.fbi.gov/homepage.htm

FBI Uniform Crime Reports
http://www.fbi.gov/ucr/ucr.htm

Federal Bureau of Prisons
http://www.bop.gov/fact0598.html

Federal Crime Statistics Organizational Index
http://www.crime.org/fed-index.html

National Criminal Justice Reference Service
http://www.ncjrs.org

Crime Mapping Research Center
http://www.ojp.usdoj.gov/cmrc/

Office of Juvenile Justice and Delinquency Prevention
http://ojjdp.ncjrs.org/

Companion Web Site for This Book
Virtual Society: The Wadsworth Sociology Resource Center
Visit **http://sociology.wadsworth.com** and click on the page for Kendall, *Sociology in Our Times: The Essentials*, Fourth Edition, to access a wide range of enrichment material to aid in your study of sociology. Click on the Student Resources section of the web site. Next, select from the pull-down menu the chapter that you are presently studying. Among the useful options for self-study are chapter objectives, flashcards, practical study tips, and practice tests for each chapter.

MicroCase Online
From the Virtual Society home page, click on MicroCase Online to access book-specific MicroCase exercises that allow you to further explore sociological issues and principles presented in the text.

InfoTrac College Edition
Another unique option available to you at the Student Resources section of the companion web site is InfoTrac College Edition, an online library with access to hundreds of scholarly and popular periodicals. Below are suggested search terms for this chapter. Results from these and other searches are found at the site.

- Search keywords: *juvenile crime*. Locate articles that deal with the prevention of juvenile crime. How many articles did your search produce? What suggestions and solutions did the articles put forth for combating juvenile crime?
- Search keywords: *death penalty*. The death penalty has been and continues to be a subject of much controversy. Examine recent articles on the debate over the death penalty. What are the main arguments made by those for the death penalty? Against?

Virtual Explorations CD-ROM
Go to the Virtual Explorations CD-ROM to begin the interactive exercise for this chapter. This Virtual Exploration will introduce you to some of the exciting resources for sociology on the World Wide Web. You will be guided through an exercise that employs related web sites on deviance and crime. Answer the questions, and e-mail your responses to your instructor.

Steve Dininno/David Goldman Agency/SIS

Global Stratification

7

What Is Social Stratification?

Global Systems of Stratification
Slavery
The Caste System
The Class System

Wealth and Poverty in Global Perspective

Problems in Studying Global Inequality
The "Three Worlds" Approach
The Levels of Development Approach

Classification of Economies by Income
Low-Income Economies
Middle-Income Economies
High-Income Economies

Measuring Global Wealth and Poverty
Absolute, Relative, and Subjective Poverty
The Gini Coefficient and Global Quality of Life Issues

Global Poverty and Human Development Issues
Life Expectancy
Health
Education and Literacy
Persistent Gaps in Human Development

Theories of Global Inequality
Development and Modernization Theory
Dependency Theory
World Systems Theory
The New International Division of Labor Theory

Global Inequality in the Future

I would watch the women turn on the water, and sometimes they would cry, but always it would be with happiness. I cried, because how can you not? You see hope in the eyes of people. You see women who have been wracked with arthritis [from carrying heavy buckets of water up the hill from the well] now able to stand up. . . . (qtd. in Stevenson, 1997a: BU1)

—James D. Wolfensohn, a former investment banker who heads the World Bank (a United Nations agency that makes loans to member nations), describing how he felt upon watching impoverished Rio de Janeirians learn that they had running water in their area of the city

When I first enquired about Taiwan's satellite factories, people told me what they had seen or experienced. There was a factory in the next building. They made caps for Kung-man [a large sportswear factory in Taiwan]. My next-door neighbor used to sew logos on the caps. Her living room was really a mess. Every time I visited her, she was on the machine. . . .

[You mean women do factory work at home?]

Oh, yeah. Many housewives do that as a subsidiary job. I have seen that. In fact, my brother's wife used to put together tiny Santa Clauses [Christmas ornaments] at home. I don't remember exactly how much she earned. Not that much, as I recall. When I visited her, I sometimes helped her. . . . (Hsiung, 1996: 65)

Michael Dwyer/Stock Boston

In some areas of the world, running water in households is taken for granted. In nations such as Mali, low-income women continue to carry water to their homes. Will the people of all nations more equally participate in economic prosperity in the twenty-first century?

—sociologist Ping-Chun Hsiung discussing the large number of small-scale subcontracting factories in Taiwan, where married women have made a significant contribution to their nation's gross domestic product (GDP) over the past three decades

Whether people live in Rio de Janeiro, Taiwan, or elsewhere in the world, social and economic inequality are pressing daily concerns. Poverty and inequality know no political boundaries or national borders. In this chapter, we examine global stratification and inequality, and various perspectives that have been developed to explain the nature and extent of this problem. Before reading on, test your knowledge of global wealth and poverty (see Box 7.1).

Questions and Issues

Chapter Focus Question: How are global stratification and gender linked?

What is global stratification, and how does it contribute to economic inequality?

How are global poverty and human development related?

What is modernization theory, and what are its stages?

How do conflict theorists explain patterns of global stratification?

What Is Social Stratification?

Social stratification is the hierarchical arrangement of large social groups based on their control over basic resources (Feagin and Feagin, 2003). Stratification involves patterns of structural inequality that are associated with membership in each of these groups, as well as the ideologies that support inequality. Sociologists examine the social groups that make up the hierarchy in a society and seek to determine how inequalities are structured and persist over time.

Max Weber's term *life chances* refers to the extent to which individuals have access to important

Box 7.1 Sociology and Everyday Life

How Much Do You Know About Global Wealth and Poverty?

True	False	
T	F	1. The world's ten richest people are U.S. citizens.
T	F	2. Although the percentage of the world's people living in absolute poverty has declined over the past decade, the total number of people living in poverty has increased.
T	F	3. The richest fifth of the world's population receives about 50 percent of the total world income.
T	F	4. The political role of governments in policing the activities of transnational corporations has expanded as companies' operations have become more globalized.
T	F	5. Most analysts agree that the World Bank was created to serve the poor of the world and their borrowing governments.
T	F	6. In low-income countries, the problem of poverty is unequally shared between men and women.
T	F	7. The assets of the 200 richest people are more than the combined income of over 40 percent of the world's population.
T	F	8. Poverty levels have declined somewhat in East Asia, the Middle East, and North Africa in recent years.
T	F	9. The majority of people with incomes below the poverty line live in rural areas of the world.
T	F	10. Poor people in low-income countries meet most of their energy needs by burning wood, dung, and agricultural wastes, which increases health hazards and environmental degradation.

Answers on page 212.

societal resources such as food, clothing, shelter, education, and health care. According to sociologists, more-affluent people typically have better life chances than the less-affluent because they have greater access to quality education, safe neighborhoods, high-quality nutrition and health care, police and private security protection, and an extensive array of other goods and services. In contrast, persons with low- and poverty-level incomes tend to have limited access to these resources. *Resources* are anything valued in a society, ranging from money and property to medical care and education; they are considered to be scarce because of their unequal distribution among social categories. If we think about the valued resources available in various nations, for example, the differences in life chances are readily apparent. As one analyst suggested, "Poverty narrows and closes life chances. The victims of poverty experience a kind of arteriosclerosis of opportunity. Being poor not only means economic in-

security, it also wreaks havoc on one's mental and physical health" (Ropers, 1991: 25).

Throughout the world, poverty is a major social concern, and it is often related to people's access to scarce resources as well as their nationality, perceived race or ethnicity, gender, age, religion, and many other factors that are used to distinguish between— and sometimes favor or disadvantage—individuals and groups that are believed to fit into certain categories. For example, all societies distinguish among

Social stratification the hierarchical arrangement of large social groups based on their control over basic resources.

Life chances Max Weber's term for the extent to which individuals have access to important societal resources such as food, clothing, shelter, education, and health care.

Box 7.1

Answers to the Sociology Quiz on Global Wealth and Poverty

1. **False.** In 2001, six of the ten richest people were U.S. citizens. These included Bill Gates, Warren Buffett, Paul G. Allen, and Michael Dell (*Forbes,* 2001).

2. **True.** Data from the World Bank indicate that the percentage of the world's people living in absolute poverty has declined since the mid-1980s, particularly in Asia. However, other regions have not reduced the incidence of poverty to the same degree, and the total number of people living in poverty rose to approximately 1.4 billion in the mid-1990s (World Bank, 2001).

3. **False.** According to the Human Development Report published by the United Nations Development Programme (1999), the richest fifth receives more than 80 percent of the total world income. This ratio doubled between the 1950s and the 1990s.

4. **False.** As companies have globalized their operations, governmental restrictions have become less effective in controlling their activities. Transnational corporations have very little difficulty sidestepping governmental restrictions based on old assumptions about national economies and foreign policy. For example, Honda is able to circumvent import restrictions that the governments of Taiwan, South Korea, and Israel have placed on its vehicles by shipping vehicles made in Ohio to those locations (Korten, 1996).

5. **False.** Some analysts point out the linkages between the World Bank and the transnational corporate sector on both the borrowing and lending ends of the bank's operation. Although the bank is supposedly owned by its members' governments and lends money only to governments, many of its projects involve vast financial dealings with transnational construction companies, consulting firms, and procurement contractors (see Korten, 1996).

6. **True.** In almost all low-income countries (as well as middle- and high-income countries), poverty is a more chronic problem for women due to sexual discrimination, resulting in a lack of educational and employment opportunities (Hauchler and Kennedy, 1994).

7. **True.** Assets of the 200 richest people are more than the combined income of 41 percent of the world's population (United Nations Development Programme, 1999).

8. **True.** These have been the primary regions in which poverty has decreased somewhat and infant mortality rates have fallen. Factors such as economic growth, oil production, foreign investment, and overall development have been credited for the decrease in poverty in East Asia, the Middle East, and North Africa (United Nations DPCSD, 1997).

9. **True.** The majority of people with incomes below the poverty line live in rural areas of the world; however, the number of poor people residing in urban areas is growing rapidly. In fact, most people living in poverty in Latin America are urban dwellers (United Nations DPCSD, 1997).

10. **True.** Although these fuels are inefficient and harmful to health, many low-income people cannot afford appliances, connection charges, and so forth. In some areas, electric hookups are not available (United Nations DPCSD, 1997).

Ed Kashi/Aurora Photos

People's life chances are enhanced by access to important societal resources such as education. How will the life chances of students who have the opportunity to pursue a college degree differ from those of young people who do not have the chance to go to college?

people by age. Young children typically have less authority and responsibility than older persons. Older persons, especially those without wealth or power, may find themselves at the bottom of the social hierarchy. Similarly, all societies differentiate between females and males: Women are often treated as subordinate to men. From society to society, people are treated differently as a result of their religion, race/ethnicity, appearance, physical strength, disabilities, or other distinguishing characteristics. All of these differentiations result in inequality. However, systems of stratification are also linked to the specific economic and social structure of a society and to a nation's position in the system of global stratification.

Global Systems of Stratification

Around the globe, one of the most important characteristics of systems of stratification is their degree of flexibility. Sociologists distinguish among such systems based on the extent to which they are open or

✓Checkpoints

Checkpoint 7.1

1. What is social stratification?
2. What are the key factors involved in life chances?

closed. In an *open system*, the boundaries between levels in the hierarchies are more flexible and may be influenced (positively or negatively) by people's achieved statuses. Open systems are assumed to have some degree of social mobility. **Social mobility is the movement of individuals or groups from one level in a stratification system to another** (Rothman, 1993). This movement can be either upward or downward. **Intergenerational mobility is the social movement experienced by family members from one generation to the next.** For example, Sarah's father is a carpenter who makes good wages in good economic times but is often unemployed when the construction industry slows to a standstill. Sarah becomes a neurologist, earning $350,000 a year, and moves from the working class to the upper-middle class. Between her father's generation and her own, Sarah has experienced upward social mobility.

By contrast, **intragenerational mobility is the social movement of individuals within their own lifetime.** Consider, for example, RaShandra, who began her career as a high-tech factory worker and through increased experience and taking specialized courses in her field became an entrepreneur, starting her own highly successful "dot.com" business. RaShandra's advancement is an example of upward intragenerational social mobility. However, both intragenerational mobility and intergenerational mobility may be downward as well as upward.

In a *closed system*, the boundaries between levels in the hierarchies of social stratification are rigid, and people's positions are set by ascribed status. Open and closed systems are ideal-type constructs; no actual stratification system is completely open or closed. The systems of stratification that we will examine—slavery, caste, and class—are characterized by different hierarchical structures

Social mobility the movement of individuals or groups from one level in a stratification system to another.

Intergenerational mobility the social movement (upward or downward) experienced by family members from one generation to the next.

Intragenerational mobility the social movement (upward or downward) of individuals within their own lifetime.

and varying degrees of mobility. Let's examine these three systems of stratification to determine how people acquire their positions in each and what potential for social movement they have.

Slavery

Slavery **is an extreme form of stratification in which some people are owned by others.** It is a closed system in which people designated as "slaves" are treated as property and have little or no control over their lives. According to some social analysts, throughout recorded history only five societies have been slave societies—those in which the social and economic impact of slavery was extensive: ancient Greece, the Roman Empire, the United States, the Caribbean, and Brazil (Finley, 1980). Others suggest that slavery also existed in the Americas prior to European settlement, and throughout Africa and Asia (Engerman, 1995).

Those of us living in the United States are most aware of the legacy of slavery in our own country. Beginning in the 1600s, slaves were forcibly imported to the United States as a source of cheap labor. Slavery was defined in law and custom by the 1750s, making it possible for one person to own another person (Healey, 1995). In fact, early U.S. presidents including George Washington, James Madison, and Thomas Jefferson owned slaves. As practiced in the United States, slavery had four primary characteristics: (1) it was for life and was inherited (children of slaves were considered to be slaves); (2) slaves were considered property, not human beings; (3) slaves were denied rights; and (4) coercion was used to keep slaves "in their place" (Noel, 1972). Although most slaves were powerless to bring about change, some were able to challenge slavery—or at least their position in the system—by engaging in activities such as sabotage, intentional carelessness, work slowdowns, or running away from owners and working for the abolition of slavery (Healey, 1995). Despite the fact that slavery officially ended many years ago in this country, sociologists such as Patricia Hill Collins (1990) believe that its legacy is deeply embedded in current patterns of prejudice and discrimination against African Americans.

Slavery is not simply an unfortunate historical legacy. Although slavery is illegal everywhere, and there is no more *legal* ownership of people, the economist Stanley L. Engerman (1995: 175) believes that the world will not be completely free of slavery as long as there are "debt bondage, child labor, contract labor, and other varieties of coerced work for limited periods of time, with limited opportunities for mobility, and with limited political and economic power." As the scholar Kevin Bales (1999) states in *Disposable People: New Slavery in the Global Economy*, "When people buy slaves today they don't ask for a receipt or ownership papers, but they do gain *control*—and they use violence to maintain this control. Slaveholders have all the benefits of ownership without the legalities."

Slavery in the United States was a topic of concern for Harriet Martineau, who, as discussed in Chapter 1, was one of the first women social scientists. In *Society in America* (1962/1837), Martineau described her visit to Montgomery, Alabama, in 1835, when she visited certain plantations and observed the lives of the slaves. Seeing through the prejudice of her day, Martineau discredited negative stereotypes about the African Americans who were held in slavery and applauded the few slave owners whom she thought treated their slaves decently:

> I spent some days at a plantation a few miles from Montgomery, and heard there of an old lady who treats her slaves in a way very unusual, but quite safe, as far as appears. She gives them knowledge, which is against the law; but the law leaves her in peace and quiet. She also commits to them the entire management of her estate, requiring only that they should make her comfortable, and letting them take the rest. There is an obligation by law to keep an overseer; to obviate insurrection. How she manages about this, I omitted to inquire: but all goes on well; the cultivation of the estate is creditable, and all parties are contented. This is only a temporary ease and contentment. The old lady must die; and her slaves will either be sold to a new owner, whose temper will be an accident; or, if freed, must leave the State: but the story is satisfactory in as far as it gives evidence of the trustworthiness of the negroes.

Martineau, an Englishwoman, was a strong—but largely unheard—voice advocating the end of slavery in the United States.

How many people are held in slavery today? Bales (1999) estimates that the number of slaves worldwide stands at about 27 million people, most of whom are held in *bonded labor* in India, Pakistan, Bangladesh, and Nepal. *Bonded labor* or *debt bondage* refers to a situation in which people give themselves into slavery as security against a loan or when they inherit a debt from a relative (Bales,

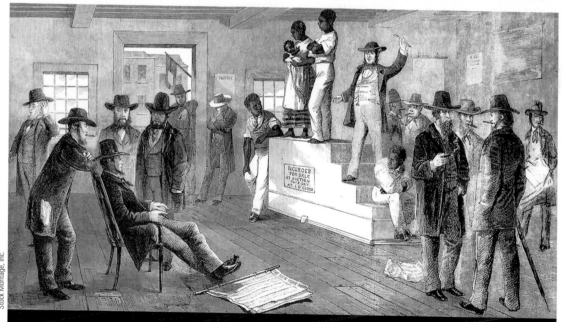

Systems of stratification include slavery, caste, and class. As shown above, the life chances of people under each of these systems differ widely.

1999). For the most part, slavery is concentrated in Southeast Asia, northern and western Africa, and parts of South America. However, Bales argues that there are slaves in almost every country of the world,

Slavery an extreme form of stratification in which some people are owned by others.

where they work in simple, nontechnical, traditional labor such as brickmaking, cloth and carpet making, domestic service, and prostitution. Although some products made with slave labor are sold at the local level, goods made by people in slavery or debt bondage are eventually sold to people around the globe. According to Bales (1999), *what slaves make is not as important as the sheer volume of work* that they can be forced to produce and the number of hours that they can be required to work. Bales believes that poverty is a more important indicator of who will be today's slaves rather than the race, ethnicity, or religion of the people involved.

As modernization and globalization have affected many previously less developed regions of the world, traditional economies and family life have been changed drastically. Subsistence farming can no longer support many people because of the shift to cash-crop agriculture, the loss of common land, and government policies that focus on the production of cheap food for cities rather than subsistence for farmers and agricultural laborers (Bales, 1999).

Various forms of slavery include child prostitution in Thailand, enslaved brickmakers in Pakistan, and domestic slaves in France; however, we will briefly examine modern-day slavery in Brazil as an example of the experiences of many who are enslaved today. Ronald, who was interviewed in a charcoal camp in Mato Grosso, is one example:

> My parents lived in a very dry area and when I got older there was no work, no work at all there . . . but one day a *gato* [labor recruiter] came and began to recruit people to work out here in Mato Grosso. The *gato* said that we would be given good food every day, and we would have good wages besides. He promised that every month his truck would bring people back [home] so that they could visit their families and bring them their pay. He even gave money to some men to give to their families before they left and to buy food to bring with them on the trip. . . .
>
> When we got to Mato Grosso we kept driving further and further into the country. This camp is almost fifty miles from anything; it is just raw *cerrado* for fifty miles before you get to even a ranch, and there is just the one road. When we reached the camp we could see it was terrible: the conditions were not good enough for animals. Standing around the camp were men with guns. And then the *gato* said, "You each owe me a lot of money: there is the cost of the trip, and all that food you ate, and the

> money I gave you for your families—so don't even think about leaving." (qtd. in Bales, 1999: 127)

Like many others who are enslaved in Brazil, Ronald was trapped by poverty, circumstances, and the threat of physical force by armed men. The *gatos* even take away the workers' state identity cards and labor cards that are the key to legal employment; consequently, workers cannot find other employment even if they are able to flee the oppressive conditions.

Today, slavery flourishes in rural and other isolated areas of Brazil—on sugarcane plantations and ranches, in gold mines, and in the charcoal industries of the Amazon. The sugarcane fields are harvested by farm workers who toil from sunup to sundown and sleep in hammocks strung in cow stalls. Employers often do not pay wages for the work: Workers instead receive script, which they can redeem for food. Because landowners need to ensure that they will have a readily available supply of cheap labor, they bind laborers by encouraging them to run up unpayable debts at company-owned stores or canteens. According to an international labor organization report, "Workers who try to escape are pursued by gunmen and returned to the estate, where they can be beaten, whipped or subjected to mutilation or sexual abuse" (Brooke, 1993b: 3).

What happens to people who are enslaved? Like people in all nations, enslaved people have hopes and dreams for their children and grandchildren. Doralice Moreira de Souza, a forty-seven-year-old woman whose hands are gnarled from years of cutting five tons of sugarcane daily in Conceicao de Macabu, Brazil, has a dream for Alan, her ten-year-old grandson: "I would like him to study, so that when he grows up, he won't end up a slave like me" (qtd. in Brooke, 1993a: Y3). However, Alan's life chances are already seriously limited because of economic conditions and labor exploitation in his country. Why does such exploitation occur? Social scientists have developed various theories to explain the persistence of extreme global inequality.

The Caste System

Like slavery, caste is a closed system of social stratification. A **caste system** is a system of social inequality in which people's status is permanently determined at birth based on their parents' as-

cribed characteristics. Vestiges of caste systems exist in contemporary India and South Africa.

In India, caste is based in part on occupation; thus, families typically perform the same type of work from generation to generation. By contrast, the caste system of South Africa was based on racial classifications and the belief of white South Africans (Afrikaners) that they were morally superior to the black majority. Until the 1990s, the Afrikaners controlled the government, the police, and the military by enforcing *apartheid*—the separation of the races. Blacks were denied full citizenship and restricted to segregated hospitals, schools, residential neighborhoods, and other facilities. Whites held almost all of the desirable jobs; blacks worked as manual laborers and servants.

In a caste system, marriage is endogamous, meaning that people are allowed to marry only within their own group. In India, parents have traditionally selected marriage partners for their children. In South Africa, interracial marriage was illegal until 1985.

Cultural beliefs and values sustain caste systems. Hinduism, the primary religion of India, reinforced the caste system by teaching that people should accept their fate in life and work hard as a moral duty. Caste systems grow weaker as societies industrialize; during that process, the values reinforcing the system break down, and people start to focus on the types of skills needed for industrialization. However, vestiges of caste systems often remain for hundreds of years beyond the period in which they are "officially" abolished. For example, some social scientists believe that past racial segregation in the United States has contributed to a caste-like condition of racial and ethnic inequality that remains in the present (Cox, 1948; Frankenberg, 1993). We will return to this issue in Chapter 9 ("Race and Ethnicity").

As we have seen, in closed systems of stratification, group membership is hereditary, and moving up within the structure is almost impossible. Custom and law frequently perpetuate privilege and ensure that higher-level positions are reserved for the children of the advantaged (Rothman, 1993).

The Class System

The *class system* is a type of stratification based on the ownership and control of resources and on the type of work people do (Rothman, 1993). At least theoretically, a class system is more open than a caste system because the boundaries between classes are less distinct than the boundaries between castes. In a class system, status comes at least partly through achievement rather than entirely by ascription.

In class systems, people may become members of a class other than that of their parents through both intergenerational mobility and intragenerational mobility, either upward or downward. Horizontal mobility occurs when people experience a gain or loss in position and/or income that does not produce a change in their place in the class structure. For example, a person may get a pay increase and a more prestigious title but still not move from one class to another. By contrast, movement up or down the class structure is *vertical mobility*. Martin, a commercial artist who owns his own firm in the United States, is an example of vertical, intergenerational mobility:

> My family came out of a lot of poverty and were eager to escape it. . . . My [mother's parents] worked in a sweatshop. My grandfather to the day he died never earned more than $14 a week. My grandmother worked in knitting mills while she had five children. . . . My father quit school when he was in eighth grade and supported his mother and his two sisters when he was twelve years old. My grandfather died when my father was four and he basically raised his sisters. He got a man's job when he was twelve and took care of the three of them. (qtd. in Newman, 1993: 65)

Martin's situation reflects upward mobility; however, as previously stated, people may also experience downward mobility, caused by any number of reasons, including a lack of jobs, low wages and employment instability, marriage to someone with fewer resources and less power than oneself, and changing social conditions (Ehrenreich, 1989; Newman, 1988, 1993).

Social classes involve much more than the upward or downward mobility of individuals: The class

Caste system   a system of social inequality in which people's status is permanently determined at birth based on their parents' ascribed characteristics.

Class system   a type of stratification based on the ownership and control of resources and on the type of work that people do.

Figure 7.1 Income Gap Between the World's Richest and Poorest People

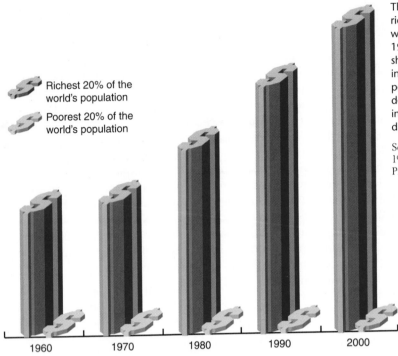

Richest 20% of the world's population

Poorest 20% of the world's population

1960   1970   1980   1990   2000

The income gap between the richest and poorest people in the world continued to grow between 1960 and 2000. As this figure shows, in 1960 the highest-income 20 percent of the world's population received $30 for each dollar received by the lowest-income 20 percent. By 2000, the disparity had increased: $74 to $1.

Source: International Monetary Fund, 1992; United Nations Development Programme, 1996, 1999.

structure of any society is shaped by the historical, economic, political, and social context in which it is embedded. For example, changes in technology contribute to the changing composition of classes and to how people are distributed within those classes. In many nations, government policies influence the economy and how economic rewards are distributed among citizens. In other words, social classes do not exist apart from the class system of which they are a component. According to many social analysts, class is a relationship that is not confined to economic divisions alone: Class relations shape all human relationships, including how parents bring up their children and how employers and employees interact with each other; thus, the life chances of individuals and their families are tied to their position in the class structure. We will examine these class structures in greater detail in Chapter 8 when we look at the U.S. class system.

Wealth and Poverty in Global Perspective

What do we mean by global stratification? *Global stratification* refers to the unequal distribution of wealth, power, and prestige on a global basis, resulting in people having vastly different lifestyles and life chances both within and among the nations of the world. Just as the United States is divided into classes, the world is divided into unequal segments characterized by extreme differences in wealth and poverty. For example, the income gap between the richest and the poorest 20 percent of the world population continues to widen (see Figure 7.1). However, when we compare social and economic inequality within other nations, we find gaps that

✓ Checkpoints

Checkpoint 7.2

3. How does intergenerational mobility differ from intragenerational mobility?
4. What is slavery?
5. Does slavery still exist today?
6. What is the caste system?
7. How does the caste system differ from the class system?

Todd Bigelow/Aurora Photos

This recent strike by service employees reflects the activism of workers throughout the years as they have sought to gain better wages and working conditions. How would Karl Marx explain the problems experienced by employees such as these?

almost 80 times the income of the poorest 20 percent (Hauchler and Kennedy, 1994; United Nations Development Programme, 1999). Income disparities *within* countries were even more pronounced.

The death of Mother Teresa in 1997 called attention to the plight of the poorest of the poor in Calcutta, India, and around the globe. Unlike many contemporary stereotypes that depict the poor as being lazy or unworthy of assistance, Mother Teresa felt strongly about people who are poor:

> The poor give us much more than we give them. They're such strong people, living day to day with no food. And they never curse, never complain. We don't have to give them pity or sympathy. We have so much to learn from them. (qtd. in Lapierre, 1997: A10)

Mother Teresa is only one of the many people who have sought to address the issue of world poverty and to determine ways in which resources can be used to meet the urgent challenge of poverty. However, not much progress has been made on this front (Lummis, 1992) despite a great deal of talk and billions of dollars in "foreign aid" flowing from high-income to low-income nations. The idea of "development" has become the primary means used in attempts to reduce social and economic inequalities and alleviate the worst effects of poverty in the less industrialized nations of the world. Often, the nations that have not been able to reduce or eliminate poverty are chastised for not making the necessary social and economic reforms to make change possible (Myrdal, 1970). Or, as another social analyst has suggested,

> The *problem* of inequality lies not in poverty, but in excess. "The problem of the world's poor," defined more accurately, turns out to be "the problem of the world's rich." This means that the solution to the problem is not a massive change in the culture of poverty so as to place it on the path of development, but a massive change in the culture of superfluity in order to place it on the path of counterdevelopment. It does not call for a new value system forcing the world's majority to feel shame at their traditionally moderate consumption habits, but for a new value system forcing the world's rich to see the shame and vulgarity of their

are more pronounced than they are in the United States. As previously defined, *high-income countries* are nations characterized by highly industrialized economies; technologically advanced industrial, administrative, and service occupations; and relatively high levels of national and per capita (per person) income. In contrast, *middle-income countries* are nations with industrializing economies, particularly in urban areas, and moderate levels of national and personal income. *Low-income countries* are primarily agrarian nations with little industrialization and low levels of national and personal income. Within some nations, the poorest one-fifth of the population has an income that is only a slight fraction of the overall average per capita income for that country. For example, in Brazil, Guatemala, and Honduras, less than 3 percent of total national income accrues to the poorest one-fifth of the population (World Bank, 1996).

Just as the differences between the richest and poorest people in the world have increased, the gap in global income differences between rich and poor countries has continued to widen over the past 50 years. In 1960, the wealthiest 20 percent of the world population had more than 30 times the income of the poorest 20 percent. By 1997, the wealthiest 20 percent of the world population had

✓**Checkpoints**

Checkpoint 7.3

8. How wide is the gap in global income differences between rich and poor countries?

9. How does social inequality contribute to the problem of economic inequality?

overconsumption habits, and the double vulgarity of standing on other people's shoulders to achieve those consumption habits. (Lummis, 1992: 50)

As this statement suggests, the increasing interdependence of all the world's nations was largely overlooked or ignored until increasing emphasis was placed on the global marketplace and the global economy. In addition, there are a number of problems inherent in studying global stratification, one of which is what terminology should be used to describe various nations.

Problems in Studying Global Inequality

One of the primary problems encountered by social scientists studying global stratification and social and economic inequality is what terminology should be used to refer to the distribution of resources in various nations. During the past forty years, major changes have occurred in the way that inequality is addressed by organizations such as the United Nations and the World Bank. Most definitions of inequality are based on comparisons of levels of income or economic development, whereby countries are identified in terms of the "three worlds" or upon their levels of economic development.

The "Three Worlds" Approach

After World War II, the terms "First World," "Second World," and "Third World" were introduced by social analysts to distinguish among nations on the basis of their levels of economic development and the standard of living of their citizens. *First World* nations were said to consist of the rich, industrialized nations that primarily had capitalist economic systems and democratic political sys-

tems. The most frequently noted First World nations were the United States, Canada, Japan, Great Britain, Australia, and New Zealand. *Second World* nations were said to be countries with at least a moderate level of economic development and a moderate standard of living. These nations included China, North Korea, Vietnam, Cuba, and portions of the former Soviet Union. According to social analysts, although the quality of life in Second World nations was not comparable to that of life in the First World, it was far greater than that of people living in the *Third World*—the poorest countries, with little or no industrialization and the lowest standards of living, shortest life expectancies, and highest rates of mortality.

The Levels of Development Approach

Among the most controversial terminology used for describing world poverty and global stratification has been the language of development. Terminology based on levels of development includes concepts such as developed nations, developing nations, less-developed nations, and underdevelopment. Let's look first at the contemporary origins of the idea of "underdevelopment" and "underdeveloped nations."

Following World War II, the concepts of *underdevelopment* and *underdeveloped nations* emerged out of the Marshall Plan (named after U.S. Secretary of State George C. Marshall), which provided massive sums of money in direct aid and loans to rebuild the European economic base destroyed during World War II. Given the Marshall Plan's success in rebuilding much of Europe, U.S. political leaders decided that the Southern Hemisphere nations that had recently been released from European colonialism could also benefit from a massive financial infusion and rapid economic development. Leaders of the developed nations argued that urgent problems such as poverty, disease, and famine could be reduced through the transfer of finance, technology, and experience from the developed nations to lesser-developed countries. From this viewpoint, economic development is the primary way to solve the poverty problem: Hadn't economic growth brought the developed nations to their own high standard of living? Moreover, "self-sustained development" in a nation would require that people in the lesser-developed nations accept the beliefs and values of people in the developed nations.

A. Perigot-Icone/The Image Works

According to the levels of development approach, economic development is the primary way to solve the poverty problem in low-income nations. Does the employment of workers such as these reduce the poverty of the poorest people in a country? What other factors are also important?

Ideas regarding *underdevelopment* were popularized by President Harry S Truman in his 1949 inaugural address. According to Truman, the nations in the Southern Hemisphere were "underdeveloped areas" because of their low gross national product, which today is referred to as *gross national income* (GNI)—a term that refers to all the goods and services produced in a country in a given year, plus the net income earned outside the country by individuals or corporations. If nations could increase their GNI, then social and economic inequality among the citizens within the country could also be reduced. Accordingly, Truman believed that it was necessary to assist the people of economically underdeveloped areas to raise their *standard of living*, by which he meant material well-being that can be measured by the quality of goods and services that may be purchased by the per capita national income (Latouche, 1992). Thus, an increase in the standard of living meant that a nation was moving toward economic development, which typically included the improved exploitation of natural resources by industrial development.

What has happened to the issue of development since the post–World War II era? After several decades of economic development fostered by organizations such as the United Nations and the World Bank, it became apparent by the 1970s that improving a country's GNI did not tend to reduce the poverty of the poorest people in that country. In fact, global poverty and inequality were increasing, and the initial optimism of a speedy end to underdevelopment faded. Although many developing countries had achieved economic growth, it was not shared by everyone in the nation. For example, the poorest 40 percent of the Brazilian population receives less than 7 percent of the total national income, whereas the richest 20 percent of the population receives more than 65 percent (Elliott, 1994).

Why did inequality increase even with greater economic development? Some analysts in the developed nations began to link growing social and economic inequality on a global basis to relatively high rates of population growth taking place in the underdeveloped nations. Organizations such as the United Nations and the World Health Organization stepped up their efforts to provide family planning services to the populations so that they could control their own fertility. More recently, however, population researchers have become aware that issues such as population growth, economic development, and environmental problems must be seen as interdependent concerns. This changing perception culminated in the U.N. Conference on Environment and Development in Rio de Janeiro, Brazil (the "Earth Summit"), in 1992; as a result, terms such as *underdevelopment* have largely been dropped in favor of measurements such as sustainable development, and economies are now classified by their levels of income (Elliott, 1994).

✓Checkpoints

Checkpoint 7.4

10. What is the "three worlds" approach to studying global inequality?
11. What levels of development are used in studying global inequality?

Map 7.1 High-, Middle-, and Low-Income Economies in Global Perspective

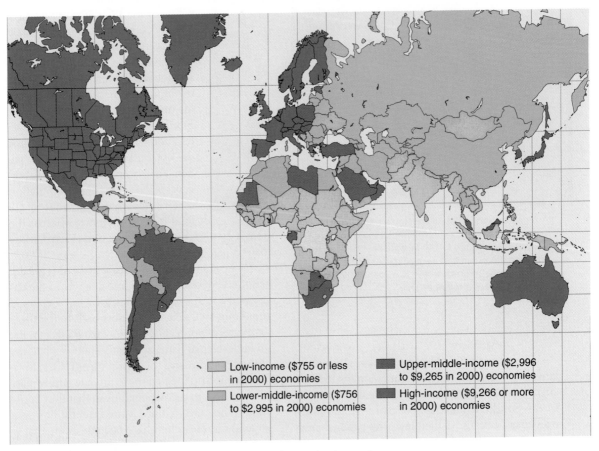

Low-income ($755 or less in 2000) economies

Lower-middle-income ($756 to $2,995 in 2000) economies

Upper-middle-income ($2,996 to $9,265 in 2000) economies

High-income ($9,266 or more in 2000) economies

How does the United States compare with other nations with regard to income?

Source: World Bank, 2001.

Classification of Economies by Income

Today, the World Bank (2001) focuses on three development themes: people, the environment, and the economy. Since the World Bank's primary business is providing loans and policy advice to low- and middle-income member countries, it classifies nations into three economic categories: *low-income economies* (a GNI per capita of $755 or less in 2000), *middle-income economies* (a GNI per capita between $756 and $9,265 in 2000), and *high-income economies* (a GNI per capita of more than $9,266 in 2000).

Reflections

Reflections 7.1

Why is it more difficult for many subordinate-group members to overcome social inequality and poverty?

Low-Income Economies

About half the world's population lives in the 63 low-income economies, where most people engage in agricultural pursuits, reside in nonurban areas, and are impoverished (World Bank, 2001). As shown in Map 7.1, low-income economies are primarily found in countries in Asia and Africa, where half of the world's population resides. Included are nations such as Rwanda, Mozambique, Ethiopia, Nigeria, Cambodia, Vietnam, Afghanistan, and Bangladesh. Eastern European countries such as Armenia and Georgia are among the low-income economies in Eastern Europe.

Among those most affected by poverty in low-income economies are women and children. Mayra Buvinić, who has served as chief of the women in development program unit at the Inter-American Development Bank, describes the plight of one Nigerian woman as an example:

> On the outskirts of Ibadan, Nigeria, Ade cultivates a small, sparsely planted plot with a baby on her back and other visibly undernourished children nearby. Her efforts to grow an improved soybean variety, which could have improved her children's diet, failed because she lacked the extra time to tend the new crop, did not have a spouse who would help her, and could not afford hired labor. (Buvinić, 1997: 38)

According to Buvinić, Ade's life is typical of many women worldwide who face obstacles to increasing their economic power because they do not have the time to invest in the additional work that could bring in more income. Also, many poor women worldwide do not have access to commercial credit and have been trained only in traditionally female skills that produce low wages. These factors have contributed to the *global feminization of poverty*, whereby women around the world tend to be more impoverished than men (Durning, 1993).

Despite the fact that women have made some gains in terms of well-being, the income gap between men and women continues to grow wider in the low-income, developing nations as well as in the high-income, developed nations such as the United States (Buvinić, 1997). In some low-income nations, children are not only affected by poverty but also by child labor practices. In 1997 the United States banned imports of goods made by children in bondage, including handmade imported rugs made by child labor in rural-based economies. Although accurate numbers do not exist for the number of children who work in servitude making rugs alone, estimates range from 300,000 to 1,000,000 children (Iovine, 1997). Since production is often spread out in homes, it is very difficult to count or monitor labor practices. Moreover, some people believe that it is unfair to

John Maier, Jr./The Image Works

Vast inequalities in income and lifestyle are shown in this photo of slums and nearby upper-class housing in Rio de Janeiro, Brazil. Do similar patterns of economic inequality exist in other nations?

evaluate the labor practices of other cultures in terms of Western values and labor standards and to take away the one means of earning income that these people may have. According to a museum curator who specializes in handmade rugs from low-income economies,

> In many cultures the economies are very different from one's own, many are family-based and the set-ups are very different from documentable child abuse. Rug weaving is one of the most perfect examples of a sustainable economy in developing countries. In Turkey virtually every living room will have a loom in it. There is pride and delight felt by the whole family with their rugs. To ban all that could have a devastating effect. (qtd. in Iovine, 1997: B9)

Middle-Income Economies

About one-third of the world's population resides in the 92 nations with middle-income economies (a GNI per capita between $756 and $9,265 in 2000). The World Bank divides middle-income economies into lower-middle-income ($756 to $2,995) and upper-middle-income ($2,996 to $9,265). Countries classified as lower-middle-income include the Latin American nations of Bolivia, Colombia, Guatemala, and El Salvador. However, even though these countries are referred to as "middle-income," more than half of the people residing in countries such as Bolivia, Guatemala, and Honduras live in poverty, defined as below $60 per month in 1994 U.S. dollars (World Bank, 2001).

Other lower-middle-income economies include Russia, Romania, and Kazakhstan. These nations had centrally planned (i.e., socialist) economies until dramatic political and economic changes occurred in the late 1980s and early 1990s. Since then, these nations have been going through a transition to a market economy. According to the World Bank (2001), some nations have been more successful than others in implementing key elements of change and bringing about a higher standard of living for their citizens. Among other factors, high rates of inflation, the growing gap between the rich and the poor, low life-expectancy rates, and homeless children have been visible signs of problems in the transition toward a free market economy in countries such as Russia.

As compared with lower-middle-income economies, nations having upper-middle-income economies typically have a somewhat higher standard of living and export diverse goods and services, ranging from manufactured goods to raw materials and fuels. Nations with upper-middle-income economies include Brazil, South Africa, Chile, Hungary, and Mexico. Although these nations are referred to as middle-income economies, they have extremely high levels of indebtedness, leaving them with few resources for fighting poverty. According to one analyst, "A major cause of debt accumulation was investment in ill-considered, ill-conceived projects, many involving bloated capital costs and healthy doses of graft" (George, 1993: 88). Additional debts also accrued through military spending, even though a sizable portion of some populations is living in hunger and misery (George, 1993). The World Bank has sought to set up funds to reduce the debts of some nations that have debts in excess of 200 to 250 percent of their annual export earnings. Among the nations in this position are Mozambique, Nicaragua, Congo, Bolivia, Ethiopia, and Tanzania. Some high-income nations believed that

such an action on the part of the World Bank might set a dangerous precedent, so the debt-reduction process was limited in scope (Lewis, 1996).

As middle-income nations have been required to make payments on their debts, the requisite structural adjustments have necessitated that the countries make spending cuts in areas that formerly helped some of the poor, including subsidized food, education, and health care. These structural adjustments have been required in agreements with lending agencies such as the World Bank and the International Monetary Fund. Thus, there is a high rate of poverty in many countries that are classified as middle-income economies. The United Nations children's relief agency, UNICEF, decries the extent to which paying off debt to these international lending agencies affects the young in Latin America and Africa:

> Debt, in particular, still shackles many developing nations, claiming a large proportion of the resources which might otherwise have been available for human progress. With falling family incomes, and cuts in public spending on services such as health and education, many African and Latin American children are still paying for their opportunity for normal growth, their opportunity to be educated, and often their lives. With no less urgency than at any time in the last five years, UNICEF must again say that it is the antithesis of civilization that so many millions of children should be continuing to pay such a price. (UNICEF, 1991; qtd. in Hauchler and Kennedy, 1994: 62)

High-Income Economies

High-income economies are found in 52 nations, including the United States, Canada, Japan, Australia, Portugal, Ireland, Israel, Italy, Norway, and Germany. According to the World Bank, people in high-income economies typically have a higher standard of living than those in low- and middle-income economies. Nations with high-income economies continue to dominate the world economy, despite the fact that shifts in the global marketplace have affected some workers who have found themselves without work due to *capital flight*—the movement of jobs and economic resources from one nation to another—and *deindustrialization*—the closing of plants and factories because of their obsolescence or the fact that work-

Figure 7.2 Percentage of Population Below the Poverty Line in East Asia

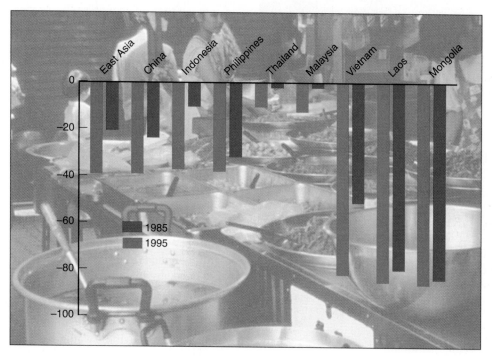

Despite economic growth, the East Asian region still has large numbers of people who live below the poverty line.

Source: World Bank, 2001.

ers in other nations are being hired to do the work more cheaply.

Some of the nations that have been in the high-income category do not have as high a rate of annual economic growth as the newly industrializing nations, particularly in East Asia and Latin America, that are still in the process of development. The only significant group of middle- and lower-income economies to close the gap with the high-income, industrialized economies over the past few decades has been the nations of East Asia. China has experienced a 270-percent increase in per capita income over the past 17 years, but some analysts wonder whether the East Asian "miracle" is over (Crossette, 1997).

Despite economic growth, the East Asian region remains home to approximately 350 million poor people (see Figure 7.2). And from all signs, it appears that nations such as India will continue to have an extremely large population and rapid annual growth rates.

✓ Checkpoints

Checkpoint 7.5

12. What are the key characteristics of low-income economies?
13. How do high-income economies compare with middle-income and low-income economies?

Measuring Global Wealth and Poverty

On a global basis, measuring wealth and poverty is a difficult task because of conceptual problems and problems in acquiring comparable data from various nations. Although gross national income continues to be one of the most widely used measures

Macduff Everton/The Image Works

Workers in this toy factory in Zhuhal, China, make toys that are often marketed in high-income nations such as the United States.

of national income, in recent years the United Nations and the World Bank have begun to use the gross domestic product (GDP). The *gross domestic product* is all the goods and services produced *within* a country's economy during a given year. Unlike the GNI, the GDP does not include any income earned by individuals or corporations if the revenue comes from sources outside of the country. For example, using GDP as a measure of economic growth, a recent World Bank report concluded that nations such as China, India, Indonesia, Brazil, and Russia were well on their way to becoming "economic powerhouses" in the next twenty-five years (Stevenson, 1997b). The report also suggested that sub-Saharan Africa, one of the poorest regions of the world, will also experience a surge of development that will help alleviate its poverty (Stevenson, 1997b). According to World Bank officials, these changes in GDP can be attributed to several factors: (1) stable economic conditions worldwide, (2) low inflation and interest rates, and (3) surging flows of foreign capital and expertise into emerging markets (Stevenson, 1997b). However, some scholars criticize the assumption that economic development always benefits low-income nations and that poor

nations *can* and *should* play "catch-up" with the wealthy, industrialized nations (Lummis, 1992).

Absolute, Relative, and Subjective Poverty

How is poverty defined on a global basis? Isn't it more a matter of comparison than an absolute standard? According to social scientists, defining poverty involves more than comparisons of personal or household income: It also involves social judgments made by researchers. From this point of view, *absolute poverty*—previously defined as a condition in which people do not have the means to secure the most basic necessities of life—would be measured by comparing personal or household income or expenses with the cost of buying a given quantity of goods and services. The World Bank has defined absolute poverty as living on less than a dollar a day. Similarly, *relative poverty*—which exists when people may be able to afford basic necessities but are still unable to maintain an average standard of living—would be measured by comparing one person's income with the incomes of others. Finally, *subjective poverty* would be measured by comparing the actual income against the income earner's expectations and perceptions (World Bank, 2001). However, for low-income nations in a state of economic transition, data on income and levels of consumption are typically difficult to obtain and are often ambiguous when they are available. Defining levels of poverty involves several dimensions: (1) how many people are poor, (2) how far below the poverty line people's incomes fall, and (3) how long they have been poor (is the poverty temporary or long term?) (World Bank, 2001).

The Gini Coefficient and Global Quality of Life Issues

The World Bank uses as its measure of income inequality what is known as the *Gini coefficient*, which ranges from zero (meaning that everyone has the same income) to 100 (one person receives all the income). Using this measure, the World Bank (2001) has concluded that inequality has increased in nations such as Bulgaria, the Baltic countries, and the Slavic countries of the former Soviet Union to levels similar to those in the less-equal industrial market economies, such as the United States. In fact, inequality in Russia rose sharply in the 1990s: By 1998, the top 20 percent of the population received a much larger percentage of the total income than it had in 1993. Stark contrasts also exist in countries such as India, where abject poverty still exists side by side with lavish opulence in Calcutta. Note the sharp contrast in lifestyle on one Calcutta street:

> On one side is the Tollygunge Club, 40 hectares of landscaped serenity with an 18-hole golf course, a driving range, riding stables, tennis courts, and two covered swimming pools. . . . On the other side stands the M.R. Bangur Hospital, a sooty building with a morgue [that sometimes takes] the corpses of paupers who die in Calcutta's streets. (Watson, 1997: F1)

This street is symbolic of the sharp chasm that divides the 11 million people who live in Calcutta. In fact, some analysts believe that the scale of poverty in south and east Asia is most visible in the heavily populated states of India and China. It is estimated that 652 million of Asia's people are poor and that 448 million of them live in India alone, where there are more than twice as many poor people as in sub-Saharan Africa (Watson, 1997).

But similar disparities between the rich and the poor can be seen in nations throughout the world. For example, in Montrouis, Haiti, Club Med runs a resort featuring pristine beaches, an Olympic-size swimming pool, and all the amenities that affluent tourists expect from a luxury resort. However, just a short walk away from Club Med, open markets have raw meat crawling with flies, and homeless, malnourished people sleep nearby on the ground. But Club Med staff members and tourists at the resort seldom see the other side of Montrouis: Most never leave the compound except when they are going to and from the airport (Emling, 1997b). In other regions of Haiti as well, starvation and disease are a way of life for most inhabitants. As the poorest nation in the Western Hemisphere, Haiti's ability to feed people has been further reduced by recent droughts. It is estimated that 40 percent of Haitian children are chronically malnourished, and an estimated 80 percent of all Haitians eat fewer than 2,200 calories a day (Emling, 1997a).

✓ Checkpoints

Checkpoint 7.6

14. How is poverty defined on a global basis?
15. What is the Gini coefficient?
16. Why are global quality of life issues important?

Global Poverty and Human Development Issues

As the example of Haiti shows, income disparities are not the only factor that defines poverty and its effect on people. Although the average income per person in lower-income countries has doubled in the past 30 years and for many years economic growth has been seen as the primary way to achieve development in low-income economies, the United Nations since the 1970s has more actively focused on human development as a crucial factor in fighting poverty. In 1990 the United Nations Development Program introduced the Human Development Index (HDI), establishing three new criteria—in addition to GDP—for measuring the level of development in a country: life expectancy, education, and living standards. According to the United Nations, human development is "the process of increasing people's options to lead a long and healthy life, to acquire knowledge, and to find access to the assets needed for a decent standard of living" (qtd. in Pietilä and Vickers, 1994: 45). Table 7.1 compares indicators such as life expectancy, infant mortality rate, proportion of underweight children under age five (a measure of nourishment and health), and adult literacy rate for low-income, middle-income, and high-income countries. The World Bank calculates that 37 percent of the population in low- and middle-income countries (1.6 billion people) lack the essentials of well-being, whereas only 21 percent (900 million people) are "income poor," as defined by the bank's poverty line. Women account for most of the 700 million "income-poor" people (Buvinić, 1997: 40).

Life Expectancy

Although some advances have been made in middle- and low-income countries regarding life expectancy, major problems still exist. On the plus

Table 7.1 Indicators of Human Development

Indicator		Least-Developed Countries	All Developing Countries	Industrialized Countries[1]
Life Expectancy	1960	39	46	69
	1963	51	62	74
Infant Mortality Rate[2]	1960	173	150	
	1963	110	70	13
Underweight Children	1975	51	40	
Under Age 5 (Percent)	1985–95	45	30	4
Daily Calorie Supply	1979	2,060	2,140	3,190
per Capita	1992	2,040	2,520	3,350
Adult Literacy Rate[3]	1970	28	43	
	1993	47	71	98
Real GDP per Capita[4]	1960	561	915	
	1993	894	2,709	15,211

[1]Data for industrialized countries for previous year not available in some categories.
[2]Deaths per thousand live births.
[3]Percent.
[4]Gross domestic product per person.

Source: United Nations Development Programme, 1997.

side, average life expectancy has increased by about a third in the past three decades and is now more than 70 years in 84 countries (United Nations Development Programme, 1999). Although no country has reached a life expectancy for men of 80 years, 19 countries have now reached an expectancy of 80 years or more for women. On a less positive note, the average life expectancy at birth of people in middle-income countries remains about 12 years less than that of people in high-income countries. Moreover, the life expectancy of people in low-income nations is as much as 23 years less than that of people in high-income nations. Especially striking are the differences in life expectancies in high-income economies and low-income economies such as sub-Saharan Africa, where estimated life expectancy has dropped significantly in Uganda (from 48 to 43 for women and from 46 to 43 for men) over the past two decades.

One major cause of shorter life expectancy in low-income nations is the high rate of infant mor-

tality. Note in Table 7.1 that the infant mortality rate (deaths per thousand live births) is more than eight times higher in low-income countries than in high-income countries (United Nations, 1997). Low-income countries typically have higher rates of illness and disease, and they do not have adequate health care facilities. Malnutrition is a common problem among children, many of whom are underweight, stunted, and have anemia — a nutritional deficiency with serious consequences for child mortality. Consider this journalist's description of a child she saw in Haiti:

Like any baby, Wisly Dorvil is easy to love. Unlike others, this 13-month-old is hard to hold.

That's because his 10-pound frame is so fragile that even the most minimal of movements can dislocate his shoulders.

As lifeless as a rag doll, Dorvil is starving. He has large, brown eyes and a feeble smile, but a stomach so tender that he suffers from ongoing bouts of vomiting and diarrhea.

Fortunately, though, Dorvil recently came to the attention of U.S. aid workers. With round-the-clock feeding, he is expected to survive.

Others are not so lucky. (Emling, 1997a: A17)

Among adults and children alike, life expectancies are strongly affected by hunger and malnutrition. It is estimated that people in the United States spend more than $5 billion each year on diet products to lower their calorie consumption, whereas the world's poorest 600 million people suffer from chronic malnutrition, and over 40 million people die each year from hunger-related diseases (Kidron and Segal, 1995). To put this figure in perspective, the number of people worldwide dying from hunger-related diseases is the equivalent of more than 300 jumbo-jet crashes per day with no survivors, and half the passengers children (Kidron and Segal, 1995).

In an effort to reduce poverty, some nations have developed adult literacy programs so that people can gain an education that will help lift them out of poverty. In regions such as Iraqi Kurdistan, women's literacy is a particularly crucial issue.

Health

Health is defined in the Constitution of the World Health Organization as "a state of complete physical, mental and social well-being and not merely the absence of disease or infirmity" (Smyke, 1991: 2). Many people in low-income nations are far from having physical, mental, and social well-being. In fact, about 17 million people die each year from diarrhea, malaria, tuberculosis, and other infectious and parasitic illnesses (Hauchler and Kennedy, 1994). According to the World Health Organization, infectious diseases are far from under control in many nations due to such factors as unsanitary or overcrowded living conditions. Despite the possible eradication of diseases such as poliomyelitis, leprosy, guinea-worm disease, and neonatal tetanus in the near future, at least 30 new diseases—for which there is no treatment or vaccine—have recently emerged.

Some middle-income countries are experiencing rapid growth in degenerative diseases such as cancer and coronary heart disease, and many more deaths are expected from smoking-related diseases. Despite the decrease in tobacco smoking in high-income countries, there has been an increase in per capita consumption of tobacco in low- and middle-income countries, many of which have been targeted for free samples and promotional advertising by U.S. tobacco companies (United Nations Development Programme, 1997).

Education and Literacy

According to the Human Development Report (United Nations Development Programme, 1999), education is fundamental to reducing both individual and national poverty. As a result, school enrollment is used as one measure of human development. Although school enrollment has increased at the primary and secondary levels in about two-thirds of the lower-income regions, enrollment is not always a good measure of educational achievement because many students drop out of school during their elementary school years.

What is literacy, and why is it important for human development? The United Nations Educational, Scientific and Cultural Organization (UNESCO) defines a literate person as "someone who can, with understanding, both read and write a short, simple statement on their everyday life" (United Nations, 1997: 89). Based on this definition, people who can write only with figures, their name, or a memorized phrase are not considered literate.

✓**Checkpoints**

Checkpoint 7.7

17. How does life expectancy differ based on the economic level of societies?
18. What is the relationship between health and education and the various income levels within a society?

Reflections

Reflections 7.2

What is the relationship between global wealth and poverty?

The adult literacy rate in the low-income countries is about half that of the high-income countries, and for women the rate is even lower (United Nations, 1997). Women constitute about two-thirds of those who are illiterate: There are approximately 74 literate women for every 100 literate men (United Nations, 1997). Literacy is crucial for women because it has been closely linked to decreases in fertility, improved child health, and increased earnings potential (Hauchler and Kennedy, 1994).

Persistent Gaps in Human Development

Some middle- and lower-income countries have made progress in certain indicators of human development. The gap between some richer and middle- or lower-income nations has narrowed significantly for life expectancy, adult literacy, and daily calorie supply; however, the overall picture for the world's poorest people remains dismal. The gap between the poorest nations and the middle-income nations has continued to widen. Poverty, food shortages, hunger, and rapidly growing populations are pressing problems for at least 1.3 billion people, most of them women and children living in a state of absolute poverty. Although more women around the globe have paid employment than in the past, more and more women are still finding themselves in poverty because of increases in single-person and single-parent households headed by women and the fact that low-wage work is often the only source of livelihood available to them.

Theories of Global Inequality

Why is the majority of the world's population growing richer while the poorest 20 percent—over 1 billion people—are so poor that they are effectively excluded from even a moderate standard of living? Social scientists have developed a variety of theories that view the causes and consequences of global inequality somewhat differently. We will examine the development approach and modernization theory, dependency theory, and world systems theory.

Development and Modernization Theory

According to some social scientists, global wealth and poverty are linked to the level of industrialization and economic development in a given society. Although the process by which a nation industrializes may vary somewhat, industrialization almost inevitably brings with it a higher standard of living in a nation and some degree of social mobility for individual participants in the society. Specifically, the traditional caste system becomes obsolete as industrialization progresses. Family status, race/ethnicity, and gender are said to become less significant in industrialized nations than in agrarian-based societies. As societies industrialize, they also urbanize as workers locate their residences near factories, offices, and other places of work. Consequently, urban values and folkways overshadow the beliefs and practices of the rural areas. Analysts using a development framework typically view industrialization and economic development as essential steps that nations must go through in order to reduce poverty and increase life chances for their citizens.

Earlier in the chapter, we discussed the post-World War II Marshall Plan, under which massive financial aid was provided to the European nations to help rebuild infrastructure lost in the war. Based on the success of this infusion of cash in bringing about modernization, President Truman and many other politicians and leaders in the business community believed that it should be possible to help so-called underdeveloped nations modernize in the same manner.

The most widely known development theory is *modernization theory*—**a perspective that links global inequality to different levels of economic development and suggests that low-income economies can move to middle- and high-income economies by achieving self-sustained economic growth.** According to modernization theory, the low-income, less-developed nations can improve their standard of living only by a period of intensive economic growth and accompanying changes in people's beliefs, values, and attitudes toward work. As a result of modernization, the values of people in developing countries supposedly become more similar to those of people in high-income nations. The number of hours that people work at their jobs each week is one measure of the extent to which individuals subscribe to the *work ethic*, a core value widely believed to be of great significance in the modernization process.

Perhaps the best-known modernization theory is that of Walt W. Rostow (1971, 1978), who, as an economic advisor to U.S. president John F. Kennedy, was highly instrumental in shaping U.S. foreign policy toward Latin America in the 1960s. To Rostow, one of the largest barriers to development in low-income nations was the traditional cultural values held by people, particularly beliefs that are fatalistic, such as viewing extreme hardship and economic deprivation as inevitable and unavoidable facts of life. In cases of fatalism, people do not see any need to work in order to improve their lot in life: If it is predetermined for them, why bother? Based on modernization theory, poverty can be attributed to people's cultural failings, which are further reinforced by governmental policies interfering with the smooth operation of the economy.

Rostow suggested that all countries go through four stages of economic development, with identical content, regardless of when these nations started the process of industrialization. He compared the stages of economic development to an airplane ride. The first stage is the *traditional stage*, in which very little social change takes place, and people do not think much about changing their current circumstances. According to Rostow, societies in this stage are slow to change because the people hold a fatalistic value system, do not subscribe to the work ethic, and save very little money. The second stage is the *take-off stage*—a period of economic growth accompanied by a growing belief in individualism,

competition, and achievement. During this stage, people start to look toward the future, to save and invest money, and to discard traditional values. According to Rostow's modernization theory, the development of capitalism is essential for the transformation from a traditional, simple society to a modern, complex one. With the financial help and advice of the high-income countries, low-income countries will eventually be able to "fly" and enter the third stage of economic development. (Box 7.2 discusses some of the problems that low-income nations may encounter in completing this step today.) In the third stage, the country moves toward *technological maturity*. At this point, the country would improve its technology, reinvest in new industries, and embrace the beliefs, values, and social institutions of the high-income, developed nations. In the fourth and final stage, the country reaches the phase of *high mass consumption* and a correspondingly high standard of living.

Modernization theory has had both its advocates and its critics. According to proponents of this approach, studies have supported the assertion that economic development occurs more rapidly in a capitalist economy. In fact, the countries that have been most successful in moving from low- to middle-income status typically have been those that are most centrally involved in the global capitalist economy. For example, the nations of East Asia have successfully made the transition from low-income to higher-income economies through factors such as a high rate of savings, an aggressive work ethic among employers and employees, and the fostering of a market economy.

Critics of modernization theory point out that it tends to be Eurocentric in its analysis of low-income countries, which it implicitly labels as backward (see Evans and Stephens, 1988). In particular, modernization theory does not take into account the possibility that all nations do not industrialize in the same manner. In contrast, some analysts have suggested that modernization of low-income

Modernization theory a perspective that links global inequality to different levels of economic development and suggests that low-income economies can move to middle- and high-income economies by achieving self-sustained economic growth.

Box 7.2 Changing Times: Media and Technology
A Wired World?

- I get angry that we had to wait so long. But now the Internet is all that my friends and I talk about. There are so many things in this world.

  —Hisham A. Turkistani, age 25, who lives in Riyadh, Saudi Arabia (qtd. in Jehl, 1999: A4)

- A forty page document can be sent from Madagascar to Côte d'Ivoire by five-day courier for $75, by 30-minute fax for $45 or by two-minute email for less than 20 cents—and the email can go to hundreds of people at no extra cost. The choice is easy, if the choice is there. (United Nations Development Programme, 1999: 58)

The Internet and the creation of the World Wide Web have rapidly transformed communications for many people in some areas of the world. The number of Internet hosts—computers with a direct connection to the Internet—rose from under 100,000 in 1988 to over 40 million in 2000. However, as some analysts have suggested, "The collapse of space, time and borders may be creating a global village, but not everyone can be a citizen. The global, professional elite now faces low borders, but billions of others find borders as high as ever" (United Nations Development Programme, 1999: 3). At a time when

many global businesses and political leaders are emphasizing the importance of global Internet connections, some have forgotten one basic fact: Many people in lower-income, less-industrialized nations do not have toilets, telephones, radios, or television sets, much less computers, modems, and other equipment necessary for becoming "wired." Nor can they afford it.

Why is rapid global communication important? According to United Nations experts, people who live in lower-income, developing nations need greater access to all kinds of information, particularly information related to health care issues. In lower-income nations, many people experience the world's most virulent and infectious diseases, yet they often have the least access to information for preventing or combating illnesses. Likewise, educational opportunities are extremely limited, and access to distance learning might provide people with needed information even though they live in an "information-poor" nation. Communications technology also opens up new opportunities for people to become part of the global economy and the political arena (United Nations Development Programme, 1999).

Who currently has access to the Internet? Although there has been a significant increase in Internet use worldwide, only about 10 percent of the world's inhabitants have such access—but that figure is up from about 3 percent in 1999. Access varies widely by region, educational level, gender, income, age, and rural versus urban location. Some of the most underrepresented areas are in Africa, where less than 1 percent of the population in nations such as Cameroon, Chad, Ethiopia, Ghana, and Nigeria have Internet access.

Is it possible to make global communications truly global? According to United Nations studies, global communications will not be possible for everyone unless a conscious effort is made to increase connectivity. From this viewpoint, access to the global communications network should be free from commercialization and profitability. Instead, the focus should be community and collaboration on the "information superhighway."

Do you believe that more people living in the most densely populated regions of the world will gain access to the Internet? What ideas can you suggest for enhancing global communications for a more diverse group of people?

Louise Gubb/The Image Works

Figure 7.3 Approaches to Studying Global Inequality

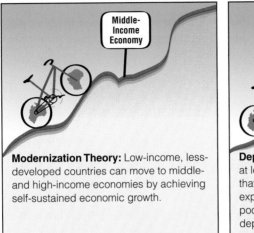

Modernization Theory: Low-income, less-developed countries can move to middle- and high-income economies by achieving self-sustained economic growth.

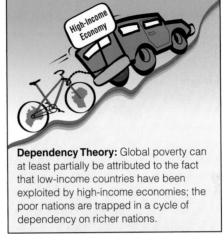

Dependency Theory: Global poverty can at least partially be attributed to the fact that low-income countries have been exploited by high-income economies; the poor nations are trapped in a cycle of dependency on richer nations.

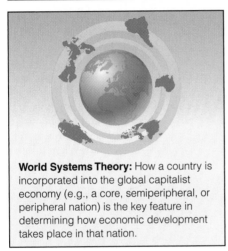

World Systems Theory: How a country is incorporated into the global capitalist economy (e.g., a core, semiperipheral, or peripheral nation) is the key feature in determining how economic development takes place in that nation.

The New International Division of Labor Theory: Commodity production is split into fragments, each of which can be moved (e.g., by a transnational corporation) to whichever part of the world can provide the best combination of capital and labor.

What causes global inequality? Social scientists have developed a variety of explanations, including the four theories shown here.

nations today will require novel policies, sequences, and ideologies that are not accounted for by Rostow's approach (see Gerschenkron, 1962).

Which sociological perspective is most closely associated with the development approach? Modernization theory is based on a market-oriented perspective, which assumes that "pure" capitalism is good and that the best economic outcomes occur when governments follow the policy of laissez-faire (or "hands-off") business, which gives capitalists the opportunity to make the "best" economic decisions, unfettered by government restraints or cumbersome rules and regulations (see Chapter 13, "Politics and the Economy in Global Perspective"). In today's

global economy, however, many analysts believe that national governments are no longer central corporate decision makers and that transnational corporations determine global economic expansion and contraction. Therefore, corporate decisions to relocate manufacturing processes around the world make the rules and regulations of any one nation irrelevant and national boundaries obsolete (Gereffi, 1994). Just as modernization theory most closely approximates a functionalist approach to explaining inequality, dependency theory, world systems theory, and the new international division of labor theory are perspectives rooted in the conflict approach. All four of these approaches are depicted in Figure 7.3.

Dependency Theory

Dependency theory states that global poverty can at least partially be attributed to the fact that the low-income countries have been exploited by the high-income countries. Analyzing events as part of a particular historical process—the expansion of global capitalism—dependency theorists see the greed of the rich countries as a source of increasing impoverishment of the poorer nations and their people. Dependency theory disputes the notion of the development approach, and modernization theory specifically, that economic growth is the key to meeting important human needs in societies. In contrast, the poorer nations are trapped in a cycle of structural dependency on the richer nations due to their need for infusions of foreign capital and external markets for their raw materials, making it impossible for the poorer nations to pursue their own economic and human development agendas. For this reason, dependency theorists believe that countries such as Brazil, Nigeria, India, and Kenya cannot reach the sustained economic growth patterns of the more-advanced capitalist economies.

Dependency theory has been most often applied to the newly industrializing countries (NICs) of Latin America, whereas scholars examining the NICs of East Asia found that dependency theory had little or no relevance to economic growth and development in that part of the world. Therefore, dependency theory had to be expanded to encompass transnational economic linkages that affect developing countries, including foreign aid, foreign trade, foreign direct investment, and foreign loans. On the one hand, in Latin America and sub-Saharan Africa, transnational linkages such as foreign aid, investments by transnational corporations, foreign debt, and export trade have been significant impediments to development within a country. On the other hand, East Asian countries such as Hong Kong, Taiwan, South Korea, and Singapore have historically also had high rates of dependency on foreign aid, foreign trade, and interdependence with transnational corporations but have still experienced high rates of economic growth despite dependency. According to the sociologist Gary Gereffi (1994), differences in outcome are probably associated with differences in the timing and sequencing of a nation's relationship with external entities such as foreign governments and trans-

national corporations. However, in her study of the satellite factory system in Taiwan, the sociologist Ping-Chun Hsiung found that managers in the plants use the norms and behavior patterns of Chinese society to derive high profits for the capitalists. As one manager explained to Hsiung,

> We managers are the mediators between the boss and the workers. We have to communicate [the workers'] opinion to the boss, while at the same time watching out for the boss's pocket. . . . It isn't really a bad thing if the workers are not happy with their wages. It implies that they have the potential to be worth more. How to manipulate them all depends on us, the managers. . . . Chinese are humble. We seldom talk about how good we are, not to mention boast. When things come down to wage conflicts, I always turn the issue around by asking the workers to give me a figure. That is, I ask them to tell me exactly how much more they believe their labor is worth. If they can't come out with a concrete price, then, they have to listen to me. I may decide to give them a raise. I may not. It's all up to me. Even if they do give me a price, the chances are it will always be lower than the real value of their labor. For example, if what they really want is a one-hundred-dollar raise, as Chinese they will only say that their work is worth eighty dollars more, at the most. When this happens, I can really cut it to fifty dollars. . . . By handling them this way, their productivity will go up because I do show them that I did recognize their unrest [and give them a raise]. . . . It is a win–win battle for the company when things come down to wage conflict, you know. (qtd. in Hsiung, 1996: 62)

Dependency theory makes a positive contribution to our understanding of global poverty by noting that "underdevelopment" is not necessarily the cause of inequality. Rather, it points out that exploitation not only of one country by another but of countries by transnational corporations may limit or retard economic growth and human development in some nations. However, what remains unexplained is how East Asia and India had successful "dependency management" whereas many Latin American countries did not (Gereffi, 1994). In fact, over the past decade, the annual economic growth in East Asia (excluding Japan) has averaged 8.5 percent, which is four times the rate of the West (Tanzer, 1996). Currently, the World Bank forecasts an 8-percent annual compound growth rate in the region during the next decade if temporary economic setbacks in the late 1990s and early twenty-first century do not have an averse impact. If we compare this rate with the Industrial Revolution, we find that it took Britain nearly 60 years to double its per capita output and the United States nearly 50 years to do the same. In contrast, East Asia doubles its per capita outcome every decade (Tanzer, 1996). Other conflict explanations have sought to explain global inequality by using the framework of world systems theory and the new international division of labor theory.

World Systems Theory

World systems theory suggests that what exists under capitalism is a truly global system that is held together by economic ties. From this approach, global inequality does not emerge solely as a result of the exploitation of one country by another. Instead, economic domination involves a complex world system in which the industrialized, high-income nations benefit from other nations and exploit their citizens. This theory is most closely associated with the sociologist Immanuel Wallerstein (1979, 1984), who believed that a country's mode of incorporation into the capitalist work economy is the key feature in determining how economic development takes place in that nation. According to world systems theory, the capitalist world economy is a global system divided into a hierarchy of three major types of nations—core, semiperipheral, and peripheral—in which upward or downward mobility is conditioned by the resources and obstacles that characterize the international system. **Core nations are dominant capitalist centers characterized by high levels of industrialization and urbanization.** Core nations such as the United States, Japan, and Germany possess most of the world's capital and technology. Even more importantly for their position of domination, they exert massive control over world trade and

James Marshall/The Image Works

A variety of factors—such as foreign investment and the presence of transnational corporations—have contributed to the economic growth of nations such as Singapore.

economic agreements across national boundaries. Some cities in core nations are referred to as *global cities* because they serve as international centers for political, economic, and cultural concerns. New York, Tokyo, and London are the largest global cities, and they are often referred to as the "command posts" for the world economy (Sassen, 1991, 1995).

Semiperipheral nations are more developed than peripheral nations but less developed than core nations. Nations in this category typically provide labor and raw materials to core nations within the world system. These nations constitute a midpoint between the core and peripheral nations that promotes the stability and legitimacy of

Reflections

Reflections 7.3

How do life chances and opportunities differ for people in core, semiperipheral, and peripheral nations?

Dependency theory the belief that global poverty can at least partially be attributed to the fact that the low-income countries have been exploited by the high-income countries.

Core nations according to world systems theory, dominant capitalist centers characterized by high levels of industrialization and urbanization.

Semiperipheral nations according to world systems theory, nations that are more developed than peripheral nations but less developed than core nations.

Figure 7.4 Maquiladora Plants

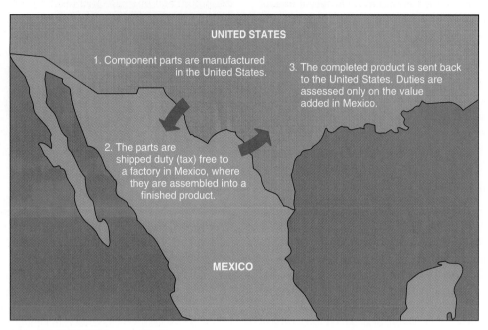

Here is the process by which transnational corporations establish plants in Mexico so that profits can be increased by using low-wage workers there to assemble products that are then brought into the United States for sale.

the three-tiered world economy. These nations include South Korea and Taiwan in East Asia, Mexico and Brazil in Latin America, India in South Asia, and Nigeria and South Africa in Africa. Only two global cities are located in semiperipheral nations: São Paulo, Brazil, which is the center of the Brazilian economy, and Singapore, which is the economic center of a multi-country region in Southeast Asia (Friedmann, 1995). According to Wallerstein, semiperipheral nations exploit peripheral nations, just as the core nations exploit both the semiperipheral and the peripheral.

Most low-income countries in Africa, South America, and the Caribbean are *peripheral nations*—nations that are dependent on core nations for capital, have little or no industrialization (other than what may be brought in by core nations), and have uneven patterns of urbanization. According to Wallerstein (1979, 1984), the wealthy in peripheral nations benefit from the labor of poor workers and from their own economic relations with core-nation capitalists, whom they uphold in order to maintain their own wealth and position. At a

global level, uneven economic growth results from capital investment by core nations; disparity between the rich and the poor within the major cities in these nations is increased in the process. The U.S./Mexican border is an example of disparity and urban growth: Transnational corporations have built *maquiladora* plants so that goods can be assembled by low-wage workers to keep production costs down. Figure 7.4 describes this process. In 2001, there were almost 3,800 maquiladora plants in Mexico, employing 1.3 million people (*Austin American-Statesman*, 2001). Because of a demand for a large supply of low-wage workers, thousands of people have moved from the rural regions of Mexico to urban areas along the border in hope of earning a higher wage. This influx has pushed already overcrowded cities far beyond their capacity. Many people live on the edge of the city in *shantytowns* made from discarded materials or in low-cost rental housing in central-city slums because their wages are low and affordable housing is nonexistent (Flanagan, 1995). In fact, housing shortages are among the most pressing problems in many peripheral nations.

Not all social analysts agree with Wallerstein's perspective on the hierarchical position of nations in the global economy. However, most scholars acknowledge that nations throughout the world are influenced by a relatively small number of cities and transnational corporations that have prompted a shift from an international to a more global economy (see Knox and Taylor, 1995; Wilson, 1997). Even Wallerstein (1991) acknowledges that world systems theory is an "incomplete, unfinished critique" of long-term, large-scale social change that influences global inequality.

The New International Division of Labor Theory

Although the term *world trade* has long implied that there is a division of labor among societies, the nature and extent of this division have recently been reassessed based on the changing nature of the world economy. According to the *new international division of labor theory*, commodity production is being split into fragments that can be assigned to whichever part of the world can provide the most profitable combination of capital and labor. Consequently, the new international division of labor has changed the pattern of geographic specialization between countries, whereby high-income countries have now become dependent on low-income countries for labor. The low-income countries provide transnational corporations with a situation in which they can pay lower wages and taxes, and face fewer regulations regarding workplace conditions and environmental protection (Waters, 1995). Overall, a global manufacturing system has emerged in which transnational corporations establish labor-intensive, assembly-oriented export production, ranging from textiles and clothing to technologically sophisticated exports such as computers, in middle- and lower-income nations (Gereffi, 1994). At the same time, manufacturing technologies are shifting from the large-scale, mass-production assembly lines of the past toward a more flexible production process involving microelectronic technologies. Even service industries—such as processing insurance claims forms—that were formerly thought to be less mobile have become exportable through electronic transmission and the Internet. The global nature of these activities has been referred to as *global commodity*

chains, a complex pattern of international labor and production processes that results in a finished commodity ready for sale in the marketplace.

Some commodity chains are producer-driven, whereas others are buyer-driven. *Producer-driven commodity chains* is the term used to describe industries in which transnational corporations play a central part in controlling the production process. Industries that produce automobiles, computers, and other capital- and technology-intensive products are typically producer-driven. In contrast, *buyer-driven commodity chains* is the term used to refer to industries in which large retailers, brand-name merchandisers, and trading companies set up decentralized production networks in various middle- and low-income countries. This type of chain is most common in labor-intensive, consumer-goods industries such as toys, garments, and footwear (Gereffi, 1994). Athletic footwear companies such as Nike and Reebok and clothing companies like The Gap and Liz Claiborne are examples of the buyer-driven model. Since these products tend to be labor intensive at the manufacturing stage, the typical factory system is very competitive and globally decentralized. Workers in buyer-driven commodity chains are often exploited by low wages, long hours, and poor working conditions. In fact, most workers cannot afford the products they make. Tini Heyun Alwi, who works on the assembly line of the shoe factory in Indonesia that makes Reebok sneakers, is an example: "I think maybe I could work for a month and still not be able to buy one pair" (qtd. in Goodman, 1996: F1). Since Tini earns only 2,600 Indonesian rupiah ($1.28) per day working a ten-hour shift six days a week, her monthly income would fall short of the retail price of the athletic shoes (Goodman, 1996). Sociologist Gary Gereffi (1994: 225) explains the problem with studying the new global patterns as follows:

> The difficulty may lie in the fact that today we face a situation where (1) the political unit is *national*, (2) industrial production is *regional*, and (3) capital movements are *international*. The rise of Japan

Peripheral nations   according to world systems theory, nations that are dependent on core nations for capital, have little or no industrialization (other than what may be brought in by core nations), and have uneven patterns of urbanization.

✓ Checkpoints

Checkpoint 7.8

19. What are the four stages of economic development according to modernization theory?
20. What is dependency theory, and when is it most useful in explaining inequality?
21. How do core nations differ from semiperipheral and peripheral nations?
22. What is the new international division of labor theory?
23. How does the global economy affect our everyday lives?

and the East Asian [newly industrializing countries] in the 1960s and 1970s is the flip side of the "deindustrialization" that occurred in the United States and much of Europe. Declining industries in North America have been the growth industries in East Asia.

As other analysts suggest, such changes have been a mixed bag for people residing in these countries. For example, Indonesia has been able to woo foreign business into the country, but workers have experienced poverty despite working full time in factories making such consumer goods as Nike tennis shoes (Gargan, 1996). As employers feel pressure from workers to raise wages, clashes erupt between the workers and managers or owners. Similarly, the governments in these countries fear that rising wages and labor strife will drive away the businesses, sometimes leaving behind workers who have no other hopes for employment and become more impoverished than they previously were. Moreover, in situations where government officials are benefiting from the presence of the companies, they also become losers if the workers rebel against their pay or working conditions. What will be the future of global inequality given this current set of conditions in countries such as Indonesia?

Global Inequality in the Future

As we have seen, social inequality is vast both within and among the countries of the world. Even in high-income nations where wealth is highly concentrated, many poor people coexist with the affluent. In middle- and low-income countries, there are small pockets of wealth in the midst of poverty and despair.

What are the future prospects for greater equality across and within nations? Not all social scientists agree on the answer to this question. Depending on the theoretical framework they apply in studying global inequality, social analysts may describe either an optimistic or a pessimistic scenario for the future. Moreover, some analysts highlight the human rights issues embedded in global inequality, whereas others focus primarily on an economic framework.

In some regions, persistent and growing poverty continues to undermine human development and future possibilities for socioeconomic change. Gross inequality has high financial and quality-of-life costs for people, even among those who are not the poorest of the poor. In the future, continued population growth, urbanization, environmental degradation, and violent conflict threaten even the meager living conditions of those residing in low-income nations. From this approach, the future looks dim not only for people in low-income and middle-income countries but also for those in high-income countries, who will see their quality of life diminish as natural resources are depleted, the environment is polluted, and high rates of immigration and global political unrest threaten the high standard of living that many people have previously enjoyed. According to some social analysts, transnational corporations and financial institutions such as the World Bank and the International Monetary Fund will further solidify and control a globalized economy, which will transfer the power to make significant choices to these organizations and away from the people and their governments. As a result, further loss of resources and means of livelihood will affect people and countries around the globe.

Because of global corporate domination, there could be a leveling out of average income around the world, with wages falling in the high-income countries and wages increasing significantly in low- and middle-income countries. If this pessimistic scenario occurs, there is likely to be greater polarization of the rich and the poor and more potential for ethnic and national conflicts over such issues as worsening environmental degradation and who has the right to natural resources. For example, pulp-and-paper companies in Indonesia, along with palm-oil-plantation owners, have continued clearing land for crops by burning off vast tracts of jungle, producing high levels of smog and pollution across seven Southeast Asian nations and creating havoc for millions of people (Mydans, 1997). Whose rights should prevail in situations such as this? The smoke from the Indonesian fires has affected agriculture; food shortages and rising prices are inevitable. Reduced sunlight is slowing the growth of crops, and people in record numbers are suffering from respiratory illnesses. The possibility remains that severe, long-term illnesses may also develop over time, especially among the young, the old, and people with respiratory problems (Mydans, 1997). This could be a future scenario if "business as usual" continues to take place around the world.

On the other hand, a more optimistic scenario is also possible. With modern technology and worldwide economic growth, it might be possible to reduce absolute poverty and to increase people's opportunities. Among the trends cited by the Human Development Report (United Nations Development Programme, 1996) that have the potential to bring about more sustainable patterns of development are the socioeconomic progress made in many low- and middle-income countries over the past thirty years as technological, social, and environmental improvements have occurred. For example, technological innovation continues to improve living standards for some people. Fertility rates are declining in some regions (but remain high in others, where there remains grave cause for concern about the availability of adequate natural resources for the future). Finally, health and education may continue to improve in lower-income countries. According to the Human Development Report (United Nations Development Programme, 1996), healthy, educated populations are crucial for the future in order to reduce global poverty. The education of women is of primary importance in the future if global inequality is to be reduced. As one analyst stated, "If you educate a boy, you educate a human being. If you educate a girl, you educate generations" (Buvinić, 1997: 49). All aspects of schooling and training are crucial for the future, including agricultural extension services in rural areas to help women farmers in regions such as western Kenya produce more crops to feed their families. As we saw at the beginning of this chapter, easier access to water can make a crucial difference in people's lives. Mayra Buvinić, of the Inter-American Development Bank, puts global poverty in perspec-

tive for people living in high-income countries by pointing out that their problems are our problems. She provides the following example:

> Reina is a former guerilla fighter in El Salvador who is being taught how to bake bread under a post-civil war reconstruction program. But as she says, "the only thing I have is this training and I don't want to be a baker. I have other dreams for my life."
>
> Once upon a time, women like Reina . . . only migrated [to the United States] to follow or find a husband. This is no longer the case. It is likely that Reina, with few opportunities in her own country, will sooner or later join the rising number of female migrants who leave families and children behind to seek better paying work in the United States and other industrial countries. Wisely spent foreign aid can give Reina the chance to realize her dreams in her *own* country. (Buvinić, 1997: 38, 52)

From this viewpoint, we can enjoy prosperity only by ensuring that other people have the opportunity to survive and thrive in their own surroundings (see Box 7.3). The problems associated with global poverty are therefore of interest to a wide-ranging set of countries and people.

We will continue to focus in subsequent chapters on issues pertaining to inequality as we examine social class in the United States, including topics such as why inequalities persist and how the distribution of income and wealth in our nation produces differential access to goods and services, health and nutrition, and educational opportunities.

Chapter Review

What is stratification, and how does it affect our daily life?

Stratification is the hierarchical arrangement of large social groups based on their control over basic resources. People are treated differently based on where they are positioned within the social hierarchies of class, race, gender, and age.

What are the major systems of stratification?

Stratification systems include slavery, caste, and class. Slavery, an extreme form of stratification in which people are owned by others, is a closed system. The caste system is also a closed one in which people's status is determined at birth, based on their parents' position in society. The class system, which exists in the United States, is a type of stratification based on ownership of resources and on the type of work that people do. Class systems are characterized by unequal distribution of

Box 7.3 You Can Make a Difference
Global Networking to Reduce World Hunger and Poverty

We, the people of the world, will mobilize the forces of transnational civil society behind a widely shared agenda that binds our many social movements in pursuit of just, sustainable, and participatory human societies. In so doing we are forging our own instruments and processes for redefining the nature and meaning of human progress and for transforming those institutions that no longer respond to our needs. We welcome to our cause all people who share our commitment to peaceful and democratic change in the interest of our living planet and the human societies it sustains.
—International NGO Forum, United Nations Conference on Environment and Development, Rio de Janeiro, Brazil, June 12, 1992 (qtd. in Korten, 1996: 333)

If everyone lit just one little candle, what a bright world this would be.
—line from the 1950s theme song for Bishop Fulton J. Sheen's television series *Life Is Worth Living* (Sheen, 1995: 245)

When many of us think about problems such as world poverty, we tend to see ourselves as powerless to bring about change in so vast an issue. However, a recurring message from social activists and religious leaders is that each person can contribute something to the betterment of other people and sometimes the entire world.

An initial way for each of us to become involved is to become better informed about global issues and to learn how we can contribute time and resources to organizations seeking to address social issues such as illiteracy and hunger. We can also find out about meetings and activities of organizations and participate in online discussion forums where we can express our opinions, ask questions, share information, and interact with other people interested in topics such as international relief and development. At first, it may not feel like you are doing much to address global problems; however, information and education are the first steps to promoting greater understanding of social problems and of the world's people, whether they reside in high-, middle-, or low-income countries and regardless of their individual socioeconomic position. Likewise, it is important to help our own nation's children understand that they can make a difference in ending hunger in the United States and other nations.

Would you like to function as a catalyst for change? You can learn how to proceed by gathering information from organizations that seek to reduce problems such as poverty and to provide forums for interacting with other people. Here are a few starting points for your information search:

- CARE International is a confederation of ten national members in North America, Europe, Japan, and Australia. CARE assists the world's poor in their efforts to achieve social and economic well-being. Its work reaches 25 million people in 53 nations in Africa, Asia, Latin America, and Eastern Europe. Programs include emergency relief, education, health and population, children's health, reproductive health, water and sanitation, small economic activity development, agriculture, community development, and environment. Contact CARE at 151 Ellis Street, NE, Atlanta, GA 30303-2439. On the Internet:

http://www.care.org

Other organizations fighting world hunger and health problems include the following:

- World Hunger Year ("WHY"):

http://www.worldhungeryear.org

- "Kids Can Make a Difference," an innovative guide developed by WHY:

http://www.kids.maine/org/cando.htm

- World Health Organization:

http://www.who.org

---

resources and by movement up and down the class structure through social mobility.

What is global stratification, and how does it contribute to economic inequality?

Global stratification refers to the unequal distribution of wealth, power, and prestige on a global basis, which re-sults in people having vastly different lifestyles and life chances both within and among the nations of the world. Today, the income gap between the richest and the poorest 20 percent of the world population continues to widen, and within some nations the poorest one-fifth of the population has an income that is only a slight fraction of the overall average per capita income for that country.

How are global poverty and human development related?
Income disparities are not the only factor that defines poverty and its effect on people. The United Nations' Human Development Index measures the level of development in a country through indicators such as life expectancy, infant mortality rate, proportion of underweight children under age five (a measure of nourishment and health), and adult literacy rate for low-income, middle-income, and high-income countries.

What is modernization theory, and what stages did Rostow believe all societies go through?
Modernization theory is a perspective that links global inequality to different levels of economic development and suggests that low-income economies can move to middle- and high-income economies by achieving self-sustained economic growth. According to Rostow, all countries go through four stages of economic development: (1) the traditional stage, in which very little social change takes place; (2) the take-off stage, a period of economic growth accompanied by a growing belief in individualism, competition, and achievement; (3) technological maturity, a period of improving technology, reinvesting in new industries, and embracing the beliefs, values, and social institutions of the high-income, developed nations; and (4) the phase of high mass consumption, which is accompanied by a high standard of living.

How does dependency theory differ from modernization theory?
Dependency theory states that global poverty can at least partially be attributed to the fact that the low-income countries have been exploited by the high-income countries. Whereas modernization theory focuses on how societies can reduce inequality through industrialization and economic development, dependency theorists see the greed of the rich countries as a source of increasing impoverishment of the poorer nations and their people.

What is world systems theory, and how does it view the global economy?
According to world systems theory, the capitalist world economy is a global system divided into a hierarchy of three major types of nations: Core nations are dominant capitalist centers characterized by high levels of industrialization and urbanization, semiperipheral nations are more developed than peripheral nations but less developed than core nations, and peripheral nations are those countries that are dependent on core nations for capital, have little or no industrialization (other than what may be brought in by core nations), and have uneven patterns of urbanization.

What is the new international division of labor theory?
The new international division of labor theory is based on the assumption that commodity production is split into fragments that can be assigned to whichever part of the world can provide the most profitable combination of capital and labor. This division of labor has changed the pattern of geographic specialization between countries, whereby high-income countries have become dependent on low-income countries for labor. The low-income countries provide transnational corporations with a situation in which they can pay lower wages and taxes, and face fewer regulations regarding workplace conditions and environmental protection.

Key Terms

caste system 216
class system 217
core nations 235
dependency theory 234
intergenerational mobility 213
intragenerational mobility 213
life chances 210
modernization theory 231
peripheral nations 236
semiperipheral nations 235
slavery 214
social mobility 213
social stratification 210

Questions for Critical Thinking

1. You have decided to study global wealth and poverty. How would you approach your study? What research methods would provide the best data for analysis? What might you find if you compared your research data with popular presentations—such as films and advertising—of everyday life in low- and middle-income countries?

2. How would you compare the lives of poor people living in the low-income nations of the world with those in central cities and rural areas of the United States? In what ways are their lives similar? In what ways are they different?

3. Should U.S. foreign policy include provisions for reducing poverty in other nations of the world? Should U.S. domestic policy include provisions for reducing poverty in the United States? How are these issues similar? How are they different?

4. Using the theories discussed in this chapter, devise a plan to alleviate global poverty. Assume that you

have the necessary wealth, political power, and other resources necessary to reduce the problem. Share your plan with others in your class, and create a consolidated plan that represents the best ideas and suggestions presented.

 **Resources on the Internet**

Chapter-Related Web Sites
The following web sites have been selected for their relevance to the topics in this chapter. These sites are among the more stable, but please note that web site addresses change frequently.

Causes of Poverty
http://www.globalissues.org/TradeRelated/Poverty.asp
This site explores the relationship among inequality, poverty, and conflict from a global perspective.

Global Stratification
http://www.sdsmt.edu/online-courses/is/soc100/Glob_Strat.htm
This page links you to a variety of resources and data that explore global poverty.

Global Poverty
http://www.mtholyoke.edu/acad/intrel/poverty.htm
With links to dozens of sites related to data on global poverty, this site is a great resource for writing papers on inequality and global poverty.

Global Stratification and Conflict
http://www.soc.ucsb.edu/honors/strat/GlobalConflict.html
The links on this page relate global poverty to a number of other social categories, such as religion and race.

The World Bank Group: Global Poverty Monitoring
http://www.worldbank.org/research/povmonitor/
The World Bank sponsors this site, which keeps tabs on global poverty and can link the student to information from various regions across the globe.

Moreover Showcase: Global Poverty News Headlines
http://www.moreover.com/cgi-local/page?o=portal&feed=264
This site displays the most recent news articles about global poverty from around the world.

The United States in Global Context
http://www.census.gov/prod/2002pubs/c2kbr01-11.pdf
This is the Census Bureau's detailed report on the U.S. economy in global perspective.

Companion Web Site for This Book
Virtual Society: The Wadsworth Sociology Resource Center
Visit http://sociology.wadsworth.com and click on the page for Kendall, *Sociology in Our Times: The Essentials*, Fourth Edition, to access a wide range of enrichment material to aid in your study of sociology. Click on the Student Resources section of the web site. Next, select from the pull-down menu the chapter that you are presently studying. Among the useful options for self-study are chapter objectives, flashcards, practical study tips, and practice tests for each chapter.

MicroCase Online
From the Virtual Society home page, click on MicroCase Online to access book-specific MicroCase exercises that allow you to further explore sociological issues and principles presented in the text.

InfoTrac College Edition
Another unique option available to you at the Student Resources section of the companion web site is InfoTrac College Edition, an online library with access to hundreds of scholarly and popular periodicals. Below are suggested search terms for this chapter. Results from these and other searches are found at the site.

- Search keywords: *world poverty*. World poverty has been a persistent problem. Conduct a search to find articles that focus on the social consequences of this topic.
- Search keywords: *human rights violations*. Look for articles that discuss human rights violations in such countries as Russia, China, Saudi Arabia, and Sierra Leone.

Virtual Explorations CD-ROM
Go to the Virtual Explorations CD-ROM to begin the interactive exercise for this chapter. This Virtual Exploration will introduce you to some of the exciting resources for sociology on the World Wide Web. You will be guided through an exercise that employs related web sites on global stratification. Answer the questions, and e-mail your responses to your instructor.

Paul Anderson/SIS

Social Class in the United States

**Income and Wealth Differences
  in the United States**
Comparing Income and Wealth
Distribution of Income and Wealth

Classical Perspectives on Social Class
Karl Marx: Relationship to the Means of Production
Max Weber: Wealth, Prestige, and Power

**Contemporary Sociological Models
  of the U.S. Class Structure**
The Weberian Model of the U.S. Class Structure
The Marxian Model of the U.S. Class Structure

Consequences of Inequality
Physical Health, Mental Health, and Nutrition
Housing
Education

Poverty in the United States
Who Are the Poor?
Economic and Structural Sources of Poverty
Solving the Poverty Problem

**Sociological Explanations of Social Inequality
  in the United States**
Functionalist Perspectives
Conflict Perspectives
Symbolic Interactionist Perspectives

U.S. Stratification in the Future

243

As a prep cook in a Mexican restaurant, I earned minimum wage. I learned to appreciate jobs that don't require long hours on your feet or work in front of a hot oven. My sister Victoria—who is my business partner now—and I walked to work and would come home smelling like tacos. As you might guess, this attracted half the dogs in the neighborhood. Fortunately, the restaurant was near our home. (qtd. in Russell, 1997: 100)

—Annette Quintana, who now owns one of the top 500 "Hispanic" businesses in the United States, describing how grateful she was at one time to have any job that produced a paycheck

[A]t the age of eight, I dreamed that I would one day own the best restaurant in the world. All of the

customers would love my food, and all of my employees would do everything they were supposed to do. But, most important, everyone would think that I was a good boss, and every day when I walked into the restaurant, people would be glad to see me. . . . Today, people seem glad to see me in about four thousand Wendy's restaurants [, and I have] the biggest single stake in one of America's largest restaurant chains, making me a millionaire many times over. I never expected it to be that way.

—the late Dave Thomas (1992: 3), founder of Wendy's International, describing how he dreamed, even as a young child living in a low-income, adoptive family, that he would be able to achieve the American Dream

As the statements of Annette and Dave suggest, the United States is viewed by many as the land of opportunity—of the "American Dream." What is the American Dream? Simply stated, the American Dream is the belief that if people work hard and play by the rules, they will have a chance to get ahead (see Hochschild, 1995). Moreover, each generation will be able to have a higher standard of living than that of its parents (Danziger and Gottschalk, 1995). The American Dream is based on the assumption that people in the United States have equality of opportunity regardless of their race, creed, color, national origin, gender, or religion.

For middle- and upper-income people, the American Dream typically means that each subsequent generation will be able to acquire more material possessions and wealth than people in the preceding generations. To some people, achieving the American Dream means having a secure job, owning a home, and getting a good education for their children. To others, it is the promise that anyone may rise from poverty to wealth (from "rags to

riches") if he or she works hard enough. In this chapter, we examine the U.S. class structure to see how people's opportunities are affected by their position in that structure. However, before we explore class and stratification in the United States, test your knowledge of wealth, poverty, and the American Dream by taking the quiz in Box 8.1.

Questions and Issues

Chapter Focus Question: How is the American Dream influenced by social stratification?

How do prestige, power, and wealth determine social class?

What role does occupational structure play in a functionalist perspective on class structure?

What role does ownership of resources play in a conflict perspective on class structure?

How are social stratification and poverty linked?

Income and Wealth Differences in the United States

Throughout human history, people have argued about the distribution of scarce resources in society. Disagreements often concentrate on whether the share we get is a fair reward for our effort and hard work (Braun, 1991). Recently, social analysts have pointed out that the old maxim "the rich get

richer" continues to be valid in the United States. To understand how this happens, we must take a closer look at income and wealth inequality in this country.

Comparing Income and Wealth

When many people discuss financial well-being—and, thus, a person's place in the U.S. class structure—they are usually referring to income. However,

Box 8.1 Sociology and Everyday Life

How Much Do You Know About Wealth, Poverty, and the American Dream?

True	False		
T	F	1.	People no longer believe in the American Dream.
T	F	2.	Individuals over age sixty-five have the highest rate of poverty.
T	F	3.	Men account for two out of three impoverished adults in the United States.
T	F	4.	About one in twenty U.S. people lives in a household whose members sometimes do not get enough to eat.
T	F	5.	With the economic slowdown of 2000 and the terrorist attacks on the United States in 2001, the wealthiest people in this country lost about 50 percent of their fortunes.
T	F	6.	Income is more unevenly distributed than wealth.
T	F	7.	People who are poor usually have personal attributes that contribute to their impoverishment.
T	F	8.	A number of people living below the official poverty line have full-time jobs.
T	F	9.	Federal welfare law (Temporary Assistance for Needy Families) does not require recipients with young children to find work.
T	F	10.	One in three U.S. children will be poor at some point of their childhood.

Answers on page 246.

it is important to distinguish between income and wealth. *Income* **is the economic gain derived from wages, salaries, income transfers (governmental aid), and ownership of property** (Beeghley, 1996). Or, to put it another way, "income refers to money, wages, and payments that periodically are received as returns from an occupation or investment" (Kerbo, 2000: 19). For most of us living in a class system, our major sources of income are a wage or salary, although some people live entirely off the earnings on their investments. But income is only one aspect of wealth. *Wealth* **is the value of all of a person's or family's economic assets, including income, personal property, and income-producing property.** As this definition suggests, wealth refers to accumulated assets in the form of various types of valued goods, including property such as buildings, land, farms, houses, factories, and cars, as well as other assets such as bank accounts, corporate stocks, bonds, and insur-

ance policies. Wealth can be used to generate income that is used to purchase necessities (such as food, clothing, and shelter) or luxuries (such as a diamond ring or a yacht). It has been said that "wealth generates more wealth" (Keister, 2000: 7), meaning that wealthy people may acquire even more wealth as a result of their investments. Wealth makes it possible for people to "buy leisure," in that the owner of wealth can decide whether he or she will work or not. Similarly, wealth can be used to help people gain an

Income the economic gain derived from wages, salaries, income transfers (governmental aid), and ownership of property.

Wealth the value of all of a person's or family's economic assets, including income, personal property, and income-producing property.

Answers to the Sociology Quiz on Wealth, Poverty, and the American Dream

1. **False.** The American Dream appears to be alive and well. U.S. culture places a strong emphasis on the goal of monetary success, and many people use legal or illegal means to attempt to achieve that goal.

2. **False.** As a group, children have a higher rate of poverty than the elderly. Government programs such as Social Security have been indexed for inflation whereas many of the programs for the young have been scaled back or eliminated. However, many elderly individuals still live in poverty.

3. **False.** Women, not men, account for two out of three impoverished adults in the United States. Reasons include the lack of job opportunities for women, lower pay than men for comparable jobs, lack of affordable day care for children, sexism in the workplace, and a number of other factors.

4. **True.** It is estimated that about 5 percent of the U.S. population (one in twenty people) resides in household units where members do not get enough to eat.

5. **False.** As calculated by *Forbes* magazine, which keeps detailed accounts of the wealthiest people's assets, the 400 wealthiest United States citizens lost about 20 percent of their fortunes. For example, Bill Gates, the wealthiest person in the nation, saw his net worth decline about 14 percent, to $54 billion.

6. **False.** Wealth is more unevenly distributed among the U.S. population than is income. However, both wealth and income are concentrated in very few hands compared with the size of the overall population.

7. **False.** According to one widely held stereotype, the poor are lazy and do not want to work. Rather than looking at structural characteristics of the society, people cite the alleged personal attributes of the poor as the reason for their plight.

8. **True.** Many of those who fall below the official poverty line are referred to as the "working poor" because they work full time but earn such low wages that they are still considered to be impoverished.

9. **False.** The latest federal welfare guidelines require recipients with young children to find work. This change has resulted in an increased demand for both child-care services and jobs for many women with little education or job training. In some states, welfare recipients have been employed by child-care centers that offer them on-site training in child care.

10. **True.** According to recent data from the Children's Defense Fund, one in three U.S. children will live in a family that is below the official poverty line at some point in their childhood. For some of the children, poverty will be a persistent problem throughout their childhood and youth.

Sources: Based on Children's Defense Fund, 2001; *Forbes*, 2001; Gilbert, 2003; and U.S. Census Bureau, 2001e.

William Strode/Woodfin, Camp & Associates

Tim Carlson/Stock Boston

Wealth and poverty influence all aspects of our daily lives, from where we live to how we think about opportunities for social mobility. Children in wealthy families typically have many opportunities that children in poverty-level families do not. What are the long-term consequences of such inequality for individuals and society?

advantage that they otherwise would not have. Such advantages include, but are not limited to, gaining high social prestige, political influence, improved opportunities and greater safety for oneself and one's family, high-quality health care, and enhanced life chances (Keister, 2000).

The idea of the American Dream is linked to both wealth and income, and this dream is therefore threatened in periods of economic problems and political unrest, particularly for those persons who must rely on income alone for their family's economic survival. Even in "good" economic times, however, wealth is much more unequally distributed than is income, and we now turn to that topic.

Distribution of Income and Wealth

Money—in the form of both income and wealth—is very unevenly distributed in the United States. Money is essential for acquiring goods and services. People without money cannot purchase food, shelter, clothing, medical care, legal aid, education, and the other things they need or desire. Median household income varies widely from one state to another, for example (see Map 8.1). Among the prosperous

AP/Wide World Photos

Factory workers are shown here while leaving for the day after learning that their plant will soon be closed and that they will soon be unemployed. How is the loss of hundreds of thousands of jobs in the United States due to changing economic conditions affecting people's opportunities for achieving the American Dream?

Map 8.1 Median Income by State

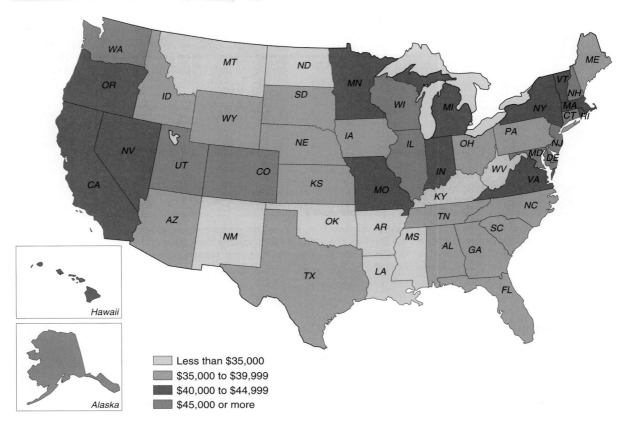

Less than $35,000
$35,000 to $39,999
$40,000 to $44,999
$45,000 or more

What factors contribute to the uneven distribution of income in the United States?

Source: U.S. Census Bureau, 2000b.

nations, the United States is number one in inequality of income distribution (Rothchild, 1995).

INCOME INEQUALITY In regard to income inequality in the United States, the economist Paul Samuelson stated that "If we made an income pyramid out of a child's blocks, with each layer portraying $500 of income, the peak would be far higher than Mount Everest, but most people would be within a few feet of the ground" (qtd. in Samuelson and Nordhaus, 1989: 644).

Similarly, sociologist Dennis Gilbert (2003) compares the distribution of income to a national pie that has been sliced into portions, ranging in size from stingy to generous, for distribution among segments of the population. As shown in Figure 8.1, in 1999 the wealthiest 20 percent of households received almost 50 percent of the total income "pie" while the poorest 20 percent of households received less than 4 percent of all income. The top 5 percent *alone* received more than 20 percent of all income—an amount greater than that received by the bottom 40 percent of all households (U.S. Census Bureau, 2001g).

Reflections

Reflections 8.1

Why are both wealth and income important to the idea of attaining the American Dream?

Figure 8.1 Distribution of Pretax Income in the United States

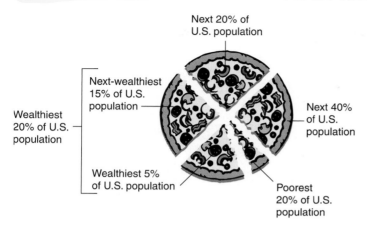

Next 20% of U.S. population

Next-wealthiest 15% of U.S. population

Wealthiest 20% of U.S. population

Next 40% of U.S. population

Wealthiest 5% of U.S. population

Poorest 20% of U.S. population

Thinking of personal income in the United States (before taxes) as a large pizza helps us to see which segments of the population receive the largest and smallest portions. What part do taxes play in redistributing parts of the pizza?

Source: U.S. Census Bureau, 2000c.

In the last two decades of the twentieth century, the gulf between the rich and the poor widened in the United States. Since the early 1990s, the poor have been more likely to stay poor, and the affluent have been more likely to stay affluent. Between 1979 and 1999, the income of the top one-fifth of U.S. families increased by 36 percent; during that same period of time, the income of the bottom one-fifth of families fell from $13,504 to $13,320 in 1999 dollars (Children's Defense Fund, 2001) (see Figure 8.2).

Income distribution varies by race/ethnicity as well as class. Figure 8.3 compares median income by race/ethnicity, showing not only the disparity among groups, but also the consistency of that disparity. Over half of African American (61.5 percent) and Latino/a (63.3 percent) households fall within the lowest two income categories; slightly over one-third (37.8 percent) of whites are in these categories (U.S. Census Bureau, 2001g). In all categories, differences in the median income of married couples and female-headed households are striking.

WEALTH INEQUALITY To see how wealth inequality has increased in recent decades, let's compare two studies. A study by the Joint Economic Committee of Congress divided the population into four categories: (1) the super-rich (0.5 percent

Figure 8.2 Average After-Tax Family Income in the United States

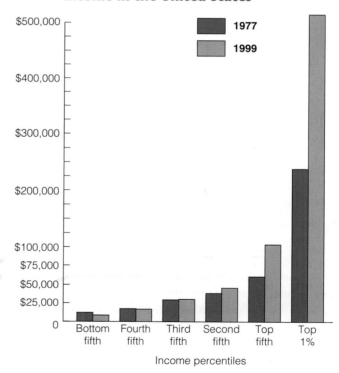

Income percentiles

As contrasted with Figure 8.1, this chart shows the distribution of after-tax family income in the United States. Notice the dramatic increase in income for the top 1 percent of U.S. families. During the past decade, the difference in income between the richest and poorest people in this nation has become even more pronounced.

Source: Johnston, 1999.

Figure 8.3 Median Household Income by Race/Ethnicity in the United States

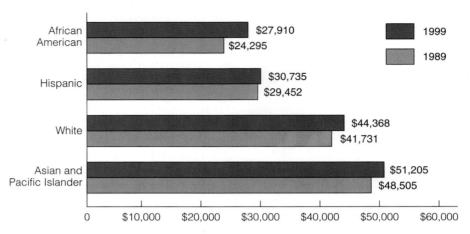

Note: Amounts shown in constant dollars.

Source: U.S. Census Bureau, 2000b.

of households), who own 35 percent of the nation's wealth, with net assets averaging almost $9 million; (2) the very rich (the next 0.5 percent of households), who own about 7 percent of the nation's wealth, with net assets ranging from $1.4 million to $2.5 million; (3) the rich (9 percent of households), who own 30 percent of the wealth, with net assets of a little over $400,000; and (4) everybody else (the bottom 90 percent), who own about 28 percent of the nation's wealth. However, by 1995, another study indicated that the holdings of super-rich households had risen from 35 percent to almost 40 percent of all assets in the nation (stocks, bonds, cash, life insurance policies, paintings, jewelry, and other tangible assets) (Rothchild, 1995).

For the upper class, wealth often comes from inheritance. According to the sociologist Michael Patrick Allen (1987), many of the wealthiest people in the United States have inherited wealth, and inheritors often are three or four generations removed from individuals who amassed the original fortune (Odendahl, 1990). After inheriting a fortune, John D. Rockefeller, Jr., stated that "I was born into [wealth] and there was nothing I could do about it. It was there, like air or food or any other element. The only question with wealth is what to do with it" (qtd. in Glastris, 1990: 26).

Disparities in wealth are more pronounced when compared across racial and ethnic categories. For example, one survey of wealth found that the net worth of the average white household was twelve times that of the average African American household. Whereas 44 percent of whites had assets of $50,000 or more, only 13 percent of African Americans were in this category (Feagin and Sikes, 1994). Married couples have a higher net worth than the unmarried, and households headed by people age 55 and older are wealthier than those headed by younger persons (U.S. Census Bureau, 2001g).

Classical Perspectives on Social Class

The issue of economic inequality is not a new concern. For many years, social scientists have developed theories to explain social class differences

✓**Checkpoints**

Checkpoint 8.1

1. How does income differ from wealth?
2. How equal is income distribution in the United States?
3. What are the key factors that produce inequalities in wealth in the United States?

Figure 8.4 Marx's View of Stratification

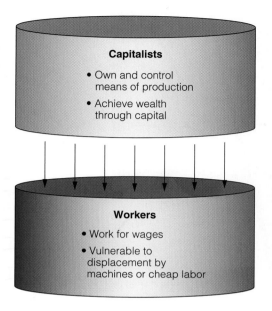

and how these differences are related to people's resources and opportunities. Early social thinkers such as Karl Marx and Max Weber identified class as an important determinant of social inequality and social change, and their works have had a profound influence on how we view the U.S. class system today.

Karl Marx: Relationship to the Means of Production

According to Karl Marx, class position and the extent of our income and wealth are determined by our work situation, or our relationship to the means of production. As we have previously seen, Marx stated that capitalistic societies consist of two classes—the capitalists and the workers. The *capitalist class (bourgeoisie)* **consists of those who own the means of production**—the land and capital necessary for factories and mines, for example. The *working class (proletariat)* **consists of those who must sell their labor to the owners in order to earn enough money to survive** (see Figure 8.4).

According to Marx, class relationships involve inequality and exploitation. The workers are exploited as capitalists maximize their profits by paying workers less than the resale value of what they produce but do not own. This exploitation results in workers' *alienation*—**a feeling of powerlessness and estrangement from other people and from oneself.** As mechanization reduced the cost of producing products and machines replaced many workers, the newly created surplus of workers became a "reserve army"—a readily available source of cheap labor that could be used by capitalists as a "weapon" against employees who demanded pay raises or better working conditions. The presence of the reserve army keeps wages low and creates even greater profits for members of the capitalist class.

In Marx's view, the capitalist class maintains its position at the top of the class structure by control of the society's *superstructure*, which is composed of the government, schools, churches, and other social institutions that produce and disseminate ideas perpetuating the existing system of exploitation. Marx predicted that the exploitation of workers by the capitalist class would ultimately lead to *class conflict*—**the struggle between the capitalist class and the working class.** According to Marx, when the workers realized that capitalists were the source of their oppression, they would overthrow the capitalists and their agents of social control, leading to the end of capitalism. The workers would then take over the government and create a more egalitarian society.

Why has no workers' revolution occurred? According to the sociologist Ralf Dahrendorf (1959), capitalism may have persisted because it has changed significantly since Marx's time. Individual capitalists no longer own and control factories and other means of production; today, ownership and control have largely been separated. For example, contemporary transnational corporations are owned

Capitalist class (or bourgeoisie) Karl Marx's term for the class that consists of those who own and control the means of production.

Working class (or proletariat) those who must sell their labor to the owners in order to earn enough money to survive.

Alienation a feeling of powerlessness and estrangement from other people and from oneself.

Class conflict Karl Marx's term for the struggle between the capitalist class and the working class.

AP/Wide World Photos

This recent strike by service employees reflects the activism of workers throughout the years as they have sought to gain better wages and working conditions. How would Karl Marx explain the problems experienced by employees such as these?

by a multitude of stockholders but run by paid officers and managers. Similarly, many (but by no means all) workers have experienced a rising standard of living, which may have contributed to a feeling of complacency. Moreover, some people have become so engrossed in the process of consumption—including acquiring more material possessions and going on outings to shopping malls, movie theaters, and amusement parks such as Disney World—that they are less likely to engage in workers' rebellions against the system that has brought them a relatively high standard of living (Gottdiener, 1997). During the twentieth century, workers pressed for salary increases and improvements in the workplace through their activism and labor union membership. They also gained more legal protection in the form of workers' rights and benefits such as workers' compensation insurance for job-related injuries and disabilities (Dahrendorf, 1959). For these reasons, and because of a myriad of other complex factors, the workers' revolution predicted by Marx never came to pass. However, the failure of his prediction does not mean that his analysis of capitalism and his theoretical contributions to sociology are without validity.

Marx had a number of important insights into capitalist societies. First, he recognized the economic basis of class systems (Gilbert, 2003). Second, he noted the relationship between people's social location in the class structure and their values, beliefs, and behavior. Finally, he acknowledged that classes may have opposing (rather than complementary) interests. For example, capitalists' best interests are served by a decrease in labor costs and other expenses and a corresponding increase in profits; workers' best interests are served by well-paid jobs, safe working conditions, and job security.

Max Weber: Wealth, Prestige, and Power

Max Weber's analysis of class builds upon earlier theories of capitalism (particularly those by Marx) and of money (particularly those by Simmel, as discussed in Chapter 1). Living in the late nineteenth and early twentieth centuries, Weber was in a unique position to see the transformation that occurred as individual, competitive, entrepreneurial capitalism went through the process of shifting to bureaucratic, industrial, corporate capitalism.

As a result, Weber had more opportunity than Marx to see how capitalism changed over time.

Weber agreed with Marx's assertion that economic factors are important in understanding individual and group behavior. However, Weber emphasized that no single factor (such as economic divisions between capitalists and workers) was sufficient for defining the location of categories of people within the class structure. According to Weber, the access that people have to important societal resources (such as economic, social, and political power) is crucial in determining people's life chances. To highlight the importance of life chances for categories of people, Weber developed a multidimensional approach to social stratification that reflects the interplay among wealth, prestige, and power. In his analysis of these dimensions of class structure, Weber viewed the concept of "class" as an *ideal type* (that can be used to compare and contrast various societies) rather than as a specific social category of "real" people (Bourdieu, 1984).

As previously defined, *wealth* is the value of all of a person's or family's economic assets, including income, personal property, and income-producing property. Weber placed categories of people who have a similar level of wealth and income in the same class. For example, he identified a privileged commercial class of *entrepreneurs* — wealthy bankers, ship owners, professionals, and merchants who possess similar financial resources. He also described a class of *rentiers* — wealthy individuals who live off their investments and do not have to work. According to Weber, entrepreneurs and rentiers have much in common. Both are able to purchase expensive consumer goods, control other people's opportunities to acquire wealth and property, and monopolize costly status privileges (such as education) that provide contacts and skills for their children.

Weber divided those who work for wages into two classes: the middle class and the working class. The middle class consists of white-collar workers, public officials, managers, and professionals. The working class consists of skilled, semiskilled, and unskilled workers.

The second dimension of Weber's system of stratification is *prestige* — **the respect or regard with which a person or status position is regarded by others.** Fame, respect, honor, and esteem are the most common forms of prestige. A person who has a high level of prestige is assumed to receive deferential and respectful treatment from others. Weber suggested that individuals who share a common level of social prestige belong to the same status group regardless of their level of wealth. They tend to socialize with one another, marry within their own group of social equals, spend their leisure time together, and safeguard their status by restricting outsiders' opportunities to join their ranks (Beeghley, 1996). Style of life, formal education, and occupation are often significant factors in establishing and maintaining prestige in industrial and postindustrial societies.

The other dimension of Weber's system is *power* — **the ability of people or groups to achieve their goals despite opposition from others.** The powerful can shape society in accordance with their own interests and direct the actions of others (Tumin, 1953). According to Weber, social power in modern societies is held by bureaucracies; individual power depends on a person's position within the bureaucracy. Weber suggested that the power of modern bureaucracies was so strong that even a workers' revolution (as predicted by Marx) would not lessen social inequality (Hurst, 1998).

Weber stated that wealth, prestige, and power are separate continuums on which people can be ranked from high to low, as shown in Figure 8.5. Individuals may be high on one dimension while being low on another. For example, people may be very wealthy but have little political power (for example, a recluse who has inherited a large sum of money). They may also have prestige but not wealth (for instance, a college professor who receives teaching excellence awards but lives on a relatively low income). In Weber's multidimensional approach, people are ranked on all three dimensions. Sociologists often use the term *socioeconomic status* **(SES) to refer to a combined measure that attempts to classify individuals, families, or households in**

Prestige the respect or regard with which a person or status position is regarded by others.

Power according to Max Weber, the ability of people or groups to achieve their goals despite opposition from others.

Socioeconomic status (SES) a combined measure that attempts to classify individuals, families, or households in terms of factors such as income, occupation, and education to determine class location.

Figure 8.5 Weber's Multidimensional Approach to Social Stratification

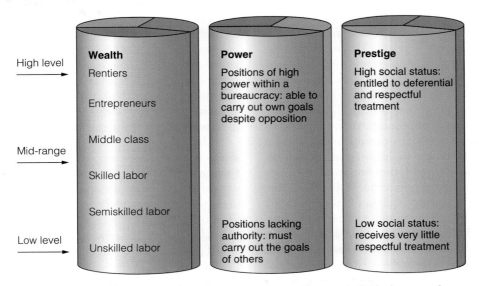

According to Max Weber, wealth, power, and prestige are separate continuums. Individuals may rank high in one dimension and low in another, or they may rank high or low in more than one dimension. Also, individuals may use their high rank in one dimension to achieve a comparable rank in another. How does Weber's model compare with Marx's approach as shown in Figure 8.4?

terms of factors such as income, occupation, and education to determine class location.

Weber's analysis of social stratification contributes to our understanding by emphasizing that people behave according to both their economic interests and their values. He also added to Marx's insights by developing a multidimensional explanation of the class structure and by identifying additional classes. Both Marx and Weber emphasized that capitalists and workers are the primary players in a class society, and both noted the importance of class to people's life chances. However, they saw different futures for capitalism and the social system. Marx saw these structures being overthrown; Weber saw the increasing bureaucratization of life even without capitalism.

Reflections

Reflections 8.2

How do the theories of Karl Marx and Max Weber help us understand why people may or may not achieve the American Dream?

✓ Checkpoints

Checkpoint 8.2

4. What are the two classes identified by Karl Marx?
5. What is class conflict?
6. Why did Max Weber believe that wealth, prestige, and power were important dimensions of the class structure?
7. What is socioeconomic status?

Contemporary Sociological Models of the U.S. Class Structure

How many social classes exist in the United States today? What criteria are used for determining class membership? No broad consensus exists about how to characterize the class structure in this country. In fact, many people deny that class distinctions exist (see Eisler, 1983; Parenti, 1994). Most people like to think of themselves as middle class; it puts them in a comfortable middle position—neither rich nor poor. Sociologists have developed several models of the class structure: One is broadly based on a Weberian approach, the second on a Marxian approach, and the third on a study of society today. We will examine all three models briefly.

The Weberian Model of the U.S. Class Structure

Expanding on Weber's analysis of the class structure, the sociologists Dennis Gilbert (2003) and Joseph A. Kahl developed a widely used model of social classes based on three elements: (1) education, (2) occupation of family head, and (3) family income (see Figure 8.6).

THE UPPER (CAPITALIST) CLASS The upper class is the wealthiest and most powerful class in the United States. About 1 percent of the population is included in this class, whose members own substantial income-producing assets and operate on both the national and international level. According to Gilbert (2003), people in this class have an influence on the economy and society far beyond their numbers.

Some models further divide the upper class into upper-upper ("old money") and lower-upper ("new money") categories (Warner and Lunt, 1941; Coleman and Rainwater, 1978). Members of the upper-upper class come from prominent families which possess great wealth that they have held for several generations. Family names—such as Rockefeller, Mellon, Du Pont, and Kennedy— are well-known and often held in high esteem. Persons in the upper-upper class tend to have strong feelings of ingroup solidarity. They belong to the same exclusive clubs and support high culture (such as the opera, symphony orchestras, ballet, and art museums). Children are educated in prestigious private schools and Ivy League universities; many acquire strong feelings of privilege from birth, as upper-class author Lewis H. Lapham (1988: 14) states:

> Together with my classmates and peers, I was given to understand that it was sufficient accomplishment merely to have been born. Not that anybody ever said precisely that in so many words, but the assumption was plain enough, and I could confirm it by observing the mechanics of the local society. A man might become a drunkard, a concert pianist or an owner of companies, but none of these occu-

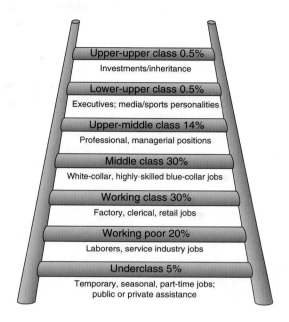

Figure 8.6 Stratification Based on Education, Occupation, and Income

- Upper-upper class 0.5%
  Investments/inheritance
- Lower-upper class 0.5%
  Executives; media/sports personalities
- Upper-middle class 14%
  Professional, managerial positions
- Middle class 30%
  White-collar, highly-skilled blue-collar jobs
- Working class 30%
  Factory, clerical, retail jobs
- Working poor 20%
  Laborers, service industry jobs
- Underclass 5%
  Temporary, seasonal, part-time jobs; public or private assistance

pations would have an important bearing on his social rank.

Children of the upper class are socialized to view themselves as different from others; they also learn that they are expected to marry within their own class (Warner and Lunt, 1941; Mills, 1959a; Domhoff, 1983).

Members of the lower-upper class may be extremely wealthy but not have attained as much prestige as members of the upper-upper class. The "new rich" have earned most of their money in their own lifetime as entrepreneurs, presidents of major corporations, sports or entertainment celebrities, or top-level professionals. For some members of the lower-upper class, the American Dream has become a reality. Others still desire the respect of members of the upper-upper class.

THE UPPER-MIDDLE CLASS Persons in the upper-middle class are often highly educated professionals who have built careers as physicians, attorneys, stockbrokers, or corporate managers. Others derive their income from family-owned businesses. According to Gilbert (2003), about 14 percent of the U.S. population is in this category.

A combination of three factors qualifies people for the upper-middle class: university degrees, authority and independence on the job, and high income. Of all the class categories, the upper-middle class is the one that is most shaped by formal education. Over the past fifty years, Asian Americans, Latinos/as, and African Americans have placed great importance on education as a means of attaining the American Dream. Many people of color have moved into the upper-middle class by acquiring higher levels of education. James P. Comer, an African American child psychiatrist, describes how his mother, who grew up in abject poverty in the rural South, encouraged her children to get a good education:

> My mother, Maggie, believed that education was the way to achieve her American Dream. When she was denied the opportunity herself, she declared that all her children would be educated. Mom and Dad together gave all five of us the support needed to acquire thirteen college degrees. (Comer, 1988: xxii)

Across racial–ethnic and class lines, children are encouraged to acquire an education suitable for upper-middle-class occupations, which typically have high prestige in the community. However, many social analysts believe that racism still diminishes the life chances for people of color even when they achieve a high income and a prestigious career (see Cose, 1993; Takaki, 1993; Feagin and Sikes, 1994).

THE MIDDLE CLASS In past decades, a high school diploma was necessary to qualify for most middle-class jobs. Today, two-year or four-year college degrees have replaced the high school diploma as an entry-level requirement for employment in many middle-class occupations, including medical technicians, nurses, legal and medical assis-

tants, lower-level managers, semiprofessionals, and nonretail sales workers. An estimated 30 percent of the U.S. population is in the middle class (Gilbert, 2003).

Traditionally, most middle-class occupations have been relatively secure and have provided more opportunities for advancement (especially with increasing levels of education and experience) than working-class positions. Recently, however, four factors have eroded the American Dream for this class: (1) escalating housing prices, (2) occupational insecurity, (3) blocked mobility on the job, and (4) the cost-of-living squeeze that has penalized younger workers, even when they have more education and better jobs than their parents (Newman, 1993). Consider, for example, Brenda and Amancio Irizarry of New York, who are struggling to get by on a combined pretax income of $38,000 per year. Although they make more money than their parents did, the Irizzarrys cannot afford to buy a home, have no savings, and have never taken a vacation together. They are working hard to send their daughter, Michelle, to college, as Mrs. Irizarry explains:

> We didn't have a college fund for Michelle. You see commercials on television saying to start [saving] when the baby is born. But because of the unforeseen things that happen in life, it's impossible to save money. (qtd. in Jones, 1995: 9)

Michelle appreciates her parents' efforts but eventually wants much more for herself:

> I don't want to live like this the rest of my life. . . . I mean, they have morals and are trying to make it. But unless they hit the Lotto, this is how it's going to be the rest of their lives. And that's sad, to think that's it. This is as good as it's going to get. (qtd. in Jones, 1995: 9)

But Michelle has not given up on the American Dream; she is determined to become a social worker and have a nice place to live (Jones, 1995). As this example suggests, class distinctions between the middle and working classes are sometimes blurred due to overlapping characteristics (Gilbert, 2003).

THE WORKING CLASS An estimated 30 percent of the U.S. population is in the working class. The core of this class is made up of semiskilled ma-

Reflections

Reflections 8.3

Many people see education as a way to achieve the American Dream. What are the strengths and weaknesses of this belief?

chine operators who work in factories and elsewhere. Members of the working class also include some workers in the service sector, as well as clerks and salespeople whose job responsibilities involve routine, mechanized tasks requiring little skill beyond basic literacy and a brief period of on-the-job training (Gilbert, 2003). Some people in the working class are employed in *pink-collar occupations* — **relatively low-paying, nonmanual, semiskilled positions primarily held by women, such as day-care workers, checkout clerks, cashiers, and waitpersons.**

How does life in the working-class family compare with that of individuals in middle-class families? According to sociologists, working-class families not only earn less than middle-class families, but they also have less financial security, particularly because of high rates of layoffs and plant closings in some regions of the country. Few people in the working class have more than a high school diploma, and many have less, which makes job opportunities scarce for them in a "high-tech" society (Gilbert, 2003). Others find themselves in low-paying jobs in the service sector of the economy, particularly fast-food restaurants, a condition that often places them among the working poor.

THE WORKING POOR The working poor account for about 20 percent of the U.S. population. Members of the working-poor class live from just above to just below the poverty line; they typically hold unskilled jobs, seasonal migrant jobs in agriculture, lower-paid factory jobs, and service jobs (such as counter help at restaurants). Employed single mothers often belong to this class; consequently, children are overrepresented in this category. African Americans and other people of color are also overrepresented among the working poor. To cite only one example, in the United States today, there are two white hospital orderlies to every one white physician, whereas there are twenty-five African American orderlies to every one African American physician (Gilbert, 2003). For the working poor, living from paycheck to paycheck makes it impossible to save money for emergencies such as periodic or seasonal unemployment, which is a constant threat to any economic stability they may have.

Social critic and journalist Barbara Ehrenreich (2001) left her upper-middle-class lifestyle for a period of time to see if it was possible for the working poor to live on the wages that they were being paid as restaurant servers, sales clerks at discount department stores, aides at nursing homes, house cleaners for franchise maid services, or similar jobs. She conducted her research by actually holding those jobs for periods of time and seeing if she could live on the wages that she received. Through her research, Ehrenreich persuasively demonstrated that people who work full time, year round, for poverty-level wages must develop survival strategies that include such things as help from relatives or constantly moving from one residence to another in order to have a place to live. Like many other researchers, Ehrenreich found that minimum-wage jobs cannot cover the full cost of living, such as rent, food, and the rest of an adult's monthly needs, even without taking into consideration the needs of children or other family members (see also Newman, 1999). According to Ehrenreich (2001: 221),

> The "working poor," as they are approvingly termed, are in fact the major philanthropists of our society. They neglect their own children so that the children of others will be cared for; they live in substandard housing so that other homes will be shiny and perfect; they endure privation so that inflation will be low and stock prices high. To be a member of the working poor is to be an anonymous donor, a nameless benefactor, to everyone else.

When the children of the working poor become successful, often through entertainment or sports careers or gaining high levels of education, their successes may receive a great deal of media attention (see Box 8.2).

THE UNDERCLASS According to Gilbert (2003), people in the underclass are poor, seldom employed, and caught in long-term deprivation that results from low levels of education and income and high rates of unemployment. Some are unable to work because of age or disability; others

Pink-collar occupations relatively low-paying, nonmanual, semiskilled positions primarily held by women, such as day-care workers, checkout clerks, cashiers, and waitpersons.

Box 8.2 Changing Times: Media and Technology

Media Stars and the American Dream

I'll never forget finding this Tonka truck. It was missing a wheel, but it was the best toy I ever had.

—Marc Anthony, in an interview about the success of his pop album in English and a salsa album in Spanish, recalling the times as a child when he had gone through "rich people's trash" looking for things that his parents—Puerto Rican émigrés in East Harlem, New York—could not afford to purchase for him (qtd. in Buia, 2001)

Today, Marc Anthony is the prototype of the American Dream; his success in the music industry and as an actor has brought him wealth, a 10,000-square-foot mansion, and a 73-foot yacht, among other possessions (Buia, 2001).

Anthony's achievement of the American Dream is an example of how increasing globalization of the music industry has brought new markets as listeners around the world enjoy the music of artists from other cultures and nations. For some, globalization of culture is a hotly contested issue; for others, it is a means of gaining wealth and worldwide recognition. Although the term *crossover* is frequently used for entertainers such as Ricky Martin and Jennifer Lopez, who have widespread appeal in multiple cultures, Marc Anthony does not believe that his success should be described in that manner. According to Anthony, "What did I cross over from? I'm as American as anybody. I was born in your backyard" (qtd. in Buia, 2001: 12).

Although most young people will not become a household name and earn large sums of money

Ethan Miller/Reuters—Getty Images

Some people believe that singer Marc Anthony reflects the American Dream because he has achieved great wealth after a childhood of poverty.

the way Marc Anthony has, many individuals hope that they, too, will hit it big in the land of opportunity—the United States—and move into the upper class of our society. What are the actual chances that most people will strike it rich through music, entertainment, sports, a great entrepreneurial idea, or a winning lottery ticket? What produces most of the social mobility in our society? Why?

experience discrimination based on race/ethnicity. Single mothers are overrepresented in this class because of the lack of jobs, lack of affordable child care, and many other impediments to the mother's future and that of her children. People without a "living wage" often must rely on public or private assistance programs for their survival. About 3 to 5 percent of the U.S. population is in this category, and the chances of their children moving out of poverty are about fifty-fifty (Gilbert, 2003).

Studies by various social scientists have found that meaningful employment opportunities are the critical missing link for people on the lowest rungs of the class ladder. According to these analysts, job creation is essential in order for people to have the opportunity to earn a decent wage; have medical

coverage; live meaningful, productive lives; and raise their children in a safe environment (see Fine and Weis, 1998; Nelson and Smith, 1999; Newman, 1999; Wilson, 1996). These issues are closely tied to the American Dream we have been discussing in this chapter.

The Marxian Model of the U.S. Class Structure

The earliest Marxian model of class structure identified ownership or nonownership of the means of production as the distinguishing feature of classes. From this perspective, classes are social groups organized around property ownership, and social stratification is created and maintained by one group in order to protect and enhance its own economic interests. Moreover, societies are organized around classes in conflict over scarce resources. Inequality results from the more powerful exploiting the less powerful.

Contemporary Marxian (or conflict) models examine class in terms of people's relationship to others in the production process. For example, conflict theorists attempt to determine what degree of control that workers have over the decision-making process and the extent to which they are able to plan and implement their own work. They also analyze the type of supervisory authority, if any, that a worker has over other workers. According to this approach, most employees are a part of the working class because they do not control either their own labor or that of others.

Erik Olin Wright (1979, 1985, 1997), one of the leading stratification theorists to examine social class from a Marxian perspective, has concluded that Marx's definition of "workers" does not fit the occupations found in advanced capitalist societies. For example, many top executives, managers, and supervisors who do not own the means of production (and thus would be "workers" in Marx's model) act like capitalists in their zeal to control workers and maximize profits. Likewise, some experts hold positions in which they have control over money and the use of their own time even though they are not owners. Wright views Marx's category of "capitalist" as being too broad as well. For instance, small-business owners might be viewed as capitalists because they own their own tools and have a few people working for them, but they have little in common with large-scale capitalists and do not share the interests of factory workers.

Wright (1979) argues that classes in modern capitalism cannot be defined simply in terms of different levels of wealth, power, and prestige, as in the Weberian model. Consequently, he outlines four criteria for placement in the class structure: (1) ownership of the means of production, (2) purchase of the labor of others (employing others), (3) control of the labor of others (supervising others on the job), and (4) sale of one's own labor (being employed by someone else). Wright (1978) assumes that these criteria can be used to determine the class placement of all workers, regardless of race/ethnicity, in a capitalist society. Let's take a brief look at Wright's (1979, 1985) four classes—(1) the capitalist class, (2) the manage-

AP/Wide World Photos

Bill Gates (co-founder of Microsoft Corp, and the world's wealthiest person) is representative of twenty-first century capitalism, which increasingly is based on information technology.

rial class, (3) the small-business class, and (4) the working class—so that you can compare them to those found in the Weberian model.

THE CAPITALIST CLASS According to Wright, this class holds most of the wealth and power in society through ownership of capital—for example, banks, corporations, factories, mines, news and entertainment industries, and agribusiness firms. The "ruling elites," or "ruling class," within the capitalist class hold political power and are often elected or appointed to influential political and regulatory positions (Parenti, 1994).

This class is composed of individuals who have inherited fortunes, own major corporations, or are top corporate executives with extensive stock holdings or control of company investments. Even though many top executives have only limited *legal ownership* of their corporations, they have substantial economic ownership and exert extensive control over investments, distribution of profits, and management of resources. The major sources of income for the capitalist class are profits, interest, and very high salaries. Members of this class make important decisions about the workplace, including which products and services to make available to consumers and how many workers to hire or fire (Feagin and Feagin, 2003).

According to *Forbes* magazine's 2001 list of the richest people in the world, Bill Gates (co-founder of Microsoft Corporation, the world's largest microcomputer software company) was the wealthiest capitalist, with a net worth of $54 billion, down almost 15 percent from his $63-billion figure in 2000 as a result of steep decreases in the stock market during 2001 (*Forbes*, 2001). Investor Warren Buffet came in second with $33.2 billion. However, the number of billion-dollar fortunes in the United States rose from 129 in 1995 to 274 in 2000 before falling to 236 in 2001 (*Forbes*, 1996, 2001). Although some of the men who made the *Forbes* list of the wealthiest people have gained their fortunes through entrepreneurship or being CEOs of large corporations, women who made the list have acquired their wealth typically through inheritance, marriage, or both. Perhaps this difference is partially due to the fact that women tend to experience more problems beginning their own businesses, but few women have headed major corporations over the past two decades, and most of those headed up family-owned business enterprises (Lorber, 1994). In 2001, only six women were heads of Fortune 500 companies (*Fortune*, 2001).

THE MANAGERIAL CLASS   People in the managerial class have substantial control over the means of production and over workers. However, these upper-level managers, supervisors, and professionals typically do not participate in key corporate decisions such as how to invest profits. Lower-level managers may have some control over employment practices, including the hiring and firing of some workers (Vanneman and Cannon, 1987).

Top professionals such as physicians, attorneys, accountants, and engineers may control the structure of their own work; however, they typically do not own the means of production and may not have supervisory authority over more than a few people. Even so, they may influence the organization of work and the treatment of other workers. Members of the capitalist class often depend on these professionals for their specialized knowledge.

As previously discussed, members of the managerial class occupy a contradictory class location between the capitalist and working classes (Wright, 1979). Like members of the working class, persons in the managerial class do not own the means of production, and they usually earn a regular salary. However, they typically have control over the work of others and may exercise considerable authority over the organization of production. They also may invest in corporate stock and gain significant unearned income. As a result, they tend to align themselves with the basic interests of the capitalist class. For people of color and white women in the managerial class, additional contradictions exist based on race and/or gender.

THE SMALL-BUSINESS CLASS   This class consists of small-business owners and craftspeople who may hire a small number of employees but largely do their own work. Some members own businesses such as "mom and pop" grocery stores, retail clothing stores, and jewelry stores. Others are doctors and lawyers who receive relatively high incomes from selling their own services. Some of these professionals now share attributes with members of the capitalist class because they have formed corporations that hire and control the employees who produce profits for the professionals.

It is in the small-business class that we find many people's hopes of achieving the American Dream. Seventeen percent of those surveyed in a Roper opinion poll thought "owning your own business" was the essence of this dream (Feagin and Sikes, 1994). However, Wright (1985) estimated that less than 5 percent of the population is in the small-business class.

Several leading scholars have argued that starting a business is a solution for the economic problems of people of color (see Sowell, 1981). However, this is not a new idea. Sociologist John Butler (1991) has demonstrated that African American-owned small businesses have existed in the United States since the 1700s. Today, African Americans own nearly 4 percent of small businesses with receipts of less than $10,000 annually, but less than 1 percent of those with receipts of $1 million or more annually. Similarly, although Latinos/as constitute 12.5 percent of the U.S. population, they own only 6 percent of all businesses (U.S. Census Bureau, 2001h). However, in recent years increasing numbers of small businesses have been formed by subordinate-group members (Shorris, 1992).

Billy Hustace/Stone-Getty Images

In which segment of the class structure would sociologists place clerical workers such as those in this office mail room? What are the key elements of that social class?

THE WORKING CLASS The working class is made up of a number of subgroups, one of which is blue-collar workers, some of whom are highly skilled and well paid and others of whom are unskilled and poorly paid. Skilled blue-collar workers include electricians, plumbers, and carpenters; unskilled blue-collar workers include janitors and gardeners.

White-collar workers are another subgroup of the working class. Referred to by some as a "new middle class," these workers are actually members of the working class because they do not own the means of production, do not control the work of others, and are relatively powerless in the workplace. Secretaries, other clerical workers, and sales workers

are members of the white-collar faction of the working class. They take orders from others and tend to work under constant supervision. Thus, these workers are at the bottom of the class structure in terms of domination and control in the workplace. The working class contains about half of all employees in the United States.

Although Marxian and Weberian models of the U.S. class structure show differences in people's occupations and access to valued resources, neither fully reflects the nature and extent of inequality in the United States. In the next section, we will take a closer look at the unequal distribution of income and wealth in the United States and the effects of inequality on people's opportunities and life chances.

Consequences of Inequality

Income and wealth are not simply statistics; they are intricately related to the American Dream and our individual life chances. Persons with a high income or substantial wealth have more control over their own lives. They have greater access to goods and services; they can afford better housing, more education, and a wider range of medical services. Persons with less income, especially those living in poverty, must spend their limited resources to acquire the basic necessities of life.

Physical Health, Mental Health, and Nutrition

People who are wealthy and well educated and who have high-paying jobs are much more likely to be healthy than are poor people. As people's economic status increases, so does their health status. The poor have shorter life expectancies and are at greater risk for chronic illnesses such as diabetes, heart disease, and cancer, as well as infectious diseases such as tuberculosis.

Children born into poor families are at much greater risk of dying during their first year of life. Some die from disease, accidents, or violence. Others are unable to survive because they are born with

✓ Checkpoints

Checkpoint 8.3

8. What social classes are identified in the Weberian model of the U.S. class structure?
9. What are the key classes identified in the Marxian model of the U.S. class structure?
10. What are the four criteria for placement in the class structure as identified by Erik Olin Wright?

low birth weight, a condition linked to birth defects and increased probability of infant mortality (Rogers, 1986). Low birth weight in infants is attributed, at least in part, to the inadequate nutrition received by many low-income pregnant women. Most of the poor do not receive preventive medical and dental checkups; many do not receive adequate medical care after they experience illness or injury.

Many high-poverty areas lack an adequate supply of doctors and medical facilities. Even in areas where such services are available, the inability to pay often prevents people from seeking medical care when it is needed. Some "charity" clinics and hospitals may provide indigent patients (those who cannot pay) with minimal emergency care but make them feel stigmatized in the process. For many of the working poor, medical insurance is out of the question (Schwarz and Volgy, 1992). An estimated 38.7 million people in the United States were without health insurance coverage in 2000 (Mills, 2001). Many people rely on their employers for health coverage; however, some employers are cutting back on health coverage, particularly for employees' family members. Despite passage of the 1996 Kassenbaum–Kennedy bill by Congress, which makes insurance more readily available for millions of people who change their jobs or lose them, many unemployed workers and their families remain without medical coverage. However, the uninsured are a changing group in the United States—not everyone who becomes uninsured for a month or more remains uninsured throughout a given year (Zaldivar, 1996). Of all age groups, persons aged eighteen to twenty-four are the most likely to be uninsured; Medicare and other benefit programs provide medical care to most persons sixty-five and over (U.S. Census Bureau, 2001c). As shown in Figure 8.7, a high percentage of poor persons do not have health insurance.

Many lower-paying jobs are often the most dangerous and have the greatest health hazards. Black lung disease, cancer caused by asbestos, and other environmental hazards found in the workplace are more likely to affect manual laborers and low-income workers, as are job-related accidents.

Although the precise relationship between class and health is not known, analysts suggest that people with higher income and wealth tend to smoke less, exercise more, maintain a healthy body weight, and eat nutritious meals. As a category, more-affluent persons tend to be less depressed and

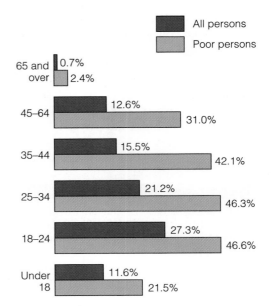

Figure 8.7 Percentage of U.S. Population Without Health Insurance, 2000

Source: U.S. Census Bureau, 2001c.

face less psychological stress, conditions that tend to be directly proportional to income, education, and job status (*Mental Medicine*, 1994).

Good health is basic to good life chances; in turn, adequate nutrition is essential for good health. Hunger is related to class position and income inequality. Recent surveys estimate that 13 percent of children under age 12 are hungry or at risk of being hungry. Among the working poor, almost 75 percent of the children are thought to be in this category. After spending 60 percent of their income on housing, low-income families are unable to provide adequate food for their children. Between one-third and one-half of all children living in poverty consume significantly less than the federally recommended guidelines for caloric and nutritional intake (Children's Defense Fund, 2001). Lack of adequate nutrition has been linked to children's problems in school.

Housing

As discussed in Chapter 4 ("Social Structure and Interaction in Everyday Life"), homelessness is a major problem in the United States. The lack of affordable housing is a pressing concern for many low-income individuals and families. With the economic prosperity of the 1990s, low-cost housing units in many cities were replaced with expensive condominiums and luxury single-family residences for affluent people. As unemployment rose dramatically beginning in 2001, partly due to terrorism and a faltering economy, housing costs remained high compared to many families' ability to pay for food, shelter, clothing, and other necessities.

Lack of *affordable* housing is one central problem brought about by economic inequality. Another concern is *substandard* housing, which refers to facilities that have inadequate heating, air conditioning, plumbing, electricity, or structural durability. Structural problems—due to faulty construction or lack of adequate maintenance—exacerbate the potential for other problems such as damage from fire, falling objects, or floors and stairways collapsing.

Housing is a problem that is directly related to inequality in both the urban and rural areas of the United States. Due to the importance of adequate housing for even a subsistence level of existence, the Census Bureau includes questions about housing in the questions that it asks during the decennial census (see "Census Profiles: Housing Characteristics as a Social Concern"). In some areas of the United States, people live in substandard housing that cannot withstand local weather conditions. An example is the problem of living in trailers and manufactured housing in areas where tornadoes and hurricanes are a possibility. Another example is the housing for migrant farm-worker families in rural areas of Arizona, California, New Mexico, Texas, and other states, where the housing that is provided may be without safe electricity, running water, and indoor toilets.

Education

Educational opportunities and life chances are directly linked. Some functionalist theorists view education as the "elevator" to social mobility. Improvements in the educational achievement levels (measured in number of years of schooling completed) of the poor, people of color, and white women have been cited as evidence that students' abilities are now more important than their class, race, or gender. From this perspective, inequality in education is declining, and students have an opportunity to achieve upward mobility through

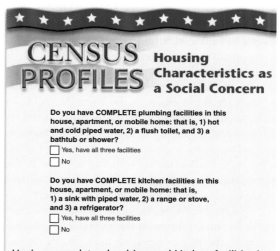

CENSUS PROFILES Housing Characteristics as a Social Concern

Do you have COMPLETE plumbing facilities in this house, apartment, or mobile home: that is, 1) hot and cold piped water, 2) a flush toilet, and 3) a bathtub or shower?
☐ Yes, have all three facilities
☐ No

Do you have COMPLETE kitchen facilities in this house, apartment, or mobile home: that is, 1) a sink with piped water, 2) a range or stove, and 3) a refrigerator?
☐ Yes, have all three facilities
☐ No

Having complete plumbing and kitchen facilities is a factor in determining whether or not a residence is substandard. Therefore, the U.S. Census Bureau asked the two questions displayed above when taking Census 2000. As shown below, the answer from approximately one out of every twenty households was that they either did not have complete plumbing facilities or did not have complete kitchen facilities—or that they had neither.

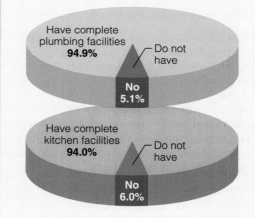

Have complete plumbing facilities **94.9%** — Do not have
No **5.1%**

Have complete kitchen facilities **94.0%** — Do not have
No **6.0%**

Although the vast majority of households in this nation reported that they *do* have complete plumbing and kitchen facilities, it is a sad fact that, even in the "land of plenty," more than half a million housing units lack facilities that most of us would consider essential to health and well-being. What other problems are associated with this pressing concern?

Source: U.S. Census Bureau, 2001b.

achievements at school (see Hauser and Featherman, 1976). Functionalists generally see the education system as flexible, allowing most students the opportunity to attend college if they apply themselves (Ballantine, 2000).

In contrast, most conflict theorists stress that schools are agencies for reproducing the capitalist class system and perpetuating inequality in society (Bowles and Gintis, 1976; Bowles, 1977). From this perspective, education perpetuates poverty. Parents with limited income are not able to provide the same educational opportunities for their children as are families with greater financial resources.

Today, great disparities exist in the distribution of educational resources. Because funding for education primarily comes from local property taxes, school districts in wealthy suburban areas generally pay higher teachers' salaries, have newer buildings, and provide state-of-the-art equipment. By contrast, schools in poorer areas have a limited funding base. Students in central-city schools and poverty-stricken rural areas often attend dilapidated schools that lack essential equipment and teaching materials. Author Jonathan Kozol (1991, qtd. in Feagin and Feagin, 1994: 191) documented the effect of a two-tiered system on students:

> Kindergartners are so full of hope, cheerfulness, high expectations. By the time they get into fourth grade, many begin to lose heart. They see the score, understanding they're not getting what others are getting. . . . They see suburban schools on television. . . . They begin to get the point that they are not valued much in our society. By the time they are in junior high, they understand it. "We have eyes and we can see; we have hearts and we can feel. . . . We know the difference."

Poverty extracts such a toll that many young people will not have the opportunity to finish high school, much less enter college.

✓ Checkpoints

Checkpoint 8.4

11. How does inequality affect physical health, mental health, and nutrition?
12. What types of housing problems are brought about by economic inequality?
13. How are education and life chances related?

Poverty in the United States

When many people think about poverty, they think of people who are unemployed or on welfare. However, many hardworking people with full-time jobs live in poverty. The U.S. Social Security Administration has established an *official poverty line*, which is based on what is considered to be the minimum amount of money required for living at a subsistence level. The poverty level is computed by determining the cost of a minimally nutritious diet (a low-cost food budget on which a family could survive nutritionally on a short-term, emergency basis) and multiplying this figure by three to allow for nonfood costs. In 2000, over 28 million people lived below the official government poverty level of $17,603 for a family of four (U.S. Census Bureau, 2001e). This constituted some improvement over previous years: In 1997, over 36 million (13.8 percent of the U.S. population) lived below the poverty level, which was $15,141 for a family of four (U.S. Census Bureau, 1998a).

When sociologists define poverty, they distinguish between absolute and relative poverty. **Absolute poverty exists when people do not have the means to secure the most basic necessities of life.** This definition comes closest to that used by the federal government. Absolute poverty often has life-threatening consequences, such as when a homeless person freezes to death on a park bench. By comparison, *relative poverty* **exists when people may be able to afford basic necessities but are still unable to maintain an average standard of living** (Ropers, 1991). A family must have income substantially above the official poverty line in order to afford the basic necessities, even when these are purchased at the lowest possible cost. At about 155 percent of the official poverty line, families could live on an economy budget. What is it like to live on the economy budget? John Schwarz and Thomas Volgy (1992: 43) offer the following distressing description:

> Members of families existing on the economy budget never go out to eat, for it is not included in the food budget; they never go out to a movie, concert, or ball game or indeed to any public or private establishment that charges admission, for there is no entertainment budget; they have no cable television, for the same reason; they never purchase alcohol or cigarettes; never take a vacation or holiday that involves any motel or hotel or, again, any

Table 8.1 Percentage Distribution of Poverty in the United States

	All Races[a]	White	African American	Hispanic[b]
By Age				
Under 18 years	16.2	13.0	30.9	28.0
18–24 years	14.4	12.6	23.6	21.5
25–44 years	9.3	8.1	16.4	17.3
45–64 years	8.5	7.3	17.0	15.7
65 and above	10.2	8.9	22.3	18.8
By Education				
No high school diploma	22.2	20.1	33.1	25.3
4 years of high school	9.2	7.7	18.0	12.9
Some college (no degree)	5.9	5.1	10.0	7.3
College degree or more	3.2	2.8	5.0	6.6

[a]Includes other races/ethnicities not shown separately.
[b]Includes Hispanic persons of any race.

Source: U.S. Census Bureau, 2001e.

meals out; never hire a baby-sitter or have any other paid child care; never give an allowance or other spending money to the children; never purchase any lessons or home-learning tools for the children; never buy books or records for the adults or children, or any toys, except in the small amounts available for birthday or Christmas presents ($50 per person over the year); never pay for a haircut; never buy a magazine; have no money for the feeding or veterinary care of any pets; and, never spend any money for preschool for the children, or educational trips for them away from home, or any summer camp or other activity with a fee.

Who Are the Poor?

Poverty in the United States is not randomly distributed, but rather is highly concentrated according to age and race/ethnicity, as indicated in Table 8.1, as well as gender.

AGE Today, children are at a much greater risk of living in poverty than are older persons. A generation ago, persons over age 65 were at the greatest risk of being poor; however, government programs such as Social Security and pension plans have been indexed for inflation and thus provide for something closer to an adequate standard of living than do other social welfare programs. Even so, older women are twice as

likely to be poor as older men; older African Americans and Latinos/as are much more likely to live below the poverty line than are non-Latino/a whites.

The age category most vulnerable to poverty today is the very young. One out of every three persons below the poverty line are under 18 years of age, and a large number of children hover just above the official poverty line. The precarious position of African American and Latino/a children is even more striking. In 2000, about 31 percent of all African Americans under age 18 lived in poverty; 28 percent of Latino/a children were also poor, as compared with about 10 percent of non-Latino/a white children (U.S. Census Bureau, 2001e).

What do such statistics indicate about the future of our society? Children as a group are poorer now than they were at the beginning of the 1980s, whether they live in one- or two-parent families. The majority live in two-parent families in which

Absolute poverty a level of economic deprivation that exists when people do not have the means to secure the most basic necessities of life.

Relative poverty a condition that exists when people may be able to afford basic necessities but are still unable to maintain an average standard of living.

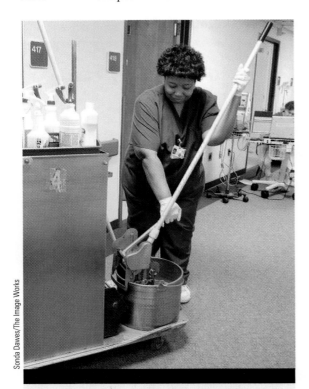

Sonda Dawes/The Image Works

Many women—men—are among the "working poor," who, although employed full time, have jobs in service occupations that are typically lower paying and less secure than jobs in other sectors of the labor market. Does the nature of women's work contribute to the feminization of poverty in the United States?

one or both parents are employed. However, children in single-parent households headed by women have a much greater likelihood of living in poverty: In 2000, approximately 33 percent of white (non-Latino/a) children under age 18 in female-headed households lived below the poverty line, as sharply contrasted with about 49 percent of Latina/o and 50 percent of African American children in the same category (U.S. Census Bureau, 2001e). Nor does the future look bright: Many governmental programs established to alleviate childhood poverty and malnutrition have been seriously cut back or eliminated altogether.

GENDER   About two-thirds of all adults living in poverty are women. In 2000, single-parent families headed by women had a 35-percent poverty rate as compared with a 10-percent rate for two-parent

families. Sociologist Diana Pearce (1978) coined a term to describe this problem: The *feminization of poverty* **refers to the trend in which women are disproportionately represented among individuals living in poverty.** According to Pearce (1978), women have a higher risk of being poor because they bear the major economic and emotional burdens of raising children when they are single heads of households but earn between 70 and 80 cents for every dollar a male worker earns—and the gap between men's and women's earnings has been increasing since 1994 (Lewin, 1997). More women than men are unable to obtain regular, full-time, year-round employment, and lack of adequate, affordable day care exacerbates this problem.

Does the feminization of poverty explain poverty in the United States today? Is poverty primarily a women's issue? On the one hand, this thesis highlights a genuine problem—the link between gender and poverty. On the other hand, several major problems exist with this argument. First, women's poverty is not a new phenomenon. Women have always been more vulnerable to poverty (see Katz, 1989). Second, all women are not equally vulnerable to poverty. Many in the upper and upper-middle classes have the financial resources, education, and skills to support themselves regardless of the presence of a man in the household. Third, event-driven poverty does not explain the realities of poverty for many women of color, who instead may experience "reshuffled poverty"—a condition of deprivation that follows them regardless of their marital status or the type of family in which they live. Research by Mary Jo Bane (1986; Bane and Ellwood, 1994) demonstrates that two out of three African American families headed by a woman were poor before the family event that made the woman a single mother. In addition, the poverty risk for a two-parent African American family is more than twice that for a white two-parent family.

Finally, poverty is everyone's problem, not just women's. When women are impoverished, so are their children. Moreover, many of the poor in our society are men, especially the chronically unemployed, older persons, the homeless, persons with disabilities, and men of color who have spent their adult lives without hope of finding work (Sidel, 1986).

RACE/ETHNICITY   According to some stereotypes, most of the poor and virtually all welfare recipients are people of color. However, this stereotype is false; white Americans (non-Latinos/as) account for approximately two-thirds of those below the official poverty line. However, such stereotypes are perpetuated because a disproportionate percentage of the impoverished in the United States are made up of African Americans, Latinos/as, and Native Americans. About 22 percent of African Americans and Latinas/os were among the officially poor in 2000, as compared with about 11 percent of non-Latino/a whites (U.S. Census Bureau, 2001e).

Native Americans are among the most severely disadvantaged persons in the United States. About one-third live below the poverty line, and some of these individuals live in conditions of extreme poverty. For example, a congressional study found that 70 percent of the Navajo households in Arizona, New Mexico, and Utah lived in houses without running water, sewer facilities, or electricity (Benokraitis, 1997). Some Native Americans receive governmental assistance; many others do not fall within the officially defined criteria for social welfare assistance.

Economic and Structural Sources of Poverty

Social inequality and poverty have both economic and structural sources. The low wages paid for many jobs is the major cause: Half of all families living in poverty are headed by someone who is employed, and one-third of those family heads work full time. A person with full-time employment in a minimum-wage job cannot keep a family of four from sinking below the official poverty line.

Structural problems contribute to both unemployment and underemployment. Corporations have been disinvesting in the United States, displacing millions of people from their jobs. Economists refer to this displacement as the *deindustrialization* of America (Bluestone and Harrison, 1982). Even as they have closed their U.S. factories and plants, many corporations have opened new facilities in other countries where "cheap labor" exists because people will, of necessity, work for lower wages. *Job deskilling*—a reduction in the proficiency needed to perform a specific job that leads to a corresponding reduction in the wages for that job—has resulted from the introduction of computers and other technology (Hodson and Parker, 1988). The shift from manufacturing to service occupations has resulted in the loss of higher-paying positions and

their replacement with lower-paying and less secure positions that do not offer the wages, job stability, or advancement potential of the disappearing manufacturing jobs. Many of the new jobs are located in the suburbs, thus making them inaccessible to central-city residents.

The problems of unemployment, underemployment, and poverty-level wages are even greater for people of color and young people in declining central cities (Collins, 1990). The unemployment rate for African Americans is 150 percent that of whites (U.S. Bureau of Labor Statistics, 2000). African Americans have also experienced gender differences in employment that may produce different types of economic vulnerability (Collins, 1990). African American men who find employment typically earn more than African American women; however, the men's employment is often less secure. In the past decade, African American men have been more likely to lose their jobs because of declining employment in the manufacturing sector (Bane, 1986; Collins, 1990; Bane and Ellwood, 1994).

Solving the Poverty Problem

The United States has attempted to solve the poverty problem in several ways. One of the most enduring is referred to as social welfare. When most people think of "welfare," they think of food stamps and programs such as Temporary Assistance for Needy Families (TANF) or the earlier program it replaced, Aid to Families with Dependent Children (AFDC). However, the primary beneficiaries of social welfare programs are not poor. Some analysts estimate that approximately 80 percent of all social welfare benefits are paid to people who do not qualify as "poor." For example, many recipients of Social Security are older people in middle- and upper-income categories (Barlett and Steele, 1994).

When older persons, including members of Congress, accept Social Security payments, they are not stigmatized. Similarly, veterans who receive

Feminization of poverty   the trend in which women are disproportionately represented among individuals living in poverty.

Job deskilling   a reduction in the proficiency needed to perform a specific job that leads to a corresponding reduction in the wages for that job.

Steve Rubin/The Image Works

According to a functional perspective, people such as these Harvard Law School graduates attain high positions in society because they are the most qualified and they work the hardest. Is our society a meritocracy? How would conflict theorists answer this question?

benefits from the Veterans Benefits Administration are not viewed as "slackers," and farmers who profit because of price supports are not considered to be lazy and unwilling to work. Unemployed workers who receive unemployment compensation are viewed with sympathy because of the financial plight of their families. By contrast, poor women

✓ Checkpoints

Checkpoint 8.5

14. How does absolute poverty differ from relative poverty?
15. Which age category in the United States is most vulnerable to poverty today?
16. What is the feminization of poverty?
17. Based on race/ethnicity, who is most likely to be below the official U.S. poverty line today?
18. What are the major structural sources of poverty?

and children who receive minimal benefits from welfare programs tend to be stigmatized and sometimes humiliated.

Recent thinking regarding overcoming poverty suggests that people with training and education are in the best position to "work their way out of poverty."

Sociological Explanations of Social Inequality in the United States

Obviously, some people are disadvantaged as a result of social inequality. Therefore, is inequality always harmful to society?

Functionalist Perspectives

According to the sociologists Kingsley Davis and Wilbert Moore (1945), inequality is not only inevitable but also necessary for the smooth functioning of society. The Davis–Moore thesis, which has become the definitive functionalist explanation for social inequality, can be summarized as follows:

1. All societies have important tasks that must be accomplished and certain positions that must be filled.
2. Some positions are more important for the survival of society than others.
3. The most important positions must be filled by the most qualified people.
4. The positions that are the most important for society and that require scarce talent, extensive training, or both must be the most highly rewarded.
5. The most highly rewarded positions should be those that are functionally unique (no other position can perform the same function) and on which other positions rely for expertise, direction, or financing.

Davis and Moore use the physician as an example of a functionally unique position. Doctors are very important to society and require extensive training, but individuals would not be motivated to go through

Reflections

Reflections 8.4

If the Davis–Moore thesis is correct, and inequality is necessary for the smooth functioning of society, does that mean that the American Dream simply isn't reasonably attainable for many people?

years of costly and stressful medical training without incentives to do so. The Davis–Moore thesis assumes that social stratification results in *meritocracy*—**a hierarchy in which all positions are rewarded based on people's ability and credentials.**

Critics have suggested that the Davis–Moore thesis ignores inequalities based on inherited wealth and intergenerational family status (Rossides, 1986). The thesis assumes that economic rewards and prestige are the only effective motivators for people and fails to take into account other intrinsic aspects of work, such as self-fulfillment (Tumin, 1953). It also does not adequately explain how such a reward system guarantees that the most qualified people will gain access to the most highly rewarded positions.

Conflict Perspectives

From a conflict perspective, people with economic and political power are able to shape and distribute the rewards, resources, privileges, and opportunities in society for their own benefit. Conflict theorists do not believe that inequality serves as a motivating force for people; they argue that powerful individuals and groups use ideology to maintain their favored positions at the expense of others. Core values in the United States emphasize the importance of material possessions, hard work, and individual initiative to get ahead, and behavior that supports the existing social structure. These same values support the prevailing resource distribution system and contribute to social inequality.

Are wealthy people smarter than others? According to conflict theorists, certain stereotypes suggest that this is the case; however, the wealthy may actually be "smarter" than others only in the sense of having "chosen" to be born to wealthy par-

ents from whom they could inherit assets. Conflict theorists also note that laws and informal social norms support inequality in the United States. For the first half of the twentieth century, both legalized and institutionalized segregation and discrimination reinforced employment discrimination and produced higher levels of economic inequality. Although laws have been passed to make these overt acts of discrimination illegal, many forms of discrimination still exist in educational and employment opportunities.

Symbolic Interactionist Perspectives

Symbolic interactionists focus on microlevel concerns and usually do not analyze larger structural factors that contribute to inequality and poverty. However, many significant insights on the effects of wealth and poverty on people's lives and social interactions can be derived from applying a symbolic interactionist approach. Using qualitative research methods and influenced by a symbolic interactionist approach, researchers have collected the personal narratives of people across all social classes, ranging from the wealthiest to the poorest people in the United States.

Microlevel studies of the wealthy have examined issues such as the social and psychological factors that influence members of the upper class to contribute vast sums of money to charitable and arts organizations (Odendahl, 1990; Ostrower, 1997). Other studies have focused on the social interactions involved in "male bonding" rituals of elite men's organizations such as the Bohemian Club, which has annual two-week retreats at the "Bohemian Grove" in Northern California. Several decades ago, the sociologist G. William Domhoff (1974) engaged in participant observation research (serving as a waiter) at the Bohemian Grove to examine how the interactions of wealthy business executives and powerful political leaders partaking in games, rituals, and other activities helped bind these male members of the upper class into a socially cohesive group. Domhoff (1974) concluded that one of the ways the upper class perpetuates its

Meritocracy a hierarchy in which all positions are rewarded based on people's ability and credentials.

Concept Table 8.A Sociological Explanations of Social Inequality in the United States

Functionalist perspectives	Some degree of social inequality is necessary for the smooth functioning of society (in order to fill the most important functions) and thus is inevitable.
Conflict perspectives	Powerful individuals and groups use ideology to maintain their favored positions in society at the expense of others, and wealth is not necessary in order to motivate people.
Symbolic interactionist perspectives	The beliefs and actions of people reflect their class location in society.

position is through social cohesion, which is furthered by a wide variety of small groups that encourage face-to-face interaction and ensure status and security for members. Similarly, other social scientists have looked at the social and psychological aspects of life in the middle class (Newman, 1988, 1993; Rubin, 1994) or the working and lower classes (Anderson, 1999; Fine and Weis, 1998; Nelson and Smith, 1999; Newman, 1999).

A few studies provide rare insights into the social interactions between people from vastly divergent class locations. Sociologist Judith Rollins's (1985) study of the relationship between household workers and their employers is one example. Based on in-depth interviews and participant observation, Rollins examined rituals of deference that were often demanded by elite white women of their domestic workers, who were frequently women of color. According to the sociologist Erving Goffman (1967), *deference* is a type of ceremonial activity that functions as a symbolic means whereby appreciation is regularly conveyed to a recipient. In fact, deferential behavior between nonequals (such as employers and employees) confirms the inequality of the relationship and each party's position in the relationship relative to the other. Rollins identified three types of linguistic deference between domestic workers and their employers: use of the first names of the workers, contrasted with titles and last names (Mrs. Adams, for example) of the employers; use of the term *girls* to refer to female household workers regardless of their age; and deferential references to employers, such as "Yes, ma'am." Spatial demeanor, including touching and how close one person stands to another, is an additional factor in deference rituals across class lines. Rollins (1985: 232) concludes that

✓**Checkpoints**

Checkpoint 8.6

19. What is the functionalist view on social inequality?
20. Why do conflict theorists believe that inequality is increasing in the United States?
21. What does the symbolic interactionist perspective contribute to our knowledge about U.S. social stratification?
22. Why is it important to study inequality from both the microlevel and the macrolevel perspectives?

The employer, in her more powerful position, sets the essential tone of the relationship; and that tone . . . is one that functions to reinforce the inequality of the relationship, to strengthen the employer's belief in the rightness of her advantaged class and racial position, and to provide her with justification for the inegalitarian social system.

Many concepts introduced by the sociologist Erving Goffman (1959, 1967) could be used as springboards for examining microlevel relationships between inequality and people's everyday interactions. What could you learn about class-based inequality in the United States by using a symbolic interactionist approach to examine a setting with which you are familiar?

Concept Table 8.A summarizes the three major perspectives on social inequality in the United States.

Box 8.3 You Can Make a Difference
Feeding the Hungry

The great fear among us all is that we are going to have to feed even more people. . . . It's not enough to just hand food out anymore.
—Robert Egger, director of the nonprofit Central Kitchen in Washington, D.C. (qtd. in Clines, 1996)

Egger is one of the people responsible for an innovative chef's training program that feeds hope as well as hunger. At the Central Kitchen, located in the nation's capital, staff and guest chefs annually train around 48 homeless persons in three-month-long kitchen-arts courses. While the trainees are learning about food preparation, which will help them get starting jobs in the restaurant industry, they are helping feed about 3,000 homeless persons each day. Much of the food is prepared using donated goods such as turkeys that people have received as gifts at office parties and given to the kitchen, and leftover food from grocery stores including 7-Eleven stores, restaurants such as Pizza Hut, hotel food services, and college cafeterias. Capital Kitchen got its start using leftovers from President George Bush's inaugural banquet in the late 1980s (Clines, 1996). Recently, donated food has gotten a boost from the Good Samaritan law passed in 1996, which exempts nonprofit organizations and *gleaners*—volunteers who collect what is

left in the field after harvesting—from liability for problems with food that they contribute in good faith (see Burros, 1996).

Can you think of ways that leftover food could be recovered from places where you eat so the food could be redistributed to persons in need? Have you thought about suggesting that members of an organization to which you belong might donate their time to help the Salvation Army, Red Cross, or other voluntary organization to collect, prepare, and serve food to others? If you would like to know more,

- "A Citizens Guide to Food Recovery" is available to help individuals participate in food recovery. Call 800-GLEAN-IT, or call the Salvation Army in your community. On the Internet:

http://www.salvationarmy.org

- World Hunger Year has projects such as Reinvesting in America that try to end hunger:

http://www.worldhungeryear.org

- Contact the American Red Cross:

http://www.redcross.org

U.S. Stratification in the Future

Will social inequality in the United States increase, decrease, or remain the same in the future? Many social scientists believe that existing trends point to an increase. First, the purchasing power of the dollar has stagnated or declined since the early 1970s. As families started to lose ground financially, more family members (especially women) entered the labor force in an attempt to support themselves and their families (Gilbert, 2003). Economist and former Secretary of Labor Robert Reich (1993) has noted that in recent years the employed have been traveling on two escalators—one going up and the other going down. The gap between the earnings of workers and the income of managers and top executives has widened (Feagin and Feagin, 2003).

Second, wealth continues to become more concentrated at the top of the U.S. class structure. As the rich have grown richer, more people have found themselves among the ranks of the poor. Third, federal tax laws in recent years have benefited corporations and wealthy families at the expense of middle- and lower-income families. Finally, structural sources of upward mobility are shrinking whereas the rate of downward mobility has increased.

Are we sabotaging our future if we do not work constructively to eliminate poverty? It has been said that a chain is no stronger than its weakest link. If we apply this idea to the problem of poverty, then it is to our advantage to see that those who cannot find work or do not have a job that provides a living wage receive adequate training and employment. Innovative programs can combine job training with producing something useful to meet the immediate needs of people living in poverty. Children of today—the adults of tomorrow—need nutrition, education, health care, and safety as they grow up (see Box 8.3).

Some social analysts believe that the United States will become a better nation if it attempts to regain the American Dream by attacking poverty. According to the sociologist Michael Harrington (1985: 13), if we join in solidarity with the poor, we will "rediscover our own best selves . . . we will regain the vision of America."

Chapter Review

How does income differ from wealth?
Whereas income is the economic gain derived from wages, salaries, income transfers (governmental aid), and ownership of property, wealth is the value of all of a person's or family's economic assets, including income, personal property, and income-producing property.

How did classical sociologists such as Karl Marx and Max Weber view social class?
Karl Marx and Max Weber acknowledged social class as a key determinant of social inequality and social change. For Marx, people's relationship to the means of production determines their class position. Weber developed a multidimensional concept of stratification that focuses on the interplay of wealth, prestige, and power.

What are some of the consequences of inequality in the United States?
The stratification of society into different social groups results in wide discrepancies in income and wealth and in variable access to available goods and services. People with high income or wealth have greater opportunity to control their own lives. People with less income have fewer life chances and must spend their limited resources to acquire basic necessities.

How do sociologists view poverty?
Sociologists distinguish between absolute poverty and relative poverty. Absolute poverty exists when people do not have the means to secure the basic necessities of life. Relative poverty exists when people may be able to afford basic necessities but are still unable to maintain an average standard of living.

Who are the poor?
Age, gender, and race tend to be factors in poverty. Children have a greater risk of being poor than do the elderly; women have a higher rate of poverty than do men. Although whites account for approximately two-thirds of those below the poverty line, people of color account for a disproportionate share of the impoverished in the United States. As the gap between rich and poor and be-

tween employed and unemployed widens, social inequality will clearly continue to increase if we do nothing.

What is the functionalist view on class?
Functionalist perspectives view classes as broad groupings of people who share similar levels of privilege on the basis of their roles in the occupational structure. According to the Davis–Moore thesis, stratification exists in all societies, and some inequality is not only inevitable but also necessary for the ongoing functioning of society. The positions that are most important within society and that require the most talent and training must be highly rewarded.

What is the conflict view on class?
Conflict perspectives on class are based on the assumption that social stratification is created and maintained by one group in order to enhance and protect its own economic interests. Conflict theorists measure class according to people's relationships with others in the production process. Erik Olin Wright identified four criteria for placement in a capitalist class structure: (1) ownership of the means of production, (2) purchase of the labor of others, (3) control of the labor of others, and (4) sale of one's own labor.

Key Terms

absolute poverty 264
alienation 251
capitalist class (bourgeoisie) 251
class conflict 251
feminization of poverty 266
income 245
job deskilling 267
meritocracy 269
pink-collar occupations 257
power 253
prestige 253
relative poverty 264
socioeconomic status (SES) 253
wealth 245
working class (proletariat) 251

Questions for Critical Thinking

1. Based on the Weberian and Marxian models of class structure, what is the class location of each of your ten closest friends or acquaintances? What is their location relative to yours? To one another's? What does their location tell you about friendship and social class?

2. Should employment be based on meritocracy, need, or affirmative action policies?
3. What might happen in the United States if the gap between rich and poor continues to widen?

 Resources on the Internet

Chapter-Related Web Sites

The following web sites have been selected for their relevance to the topics in this chapter. These sites are among the more stable, but please note that web site addresses change frequently.

U.S. Census Bureau: Poverty
http://www.census.gov/hhes/www/poverty.html
From this page you can access a number of census findings related to poverty in the United States.

U.S. Census Bureau: "Depth of Poverty" Measures
http://www.census.gov/hhes/poverty/depthpov.html
This page explains how the Census Bureau calculates poverty rates for each census.

U.S. Census Bureau: Who Are the Poor?
http://www.census.gov/hhes/poverty/povmeas/topicpg2.html
Here are a number of census papers detailing the various segments of the population that experience poverty.

Explorations in Social Inequality
http://www.trinity.edu/mkearl/strat.html
This site is an elaborate resource for learning about social class. There are a number of links to related articles and data sets.

Social Stratification
http://www.sdsmt.edu/online-courses/is/soc100/Soc_Strat.htm
A page with a number of links to important resources for learning about stratification in the United States.

Karl Marx, 1818–1883
http://www.historyguide.org/intellect/marx.html
This page contains a brief biography and references to several of Marx's important ideas.

Karl Marx's View of Class Differentiation
http://sharh.freenet.uz/lib/english/en_3.htm
In Marx's view, social relations during any period of history depend on who controls the primary mode of economic production. His analysis centered on how the relationships between various groups were shaped by differential access to scarce resources.

Inequality.org
http://www.inequality.org/
The site is a comprehensive source of information, statistics, stories, news, opinion, and academic articles on social and economic inequality.

Companion Web Site for This Book
Virtual Society: The Wadsworth Sociology Resource Center
Visit http://sociology.wadsworth.com and click on the page for Kendall, *Sociology in Our Times: The Essentials*, Fourth Edition, to access a wide range of enrichment material to aid in your study of sociology. Click on the Student Resources section of the web site. Next, select from the pull-down menu the chapter that you are presently studying. Among the useful options for self-study are chapter objectives, flashcards, practical study tips, and practice tests for each chapter.

MicroCase Online
From the Virtual Society home page, click on MicroCase Online to access book-specific MicroCase exercises that allow you to further explore sociological issues and principles presented in the text.

InfoTrac College Edition
Another unique option available to you at the Student Resources section of the companion web site is InfoTrac College Edition, an online library with access to hundreds of scholarly and popular periodicals. Below are suggested search terms for this chapter. Results from these and other searches are found at the site.

- Search keywords: *income distribution*. Locate articles that focus on income inequality in contemporary U.S. society. What types of race, class, or gender issues are discussed in the articles you found?
- Search keywords: *welfare reform*. Look for articles that evaluate the impact of recent welfare reform legislation. To narrow the range of articles, do a limit search and enter your state of residence to see how many articles discuss welfare reform within your state.

Virtual Explorations CD-ROM
Go to the Virtual Explorations CD-ROM to begin the interactive exercise for this chapter. This Virtual Exploration will introduce you to some of the exciting resources for sociology on the World Wide Web. You will be guided through an exercise that employs related web sites on social class in the United States. Answer the questions, and e-mail your responses to your instructor.

Race and Ethnicity

Will Terry/SIS

Race and Ethnicity
The Social Significance of Race and Ethnicity
Racial Classifications and the Meaning of Race
Dominant and Subordinate Groups

Prejudice
Stereotypes
Racism
Theories of Prejudice

Discrimination

**Sociological Perspectives on
   Race and Ethnic Relations**
Symbolic Interactionist Perspectives
Functionalist Perspectives
Conflict Perspectives
An Alternative Perspective: Critical Race Theory

Racial and Ethnic Groups in the United States
Native Americans
White Anglo-Saxon Protestants (British Americans)
African Americans
White Ethnic Americans
Asian Americans
Latinos/as (Hispanic Americans)
Middle Eastern Americans

Global Racial and Ethnic Inequality in the Future
Worldwide Racial and Ethnic Struggles
Growing Racial and Ethnic Diversity in the United States

9

REPORTER: Mr. Ashe, I guess [having AIDS] must be the heaviest burden you have ever had to bear, isn't it?

ARTHUR ASHE: No, it isn't. It's a burden, all right. But AIDS isn't the heaviest burden I have had to bear. . . . Being black is the greatest burden I've had to bear. . . . No question about it. Race has always been my biggest burden. Having to live as a minority in America. Even now it continues to feel like an extra weight tied around me. . . .

—tennis star Arthur Ashe, a Wimbledon champion in the 1970s and one of the early African American players to break the "color line," comparing racial discrimination with having AIDS (which he had contracted through a blood transfusion), a few months before his death (Ashe and Rampersad, 1994: 139–141)

Reuters NewMedia Inc./CORBIS

Were it not for early trailblazers such as African American tennis champion Arthur Ashe, top athletes today—including Venus and Serena Williams—might not have the opportunities they do to compete in tournaments and other sporting matches.

Would tennis stars such as the African American sisters Serena and Venus Williams—the 2002 Wimbledon women's singles champion and runner-up, respectively, and the finalists in 2001, when Venus won and Serena was runner-up—have always been welcomed on tennis courts and in matches around the globe? As strange as it may seem, some of the athletes who today have invigorated sports such as tennis and golf would not have been permitted to play the game a few decades ago because of their race. Historically, sports played in "private club" settings (such as tennis, golf, swimming, sailing, and polo) have been less accessible for people of color because of the exclusionary policies of some of the most prestigious clubs, where matches, tournaments, and regattas are likely to be held (Symonds, 1990). In this chapter, sports is used as an example of the effects of race and ethnicity on people's lives—and to examine whether those effects are changing. Before reading on, test your knowledge about race, ethnicity, and sports by taking the quiz in Box 9.1.

Questions and Issues

Chapter Focus Question: How is sports a reflection of racial and ethnic relations in the United States and other nations?

How do race and ethnicity differ?

How do discrimination and prejudice differ?

How are racial and ethnic relations analyzed according to sociological perspectives?

What are the unique experiences of racial and ethnic groups in the United States?

Race and Ethnicity

What is race? Some people think it refers to skin color (the Caucasian "race"); others use it to refer to a religion (the Jewish "race"), nationality (the British "race"), or the entire human species (the human "race") (Marger, 2000). Popular usages of race have been based on the assumption that a race is a grouping or classification based on *genetic* variations in physical appearance, particularly skin color. However, social scientists and biologists dispute the idea that biological race is a meaningful concept (Johnson, 1995). In fact, the idea of race has little meaning in a biological sense because of the enormous amount of interbreeding that has taken place within the human population. For these reasons, sociologists sometimes place "race" in quotation marks to show that categorizing individuals and population groups on biological characteristics is neither accurate nor based on valid distinctions

Box 9.1 Sociology and Everyday Life

How Much Do You Know About Race, Ethnicity, and Sports?

True	False	
T	F	1. Because sports are competitive and fans, coaches, and players want to win, the color of the players has not been a factor, only their performance.
T	F	2. In the late 1800s and early 1900s, boxing provided social mobility for some Irish, Jewish, and Italian immigrants.
T	F	3. African Americans who competed in boxing matches in the late 1800s often had to agree to lose before they could obtain a match.
T	F	4. Racially linked genetic traits explain many of the differences among athletes.
T	F	5. All racial and ethnic groups have viewed sports as a means to become a part of the mainstream.
T	F	6. In the National Football League, the positions of quarterback and kicker have been held almost exclusively by white players.
T	F	7. In recent years, players of color have moved into coaching, management, and ownership positions in professional sports.
T	F	8. The odds are good that many outstanding high school and college athletes will make the pros if they do not get injured.
T	F	9. Racism and sexism appear to be on the decline in sports in the United States.

Answers on page 278.

between the genetic makeup of differently identified "races" (Marshall, 1998). Today, sociologists emphasize that race is a *socially constructed reality*, not a biological one (Feagin and Feagin, 2003). From this approach, the social significance that people accord to race is more significant than any biological differences that might exist among people who are placed in arbitrary categories.

A *race* **is a category of people who have been singled out as inferior or superior, often on the basis of real or alleged physical characteristics such as skin color, hair texture, eye shape, or other subjectively selected attributes** (Feagin and Feagin, 2003). Categories of people frequently thought of as racial groups include Native Americans, Mexican Americans, African Americans, and Asian Americans.

As compared with race, *ethnicity* defines individuals who are believed to share common characteristics that differentiate them from the other collectivities in a society. An ***ethnic group* is a col-**

lection of people distinguished, by others or by themselves, primarily on the basis of cultural or nationality characteristics (Feagin and Feagin, 2003). Examples of ethnic groups include Jewish Americans, Irish Americans, and Italian Americans. Ethnic groups share five main characteristics: (1) *unique cultural traits*, such as language, clothing, holidays, or religious practices; (2) *a sense of community*; (3) *a feeling of ethnocentrism*; (4) *ascribed membership from birth*; and (5) *territoriality*, or the tendency to occupy a distinct geographic area (such as Little Italy or Little Havana) by choice and/or for self-protection. Although some people do not identify with any ethnic group, others participate in social interaction with individuals in their ethnic group and feel a sense of common identity based on cultural characteristics such as language, religion, or politics. However, ethnic groups are not only influenced by their own past history but also by patterns of ethnic domination and subordination in societies (Marshall, 1998).

The Social Significance of Race and Ethnicity

Race and ethnicity take on great social significance because how people act in regard to these terms drastically affects other people's lives, including what opportunities they have, how they are treated, and even how long they live. According to the sociologists Michael Omi and Howard Winant (1994: 158), race "permeates every institution, every relationship, and every individual" in the United States:

> As we . . . compare real estate prices in different neighborhoods, select a radio channel to enjoy while we drive to work, size up a potential client, customer, neighbor, or teacher, stand in line at the unemployment office, or carry out a thousand other normal tasks, we are compelled to think racially, to use the racial categories and meaning systems into which we have been socialized. (Omi and Winant, 1994: 158)

Historically, stratification based on race and ethnicity has pervaded all aspects of political, economic, and social life. Consider sports as an example. Throughout the early history of the game of baseball, many African Americans had outstanding skills as players but were categorically excluded from Major League teams because of their skin color. Even after Jackie Robinson broke the "color line" to become the first nonwhite in the Major Leagues in 1947, his experience was marred by racial slurs, hate letters, death threats against his infant son, and assaults on his wife (Ashe, 1988; Peterson, 1992). With some professional athletes from diverse racial–ethnic categories having multimillion-dollar contracts and lucrative endorsement deals, it is easy to assume that racism in sports is a thing of the past. However, this *commercialization* of sports does not mean that racial prejudice and discrimination no longer exist (Coakley, 1998).

Racial Classifications and the Meaning of Race

If we examine racial classifications throughout history, we find that in ancient Greece and Rome a person's race was the group to which she or he belonged, associated with an ancestral place and culture. From the Middle Ages until about the eighteenth century, a person's race was based on

family and ancestral ties, in the sense of a *line*, or ties to a national group. During the eighteenth century, physical differences such as the darker skin hues of Africans became associated with race, but racial divisions were typically based on differences in religion and cultural tradition rather than on human biology. With the intense (though misguided) efforts that surrounded the attempt to justify black slavery and white dominance in all areas of life during the second half of the nineteenth century, *races* came to be defined as distinct biological categories of people who were not all members of the same family but who shared inherited physical and cultural traits that were alleged to be different from those traits shared by people in other races. Hierarchies of races were established, placing the "white race" at the top, the "black race" at the bottom, and others in between.

Second, racial purity is assumed to exist. Prior to the 2000 census, for example, the true diversity of the U.S. population was not revealed in census data, since multiracial individuals were forced to either select a single race as being their "race" or to select the vague category of "other." Professional golfer Tiger Woods is an example of how people often have a mixed racial heritage. Woods describes himself as one-half Asian American (one-fourth Thai and one-fourth Chinese), one-eighth white, one-eighth Native American, and one-fourth African American (White, 1997). Census 2000 made it possible—for the first time—for individuals to classify themselves as being of more than one race (see "Census Profiles: Percent Distribution of Persons Reporting Two or More Races").

Third, categories of official racial classifications may (over time) create a sense of group membership or "consciousness of kind" for people within a somewhat arbitrary classification. When people of European descent were classified as

Race a category of people who have been singled out as inferior or superior, often on the basis of physical characteristics such as skin color, hair texture, and eye shape.

Ethnic group a collection of people distinguished, by others or by themselves, primarily on the basis of cultural or nationality characteristics.

Box 9.1

Answers to the Sociology Quiz on Race, Ethnicity, and Sports

1. **False.** Discrimination has been pervasive throughout the history of sports in the United States. For example, African American athletes, regardless of their abilities, were excluded from white teams for many years.

2. **True.** Irish Americans were the first to dominate boxing, followed by Jewish Americans and then Italian Americans. Boxing, like other sports, was a source of social mobility for some immigrants.

3. **True.** Promoters, who often set up boxing matches that pitted fighters by race, assumed that white fans were more likely to buy tickets if the white fighters frequently won.

4. **False.** Although some scholars and journalists have used biological or genetic factors to explain the achievements of athletes, sociologists view these explanations as being based on the inherently racist assumption that people have "natural" abilities (or disabilities) because of their race or ethnicity.

5. **False.** Some racial and ethnic groups—including Chinese Americans and Japanese Americans—have not viewed sports as a means of social mobility.

6. **True.** In the 1990s, whites accounted for about 90 percent of the quarterbacks and kickers on NFL teams. Different reasons have been given for this overrepresentation; however, some sociologists believe that a discriminatory practice known as stacking has been responsible for the racial distribution of players in football and baseball.

7. **False.** Although more African American players are employed by these teams (especially in basketball), their numbers have not increased significantly in coaching and management positions. Only one professional team is currently owned by an African American.

8. **False.** The odds of becoming a professional athlete are very low. The percentage of high school football players who make it to the pros is estimated at about .14 percent; for basketball, about .06 percent; and for baseball, about .18 percent. The percentages of college athletes who make it to the pros are a little higher: 3.5 percent for football, 2.6 percent for basketball, and 4.3 percent for baseball.

9. **False.** Even as people of color and white women have made gains on collegiate and professional teams, scholars have documented the continuing significance of racial and gender discrimination in sports.

Sources: Based on Coakley, 1998; Messner, Duncan, and Jensen, 1993; Hight, 1994; and Nelson, 1994.

"white," some began to see themselves as different from "nonwhite." Consequently, Jewish, Italian, and Irish immigrants may have felt more a part of the Northern European white mainstream in the late eighteenth and early nineteenth centuries. Whether Chinese Americans, Japanese Americans, Korean Americans, and Filipino Americans come to think of themselves collectively as "Asian Americans" because of official classifications remains to be seen (S. Lee, 1993).

In the future, increasing numbers of children in the United States will be likely to have a mixed racial or ethnic heritage such as this student describes:

> I am part French, part Cherokee Indian, part Filipino, and part black. Our family taught us to be aware of all these groups, and just to be ourselves. But I have never known what I am. People have asked if I am a Gypsy, or a Portuguese, or a Mexican, or lots of other things. It seems to make people curious, uneasy, and sometimes belligerent. Stu-

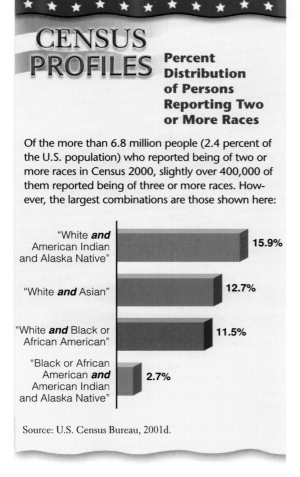

CENSUS PROFILES

Percent Distribution of Persons Reporting Two or More Races

Of the more than 6.8 million people (2.4 percent of the U.S. population) who reported being of two or more races in Census 2000, slightly over 400,000 of them reported being of three or more races. However, the largest combinations are those shown here:

"White **and** American Indian and Alaska Native"	15.9%
"White **and** Asian"	12.7%
"White **and** Black or African American"	11.5%
"Black or African American **and** American Indian and Alaska Native"	2.7%

Source: U.S. Census Bureau, 2001d.

dents I don't even know stop me on campus and ask, "What are you anyway?" (qtd. in Davis, 1991: 133)

How people are classified remains important because such classifications affect their access to employment, housing, social services, federal aid, and many other "publicly or privately valued goods" (Omi and Winant, 1994: 3). However, recent publicity about celebrities of mixed-race heritage, such as the golfer Tiger Woods, together with the change in the way that the Census Bureau asks about race, may lead more people to identify themselves as having more than one racial heritage.

Dominant and Subordinate Groups

The terms *majority group* and *minority group* are widely used, but their meanings are less clear as the composition of the U.S. population continues to change. Accordingly, many sociologists prefer the

Checkpoint 9.1

1. Why is race a socially constructed reality?
2. What are the main characteristics that members of an ethnic group share?
3. Why do sociologists prefer the terms *dominant group* and *subordinate group* to the terms *majority group* and *minority group* with regard to racial and ethnic groups in the United States?

terms *dominant* and *subordinate* to identify power relationships that are based on perceived racial, ethnic, or other attributes and identities. To sociologists, a **dominant group is one that is advantaged and has superior resources and rights in a society** (Feagin and Feagin, 2003). In the United States, whites with Northern European ancestry (often referred to as Euro-Americans, white Anglo-Saxon Protestants, or WASPs) are considered to be the dominant group. A **subordinate group is one whose members, because of physical or cultural characteristics, are disadvantaged and subjected to unequal treatment by the dominant group and who regard themselves as objects of collective discrimination** (Wirth, 1945). Persons of color and white women are considered to be subordinate-group members in the United States, particularly when these individuals are from lower-income categories.

In the sociological sense, the word *group* as used in these two terms is misleading because people who merely share ascribed racial or ethnic characteristics do not constitute a group. However, the terms *dominant group* and *subordinate group* do give us a way to describe relationships of advantage/disadvantage and power/exploitation that exist in contemporary nations.

Majority (dominant) group a group that is advantaged and has superior resources and rights in a society.

Minority (subordinate) group a group whose members, because of physical or cultural characteristics, are disadvantaged and subjected to unequal treatment by the dominant group and who regard themselves as objects of collective discrimination.

Prejudice

Prejudice **is a negative attitude based on faulty generalizations about members of selected racial, ethnic, or other groups** (Allport, 1958; Feagin and Feagin, 2003). The term *prejudice* is from the Latin words *prae* ("before") and *judicium* ("judgment"), which means that people may be biased either for or against members of other groups even before they have had any contact with them (Cashmore, 1996). Although prejudice can be either *positive* (bias in favor of a group—often our own) or *negative* (bias against a group—one we deem less worthy than our own), it most often refers to the negative attitudes that people may have about members of other racial or ethnic groups.

Stereotypes

Prejudice is rooted in ethnocentrism and stereotypes. When used in the context of racial and ethnic relations, *ethnocentrism* refers to the tendency to regard one's own culture and group as the standard—and thus superior—whereas all other groups are seen as inferior. Ethnocentrism is maintained and perpetuated by *stereotypes* **—overgeneralizations about the appearance, behavior, or other characteristics of members of particular categories.** Although stereotypes can be either positive or negative, examples of negative stereotyping abound in sports. Think about the Native American names, images, and mascots used by sports teams such as the Atlanta Braves, Cleveland Indians, Golden State Warriors, Washington Redskins, and Kansas City Chiefs. Members of Native American groups have been actively working to eliminate the use of stereotypic mascots (with feathers, buckskins, beads, spears, and "warpaint"), "Indian chants," and gestures (like the "tomahawk chop"), which they claim trivialize and exploit Native American culture (Churchill, 1994; Spindel, 2000). According to sociologist Jay Coakley (1998), the use of stereotypes and words such as *Redskin* symbolize a lack of understanding of the culture and heritage of native peoples and are offensive to many Native Ameri-

AP/Wide World Photos

Contemporary prejudice and discrimination cannot be understood without taking into account the historical background. School integration in the 1950s was accomplished despite white resistance. Today, integration in education, housing, and many other areas of social life remains a pressing social issue.

cans. Although some people see the use of these names and activities as "innocent fun," others view them as a form of racism.

Racism

What is racism? *Racism* **is a set of attitudes, beliefs, and practices that is used to justify the superior treatment of one racial or ethnic group and the inferior treatment of another racial or ethnic group.** The world has seen a long history of racism: It can be traced from the earliest civilizations. At various times throughout U.S. history, various categories of people, including the Irish Americans, Italian Americans, Jewish Americans, African Americans, and Latinos/as, have been the objects of racist ideology. However, not everyone is equally racist. Recent studies have shown that the underlying reasoning behind racism differs according to factors such as gender, age, class, and geography (see Cashmore, 1996; Feagin and Vera, 1995).

According to the frustration–aggression hypothesis, members of white supremacy groups such as the Ku Klux Klan often use members of subordinate racial and ethnic groups as scapegoats for societal problems over which they have no control.

Michael Greenlar/The Image Works

Racism may be overt or subtle. Overt racism is more blatant and may take the form of public statements about the "inferiority" of members of a racial or ethnic group. In sports, for example, the late Al Campanis (former Los Angeles Dodgers general manager) repeatedly stated on a nationwide television show that African Americans "lacked the necessities" to handle coaching and management positions. More recently, pitcher John Rocker (who at the time was with the Atlanta Braves) was widely criticized for referring to a subordinate-group teammate as "a fat monkey" during a *Sports Illustrated* interview and stating his dislike of all of the "foreigners" in this country.

Comments such as these are blatant, but subtle forms of racism are often hidden from sight and more difficult to prove. Examples of subtle racism in sports include those descriptions of African American athletes which suggest that they have "natural" abilities and are better suited for team positions requiring speed and agility. By contrast, whites are described as having the intelligence, dependability, and leadership and decision-making skills needed in positions requiring higher levels of responsibility and control.

Theories of Prejudice

Are some people more prejudiced than others? To answer this question, some theories focus on how individuals may transfer their internal psychological problem onto an external object or person (Feagin and Feagin, 2003). Others look at factors such as social learning and personality types.

The *frustration–aggression hypothesis* states that people who are frustrated in their efforts to achieve a highly desired goal will respond with a pattern of aggression toward others (Dollard et al., 1939). The object of their aggression becomes the *scapegoat*—a person or group that is incapable of offering resistance to the hostility or aggression of others (Marger, 2000). Scapegoats are often used as substitutes for the actual source of the frustration. For example, members of subordinate racial and ethnic groups are often blamed for societal problems (such as unemployment or an economic recession) over which they have no control.

According to some symbolic interactionists, prejudice results from social learning; in other words, it is learned from observing and imitating significant others, such as parents and peers. Initially,

Prejudice a negative attitude based on faulty generalizations about members of selected racial, ethnic, and other groups.

Stereotypes overgeneralizations about the appearance, behavior, or other characteristics of members of particular groups.

Racism a set of attitudes, beliefs, and practices that is used to justify the superior treatment of one racial or ethnic group and the inferior treatment of another racial or ethnic group.

Scapegoat a person or group that is incapable of offering resistance to the hostility or aggression of others.

✓**Checkpoints**

Checkpoint 9.2

4. What is prejudice?
5. What part do stereotypes play in prejudice?
6. What is racism, and what are its major forms?
7. How does the frustration–aggression hypothesis explain prejudice?
8. What is the authoritarian personality?

children do not have a frame of reference from which to question the prejudices of their relatives and friends. When they are rewarded with smiles or laughs for telling derogatory jokes or making negative comments about outgroup members, children's prejudiced attitudes may be reinforced.

Psychologist Theodor W. Adorno and his colleagues (1950) concluded that highly prejudiced individuals tend to have an *authoritarian personality*, which is characterized by excessive conformity, submissiveness to authority, intolerance, insecurity, a high level of superstition, and rigid, stereotypic thinking (Adorno et al., 1950). This type of personality is most likely to develop in a family environment in which dominating parents who are anxious about status use physical discipline but show very little love in raising their children (Adorno et al., 1950). Other scholars have linked prejudiced attitudes to traits such as submissiveness to authority, extreme anger toward outgroups, and conservative religious and political beliefs (Altemeyer, 1981, 1988; Weigel and Howes, 1985).

Discrimination

Whereas prejudice is an attitude, *discrimination* involves actions or practices of dominant-group members (or their representatives) that have a harmful impact on members of a subordinate group (Feagin and Feagin, 2003). Prejudiced attitudes do not always lead to discriminatory behavior. As shown in Figure 9.1, the sociologist Robert Merton (1949) identified four combinations of attitudes and responses. Unprejudiced nondiscriminators are not personally prejudiced and do not discriminate against others. For example, two players on a professional sports team may be best friends although they

Figure 9.1 Merton's Typology of Prejudice and Discrimination

	Prejudiced attitude?	Discriminatory behavior?
Unprejudiced nondiscriminator	No	No
Unprejudiced discriminator	No	Yes
Prejudiced nondiscriminator	Yes	No
Prejudiced discriminator	Yes	Yes

Merton's typology shows that some people may be prejudiced but not discriminate against others. Do you think that it is possible for a person to discriminate against some people without holding a prejudiced attitude toward them? Why or why not?

are of different races. Unprejudiced discriminators may have no personal prejudice but still engage in discriminatory behavior because of peer-group pressure or economic, political, or social interests. For example, a coach may feel no prejudice toward African American players but believe that white fans will accept only a certain percentage of people of color on the team (Coakley, 1998). Prejudiced nondiscriminators hold personal prejudices but do not discriminate due to peer pressure, legal demands, or a desire for profits. For example, a coach with prejudiced beliefs may hire an African American player to enhance the team's ability to win (Coakley, 1998). Finally, prejudiced discriminators hold personal prejudices and actively discriminate against others. For example, a baseball umpire who is personally prejudiced against African Americans may intentionally call a play incorrectly based on that prejudice.

Discriminatory actions vary in severity from the use of derogatory labels to violence against individuals and groups. The ultimate form of discrimination occurs when people are considered to be unworthy to live because of their race or ethnicity. *Genocide* is the deliberate, systematic

killing of an entire people or nation (Schaefer, 2000). Examples of genocide include the killing of thousands of Native Americans by white settlers in North America and the extermination of six million European Jews by Nazi Germany. More recently, the term *ethnic cleansing* has been used to define a policy of "cleansing" geographic areas (such as in Yugoslavia) by forcing persons of other races or religions to flee—or die (Schaefer, 1995).

Discrimination also varies in how it is carried out. Individuals may act on their own, or they may operate within the context of large-scale organizations and institutions, such as schools, churches, corporations, and governmental agencies. How does individual discrimination differ from institutional discrimination? ***Individual discrimination consists of one-on-one acts by members of the dominant group that harm members of the subordinate group or their property*** (Carmichael and Hamilton, 1967; Feagin and Feagin, 2003). For example, a person may decide not to rent an apartment to someone of a different race. By contrast, ***institutional discrimination consists of the day-to-day practices of organizations and institutions that have a harmful impact on members of subordinate groups*** (Feagin and Feagin, 2003). For example, a bank might consistently deny loans to people of a certain race. Institutional discrimination is carried out by the individuals who implement the policies and procedures of organizations.

Sociologist Joe R. Feagin has identified four major types of discrimination:

1. *Isolate discrimination* is harmful action intentionally taken by a dominant-group member against a member of a subordinate group. This type of discrimination occurs without the support of other members of the dominant group in the immediate social or community context. For example, a prejudiced judge may give harsher sentences to all African American defendants but may not be supported by the judicial system in that action.

2. *Small-group discrimination* is harmful action intentionally taken by a limited number of dominant-group members against members of subordinate groups. This type of discrimination is not supported by existing norms or other dominant-group members in the immediate social or community context. For example, a small group of white students may deface a professor's office with racist epithets without the support of other students or faculty members.

3. *Direct institutionalized discrimination* is organizationally prescribed or community-prescribed action that intentionally has a differential and negative impact on members of subordinate groups. These actions are routinely carried out by a number of dominant-group members based on the norms of the immediate organization or community (Feagin and Feagin, 2003). Intentional exclusion of people of color from public accommodations is an example of this type of discrimination.

4. *Indirect institutionalized discrimination* refers to practices that have a harmful impact on subordinate-group members even though the organizationally or community-prescribed norms or regulations guiding these actions were initially established with no intent to harm. For example, special education classes were originally intended to provide extra educational opportunities for children with various types of disabilities. However, critics claim that these programs have amounted to racial segregation in many school districts. Over a period of time, special education classes may have contributed to racial stereotyping that tells young people of color that "they will not make it in life, so why even try" (Richardson, 1994: B8).

Authoritarian personality a personality type characterized by excessive conformity, submissiveness to authority, intolerance, insecurity, a high level of superstition, and rigid, stereotypic thinking.

Discrimination actions or practices of dominant-group members (or their representatives) that have a harmful effect on members of a subordinate group.

Genocide the deliberate, systematic killing of an entire people or nation.

Individual discrimination behavior consisting of one-on-one acts by members of the dominant group that harm members of the subordinate group or their property.

Institutional discrimination the day-to-day practices of organizations and institutions that have a harmful impact on members of subordinate groups.

✓**Checkpoints**

Checkpoint 9.3

9. What is discrimination, and how is it related to prejudice?
10. What is genocide?
11. What are the four major types of discrimination?

Sociological Perspectives on Race and Ethnic Relations

Symbolic interactionist, functionalist, and conflict analysts examine race and ethnic relations in different ways. Functionalists focus on the macrolevel intergroup processes that occur between members of dominant and subordinate groups in society. Conflict theorists analyze power and economic differentials between the dominant group and subordinate groups. Symbolic interactionists examine how microlevel contacts between people may produce either greater racial tolerance or increased levels of hostility.

Symbolic Interactionist Perspectives

What happens when people from different racial and ethnic groups come into contact with one another? In the *contact hypothesis*, symbolic interactionists point out that contact between people from divergent groups should lead to favorable attitudes and behavior when certain factors are present. Members of each group must (1) have equal status, (2) pursue the same goals, (3) cooperate with one another to achieve their goals, and (4) receive positive feedback when they interact with one another in positive, nondiscriminatory ways (Allport, 1958; Coakley, 1998).

What happens when individuals meet someone who does not conform to their existing stereotype? Frequently, they ignore anything that contradicts the stereotype, or they interpret the situation to support their prejudices (Coakley, 1998). For example, a person who does not fit the stereotype may be seen as an exception—"You're not like other [persons of a particular race]."

When a person is seen as conforming to a stereotype, he or she may be treated simply as one

of "you people." Former Los Angeles Lakers basketball star Earvin "Magic" Johnson (1992: 31–32) described how he was categorized along with all other African Americans when he was bused to a predominantly white school:

> On the first day of [basketball] practice, my teammates froze me out. Time after time I was wide open, but nobody threw me the ball. At first I thought they just didn't see me. But I woke up after a kid named Danny Parks looked right at me and then took a long jumper. Which he missed.
>
> I was furious, but I didn't say a word. Shortly after that, I grabbed a defensive rebound and took the ball all the way down for a basket. I did it again and a third time, too.
>
> Finally Parks got angry and said, "Hey, pass the [bleeping] ball."
>
> That did it. I slammed down the ball and glared at him. Then I exploded. "I *knew* this would happen!" I said. "That's why I didn't want to come to this [bleeping] school in the first place!"
>
> "Oh, yeah? Well, you people are all the same," he said. "You think you're gonna come in here and do whatever you want? Look, hotshot, your job is to get the rebound. Let us do the shooting."

The interaction between Johnson and Parks demonstrates that when people from different racial and ethnic groups come into contact with one another, they may treat one another as stereotypes, not as individuals. Eventually, Johnson and Parks were able to work out most of their differences. "There's nothing like winning to help people get along," Johnson explained (1992: 32).

Symbolic interactionist perspectives make us aware of the importance of intergroup contact and the fact that it may either intensify or reduce racial and ethnic stereotyping and prejudice.

Functionalist Perspectives

How do members of subordinate racial and ethnic groups become a part of the dominant group? To answer this question, early functionalists studied immigration and patterns of majority- and minority-group interactions.

ASSIMILATION *Assimilation* **is a process by which members of subordinate racial and ethnic groups become absorbed into the dominant culture.** To some analysts, assimilation is functional because it contributes to the stability of society by minimizing group differences that might otherwise result in hostility and violence (Gordon, 1964).

Assimilation occurs at several distinct levels, including the cultural, structural, biological, and psychological stages. *Cultural assimilation*, or *acculturation*, occurs when members of an ethnic group adopt dominant-group traits, such as language, dress, values, religion, and food preferences. Cultural assimilation in this country initially followed an "Anglo conformity" model; members of subordinate ethnic groups were expected to conform to the culture of the dominant white Anglo-Saxon population (Gordon, 1964). However, members of some groups refused to be assimilated and sought to maintain their unique cultural identity.

Structural assimilation, or *integration*, occurs when members of subordinate racial or ethnic groups gain acceptance in everyday social interaction with members of the dominant group. This type of assimilation typically starts in large, impersonal settings such as schools and workplaces, and only later (if at all) results in close friendships and intermarriage. *Biological assimilation*, or *amalgamation*, occurs when members of one group marry those of other social or ethnic groups. Biological assimilation has been more complete in some other countries, such as Mexico and Brazil, than in the United States.

Psychological assimilation involves a change in racial or ethnic self-identification on the part of an individual. Rejection by the dominant group may prevent psychological assimilation by members of some subordinate racial and ethnic groups, especially those with visible characteristics such as skin color or facial features that differ from those of the dominant group.

ETHNIC PLURALISM   Instead of complete assimilation, many groups share elements of the mainstream culture while remaining culturally distinct from both the dominant group and other social and ethnic groups. **Ethnic pluralism is the coexistence of a variety of distinct racial and ethnic groups within one society.**

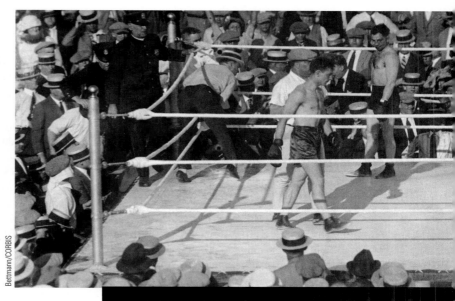

Bettmann/CORBIS

Are sports a source of upward mobility for recent immigrants and ethnic minorities, as was true for some in previous generations? For example, early-twentieth-century Jewish American and Italian American boxers not only introduced intragroup ethnic pride but also earned a livelihood in such sporting events.

Equalitarian pluralism, or *accommodation*, is a situation in which ethnic groups coexist in equality with one another. Switzerland has been described as a model of equalitarian pluralism; over six million people with French, German, and Italian cultural heritages peacefully coexist there (Simpson and Yinger, 1972). *Inequalitarian pluralism*, or *segregation*, exists when specific ethnic groups are set apart from the dominant group and have unequal access to power and privilege (Marger, 2000). **Segregation is the spatial and social separation of categories of**

Assimilation   a process by which members of subordinate racial and ethnic groups become absorbed into the dominant culture.

Ethnic pluralism   the coexistence of a variety of distinct racial and ethnic groups within one society.

Segregation   the spatial and social separation of categories of people by race, ethnicity, class, gender, and/or religion.

people by race, ethnicity, class, gender, and/or religion. Segregation may be enforced by law. *De jure segregation* refers to laws that systematically enforced the physical and social separation of African Americans in all areas of public life. An example of de jure segregation was the Jim Crow laws, which legalized the separation of the races in public accommodations (such as hotels, restaurants, transportation, hospitals, jails, schools, churches, and cemeteries) in the southern United States after the Civil War (Feagin and Feagin, 2003).

Segregation may also be enforced by custom. *De facto segregation*—racial separation and inequality enforced by custom—is more difficult to document than de jure segregation. For example, residential segregation is still prevalent in many U.S. cities; owners, landlords, real estate agents, and apartment managers often use informal mechanisms to maintain their properties for "whites only." Even middle-class African Americans find that racial polarization is fundamental to the residential layout of many cities.

Although functionalist explanations provide a description of how some early white ethnic immigrants assimilated into the cultural mainstream, they do not adequately account for the persistent racial segregation and economic inequality experienced by people of color.

Conflict Perspectives

Conflict theorists focus on economic stratification and access to power in their analyses of race and ethnic relations. Some emphasize the castelike nature of racial stratification, others analyze class-based discrimination, and still others examine internal colonialism and gendered racism.

THE CASTE PERSPECTIVE The caste perspective views racial and ethnic inequality as a permanent feature of U.S. society. According to this approach, the African American experience must be viewed as different from that of other racial or ethnic groups. African Americans were the only group to be subjected to slavery; when slavery was abolished, a caste system was instituted to maintain economic and social inequality between whites and African Americans (Dollard, 1957/1937).

The caste system was strengthened by *antimiscegenation laws*, which prohibited sexual intercourse or marriage between persons of different races. Most states had such laws, which were later expanded to include relationships between whites and Chinese, Japanese, and Filipinos. These laws were not declared unconstitutional until 1967 (Frankenberg, 1993).

Although the caste perspective points out that racial stratification may be permanent because of structural elements such as the law, it has been criticized for not examining the role of class in perpetuating racial inequality (Cox, 1948).

CLASS PERSPECTIVES Class perspectives emphasize the role of the capitalist class in racial exploitation. Based on early theories of race relations by the African American scholar W. E. B. Du Bois, the sociologist Oliver C. Cox (1948) suggested that African Americans were enslaved because they were the cheapest and best workers the owners could find for heavy labor in the mines and on plantations. Thus, the profit motive of capitalists, not skin color or racial prejudice, accounts for slavery.

More recently, sociologists have debated the relative importance of class and race in explaining the unequal life chances of African Americans. Sociologist William Julius Wilson (1996) has suggested that race, cultural factors, social psychological variables, and social class must all be taken into account in examining the life chances of "inner-city residents." His analysis focuses on how class-based economic determinants of social inequality, such as deindustrialization and the decline of the central (inner) city, have affected many African Americans, especially in the Northeast. African Americans were among the most severely affected by the loss of factory jobs because work in the manufacturing sector had previously made upward mobility possible. Wilson (1996) is not suggesting that prejudice and discrimination have been eradicated; rather, he is arguing that they may be less important than class in explaining the current status of African Americans.

How do conflict theorists view the relationship among race, class, and sports? Simply stated, sports reflects the interests of the wealthy and powerful. At all levels, sports exploits athletes (even highly paid ones) in order to gain high levels of profit and prestige for coaches, managers, and owners. African American athletes and central-city youths in particular are exploited by the message of rampant consumerism. Many are given the unrealistic expectation that sports can be a ticket out of the ghetto or barrio. If they try hard enough (and wear the right sneakers), they too can become

Reflections

Reflections 9.1

Is it possible that even highly paid athletes are the victims of racial exploitation?

wealthy and famous. However, even for those with virtually no chance of becoming professional athletes, consumerism is encouraged by media blitzes featuring the tennis shoes, clothes, and other gear worn by their favorite college and professional sports "heroes."

INTERNAL COLONIALISM Why do some racial and ethnic groups continue to experience subjugation after many years? According to the sociologist Robert Blauner (1972), groups that have been subjected to internal colonialism remain in subordinate positions longer than groups that voluntarily migrated to the United States. *Internal colonialism* **occurs when members of a racial or ethnic group are conquered or colonized and forcibly placed under the economic and political control of the dominant group.**

In the United States, indigenous populations (including groups known today as Native Americans and Mexican Americans) were colonized by Euro-Americans and others who invaded their lands and conquered them. In the process, indigenous groups lost property, political rights, aspects of their culture, and often their lives (Blauner, 1972). The capitalist class acquired cheap labor and land through this government-sanctioned racial exploitation (Blauner, 1972). The effects of past internal colonialism are reflected today in the number of Native Americans who live on government reservations and in the poverty of Mexican Americans who lost their land and had no right to vote.

The experiences of internally colonized groups are unique in three ways: (1) these groups have been forced to exist in a society other than their own; (2) they have been kept out of the economic and political mainstream, so it is difficult for them to compete with dominant-group members; and (3) they have been subjected to severe attacks on their culture, which may lead to its extinction (Blauner, 1972).

Grinding poverty is a pressing problem for families living along the border between the United States and Mexico. Economic development has been limited in areas where *colonias* such as this are are located, and the wealthy have derived far more benefit than others from recent changes in the global economy.

The internal colonialism perspective is rooted in the historical foundations of racial and ethnic inequality in the United States. However, it tends to view all voluntary immigrants as having many more opportunities than do members of colonized groups. Thus, this model does not explain the continued exploitation of some immigrant groups, such as the Chinese, Filipinos, Cubans, Vietnamese, and Haitians, and the greater acceptance of others, primarily those from Northern Europe (Cashmore, 1996).

THE SPLIT-LABOR-MARKET THEORY Who benefits from the exploitation of people of color? Dual- or split-labor-market theory states that white workers and members of the capitalist class both

Internal colonialism according to conflict theorists, a practice that occurs when members of a racial or ethnic group are conquered or colonized and forcibly placed under the economic and political control of the dominant group.

benefit from the exploitation of people of color. *Split labor market* **refers to the division of the economy into two areas of employment, a primary sector or upper tier, composed of higher-paid (usually dominant-group) workers in more-secure jobs, and a secondary sector or lower tier, composed of lower-paid (often subordinate-group) workers in jobs with little security and hazardous working conditions** (Bonacich, 1972, 1976). According to this perspective, white workers in the upper tier may use racial discrimination against nonwhites to protect their positions. These actions most often occur when upper-tier workers feel threatened by lower-tier workers hired by capitalists to reduce labor costs and maximize corporate profits. In the past, immigrants were a source of cheap labor that employers could use to break strikes and keep wages down. Throughout U.S. history, higher-paid workers have responded with racial hostility and joined movements to curtail immigration and thus do away with the source of cheap labor (Marger, 2000).

Proponents of the split-labor-market theory suggest that white workers benefit from racial and ethnic antagonisms. However, these analysts typically do not examine the interactive effects of race, class, and gender in the workplace.

RECENT PERSPECTIVES ON RACE AND GENDER The term *gendered racism* refers to the interactive effect of racism and sexism on the exploitation of women of color. According to the social psychologist Philomena Essed (1991), women's particular position must be explored within each racial or ethnic group because their experiences will not have been the same as men's in each grouping.

All workers are not equally exploited by capitalists. Gender and race or ethnicity are important in this exploitation. Historically, the high-paying primary labor market has been monopolized by white men. People of color and most white women more often hold lower-tier jobs. Below that tier is the underground sector of the economy, characterized by illegal or quasi-legal activities such as drug trafficking, prostitution, and working in sweatshops that do not meet minimum wage and safety standards. Many undocumented workers and some white women and people of color attempt to earn a living in this sector (Amott and Matthaei, 1996).

The *theory of racial formation* states that actions of the government substantially define racial and ethnic relations in the United States. Government actions range from race-related legislation to imprisonment of members of groups believed to be a threat to society. Sociologists Michael Omi and Howard Winant (1994) suggest that the U.S. government has shaped the politics of race through actions and policies that cause people to be treated differently because of their race. For example, immigration legislation reflects racial biases. The Naturalization Law of 1790 permitted only white immigrants to qualify for naturalization; the Immigration Act of 1924 favored Northern Europeans and excluded Asians and Southern and Eastern Europeans.

The government's definition of racial realities is periodically challenged by social protest movements of various racial and ethnic groups. When this social rearticulation occurs, people's understanding about race may be restructured somewhat. For example, the African American protest movements of the 1950s and 1960s helped redefine the rights of people of color in the United States.

An Alternative Perspective: Critical Race Theory

Emerging out of scholarly law studies on racial and ethnic inequality, critical race theory derives its foundation from the U.S. civil rights tradition and the writing of persons such as Martin Luther King, Jr., W. E. B. Du Bois, Malcolm X, and César Chávez. Critical race theory has several major premises, including the belief that racism is such an ingrained feature of U.S. society that it appears to be ordinary and natural to many people (Delgado, 1995). As a result, civil rights legislation and affirmative action laws (formal equality) may remedy some of the more overt, blatant forms of racial injustice but have little effect on subtle, business-as-usual forms of racism that people of color experience as they go about their everyday lives. According to this approach, the best way to document racism and ongoing inequality in society is to listen to the lived experiences of people who have experienced such discrimination. In this way, we can learn what actually happens in regard to racial oppression and the many effects it has on people, including alienation, depression, and certain physical illnesses. Central to this argument is the belief that *interest convergence* is a crucial factor in bringing about social change. According to the legal scholar Derrick Bell, white

✓**Checkpoints**

Checkpoint 9.4

12. How do symbolic interactionist perspectives explain intergroup contact?
13. What are the key functionalist perspectives on racial and ethnic relations?
14. What are the key points in conflict theories on racial and ethnic relations?
15. What is internal colonialism?
16. What is critical race theory?

elites tolerate or encourage racial advances for people of color *only* if the dominant-group members believe that their own self-interest will be served in so doing (cited in Delgado, 1995). From this approach, civil rights laws have typically benefited white Americans as much (or more) as people of color because these laws have been used as mechanisms to ensure that "racial progress occurs at just the right pace: change that is too rapid would be unsettling to society at large; change that is too slow could prove destabilizing" (Delgado, 1995: xiv).

Critical race theory is similar to postmodernist approaches in that it calls our attention to the fact that things are not always as they seem. Formal equality under the law does not necessarily equate to actual equality in society. This theory also makes us aware of the ironies and contradictions in civil rights law, which some see as self-serving laws. However, mainstream critics argue that critical race theory is unduly pessimistic because it largely ignores the strides that have been made toward racial equality in the United States. Concept Table 9.A outlines the key aspects of each sociological perspective.

Racial and Ethnic Groups in the United States

How do racial and ethnic groups come into contact with one another? How do they adjust to one another and to the dominant group over time? Sociologists have explored these questions extensively; however, a detailed historical account of the unique experiences of each group is beyond the scope of this chapter. Instead, we will look briefly at intergroup contacts. In

the process, sports is used as an example of how members of some groups have attempted to gain upward mobility and become integrated into society.

Native Americans

Native Americans are believed to have migrated to North America from Asia thousands of years ago, as shown on the time line in Figure 9.2. One of the most widely accepted beliefs about this migration is that the first groups of Mongolians made their way across a natural bridge of land called Beringia into present-day Alaska. From there, they moved to what is now Canada and the northern United States, eventually making their way as far south as the tip of South America (Cashmore, 1996).

As schoolchildren are taught, Spanish explorer Christopher Columbus first encountered the native inhabitants in 1492 and referred to them as "Indians." When European settlers (or invaders) arrived on this continent, the native inhabitants' way of life was changed forever. Experts estimate that approximately two million native inhabitants lived in North America at that time (Cashmore, 1996); however, their numbers had been reduced to less than 240,000 by 1900 (Churchill, 1994).

GENOCIDE, FORCED MIGRATION, AND FORCED ASSIMILATION Native Americans have been the victims of genocide and forced migration. Although the United States never had an official policy that set in motion a pattern of deliberate extermination, many Native Americans either were massacred or died from European diseases (such as typhoid, smallpox, and measles) and starvation (Wagner and Stearn, 1945; Cook, 1973). In battle, Native Americans were often no match for the Europeans, who had "modern" weaponry (Amott and Matthaei, 1996). Europeans justified

Split labor market a term used to describe the division of the economy into two areas of employment, a primary sector or upper tier, composed of higher-paid (usually dominant-group) workers in more-secure jobs, and a secondary sector or lower tier, composed of lower-paid (often subordinate-group) workers in jobs with little security and hazardous working conditions.

Concept Table 9.A Sociological Perspectives on Race and Ethnic Relations

	Focus	Theory/Hypothesis
Symbolic Interactionist	Microlevel contacts between individuals	Contact hypothesis

Functionalist	Macrolevel intergroup processes	1. Assimilation
		  a. cultural
		  b. biological
		  c. structural
		  d. psychological
		2. Ethnic pluralism
		  a. equalitarian pluralism
		  b. inequalitarian pluralism (segregation)

Conflict	Power/economic differentials between dominant and subordiante groups	1. Caste perspective 2. Class perspective 3. Internal colonialism 4. Split labor market 5. Gendered racism 6. Racial formation
Critical Race Theory	Racism as an ingrained feature of society that affects everyone's daily life	Laws may remedy overt discrimination but have little effect on subtle racism. Interest convergence is required for social change.

their aggression by stereotyping the Native Americans as "savages" and "heathens" (Takaki, 1993).

After the Revolutionary War, the federal government offered treaties to the Native Americans so that more of their land could be acquired for the growing white population. Scholars note that the government broke treaty after treaty as it engaged in a policy of wholesale removal of indigenous nations in order to clear the land for settlement by Anglo-Saxon "pioneers" (Green, 1977; Churchill, 1994). Entire nations were forced to move in order

to accommodate the white settlers. The "Trail of Tears" was one of the most disastrous of the forced migrations. In the coldest part of the winter of 1832, over half of the Cherokee Nation died during or as a result of their forced relocation from the southeastern United States to the Indian Territory in Oklahoma (Thornton, 1984).

Native Americans were made wards of the government (meaning they had a legal status similar to that of minors and incompetents) and were subjected to forced assimilation on the reservations

Figure 9.2 Time Line of Racial and Ethnic Groups in the United States

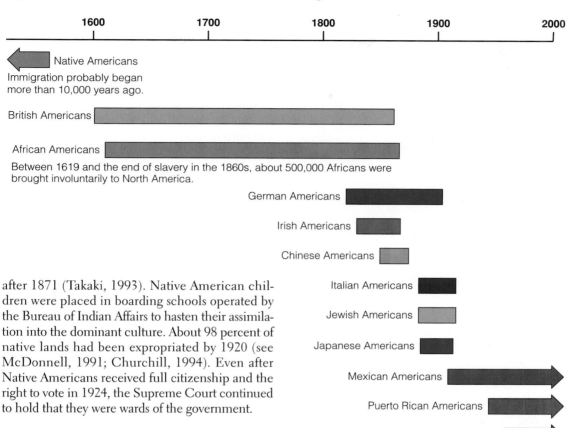

after 1871 (Takaki, 1993). Native American children were placed in boarding schools operated by the Bureau of Indian Affairs to hasten their assimilation into the dominant culture. About 98 percent of native lands had been expropriated by 1920 (see McDonnell, 1991; Churchill, 1994). Even after Native Americans received full citizenship and the right to vote in 1924, the Supreme Court continued to hold that they were wards of the government.

NATIVE AMERICANS TODAY Currently, about two million Native Americans live in the United States, including Aleuts, Inuit (Eskimos), Cherokee, Navajo, Chippewa, Sioux, and over five hundred other nations of varying sizes and different locales. Most are concentrated in the Southwest, and about one-third live on reservations. Native Americans are the most disadvantaged racial or ethnic group in the United States in terms of income, employment, housing, nutrition, and health. The life chances of Native Americans who live on reservations are especially limited. They have the highest rates of infant mortality and death by exposure and malnutrition. They also have high rates of tuberculosis, alcoholism, and suicide (Bachman, 1992). Reservation-based Native American men have an average life expectancy of less than forty-five years; for women, it is less than forty-eight years (see Churchill, 1994). Native Americans have had very limited educational opportunities and have a very high rate of unemployment. In recent years, however, a network of tribal colleges has been successful in providing some

Native Americans with the education they need to move into the ranks of the skilled working class and beyond (Bordewich, 1996).

In spite of the odds against them, many Native Americans resist oppression. The American Indian Movement, Women of All Red Nations, and other groups have demanded the recovery of Native American lands and reparation for past losses (Serrill, 1992). Native American women have publicized the harmful conditions (including radiation sickness and the forced sterilization of women) that exist on reservations. The American Indian Anti-Defamation Council advocates doing away with the word *tribe* because it demeans Native Americans by equating their level of cultural attainment with "primitivism" or "barbarism" (see Churchill, 1994). Moreover, reinterpretation of federal law in the 1990s has made it possible for Native American nations to open lucrative cigarette shops, bingo

David Butow/Black Star

Life chances are extremely limited for Native Americans who live on reservations. Although a few Native Americans early in the twentieth century were well-known athletes, Native Americans today have little opportunity to compete in sports at the college, professional, or Olympic level.

halls, and casino gambling operations on reservations (Cashmore, 1996).

NATIVE AMERICANS AND SPORTS Early in the twentieth century, Native Americans such as Jim Thorpe gained national visibility as athletes in football, baseball, and track and field. Teams at boarding schools such as the Carlisle Indian Industrial School in Pennsylvania and the Haskell Institute in Kansas were well-known. However, after the first three decades of the twentieth century, Native Americans became much less prominent in sports. Although some Navajo athletes have been very successful in basketball and some Choctaws have excelled in baseball, Native Americans have seldom been able to compete at the college, professional, or Olympic level (Blanchard, 1980; Oxendine, 1995). Native American scholar Joseph B. Oxendine (1995) attributes the lack of athletic participation to these factors: (1) reduction in opportunities for developing sports skills, (2) restricted opportunities for participation, and (3) a lessening of Native Americans' interest in competing with and against non-Native Americans.

White Anglo-Saxon Protestants (British Americans)

Whereas Native Americans have been among the most disadvantaged peoples, white Anglo-Saxon Protestants (WASPs) have been the most privileged group in this country. Although many English settlers initially came to North America as indentured servants or as prisoners, they quickly emerged as the dominant group, creating a core culture (including language, laws, and holidays) to which all other groups were expected to adapt. Most of the WASP immigrants arriving from northern Europe were advantaged over later immigrants because they were highly skilled and did not experience high levels of prejudice and discrimination.

CLASS, GENDER, AND WASPS Like members of other racial and ethnic groups, not all WASPs are alike. Social class and gender affect their life chances and opportunities. For example, members of the working class and the poor do not have political and economic power; men in the capitalist class do. WASPs constitute the majority of the upper class and maintain cohesion through listings such as the *Social Register* and interactions with one another in elite settings such as private schools and country clubs (Higley, 1997). However, WASP women do not always have the same rights as the men of their group. Although WASP women have the privilege of a dominant racial position, they do not have the gender-related privileges of men (Amott and Matthaei, 1996).

WASPS AND SPORTS Family background, social class, and gender play an important role in the sports participation of WASPs. Contemporary North American football was invented at the Ivy League colleges and was dominated by young, affluent WASPs who had the time and money to attend college and participate in sports activities. As Table 9.1 shows, whites are more likely than any other racial or ethnic group to become professional athletes in all sports except football and basketball.

Affluent WASP women participated in intercollegiate women's basketball in the late 1800s, and various other sporting events were used as a

Table 9.1 Odds of Becoming a Professional Athlete by Race/Ethnicity and Sport

Many young people dream of becoming a professional athlete; however, as this table shows, the odds of actually becoming one are very small.

	Race/Ethnicity			
	White	African American	Latino/a	Asian American
Football	1 in 62,500	1 in 47,600	1 in 2,500,000	1 in 5,000,000
Baseball	1 in 83,300	1 in 333,300	1 in 500,000	1 in 50,000,000
Basketball	1 in 357,100	1 in 153,800	1 in 33,300,000	—
Hockey	1 in 66,700	—	—	—
Golf				
Men's	1 in 312,500	1 in 12,500,000	1 in 33,300,000	1 in 20,000,000
Women's	1 in 526,300	—	1 in 33,300,000	1 in 3,300,000
Tennis				
Men's	1 in 285,700	1 in 2,000,000	1 in 3,300,000	—
Women's	1 in 434,800	1 in 20,000,000	1 in 20,000,000	—

Note: The odds of Native Americans participating in professional football are 1 in 12,500,000; they are not represented in other professional sports.

Source: Leonard and Reyman, 1988: 162–169.

means to break free of restrictive codes of femininity (Nelson, 1994). Until recently, however, most women have had little chance for any involvement in college and professional sports.

African Americans

The African American (black) experience has been one uniquely marked by slavery, segregation, and persistent discrimination. There is a lack of consensus about whether *African American* or *black* is the most appropriate term to refer to the 35 million Americans of African descent who live in the United States today. Those who prefer the term *black* point out that it incorporates many African-descent groups living in this country that do not use *African American* as a racial or ethnic self-description. For example, people who trace their origins to Haiti, Puerto Rico, or Jamaica typically identify themselves as "black" but not as "African American" (Cashmore, 1996).

Although the earliest African Americans probably arrived in North America with the Spanish conquerors in the fifteenth century, most historians trace their arrival to about 1619, when the first groups of indentured servants were brought to the colony of Virginia. However, by the 1660s, indentured servanthood had turned into full-fledged slavery with the enactment of laws in states such as Virginia that sanctioned the enslavement of African Americans. Although the initial status of persons of African descent in this country may not have been too different from that of the English indentured servants, all of that changed with the passage of laws turning human beings into property and making slavery a status from which neither individuals nor their children could escape (Franklin, 1980). Slavery laws dehumanized not only the people who were labeled as slaves but also the slave owners and political leaders who created and maintained this so-called "peculiar institution" (Feagin and Feagin, 2003).

Between 1619 and the 1860s, about 500,000 Africans were forcibly brought to North America, primarily to work on southern plantations, and these actions were justified by the devaluation and

Box 9.2 Sociology and Social Policy

The Continuing Debate Over Affirmative Action in Colleges and Universities

Education was one of the earliest targets of social policy pertaining to civil rights in the United States. Increased educational opportunity has been a goal of many subordinate-group members because of the widely held belief that education is the key to economic and social advancement. However, virtually all social policies seeking to bring about gains for excluded or disadvantaged people have been met with political debates that are fought out in the courts, the media, and other public opinion venues. Affirmative action programs in colleges and universities are no exception.

Even after passage of the Civil Rights Act of 1964 and other initiatives to promote equality of employment opportunity, regardless of race, religion, national origin, or gender, educational institutions—particularly colleges and graduate schools—remained predominantly white and male.

In the 1970s, some schools began to establish affirmative action programs to attract a more diverse student population and provide more equal educational opportunities to all students in a state or region. The 1978 U.S. Supreme Court decision in

Bakke v. The University of California at Davis inadvertently set up the parameters by which affirmative action programs have been evaluated in recent decades. In that case, Allan Bakke, a white male, was denied admission to the University of California at Davis medical school even though his grade point average and Medical College Admissions Test score were higher than those of some students of color who gained admission under a special program. Although the Supreme Court ruled that Bakke should be admitted to the medical school, the court did not prohibit colleges and universities from implementing plans to increase diversity in their student populations. During the 1980s and 1990s, most schools developed guidelines for admissions, financial aid, scholarships, and faculty hiring that took race, ethnicity, and gender into account.

The debate over affirmative action intensified with the 1996 U.S. Fifth Circuit Court of Appeals ruling in *Hopwood v. State of Texas*. This case is named for Cheryl Hopwood, one of four white students denied admission to the University of Texas law school who claimed that they were the victims of reverse discrimination because they had higher

---

stereotyping of African Americans. Some analysts believe that the central factor associated with the development of slavery in this country was the plantation system, which was heavily dependent on cheap and dependable manual labor. Slavery was primarily beneficial to the wealthy southern plantation owners, but many of the stereotypes used to justify slavery were eventually institutionalized in southern custom and practice (Wilson, 1978). However, some slaves and whites engaged in active resistance against slavery and its barbaric practices, eventually resulting in slavery being outlawed in the northern states by the late 1700s. Slavery continued in the South until 1863, when it was abolished by the Emancipation Proclamation (Takaki, 1993).

SEGREGATION AND LYNCHING Gaining freedom did not give African Americans equality with whites. African Americans were subjected to

many indignities because of race. Through informal practices in the North and *Jim Crow laws* in the South, African Americans experienced segregation in housing, employment, education, and all public accommodations. (These laws were referred to as Jim Crow laws after a derogatory song about a black man.) African Americans who did not stay in their "place" were often the victims of violent attacks and lynch mobs (Franklin, 1980). *Lynching* is a killing carried out by a group of vigilantes seeking revenge for an actual or imagined crime by the victim (Feagin and Feagin, 2003). Lynchings were used by whites to intimidate African Americans into staying "in their place." It is estimated that as many as 6,000 lynchings occurred between 1892 and 1921 (Feagin and Feagin, 2003). In spite of all odds, many African American women and men resisted oppression and did not give up in their struggle for equality (Amott and Matthaei, 1996).

scores than some subordinate-group students who gained admission. In order to promote diversity, for a number of years the law school had been admitting some African Americans and other students of color despite the fact that they had lower scores on the law school admissions test. According to the appeals court, the law school's affirmative action policies discriminated against whites. The court ruled that universities in Texas, Louisiana, and Mississippi (the states in the jurisdiction of the Fifth Circuit) could no longer use race as a factor in determining which applicants to admit. The Eleventh Circuit Court of Appeals (which has jurisdiction in the states of Alabama, Florida, and Georgia) also struck down race-conscious admissions criteria. As a result, colleges and universities in these states largely dismantled existing affirmative action programs, and a dramatic drop in minority enrollment took place in many schools until some of them reassessed admissions criteria, such as whether a student was in the top ten percent of his or her high school class or had other unique attributes or skills that enhanced the overall quality of the student population.

In 2002, however, the Sixth Circuit Court of Appeals (which has jurisdiction in Kentucky, Michigan, Ohio, and Tennessee) ruled in *Grutter v. Bollinger* that a law school may take race into account in its admissions policies as long as such a policy is not used to establish a quota system. The majority of

the court in *Grutter* held that race and ethnicity could be considered "potential 'plus' factors in a particular applicant's file [as long as they did not] insulate an applicant from competition or act to foreclose competition" from other applicants. Such an affirmative action policy, according to the court, was essential for educational diversity—making sure that students were exposed to other students from different backgrounds and with different views.

In light of the different rulings by the circuit courts in *Hopwood* and *Grutter,* it is likely that the U.S. Supreme Court will review this latest decision. In the meantime, however, there are more questions than answers regarding how to achieve greater representation of all racial and ethnic groups and of both women and men among students, faculty, and administrators. Critics of affirmative action argue that such policies allow some subordinate-group students with lower admissions scores to enroll in colleges and graduate schools at the expense of other students with higher scores. But are test scores the most important criteria for determining academic success? According to one journalist, "The fact is that . . . values apart from test scores do figure in university admission policies—and should. A society that has prospered in diversity cannot want monochrome higher education" (Lewis, 1997: A23).

Sources: Based on Jayson, 2000; Kendall, 1998; Lewis, 1997; and Schmidt, 2002.

DISCRIMINATION In the twentieth century, the lives of many African Americans were changed by industrialization and two world wars. When factories were built in the northern United States, many African American families left the rural South in hopes of finding jobs and a better life.

During World Wars I and II, African Americans were a vital source of labor in war production industries; however, racial discrimination continued both on and off the job. In World War II, many African Americans fought for their country in segregated units in the military; after the war, they sought—and were denied—equal opportunities in the country for which they had risked their lives.

African Americans began to demand sweeping societal changes in the 1950s. Initially, the Reverend Dr. Martin Luther King, Jr., and the civil rights movement used *civil disobedience*—nonviolent action seeking to change a policy or

law by refusing to comply with it—to call attention to racial inequality and to demand greater inclusion of African Americans in all areas of public life. Subsequently, leaders of the Black Power movement, including Malcolm X and Marcus Garvey, advocated black pride and racial awareness among African Americans. Gradually, racial segregation was outlawed by the courts and the federal government. For example, the Civil Rights Acts of 1964 and 1965 sought to do away with discrimination in education, housing, employment, and health care. Affirmative action programs were instituted in both public-sector and private-sector organizations in an effort to bring about greater opportunities for African Americans and other previously excluded groups. However, as discussed in Box 9.2, issues such as affirmative action remain highly controversial in the twenty-first century.

Lawrence Migdale/Stock Boston

As more African Americans have made gains in education and employment, many of them have also made a conscious effort to increase awareness of African culture and to develop a sense of unity, cooperation, and self-determination. The seven-day celebration of Kwanzaa in late December and early January exemplifies this desire to maintain a distinct cultural identity.

AFRICAN AMERICANS TODAY African Americans make up about 13 percent of the U.S. population. Some are descendants of families that have been in this country for many generations; others are recent immigrants from Africa and the Caribbean. Black Haitians make up the largest group of recent Caribbean immigrants; others come from Jamaica, Trinidad, and Tobago. Recent African immigrants are primarily from Nigeria, Ethiopia, Ghana, and Kenya. They have been simultaneously "pushed" out of their countries of origin by severe economic and political turmoil and "pulled" by perceived opportunities for a better life in the United States. Recent immigrants are often victimized by the same racism that has plagued African Americans as a people for centuries.

Since the 1960s, many African Americans have made significant gains in politics, education, employment, and income. Between 1964 and 1994, the number of African Americans elected to political office increased from about 100 to almost 8,000 nationwide. African Americans have won mayoral elections in major cities that have large African American populations, such as Washing-

ton, D.C., Los Angeles, New York, and Atlanta. Despite these political gains, African Americans still represent less than 3 percent of all elected officials in the United States.

Some African Americans have made impressive occupational gains and joined the ranks of professionals in the upper middle class. Others have achieved great wealth and fame as entertainers, professional athletes, and entrepreneurs. African Americans head three of the "Fortune 500" list of the nation's largest companies (Daniels, 2002). However, even those who make millions of dollars per year and live in affluent neighborhoods are not always exempt from racial prejudice and discrimination. And although some African Americans have made substantial occupational and educational gains, many more have not. The African American unemployment rate remains twice as high as that of whites (Feagin and Feagin, 2003).

AFRICAN AMERICANS AND SPORTS In recent decades, many African Americans have seen sports as a possible source of upward mobility because other means have been unavailable. However, their achievements in sports have often been attributed to "natural ability" and not determination and hard work. Sociologists have rejected such biological explanations for African Americans' success in sports and have focused instead on explanations rooted in the social structure of society.

During the slavery era, a few African Americans gained better treatment and, occasionally, freedom by winning boxing matches on which their owners had bet large sums of money (McPherson, Curtis, and Loy, 1989). After emancipation, some African Americans found jobs in horse racing and baseball. For example, fourteen of the fifteen jockeys in the first Kentucky Derby (in 1875) were African Americans. A number of African Americans played on baseball teams; a few played in the Major Leagues until the Jim Crow laws forced them out. Then they formed their own "Negro" baseball and basketball leagues (Peterson, 1992/1970).

Since Jackie Robinson broke baseball's "color line" in 1947, many African American athletes have played collegiate and professional sports. Even now, however, persistent class inequalities between whites and African Americans are reflected in the fact that, until recently, African Americans have primarily excelled in sports (such as basketball or football) that do not require much expensive equipment and specialized facilities in order to develop athletic skills (Coakley, 1998). According to one sports analyst, African Americans typically participate in certain sports and not others because of the *sports opportunity structure* — the availability of facilities, coaching, and competition in the schools and community recreation programs in their area (Phillips, 1993). Regardless of the sport in which they participate, African American men athletes continue to experience inequalities in assignment of playing positions, rewards and authority structures, and management and ownership opportunities in professional sports (Eitzen and Sage, 1997). For example, in 1999 only 3 of the 31 National Football League head coaches and only 5 of the 114 Division I–A head coaches in college football were African Americans (Bohls, 1999). Today, African Americans remain significantly underrepresented in other sports, including hockey, skiing, figure skating, golf, volleyball, softball, swimming, gymnastics, sailing, soccer, bowling, cycling, and tennis (Coakley, 1998); however, the success of some African American athletes — such as Serena and Venus Williams and Tiger Woods — has en-

Reflections

Reflections 9.2

In what ways has "breaking the color line" in sports opened other avenues of opportunity for African Americans, and in what ways has it failed to do so?

couraged more young people of color to participate in sports such as tennis and golf today than in the past (Woodward, 2002).

Although African American women athletes tend to experience dual discrimination based on both race and gender, they have excelled in college and professional sports and in international competitions such as the Olympic Games.

White Ethnic Americans

The American Dream initially brought many white ethnics to the United States. The term *white ethnic Americans* is applied to a wide diversity of immigrants who trace their origins to Ireland and to Eastern and Southern European countries such as Poland, Italy, Greece, Germany, Yugoslavia, and Russia and other former Soviet republics. Unlike the WASPs, who immigrated primarily from Northern Europe and assumed a dominant cultural position in society, white ethnic Americans arrived late in the nineteenth century and early in the twentieth century to find relatively high levels of prejudice and discrimination directed at them by nativist organizations that hoped to curb the entry of non-WASP European immigrants. Because many of the people in white ethnic American categories were not Protestant, they experienced discrimination because they were Catholic, Jewish, or members of other religious bodies, such as the Eastern Orthodox churches (Farley, 2000).

DISCRIMINATION AGAINST WHITE ETHNICS

Many white ethnic immigrants entered the United States between 1830 and 1924. Irish Catholics were among the first to arrive, with more than four million Irish fleeing the potato famine and economic crisis in Ireland and seeking jobs in the United States (Feagin and Feagin, 2003). When they arrived, they found that British Americans controlled the major institutions of society. The next arrivals were Italians who had been recruited for low-wage industrial and construction jobs. British Americans viewed Irish and Italian immigrants as "foreigners": The Irish were stereotyped as ape-like, filthy, bad-tempered, and heavy drinkers, and the Italians were depicted as lawless, knife-wielding thugs looking for a fight, "dagos," and "wops" (short for "without papers") (Feagin and Feagin, 2003: 79–81, 92–94).

Both Irish Americans and Italian Americans were subjected to institutionalized discrimination in employment. Employment ads read "Help Wanted—No Irish Need Apply" and listed daily wages at $1.30–$1.50 for "whites" and $1.15–$1.25 for "Italians" (Gambino, 1975: 77). In spite of discrimination, white ethnics worked hard to establish themselves in the United States, often establishing mutual self-help organizations and becoming politically active (Mangione and Morreale, 1992).

Between 1880 and 1920, over two million Jewish immigrants arrived in the United States and settled in the Northeast. Jewish Americans differ from other white ethnic groups in that some focus their identity primarily on their religion whereas others define their Jewishness in terms of ethnic group membership (Feagin and Feagin, 2003). In any case, Jews continued to be the victims of *anti-Semitism*—prejudice, hostile attitudes, and discriminatory behavior targeted at Jews. For example, signs in hotels read "No Jews Allowed," and some "help wanted" ads stated "Christians Only" (Levine, 1992: 55). In spite of persistent discrimination, Jewish Americans achieved substantial success in many areas, including business, education, the arts and sciences, law, and medicine.

WHITE ETHNICS AND SPORTS   Sports provided a pathway to assimilation for many white ethnics. The earliest collegiate football players who were not white Anglo-Saxon Protestants were of Irish, Italian, and Jewish ancestry. Sports participation provided educational opportunities that some white ethnics would not have had otherwise.

Boxing became a way to make a living for white ethnics who did not participate in collegiate sports. Boxing promoters encouraged ethnic rivalries to increase their profits, pitting Italians against Irish or Jews, and whites against African Americans (Levine, 1992; Mangione and Morreale, 1992).

Eventually, Italian Americans graduated from boxing into baseball and football. Jewish Americans found that sports lessened the shock of assimilation and gave them an opportunity to refute stereotypes about their physical weaknesses and counter anti-Semitic charges that they were "unfit to become Americans" (Levine, 1992: 272).

Today, assimilation is so complete that little attention is paid to the origins of white ethnic athletes. As former Pittsburgh Steeler running back Franco Harris stated, "I didn't know I was part Italian until I became famous" (qtd. in Mangione and Morreale, 1992: 384).

Asian Americans

The U.S. Census Bureau uses the term *Asian Americans* to designate the many diverse groups with roots in Asia. Chinese and Japanese immigrants were among the earliest Asian Americans. Many Filipinos, Asian Indians, Koreans, Vietnamese, Cambodians, Pakistani, and Indonesians have arrived more recently. Today, Asian Americans belong to the fastest-growing ethnic minority group in the United States.

CHINESE AMERICANS   The initial wave of Chinese immigration occurred between 1850 and 1880, when over 200,000 Chinese men were "pushed" from China by harsh economic conditions and "pulled" to the United States by the promise of gold in California and employment opportunities in the construction of transcontinental railroads. Far fewer Chinese women immigrated; however, many of them were brought to the United States against their will and forced into prostitution, where they were treated like slaves (Takaki, 1993).

Chinese Americans were subjected to extreme prejudice and stereotyped as "coolies," "heathens," and "Chinks." Some Asian immigrants were attacked and even lynched by working-class whites who feared that they were losing their jobs to the immigrants. Passage of the Chinese Exclusion Act of 1882 brought Chinese immigration to a halt. The Exclusion Act was not repealed until World War II, when Chinese Americans who were contributing to the war effort by working in defense plants pushed for its repeal (Takaki, 1993). After immigration laws were further relaxed in the 1960s, the second and largest wave of Chinese im-

AP/Wide World Photos

Historically, Chinatowns in major U.S. cities have provided a safe haven and an economic enclave for many Asian immigrants. Some contemporary Chinese Americans reside in these neighborhoods, whereas others visit to celebrate cultural diversity and ethnic pride.

migration occurred, with immigrants coming primarily from Hong Kong and Taiwan. These recent immigrants have had more education and workplace skills than earlier arrivals, and brought families and capital with them to pursue the American Dream (Chen, 1992).

Today, many Chinese Americans live in large urban enclaves in California, New York, Hawaii, Illinois, and Texas. As a group, they have enjoyed considerable upward mobility. Some own laundries, restaurants, and other businesses; others have professional careers (Chen, 1992). However, many Chinese Americans, particularly recent immigrants, remain in the lower tier of the working class—providing low-wage labor in garment and knitting factories and Chinese restaurants.

JAPANESE AMERICANS Most of the early Japanese immigrants were men who worked on sugar plantations in the Hawaiian Islands in the 1860s. Like Chinese immigrants, the Japanese American workers were viewed as a threat by white workers, and immigration of Japanese men was curbed in 1908. However, Japanese women were permitted to enter the United States for several years thereafter because of the shortage of women on the West Coast. Although some Japanese women married white men, this practice was stopped by laws prohibiting interracial marriage.

With the exception of the enslavement of African Americans, Japanese Americans experienced one of the most vicious forms of discrimination ever sanctioned by U.S. laws. During World War II, when the United States was at war with Japan, nearly 120,000 Japanese Americans were placed in internment camps, where they remained for more than two years despite the total lack of evidence that they posed a security threat to this country (Takaki, 1993). This action was a direct violation of the citizenship rights of many *Nisei* (second-generation Japanese Americans) who were born in the United States (see Daniels, 1993). Ironically, only Japanese Americans were singled out for such harsh treatment; German Americans avoided this fate even though the United States was also at war with Germany. Four decades later, the U.S. government issued an apology for its actions and eventually paid $20,000 each to some of those who had been placed in internment camps (Daniels, 1993; Takaki, 1993).

Since World War II, many Japanese Americans have been very successful. The median income of Japanese Americans is more than 30 percent above the national average. However, most Japanese Americans (and other Asian Americans) live in states that not only have higher incomes but also higher costs of living than the national average. In addition, many Asian American families have more persons in the paid labor force than do other families (Takaki, 1993).

KOREAN AMERICANS The first wave of Korean immigrants were male workers who arrived in Hawaii between 1903 and 1910. The second wave came to the U.S. mainland following the Korean War in 1954 and was made up primarily of the wives of servicemen and Korean children who had lost their parents in the war. The third wave arrived after the Immigration Act of 1965 permitted well-educated professionals to migrate to the United States. Korean Americans have helped one another open small businesses by pooling money through the *kye*—an association that grants members money on a rotating basis to gain access to more capital. According to Takaki (1989), Korean Americans were a hidden minority before 1965 because so few lived in the United States. After that time, however, Korean Americans have become a very visible group in this country.

Today, many Korean Americans live in California and New York, where there is a concentration of Korean-owned grocery stores, businesses, and churches. Unlike earlier Korean immigrants, more-recent arrivals have come as settlers and have brought their families with them. However, their experiences with other subordinate racial and ethnic groups have not always been harmonious. Ongoing discord has existed between African Americans and Korean Americans in New York and among African Americans, Latinos, and Korean Americans in California.

FILIPINO AMERICANS   Today, Filipino Americans constitute the second largest category of Asian Americans, with over a million population in the United States. To understand the status of Filipino Americans, it is important to look at the complex relationship between the Philippine Islands and the United States government. After Spain lost the Spanish–American War, the United States established colonial rule over the islands, a rule that lasted from 1898 to 1946 (Feagin and Feagin, 2003). Despite control by the United States, Filipinos were not granted U.S. citizenship. But, like the Chinese and the Japanese, male Filipinos were allowed to migrate to Hawaii and the U.S. mainland to work in agriculture and in fish canneries in Seattle and Alaska. Like other Asian Americans, Filipino Americans were accused of taking jobs away from white workers and suppressing wages, and Congress restricted Filipino immigration to fifty people per year between the Great Depression and the aftermath of World War II.

The second wave of Filipino immigrants came following the Immigration Act of 1965, when large numbers of physicians, nurses, technical workers, and other professionals moved to the U.S. mainland. Most Filipinos have not had the start-up capital necessary to open their own businesses, and many have been employed in the low-wage sector of the service economy. However, the average household income of Filipino American families is relatively high because about 75 percent of Filipino American women are employed, and nearly half have a four-year college degree (Espiritu, 1995).

INDOCHINESE AMERICANS   Indochinese Americans include people from Vietnam, Cambodia, Thailand, and Laos, most of whom have come to the United States in the past three decades. Vietnamese refugees who had the resources to flee at the beginning of the Vietnam War were the first to arrive. Next came Cambodians and lowland Laotians, referred to as "boat people" by the media. Many who tried to immigrate did not survive at sea; others were turned back when they reached this country or were kept in refugee camps for long periods of time. When they arrived in the United States, inflation was high, the country was in a recession, and many native-born citizens feared that they would lose their jobs to these new refugees, who were willing to work very hard for low wages. The frustrations of many Indochinese American immigrants were expressed by a Hmong refugee from Laos:

> In our old country, whatever we had was made or brought in by our own hands; we never had any doubts that we would not have enough for our mouths. But from now on to the future, that time is over. We are so afraid and worried that there will be one day when we will not have anything for eating or paying the rent, and these days these things are always in our minds. . . . Don't know how to read or write, don't know how to speak the language. My life is only to live day by day until the last day I live, and maybe that is the time when my problems will be solved. (qtd. in Portes and Rumbaut, 1996: 155)

Today, many Indochinese Americans are foreign born; about half live in the western states, especially California. Even though most Indochinese immigrants spoke no English when they arrived in this country, some of their children have done very well in school and have been stereotyped as "brains."

ASIAN AMERICANS AND SPORTS   Until recently, Asian Americans received little recognition in sports. However, as women's athletic events, including ice skating and gymnastics, have garnered more media coverage, names of persons such as Kristi Yamaguchi, Michelle Kwan, and Amy Chow have become widely known and often idolized by fans. As one sports analyst stated, "[These athletes] are of Asian descent, but more importantly they are Asian Americans whose actions reflect upon the United States. . . . As role models,

Until recently, few Asian Americans played on college or professional sports teams. After playing college football for Texas A&M University, linebacker Dat Nguyen became the first Vietnamese American signed by an NFL team, the Dallas Cowboys.

AP/Wide World Photos

particularly for the Asian-American community, they exemplify success, integrity, discipline and a dedicated work ethic" (Shum, 1997).

One Asian American athlete who has captured extensive media attention is football player Dat Nguyen (pronounced Win), who won numerous awards at Texas A&M and joined the Dallas Cowboys football team. Nguyen is the son of Vietnamese American parents who fled a war-torn Vietnam with five children and a sixth child (Dat) on the way and entered the United States as refugees in 1975. When asked by a sports reporter whether it was harder to overcome his lack of "football size" (he is 5'11" and weighed about 230 pounds at the time) or the fact that he was an Asian American in "big-time college football," Dat Nguyen replied: "That's hard to answer. It's probably been overcoming a lot of adversity and not just on the football field. People weren't expecting me to be where I am now and what I've accomplished" (qtd. in Silversten, 1998: 1). In the future, there will, no doubt, be many more well-known Asian American men and women participating in college and professional sporting events.

Latinos/as (Hispanic Americans)

The terms *Latino* (for males), *Latina* (for females), and *Hispanic* are used interchangeably to refer to people who trace their origins to Spanish-speaking Latin America and the Iberian peninsula (Cashmore,

1996). However, as racial–ethnic scholars have pointed out, the label *Hispanic* was first used by the U.S. government to designate people of Latin American and Spanish descent living in the United States (Oboler, 1995), and it has not been fully accepted as a source of identity by the more than 35 million Latinos/as who live in the United States today (Oboler, 1995; Romero, 1997). Instead, many of the people who trace their roots to Spanish-speaking countries think of themselves as Mexican Americans, Chicanos/as, Puerto Ricans, Cuban Americans, Salvadorans, Guatemalans, Nicaraguans, Costa Ricans, Argentines, Hondurans, Dominicans, or members of other categories. Many also think of themselves as having a combination of Spanish, African, and Native American ancestry.

MEXICAN AMERICANS OR CHICANOS/AS

Mexican Americans—including both native- and foreign-born people of Mexican origin—are the largest segment (approximately two-thirds) of the Latino/a population in the United States. Most Mexican Americans live in the southwestern region of the United States, although more have recently moved to the Midwest and the Washington, D.C., metropolitan area (Cashmore, 1996).

Immigration from Mexico is the primary vehicle by which the Mexican American population grew in this country. Initially, Mexican-origin workers came to work in agriculture, where they were viewed as a readily available cheap and seasonal labor force. Many initially entered the United States as undocumented workers ("illegal aliens"); however, they were more vulnerable to deportation than other illegal immigrants because of their visibility and the proximity of their country of origin. For more than a century, there has been a "revolving door" between the United States and Mexico that has been open when workers were needed and closed during periods of economic recession and high rates of U.S. unemployment.

Mexican Americans have long been seen as a source of cheap labor, while—ironically—at the same time, they have been stereotyped as lazy and unwilling to work. As has been true of other groups, when white workers viewed Chicanos/as a threat to their jobs, they demanded that the "illegal aliens" be sent back to Mexico. Consequently, U.S. citizens who happen to be Mexican American have been asked for proof of their citizenship, especially when anti-immigration sentiments are running high. Many Mexican American families have lived in the United States for four or five generations—they have fought in wars, made educational and political gains, and consider themselves to be solid U.S. citizens. Thus, it is a great source of frustration for them to be viewed as illegal immigrants or to be asked "How long have you been in this country?"

PUERTO RICANS When Puerto Rico became a possession of the United States in 1917, Puerto Ricans acquired U.S. citizenship and the right to move freely to and from the mainland. In the 1950s, many migrated to the mainland when the Puerto Rican sugar industry collapsed, settling primarily in New York and New Jersey. Although living conditions have improved substantially for some Puerto Ricans, life has been difficult for the many living in poverty in Spanish Harlem and other barrios. Nevertheless, in recent years Puerto Ricans have made dramatic advances in education, the arts, and politics. Increasing numbers have become lawyers, physicians, and college professors (see Rodriguez, 1989).

CUBAN AMERICANS Cuban Americans live primarily in the Southeast, especially Florida. As a group, they have fared somewhat better than other Latinos/as because many Cuban immigrants were affluent professionals and businesspeople who fled

AP/Wide World Photos

Sammy Sosa of the Chicago Cubs is one of the most visible Latino professional athletes. Latino and Latina sports figures have gained prominence in a wide variety of sports, including boxing, baseball, basketball, golf, and tennis.

Cuba after Fidel Castro's 1959 Marxist revolution. This early wave of Cuban immigrants has median incomes well above those of other Latinos/as; however, this group is still below the national average. The second wave of Cuban Americans, arriving in the 1970s, has fared worse. Many had been released from prisons and mental hospitals in Cuba, and their arrival fueled an upsurge in prejudice against all Cuban Americans. The more recent arrivals have developed their own ethnic and economic enclaves in Miami's Little Havana, and many of the earlier immigrants have become mainstream professionals and entrepreneurs.

LATINOS/AS AND SPORTS For most of the twentieth century, Latinos have played Major League Baseball. Originally, Cubans, Puerto Ricans, and Venezuelans were selected for their light skin as well as for their skill as players (Hoose, 1989). Today, Latinos represent more than 20 percent of all major leaguers. If not for a 1974 U.S. Labor Department quota limiting how many foreign-born players can play professional baseball, this number might be even larger (Hoose, 1989).

Recently, Latinos in sports have gained more recognition as books and web sites have been created to describe their accomplishments. For example, the web site "Latino Legends in Sports" was created in 1999 to inform people about the contributions of Latino and Latina athletes (see http://www.latinosportslegends.com).

Education is a crucial issue for Latinos/as. Because of past discrimination and unequal educational opportunities, many Latinos/as currently have low levels of educational attainment. Many are unable to attend college or participate in collegiate sports, which is essential for being drafted in professional sports other than baseball. Consequently, the overall number of Latinas/os in college and professional sports is low compared to the rest of the U.S. population who are in this age bracket.

Middle Eastern Americans

Since 1970, many immigrants have arrived in the United States from countries located in the "Middle East," which is the geographic region from Afghanistan to Libya and includes Arabia, Cyprus, and Asiatic Turkey. Placing people in the "Middle Eastern" American category is somewhat like placing wide diversities of people in the categories of Asian American or Latino/a; some U.S. residents trace their origins to countries such as Bahrain, Egypt, Iran, Iraq, Kuwait, Lebanon, Oman, Qatar, Saudi Arabia, Syria, UAE (United Arab Emirates), and Yemen. Middle Eastern Americans speak a variety of languages and have diverse religious backgrounds: Some are Muslim, some are Coptic Christian, and others are Melkite Catholic. Although some are from working-class families, Lebanese Americans, Syrian Americans, Iranian Americans, and Kuwaiti Americans primarily come from middle- and upper-income family backgrounds. For example, numerous Iranian Americans are scientists, professionals, and entrepreneurs.

In cities across the United States, Muslims have established social, economic, and ethnic enclaves. On the Internet, they have created web sites that provide information about Islamic centers, schools, and lists of businesses and services run by those who adhere to Islam, one of the fastest growing religions in this country. In cities such as Seattle, incorporation into the economic mainstream has been relatively easy for Palestinian immigrants who left their homeland in the 1980s. Some have found well-paid employment with corporations such as Microsoft because they bring educational skills and talents to the information-based economy, including the ability to translate software into Arabic for Middle Eastern markets (Ramirez, 1999). In the United States, Islamic schools and centers often bring together people from a diversity of countries such as Egypt and Pakistan. Many Muslim leaders and parents focus on how to raise children to be good Muslims and good U.S. citizens. However, recent immigrants continue to be torn between establishing roots in the United States and the continuing divisions and strife that exist in their homelands. Some Middle Eastern Americans experience prejudice and discrimination based on their speech patterns, appearance (such as the *hijabs*, or "head-to-toe covering" that leaves only the face exposed, which many girls and women wear), or the assumption that "all Middle Easterners" are somehow associated with terrorism.

Following the September 11, 2001, attacks on the United States by terrorists whose origins were traced to the Middle East, hate crimes and other forms of discrimination against people who were assumed to be Arabs, Arab Americans, or Muslims escalated in this country. With the passage of the U.S. Patriot Act—a law giving the federal government greater authority to engage in searches and surveillance with less judicial review than previously—in the aftermath of the terrorist attacks, many Arab Americans have expressed concern that this new law could be used to target people who appear to be of Middle Eastern origins. To counter this potential oppression and loss of rights, some Arab Americans have engaged in social activism to highlight their concerns (Feagin and Feagin, 2003).

MIDDLE EASTERN AMERICANS AND SPORTS
Although more Islamic schools are beginning to focus on sports, particularly for teenage boys, there has been less emphasis on competitive athletics among many Middle Eastern Americans. Based on popular sporting events in their countries of origin, some Middle Eastern Americans play golf or soccer. For example, some Iranian Americans follow the soccer careers of professional players from Iran who now play for German, Austrian, Belgian, and Greek clubs. Keeping up with global sporting events is easy with all-sports television cable channels and web sites that provide up-to-the-minute information about players and competitions. Over time, there will probably be greater participation

✓Checkpoints

Checkpoint 9.5

17. What major problems have been experienced by Native Americans that limit their life chances and opportunities?
18. Why are British Americans sometimes referred to as WASPs?
19. What effect did slavery have on African Americans?
20. What is meant by the term *white ethnic Americans*?
21. What are the major Asian American groups in the United States today?
22. What categories of people are included in the designation Latinos/as (Hispanic Americans)?
23. Who are the Middle Eastern Americans, and what unique problems have they experienced in the United States?

Reflections

Reflections 9.3

Do the Olympic games and other international sports events offer opportunities for people to demonstrate racial tolerance and global understanding, or are these competitive events more likely to produce heightened antagonisms among divergent groups?

by Middle Eastern American males in competitions such as soccer and golf; however, girls and women in Muslim families are typically not allowed to engage in athletic activities. Although little research has been done on this issue in the United States, one study of Islamic countries in the Middle East found that female athletes face strong cultural opposition to their sports participation (Dupre and Gains, 1997).

Global Racial and Ethnic Inequality in the Future

Throughout the world, many racial and ethnic groups seek *self-determination*—the right to choose their own way of life. As many nations are currently structured, however, self-determination is impossible.

Worldwide Racial and Ethnic Struggles

The cost of self-determination is the loss of life and property in ethnic warfare. In recent years, the Cold War has given way to dozens of smaller wars over ethnic dominance. In Europe, for example, ethnic violence has persisted in Yugoslavia, Spain, Britain (between the Protestant majority and the Catholic minority in Northern Ireland), Romania, Russia, Moldova, and Georgia. Ethnic violence

continues in the Middle East, Africa, Asia, and Latin America. Hundreds of thousands have died from warfare, disease (such as the cholera epidemic in war-torn Rwanda), and refugee migration.

Ethnic wars have a high price even for survivors, whose life chances can become bleaker even after the violence subsides. In ethnic conflict between Abkhazians and Georgians in the former Soviet Union, for example, as many as two thousand people have been killed and over eighty thousand displaced. Ethnic hatred also devastated the province of Kosovo, which is located in Serbia—Yugoslavia's dominant republic—and brought about the deaths of thousands of ethnic Albanians (Bennahum, 1999).

In the twenty-first century, the struggle between the Israeli government and various Palestinian factions over the future and borders of Palestine continues to make headlines. Discord in this region has heightened tensions among people not only in Israel and Palestine but also in the United States and around the world as deadly clashes continue and political leaders are apparently unable to reach a lasting solution to the decades-long strife.

Growing Racial and Ethnic Diversity in the United States

Racial and ethnic diversity is increasing in the United States. African Americans, Latinos/as, Asian Americans, and Native Americans constitute one-fourth of the U.S. population, whereas whites are a shrinking percentage of the population. In the year 2000, white Americans made up 70 percent of the population, in contrast to 80 percent in 1980. It is predicted that by 2056, the roots of the average U.S. resident will be in Africa, Asia, Hispanic countries, the Pacific islands, and the Middle East—not white Europe (Henry, 1990).

These Kosovo refugees are among the millions of people whose lives have been torn asunder by ethnic warfare.

AP/Wide World Photos

What effect will these changes have on racial and ethnic relations? Several possibilities exist. On the one hand, conflicts may become more overt and confrontational as people continue to use *sincere fictions*—personal beliefs that reflect larger societal mythologies, such as "I am not a racist" or "I have never discriminated against anyone"—even when these are inaccurate perceptions (Feagin and Vera, 1995). Interethnic tensions may increase as competition for education, jobs, and other resources continues to grow.

On the other hand, there is reason for cautious optimism. Throughout U.S. history, members of diverse racial and ethnic groups have struggled to gain the freedom and rights that were previously withheld from them. Today, minority grassroots organizations are pressing for affordable housing, job training, and educational opportunities (Feagin and Feagin, 2003). As discussed in Box 9.3, movements composed of both whites and people of color continue to oppose racism in everyday life, to seek to heal divisions among racial groups, and to teach children about racial tolerance (Rutstein, 1993). Many groups hope not only to affect their own microcosm but also to contribute to worldwide efforts to end racism (Ford, 1994).

To eliminate racial discrimination, it will be necessary to equalize opportunities in schools and workplaces. As Michael Omi and Howard Winant (1994: 158) have emphasized,

> Today more than ever, opposing racism requires that we notice race, not ignore it, that we afford it the recognition it deserves and the subtlety it embodies. By noticing race we can begin to challenge racism, with its ever-more-absurd reduction of human experience to an essence attributed to all without regard for historical or social context. . . . By noticing race we can develop the political insight and mobilization necessary to make the U.S. a more racially just and egalitarian society.

Chapter Review

How do race and ethnicity differ?

A race is a category of people who have been singled out as inferior or superior, often on the basis of physical characteristics such as skin color, hair texture, or eye shape. An ethnic group is a collection of people distinguished primarily by cultural or national characteristics, including unique cultural traits, a sense of community, a feeling of ethnocentrism, ascribed membership, and territoriality.

What are dominant and subordinate groups?

A dominant group is an advantaged group that has superior resources and rights in society. A subordinate group is a disadvantaged group whose members are subjected to unequal treatment by the dominant group. Use of the terms *dominant* and *subordinate* reflect the importance of power in relationships.

How is prejudice related to discrimination?

Prejudice is a negative attitude often based on stereotypes, which are overgeneralizations about the appearance, behavior, or other characteristics of all members of a group. Discrimination involves actions or practices of dominant-group members that have a harmful impact on members of a subordinate group. Whereas prejudice involves attitudes, discrimination involves actions. Discriminatory actions range from name-calling to violent actions; genocide is the ultimate form of discrimination.

What are the major psychological explanations of prejudice?

According to the frustration–aggression hypothesis of prejudice, people frustrated in their efforts to achieve a highly desired goal may respond with aggression toward others, who then become scapegoats. Another theory of prejudice focuses on the authoritarian personality, marked by excessive conformity, submissiveness to authority, intolerance, insecurity, superstition, and rigid thinking.

Box 9.3 You Can Make a Difference
Working for Racial Harmony

Suppose that you are talking with several friends about a series of racist incidents at your college. Having studied the sociological imagination, you decide to start an organization similar to No Time to Hate, which was started at Emory University several years ago to reduce racism on campus. In analyzing racism, your group identifies factors contributing to the problem: (1) divisiveness between different cultural and ethnic communities, (2) persistent lack of trust, (3) the fact that many people never really communicate with one another, (4) the need to bring different voices into the curriculum and college life generally, and (5) the need to learn respect for people from different backgrounds (Loeb, 1994). Your group also develops a set of questions to be answered regarding racism on campus:

- *Encouraging inclusion and acceptance.* Do members of our group reflect the college's racial and ethnic diversity? How much do I know about other people's history and culture? How can I become more tolerant—or accepting—of people who are different from me?
- *Raising consciousness.* What is racism? What causes it? Can people participate in racist language and behavior without realizing what they are doing? What is our college or university doing to reduce racism?
- *Becoming more self-aware.* How much do I know about my own family roots and ethnic back-

ground? How do the families and communities in which we grow up affect our perceptions of racial and ethnic relations?
- *Using available resources.* What resources are available for learning more about working to reduce racism?

Here are some agencies to contact:

- ACLU (American Civil Liberties Union), 132 West 43rd Street, New York, NY 10036. (212) 944-9800.
- ADL (Anti-Defamation League of B'nai B'rith), 823 United Nations Plaza, New York, NY 10017. (212) 490-2525.
- NAACP (National Association for the Advancement of Colored People), 4805 Mt. Hope Drive, Baltimore, MD 21215. (410) 358-8900.

On the Internet:

- The American Studies Web/Race and Ethnicity provides links to related web sites on race and ethnicity in the United States. It also includes information on African Americans, Native Americans, Latinos/as and Chicanas/os, and Asian Americans:

http://www.georgetown.edu/crossroads/asw

What additional items would you add to the list of problem areas on your campus? How might your group's objective be reached? Over time, many colleges and universities have been changed as a result of involvement by students like you!

How do individual discrimination and institutional discrimination differ?

Individual discrimination involves actions by individual members of the dominant group that harm members of subordinate groups or their property. Institutional discrimination involves day-to-day practices of organizations and institutions that have a harmful impact on members of subordinate groups.

How do sociologists view racial and ethnic group relations?

Symbolic interactionists suggest that increased contact between people from divergent groups should lead to favorable attitudes and behavior when members of each

group (1) have equal status, (2) pursue the same goals, (3) cooperate with one another to achieve goals, and (4) receive positive feedback when they interact with one another. Functionalists stress that members of subordinate groups become a part of the mainstream through assimilation, the process by which members of subordinate groups become absorbed into the dominant culture. Conflict theorists focus on economic stratification and access to power in race and ethnic relations. The caste perspective views inequality as a permanent feature of society, whereas class perspectives focus on the link between capitalism and racial exploitation. According to racial formation theory, the actions of the U.S. government substantially define racial and ethnic relations.

How have the experiences of various racial–ethnic groups differed in the United States?

Native Americans suffered greatly from the actions of European settlers, who seized their lands and made them victims of forced migration and genocide. Today, they lead lives characterized by poverty and lack of opportunity. White Anglo-Saxon Protestants are the most privileged group in the United States, although social class and gender affect their life chances. White ethnic Americans, whose ancestors migrated from Southern and Eastern European countries, have gradually made their way into the mainstream of U.S. society. Following the abolishment of slavery through the Emancipation Proclamation in 1863, African Americans were still subjected to segregation, discrimination, and lynchings. More recently, despite civil rights legislation and economic and political gains by many African Americans, racial prejudice and discrimination still exist. Asian American immigrants as a group have enjoyed considerable upward mobility in U.S. society in recent decades, but many Asian Americans still struggle to survive by working at low-paying jobs and living in urban ethnic enclaves. Although some Latinos/as have made substantial political, economic, and professional gains in U.S. society, as a group they still are subjected to anti-immigration sentiments. Middle Eastern immigrants to the United States speak a variety of languages and have diverse religious backgrounds. Because they generally come from middle-class backgrounds, they have made inroads into mainstream U.S. society.

Key Terms

assimilation 284
authoritarian personality 282
discrimination 282
ethnic group 276
ethnic pluralism 285
genocide 282
individual discrimination 283
institutional discrimination 283
internal colonialism 287
majority (dominant) group 279
minority (subordinate) group 279
prejudice 280
race 276
racism 280
scapegoat 281
segregation 285
split labor market 288
stereotypes 280

Questions for Critical Thinking

1. Do you consider yourself defined more strongly by your race or by your ethnicity? How so?
2. Given that subordinate groups have some common experiences, why is there such deep conflict between some of these groups?
3. What would need to happen in the United States, both individually and institutionally, for a positive form of ethnic pluralism to flourish in the twenty-first century?

 # Resources on the Internet

Chapter-Related Web Sites
The following web sites have been selected for their relevance to the topics in this chapter. These sites are among the more stable, but please note that web site addresses change frequently.

U.S. Census Bureau: Rankings and Comparisons: Population and Housing Tables
http://www.census.gov/population/www/cen2000/tablist.html
This is the entrance page for a number of statistical reports related to race and ethnicity.

Current Population Survey: Definitions and Explanations
http://www.census.gov/population/www/cps/cpsdef.html
Here are definitions and explanations of various terms used by the Census Bureau, including those regarding data on race and Hispanic origin.

The Sociology of Race and Ethnicity
http://www.trinity.edu/~mkearl/race.html
This page on race and ethnicity contains volumes of information and links to help with a deeper investigation of these important social categories.

"Shelby Steele's Race Problem and Mine"
http://www.raleightavern.org/steele.html
This is an essay by Tom Lovell on the work of Shelby Steele. Specific reference is made to Steele's term "race holding."

The Center for the Study of Race and Ethnicity in America
http://www.brown.edu/Departments/Race_Ethnicity/mission/history/
The CSREA was established in 1988 with the premise that it is crucial to understand race as a historical and

sociological reality in America and to understand the implications of race and ethnicity as historical, social, and analytical categories for multidisciplinary studies and multiple modes of discourse.

Race and Ethnicity Today: Some Statistics
http://smccd.net/accounts/helton/racefacts.htm
Although not as current as the 2000 Census, this site offers a concise composition of important statistical data about racial and ethnic groups in America.

MLK Page
http://www.wmich.edu/politics/mlk/
Sponsored by Western Michigan State University, this site contains a great history of the civil rights movement, concentrating on the efforts of Dr. Martin Luther King, Jr.

The Civil Rights Movement 1955–1965: Introduction
http://www.watson.org/~lisa/blackhistory/civilrights-55-65/
This page contains several important links to historical data necessary to put a study of race and ethnicity in America in context.

Companion Web Site for This Book
Virtual Society: The Wadsworth Sociology Resource Center
Visit **http://sociology.wadsworth.com** and click on the page for Kendall, *Sociology in Our Times: The Essentials*, Fourth Edition, to access a wide range of enrichment material to aid in your study of sociology. Click on the Student Resources section of the web site. Next, select from the pull-down menu the chapter that you are presently studying. Among the useful options

for self-study are chapter objectives, flashcards, practical study tips, and practice tests for each chapter.

MicroCase Online
From the Virtual Society home page, click on MicroCase Online to access book-specific MicroCase exercises that allow you to further explore sociological issues and principles presented in the text.

InfoTrac College Edition
Another unique option available to you at the Student Resources section of the companion web site is InfoTrac College Edition, an online library with access to hundreds of scholarly and popular periodicals. Below are suggested search terms for this chapter. Results from these and other searches are found at the site.

- Search keywords: *civil rights movement*. Locate articles that address civil rights issues for subordinate groups in the United States.
- Search keyword: *prejudice*. Conduct a limit search on the initial list of articles that this keyword produces. Type *sociology* into the journal block. What kinds of prejudices do the articles from this search discuss?

Virtual Explorations CD-ROM
Go to the Virtual Explorations CD-ROM to begin the interactive exercise for this chapter. This Virtual Exploration will introduce you to some of the exciting resources for sociology on the World Wide Web. You will be guided through an exercise that employs related web sites on race and ethnicity. Answer the questions, and e-mail your responses to your instructor.

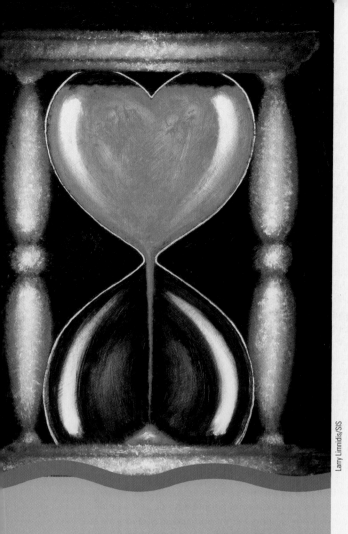

Larry Limnidis/SIS

Sex and Gender

10

Sex: The Biological Dimension
Hermaphrodites/Transsexuals
Sexual Orientation

Gender: The Cultural Dimension
The Social Significance of Gender
Sexism

Gender Stratification in Historical and Contemporary Perspective
Hunting and Gathering Societies
Horticultural and Pastoral Societies
Agrarian Societies
Industrial Societies
Postindustrial Societies

Gender and Socialization
Parents and Gender Socialization
Peers and Gender Socialization
Teachers, Schools, and Gender Socialization
Sports and Gender Socialization
Mass Media and Gender Socialization
Adult Gender Socialization

Contemporary Gender Inequality
Gendered Division of Paid Work
Pay Equity (Comparable Worth)
Paid Work and Family Work

Perspectives on Gender Stratification
Functionalist and Neoclassical Economic Perspectives
Conflict Perspectives
Feminist Perspectives

Gender Issues in the Future

Today was my first appointment with Dr. Gold [a psychiatrist]. . . . Dr. Gold asked if I thought the girls at school who diet are overweight. It was such a stupid question. . . . "*Of course* they aren't overweight, didn't I already say that they were *popular*?" . . . Dr. Gold . . . gave me [some] paper and asked me to draw my "ideal" of what I wanted to look like. . . . So I picked up the pencil and drew a girl I want to look like. She was tall and skinny, but she had my face and hair. When Dr. Gold took the drawing back, he didn't nod. "This is a stick figure," he said, like I didn't understand the assignment the first time. "Try to draw a *realistic* picture of how you'd like to look. Don't worry if you aren't very good at art." He must have thought I was terrible at art. I tried explaining how that was *exactly* the way I wanted to look, but Dr. Gold said I wouldn't be alive if I looked like that drawing. . . .

[During another appointment, Dr. Gold decided that I had to be hospitalized so that I would gain weight. At the hospital, I had to follow a bunch of rules, like getting weighed every morning, being watched by a nurse during meals, not being allowed to exercise, and not being allowed to look in a full-length mirror. After I began eating more, I started being allowed more privileges, however, one of which was being allowed out of the hospital to attend my brother's graduation.]

After the ceremony, we all went out to dinner at this fancy prime rib place, [and] my stomach started hurting because I wasn't used to eating so much at the hospital. I had to go to the bathroom pretty bad. . . . The first thing I saw when I walked into the bathroom was a really skinny girl who looked just like the pictures in Dr. Katz's book. Whoever decorated the ladies' bathroom was madly in love with mirrors, so I kept seeing the skinny girl everywhere. *She* was a stick figure, . . . but when I turned around to look, I was the only person in the bathroom. Except the girl couldn't be me because she was so skinny. . . . So I walked closer to the sink to get a better look, but the skinny legs in the mirror kept moving at the same time as mine. Then I smiled, and so did the mouth in the mirror. . . . Finally I turned on the hot water and washed all [the] makeup off my face. That's when I knew for sure it was me. I couldn't believe it! I looked disgusting. I used to want to be a stick figure, but now I'm not so sure.

—Lori Gottlieb, who weighed only 55 pounds at age 11, describing her experience with anorexia nervosa (Gottlieb, 2000: 112–114, 151–152, 204–208)

Is Lori's situation an isolated case of an eating disorder? Her comments are similar to those found in numerous sociological studies, including research by the sociologist Sharlene Hesse-Biber (1996), who found that many college women were extremely critical of their weight and shape. As Lori and many other women are aware, people who deviate significantly from existing weight and appearance norms are often devalued and objectified by others. *Objectification* is the process of treating people as if they were objects or things, not human beings. We objectify people when we judge them on the basis of their physical appearance rather than on the basis of their individual qualities or actions (Schur, 1983). Although men may be objectified, objectification of women is especially common in the United States and other nations (see Table 10.1).

According to Hesse-Biber, college women—more than men—express dissatisfaction with their body weight and shape. She discovered several distinct differences between women and men in perception of body weight. For example, women overestimated their weights and thought their bodies were heavier than the medically desirable weights, whereas men judged their weights more accurately. Why do women and men feel differently about their bodies? Cultural differences in appearance norms may explain women's greater concern; they tend to be judged more harshly, and they know it (Schur, 1983). Throughout life, men and women receive different cultural messages about body image, food, and eating. Men are encouraged to eat; women are made to feel guilty about eating (Basow, 1992). Hesse-Biber (1996: 11) refers to this

Marjorie Farrell/The Image Works

No wonder many women are extremely concerned about body image; even billboards communicate cultural messages about women's appearance.

Table 10.1 The Objectification of Women

General Aspects of Objectification	Objectification Based on Cultural Preoccupation with "Looks"
Women are responded to primarily as "females," whereas their personal qualities and accomplishments are of secondary importance.	Women are often seen as the objects of sexual attraction, not full human beings—for example, when they are stared at.
Women are seen as "all alike."	Women are seen by some as depersonalized body parts—for example, "a piece of ass."
Women are seen as being subordinate and passive, so things can easily be "done to a woman"—for example, discrimination, harassment, and violence.	Depersonalized female sexuality is used for cultural and economic purposes—such as in the media, advertising, fashion and cosmetics industries, and pornography.
Women are seen as easily ignored, or trivialized.	Women are seen as being "decorative" and status-conferring objects to be bought (sometimes collected) and displayed by men and sometimes by other women.
	Women are evaluated according to prevailing, narrow "beauty" standards and often feel pressure to conform to appearance norms.

Source: Schur, 1983.

phenomenon as the *cult of thinness,* in which people worship the "perfect" body and engage in rituals such as dieting and exercising with "obsessive attention to monitoring progress—weighing the body at least once a day and constantly checking calories." The main criterion for joining the cult of thinness is being female (Hesse-Biber, 1996).

Body image is only one example of the many socially constructed differences between men and women—differences that relate to gender (a social concept) rather than to a person's biological makeup, or sex. In this chapter, we examine the issue of gender: what it is and how it affects us. Before reading on, test your knowledge about body image and gender by taking the quiz in Box 10.1.

Questions and Issues

Chapter Focus Question: How do expectations about female and male appearance, and especially weight, reflect gender inequality?

How do a society's resources and economic structure influence gender stratification?

What are the primary agents of gender socialization?

How does the contemporary workplace reflect gender stratification?

How do functionalist, conflict, and feminist perspectives on gender stratification differ?

Sex: The Biological Dimension

Whereas the word *gender* is often used to refer to the distinctive qualities of men and women (masculinity and femininity) that are culturally created, **sex refers to the biological and anatomical differences between females and males** (Epstein, 1988; Marshall, 1994). At the core of these differences is the chro-

mosomal information transmitted at the moment a child is conceived. The mother contributes an X chromosome and the father either an X (which produces a female embryo) or a Y (which produces a

Sex the biological and anatomical differences between females and males.

Box 10.1 Sociology and Everyday Life

How Much Do You Know About Body Image and Gender?

True	False	
T	F	1. Most people have an accurate perception of their physical appearance.
T	F	2. Recent studies show that up to 95 percent of men express dissatisfaction with some aspect of their bodies.
T	F	3. Many young girls and women believe that being even slightly "overweight" makes them less "feminine."
T	F	4. Physical attractiveness is a more central part of self-concept for women than for men.
T	F	5. Virtually no men have eating problems such as anorexia and bulimia.
T	F	6. Thinness has always been the "ideal" body image for women.
T	F	7. Women bodybuilders have gained full acceptance in society.
T	F	8. Young girls and women very rarely die as a result of anorexia or bulimia.

Answers on page 314.

male embryo). At birth, male and female infants are distinguished by *primary sex characteristics:* **the genitalia used in the reproductive process.** At puberty, an increased production of hormones results in the development of *secondary sex characteristics:* **the physical traits (other than reproductive organs) that identify an individual's sex.** For women, these include larger breasts, wider hips, and narrower shoulders; a layer of fatty tissue throughout the body; and menstruation. For men, they include development of enlarged genitals, a deeper voice, greater height, a more muscular build, and more body and facial hair (see Lott, 1994).

Hermaphrodites/Transsexuals

Sex is not always clear-cut. Occasionally, a hormone imbalance before birth produces a *hermaphrodite*—**a person in whom sexual differentiation is ambiguous or incomplete.** Hermaphrodites tend to have some combination of male and female genitalia. In one case, for example, a chromosomally normal (XY) male was born with a penis just one centimeter long and a urinary opening similar to that of a female (Money and Ehrhardt, 1972). Some people may be genetically of one sex but have a gender identity of the other. That is true for a *transsexual,* **a person in**

whom the sex-related structures of the brain that define gender identity are opposite from the physical sex organs of the person's body. Consequently, transsexuals often feel that they are the opposite sex from that of their sex organs. Some transsexuals are aware of this conflict between gender identity and physical sex as early as the preschool years. Some transsexuals take hormone treatments or have a sex change operation to alter their genitalia in order to achieve a body congruent with their sense of sexual identity (Basow, 1992). Many transsexuals who receive hormone treatments or undergo surgical procedures go on to lead lives that they view as being compatible with their true sexual identity.

Western societies acknowledge the existence of only two sexes; some other societies recognize three—men, women, and *berdaches* (or *hijras* or *xaniths*), biological males who behave, dress, work, and are treated in most respects as women. The closest approximation of a third sex in Western societies is a *transvestite,* **a male who lives as a woman or a female who lives as a man but does not alter the genitalia.** Although transvestites are not treated as a third sex, they often "pass" for members of that sex because their appearance and mannerisms fall within the range of what is expected from members of the other sex (Lorber, 1994).

Transsexuality may occur in conjunction with homosexuality, but this is frequently not the case. Some researchers believe that both transsexuality and homosexuality have a common prenatal cause such as a critically timed hormonal release due to stress in the mother or the presence of certain hormone-mimicking chemicals during critical steps of fetal development. Researchers continue to examine this issue and debate the origins of transsexuality and homosexuality.

Sexual Orientation

Sexual orientation **refers to an individual's preference for emotional–sexual relationships with members of the opposite sex (heterosexuality), the same sex (homosexuality), or both (bisexuality)** (Lips, 1997). Some scholars believe that sexual orientation is rooted in biological factors that are present at birth (Pillard and Weinrich, 1986); others believe that sexuality has both biological and social components and is not preordained at birth (Golden, 1987).

The terms *homosexual* and *gay* are most often used in association with males who prefer same-sex relationships; the term *lesbian* is used in association with females who prefer same-sex relationships. Heterosexual individuals, who prefer opposite-sex relationships, are sometimes referred to as *straight*. However, it is important to note that heterosexual people are much less likely to be labeled by their sexual orientation than are people who are gay, lesbian, or bisexual.

What criteria do social scientists use to classify individuals as gay, lesbian, or homosexual? In a definitive study of sexuality in the mid-1990s, researchers at the University of Chicago established three criteria for identifying people as homosexual or bisexual: (1) *sexual attraction* to persons of one's own gender, (2) *sexual involvement* with one or more persons of one's own gender, and (3) *self-identification* as a gay, lesbian, or bisexual (Michael et al., 1994). According to these criteria, then, having engaged in a homosexual act does not necessarily classify a person as homosexual. In fact, many respondents in the University of Chicago study indicated that although they had at least one homosexual encounter when they were younger, they were no longer involved in homosexual conduct and never identified themselves as gay, lesbian, or bisexual.

Other studies have examined how sexual orientation is linked to identity. For example, the sociologist Kristin G. Esterberg (1997) interviewed lesbian and bisexual women to determine how they "perform" lesbian or bisexual identity through daily activities such as choice of clothing and hairstyles, as well as how they use body language and talk. According to Esterberg (1997), some of the women viewed themselves as being "lesbian from birth" whereas others had experienced shifts in their identities, depending on social surroundings, age, and political conditions at specific periods in their lives.

In the 1990s, the term *transgender* was created to describe individuals whose appearance, behavior, or self-identification does not conform to common social rules of gender expression. Transgenderism is sometimes used to refer to those who cross-dress, to transsexuals, and to others outside mainstream categories. Although some gay and lesbian advocacy groups oppose the concept of transgender as being somewhat meaningless, others applaud the term as one that might help unify diverse categories of people based on sexual identity. Various organizations of gays, lesbians, and transgendered persons have been unified in their desire to reduce hate crimes and other forms of *homophobia*—**extreme prejudice directed at gays, lesbians, bisexuals, and others who are perceived as not being heterosexual.**

Primary sex characteristics   the genitalia used in the reproductive process.

Secondary sex characteristics   the physical traits (other than reproductive organs) that identify an individual's sex.

Hermaphrodite   a person in whom sexual differentiation is ambiguous or incomplete.

Transsexual   a person who believes that he or she was born with the body of the wrong sex.

Transvestite   a male who lives as a woman or a female who lives as a man but does not alter the genitalia.

Sexual orientation   a person's preference for emotional–sexual relationships with members of the opposite sex (heterosexuality), the same sex (homosexuality), or both (bisexuality).

Homophobia   extreme prejudice directed at gays, lesbians, bisexuals, and others who are perceived as not being heterosexual.

Box 10.1

Answers to the Sociology Quiz on Body Image and Gender

1. **False.** Many people do not have a very accurate perception of their bodies. For example, many young girls and women think of themselves as "fat" when they are not. Some young boys and men tend to believe that they need a well-developed chest and arm muscles, broad shoulders, and a narrow waist.

2. **True.** In recent studies, up to 95 percent of men believed they needed to improve some aspect of their bodies.

3. **True.** More than half of all adult women in the United States are currently dieting, and over three-fourths of normal-weight women think they are "too fat." Recently, very young girls have developed similar concerns. For example, 80 percent of fourth-grade girls in one study were watching their weight.

4. **True.** Women have been socialized to believe that being physically attractive is very important. Studies have found that weight and body shape are the central determinants of women's perception of their physical attractiveness.

5. **False.** Some men do have eating problems such as anorexia and bulimia. These problems have been found especially among gay men and male fashion models and dancers.

6. **False.** The "ideal" body image for women has changed a number of times. A positive view of body fat has prevailed for most of human history; however, in the twentieth century in the United States, this view gave way to "fat aversion."

7. **False.** Although bodybuilding among women has gained some degree of acceptance, women bodybuilders are still expected to be very "feminine" and not to overdevelop themselves.

8. **False.** Although the exact number is not known, many young girls and women do die as a result of starvation, malnutrition, and other problems associated with anorexia and bulimia. These are considered to be "life-threatening" behaviors by many in the medical profession.

Sources: Based on Fallon, Katzman, and Wooley, 1994; Kilbourne, 1999; Lips, 1997; and Seid, 1994.

✓ Checkpoints

Checkpoint 10.1

1. How do sex and gender differ?
2. What is meant by each of the following terms: (a) hermaphrodite, (b) transsexual, and (c) transvestite?
3. What is sexual orientation?
4. What have recent studies shown about the relationship between sexual orientation and identity?

Gender: The Cultural Dimension

Gender refers to the culturally and socially constructed differences between females and males found in the meanings, beliefs, and practices associated with "femininity" and "masculinity." Although biological differences between women and men are very important, in reality most "sex differences" are socially constructed "gender differences" (Gailey, 1987). According to sociologists, social and cultural processes, not biological "givens," are most important in defining what

Bob Daemmrich/The Image Works

In nations around the world, women are often the objects of the male gaze. How might the behavior of others influence our own perceptions of body consciousness?

females and males are, what they should do, and what sorts of relations do or should exist between them (Ortner and Whitehead, 1981; Lott, 1994). Sociologist Judith Lorber (1994: 6) summarizes the importance of gender:

> Gender is a human invention, like language, kinship, religion, and technology; like them, gender organizes human social life in culturally patterned ways. Gender organizes social relations in everyday life as well as in the major social structures, such as social class and the hierarchies of bureaucratic organizations.

Virtually everything social in our lives is *gendered:* People continually distinguish between males and females and evaluate them differentially. Gender is an integral part of the daily experiences of both women and men (Kimmel and Messner, 1998).

A microlevel analysis of gender focuses on how individuals learn gender roles and acquire a gender identity. *Gender role* **refers to the attitudes, behavior, and activities that are socially defined as appropriate for each sex and are learned through the socialization process** (Lips, 1997). For example, in U.S. society, males are traditionally expected to demonstrate aggressiveness and toughness whereas females are expected to be passive and nurturing. *Gender identity* **is a person's perception of the self as female or male.** Typically established between eighteen months and three years of age, gender identity is a powerful aspect of our self-concept (Cahill, 1986; Lips, 1997). Although this identity is an indi-

vidual perception, it is developed through interaction with others. As a result, most people form a gender identity that matches their biological sex: Most biological females think of themselves as female, and most biological males think of themselves as male. Body consciousness is a part of gender identity (Basow, 1992). *Body consciousness* **is how a person perceives and feels about his or her body;** it also includes an awareness of social conditions in society that contribute to this self-knowledge (Thompson, 1994). Consider, for example, these comments by Steve Michalik, a former Mr. Universe:

> I was small and weak, and my brother Anthony was big and graceful, and my old man made no bones about loving him and hating me. . . . The minute I walked in from school, it was, "You worthless little s--t, what are you doing home so early?" His favorite way to torture me was to tell me he was going to put me in a home. We'd be driving along in Brooklyn somewhere, and we'd pass a building with iron bars on the windows, and he'd stop the car and say to me, "Get out. This is the home we're putting you in." I'd be standing there sobbing on the curb—I was maybe eight or nine at the time. (qtd. in Klein, 1993: 273)

As we grow up, we become aware, as Michalik did, that the physical shape of our bodies subjects us to the approval or disapproval of others. Being small and weak may be considered positive attributes for women, but they are considered negative characteristics for "true men."

A macrolevel analysis of gender examines structural features, external to the individual, that perpetuate gender inequality. These structures have been referred to as *gendered institutions,*

Gender the culturally and socially constructed differences between females and males found in the meanings, beliefs, and practices associated with "femininity" and "masculinity."

Gender role the attitudes, behavior, and activities that are socially defined as appropriate for each sex and are learned through the socialization process.

Gender identity a person's perception of the self as female or male.

Body consciousness a term that describes how a person perceives and feels about his or her body.

Reflections

Reflections 10.1

How does the issue of body consciousness differ for many women and men? What part do popular culture and the media play in our perceptions about body consciousness?

meaning that gender is one of the major ways by which social life is organized in all sectors of society. Gender is embedded in the images, ideas, and language of a society and is used as a means to divide up work, allocate resources, and distribute power. For example, every society uses gender to assign certain tasks—ranging from child rearing to warfare—to females and to males, and differentially rewards those who perform these duties.

These institutions are reinforced by a *gender belief system* which includes all the ideas regarding masculine and feminine attributes that are held to be valid in a society. This belief system is legitimated by religion, science, law, and other societal values (Lorber, 1994). For example, gendered belief systems may change over time as gender roles change. Many fathers take care of young children today, and there is a much greater acceptance of this change in roles. However, popular stereotypes about men and women, as well as cultural norms about gender-appropriate appearance and behavior, serve to reinforce gendered institutions in society (Deaux and Kite, 1987).

The Social Significance of Gender

Gender is a social construction with important consequences in everyday life. Just as stereotypes regarding race/ethnicity have built-in notions of superiority and inferiority, gender stereotypes hold that men and women are inherently different in attributes, behavior, and aspirations. Stereotypes define men as strong, rational, dominant, independent, and less concerned with their appearance. Women are stereotyped as weak, emotional, nurturing, dependent, and anxious about their appearance.

The social significance of gender stereotypes is illustrated by eating problems. The three most common eating problems are anorexia, bulimia, and obesity. With *anorexia*, a person has lost at least

25 percent of body weight due to a compulsive fear of becoming fat (Lott, 1994). With *bulimia*, a person binges by consuming large quantities of food and then purges the food by induced vomiting, excessive exercise, laxatives, or fasting. With *obesity*, individuals are 20 percent or more above their desirable weight, as established by the medical profession. For a 5-foot-4-inch woman, that is about twenty-five pounds; for a 5-foot-10-inch man, it is about thirty pounds (Burros, 1994: 1).

Sociologist Becky W. Thompson argues that, based on stereotypes, the primary victims of eating problems are presumed to be white, middle-class, heterosexual women. However, such problems also exist among women of color, working-class women, lesbians, and some men. According to Thompson, explanations regarding the relationship between gender and eating problems must take into account a complex array of social factors, including gender socialization and women's responses to problems such as racism and emotional, physical, and sexual abuse (Thompson, 1994; see also Wooley, 1994).

Bodybuilding is another gendered experience. *Bodybuilding* is the process of deliberately cultivating an increase in mass and strength of the skeletal muscles by means of lifting and pushing weights (Mansfield and McGinn, 1993). In the past, bodybuilding was predominantly a male activity; musculature connoted power, domination, and virility (Klein, 1993). Today, an increasing number of women engage in this activity. As gendered experiences, eating problems and bodybuilding have more in common than we might think. Historian Susan Bordo (1993) has noted that the anorexic body and the muscled body are not opposites; instead, they exist on a continuum because they are united against a "common platoon of enemies: the soft, the loose; unsolid, excess flesh." The *body* is objectified in both compulsive dieting and bodybuilding (Mansfield and McGinn, 1993: 53).

Sexism

Sexism **is the subordination of one sex, usually female, based on the assumed superiority of the other sex.** Sexism directed at women has three components: (1) negative attitudes toward women; (2) stereotypical beliefs that reinforce, complement, or justify the prejudice; and (3) discrimination—acts that exclude, distance, or keep women separate (Lott, 1994).

Can men be victims of sexism? Although women are more often the target of sexist remarks and practices, men can be victims of sexist assumptions. As the social psychologist Hilary M. Lips (1993: 11) notes, "Sexism cuts both ways; for example, the other side of the prejudiced attitude that [usually bars] women from combat positions in the military is the attitude that it is somehow less upsetting to have male soldiers killed than to have female soldiers killed."

Like racism, sexism is used to justify discriminatory treatment. When women participate in what are considered gender-inappropriate endeavors in the workplace, at home, or in leisure activities, they often find that they are the targets of prejudice and discrimination. Obvious manifestations of sexism are found in the undervaluing of women's work, in hiring and promotion practices that effectively exclude women from an organization or confine them to the bottom of the organizational hierarchy, and in the denial of equal access for women to educational opportunities (Lips, 1997). Women who attempt to enter nontraditional occupations (such as firefighting and welding) or professions (such as dentistry and architecture) often encounter hurdles that men do not face. Even in leisure activities such as bodybuilding, women may experience discrimination because they are perceived to be "out of place" (Klein, 1993).

Sexism is interwoven with *patriarchy*—**a hierarchical system of social organization in which cultural, political, and economic structures are controlled by men.** By contrast, *matriarchy* **is a hierarchical system of social organization in which cultural, political, and economic structures are controlled by women;** however, few (if any) societies have been organized in this manner (Lengermann and Wallace, 1985). Patriarchy is reflected in the way that men may think of their position as men as a given whereas women may deliberate on what their position in society should be. As the sociologist Virginia Cyrus (1993: 6) explains, "Under patriarchy, men are seen as 'natural' heads of households, Presidential candidates, corporate executives, college presidents, etc. Women, on the other hand, are men's subordinates, playing such supportive roles as housewife, mother, nurse, and secretary." Gender inequality and a division of labor based on male dominance are nearly universal, as we will see in the following discussion on the origins of gender-based stratification.

✓ Checkpoints

Checkpoint 10.2

5. What key concerns would sociologists using a microlevel analysis of gender examine?
6. What key concerns would sociologists using a macrolevel analysis of gender examine?
7. What is sexism?
8. How is sexism related to patriarchy?

Gender Stratification in Historical and Contemporary Perspective

How do tasks in a society come to be defined as "men's work" or "women's work"? Three factors are important in determining the gendered division of labor in a society: (1) the type of subsistence base, (2) the supply of and demand for labor, and (3) the extent to which women's child-rearing activities are compatible with certain types of work. *Subsistence* refers to the means by which a society gains the basic necessities of life, including food, shelter, and clothing (Nielsen, 1990). The three factors vary according to a society's *technoeconomic base*—the level of technology and the organization of the economy in a given society. Five such bases have been identified: hunting and gathering societies, horticultural and pastoral societies, agrarian societies, industrial societies, and postindustrial societies, as shown in Table 10.2.

Hunting and Gathering Societies

The earliest known division of labor between women and men is in hunting and gathering societies. While the men hunt for wild game, women gather roots and berries (Nielsen, 1990). A relatively equitable relationship exists because neither sex has

Sexism the subordination of one sex, usually female, based on the assumed superiority of the other sex.

Patriarchy a hierarchical system of social organization in which cultural, political, and economic structures are controlled by men.

Matriarchy a hierarchical system of social organization in which cultural, political, and economic structures are controlled by women.

Table 10.2 Technoeconomic Bases of Society

	Hunting and Gathering	Horticultural and Pastoral	Agrarian
Change from Prior Society	—	Use of hand tools, such as digging stick and hoe	Use of animal-drawn plows and equipment
Economic Characteristics	Hunting game, gathering roots and berries	Planting crops, domestication of animals for food	Labor-intensive farming
Control of Surplus	None	Men, who begin to control societies	Men who own land or herds
Inheritance	None	Shared—patrilineal and matrilineal	Patrilineal
Control Over Procreation	None	Increasingly by men	Men—to ensure legitimacy of heirs
Women's Status	Relative equality	Decreasing in move to pastoralism	Low

Holton Collection/SuperStock

Alison Wright/The Image Works

Source: Adapted from Lorber, 1994, 140.

the ability to provide all the food necessary for survival. When wild game is nearby, both men and women may hunt (Basow, 1992). When it is far away, hunting becomes incompatible with child rearing (which women tend to do because they breast-feed their young), and women are placed at a disadvantage in terms of contributing to the food supply (Lorber, 1994). In most hunting and gathering societies, women are full economic partners with men; relations between them tend to be cooperative and relatively egalitarian (Chafetz, 1984). Little social stratification of any kind is found because people do not acquire a food surplus.

A few hunting and gathering societies remain, including the Bushmen of Africa, the aborigines of Australia, the Kaska Indians of Canada, and the Yanomami of South America. However, some analysts predict that these groups will cease to exist within the next few years (Nolan and Lenski, 1999).

Horticultural and Pastoral Societies

In horticultural societies, which first developed ten to twelve thousand years ago, a steady source of food becomes available. People are able to grow their own food because of hand tools, such as the

Table 10.2 continued

	Industrial	Postindustrial
Change from Prior Society	Invention of steam engine	Invention of computer and development of "high-tech" society
Economic Characteristics	Mechanized production of goods	Information and service economy
Control of Surplus	Men who own means of production	Corporate shareholders and high-tech entrepreneurs
Inheritance	Bilateral	Bilateral
Control Over Procreation	Men—but less so in later stages	Mixed
Women's Status	Low	Varies by class, race, and age

Andy Sacks/Stone—Getty Images

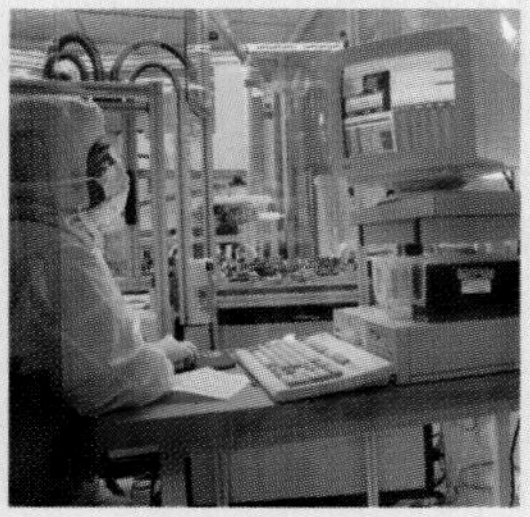

Lonnie Duka/Stone—Getty Images

digging stick and the hoe. Women make an important contribution to food production because hoe cultivation is compatible with child care. A fairly high degree of gender equality exists because neither sex controls the food supply (Basow, 1992).

When inadequate moisture in an area makes planting crops impossible, *pastoralism*—the domestication of large animals to provide food— develops. Herding is primarily done by men, and women contribute relatively little to subsistence production in such societies. In some herding societies, women have relatively low status; their primary value is their ability to produce male offspring

so that the family lineage can be preserved and enough males will exist to protect the group against attack (Nielsen, 1990).

Social practices contribute to gender inequality in horticultural and pastoral societies. Male dominance is promoted by practices such as polygyny, menstrual taboos, and bridewealth (Nielsen, 1990). *Polygyny*—the marriage of one man to multiple wives—contributes to power differences between women and men. A man with multiple wives can produce many children, who will enhance his resources, take care of him in his "old age," and become heirs to his property (Nielsen,

1990). *Menstrual taboos* place women in a subordinate position by segregating them into menstrual huts for the duration of their monthly flow. Even when women are not officially segregated, they are defined as "unclean." *Bridewealth*—the payment of a price by a man for a wife—turns women into property that can be bought and sold. The man gives the bride's family material goods or services in exchange for their daughter's exclusive sexual services and his sole claim to their offspring.

In contemporary horticultural societies, women do most of the farming while men hunt game, clear land, work with arts and crafts, make tools, participate in religious and ceremonial activities, and engage in war (Nielsen, 1990). A combination of horticultural and pastoral activities is found in some contemporary societies in Asia, Africa, the Middle East, and South America. These societies are characterized by more gender inequality than in hunting and gathering societies but less than in agrarian societies (Nielsen, 1990).

Agrarian Societies

In agrarian societies, which first developed about eight to ten thousand years ago, gender inequality and male dominance become institutionalized. The most extreme form of gender inequality developed about five thousand years ago in societies in the fertile crescent around the Mediterranean Sea (Lorber, 1994). Agrarian societies rely on agriculture—farming done by animal-drawn or mechanically powered plows and equipment. Because agrarian tasks require more labor and greater physical strength than do horticultural ones, men become more involved in food production. It has been suggested that women are excluded from these tasks because they are viewed as too weak for the work and because child-care responsibilities are considered incompatible with the full-time labor that the tasks require (Nielsen, 1990).

Why does gender inequality increase in agrarian societies? Scholars cannot agree on an answer; some suggest that it results from private ownership of property. When people no longer have to move continually in search of food, they can acquire a surplus. Men gain control over the disposition of the surplus and the kinship system, and this control serves men's interests (Lorber, 1994). The importance of producing "legitimate" heirs to inherit the surplus increases significantly, and women's lives become more secluded and restricted as men attempt to ensure the legitimacy of their children. Premarital virginity and marital fidelity are required; indiscretions are punished (Nielsen, 1990). However, some scholars argue that male dominance existed before the private ownership of property (Firestone, 1970; Lerner, 1986).

Two practices in agrarian societies contribute to subordination of women. *Purdah*, found primarily among Hindus and Muslims, requires the seclusion of women, extreme modesty in apparel, and the visible subordination of women to men. Women must show deference to men by walking behind them, speaking only when spoken to, and eating only after the men have finished a meal (Nielsen, 1990). *Genital mutilation* is a surgical procedure performed on young girls as a method of sexual control (Nielsen, 1990). The mutilation involves cutting off all or part of a girl's clitoris and labia, and in some cases stitching her vagina closed until marriage (Simons, 1993). Often justified on the erroneous belief that the Qur'an commands it, these procedures are supposed to ensure that women are chaste before marriage and have no extramarital affairs after marriage. Genital mutilation has resulted in the maiming of many females, some of whom died as a result of hemorrhage, infection, or other complications. It is still practiced in more than twenty-five countries.

In sum, male dominance is very strong in agrarian societies. Women are secluded, subordinated, and mutilated as a means of regulating their sexuality and protecting paternity. Most of the world's population currently lives in agrarian societies in various stages of industrialization. Issues concerning the rights of women in some of these societies became the subject of media attention after the 2001 terrorist attacks on the United States were linked to an organization with ties to the repressive Taliban regime in Afghanistan (see Box 10.2).

Industrial Societies

An *industrial society* is one in which factory or mechanized production has replaced agriculture as the major form of economic activity (Nielsen, 1990). As societies industrialize, the status of women tends to decline further. Industrialization in the United States created a gap between the

Box 10.2 Sociology in Global Perspective

Oppression, Resistance, and the Women of Afghanistan

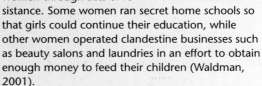

It was a very emotional moment. After years the women of Afghanistan came out in the open. Under the Taliban we all wore burkas and did not know each other. Now we all know each other's faces. . . . I just want to tell the world that women should be able to speak out about their own problems.

—Soraya Parlika, a prominent Afghan activist, describing how she felt when many women removed their veils after the overthrow of the Taliban regime in her country (qtd. in McCarthy, 2001: 46)

Although the plight of women and girls in Afghanistan was already a topic of concern for some international women's activist groups prior to the September 11, 2001, terrorist attacks, it was not until after this date that U.S. political leaders and the media highlighted the problems that many Afghan women had experienced for nearly a half decade under the Taliban. After the Taliban's takeover of that country in 1996, Afghan women were prohibited from most employment, walking alone on the streets, and teaching or studying in the schools. Residents were required to paint their windows black so that passersby could not see the face of any woman in the home (Lacayo, 2001). During the Taliban era, women were virtually imprisoned in their homes.

However, women had not always been so oppressed in Afghanistan. Prior to the Taliban takeover, Afghan women comprised 60 percent of the teachers and 50 percent of students at Kabul University; they had also been 40 percent of the doctors and 70 percent of the schoolteachers. After the takeover, many schools were closed because of a shortage of teachers and students, and women and girls were forced to wear a head-to-toe covering called a "burqa" (or "burka"), which has only a small mesh opening through which the person can see and breathe (NOW, 2002).

Although women lost virtually all of their rights under the repressive Taliban regime, journalists and other social analysts have documented how, even under these adverse conditions, some Afghan women subtly fought against the Taliban's oppression of women through acts of resistance. Some women ran secret home schools so that girls could continue their education, while other women operated clandestine businesses such as beauty salons and laundries in an effort to obtain enough money to feed their children (Waldman, 2001).

What rights will emerge for the women of Afghanistan? It has been suggested by some analysts that people who live in the United States are not able to fully understand the many factors that are involved in the struggle for women's rights in other countries. According to these observers, U.S. beliefs about women's rights are linked to Western thinking and the ideas of feminist movements in Western Europe and the United States, which are not applicable to people in other nations and cultures. According to Rina Amiri (2001: A21), senior associate for research with the Women Waging Peace Initiative at Harvard's Kennedy School of Government,

> It has come to be assumed in much of the Muslim world that to be a proponent of women's rights is to be pro-Western. This enmeshing of gender and geopolitics has robbed Muslim women of their ability to develop a discourse on their rights independent of a cultural debate between the Western and Muslim worlds.

Many social scientists who specialize in international relations hope that such a discourse—free of cultural bias—will develop so that girls and women will not continue to experience the high levels of subordination that exist in many areas of the world today. Although not as oppressive as the rules established by the Taliban, these patterns of subordination may limit the human potential of more than half of the population of these regions.

---

nonpaid work performed by middle- and upper-class women at home and the paid work that was increasingly performed by men and unmarried girls (Amott and Matthaei, 1996). Husbands were responsible for being "breadwinners"; wives were seen as "homemakers." In this new "cult of domesticity" (also referred to as the "cult of true womanhood"), the home became a private, personal sphere in which women created a haven for the family (Amott and Matthaei, 1996). The "breadwinner" role placed enormous pressures on men to support their families—being a good provider was considered to be a sign of manhood (Amott and Matthaei, 1996).

Reflections

Reflections 10.2

Do perceptions about women's appearance change as societies move through various stages of technoeconomic development? If so, how?

This gendered division of labor increased the economic and political subordination of women. It also became a source of discrimination against women of color based on both their race and the fact that many of them had to work in order to survive. In the late 1800s and early 1900s, many African American women were employed as domestic servants in affluent white households (Lorber, 1994).

As people moved from a rural, agricultural lifestyle to an urban existence, body consciousness increased. People who worked in offices often became sedentary and exhibited physical deterioration from their lack of activity. As gymnasiums were built to fight this lack of physical fitness, images of masculinity shifted from the "burly farmer" or "robust workman" to the middle-class man who exercised and lifted weights (Klein, 1993). As industrialization progressed and food became more plentiful, the social symbolism of women's body weight and size also changed, and middle-class women became more preoccupied with body fitness (Bordo, 1993; Seid, 1994). Today, women's bodies (even in bodybuilding programs) are supposed to be "inviting, available, and welcoming" whereas men's bodies should be "self-contained, active and invasive" (MacSween, 1993: 156).

Postindustrial Societies

Chapter 4 defines *postindustrial societies* as ones in which technology supports a service- and information-based economy. In such societies, the division of labor in paid employment is increasingly based on whether people provide or apply information or are employed in service jobs such as fast-food restaurant counter help or health care workers. For both women and men in the labor force, formal education is increasingly crucial for economic and social success. However, as some women have moved into entrepreneurial, managerial, and professional occupations, many others have remained in the low-paying service sector, which affords few opportunities for upward advancement.

Will technology change the gendered division of labor in postindustrial societies? Scholars do not agree on the effects of computers, the Internet, the World Wide Web, cellular phones, and many newer forms of communications technology on the role of women in society. For example, some feminist writers had a pessimistic view of the impact of computers and monitors on women's health and safety, predicting that women in secretarial and administrative roles would experience an increase in eyestrain, headaches, risks of radiation, and problems such as carpal tunnel syndrome. However, some medical experts now believe that such problems extend to both men and women, as computers have become omnipresent in more people's lives. The term "24-7" has come to mean that a person is available "twenty-four hours a day, seven days a week" via cell phones, pocket pagers, fax machines, e-mail, and other means of communication, whether the individual is at the office or four thousand miles away on "vacation." Currently, some phone companies are offering individuals a "universal phone number" at which they can be reached anywhere in the world.

How do new technologies influence gender relations in the workplace? Although some analysts presumed that technological developments would reduce the boundaries between women's and men's work, researchers have found that the gender stereotyping associated with specific jobs has remained remarkably stable even when the nature of work and the skills required to perform it have been radically transformed. Today, men and women continue to be segregated into different occupations, and this segregation is particularly visible within individual workplaces (as discussed later in the chapter).

How does the division of labor change in families in postindustrial societies? For a variety of reasons, more households are headed by women with no adult male present. As shown in the Census Profiles feature, the percentage of U.S. households headed by a single mother with children under eighteen has increased. Chapter 11 ("Families and Intimate Relationships") discusses a number of reasons why the current division of labor in household chores in some families is between a woman

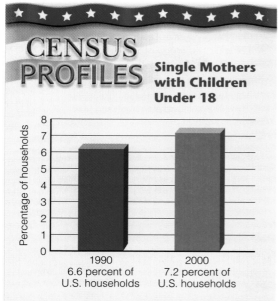

CENSUS PROFILES

Single Mothers with Children Under 18

As shown above, between 1990 and 2000, the number of U.S. families headed by single mothers increased by about 25 percent, to more than 7.5 million households. For census purposes, a single mother is identified as a woman who is widowed, divorced, separated, or never married and who has children under 18 living at home. Not only does this increase mark a change in the roles of many women, but it may also indicate that "traditional" households are in decline in this country.

Sources: U.S. Census Bureau, 2000c, 2001e.

and her children rather than between women and men. Consider, for example, that more than one-fourth (27 percent) of all U.S. children live with their mother only (as contrasted with just 5 percent who reside with their father only); among African American children, 58 percent live with their mother only (U.S. Census Bureau, 2001g). This means that women in these households truly have a double burden, both from family responsibilities and from the necessity of holding gainful employment in the labor force.

Even in single-person or two-parent households, programming "labor-saving" devices (if they can be afforded) often means that a person must have some leisure time to learn how to do the programming. According to analysts, leisure is deeply divided along gender lines, and women have less time to "play in the house" than do men and boys.

✓ Checkpoints

Checkpoint 10.3

9. How does men's work and women's work differ in hunting and gathering societies? In horticultural and pastoral societies?
10. How does gender inequality become institutionalized in agrarian societies?
11. Why does the gendered division of labor in industrial societies contribute to the subordination of women?
12. Does the gendered division of labor change in postindustrial societies?

Some web sites seek to appeal to women who have economic resources but are short on time, making it possible for them to shop, gather information, "telebank," and communicate with others at all hours of the day and night.

In postindustrial societies such as the United States, more than 60 percent of adult women are in the labor force, meaning that finding time to care for children, help aging parents, and meet the demands of the workplace will continue to place a heavy burden on women, despite living in an information- and service-oriented economy.

How people accept new technologies and the effect these technologies have on gender stratification are related to how people are socialized into gender roles. However, gender-based stratification remains rooted in the larger social structures of society, which individuals have little ability to control.

Gender and Socialization

We learn gender-appropriate behavior through the socialization process. Our parents, teachers, friends, and the media all serve as gendered institutions that communicate to us our earliest, and often most lasting, beliefs about the social meanings of being male or female and thinking and behaving in masculine or feminine ways. Some gender roles have changed dramatically in recent years; others remain largely unchanged over time.

Many parents prefer boys to girls because of stereotypical ideas about the relative importance of males and females to the future of the family

Bob Thomas/Getty Images

Bob Thomas/Stone—Getty Images

Are children's toys a reflection of their own preferences and choices? How do toys reflect gender socialization by parents and other adults?

and society (Basow, 1992). Although some parents prefer boys to girls because these parents believe old myths about the biological inferiority of females, research suggests that social expectations also play a major role in this preference. We are socialized to believe that it is important to have a son, especially for a first or only child. For many years, it was assumed that only a male child could support his parents in their later years and carry on the family name.

Across cultures, boys are preferred to girls, especially when the number of children that parents can have is limited by law or economic conditions. For example, in China, which strictly regulates the allowable number of children to one per family, a disproportionate number of female fetuses are aborted (Basow, 1992). In India, the practice of aborting female fetuses is widespread, and female infanticide occurs frequently (Burns, 1994). As a result, both India and China have a growing sur-

plus of young men who will face a shortage of women their own age (Shenon, 1994).

Parents and Gender Socialization

From birth, parents act toward children on the basis of the child's sex. Baby boys are perceived to be less fragile than girls and tend to be treated more roughly by their parents. Girl babies are thought to be "cute, sweet, and cuddly" and receive more gentle treatment (MacDonald and Parke, 1986). Parents strongly influence the gender-role development of children by passing on—both overtly and covertly—their own beliefs about gender (Witt, 1997). When girl babies cry, parents respond to them more quickly, and parents are more prone to talk and sing to girl babies (Basow, 1992). However, one study ("The Favored Infants," 1976) found that African American mothers "rubbed, patted, rocked, touched, kissed and talked to" boy babies more than girl babies.

Children's toys reflect their parents' gender expectations (Thorne, 1993). Gender-appropriate toys for boys include computer games, trucks and other vehicles, sports equipment, and war toys such as guns and soldiers. Girls' toys include "Barbie" dolls, play makeup, and homemaking items. Parents' choices of toys for their children are not likely to change in the near future. A group of college students in one study were shown slides of toys and asked to decide which ones they would buy for girls and boys. Most said they would buy guns, soldiers, jeeps, carpenter tools, and red bicycles for boys; girls would get baby dolls, dishes, sewing kits, jewelry boxes, and pink bicycles (Fisher-Thompson, 1990).

When children are old enough to help with household chores, they are often assigned different tasks. Maintenance chores (such as mowing the lawn) are assigned to boys whereas domestic chores (such as shopping, cooking, and clearing the table) are assigned to girls. Chores may also become linked with future occupational choices and personal characteristics. Girls who are responsible for domestic chores such as caring for younger brothers and sisters may learn nurturing behaviors that later translate into employment as a nurse or schoolteacher. Boys may learn about computers and other types of technology that lead to different career options.

Because most studies of gender socialization by parents focus on white, middle-class families, it is probably inappropriate to assume that the patterns of differential parental treatment of girls and boys hold true in all respects for other class and racial groups (Lips, 1997). For example, children from middle- and upper-income families are less likely to be assigned gender-linked chores than children from lower-income backgrounds. In addition, analysts have found that gender-linked chore assignments occur less frequently in African American families, where both sons and daughters tend to be socialized toward independence, employment, and child care (Bardwell, Cochran, and Walker, 1986; Hale-Benson, 1986). Sociologist Patricia Hill Collins (1991) suggests that African American mothers are less likely to socialize their daughters into roles as subordinates; instead, they are likely to teach them a critical posture that allows them to cope with contradictions.

Gender socialization by parents may become intertwined with appearance norms that produce dysfunctional behavior, such as eating problems (Thompson, 1994). However, many parents are

Barbara Campbell/Liaison–Getty Images

Not only parents and peers but also the larger society influences our perceptions about gender-appropriate behavior.

aware of the effect that gender socialization has on their children and make a conscientious effort to provide nonsexist experiences for them (Brooks-Gunn, 1986).

Peers and Gender Socialization

Peers help children learn prevailing gender-role stereotypes, as well as gender-appropriate and gender-inappropriate behavior (Hibbard and Buhrmester, 1998). During the preschool years, same-sex peers have a powerful effect on how children see their gender roles (Maccoby and Jacklin, 1987); children are more socially acceptable to their peers when they conform to implicit societal norms governing the "appropriate" ways that girls and boys should act in social situations and what prohibitions exist in such cases (Martin, 1989).

Male peer groups place more pressure on boys to do "masculine" things than female peer groups

Teachers often use competition between boys and girls because they hope to make a learning activity more interesting. What are the advantages and disadvantages of gender-based competition in classroom settings?

Hazel Hankin/Stock, Boston

place on girls to do "feminine" things (Fagot, 1984). For example, girls wear jeans and other "boy" clothes, play soccer and softball, and engage in other activities traditionally associated with males. By contrast, if a boy wears a dress, plays hopscotch with girls, and engages in other activities associated with being female, he will be ridiculed by his peers. This distinction between the relative value of boys' and girls' behaviors strengthens the cultural message that masculine activities and behavior are more important and more acceptable (Wood, 1999).

During adolescence, peers are often stronger and more effective agents of gender socialization than adults (Hibbard and Buhrmester, 1998). Peers are thought to be especially important in boys' development of gender identity (Maccoby and Jacklin, 1987). Male bonding that occurs during adolescence is believed to reinforce masculine identity (Gaylin, 1992) and to encourage gender-stereotypical attitudes and behavior (Huston, 1985; Martin, 1989). For example, male peers have a tendency to ridicule and bully others about their appearance, size, and weight. Aleta Walker painfully recalls walking down the halls at school when boys would flatten themselves against the lockers and cry, "Wide load!" At lunchtime, the boys made a production of watching her eat lunch and frequently made sounds like grunts or moos (Kolata, 1993). Because peer acceptance is so important for both males and females

during their first two decades, such actions can have very harmful consequences for the victims.

As young adults, men and women still receive many gender-related messages from peers. Among college students, for example, peer groups are organized largely around gender relations and play an important role in career choices and the establishment of long-term, intimate relationships (Holland and Eisenhart, 1990). In a study of women college students at two universities (one primarily white, the other predominantly African American), anthropologists Dorothy C. Holland and Margaret A. Eisenhart (1990) found that the peer system propelled women into a world of romance in which their attractiveness to men counted most. Although peers initially did not influence the women's choices of majors and careers, they did influence whether the women continued to pursue their original goals, changed their course of action, or were "derailed" (Holland and Eisenhart, 1981, 1990).

Teachers, Schools, and Gender Socialization

From kindergarten through college, schools operate as a gendered institution. Teachers provide important messages about gender through both the formal content of classroom assignments and informal interactions with students. Sometimes, gender-related messages from teachers and other students reinforce gender roles that have been taught at home; however, teachers may also contradict parental socialization. During the early years of a child's schooling, teachers' influence is very powerful; many children spend more hours per day with their teachers than they do with their own parents.

According to some researchers, the quantity and quality of teacher–student interactions often vary between the education of girls and that of boys (Wellhousen and Yin, 1997). One of the messages that teachers may communicate to students is that boys are more important than girls. Research spanning the past thirty years shows that unintentional gender bias occurs in virtually all educational settings. *Gender bias* **consists of showing favoritism toward one gender over the other.** Researchers consistently find that teachers devote more time, effort, and attention to boys than to girls (Sadker and Sadker, 1994). Males receive more praise for their contributions and are called on more frequently in class, even when they do not volunteer. Very often, boys receive attention because they call out in class, demand help, and sometimes engage in disruptive behavior (Sadker and Sadker, 1994). Teachers who do not negatively sanction such behavior may unintentionally encourage it. Boys learn that when they yell out an answer without being called on, their answer will be accepted by the teacher; girls learn that they will be praised when they are compliant and wait for the teacher to call on them. If they call out an answer, they may be corrected with comments such as "Please raise your hand if you want to speak" (Sadker and Sadker, 1994).

Teacher–student interactions influence not only students' learning but also their self-esteem (Sadker and Sadker, 1985, 1986, 1994). A comprehensive study of gender bias in schools suggested that girls' self-esteem is undermined in school through such experiences as (1) a relative lack of attention from teachers; (2) sexual harassment by male peers; (3) the stereotyping and invisibility of females in textbooks, especially in science and math texts; and (4) test bias based on assumptions about the relative importance of quantitative and visual–spatial ability, as compared with verbal ability, that restricts some girls' chances of being admitted to the most prestigious colleges and being awarded scholarships. White males may have better self-esteem because they receive more teacher attention than all other students (Sadker and Sad-

ker, 1994). Although African American girls may start out with very positive behaviors, they receive little teacher feedback or academic encouragement (Irvine, 1986). Of all groups, African American males may receive the most unfavorable treatment, as well as the most frequent teacher referrals to special education programs (McIntyre and Pernell, 1985). Teachers may inadvertently subscribe to racist and sexist stereotypes when they perceive Asian American students as the best students and African American males as the worst (Basow, 1992).

Teachers also influence how students treat one another during school hours. Many teachers use sex segregation as a way to organize students, resulting in unnecessary competition between females and males (Basow, 1992). In addition, teachers may take a "boys will be boys" attitude when girls complain of sexual harassment. Even though sexual harassment is prohibited by law, and teachers and administrators are obligated to investigate such incidents, the complaints may be dealt with superficially. If that happens, the school setting can become a hostile environment rather than a site for learning (Sadker and Sadker, 1994).

Most problems that exist in pre-kindergarten through high school are also found in colleges and universities. Studies in the 1980s showed that college instructors often paid more attention to men than women in their classes (Hall and Sandler, 1982, 1984) and that women were called on less frequently, received less encouragement, and were often interrupted, ignored, or devalued (Fox, 1989). With recent gains in women's undergraduate and graduate school enrollment, these problems may be less severe in certain academic fields. There are more women than men in some classes, making it more difficult for professors to overlook women's potential contributions in the classroom. However, much remains to be done to produce a positive learning environment for all students.

Today, women are more likely than men to earn a college degree. Although men still constitute the majority of majors in architecture, engineering, computer technology, and physical sciences, women

Gender bias   behavior that shows favoritism toward one gender over the other.

Reuters/Fabrizio Bensch/Archive Photos–Getty Images

In recent years, women have expanded their involvement in professional sports through organizations such as the WNBA; however, most sports remain rigidly divided into female events and male events. Do you think that media coverage of women's and men's college and professional sporting events differs?

are more likely to earn undergraduate degrees in the health professions, education, English, and psychology. Even so, men make up more than 60 percent of faculty members at four-year colleges and universities, and are more likely than women faculty members to hold full-time, tenured or tenure-track positions (*Chronicle of Higher Education*, 2001).

Sports and Gender Socialization

Children spend more than half of their nonschool time in play and games, but the type of games played differs with the child's sex. Studies indicate that boys are socialized to participate in highly competitive, rule-oriented games with a larger number of participants than games played by girls. Girls have been socialized to play exclusively with others of their own age, in groups of two or three, in activities such as hopscotch and jump rope that

involve a minimum of competitiveness (Lever, 1978; Ignico and Mead, 1990). Other research shows that boys express much more favorable attitudes toward physical exertion and exercise than girls do. Some analysts believe this difference in attitude is linked to ideas about what is gender-appropriate behavior for boys and girls (Brustad, 1996). For males, competitive sport becomes a means of "constructing a masculine identity, a legitimated outlet for violence and aggression, and an avenue for upward mobility" (Lorber, 1994: 43). Recently, more girls have started to play soccer and softball and to participate in sports formerly regarded as exclusively "male" activities. However, even with these changes, many women athletes believe that they have to manage the contradictory statuses of being both "women" and "athletes." One study found that women college basketball players dealt with this contradiction by dividing their lives into segments. On the basketball court, the women "did athlete": They pushed, shoved, fouled, ran hard, sweated, and cursed. Off the court, they "did woman": After the game, they showered, dressed, applied makeup, and styled their hair, even if they were only getting in a van for a long ride home (Watson, 1987).

Most sports are rigidly divided into female and male events. Assumptions about male and female physiology and athletic capabilities influence the types of sports in which members of each sex are encouraged to participate. For example, women who engage in activities that are assumed to be "masculine" (such as bodybuilding) may either ignore their critics or attempt to redefine the activity or its result as "feminine" or womanly (Duff and Hong, 1984; Klein, 1993). Some women bodybuilders do not want their bodies to get "overbuilt." They have learned that they are more likely to win women's bodybuilding competitions if they look and pose "more or less along the lines of fashion models" (Klein, 1993: 179).

Mass Media and Gender Socialization

The media are a powerful source of gender stereotyping. Some critics argue that the media simply reflect existing gender roles in society, but others point out that the media have a uniquely persuasive ability to shape ideas. From children's cartoons to adult shows, television programs are sex-typed, and many are male oriented. More male than female roles are shown, and male characters act strikingly different from female ones. Typically, males are more aggressive, constructive, and direct and are rewarded for their actions. By contrast, females are depicted as acting deferential toward other people or as manipulating them through helplessness or seductiveness to get their way (Basow, 1992). Because advertisers hope to appeal to boys, who constitute more than half of the viewing audience for some shows, many programs feature lively adventure and lots of loud noise and violence.

In prime-time television, a number of significant changes in the past three decades have reduced gender stereotyping; however, men still outnumber women as leading characters. Men, for the most part, have been police officers, detectives, attorneys, doctors, and businessmen. In recent years, women in professional careers have been overrepresented, which may give an erroneous impression that most women in the work force are in executive, managerial, and professional positions such as lawyers; in the "real world," most employed women work in low-paying, low-status jobs (Basow, 1992). In most programs, women's appearance is considered very important and is frequently a topic of discussion in the program itself.

Advertising—whether on television and billboards or in magazines and newspapers—can be very persuasive. The intended message is clear to many people: If they embrace traditional notions of masculinity and femininity, their personal and social success is assured; if they purchase the right products and services, they can enhance their appearance and gain power over other people. For example, one study of television commercials reflects that men's roles are portrayed differently from women's roles (Kaufman, 1999). Men are more likely to be away from home or are more likely to be shown working or playing outside the house rather than inside, where women are more likely to be located. There is also a far greater likelihood that women will be doing domestic tasks such as cooking, cleaning, or shopping, whereas men are more likely to be taking care of cars or yards or playing games. As such, television commercials may act as agents of socialization, showing children and others what women's and men's designated activities are (Kaufman, 1999).

A study by the sociologist Anthony J. Cortese (1999) found that women—regardless of what they were doing in a particular ad—were frequently

Reflections

Reflections 10.3

What part does gender socialization play in creating our ideas about our physical appearance?

shown in advertising as being young, beautiful, and seductive. Although such depictions may sell products, they may also have the effect of influencing how we perceive ourselves and others with regard to issues of power and subordination.

Adult Gender Socialization

Gender socialization continues as women and men complete their training or education and join the work force. Men and women are taught the "appropriate" type of conduct for persons of their sex in a particular job or occupation—both by their employers and by co-workers. However, men's socialization usually does not include a measure of whether their work can be successfully combined with having a family; it is often assumed that men can and will do both. Even today, the reason given for women not entering some careers and professions is that this kind of work is not suitable for women because of their physical capabilities or assumed child-care responsibilities.

Different gender socialization may occur as people reach their forties and enter "middle age." A double standard of aging exists that affects women more than men. Often, men are considered to be at the height of their success as their hair turns gray and their face gains a few wrinkles. By contrast, not only do other people in society make middle-aged women feel as if they are "over the hill," but multimillion-dollar advertising campaigns continually call attention to women's every weakness, every pound gained, and every bit of flabby flesh, wrinkle, or gray hair. Increasingly, both women and men have turned to "miracle" products, and sometimes to cosmetic surgery, to reduce the visible signs of aging.

A knowledge of how we develop a gender-related self-concept and learn to feel, think, and act in feminine or masculine ways is important for an understanding of ourselves. Examining gender socialization makes us aware of the impact of our

✓ **Checkpoints**

Checkpoint 10.4

13. What part do parents play in the gender socialization of children?
14. Why are peers especially important in the gender socialization process?
15. What is gender bias in education, and how does it affect students?
16. What part do sports and mass media play in gender socialization?

parents, siblings, teachers, friends, and the media on our perspectives about gender. However, the gender socialization perspective has been criticized on several accounts. Childhood gender-role socialization may not affect people as much as some analysts have suggested. For example, the types of jobs that people take as adults may have less to do with how they were socialized in childhood than it does with how they are treated in the workplace. From this perspective, women and men will act in ways that bring them the most rewards and produce the fewest punishments (Reskin and Padavic, 1994). Also, gender socialization theories can be used to blame women for their own subordination by not taking into account structural barriers that perpetuate gender inequality. We will now examine a few of those structural forces.

Contemporary Gender Inequality

According to feminist scholars, women experience gender inequality as a result of past and present economic, political, and educational discrimination (Hacker, 1951, 1974; Hochschild, 1973; Dworkin, 1982). Women's position in the U.S. work force reflects the years of subordination that they have experienced in society.

Gendered Division of Paid Work

Where people are located in the occupational structure of the labor market has a major impact on their earnings (Kemp, 1994). The workplace is another example of a gendered institution. In industrialized countries, most jobs are segregated by gender and by race/ethnicity. Sociologist Judith Lorber (1994: 194) gives this example:

> In a workplace in New York City—for instance, a handbag factory—a walk through the various departments might reveal that the owners and managers are white men; their secretaries and bookkeepers are white and Asian women; the order takers and data processors are African American women; the factory hands are [Latinos] cutting pieces and [Latinas] sewing them together; African American men are packing and loading the finished product; and non-English-speaking Eastern European women are cleaning up after everyone. The workplace as a whole seems integrated by race, ethnic group, and gender, but the individual jobs are markedly segregated according to social characteristics.

Lorber notes that in most workplaces, employees are either gender segregated or all of the same gender. *Gender-segregated work* refers to the concentration of women and men in different occupations, jobs, and places of work (Reskin and Padavic, 1994). In 2000, for example, 98 percent of all secretaries in the United States were women; 90 percent of all engineers were men (U.S. Census Bureau, 2001g). To eliminate gender-segregated jobs in the United States, more than half of all men or all women workers would have to change occupations (Reskin and Hartmann, 1986). Moreover, women are severely underrepresented at the top of U.S. corporations. Only about 10 percent of the executive jobs at Fortune 500 companies are held by women, and only six women are the CEO of such a company (*Fortune*, 2001).

Although the degree of gender segregation in the professional labor market (including physicians, dentists, lawyers, accountants, and managers) has declined since the 1970s, racial–ethnic segregation has remained deeply embedded in the social structure. As the sociologist Elizabeth Higginbotham (1994) points out, African American professional women for many years found themselves limited to employment in certain sectors of the labor market. Although some change has occurred in recent years, women of color are more likely than their white counterparts to be concentrated in public-sector employment (as public schoolteachers, welfare workers, librarians, public defenders, and faculty members at public colleges, for example) rather than in the private sector (for example, in large corporations, major law firms, and private educational

Table 10.3 Percentage of Work Force Represented by Women, African Americans, and Hispanics in Selected Occupations

The U.S. Census Bureau accumulates data that show what percentage of the total work force is made up of women, African Americans, and Hispanics. As used in this table, *women* refers to females in all racial–ethnic categories, whereas *African Americans* and *Hispanics* refer to both women and men. Based on this table, in which occupations are white men most likely to be employed? In which occupations are they least likely to be employed?

	Women	African Americans	Hispanics
All occupations	46.2	11.1	10.1
Managerial and professional specialty (all)	49.0	7.6	5.0
Executive, administrative, and managerial	44.4	7.2	5.4
Professional specialty	53.3	7.9	4.6
Technical, sales, and administrative support (all)	64.2	11.1	8.3
Technicians and related support	53.6	10.4	6.6
Sales occupations	50.3	8.9	7.9
Service occupations (all)	59.5	17.6	15.0
Private household workers	94.6	13.7	30.9
Cleaning and building service workers	44.5	22.8	20.5
Health service assistants, aides, nurses, attendants	88.5	30.1	10.0
Operators, fabricators, and laborers	24.6	15.7	16.0

Source: U.S. Census Bureau, 2000c.

institutions). Across all categories of occupations, white women and all people of color are not evenly represented, as shown in Table 10.3.

Labor market segmentation—the division of jobs into categories with distinct working conditions—results in women having separate and unequal jobs (Amott and Matthaei, 1996; Lorber, 1994). The pay gap between men and women is the best-documented consequence of gender-segregated work (Reskin and Padavic, 1994). Most women work in lower-paying, less-prestigious jobs, with little opportunity for advancement. Because many employers assume that men are the breadwinners, men are expected to make more money than women in order to support their families. For many years, women have been viewed as supplemental wage earners in a male-headed household, regardless of the women's marital status. Consequently, women have not been seen as legitimate workers but mainly as wives and mothers (Lorber, 1994). Such thinking has especially harmful consequences for the many women who are

the only breadwinner in their family due to divorce, widowhood, unmarried status, or other reasons.

Gender-segregated work affects both men and women. Men are often kept out of certain types of jobs. Those who enter female-dominated occupations often have to justify themselves and prove that they are "real men." They have to fight stereotypes ("Is he gay? Lazy?") about why they are interested in such work (Williams, 1993: 3). Even if these assumptions do not push men out of female-dominated occupations, they affect how the men manage their gender identity at work. For example, men in occupations such as nursing tend to emphasize their masculinity, attempt to distance themselves from female colleagues, and try to move quickly into management and supervisory positions (Williams, 1989, 1993).

Occupational gender segregation contributes to stratification in society. Job segregation is structural; it does not occur simply because individual workers have different abilities, motivations, and material

Figure 10.1 The Wage Gap

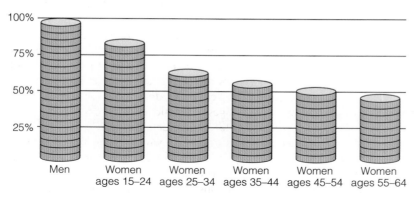

Although men's average wages vary depending upon age, women's average wages are always lower than those of men in the same age group, and the older women get, the greater the gap. Women ages 15–24 earn 80 cents for every dollar earned by men the same age, but women ages 55–64 earn only 45 cents for every dollar earned by men ages 55–64.

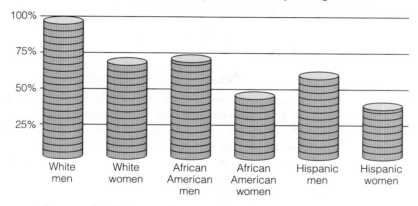

Across racial–ethnic groupings, men's earnings remain higher than the earnings of women in the same category.

Regardless of occupation, women on average receive lower wages.

Source: U.S. Census Bureau, 2000g.

needs. As a result of gender and racial segregation, employers are able to pay many men of color and all women less money, promote them less often, and provide fewer benefits.

Pay Equity (Comparable Worth)

Occupational segregation contributes to a *pay gap*—**the disparity between women's and men's earnings.** It is calculated by dividing women's earnings by men's to yield a percentage, also known as the earnings ratio (Reskin and Padavic, 1994). As Figure 10.1 shows, women at all levels of educational attainment receive less pay than men with the same levels of education. Across different categories of employment, data continue to demonstrate this disparity (Lott, 1994). Moreover, the pay gap is even greater for women of color. Although white women in 2000 earned 73 percent as much as white men, African American women earned only 48 percent and Latinas 39 percent of what white male workers earned (U.S. Census Bureau, 2001g). The gap increases as a person's salary grows: Whereas 48 percent of people earning $20,000 to $25,000 per year are women, women constitute less than 10 percent of people earning over $200,000 per year (Johnston, 2002). Likewise, among older workers, the pay gap between men and women is larger. Women's earnings tend to rise more slowly than men's when they are younger and then drop off when women reach their late thirties and early forties, whereas the wages of men tend to increase as they age (Reskin and Padavic, 1994).

Pay equity or *comparable worth* **is the belief that wages ought to reflect the worth of a job, not the gender or race of the worker** (Kemp, 1994). How can the comparable worth of different kinds of jobs be determined? One way is to compare the actual work of women's and men's jobs and see if there is a disparity in the salaries paid for each. To do this, analysts break a job into components—such as the education, training, and skills required, the extent of responsibility for others' work, and the working conditions—and then allocate points for each (Lorber, 1994). For pay equity to exist, men and women in occupations that receive the same number of points should be paid the same. However, pay equity exists for very few jobs. For example, for every dollar earned by men, women in the same occupation earned 73 cents as bookkeepers,

80 cents as computer programmers, 76 cents as cooks, 75 cents as lawyers, and 61 cents as office managers (Colatosi, 1992).

Comparable worth is important for all workers, not just women. Both men and women suffer an economic penalty when they work in jobs that employ mostly women. Compared to what they could make in male-dominated jobs, men suffer a loss of wages in female-dominated jobs such as nurse, secretary, and elementary schoolteacher. Perhaps this helps explain why men are less likely to seek employment in female-dominated fields than women are to seek employment in male-dominated fields (Jacobs, 1993). In their study of men in female-dominated jobs, the sociologists Paula England and Melissa Herbert (1993) found that the differences in pay are due to the cultural devaluation of women. If women were compensated fairly, an employer could not undercut men's wages by hiring women at a cheaper rate (Kessler-Harris, 1990).

Paid Work and Family Work

As previously discussed, the first big change in the relationship between family and work occurred with the Industrial Revolution and the rise of capitalism. The cult of domesticity kept many middle- and upper-class women out of the work force during this period. Primarily, working-class and poor women were the ones who had to deal with the work/family conflict. Today, however, the issue spans the entire economic spectrum (Reskin and Padavic, 1994). The typical married woman in the United States combines paid work in the labor force and family work as a homemaker. Although this change has occurred at the societal level, individual women bear the brunt of the problem.

Even with dramatic changes in women's work force participation, the sexual division of labor in the family remains essentially unchanged. Most married women now share responsibility for the breadwinner role, yet many men do not accept their share of

Pay gap   a term used to describe the disparity between women's and men's earnings.

Comparable worth (or **Pay equity**)   the belief that wages ought to reflect the worth of a job, not the gender or race of the worker.

What stereotypes are associated with men in female-oriented occupations? With women in male-oriented occupations? Do you think such stereotypes will change in the near future?

domestic responsibilities (Reskin and Padavic, 1994). Consequently, many women have a "double day" or "second shift" because of their dual responsibilities for paid and unpaid work (Hochschild, 1989). Working women have less time to spend on housework; if husbands do not participate in routine domestic chores, some chores simply do not get done or get done less often. Although the income that many women earn is essential to the economic survival of their families, they still must spend part of their earnings on family maintenance, such as daycare centers, fast-food restaurants, and laundries, in an attempt to keep up with their obligations (Bergmann, 1986).

Especially in families with young children, domestic responsibilities consume a great deal of time and energy. Although some kinds of housework can be put off, the needs of children often cannot be ignored or delayed. When children are ill or school events cannot be scheduled around work, parents (especially mothers) may experience stressful role conflicts ("Shall I be a good employee or a good mother?"). Many working women care not only for themselves, their husbands, and their children but also for elderly parents or in-laws. Some analysts refer to these women as "the sandwich generation"—caught between the needs of their young children and of their elderly relatives. Many women try to solve their time crunch by forgoing leisure time and sleep. When Arlie Hochschild interviewed working mothers, she found that they talked about sleep "the way a hungry person talks about food" (1989: 9). Perhaps this is one reason that, in later research, Hochschild (1997) learned that some married women with children found more fulfillment at work and that they worked longer hours because they liked work better than facing the pressures of home.

✓ Checkpoints

Checkpoint 10.5

17. How does gender segregation in the workplace affect people's opportunities and economic conditions?
18. What is pay equity, and what are the central arguments in the comparable worth debate?
19. How has the nature of paid work and family work changed in the United States since the Industrial Revolution?

Perspectives on Gender Stratification

Sociological perspectives on gender stratification vary in their approach to examining gender roles and power relationships in society. Some focus on the roles of women and men in the domestic sphere; others note the inequalities arising from a gendered division of labor in the workplace. Still others attempt to integrate both the public and private spheres into their analyses.

Functionalist and Neoclassical Economic Perspectives

As seen earlier, functionalist theory views men and women as having distinct roles that are important for the survival of the family and society. The most basic division of labor is biological: Men are physically stronger, and women are the only ones able to bear and nurse children. Gendered belief systems foster assumptions about appropriate behavior for men and women and may have an impact on the types of work that women and men perform.

THE IMPORTANCE OF TRADITIONAL GENDER ROLES According to functional analysts such as Talcott Parsons (1955), women's roles as nurturers and caregivers are even more pronounced in contemporary industrialized societies. While the husband performs the *instrumental* tasks of providing economic support and making decisions, the wife assumes the *expressive* tasks of providing affection and emotional support for the family. This division of family labor ensures that

According to the human capital model, women may earn less in the labor market because of their child-rearing responsibilities. What other sociological explanations are offered for the lower wage that women receive?

important societal tasks will be fulfilled; it also provides stability for family members.

This view has been adopted by a number of politically conservative analysts who assert that relationships between men and women are damaged when changes in gender roles occur, and family life suffers as a consequence (Gilder, 1986). From this perspective, the traditional division of labor between men and women is the natural order of the universe (Kemp, 1994).

THE HUMAN CAPITAL MODEL Functionalist explanations of occupational gender segregation are similar to neoclassical economic perspectives, such as the human capital model (Horan, 1978; Kemp, 1994). According to this model, individuals vary widely in the amount of human capital they bring to the labor market. *Human capital* is acquired by education and job training; it is the source of a person's productivity and can be measured in terms of the return on the investment (wages) and the cost (schooling or training) (Stevenson, 1988; Kemp, 1994).

From this perspective, what individuals earn is the result of their own choices (the kinds of training, education, and experience they accumulate, for example) and of the labor market need (demand) for and availability (supply) of certain kinds of workers at specific points in time. For example, human capital analysts argue that women diminish their human capital when they leave the labor force to engage in childbearing and child-care activities. While women are out of the labor force, their human capital deteriorates from nonuse. When they return to work, women earn lower wages than men because they have fewer years of work experience and have "atrophied human capital" because their education and training may have become obsolete (Kemp, 1994: 70).

EVALUATION OF FUNCTIONALIST AND NEOCLASSICAL ECONOMIC PERSPECTIVES
Although Parsons and other functionalists did not specifically endorse the gendered division of labor, their analysis suggests that it is natural and perhaps inevitable. However, critics argue that problems inherent in traditional gender roles, including the personal role strains of men and women and the social costs to society, are minimized by this approach. For example, men are assumed to be "money machines" for their families when they might prefer to spend more time in child-rearing activities. Also, the woman's place is assumed to be in the home, an assumption that ignores the fact that many women hold jobs due to economic necessity.

In addition, the functionalist approach does not take a critical look at the structure of society (especially the economic inequalities) that makes educational and occupational opportunities more available to some than to others. Furthermore, it fails to examine the underlying power relations between men and women or to consider the fact that the tasks assigned to women and to men are unequally valued by society (Kemp, 1994). Similarly, the human capital model is rooted in the premise that individuals are evaluated based on their human capital in an open, competitive market where education, training, and other job-enhancing characteristics are taken into account. From this perspective, those who make less money (often men of color and all women) have no one to blame but themselves.

Critics note that instead of blaming people for their choices, we must acknowledge other realities. Wage discrimination occurs in two ways: (1) the wages are higher in male-dominated jobs, occupations, and segments of the labor market, regardless of whether women take time for family duties, and (2) in any job, women and people of color will be paid less (Lorber, 1994).

Conflict Perspectives

According to many conflict analysts, the gendered division of labor within families and in the workplace results from male control of and dominance over women and resources. Differentials between men and women may exist in terms of economic, political, physical, and/or interpersonal power. The importance of a male monopoly in any of these arenas depends on the significance of that type of power in a society (Richardson, 1993). In hunting and gathering and horticultural societies, male dominance over women is limited because all members of the society must work in order to survive (Collins, 1971; Nielsen, 1990). In agrarian societies, however, male sexual dominance is at its peak. Male heads of household gain a monopoly not only on physical power but also on economic power, and women become sexual property.

Although men's ability to use physical power to control women diminishes in industrial societies, men still remain the head of household and control the property. In addition, men gain more power through their predominance in the most highly paid and prestigious occupations and the highest elected offices. By contrast, women have the ability to trade their sexual resources, companionship, and emotional support in the marriage market for men's financial support and social status. As a result, women as a group remain subordinate to men (Collins, 1971; Nielsen, 1990).

All men are not equally privileged; some analysts argue that women and men in the upper classes are more privileged, because of their economic power, than men in lower-class positions and all people of color (Lorber, 1994). In industrialized societies, persons who occupy elite positions in corporations, universities, the mass media, and government or who have great wealth have the most power (Richardson, 1993). Most of these are men, however.

Conflict theorists in the Marxist tradition assert that gender stratification results from private ownership of the means of production; some men not only gain control over property and the distribution of goods but also gain power over women. According to Friedrich Engels and Karl Marx, marriage serves to enforce male dominance. Men of the capitalist class instituted monogamous marriage (a gendered institution) so that they could be

Box 10.3 You Can Make a Difference

Joining Organizations to Overcome Sexism and Gender Inequality

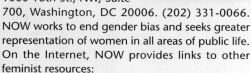

Over the past four decades, many college students—like yourself—have been actively involved in organizations seeking to dismantle sexism and reduce gender-based inequalities in the United States and other nations. Two people speak about the importance of feminist advocacy for women and men:

> I was raised on a pure, unadulterated feminist ethic. . . . [But] the feminism I was raised on was very cerebral. It forced a world full of people to change the way they think about women. I want more than their minds. I want to see them do it. . . . I know that sitting on the sidelines will not get me what I want from my movement. And it is mine. . . . Don't be fooled into thinking that feminism is old-fashioned. The movement is ours and we need it. . . . The next generation is coming. (Neuborne, 1995: 29–35)
> —Ellen Neuborne, a reporter for *USA Today*

> For men who are messed up (that is, facing problems related to their emotional lives, sexuality, their place in society, and gender politics—in other words, me and virtually every other man I have ever met) feminism offers the best route to understanding the politics of such personal problems and coming to terms with those problems. (Jensen, 1995: 111)
> —Robert Jensen, a journalism professor at the University of Texas at Austin

If you are interested in joining an organization that deals with the problem of sexism or organizes activities such as Take Back the Night, an annual event that promotes awareness of violence against women, contact your college's student activities office.

Opportunities for involvement in local, state, and national organizations that promote women's and/or men's rights are available. Here are a few examples:

- The National Organization for Women (NOW), 1000 16th St., NW, Suite 700, Washington, DC 20006. (202) 331-0066. NOW works to end gender bias and seeks greater representation of women in all areas of public life. On the Internet, NOW provides links to other feminist resources:

  http://www.now.org/now/home.html

- Feminist Activist Resources provides information on equal rights advocacy, a calendar of activities, and links to a gender-related electronic forum where people chat about women's issues:

  http://www.igc.apc.org/women/feminist.html

- National Organization for Men Against Sexism (NOMAS), 54 Mint Street, San Francisco, CA 94103. NOMAS has a *profeminist stance* that seeks to end sexism and an *affirmative stance* on the rights of gay men and lesbians.

On the Internet:

- Men's Issue Page includes topics such as sexual harassment, domestic violence, fatherhood, and fathers' rights for single fathers. It lists men's movement organizations and links to other sites:

  http://www.vix.com:80/pub/men

- *M.E.N. Magazine* provides an e-zine to offer support and advocacy for men. It features articles on the men's movement and a national calendar of men's events, as well as links to other sites:

  http://www.vix.com/menmag

certain of the paternity of their offspring, especially sons, whom they wanted to inherit their wealth. Feminist analysts have examined this theory, among others, as they have sought to explain male domination and gender stratification.

Feminist Perspectives

Feminism—the belief that women and men are equal and should be valued equally and have equal rights—is embraced by many men as well as women. It holds in common with men's studies the view that gender is a socially constructed concept that has important consequences in the lives of all people (Craig, 1992). According to the sociologist Ben Agger (1993), men can be feminists

Feminism   the belief that all people—both women and men—are equal and that they should be valued equally and have equal rights.

Martin R. Jones/Unicorn Stock Photos

The disparity between women's and men's earnings is even greater for women of color. Feminists who analyze race, class, and gender suggest that equality will occur only when all women are treated more equitably.

and propose feminist theories; both women and men have much in common as they seek to gain a better understanding of the causes and consequences of gender inequality. Over the past three decades, many different organizations have been formed to advocate causes uniquely affecting women or men and to help people gain a better understanding of gender inequality (see Box 10.3).

Feminist theory seeks to identify ways in which norms, roles, institutions, and internalized expectations limit women's behavior. It also seeks to demonstrate how women's personal control operates even within the constraints of relative lack of power (Stewart, 1994).

LIBERAL FEMINISM   In liberal feminism, gender equality is equated with equality of opportunity. The roots of women's oppression lie in women's lack of equal civil rights and educational opportunities. Only when these constraints on women's participation are removed will women have the same chance for success as men. This approach notes the importance of gender-role socialization and suggests that changes need to be made in what children learn from their families, teach-

ers, and the media about appropriate masculine and feminine attitudes and behavior. Liberal feminists fight for better child-care options, a woman's right to choose an abortion, and the elimination of sex discrimination in the workplace.

RADICAL FEMINISM   According to radical feminists, male domination causes all forms of human oppression, including racism and classism (Tong, 1989). Radical feminists often trace the roots of patriarchy to women's childbearing and child-rearing responsibilities, which make them dependent on men (Firestone, 1970; Chafetz, 1984). In the radical feminist view, men's oppression of women is deliberate, and ideological justification for this subordination is provided by other institutions such as the media and religion. For women's condition to improve, radical feminists claim, patriarchy must be abolished. If institutions are currently gendered, alternative institutions—such as women's organizations seeking better health and day care and shelters for victims of domestic violence and rape—should be developed to meet women's needs.

SOCIALIST FEMINISM   Socialist feminists suggest that women's oppression results from their dual roles as paid *and* unpaid workers in a capitalist economy. In the workplace, women are exploited by capitalism; at home, they are exploited by patriarchy (Kemp, 1994). Women are easily exploited in both sectors; they are paid low wages and have few economic resources. Gendered job segregation is "the primary mechanism in capitalist society that maintains the superiority of men over women, because it enforces lower wages for women in the labor market" (Hartmann, 1976: 139). As a result, women must do domestic labor either to gain a better-paid man's economic support or to stretch their own wages (Lorber, 1994). According to socialist feminists, the only way to achieve gender equality is to eliminate capitalism and develop a socialist economy that would bring equal pay and rights to women.

MULTICULTURAL FEMINISM   Recently, academics and activists have been rethinking the experiences of women of color from a feminist perspective. The experiences of African American women and

Latinas/Chicanas have been of particular interest to some social analysts. Building on the civil rights and feminist movements of the late 1960s and early 1970s, some contemporary black feminists have focused on the cultural experiences of African American women. A central assumption of this analysis is that race, class, and gender are forces that simultaneously oppress African American women (Hull, Bell-Scott, and Smith, 1982). The effects of these three statuses cannot be adequately explained as "double" or "triple" jeopardy (race + class + gender = a poor African American woman) because these ascribed characteristics are not simply added to one another. Instead, they are multiplicative in nature (race × class × gender); different characteristics may be more significant in one situation than another. For example, a well-to-do white woman (class) may be in a position of privilege when compared to people of color (race) and men from lower socioeconomic positions (class), yet be in a subordinate position as compared with a white man (gender) from the capitalist class (Andersen and Collins, 1998). In order to analyze the complex relationship among these characteristics, the lived experiences of African American women and other previously "silenced people" must be heard and examined within the context of particular historical and social conditions.

Another example of multicultural feminist studies is the work of the psychologist Aida Hurtada (1996), who explored the cultural identification of Latina/Chicana women. According to Hurtada, distinct differences exist between the world views of the white (non-Latina) women who participate in the women's movement and many Chicanas, who have a strong sense of identity with their own communities. Hurtada (1996) suggests that women of color do not possess the "relational privilege" that white women have because of their proximity to white patriarchy through husbands, fathers, sons, and others. Like other multicultural feminists, Hurtada calls for a "politics of inclusion," creating social structures that lead to positive behavior and bring more people into a dialogue about how to improve social life and reduce inequalities.

EVALUATION OF CONFLICT AND FEMINIST PERSPECTIVES Conflict and feminist perspectives provide insights into the structural aspects of gender inequality in society. These approaches emphasize factors external to individuals that contribute to the oppression of white women and people of color; however, they have been criticized for emphasizing the differences between men and women without taking into account the commonalities they share. Feminist approaches have also been criticized for their emphasis on male dominance without a corresponding analysis of the ways in which some men may also be oppressed by patriarchy and capitalism.

Recently, the debate has continued about whether the feminist movement has diminished the well-being of boys and men. Sociologist William J. Goode (1982) suggests that some men have felt "under attack" by women's demands for equality because the men do not see themselves as responsible for societal conditions such as patriarchy and attribute their own achievements to hard work and intelligence, not to built-in societal patterns of male domination and female subordination.

In *Stiffed: The Betrayal of the American Man*, the author and journalist Susan Faludi (1999) argues that men are not as concerned about the possibility of the feminist movement diminishing their own importance as they are experiencing an identity crisis

Reflections

Reflections 10.4

Would greater social and economic equality reduce the pressures that many women feel regarding their appearance? Why or why not?

✓ Checkpoints

Checkpoint 10.6

20. What are instrumental tasks and expressive tasks?
21. How does the human capital model explain occupational gender segregation?
22. According to the conflict perspective, what are the origins of gender inequality?
23. What are the key feminist perspectives on gender inequality?

brought about by the current societal emphasis on wealth, power, fame, and looks (often to the exclusion of significant social values). According to Faludi, men are increasingly aware of their body image and are spending ever-increasing sums of money on men's cosmetics, health and fitness gear and classes, and cigar bars and "gentlemen's clubs." The popularity of her book, among others, suggests that issues of gender inequality and of men's and women's roles in society are far from resolved.

Gender Issues in the Future

Over the past century, women made significant progress in the labor force (Reskin and Padavic, 1994). Laws were passed to prohibit sexual discrimination in the workplace and school. Affirmative action programs helped make women more visible in education, government, and the professional world. More women entered the political arena as candidates instead as of volunteers in the campaign offices of male candidates (Lott, 1994).

Many men joined movements to raise their consciousness, realizing that what is harmful to women may also be harmful to men. For example, women's lower wages in the labor force suppress men's wages as well; in a two-paycheck family, women who are paid less contribute less to the family's finances, thus placing a greater burden on men to earn more money.

In the midst of these changes, however, many gender issues remain unresolved in the twenty-first century. In the labor force, gender segregation and the wage gap are still problems. Although women continue to narrow the pay gap, women earn slightly less than 77 cents on the dollar compared with men (Barakat, 2000), and analysts believe that the narrowing can partly be attributed to the economic boom (for some) of recent decades. Employers have had to look for employees based on merit (rather than race, class, and gender) in order to have the number and types of employees they need to meet the demands of global competition (Barakat, 2000).

Although some employers have implemented family-leave policies, these do not relieve women's domestic burden in the family. Analysts believe that the burden of the "double day" or "second shift" will probably preserve women's inequality at home and in the workplace for another generation (Reskin and Padavic, 1994).

In regard to specific problems such as the eating disorders discussed in this chapter, we have a long way to go before all people, particularly women, are valued as human beings rather than as possessions or sex objects. What do you think might be a first step toward reducing gender inequality in the United States and the world?

Chapter Review

How do sex and gender differ?
Sex refers to the biological categories and manifestations of femaleness and maleness; gender refers to the socially constructed differences between females and males. In short, sex is what we (generally) are born with; gender is what we acquire through socialization.

How do gender roles and gender identity differ from gendered institutions?
Gender role encompasses the attitudes, behaviors, and activities that are socially assigned to each sex and that are learned through socialization. Gender identity is an individual's perception of self as either female or male. By contrast, gendered institutions are those structural features that perpetuate gender inequality.

How does the nature of work affect gender equity in societies?
In most hunting and gathering societies, fairly equitable relationships exist between women and men because neither sex has the ability to provide all of the food necessary for survival. In horticultural societies, a fair degree of gender equality exists because neither sex controls the food supply. In agrarian societies, male dominance is overt; agrarian tasks require more labor and physical strength, and females are often excluded from these tasks because they are viewed as too weak or too tied to child-rearing activities. In industrialized societies, a gap exists between nonpaid work performed by women at home and paid work performed by men and women. A wage gap also exists between women and men in the marketplace.

What are the key agents of gender socialization?
Parents, peers, teachers and schools, sports, and the media are agents of socialization that tend to reinforce stereotypes of appropriate gender behavior.

What causes gender inequality in the United States?
Gender inequality results from economic, political, and educational discrimination against women. In most workplaces, jobs are either gender segregated or the majority of employees are of the same gender. Although the degree of gender segregation in the professional work-

place has declined since the 1970s, racial and ethnic segregation remains deeply embedded.

How is occupational segregation related to the pay gap?

Many women work in lower-paying, less-prestigious jobs than men. This occupational segregation leads to a disparity, or pay gap, between women's and men's earnings. Even when women are employed in the same job as men, on average they do not receive the same, or comparable, pay.

How do functionalists and conflict theorists differ in their view of division of labor by gender?

According to functionalist analysts, women's roles as caregivers in contemporary industrialized societies are crucial in ensuring that key societal tasks are fulfilled. While the husband performs the instrumental tasks of economic support and decision making, the wife assumes the expressive tasks of providing affection and emotional support for the family. According to conflict analysts, the gendered division of labor within families and the workplace—particularly in agrarian and industrial societies—results from male control and dominance over women and resources.

Key Terms

body consciousness 315
comparable worth 332
feminism 337
gender 314
gender bias 327
gender identity 315
gender role 315
hermaphrodite 312
homophobia 313
matriarchy 317
patriarchy 317
pay gap 332
primary sex characteristics 312
secondary sex characteristics 312
sex 311
sexism 316
sexual orientation 313
transsexual 312
transvestite 312

Questions for Critical Thinking

1. Do the media reflect societal attitudes on gender, or do the media determine and teach gender behavior? (As a related activity, watch television for several hours and list the roles for women and men depicted in programs and those represented in advertising.)

2. Review the concept of cultural relativism in Chapter 2. Should the U.S. State Department and human rights groups such as Amnesty International protest genital mutilation in other nations and withhold any funding or aid for those nations until they cease the practice?

3. Examine the various academic departments at your college. What is the gender breakdown of the faculty in selected departments? What is the gender breakdown of undergraduates and graduate students in those departments? Are there major differences among various academic areas of teaching and study? What hypothesis can you come up with to explain your observations?

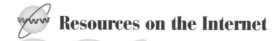

Resources on the Internet

Chapter-Related Web Sites

The following web sites have been selected for their relevance to the topics in this chapter. These sites are among the more stable, but please note that web site addresses change frequently.

Gender Inn
http://www.uni-koeln.de/phil-fak/englisch/datenbank/e_index.htm
Gender Inn is a searchable database providing access to more than 7,500 records pertaining to feminist theory, feminist literary criticism, and gender studies focusing on English and American literature.

U.S. Census Bureau Facts & Features: Women's History Month
http://www.census.gov/Press-Release/www/2002/cb02ff03.html
A wide range of information about women from the 2000 census.

U.S. Census Bureau: School Enrollment
http://www.census.gov/population/www/socdemo/school/ppl-148.html
Here are a number of reports from the 2000 census about education. A number of reports discuss education and gender.

2000 Census: Gender
http://www.census.gov/prod/2001pubs/c2kbr01-9.pdf
This report contains demographic comparisons and a detailed analysis of gender in the United States.

WomenWatch: The UN Gateway on the Advancement and Empowerment of Women
www.un.org/womenwatch
WomenWatch is a joint United Nations project to create a core Internet space on global women's issues. It was created to monitor the results of the Fourth World Conference on Women, held in Beijing in 1995. It was founded by the Division for the Advancement of Women (DAW), the United Nations Development Fund for Women (UNIFEM), and the International Research and Training Institute for the Advancement of Women (INSTRAW).

Feminism and Women's Studies
http://eserver.org/feminism/index.html
This site is an organized collection of links related to women's issues, such as education, employment, history, and health.

Feminist Theory Website
www.cddc.vt.edu/feminism/
The Feminist Theory Website provides research materials and information for students, activists, and scholars interested in women's conditions and struggles around the world.

Gender and Society: A Matter of Nature or Nurture?
http://www.trinity.edu/mkearl/gender.html
"The Sociological Tour Through Cyberspace" takes on gender and socialization.

Companion Web Site for This Book

Virtual Society: The Wadsworth Sociology Resource Center
Visit **http://sociology.wadsworth.com** and click on the page for Kendall, *Sociology in Our Times: The Essentials*, Fourth Edition, to access a wide range of enrichment material to aid in your study of sociology. Click on the Student Resources section of the web site. Next, select from the pull-down menu the chapter that you are presently studying. Among the useful options for self-study are chapter objectives, flashcards, practical study tips, and practice tests for each chapter.

MicroCase Online
From the Virtual Society home page, click on MicroCase Online to access book-specific MicroCase exercises that allow you to further explore sociological issues and principles presented in the text.

InfoTrac College Edition
Another unique option available to you at the Student Resources section of the companion web site is InfoTrac College Edition, an online library with access to hundreds of scholarly and popular periodicals. Below are suggested search terms for this chapter. Results from these and other searches are found at the site.

- Search keywords: *wage gap*. Find articles that discuss the wage gap between the genders. Is the wage gap still present? Who is most affected?
- Search keyword: *fatherhood*. Look for articles that address fatherhood in contemporary society. To what extent is the role of fatherhood changing? In what ways?

Virtual Explorations CD-ROM
Go to the Virtual Explorations CD-ROM to begin the interactive exercise for this chapter. This Virtual Exploration will introduce you to some of the exciting resources for sociology on the World Wide Web. You will be guided through an exercise that employs related web sites on sex and gender. Answer the questions, and e-mail your responses to your instructor.

Jim Dandy/SIS

Families and Intimate Relationships

Families in Global Perspective
Family Structure and Characteristics
Marriage Patterns
Patterns of Descent and Inheritance
Power and Authority in Families
Residential Patterns

Theoretical Perspectives on Families
Functionalist Perspectives
Conflict and Feminist Perspectives
Symbolic Interactionist Perspectives
Postmodernist Perspectives

**Developing Intimate Relationships
   and Establishing Families**
Love and Intimacy
Cohabitation and Domestic Partnerships
Marriage
Housework and Child-Care Responsibilities

Child-Related Family Issues and Parenting
Deciding to Have Children
Adoption
Teenage Pregnancies
Single-Parent Households
Two-Parent Households
Remaining Single

Transitions and Problems in Families
Family Transitions Based on Age and the Life Course
Family Violence
Children in Foster Care
Elder Abuse
Divorce
Remarriage

Family Issues in the Future

343

My son began commuting between his two homes at age 4. He traveled with a Hello Kitty suitcase with a pretend lock and key until it embarrassed him. . . . He graduated to a canvas backpack filled with a revolving arsenal of essential stuff: books and journals, plastic vampire teeth, "Star Trek" Micro Machines, a Walkman, CD's, a teddy bear.

The commuter flights between San Francisco and Los Angeles were the only times a parent wasn't lording over him, so he was able to order Coca-Cola, verboten at home. . . . But such benefits were insignificant when contrasted with his preflight nightmares about plane crashes. . . .

Like so many divorcing couples, we divided the china and art and our young son. . . . First he was ferried back and forth between our homes across town, and then, when his mother moved to Los Angeles, across the state. For the eight years since, he has been one of the thousands of American children with two homes, two beds, two sets of clothes and toys and two toothbrushes.

—several years ago, David Sheff (1995: 64), explaining how divorce and joint custody had affected his son, Nick Sheff

Four years later, Nick Sheff stated his own view on long-distance joint custody:

I am 16 now and I still travel back and forth, but it's mostly up to me to decide when. I've chosen to spend more time with my friends at the expense of visits with my mom. When I do go to L.A., it's like my stepdad put it: I have a cameo role in their lives. I say my lines and I'm off. It's painful.

What's the toll of this arrangement? I'm always missing somebody. When I'm in northern California, I miss my mom and stepdad. But when I'm in L.A., I miss hanging out with my friends, my other set of parents and little brothers and sisters. After all these back-and-forth flights, I've learned not to get too emotionally attached. I have to protect myself. . . . No child should be subjected to the hardship of long-distance joint custody. To prevent it, maybe there

Tony Freeman/PhotoEdit

Changing family patterns and relocations for employment opportunities have made sights such as this increasingly familiar, as children travel alone to various destinations to be reunited with a parent or other family member.

should be an addition to the marriage vows: Do you promise [that] if you ever have children and wind up divorced, [you will] stay within the same geographical area as your kids? . . . Or how about some common sense? If you move away from your children, *you* have to do the traveling to see them. (Sheff, 1999: 16)

Nick's family is one of millions around the globe experiencing major change as a result of divorce, death, or some other significant event. Although in the past, children in the United States who did not grow up with both biological parents were often considered to be the "exceptions," we cannot continue to ignore such children, whose numbers continue to increase (Coontz, 1997).

Extensive sociological research has been conducted to examine the causes of divorce and to iden-

tify the possible effects it may have on adults and children. In one study, the sociologists Frank F. Furstenberg, Jr., and Andrew J. Cherlin (1991) found that although children describe divorce and parental separation as an unsettling experience, most children eventually make a successful adaptation if two factors are present: (1) the mother does reasonably well financially and psychologically after the divorce, and (2) the divorced parents have a low level of conflict between them. Another study by the sociologist

Box 11.1 Sociology and Everyday Life

How Much Do You Know About Contemporary Trends in U.S. Family Life?

True	False	
T	F	1. Today, people in the United States are more inclined to get married than at any time in history.
T	F	2. Most U.S. family households are composed of a married couple with one or more children under age 18.
T	F	3. One in two U.S. preschoolers has a mother in the paid labor force.
T	F	4. Recent studies have found that sexual activity is more satisfying to people who are in sustained relationships such as marriage.
T	F	5. People under age 45 are more likely to cohabit than are people over age 45.
T	F	6. With the strong U.S. economy at the end of the 1990s, the number of children living in extreme poverty went down significantly.
T	F	7. Recent studies have found that adult children of divorced parents are more likely to dissolve their own marriages than they were two decades ago.
T	F	8. Teenage pregnancies have increased dramatically over the past 30 years.
T	F	9. Couples who already have children at the beginning of their marriage are less likely to divorce.
T	F	10. The number of children living with grandparents and with no parent in the home increased by more than 50 percent between 1990 and 2000.

Answers on page 346.

Demie Kurz (1995) focused on how divorce affects women and children. According to Kurz (1995), most of the women in her study were happy to see their marriages end because they found their marriages to be very difficult and in some cases abusive. By contrast, divorce offered them freedom from domination, violence, and destructive emotional circumstances.

In this chapter, we examine the increasing complexity and diversity of contemporary families in the United States and other nations. Pressing social issues such as divorce, child care, and new reproductive technologies will be used as examples of how families and intimate relationships continue to change. Before reading on, test your knowledge about some contemporary trends in U.S. family life by taking the quiz in Box 11.1.

Questions and Issues

Chapter Focus Question: Why are many family-related concerns—such as divorce and child care—viewed primarily as personal problems rather than as social concerns requiring macrolevel solutions?

Why is it difficult to define family?

How do marriage patterns vary around the world?

What are the key assumptions of functionalist, conflict/feminist, and symbolic interactionist perspectives on families?

What significant trends affect many U.S. families today?

Box 11.1

Answers to the Sociology Quiz on Contemporary Trends in U.S. Family Life

1. **False.** Census data show that the marriage rate has gone down by about one-third since 1960. In 1960, there were about 73 marriages per 1,000 unmarried women age 15 and up, whereas today the rate is about 49 per 1,000.
2. **False.** Less than 25 percent of all family households are composed of married couples with one or more children under age 18.
3. **True.** One in two preschoolers in the United States has a mother in the paid labor force, which makes affordable, high-quality child care an important issue for many families.
4. **True.** Recent research indicates that people who are married or are in a sustained relationship with their sexual partner express greater satisfaction with their sexual activity than do those who are not.
5. **True.** People under age 45 are more likely to cohabit than are older individuals.
6. **False.** Despite the strong economy prior to 2001, the number of children living in extreme poverty—with incomes below one-half of the poverty line—crept upward by almost 400,000 in recent years.
7. **False.** Adult children of divorced parents are less likely to dissolve their own marriages than they were two decades ago.
8. **False.** Teenage pregnancies have actually decreased over the past 30 years. However, the percentage of *unmarried* teenage pregnancies has increased dramatically.
9. **False.** Couples who already have children at the beginning of their marriage are *more* likely to divorce than those who do not.
10. **True.** The number of children living with grandparents and with no parent in the home grew by 51.5 percent between 1990 and 2000.

Sources: Based on Children's Defense Fund, 2001; Coontz, 1992, 1997; Fox News, 1999; Popenoe and Whitehead, 1999.

Families in Global Perspective

As the nature of family life and work has changed in high-, middle-, and low-income nations, the issue of what constitutes a "family" has been widely debated. For many years, the standard sociological definition of *family* has been a group of people who are related to one another by bonds of blood, marriage, or adoption and who live together, form an economic unit, and bear and raise children (Benokraitis, 1999). Many people believe that this definition should not be expanded—that social approval should not be extended to other relationships simply because the persons in those relationships wish to consider themselves a family. However, others challenge this definition because it simply does not match the reality of family life in contemporary society (Lamanna and Riedmann, 2003). Today's families include many types of living arrangements and relationships, including single-parent households, unmarried couples, lesbian and gay couples, and multiple generations (such as grandparent, parent, and child) living in the same household. To accurately reflect these changes in family life, some sociologists believe that we need a more encompassing definition of what constitutes a family. Accordingly, we will define *families* as **relationships in which people live together with commitment, form an economic unit and care for any young, and consider their identity to be significantly attached to the group.** Sexual expression and parent–child relationships are a part of most, but not all, family relationships (based on Benokraitis, 1999; Lamanna and Riedmann, 2003).

Bob Daemmrich/The Image Works

Steve Rubin/The Image Works

While the relationship between a husband and wife is based on legal ties, relationships between parents and children may be established either by blood ties or legal ties.

Although families differ widely around the world, they also share certain common concerns in their everyday lives. For example, women and men of all racial–ethnic categories, nationalities, and income levels face problems associated with child care. Various nongovernmental agencies of the United Nations have established international priorities to help families. Suggestions have included the development of "social infrastructures for the care and education of children of working parents in order to reduce their . . . burden" and implementation of flexible working hours so that parents will have more opportunities to spend time with their children (Pietilä and Vickers, 1994: 115).

How do sociologists approach the study of families? In our study of families, we will use our sociological imagination to see how our personal experiences are related to the larger happenings in society. At the microlevel, each of us has a "biography," based on our experience within our family; at the macrolevel, our families are embedded in a specific social context that has a major effect on them (Aulette, 1994). We will examine the institution of the family at both of these levels, starting with family structure and characteristics.

Family Structure and Characteristics

In preindustrial societies, the primary form of social organization is through kinship ties. *Kinship* **refers to a social network of people based on common ancestry, marriage, or adoption.** Through kinship networks, people cooperate so that they can acquire the basic necessities of life, including food and shelter. Kinship systems can also serve as a means by which property is transferred, goods are produced and distributed, and power is allocated.

In industrialized societies, other social institutions fulfill some of the functions previously taken care of by the kinship network. For example, political systems provide structures of social control and authority, and economic systems are responsible for the production and distribution of goods and services. Consequently, families in industrialized societies serve fewer and more-specialized purposes than do families in preindustrial societies. Contemporary families are responsible primarily for regulating sexual activity, socializing children, and providing affection and companionship for family members.

FAMILIES OF ORIENTATION AND PROCREATION During our lifetime, many of us will be members of two different types of families—a family of orientation and a family of procreation. The

Families relationships in which people live together with commitment, form an economic unit and care for any young, and consider their identity to be significantly attached to the group.

Kinship a social network of people based on common ancestry, marriage, or adoption.

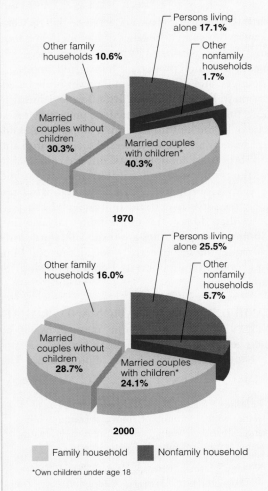

★ ★ ★ ★ ★ ★ ★ ★ ★ ★ ★ ★

CENSUS PROFILES

Household Composition, 1970 and 2000

Census 2000 quizzed respondents about their marital status and asked them to fill out an additional page of data for each other individual residing in the household. Based on this data, the Census Bureau reported that the percentage distribution of nonfamily and family households had changed during the past 30 years. The most noticeable trend is the decline in the number of married-couple households with their own children living with them, which decreased from about 40 percent of all households in 1970 to about 24 percent in 2000. The distribution shown below reflects this significant change in family structure.

1970

- Persons living alone **17.1%**
- Other nonfamily households **1.7%**
- Married couples with children* **40.3%**
- Married couples without children **30.3%**
- Other family households **10.6%**

2000

- Persons living alone **25.5%**
- Other nonfamily households **5.7%**
- Married couples with children* **24.1%**
- Married couples without children **28.7%**
- Other family households **16.0%**

Family household | Nonfamily household

*Own children under age 18

Source: Fields and Casper, 2001.

family of orientation **is the family into which a person is born and in which early socialization usually takes place.** Although most people are related to members of their family of orientation by blood ties, those who are adopted have a legal tie that is patterned after a blood relationship (Aulette, 1994). The *family of procreation* **is the family that a person forms by having or adopting children** (Benokraitis, 1999). Both legal and blood ties are found in most families of procreation. The relationship between a husband and wife is based on legal ties; however, the relationship between a parent and child may be based on either blood ties or legal ties, depending on whether the child has been adopted (Aulette, 1994).

Some sociologists have emphasized that "family of orientation" and "family of procreation" do not encompass all types of contemporary families. Instead, as the sociologist Kath Weston (1991) notes, many gay men and lesbians have *families we choose*—social arrangements that include intimate relationships between couples and close familial relationships among other couples and other adults and children. According to the sociologist Judy Root Aulette (1994), "families we choose" include blood ties and legal ties, but they also include *fictive kin*—persons who are not actually related by blood but who are accepted as family members.

EXTENDED AND NUCLEAR FAMILIES Sociologists distinguish between extended families and nuclear families based on the number of generations that live within a household. An *extended family* **is a family unit composed of relatives in addition to parents and children who live in the same household.** These families often include grandparents, uncles, aunts, or other relatives who live close to the parents and children, making it possible for family members to share resources. In horticultural and agricultural societies, extended families are extremely important; having a large number of family members participate in food production may be essential for survival. Today, extended family patterns are found in Latin America, Africa, Asia, and some parts of Eastern and Southern Europe (Busch, 1990). With the advent of industrialization and urbanization, maintaining the extended family pattern becomes more difficult. Increasingly, young people move from rural to urban areas in search of employment in the industrializing sector of the economy. At that time, the

Reflections

Reflections 11.1

Are some family problems more difficult to solve in a nuclear family than in an extended family? If so, which ones?

nuclear family typically becomes the predominant family form in the society.

A *nuclear family* **is a family composed of one or two parents and their dependent children, all of whom live apart from other relatives.** A traditional definition specifies that a nuclear family is made up of a "couple" and their dependent children; however, this definition became outdated when a significant shift occurred in the family structure. A comparison of Census Bureau data from 1970 and 2000 shows that there has been a significant decline in the percentage of U.S. households comprising a married couple with their own children under eighteen years of age (see "Census Profiles: Household Composition"). Conversely, there has been an increase in the percentage of households in which either a woman or a man lives alone.

Marriage Patterns

Across cultures, families are characterized by different forms of marriage. *Marriage* **is a legally recognized and/or socially approved arrangement between two or more individuals that carries certain rights and obligations and usually involves sexual activity.** In most societies, marriage involves a mutual commitment by each partner, and linkages between two individuals and families are publicly demonstrated.

In the United States, the only legally sanctioned form of marriage is *monogamy*—**a marriage between two partners, usually a woman and a man.** For some people, marriage is a lifelong commitment that ends only with the death of a partner. For others, marriage is a commitment of indefinite duration. Through a pattern of marriage, divorce, and remarriage, some people practice *serial monogamy*—a succession of marriages in which a person has several spouses over a lifetime but is legally married to only one person at a time.

Polygamy **is the concurrent marriage of a person of one sex with two or more members of the opposite sex** (Marshall, 1998). The most prevalent form of polygamy is *polygyny*—**the concurrent marriage of one man with two or more women.** Polygyny has been practiced in a number of societies, including parts of Europe until the Middle Ages. More recently, some marriages in Islamic societies in Africa and Asia have been polygynous; however, the cost of providing for multiple wives and numerous children makes the practice impossible for all but the wealthiest men. In addition, because roughly equal numbers of women and men live in these areas, this nearly balanced sex ratio tends to limit polygyny.

The second type of polygamy is *polyandry*—**the concurrent marriage of one woman with two or more men.** Polyandry is very rare; when it does occur, it is typically found in societies where men greatly outnumber women because of high rates of female infanticide.

Family of orientation   the family into which a person is born and in which early socialization usually takes place.

Family of procreation   the family that a person forms by having or adopting children.

Extended family   a family unit composed of relatives in addition to parents and children who live in the same household.

Nuclear family   a family composed of one or two parents and their dependent children, all of whom live apart from other relatives.

Marriage   a legally recognized and/or socially approved arrangement between two or more individuals that carries certain rights and obligations and usually involves sexual activity.

Monogamy   a marriage between two partners, usually a woman and a man.

Polygamy   the concurrent marriage of a person of one sex with two or more members of the opposite sex.

Polygyny   the concurrent marriage of one man with two or more women.

Polyandry   the concurrent marriage of one woman with two or more men.

Patterns of Descent and Inheritance

Even though a variety of marital patterns exist across cultures, virtually all forms of marriage establish a system of descent so that kinship can be determined and inheritance rights established. In preindustrial societies, kinship is usually traced through one parent (unilineally). The most common pattern of unilineal descent is *patrilineal descent*—a system of tracing descent through the father's side of the family. Patrilineal systems are set up in such a manner that a legitimate son inherits his father's property and sometimes his position upon the father's death. In nations such as India, where boys are seen as permanent patrilineal family members but girls are seen as only temporary family members, girls tend to be considered more expendable than boys (O'Connell, 1994). Recently, some scholars have concluded that cultural and racial nationalism in China is linked to the idea of patrilineal descent being crucial to the modern Chinese national identity (Dikotter, 1996).

Even with the less common pattern of *matrilineal descent*—a system of tracing descent through the mother's side of the family—women may not control property. However, inheritance of property and position is usually traced from the maternal uncle (mother's brother) to his nephew (mother's son). In some cases, mothers may pass on their property to daughters.

By contrast, kinship in industrial societies is usually traced through both parents (bilineally). The most common form is *bilateral descent*—a system of tracing descent through both the mother's and father's sides of the family. This pattern is used in the United States for the purpose of determining kinship and inheritance rights; however, children typically take the father's last name.

Power and Authority in Families

Descent and inheritance rights are intricately linked with patterns of power and authority in families. The most prevalent forms of familial power and authority are patriarchy, matriarchy, and egalitarianism. A *patriarchal family* is a family structure in which authority is held by the eldest male (usually the father). The male authority figure acts as head of the household and holds power and authority over the women and children, as well as over other males. A *matriarchal family* is a family structure in which authority is held by the eldest female (usually the mother). In this case, the female authority figure acts as head of the household. Although there has been a great deal of discussion about matriarchal families, scholars have found no historical evidence to indicate that true matriarchies ever existed.

The most prevalent pattern of power and authority in families is patriarchy. Across cultures, men are the primary (and often sole) decision makers regarding domestic, economic, and social concerns facing the family. The existence of patriarchy may give men a sense of power over their own lives, but it can also create an atmosphere in which some men feel greater freedom to abuse women and children (O'Connell, 1994).

An *egalitarian family* is a family structure in which both partners share power and authority equally. Recently, a trend toward more-egalitarian relationships has been evident in a number of countries as women have sought changes in their legal status and increased educational and employment opportunities. Some degree of economic independence makes it possible for women to delay marriage or to terminate a problematic marriage (O'Connell, 1994). Among gay and lesbian couples, power and authority issues are also important. Although some analysts suggest that legalizing marriage among gay couples would produce more-egalitarian relationships, others argue that such marriages would reproduce the inequalities in power and authority found in conventional heterosexual relationships (see Aulette, 1994).

Residential Patterns

Residential patterns are interrelated with the authority structure and method of tracing descent in families. *Patrilocal residence* refers to the custom of a married couple living in the same household (or community) as the husband's family. Across cultures, patrilocal residency is most common. One example of contemporary patrilocal residency can be found in al-Barba, a lower-middle-class neighborhood in the Jordanian city of Irbid (McCann, 1997). According to researchers, the high cost of renting an apartment or building a new home has resulted in many sons building their own living quarters onto their parents' home, resulting in multi-family households consisting of an older married couple, their unmarried children, their married sons, and their sons' wives and children. The family units often have separate dwellings within the confines of the home. By sharing resources, all members of the extended family can maintain a higher standard of living than if each family unit lived independently. Thus, socioeconomic factors may be an important factor in the establishment of contemporary patrilocal residences (McCann, 1997).

Few societies have residential patterns known as *matrilocal residence*—the custom of a married couple living in the same household (or community) as the wife's parents. In industrialized nations such as the United States, most couples hope to live in a *neolocal residence*—the custom of a married couple living in their own residence apart from both the husband's and the wife's parents. For many couples, however, economic conditions, availability of housing, and other considerations may make neolocal residency impossible, at least initially.

To this point, we have examined a variety of marriage and family patterns found around the world. Even with the diversity of these patterns, most people's behavior is shaped by cultural rules pertaining to endogamy and exogamy. *Endogamy is the practice of marrying within one's own group.* In the United States, for example, most people practice endogamy: They marry people who come from the same social class, racial–ethnic group, religious affiliation, and other categories considered important within their own social group. *Exogamy is the practice of marrying outside one's own social group or category.* Depending on the circumstances, exogamy may not be noticed at all, or it may result in a person being ridiculed or ostracized by other members of the "in" group. The three most important sources of positive or negative sanctions for intermarriage are the family, the church, and the state. Participants in these social institutions may look unfavorably on the marriage of an in-group member to an "outsider" because of the belief that it diminishes social cohesion in the group (Kalmijn, 1998). However, educational attainment is also a strong indicator of marital choice. Higher education emphasizes individual achievement, and college-educated people may be less likely than others to identify themselves with their social or cultural roots and thus more

✓Checkpoints

Checkpoint 11.1

1. How do sociologists study the institution of the family?
2. How does the family of orientation differ from the family of procreation?
3. How does the extended family differ from the nuclear family?
4. What are the two types of polygamy?
5. How do patrilineal, matrilineal, and bilateral patterns of descent differ from one another?
6. How are power and authority distributed in (a) patriarchal, (b) matriarchal, and (c) egalitarian families?
7. What are the main forms of residential patterns?

willing to marry outside their own social group or category if their potential partner shares a similar level of educational attainment (Hwang, Saenz, and Aguirre, 1995; Kalmijn, 1998).

Theoretical Perspectives on Families

The *sociology of family* is the subdiscipline of sociology that attempts to describe and explain patterns of family life and variations in family

Patrilineal descent a system of tracing descent through the father's side of the family.

Matrilineal descent a system of tracing descent through the mother's side of the family.

Bilateral descent a system of tracing descent through both the mother's and father's sides of the family.

Patriarchal family a family structure in which authority is held by the eldest male (usually the father).

Matriarchal family a family structure in which authority is held by the eldest female (usually the mother).

Egalitarian family a family structure in which both partners share power and authority equally.

Functionalist theorists believe that families serve a variety of functions that no other social institution can adequately fulfill. In contrast, conflict and feminist analysts believe that families may be a source of conflict over values, goals, and access to resources and power. Children in upper-class families have many advantages and opportunities that are not available to other children, as shown in this photo of an affluent family in front of its horse barn.

structure. Functionalist perspectives emphasize the functions that families perform at the macrolevel of society whereas conflict and feminist perspectives focus on families as a primary source of social inequality. Symbolic interactionists examine microlevel interactions that are integral to the roles of different family members.

Functionalist Perspectives

Functionalists emphasize the importance of the family in maintaining the stability of society and the well-being of individuals. According to Emile Durkheim, marriage is a microcosmic replica of the larger society; both marriage and society in-

volve a mental and moral fusion of physically distinct individuals (Lehmann, 1994). Durkheim also believed that a division of labor contributes to greater efficiency in all areas of life—including marriages and families—even though he acknowledged that this division imposes significant limitations on some people.

In the United States, Talcott Parsons was a key figure in developing a functionalist model of the family. According to Parsons (1955), the husband/father fulfills the *instrumental role* (meeting the family's economic needs, making important decisions, and providing leadership), whereas the wife/mother fulfills the *expressive role* (running the household, caring for children, and meeting the emotional needs of family members).

Contemporary functionalist perspectives on families derive their foundation from Durkheim. Division of labor makes it possible for families to fulfill a number of functions that no other institution can perform as effectively. In advanced industrial societies, families serve four key functions:

1. *Sexual regulation.* Families are expected to regulate the sexual activity of their members and thus control reproduction so that it occurs within specific boundaries. At the macrolevel, incest taboos prohibit sexual contact or marriage between certain relatives. For example, virtually all societies prohibit sexual relations between parents and their children and between brothers and sisters.

2. *Socialization.* Parents and other relatives are responsible for teaching children the necessary knowledge and skills to survive. The smallness and intimacy of families make them best suited for providing children with the initial learning experiences they need.

3. *Economic and psychological support.* Families are responsible for providing economic and psychological support for members. In preindustrial societies, families are economic production units; in industrial societies, the economic security of families is tied to the workplace and to macrolevel economic systems. In recent years, psychological support and emotional security have been increasingly important functions of the family (Chafetz, 1989).

4. *Provision of social status.* Families confer social status and reputation on their members.

These statuses include the ascribed statuses with which individuals are born, such as race/ethnicity, nationality, social class, and sometimes religious affiliation. One of the most significant and compelling forms of social placement is the family's class position and the opportunities (or lack thereof) resulting from that position. Examples of class-related opportunities are access to quality health care, higher education, and a safe place to live.

Conflict and Feminist Perspectives

Conflict and feminist analysts view functionalist perspectives on the role of the family in society as idealized and inadequate. Rather than operating harmoniously and for the benefit of all members, families are sources of social inequality and conflict over values, goals, and access to resources and power (Benokraitis, 1999).

According to some conflict theorists, families in capitalist economies are similar to the work environment of a factory. Women are dominated by men in the home in the same manner that workers are dominated by capitalists and managers in factories (Engels, 1970/1884). Although childbearing and care for family members in the home contribute to capitalism, these activities also reinforce the subordination of women through unpaid (and often devalued) labor. Other conflict analysts are concerned with the effect that class conflict has on the family. The exploitation of the lower classes by the upper classes contributes to family problems such as high rates of divorce and overall family instability.

Some feminist perspectives on inequality in families focus on patriarchy rather than class. From this viewpoint, men's domination over women existed long before capitalism and private ownership of property (Mann, 1994). Women's subordination is rooted in patriarchy and men's control over women's labor power (Hartmann, 1981). According to one scholar, "Male power in our society is expressed in economic terms even if it does not originate in property relations; women's activities in the home have been undervalued at the same time as their labor has been controlled by men" (Mann, 1994: 42). In addition, men have benefited from the privileges they derive from their status as family breadwinners (see Firestone, 1970; Daly, 1978; Goode, 1982).

Symbolic Interactionist Perspectives

Early symbolic interactionists such as Charles Horton Cooley and George Herbert Mead provided key insights on the roles we play as family members and how we modify or adapt our roles to the expectations of others—especially significant others such as parents, grandparents, siblings, and other relatives. How does the family influence the individual's self-concept and identity? Contemporary symbolic interactionists examine the roles of husbands, wives, and children as they act out their own part and react to the actions of others in order to answer questions such as this one. From such a perspective, what people think, as well as what they say and do, is very important in understanding family dynamics.

According to the sociologists Peter Berger and Hansfried Kellner (1964), interaction between marital partners contributes to a shared reality. Although newlyweds bring separate identities to a marriage, over time they construct a shared reality as a couple. In the process, the partners redefine their past identities to be consistent with new realities. Development of a shared reality is a continuous process, taking place not only in the family but in any group in which the couple participates together. Divorce is the reverse of this process; couples may start with a shared reality and, in the process of uncoupling, gradually develop separate realities (Vaughan, 1985).

Patrilocal residence the custom of a married couple living in the same household (or community) as the husband's family.

Matrilocal residence the custom of a married couple living in the same household (or community) as the wife's parents.

Neolocal residence the custom of a married couple living in their own residence apart from both the husband's and the wife's parents.

Endogamy the practice of marrying within one's own social group or category.

Exogamy the practice of marrying outside one's own social group or category.

Sociology of family the subdiscipline of sociology that attempts to describe and explain patterns of family life and variations in family structure.

Reflections

Reflections 11.2

How might a symbolic interactionist perspective be useful in reducing family problems?

Symbolic interactionists explain family relationships in terms of the subjective meanings and everyday interpretations that people give to their lives. As the sociologist Jessie Bernard (1982/1973) pointed out, women and men experience marriage differently. Although the husband may see *his* marriage very positively, the wife may feel less positive about *her* marriage, and vice versa. Researchers have found that husbands and wives may give very different accounts of the same event; their "two realities" frequently do not coincide (Safilios-Rothschild, 1969).

Postmodernist Perspectives

Although postmodern theorists disparage the idea that a universal theory can be developed to explain social life, a postmodernist perspective might provide insights on questions such as this: How is family life different in the "information age"? Social scientist David Elkind (1995) describes the postmodern family as *permeable*—capable of being diffused or invaded in such a manner that an entity's original purpose is modified or changed. According to Elkind (1995), if the nuclear family is a reflection of the age of modernity, the permeable family reflects the postmodern assumptions of difference, particularity, and irregularity. Difference is evident in the fact that the nuclear family is now only one of many family forms. Similarly, the idea of romantic love under modernity has given way to the idea of consensual love: Individuals agree to have sexual relations with others whom they have no intention of marrying or, if they marry, do not necessarily see the marriage as having permanence. Maternal love has also been transformed into shared parenting, which includes not only mothers and fathers but also caregivers who may either be relatives or nonrelatives (Elkind, 1995).

Urbanity is another characteristic of the postmodern family. The boundaries between the public sphere (the workplace) and the private sphere (the home) are becoming much more open and flexible. In fact, family life may be negatively affected by the decreasing distinction between what is work time and what is family time. As more people are becoming connected "24/7" (twenty-four hours a day, seven days a week), the boss who in the past would not call at 11:30 P.M. may send an e-mail asking for an immediate response to some question that has arisen while the person is away on vacation with family members (Leonard, 1999). According to some postmodern analysts, this is an example of the "power of the new communications technologies to integrate and control labour despite extensive dispersion and decentralization" (Haraway, 1994: 439).

Social theorist Jean Baudrillard's idea that the simulation of reality may come to be viewed as "reality" by some people can be applied to family interactions in the "Information Age." Does the ability to contact someone anywhere and any time of the day or night provide greater happiness and stability in families? Or is "reach out and touch someone" merely an ideology promulgated by the consumer society? Journalists have written about the experience of watching a family gathering at an amusement park, restaurant, mall, or other location only to see family members pick up their cell phones to receive or make calls to individuals not present, rather than spending "face time" with those family members who are present.

Concept Table 11.A summarizes sociological perspectives on the family. Taken together, these perspectives on the social institution of families reflect various ways in which familial relationships may be viewed in contemporary societies. Now we shift our focus to love, marriage, intimate relationships, and family issues in the United States.

✓ Checkpoints

Checkpoint 11.2

8. According to the functionalist perspective, what are the four key functions served by families?
9. How do conflict and feminist perspectives differ from functionalist views on families?
10. How do symbolic interactionists explain family relationships?
11. How do postmodernists explain contemporary family problems?

Concept Table 11.A Theoretical Perspectives on Families

	Focus	Key Points	Perspective on Family Pr
Functionalist	Role of families in maintaining stability of society and individuals' well-being.	In modern societies, families serve the functions of sexual regulation, socialization, economic and psychological support, and provision of social status.	Family problems are to changes in social institutions such as the economy, religion, education, and law/ government
Conflict/ Feminist	Families as sources of conflict and social inequality.	Families both mirror and help perpetuate social inequalities based on class and gender.	Family problems reflect social patterns of dominance and subordination.
Interactionist	Family dynamics, including communication patterns and the subjective meanings that people assign to events.	Interactions within families create a shared reality.	How family problems are perceived and defined depends on patterns of communication, the meanings that people give to roles and events, and individuals' interpretations of family interactions.
Postmodernist	Permeability of families.	In postmodern societies, families are diverse and fragmented. Boundaries between workplace and home are blurred.	Family problems are related to cyberspace, consumerism, and the hyperreal in an age increasingly characterized by high-tech "haves" and "have-nots."

Developing Intimate Relationships and Establishing Families

The United States has been described as a "nation of lovers"; it has been said that we are "in love with love." Why is this so? Perhaps the answer lies in the fact that our ideal culture emphasizes *romantic love*, which refers to a deep emotion, the satisfaction of significant needs, a caring for and acceptance of the person we love, and involvement in an intimate relationship (Lamanna and Riedmann, 2003).

Love and Intimacy

In the late nineteenth century, during the Industrial Revolution, people came to view work and home as separate spheres in which different feelings and emotions were appropriate (Coontz, 1992). The public sphere of work—men's sphere—emphasized self-reliance and independence. By contrast, the private sphere of the home—women's sphere—emphasized the giving of services, the exchange of gifts, and love. Accordingly, love and emotions became the domain of women, and work and rationality became the domain of men (Lamanna and Riedmann, 2003). Although the roles of women and

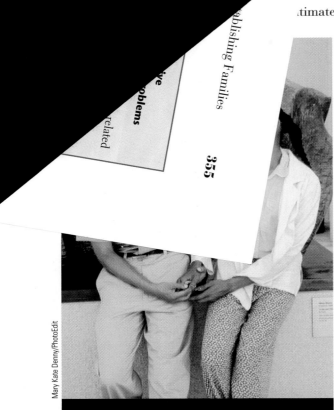

Mary Kate Denny/PhotoEdit

In the United States, the notion of romantic love is deeply intertwined with our beliefs about how and why people develop intimate relationships and establish families. Not all societies share this concern with romantic love.

men changed dramatically in the twentieth century, women and men may still not share the same perceptions about romantic love today. According to the sociologist Francesca Cancian (1990), women tend to express their feelings verbally whereas men tend to express their love through nonverbal actions, such as running an errand for someone or repairing a child's broken toy.

Love and intimacy are closely intertwined. Intimacy may be psychic ("the sharing of minds"), sexual, or both. Although sexuality is an integral part of many intimate relationships, perceptions about sexual activities vary from one culture to the next and from one time period to another. For example, kissing is found primarily in Western cultures; many African and Asian cultures view kissing negatively (Reinisch, 1990).

For many years, the work of the biologist Alfred C. Kinsey was considered to be the definitive research on human sexuality, even though some of his methodology had serious limitations. Recently, the work of Kinsey and his associates has been superseded by the National Health and Social Life Survey conducted by the National Opinion Research Center at the University of Chicago (see Laumann et al., 1994; Michael et al., 1994). Based on interviews with more than 3,400 men and women aged 18 to 59, this random survey tended to reaffirm the significance of the dominant sexual ideologies. Most respondents reported that they engaged in heterosexual relationships, although 9 percent of the men said they had had at least one homosexual encounter resulting in orgasm. Although 6.2 percent of men and 4.4 percent of women said that they were at least somewhat attracted to others of the same gender, only 2.8 percent of men and 1.4 percent of women identified themselves as gay or lesbian. According to the study, persons who engaged in extramarital sex found their activities to be more thrilling than those with a marital partner, but they also felt more guilt. Persons in sustained relationships such as marriage or cohabitation found sexual activity to be the most satisfying emotionally and physically.

Cohabitation and Domestic Partnerships

Attitudes about cohabitation have changed in the past three decades. Although cohabitation is still defined by the Census Bureau as the sharing of a household by one man and one woman who are not related to each other by kinship or marriage, many sociologists now use a more inclusive definition. For our purposes, we will define **cohabitation as two people who live together, and think of themselves as a couple, without being legally married.** It is not known how many people actually cohabit because the Census Bureau does not ask about emotional or sexual involvement between unmarried individuals sharing living quarters or between gay and lesbian couples.

Based on Census Bureau data, the people who are most likely to cohabit are under age 45, have been married before, or are older individuals who do not want to lose financial benefits (such as retirement benefits) that are contingent upon not remarrying. Among younger people, employed couples are more likely to cohabit than college students are.

Today, some people view cohabitation as a form of "trial marriage." Some people who have cohabited do eventually marry the person with whom they have been living whereas others do not. A recent study of 11,000 women found that there was a 70-percent marriage rate for women who remained in a cohabiting relationship for at least 5 years. However, of the women in that study who cohabited and then married their partner, 40 percent became divorced within a 10-year period (Bramlett and Mosher, 2001). Whether these findings will be supported by subsequent research remains to be seen. But we do know that studies over the past decade have supported the proposition that couples who cohabit before marriage do not necessarily have a stable relationship following marriage (Bumpass, Sweet, and Cherlin, 1991; London, 1991).

Among heterosexual couples, many reasons exist for cohabitation; for gay and lesbian couples, however, no alternatives to cohabitation exist in most U.S. states. For that reason, many lesbians and gays seek recognition of their *domestic partnerships*—**household partnerships in which an unmarried couple lives together in a committed, sexually intimate relationship and is granted the same rights and benefits as those accorded to married heterosexual couples** (Aulette, 1994; Gerstel and Gross, 1995). Benefits such as health and life insurance coverage are extremely important to *all* couples, as Gayle, a lesbian, points out: "It makes me angry that [heterosexuals] get insurance benefits and all the privileges, and Frances [her partner] and I take a beating financially. We both pay our insurance policies, but we don't get the discounts that other people get and that's not fair" (qtd. in Sherman, 1992: 197).

Despite laws to the contrary, many lesbian and gay couples consider themselves married and in a lifelong commitment. To make their commitment public, some couples exchange rings and vows under the auspices of churches such as the Metropolitan Community Church (the national gay church). Others, like Gayle, do not feel the need to be married:

> Within my home I feel married to Frances, but I don't consider us "married." The marriage part is still very heterosexual to me. . . . Frances is my life partner; that's how I'm accustomed to thinking of her. (qtd. in Sherman, 1992: 189–190)

Marriage

Why do people get married? Couples get married for a variety of reasons. Some do so because they are "in love," desire companionship and sex, want to have children, feel social pressure, are attempting to escape from a bad situation in their parents' home, or believe that they will have more money or other resources if they get married. These factors notwithstanding, the selection of a marital partner is actually fairly predictable. As previously discussed, most people in the United States tend to choose marriage partners who are similar to themselves. *Homogamy* **refers to the pattern of individuals marrying those who have similar characteristics, such as race/ethnicity, religious background, age, education, or social class.** However, homogamy provides only the general framework within which people select their partners; people are also influenced by other factors. For example, some researchers claim that people want partners whose personalities match their own in significant ways. Thus, people who are outgoing and friendly may be attracted to other people with those same traits. However, other researchers claim that people look for partners whose personality traits differ from but complement their own.

Housework and Child-Care Responsibilities

Today, over 50 percent of all marriages in the United States are *dual-earner marriages*—**marriages in which both spouses are in the labor**

Cohabitation a situation in which two people live together, and think of themselves as a couple, without being legally married.

Domestic partnerships household partnerships in which an unmarried couple lives together in a committed, sexually intimate relationship and is granted the same rights and benefits as those accorded to married heterosexual couples.

Homogamy the pattern of individuals marrying those who have similar characteristics, such as race/ethnicity, religious background, age, education, or social class.

Dual-earner marriages marriages in which both spouses are in the labor force.

Michael Newman/PhotoEdit

Juggling housework, child care, and a job in the paid work force are all part of the average day for many women. Why does sociologist Arlie Hochschild believe that many women work a "second shift"?

force. Over half of all employed women hold full-time, year-round jobs. Even when their children are very young, most working mothers work full time. For example, in 2000 more than 75 percent of employed mothers with children under age 6 worked full time (U.S. Census Bureau, 2001g). Moreover, as Chapter 10 points out, many married women leave their paid employment at the end of the day and go home to perform hours of housework and child care. Sociologist Arlie Hochschild (1989) refers to this as the *second shift*—the domestic work that employed women perform at home after they complete their workday on the job. Thus, many married women contribute to the economic well-being of their families and also meet many, if not all, of the domestic needs of family members by cooking, cleaning, shopping, taking care of children, and managing household routines. According to Hochschild, the unpaid housework that women do on the second shift amounts to an extra month of work each year. In households with small children or many children, the amount of housework increases (Hartmann, 1981). Across race and class, numerous studies have confirmed that domestic work remains primarily women's work (Gerstel and Gross, 1995).

In recent years, more husbands have attempted to share some of the household and child-care responsibilities, especially in families in which the wife's earnings are essential to family finances (Perry-Jenkins and Crouter, 1990). By contrast, in families in which husbands see themselves as the primary breadwinners, they are less likely to share housework with their wives. Even when husbands share some of the household responsibilities, they typically spend much less time in these activities than do their wives (see Coverman, 1989). Women and men perform different household tasks, and the deadlines for their work vary widely. Recurring tasks that have specific times for completion (such as bathing a child or cooking a meal) tend to be the women's responsibility; by contrast, men are more likely to do the periodic tasks that have no highly structured schedule (such as mowing the lawn or changing the oil in the car) (Hochschild, 1989). Some men are also more reluctant to perform undesirable tasks such as scrubbing the toilet or diapering a baby, or to give up leisure pursuits.

Couples with more-egalitarian ideas about women's and men's roles tend to share more equally in food preparation, housework, and child care (Wright et al., 1992). For some men, the shift to a more egalitarian household occurs gradually, as Wesley, whose wife works full time, explains:

> It was me taking the initiative, and also Connie pushing, saying, "Gee, there's so much that has to be done." At first I said, "But I'm supposed to be the breadwinner," not realizing she's also the breadwinner. I was being a little blind to what was going on, but I got tired of waiting for my wife to come home to start cooking, so one day I surprised the hell out [of] her and myself and the kids, and I had supper waiting on the table for her. (qtd. in Gerson, 1993: 170)

In the United States, millions of parents rely on child care so that they can work and so that their young children can benefit from early educational experiences that will help in their future school endeavors. For millions more parents, after-school care for school-age children is an urgent concern. Nearly five million children are home alone after school each week in this country. The children need productive and safe activities to engage in while their parents are working (Children's Defense Fund, 1999). Although child care is often unavailable or unaffordable for many parents,

Figure 11.1 The Child-Care Class System: Who's Watching Out for Our Children?

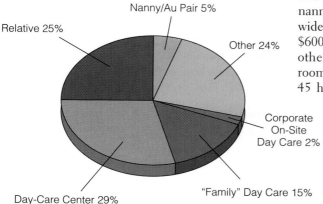

Nanny/Au Pair 5%
Relative 25%
Other 24%
Corporate On-Site Day Care 2%
Day-Care Center 29%
"Family" Day Care 15%

Source: *New York Times Magazine*, 1998.

those children who are in day care for extended hours often come to think of child-care workers and other caregivers as members of their extended families because they may spend nearly as many hours with them as they do with their own parents. For children of divorced parents and other young people living in single-parent households, the issue of child care is often a pressing concern because of the limited number of available adults and lack of financial resources.

In the United States, child-care arrangements are a reflection of the class system (see Figure 11.1). For example, in the upper-income categories, preschool-age children of working mothers are more likely to be cared for in their own homes by nannies or governesses. The "Six-Figure Nanny" is the best-paid and most-prestigious home care-

giver because this individual may earn up to $2,000 per week, plus benefits (*New York Times Magazine*, 1998). The Six-Figure Nanny is college educated and speaks several languages, whereas the typical nanny's language skills and level of experience vary widely. Typical nannies are paid between $250 and $600 per week; an au pair (a young person from another country) makes about $150 per week, plus room and board, taking care of children for about 45 hours a week. Middle-income families with working mothers typically send preschoolers to day-care centers or family day care, which takes place in the home of the caregiver. In low-income families, most preschoolers with working mothers are taken care of by nonparental relatives.

Child-Related Family Issues and Parenting

Not all couples become parents. Those who decide not to have children often consider themselves to be "child-free," whereas those who do not produce children through no choice of their own may consider themselves "childless."

Deciding to Have Children

Cultural attitudes about having children and about the ideal family size began to change in the United States in the late 1950s. Women, on average, are now having two children each. However, rates of fertility differ across racial and ethnic categories. In 2000, for example, Latinas (Hispanic women) had a total fertility rate of 2.6, which was 50 percent above that of white (non-Hispanic) women. Among Latinas, the highest rate of fertility was found among Mexican American women, whereas Puerto Rican and Cuban American women had relatively lower rates. Similarly, African American women, on average, had a higher rate of fertility—2.0 as compared to 1.8—than white (non-Hispanic) women in the 1990s (Bachu and O'Connell, 2001).

✓Checkpoints

Checkpoint 11.3

12. Why do sociologists study issues such as love and intimacy?
13. What has recent research shown about cohabitation and domestic partnerships?
14. What is homogamy?
15. How have dual-earner marriages affected U.S. family life?

Second shift Arlie Hochschild's term for the domestic work that employed women perform at home after they complete their workday on the job.

Advances in birth control techniques over the past four decades—including the birth control pill and contraceptive patches and shots—now make it possible for people to decide whether or not they want to have children, how many they wish to have, and to determine (at least somewhat) the spacing of their births. However, sociologists suggest that fertility is linked not only to reproductive technologies but also to women's beliefs that they do or do not have other opportunities in society that are viable alternatives to childbearing (Lamanna and Riedmann, 2003).

Today, the concept of reproductive freedom includes both the desire *to have* or *not to have* one or more children. According to the sociologists Leslie King and Madonna Harrington Meyer (1997), many U.S. women spend up to one-half of their life attempting to control their reproductivity. Other analysts have found that women, more often than men, are the first to choose a child-free lifestyle (Seccombe, 1991). However, the desire not to have children often comes in conflict with our society's *pronatalist bias*, which assumes that having children is the norm and can be taken for granted, whereas those who choose not to have children believe they must justify their decision to others (Lamanna and Riedmann, 2003).

However, some couples experience the condition of *involuntary infertility*, whereby they want to have a child but find that they are physically unable to do so. *Infertility* is defined as an inability to conceive after a year of unprotected sexual relations. Today, infertility affects nearly five million U.S. couples, or one in twelve couples in which the wife is between the ages of fifteen and forty-four (Gabriel, 1996). Research suggests that fertility problems originate in females in approximately 30–40 percent of the cases and with males in about 40 percent of the cases; in the other 20 percent of the cases, the cause is impossible to determine (Gabriel, 1996). A leading cause of infertility is sexually transmitted diseases, especially those cases that develop into pelvic inflammatory disease (Gold and Richards, 1994). It is estimated that about half of infertile couples who seek treatments such as fertility drugs, artificial insemination, and surgery to unblock fallopian tubes can be helped; however, some are unable to conceive despite expensive treatments such as *in vitro fertilization*, which costs as much as $11,000 per attempt (Gabriel, 1996). (See Box 11.2 for a discussion of issues arising from reproductive technology.)

According to the sociologist Charlene Miall (1986), women who are involuntarily childless engage in "information management" to combat the social stigma associated with childlessness. Their tactics range from avoiding people who make them uncomfortable to revealing their infertility so that others will not think of them as "selfish" for being childless. Some people who are involuntarily childless may choose to become parents by adopting a child.

Adoption

Adoption is a legal process through which the rights and duties of parenting are transferred from a child's biological and/or legal parents to a new legal parent or parents. This procedure gives the adopted child all the rights of a biological child. In most adoptions, a new birth certificate is issued and the child has no future contact with the biological parents; however, some states have "right-to-know" laws under which adoptive parents must grant the biological parents visitation rights.

Matching children who are available for adoption with prospective adoptive parents can be difficult. The available children have specific needs, and the prospective parents often set specifications on the type of child they want to adopt. Some adoptions are by relatives of the child; others are by infertile couples (although many fertile couples also adopt). Increasing numbers of gays, lesbians, and people who are single are adopting children. Although thousands of children are available for adoption each year in the United States, many prospective parents seek out children in developing nations such as Romania, South Korea, and India. The primary reason is that the available children in the United States are thought to be "unsuitable." They may have disabilities, or they may be sick, nonwhite (most of the prospective parents are white), or too old (Zelizer, 1985). In addition, fewer infants are available for adoption today than in the past because better means of contraception exist, abortion is more readily available, and more unmarried teenage parents decide to keep their babies.

Ironically, although many couples who would like to have a child are unable to do so, other couples conceive a child without conscious intent. Consider the fact that for the approximately 6.4 million women who become pregnant each year in the United States, about 2.8 million (44 percent) pregnancies are intended whereas about

Box 11.2 Changing Times: Media and Technology
The Reproductive Revolution: Who Are My Parents?

Consider the possibilities of in vitro fertilization, which involves fertilizing a woman's eggs in a Petri dish and then transferring them to the woman's womb. The eggs must come from the woman herself or from a female donor; the sperm can come from the woman's partner or another male donor. An ovum transfer involves moving a five-day-old embryo from one female to another, who then carries the embryo to term. When these are combined, it means that a child can have at least five parents, not counting any later changes with remarriage: a donor mother, a birth mother, a social mother (the one who raises the child), a donor father, and a social father. (Coontz, 1997: 93)

According to the sociologist Stephanie Coontz (1997), medical and technological change has created many new issues in family life. For example, in recent years biological mothers have been pitted against biological fathers in surrogate custody disputes. Ex-spouses have fought bitter court battles over custody of frozen embryos stored while the couple was still married (in the event they had decided to use in vitro fertilization).

Many infertile couples have had their hopes raised that they might be able to have a child through various forms of assisted reproductive technology. The main forms of this technology have been outlined by Gabriel (1996):

- *In vitro fertilization.* Eggs that were produced as a result of administering fertility drugs are removed from the woman's body and fertilized by sperm in a laboratory dish. The embryos that result from the process are transferred to the woman's uterus.

- *Micromanipulation.* A specialist, looking through a microscope, manipulates egg and sperm in a laboratory dish to improve the chances of a pregnancy.
- *Cryopreserved embryo transfer.* Embryos that were frozen after a previous assisted reproductive technology procedure are thawed and then transferred to the uterus.
- *Egg donation.* Eggs are removed from a donor's uterus, fertilized in a laboratory dish, and transferred to an infertile woman's uterus.
- *Surrogacy.* An embryo is implanted in the uterus of a woman who is paid to carry the fetus until birth. The egg may come from either the legal or the surrogate mother, and the sperm may come from either the legal father or a donor.

How successful are these procedures in cases of infertility? The cost of these procedures prohibits many women from trying one of them. However, among those couples who try assisted reproductive technology, it is estimated that only one in five will become parents (Gabriel, 1996).

What influence will new reproductive technology have on families and society in the future? Are questions arising from the reproductive revolution a matter of *cultural lag,* or are they linked to larger social and ethical issues in our society? What do you think?

3.6 million (56 percent) are unintended (Gold and Richards, 1994). As with women and motherhood, some men feel that their fatherhood was planned; others feel that it was thrust upon them (Gerson, 1993). Unplanned pregnancies usually result from failure to use contraceptives or from using contraceptives that do not work. Even with "planned pregnancies," it is difficult to plan exactly when conception will occur and a child will be born.

Teenage Pregnancies

Teenage pregnancies are a popular topic in the media and political discourse, and the United States has the highest rate of teen pregnancy in the Western industrialized world (National Campaign to Prevent Teen Pregnancy, 1997). However, in 2000 the total number of live births per 1,000 women aged 15 to 19 was 59.7, down from 62.1 in 1991 (Bachu and O'Connell, 2001). Perhaps teenage pregnancy is seen as a major problem because teen birth rates have not declined as rapidly as rates for older married women.

What are the primary reasons for the high rates of teenage pregnancy? At the microlevel, several issues are most important: (1) many sexually active teenagers do not use contraceptives; (2) teenagers—especially those from some low-income families and/or subordinate racial and ethnic groups—may receive little accurate information about the use of, and problems associated with, contraception; (3) some teenage males (due to a double standard

Increasingly, people who adopt children create households in which individuals from diverse racial and ethnic groups learn to live together.

based on the myth that sexual promiscuity is acceptable among males but not females) believe that females should be responsible for contraception; and (4) some teenagers view pregnancy as a sign of male prowess or as a way to gain adult status. At the macrolevel, structural factors also contribute to teenage pregnancy rates. Lack of education and employment opportunities in some central-city and rural areas may discourage young people's thoughts of upward mobility that might make early parenting appear less appealing. Likewise, religious and political opposition has resulted in issues relating to reproductive responsibility not being dealt with as openly in the United States as in some other nations. Finally, advertising, films, television programming, magazines, music, and other forms of media often flaunt the idea of being sexually active without showing the possible consequences of such behavior.

Teen pregnancies have been of concern to analysts who suggest that teenage mothers may be less skilled at parenting, are less likely to complete high school than their counterparts without children, and possess few economic and social supports other than their relatives (Maynard, 1996; Moore, Driscoll, and Lindberg, 1998). In addition, the increase in births among unmarried teenagers may have negative long-term consequences for mothers and their children, who have severely limited educational and employment opportunities and a high

likelihood of living in poverty. Moreover, the Children's Defense Fund estimates that among those who first gave birth between the ages of fifteen and nineteen, 43 percent will have a second child within three years (Benokraitis, 1999).

Teenage fathers have largely been left out of the picture. According to the sociologist Brian Robinson (1988), a number of myths exist regarding teenage fathers: (1) they are worldly wise "superstuds" who engage in sexual activity early and often, (2) they are "Don Juans" who sexually exploit unsuspecting females, (3) they have "macho" tendencies because they are psychologically inadequate and need to prove their masculinity, (4) they have few emotional feelings for the women they impregnate, and (5) they are "phantom fathers" who are rarely involved in caring for and rearing their children. However, these myths overlook the fact that some teenage males try to be good fathers; for example, Adan Chamul of Northridge, California, decided not to be like his own father, who left Adan's mother when Adan was four months old:

> [My father] didn't want to have anything to do with me until I was able to work and earn money for him. I said "No." [When my girlfriend became pregnant,] my friends were real upset. They told me, "Kick her out of the house. Dump her." But I didn't want my baby to grow up like that, without a dad. It's not the baby's fault. It was our mistake. . . . I regret having her at such a young age, but I love her and I'll never regret that she is here. We're going to grow up. All three of us. (qtd. in Gleick, Reed, and Schindehette, 1994: 55)

Many discussions of teenage pregnancy focus on the negative consequences of behavior rather than on the human faces of the young people involved.

Single-Parent Households

In recent years, there has been a significant increase in single- or one-parent households due to divorce and to births outside of marriage. Even for a person with a stable income and a network of friends and family to help with child care, raising a child alone can be an emotional and financial burden. Single-parent households headed by women have been stereotyped by some journalists, politicians, and analysts as being problematic for children. About 42 percent of all white children and 86 percent of all African American children will spend part of their childhood living in a household headed by a single mother who is divorced, separated, never married, or widowed (Garfinkel and McLanahan, 1986). According to sociologists Sara McLanahan and Karen Booth (1991), children from mother-only families are more likely than children in two-parent families to have poor academic achievement, higher school absentee and dropout rates, early marriage and parenthood, higher rates of divorce, and more drug and alcohol abuse. Does living in a one-parent family *cause* all of this? Certainly not! Many other factors — including poverty, discrimination, unsafe neighborhoods, and high crime rates — contribute to these problems.

Lesbian mothers and gay fathers are counted in some studies as single parents; however, they often share parenting responsibilities with a same-sex partner. Due to homophobia (hatred and fear of homosexuals and lesbians), lesbian mothers and gay fathers are more likely to lose custody to a heterosexual parent in divorce cases (Falk, 1989; Robson, 1992). In any case, between one million and three million gay men in the United States and Canada are fathers. Some gay men are married natural fathers, others are single gay men, and still others are part of gay couples who have adopted children. Very little research exists on gay fathers; what research does exist tends to show that noncustodial gay fathers try to maintain good relationships with their children (Bozett, 1988).

Single fathers who do not have custody of their children may play a relatively limited role in the lives of those children. Although many remain actively involved in their children's lives, others may become "Disneyland daddies" who take their children to recreational activities and buy them presents for special occasions but have a very small part in their children's day-to-day life. Sometimes, this limited role is by choice, but more often it is caused by workplace demands on time and energy, location of the ex-wife's residence, and limitations placed on visitation by custody arrangements.

Currently, men head about 17 percent of white one-parent families and 7 percent of African American one-parent families; among many of the men, a

Laura Dwight/PhotoEdit

In recent years, many more fathers and mothers alike have been confronting the unique challenges of single parenting.

pattern of "involved fatherhood" has emerged (Gerson, 1993). For example, in a study of men who became single fathers because of their wife's death, desertion, or relinquishment of custody, the sociologist Barbara Risman (1987) found that the men had very strong relationships with their children.

Two-Parent Households

Parenthood in the United States is idealized, especially for women. According to the sociologist Alice Rossi (1992), maternity is the mark of adulthood for women, whether or not they are employed. By contrast, men secure their status as adults by their employment and other activities outside the family (Hoffnung, 1995).

For families in which a couple truly shares parenting, children have two primary caregivers. Some parents share parenting responsibilities by choice; others share out of necessity because both hold full-time jobs. Some studies have found that men's taking an active part in raising the children is beneficial not only for mothers (who then have a little more time for other activities) but also for the

Figure 11.2 Marital Status of U.S. Population Aged 15 and Over by Race/Ethnicity

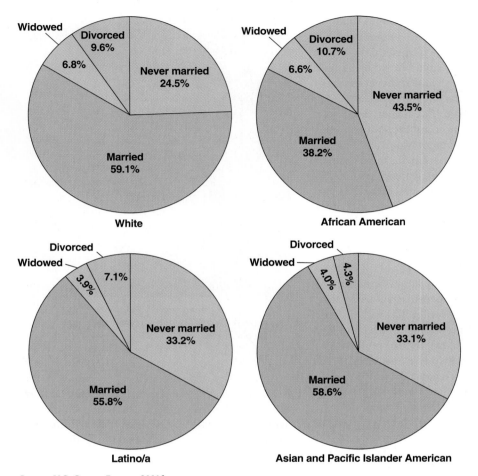

Source: U.S. Census Bureau, 2001f.

men and the children. The men benefit through increased access to children and greater opportunity to be nurturing parents (Coltrane, 1989).

Children in two-parent families are not guaranteed a happy childhood simply because both parents reside in the same household. Children whose parents argue constantly, are alcoholics, or abuse them may be worse off than children in a single-parent family in which the environment is more conducive to their well-being.

Remaining Single

Some never-married people remain single by choice. Reasons include opportunities for a career (especially for women), the availability of sexual partners without marriage, the belief that the single lifestyle is full of excitement, and the desire for self-sufficiency and freedom to change and experiment (Stein, 1976, 1981). Some scholars have concluded that individuals who prefer to remain single hold more-individualistic values and are less family-oriented than those who choose to marry. Friends and personal growth tend to be valued more highly than marriage and children (Cargan and Melko, 1982; Alwin, Converse, and Martin, 1985).

Other never-married individuals remain single out of necessity. Being single is an economic necessity for those who cannot afford to marry and set up their own household. Structural changes in the economy have limited the options of many working-class young people. Even some college graduates

✓Checkpoints

Checkpoint 11.4

16. What major factors come into play when couples decide whether or not to have children?
17. What are the current trends in adoption and teenage pregnancies?
18. How do single-parent and two-parent households compare in the United States?
19. Why do some people remain single?

have found that they cannot earn enough money to set up a household separate from that of their parents.

The proportion of singles varies significantly by racial and ethnic group, as shown in Figure 11.2. Among persons age 15 and over, 43.5 percent of African Americans have never married, compared with 33.2 percent of Latinos/as, 33.1 percent of Asian and Pacific Islander Americans, and 24.5 percent of whites. Among women age 20 and over, the difference is even more pronounced; almost twice as many African American women in this age category have never married, compared with U.S. women of the same age in general (U.S. Census Bureau, 2001f).

Although a number of studies have examined why African Americans remain single, few studies have focused on Latinos/as. Some analysts cite the diversity of experiences among Mexican Americans, Cuban Americans, and Puerto Ricans as the reason for this lack of research. Existing research attributes increased rates of singlehood among Latinos/as to several factors, including the youthful age of the Latino/a population and economic conditions experienced by many young Latinos/as (see Mindel, Habenstein, and Wright, 1988).

Transitions and Problems in Families

Families go through many transitions and experience a wide variety of problems, ranging from high rates of divorce and teen pregnancy to domestic abuse and family violence. These all-too-common experiences highlight two important facts about families: (1) for good or ill, families are central to

our existence, and (2) the reality of family life is far more complicated than the idealized image of families found in the media and in many political discussions. Moreover, as people grow older, transitions inevitably occur in family life.

Family Transitions Based on Age and the Life Course

Throughout our lives, our families play an important role in our individual development. People are assigned different roles and positions based on their age and the family structure in a particular society. Stages in the contemporary life course in industrialized nations typically include infancy and childhood, adolescence and young adulthood, middle adulthood, and late adulthood.

INFANCY AND CHILDHOOD Infancy and childhood are largely creations of industrialized nations. In agricultural societies, children are often a source of physical labor for both their families and the larger society. By contrast, children in industrialized nations are expected to attend school and learn the necessary skills for future employment rather than perform unskilled labor. In this context, families must provide for extensive periods of time in which their children are economically dependent upon them.

ADOLESCENCE AND YOUNG ADULTHOOD In contemporary industrialized countries, adolescence roughly spans the teenage years. As compared with preindustrial societies, in which adolescents are seen as adults, adolescents in industrialized nations are treated by their families, teachers, and others as neither children nor adults. Young adulthood, which follows adolescence and lasts to about age 39, is socially significant because, during this time, people are expected to establish their own families and get a job.

Although many young people leave their family of orientation as they reach adulthood, finish school, and/or get married, recent patterns show that older persons may still be actively involved in parenting and friendship relationships with their adult children. Author Betty Friedan explains how an intergenerational pattern of extended family has emerged in her family:

What can people across generations learn by taking time to talk to each other? Can the learning process flow in both directions?

With my own children . . . I have noticed an interesting development. In their twenties and thirties, they started coming back home again—Thanksgiving, Passover, summers—bringing wives, husbands, their own children now, to my house in Sag Harbor. I relished the renewed sense of family which they clearly wanted again and enjoyed. I realized that they were confident enough now, in their grown-up selves, in their busy careers and the rich texture of their marriages . . . that they did not need to defend themselves anymore against their childish need for mother. (Friedan, 1993: 297)

MIDDLE ADULTHOOD The concept of middle adulthood—people between the ages of 40 and 65—did not exist until fairly recently. For most people, middle adulthood represents the time during which their families have the highest levels of income and they may conclude child rearing but begin an era of grandparenting. Today, more grandparents have partial or full responsibility for rearing their grandchildren.

LATE ADULTHOOD Late adulthood is generally considered to begin at age 65—which for some people is "normal" retirement age. *Retirement*—the institutionalized separation of an individual from an occupational position, with continuation of income through a retirement pension based on prior years of service—means the end of a status that has long been a source of income and a means of personal identity. Loss of such status may produce family discord for a period of time as spouses become adjusted to having more time alone with each other. Increasingly, however, older adults are maintaining either full-time or part-time employment outside the home. For those older adults who are financially able, travel and other leisure-time pursuits may become the focus of their activities rather than daily attendance at a job.

Many older adults continue to live in their own homes; however, some may prefer smaller housing units or apartments if they no longer have other family members residing in the household. Those who have chronic medical conditions or need the assistance of others may rely on *assisted living* arrangements—a concept that refers to housing, support services, and health care designed to meet the varied needs of older adults who need help with daily activities such as bathing, dressing, food preparation, or taking medications. For others, living with family members remains the preferred approach. Factors involved in living arrangements of older family members include how close they live to other relatives and friends and how frequently they react with one another. Among members of some subordinate racial and ethnic groups, intergenerational family ties are very important survival mechanisms for older family members. Mary Crow Dog (Crow Dog and Erdoes, 1991: 12–13) explains the importance of extended family networks for some Native Americans:

Our people have always been known for their strong family ties, for people within one family group caring for each other, for the "helpless ones," the old folks and especially the children, the coming generation. . . . At the center of the old Sioux society was the . . . extended family group. . . . [It] was like a warm womb cradling all within it.

As the population of persons aged 65 and older continues to grow in the United States, more families will be seeking viable options for older family members.

At each stage in the life course, family life differs significantly based on the number of people and the diversity of age categories represented in the social unit at that time. Although some families provide their members with love, warmth, and satisfying emotional experiences, others may be hazardous to the individual's physical and mental well-being. Because of this dichotomy in family life, sociologists have described families as both a "haven in a heartless world" (Lasch, 1977) and a "cradle of violence" (Gelles and Straus, 1988).

Family Violence

Violence between men and women in the home is often called spouse abuse or domestic violence. *Spouse abuse* refers to any intentional act or series of acts—whether physical, emotional, or sexual—that causes injury to a female or male spouse (Wallace, 1996). According to sociologists, the term *spouse abuse* refers not only to people who are married but also to those who are cohabiting or involved in a serious relationship, as well as those individuals who are separated or living apart from their former spouse (Wallace, 1996).

How much do we know about family violence? Women, as compared with men, are more likely to be the victim of violence perpetrated by intimate partners. Recent statistics indicate that women are five times more likely than men to experience such violence and that many of these women live in households with children younger than twelve. However, we cannot know the true extent of family violence because much of it is not reported to police. For example, it is estimated that only about one-half of the intimate partner violence against women in the 1990s was reported to police (U.S. Department of Justice, 2000b). African American women were more likely than other women to report such violence, which may further skew data about who is most likely to be victimized by a domestic partner (U.S. Department of Justice, 2000b).

Although everyone in a household where family violence occurs is harmed psychologically, whether or not they are the victims of violence, children are especially affected by household violence. It is estimated that between three and ten million children witness some form of domestic violence in their homes each year, and there is evidence to suggest that domestic violence and child maltreatment often take place in the same household (Children's Defense Fund, 2001). According to some experts, domestic violence is an important indicator that child abuse and neglect are also taking place in the household.

In some situations, family violence can be reduced or eliminated through counseling, the removal of one parent from the household, or other steps that are taken either by the family or by social service or law enforcement officials. However, children who witness violence in the home may display certain emotional and behavioral problems that adversely affect their school life and communication with other people. In some families, the problems of family violence are great enough that the children are removed from the household and placed in foster care.

Children in Foster Care

Not all of the children in foster care have come from violent homes, but many foster children have been in dysfunctional homes where parents or other relatives lacked the ability to meet the children's daily needs. *Foster care* refers to institutional settings or residences where adults other than a child's own parents or biological relatives serve as caregivers. States provide financial aid to foster parents, and the intent of such programs is that the children will either return to their own families or be adopted by other families. However, this is often not the case for "difficult to place" children, particularly those who are over 10 years of age, have illnesses or disabilities, or are perceived as suffering from "behavioral problems." More than 568,000 children are in foster care at any given time (Barovick, 2001). About 60 percent of children in foster care are children of color, with about 42 percent of them being African American—almost three times the percentage of African American children in the total U.S. child population (Children's Defense Fund, 2001). Even when the number of children entering foster care for the first time remains relatively stable, the total number of children in foster care continues to increase because fewer children are leaving foster care and being adopted or placed in permanent homes (Children's Defense Fund, 2001). Many children in foster care have limited prospects for finding a permanent home; however, a few innovative programs have offered hope for children who previously had been moved from one foster care setting to another (see Box 11.3).

Problems in the family contribute to the large numbers of children who are in foster care. Such factors include parents' illness, unemployment, or

Box 11.3 You Can Make a Difference
Providing Hope and Help for Children

I take it personally when I see kids mistreated. I just think they need an advocate to fight for them. . . . For me, it's very simple: The kids' needs come first. That's the bottom line at Hope Meadows. We make decisions as if these are our own children, and when you think that way, your decisions are different than if you are just trying to work within a bureaucratic system.
—sociologist Brenda Eheart, describing why she founded Hope Meadows (qtd. in Smith, 2001: 22)

After five years of research into the adoptions of older children, the sociologist Brenda Eheart realized that foster families faced many problems when they tried to help children who had been moved from home to home. Thinking that she might be able to make a difference, Eheart developed the plan for Hope Meadows, a community established in 1994 on an abandoned Illinois air force base. Hope Meadows is made up of a three-block-long series of ranch houses that provide multigenerational and multiracial housing for foster children, their temporary families, and older adults who live and work with the children. Older adults who interact with the children receive reduced rent in exchange for at least six hours per week of volunteer work. Foster families that reside at Hope Meadows gain a feeling of community as they work together to help children who have experienced severe abuse or neglect, have been exposed to drugs and numerous foster homes, and often have physical, emotional, and behavioral problems.

Since its commencement, Hope Meadows has been largely successful in helping children get adopted. However, children are not the only beneficiaries of this community: Older residents gain the benefit of interacting with children and feeling that they can *make a positive contribution* to the lives of others (Barovick, 2001). Debbie Calhoun, a foster parent at Hope Meadows, has suggested things that chil-

dren need the most when they come there, and we can make a difference by providing the children in our lives with these same things (based on Smith, 2001):

- *Understanding.* We need to gain an awareness of how children feel and why they say and do certain things.
- *Trust.* We need to help children to see us as people they can rely on and believe in as people they can trust.
- *Love.* We must show children that they are loved and that they will still be loved even when they make mistakes.
- *Compassion.* We must show children compassion because they must experience compassion in order to be able to show it to others.
- *Time.* We must give children time to be a part of our lives, and we must also give them time to adjust and to start over when they need to do so.
- *Security.* We must help children to feel secure in their surroundings and to believe that there is stability or permanency in their living arrangements.
- *Praise.* We must tell children when they are doing well and not always be critical of them.
- *Discipline.* We must let children know what behavior is acceptable and what behavior is not, all the while showing them that we love them, even when discipline is necessary.
- *Self-Esteem.* We must help children feel good about themselves.
- *Pride.* We should provide opportunities for children to learn to take pride in their accomplishments and in themselves.

If these suggestions are beneficial for children in foster care settings, then they are certainly useful ideas for each of us to implement in our own families and communities as well. What other ideas would you add to the list? Why?

---

death; violence or abuse in the family; and high rates of divorce.

Elder Abuse

Abuse and neglect of older persons have received increasing public attention in recent years, due both to the increasing number of older people and

to the establishment of more-vocal groups to represent their concerns. *Elder abuse* refers to physical abuse, psychological abuse, financial exploitation, and medical abuse or neglect of people aged 65 or older (Hooyman and Kiyak, 1999).

According to a congressional committee on aging, more than 1.5 million older people in the United States are the victims of physical or mental abuse each year (Friedan, 1993). Just as with violence against children or women, it is difficult to determine how much abuse against older persons occurs. Many victims are understandably reluctant to talk about it. One study indicates that slightly over 2 percent of all older people experience physical abuse (Pillemer and Finkelhor, 1988; Atchley, 2000). Although this may appear to be a small percentage, it represents a large number of people. Studies have shown that elder abuse tends to be concentrated among those over age 75 (Steinmetz, 1987). Most of the victims are white, middle-to-lower-middle-class Protestant women, aged 75–85, who suffer some form of impairment (Garbarino, 1989; Benokraitis, 1997). Sons, followed by daughters, are the most frequent abusers of older persons (Friedan, 1993).

What causes elder abuse? Scholars do not agree on an answer to this question. Some believe that elder abuse may occur in families because of ageism, which you will recall is prejudice and discrimination against people on the basis of age. Other analysts link elder abuse with the physical impairment of some elderly persons (Harris, 1990); however, little evidence has been found to support that conclusion (Pedrick-Cornell and Gelles, 1982). Similarly, the sociologist Karl Pillemer found no support for the common assumption that elderly persons who are dependent on their children are more likely to be abused. To the contrary, the abusers are very likely to be *dependent on the older person* for housing and financial assistance (Pillemer, 1985).

Most states have a wide variety of laws requiring the reporting of domestic violence and elder abuse in families. But family problems, including those leading to divorce, are often so complex that legislation cannot solve the underlying concerns associated with these issues.

Divorce

Divorce is the legal process of dissolving a marriage that allows former spouses to remarry if they so choose. Most divorces today are granted on the grounds of *irreconcilable differences*, meaning that there has been a breakdown of the marital relationship for which neither partner is specifically blamed. Prior to the passage of more-lenient divorce laws, many states required that the partner

seeking the divorce prove misconduct on the part of the other spouse. Under *no-fault divorce laws*, however, proof of "blameworthiness" is generally no longer necessary.

Over the past 100 years, the U.S. divorce rate (number of divorces per 1,000 population) has varied from a low of 0.7 in 1900 to an all-time high of 5.3 in 1981; by 1995, it had stabilized at 4.4 (U.S. Census Bureau, 2001g). Although many people believe that marriage should last for a lifetime, others believe that marriage is a commitment that may change over time.

Recent studies have shown that 43 percent of first marriages end in separation or divorce within 15 years (National Centers for Disease Control, 2001). Many first marriages do not last even 15 years: One in three first marriages ends within 10 years while one in five ends within 5 years. The likelihood of divorce goes up with each subsequent marriage in the serial monogamy pattern. When divorces that terminate second or subsequent marriages are taken into account, there are about half as many divorces each year in the United States as there are marriages. These data concern researchers in agencies such as the Centers for Disease Control because separation and divorce often have adverse effects on the health and well-being of both children and adults (National Centers for Disease Control, 2001).

CAUSES OF DIVORCE Why do divorces occur? As you will recall from Chapter 1, sociologists look for correlations (relationships between two variables) in attempting to answer questions such as this. Existing research has identified a number of factors at both the macrolevel and microlevel that make some couples more or less likely to divorce. At the macrolevel, societal factors contributing to higher rates of divorce include changes in social institutions, such as religion and family. Some religions have taken a more lenient attitude toward divorce, and the social stigma associated with divorce has lessened. Further, as we have seen in this chapter, the family institution has undergone a major change that has resulted in less economic and emotional dependency among family members—thus reducing a barrier to divorce.

At the microlevel, a number of factors contribute to a couple's "statistical" likelihood of divorcing. Here are some of the primary social characteristics of those most likely to get divorced:

- Marriage at an early age (59 percent of marriages to brides under 18 end in separation or divorce within 15 years) (National Centers for Disease Control, 2001)
- A short acquaintanceship before marriage
- Disapproval of the marriage by relatives and friends
- Limited economic resources and low wages
- A high school education or less (although deferring marriage to attend college may be more of a factor than education per se)
- Parents who are divorced or have unhappy marriages
- The presence of children (depending on their gender and age) at the start of the marriage

The interrelationship of these and other factors is complicated. For example, the effect of age is intertwined with economic resources; persons from families at the low end of the income scale tend to marry earlier than those at more affluent income levels. Thus, the question becomes whether age itself is a factor or whether economic resources are more closely associated with divorce.

The relationship between divorce and factors such as race, class, and religion is another complex issue. Although African Americans are more likely than whites of European ancestry to get a divorce, other factors—such as income level and discrimination in society—must also be taken into account. Latinos/as share some of the problems faced by African Americans, but their divorce rate is only slightly higher than that of whites of European ancestry. Religion may affect the divorce rate of some people, including many Latinos/as who are Roman Catholic. However,

despite the Catholic doctrine that discourages divorce, the rate of Catholic divorces is now approximately equal to that of Protestant divorces.

CONSEQUENCES OF DIVORCE   Divorce may have a dramatic economic and emotional impact on family members. An estimated 60 percent of divorcing couples have one or more children. By age 16, about one out of every three white children and two out of every three African American children will experience divorce within their families. As a result, most of them will remain with their mothers and live in a single-parent household for a period of time. In recent years, there has been a debate over whether children who live with their same-sex parent after divorce are better off than their peers who live with an opposite-sex parent. However, sociologists have found virtually no evidence to support the belief that children are better off living with a same-sex parent (Powell and Downey, 1997).

Although divorce decrees provide for parental joint custody of approximately 100,000–200,000 children annually, this arrangement may create unique problems for some children, as the personal narratives of David and Nick Sheff at the beginning of this chapter demonstrate. Furthermore, some children experience more than one divorce during their childhood because one or both of their parents may remarry and subsequently divorce again.

Divorce changes relationships not only for the couple involved but also for other relatives. In some divorces, grandparents feel that they are the big losers. To see their grandchildren, the grandparents have to keep in touch with the parent who has custody. In-laws are less likely to be welcomed and may be seen as being on the "other side" simply because they are the parents of the ex-spouse. Recently, some grandparents have sued for custody of minor grandchildren. Generally, they have not been successful except in cases where questions existed about the emotional stability of the biological parents or the suitability of a foster care arrangement.

But divorce does not have to be always negative. For some people, divorce may be an opportunity to terminate destructive relationships (Lund, 1990; Kurz, 1995). For others, it may represent a

Reflections

Reflections 11.3

What are some of the societal factors that contribute to seemingly personal problems such as divorce and teenage pregnancies?

Michelle D. Birdwell/PhotoEdit

Remarriage and blended families create new opportunities and challenges for parents and children alike.

means to achieve personal growth by managing their lives and social relationships and establishing their own social identity (see Reissman, 1991). Still others choose to remarry one or more times.

Remarriage

Most people who divorce get remarried. In recent years, more than 40 percent of all marriages have been between previously married brides and/or grooms. Among individuals who divorce before age thirty-five, about half will remarry within three years of their first divorce (Bramlett and Mosher, 2001). Most divorced people remarry others who have

✓ **Checkpoints**

Checkpoint 11.5

20. What are the stages in the life course?
21. What is family violence, and why has our society been slow to respond to this problem?
22. What are some of the major causes of divorce?
23. What are the consequences of divorce in people's lives?
24. What are blended families?

been divorced. However, remarriage rates vary by gender and age. At all ages, a greater proportion of men than women remarry, often relatively soon after the divorce. Among women, the older a woman is at the time of divorce, the lower her likelihood of remarrying. Women who have not graduated from high school and who have young children tend to remarry relatively quickly; by contrast, women with a college degree and without children are less likely to remarry (Bramlett and Mosher, 2001).

As a result of divorce and remarriage, complex family relationships are often created. Some become stepfamilies or *blended families*, which consist of a husband and wife, children from previous marriages, and children (if any) from the new marriage. At least initially, levels of family stress may be fairly high because of rivalry among the children and hostilities directed toward stepparents or babies born into the family. In spite of these problems, however, many blended families succeed. The family that results from divorce and remarriage is typically a complex, binuclear family in which children may have a biological parent and a stepparent, biological siblings and stepsiblings, and an array of other relatives, including aunts, uncles, and cousins.

According to the sociologist Andrew Cherlin (1992), the norms governing divorce and remarriage are ambiguous. Because there are no clear-cut guidelines, people must make decisions about family life (such as whom to invite for a birthday celebration or wedding) based on their beliefs and feelings about the people involved.

Family Issues in the Future

As we have seen, families and intimate relationships changed dramatically during the twentieth century. Some people believe that the family as we know it is doomed. Others believe that a return to traditional family values will save this important social institution and create greater stability in society. However, the sociologist Lillian Rubin

(1986: 89) suggests that clinging to a traditional image of families is hypocritical in light of our society's failure to support families: "We are after all, the only advanced industrial nation that has no public policy of support for the family whether with family allowances or decent publicly-sponsored childcare facilities." Some laws even have the effect of hurting children whose families do not fit the traditional model. For example, cutting back on government programs that provide food and medical care for pregnant women and infants will result in seriously ill children rather than model families (Aulette, 1994).

According to the psychologist Bernice Lott (1994: 155), people's perceptions about what constitutes a family will continue to change in the future:

> Persons on whom one can depend for emotional support, who are available in crises and emergencies, or who provide continuing affections, concern, and companionship can be said to make up a family. Members of such a group may live together in the same household or in separate households, alone or with others. They may be related by birth, marriage, or a chosen commitment to one another that has not been legally formalized.

Some of these changes are already becoming evident. For example, many men are attempting to take an active role in raising their children and helping with household chores. Many couples terminate abusive relationships and marriages.

Regardless of problems facing families today, many people still demonstrate their faith in the future by choosing to have children. As members of the Baby Boom generation (born between 1946 and 1964) entered their thirties and forties, many decided that they wanted children. New waves of immigrants have also become parents. Thus, the "boomlet" generation (born since 1988) is far more racially and ethnically diverse than the Baby Boom generation. According to futurist Watts Wacker, the boomlet generation knows "how to take care of themselves and be their own boss, because their parents have always worked. They have an unbelievable sense of being self-sufficient and confident. . . . They feel very comfortable about handling what life's throwing at them" (qtd. in Gabriel, 1995: 15). It will be interesting to see what the families created by the boomlet generation will be like. What will your family be like?

Chapter Review

What is the family?
Today, families may be defined as relationships in which people live together with commitment, form an economic unit and care for any young, and consider their identity to be significantly attached to the group.

How does the family of orientation differ from the family of procreation?
The family of orientation is the family into which a person is born; the family of procreation is the family a person forms by having or adopting children.

What pattern of marriage is legally sanctioned in the United States?
In the United States, monogamy is the only form of marriage sanctioned by law. Monogamy is a marriage between two partners, usually a woman and a man.

What are the primary sociological perspectives on families?
Functionalists emphasize the importance of the family in maintaining the stability of society and the well-being of individuals. Conflict and feminist perspectives view the family as a source of social inequality and an arena for conflict. Symbolic interactionists explain family relationships in terms of the subjective meanings and everyday interpretations that people give to their lives. Postmodern analysts view families as being permeable, capable of being diffused or invaded so that the original purpose is modified.

What are the major stages in the family life course?
The major stages are infancy and childhood, adolescence and young adulthood, middle adulthood, and late adulthood.

How are families in the United States changing?
Families are changing dramatically in the United States. Cohabitation has increased significantly in the past three decades. With the increase in dual-earner marriages, women have increasingly been burdened by the second shift—the domestic work that employed women perform at home after they complete their workday on the job. Many single-parent families also exist today.

What is divorce, and what are some of its causes?
Divorce is the legal process of dissolving a marriage. At the macrolevel, changes in social institutions may contribute to an increase in divorce rates; at the microlevel, factors contributing to divorce include age at marriage,

length of acquaintanceship, economic resources, education level, and parental marital happiness.

Key Terms

bilateral descent 350
cohabitation 356
domestic partnerships 357
dual-earner marriages 357
egalitarian family 350
endogamy 351
exogamy 351
extended family 348
families 346
family of orientation 348
family of procreation 348
homogamy 357
kinship 347
marriage 348
matriarchal family 350
matrilineal descent 350
matrilocal residence 351
monogamy 348
neolocal residence 351
nuclear family 348
patriarchal family 350
patrilineal descent 350
patrilocal residence 350
polyandry 349
polygamy 349
polygyny 349
second shift 358
sociology of family 351

Questions for Critical Thinking

1. In your opinion, what constitutes an ideal family? How might functionalist, conflict, feminist, and symbolic interactionist perspectives describe this family?
2. Suppose that you wanted to find out about women's and men's perceptions about love and marriage. What specific issues might you examine? What would be the best way to conduct your research?
3. You have been appointed to a presidential commission on child-care problems in the United States. How to provide high-quality child care at affordable prices is a key issue for the first meeting. What kinds of suggestions would you take to the meeting? How do you think your suggestions should be funded? How does the future look for

children in high-, middle-, and low-income families in the United States?

 **Resources on the Internet**

Chapter-Related Web Sites
The following web sites have been selected for their relevance to the topics in this chapter. These sites are among the more stable, but please note that web site addresses change frequently.

Census 2000 Brief: Households and Families
http://www.census.gov/prod/2001pubs/
c2kbr01-8.pdf
This page contains demographic data on families and housing in America as well as a brief analysis of the data.

U.S. Census Bureau Facts & Features: Mother's Day, 2002
http://www.census.gov/Press-Release/www/2002/
cb02ff08.html
Brief census data on motherhood as well as links to more-detailed information.

U.S. Census Bureau Facts & Features: Father's Day, 2002
http://www.census.gov/Press-Release/www/2002/
cb02ff09.html
Brief census data on fatherhood as well as links to more-detailed information.

U.S. Census Bureau Facts & Features: Grandparent's Day, 2001
http://www.census.gov/Press-Release/www/2001/
cb01fff12.html
Brief census data on grandparents as well as links to more-detailed information.

FamilyFun
http://family.go.com
This Disney-sponsored site is a resource center for busy families. Can you guess who the primary user of this site might be?

GCSocWeb: Resources: Family
http://web.gc.cuny.edu/dept/socio/resource/family/
index.htm
This site contains gateways to a number of sites with information relative to a study of the family and children.

UNICEF: The State of the World's Children 1996
http://www.unicef.org/sowc96/contents.htm?477,233
This site's data are not current, yet the site contains important material related to the welfare of children in a global context.

The Administration for Children and Families
http://www.acf.dhhs.gov/programs/cse/extinf.htm
This is the U.S. child support enforcement site. From here, you can learn about government-sponsored efforts to protect children.

Companion Web Site for This Book

Virtual Society: The Wadsworth Sociology Resource Center
Visit **http://sociology.wadsworth.com** and click on the page for Kendall, *Sociology in Our Times: The Essentials*, Fourth Edition, to access a wide range of enrichment material to aid in your study of sociology. Click on the Student Resources section of the web site. Next, select from the pull-down menu the chapter that you are presently studying. Among the useful options for self-study are chapter objectives, flashcards, practical study tips, and practice tests for each chapter.

MicroCase Online
From the Virtual Society home page, click on MicroCase Online to access book-specific MicroCase exercises that allow you to further explore sociological issues and principles presented in the text.

InfoTrac College Edition
Another unique option available to you at the Student Resources section of the companion web site is InfoTrac College Edition, an online library with access to hundreds of scholarly and popular periodicals. Below are suggested search terms for this chapter. Results from these and other searches are found at the site.

- Search keywords: *teen pregnancy*. Locate articles that discuss the incidence of teen pregnancy in the United States. From the articles, can you determine if the rate of teen pregnancy is changing? What are some possible explanations?
- Search keywords: *domestic violence*. Domestic violence is a social problem that has been thoroughly researched in the United States. Do any of the articles that you found address domestic violence in other countries?

Virtual Explorations CD-ROM
Go to the Virtual Explorations CD-ROM to begin the interactive exercise for this chapter. This Virtual Exploration will introduce you to some of the exciting resources for sociology on the World Wide Web. You will be guided through an exercise that employs related web sites on families and intimate relationships. Answer the questions, and e-mail your responses to your instructor.

Sean Kane/SIS

Education and Religion

12

An Overview of Education and Religion

Sociological Perspectives on Education
Functionalist Perspectives on Education
Conflict Perspectives on Education
Symbolic Interactionist Perspectives on Education

Problems in Education
Unequal Funding of Public Schools
School Violence
Dropping Out
Racial Segregation and Resegregation
Class, Race, and Social Reproduction in Higher Education

Religion in Historical Perspective
Religion and the Meaning of Life
Religion and Scientific Explanations

Sociological Perspectives on Religion
Functionalist Perspectives on Religion
Conflict Perspectives on Religion
Symbolic Interactionist Perspectives on Religion

Types of Religious Organization
Ecclesia
The Church–Sect Typology
Cults

Trends in Religion in the United States

Education and Religion in the Future

I am a fundamentalist in the sense that I believe that the Bible is the inerrant word of God. . . . I am a student of the Bible as well as I am a student of biology, and my responsibility is to teach the truth in the classroom, and I believe that there is a major deception occurring in our public education system today, and that is the myth that evolutionism is a scientific theory that can be substantiated with scientific evidence. I have found that it is not. . . .

I also believe there is a conspiracy to cover up what I am attempting to expose, and that is we do have a state-sponsored religion, it's called the religion of humanism with evolutionism being the primary tenet of it, and it's being promoted in the public classroom despite the Constitution that prohibits the establishment of a particular religious view. Now of course, it comes out as if I am attempting to promote my religious view, but I believe that's a smoke screen. I believe that it is an attempt to divert the public from the real issue.

And the real issue is: Is evolutionism science or is it a religious belief system?

—John Peloza, a biology teacher at a Mission Viejo, California, public high school, explaining why he believes that he has the right to teach the theory of creationism (a belief in the divine origins of the universe and human beings based on a literal interpretation of the Bible) in addition to the theory of evolution (PBS, 1992a)

What's at stake here, it seems to me, is what kind of society are we going to have. Are we going to let religious fundamentalists of any ilk, I don't care if they are Muslim or Christian or Jewish or what they are, are we going to let religious fundamentalists of any ilk make public policy based on their religion?

—Jim Corbett, a journalism advisor at the same high school, stating why he disagrees with Peloza (PBS, 1992a)

This argument is only one in a lengthy history of debates about the appropriate relationship between public education and religion in the United States. Recently, the issue of having the words "under God" in the Pledge of Allegiance that is recited by teachers and students in public classrooms has been the topic of a protracted debate that is likely to reach the United States Supreme Court.

For many years, controversies have arisen periodically over such topics as the teaching of creationism versus evolutionism, moral education, sex education, school prayer, and the subject matter of textbooks and library books. More than seventy years ago, for example, evolutionism versus creationism was hotly debated in the famous "Scopes monkey trial," so named because of Charles Darwin's assertion that human beings had evolved from lower primates. In this case, John Thomas Scopes, a substitute high school biology teacher in Tennessee, was found guilty of teaching evolution, which denied the "divine creation of man as taught in the Bible." Although an appeals court later overturned Scopes's conviction and $100 fine (on the grounds that the fine was excessive), teaching evolution in Tennessee's public schools remained illegal until 1967 (Chalfant, Beckley, and Palmer, 1994). By contrast, recent U.S. Supreme Court rulings have looked unfavorably on the teaching of creationism in public schools, based on a provision in the Constitution

that requires a "wall of separation" between church (religion) and state (government) (Albanese, 1992). Initially, this wall of separation was erected to protect religion from the state, not the state from religion (Carter, 1994).

Who is to decide what should be taught in U.S. public schools? What is the purpose of education? Of religion? In this chapter, we examine education and religion, two social institutions that have certain commonalities both as institutions and as objects of sociological inquiry. Before reading on, test your knowledge about the impact religion has had on U.S. education by taking the quiz in Box 12.1.

Questions and Issues

Chapter Focus Question: Why are education and religion powerful and influential forces in contemporary society?

How might education become a self-fulfilling prophecy for some people?

What unique functions might religion serve that no other institution can fulfill?

How do religious groups differ in organizational structure?

What actions could be taken to resolve debates over religious values and public education?

Box 12.1 Sociology and Everyday Life

How Much Do You Know About the Impact of Religion on U.S. Education?

True	False	
T	F	1. The Constitution of the United States originally specified that religion should be taught in the public schools.
T	F	2. Virtually all sociologists have advocated the separation of moral teaching from academic subject matter.
T	F	3. The federal government has limited control over how funds are spent by school districts because most of the money comes from the state and local levels.
T	F	4. Parochial schools have decreased in enrollment as interest in religion has waned in the United States.
T	F	5. The number of children from religious backgrounds other than Christianity and Judaism has grown steadily in public schools over the past three decades.
T	F	6. Debates over textbook content focus only on elementary education because of the vulnerability of young children.
T	F	7. More parents are instructing their own children through home schooling because of their concerns about what public schools are (or are not) teaching their children.
T	F	8. The U.S. Congress has the ultimate authority over whether religious education can be included in public school curricula.

Answers on page 378.

An Overview of Education and Religion

Education and religion are powerful and influential forces in contemporary societies. Both institutions impart values, beliefs, and knowledge considered essential to the social reproduction of individual personalities and entire cultures (Bourdieu and Passeron, 1990). Education and religion both grapple with issues of societal stability and social change, reflecting society even as they attempt to shape it. Education and religion also share certain commonalities as objects of sociological study; for example, both are socializing institutions. Whereas early socialization is primarily informal and takes place within our families and friendship networks, as we grow older, socialization passes to the more formalized organizations created for the specific purposes of education and religion.

Areas of sociological inquiry that specifically focus on those institutions are (1) the *sociology of education*, which primarily examines formal education or schooling in industrial societies, and (2) the *sociology of religion*, which focuses on religious groups and organizations, on the behavior of individuals within those groups, and on ways in which religion is intertwined with other social institutions (Roberts, 1995). Let's start our examination by looking at sociological perspectives on education.

✓Checkpoints

Checkpoint 12.1

1. Why are sociologists interested in studying education and religion as social institutions?

Box 12.1

Answers to the Sociology Quiz on Religion and Education

1. **False.** Due to the diversity of religious backgrounds of the early settlers, no mention of religion was made in the original Constitution. Even the sole provision that currently exists (the establishment clause of the First Amendment) does not speak directly of the issue of religious learning in public education.

2. **False.** Obviously, contemporary sociologists hold strong beliefs and opinions on many subjects; however, most of them do not think that it is their role to advocate specific stances on a topic. Early sociologists were less inclined to believe that they had to be "value-free." For example, Durkheim strongly advocated that education should have a moral component and that schools had a responsibility to perpetuate society by teaching a commitment to the common morality.

3. **True.** Most public school revenue comes from local funding through property taxes and state funding from a variety of sources, including sales taxes, personal income taxes, and, in some states, oil revenues or proceeds from lotteries.

4. **False.** Just the opposite has happened. As parents have felt that children were not receiving the type of education they desired in public schools, parochial schools have flourished. Christian schools have grown to over five thousand in number; Jewish parochial schools have also grown rapidly over the past decade.

5. **True.** Although about 86 percent of those age eighteen and over in the forty-eight contiguous states of the United States describe their religion as some Christian denomination, there has still been a significant increase in those who either adhere to no religion (7.5 percent) or who are Jewish, Muslim/Islamic, Unitarian–Universalist, Buddhist, or Hindu.

6. **False.** Attempts to remove textbooks occur at all levels of schooling. A recent case involved the removal of Chaucer's "The Miller's Tale" and Aristophanes' *Lysistrata* from a high school curriculum.

7. **True.** Some parents choose home schooling for religious reasons; others embrace it for secular reasons, including fear for their children's safety and concerns about the quality of public schools.

8. **False.** Ultimately, issues relating to the separation of church and state, including religious instruction in public schools, are constitutional issues that are decided by the U.S. Supreme Court in the absence of a constitutional amendment.

Sources: Based on Ballantine, 2000; Gibbs, 1994; Greenberg, 1999; C. Johnson, 1994; Kosmin and Lachman, 1993; and Roof, 1993.

Sociological Perspectives on Education

Education is the social institution responsible for the systematic transmission of knowledge, skills, and cultural values within a formally organized structure. In all societies, people must acquire certain knowledge and skills in order to survive. In less-developed societies, these skills might include hunting, gathering, fishing, farming, and self-preservation. In contemporary, developed societies, knowledge and skills are often related to the requirements of a highly competitive job market.

Sociologists have divergent perspectives on the purpose of education in contemporary society. Functionalists suggest that education contributes to the maintenance of society and provides people with

Figure 12.1 Manifest and Latent Functions of Education

Manifest functions—open, stated, and intended goals or consequences of activities within an organization or institution. In education, these are:

Latent functions—hidden, unstated, and sometimes unintended consequences of activities within an organization. In education, these include:

- socialization
- transmission of culture
- social control
- social placement
- change and innovation

- matchmaking and production of social networks
- restricting some activities
- creation of a generation gap

an opportunity for self-enhancement and upward social mobility. Conflict theorists argue that education perpetuates social inequality and benefits the dominant class at the expense of all others. Symbolic interactionists focus on classroom dynamics and the effect of self-concept on grades and aspirations.

Functionalist Perspectives on Education

Functionalists view education as one of the most important components of society. According to Durkheim, education is the "influence exercised by adult generations on those that are not yet ready for social life" (Durkheim, 1956: 28). Durkheim asserted that moral values are the foundation of a cohesive social order and that schools have the responsibility of teaching a commitment to the common morality.

From this perspective, students must be taught to put the group's needs ahead of their individual desires and aspirations. Contemporary functionalists suggest that education is responsible for teaching U.S. values. According to sociologist Amitai Etzioni (1994: 258–259),

We ought to teach those values Americans share, for example, that the dignity of all persons ought to be respected, that tolerance is a virtue and discrimination abhorrent, that peaceful resolution of conflicts is superior to violence, that . . . truth telling is morally superior to lying, that democratic government is morally superior to totalitarianism and authoritarianism, that one ought to give a day's work for a day's pay, that saving for one's own and one's country's future is better than squandering one's income and relying on others to attend to one's future needs.

Etzioni suggests that "shared" values should be transmitted by schools from kindergarten through college. However, not all analysts agree on what those shared values should be or what functions education should serve in contemporary societies. In analyzing the values and functions of education, sociologists using a functionalist framework distinguish between manifest and latent functions. Manifest functions and latent functions are compared in Figure 12.1.

Education the social institution responsible for the systematic transmission of knowledge, skills, and cultural values within a formally organized structure.

Larry Martin/Little Big Horn College

Community colleges, especially those located on reservations, have increased access to higher education for some Native Americans. However, Native Americans as a category remain far below median levels of educational attainment in the United States.

MANIFEST FUNCTIONS OF EDUCATION

Some functions of education are *manifest functions*, previously defined as the open, stated, and intended goals or consequences of activities within an organization or institution. Examples of manifest functions in education include teaching specific subjects, such as science, mathematics, reading, history, and English. Education serves five major manifest functions in society:

1. *Socialization.* From kindergarten through college, schools teach students the student role, specific academic subjects, and political socialization. In primary and secondary schools, students are taught specific subject matter appropriate to their age, skill level, and previous educational experience. At the college level, students focus on more detailed knowledge of subjects they previously have studied while also being exposed to new areas of study and research.

2. *Transmission of culture.* Schools transmit cultural norms and values to each new genera-

tion and play an active part in the process of assimilation, whereby recent immigrants learn dominant cultural values, attitudes, and behavior so that they can be productive members of society.

3. *Social control.* Schools are responsible for teaching values such as discipline, respect, obedience, punctuality, and perseverance. Schools teach conformity by encouraging young people to be good students, conscientious future workers, and law-abiding citizens.

4. *Social placement.* Schools are responsible for identifying the most qualified people to fill available positions in society. As a result, students are channeled into programs based on individual ability and academic achievement. Graduates receive the appropriate credentials for entry into the paid labor force.

5. *Change and innovation.* Schools are a source of change and innovation. As student populations change over time, new programs are introduced to meet societal needs; for example, sex education, drug education, and multicultural studies have been implemented in some schools to help students learn about pressing social issues. Innovation in the form of new knowledge is required of colleges and universities. Faculty members are encouraged, and sometimes required, to engage in research and share the results with students, colleagues, and others.

LATENT FUNCTIONS OF EDUCATION
In addition to manifest functions, all social institutions, including education, have some *latent functions*, which, as you will recall, are the hidden, unstated, and sometimes unintended consequences of activities within an organization or institution. Education serves at least three latent functions:

1. *Restricting some activities.* Early in the twentieth century, all states passed *mandatory education laws* that require children to attend school until they reach a specified age or until they complete a minimum level of formal education. Out of these laws grew one latent function of education, which is to keep students off the streets and out of the full-time job market for a number of years, thus helping keep unemployment within reasonable bounds (Braverman, 1974).

Donna Day/Stone-Getty Images

What values are these schoolchildren being taught? Is there a consensus about what today's schools should teach? Why or why not?

2. *Matchmaking and production of social networks.* Because schools bring together people of similar ages, social class, and race/ethnicity, young people often meet future marriage partners and develop social networks that may last for many years.

3. *Creation of a generation gap.* Students may learn information in school that contradicts beliefs held by their parents or their religion. When education conflicts with parental attitudes and beliefs, a generation gap is created if students embrace the newly acquired perspective.

Functionalists acknowledge that education has certain dysfunctions. Some analysts argue that U.S. education is not promoting the high-level skills in

reading, writing, science, and mathematics that are needed in the workplace and the global economy. For example, mathematics education in the United States does not compare favorably with that found in many other industrialized countries. In the 1996 Third International Math and Science Study—a comprehensive study of science and math achievement by students in forty-one countries—U.S. nine-year-olds scored lower on the math exam than did students in at least eleven other nations, including Singapore, Korea, Japan, and Hong Kong (*Time*, 1997: 67). However, analysts do not agree on what these score differences mean. For example, Japanese students may outperform their U.S. counterparts because their schools are more structured and teachers focus on drill and practice (Celis, 1994). Educators David C. Berliner and Bruce J. Biddle (1995) believe that data on cross-cultural differences in educational attainment actually involve comparisons of "apples" and "oranges": They found that the International Math and Science Study compares the achievement of eighth-grade Japanese students who have already taken algebra with the achievement of U.S. students who typically take such courses a year or two later. Among U.S. students who had already completed an algebra course, most did at least as well as their Japanese counterparts on the mathematics exam (Berliner and Biddle, 1995). Are U.S. schools dysfunctional if students are not taking algebra courses sooner? According to some functionalists, limited academic demands placed upon students are a dysfunctional aspect of public schools in this country.

Conflict Perspectives on Education

In contrast with the functionalist perspective, conflict theorists argue that schools often perpetuate class, racial–ethnic, and gender inequalities as some groups seek to maintain their privileged position at the expense of others (Apple, 1980; Ballantine, 2000).

CULTURAL CAPITAL AND CLASS REPRODUCTION Although many factors—including intelligence, family income, motivation, and previous

Reflections

Reflections 12.1

Although ethics and morality are generally not courses that are taught in the schools, most children learn about these topics while attending school. Is this a manifest or a latent function of education? Might the answer vary depending on whether it is a public or a private school?

achievement—are important in determining how much education a person will attain, conflict theorists argue that access to quality education is closely related to social class. From this approach, education is a vehicle for reproducing existing class relationships. According to the French sociologist Pierre Bourdieu, the school legitimates and reinforces the social elites by engaging in specific practices that uphold the patterns of behavior and the attitudes of the dominant class. Bourdieu asserts that students from diverse class backgrounds come to school with different amounts of *cultural capital—social assets that include values, beliefs, attitudes, and competencies in language and culture* (Bourdieu and Passeron, 1990). Cultural capital involves "proper" attitudes toward education, socially approved dress and manners, and knowledge about books, art, music, and other forms of high and popular culture.

Middle- and upper-income parents endow their children with more cultural capital than do working-class and poverty-level parents. Because cultural capital is essential for acquiring an education, children with less cultural capital have fewer opportunities to succeed in school. For example, standardized tests that are used to group students by ability and to assign them to classes often measure students' cultural capital rather than their "natural" intelligence or aptitude. Thus, a circular effect occurs: Students with dominant cultural values are more highly rewarded by the educational system. In turn, the educational system teaches and reinforces those values that sustain the elite's position in society.

TRACKING AND SOCIAL INEQUALITY
Closely linked to the issue of cultural capital is how tracking in schools is related to social inequality. Conflict theorists who study ability grouping focus on how the process of tracking affects students' educational performance. Ability grouping is often used in elementary schools, based on the assumption that it is easier to teach students with similar abilities. However, class-based factors also affect which children are most likely to be placed in "high," "middle," or "low" groups, often referred to by such innocuous terms as "Blue Birds," "Red Birds," and "Yellow Birds."

In middle school/junior high and high school, most students experience *tracking—the assignment*

Syracuse Newspapers/Dick Blume/The Image Works

Computers make it possible for students with a disability to gain new educational opportunities and enhance their cultural capital. Without a special reading screen, this student with a severe vision problem would be unable to read, write, or complete his homework.

of students to specific courses and educational programs based on their test scores, previous grades, or both. Ruben Navarrette, Jr. (1997: 274–275), tells us about his experience with tracking:

> One fateful day, in the second grade, my teacher decided to teach her class more efficiently by dividing it into six groups of five students each. Each group was assigned a geometric symbol to differentiate it from the others. There were the Circles. There were the Squares. There were the Triangles and Rectangles.
>
> I remember being a Hexagon. . . . I remember something else, an odd coincidence. The Hexagons were the smartest kids in the class. These distinctions are not lost on a child of seven. . . . Even in the second grade, my classmates and I knew who was smarter than whom. And on the day on which we were assigned our respective shapes, we knew that our teacher knew, too. As Hexagons, we would wait for her to call on us, then answer by hurrying to her with books and pencils in hand. We sat

around a table in our "reading group," chattering excitedly to one another and basking in the intoxication of positive learning. We did not notice, did not care to notice, over our shoulders, the frustrated looks on the faces of Circles and Squares and Triangles who sat quietly at their desks, doodling on scratch paper or mumbling to one another. We knew also that, along with our geometric shapes, our books were different and that each group had different amounts of work to do. . . . The Circles had the easiest books and were assigned to read only a few pages at a time. . . . Not surprisingly, the Hexagons had the most difficult books of all, those with the biggest words and the fewest pictures, and we were expected to read the most pages.

The result of all of this education by separation was exactly what the teacher had imagined that it would be: Students could, and did, learn at their own pace without being encumbered by one another. Some learned faster than others. Some, I realized only [later], did not learn at all.

Numerous studies have found that ability grouping and tracking affect students' academic achievements and career choices (Oakes, 1985). Moreover, some social scientists believe that tracking is one of the most obvious mechanisms through which poor and minority students receive a diluted academic program, making it much more likely that they will fall even further behind their white, middle-class counterparts (see Miller, 1995). Instead of enhancing school performance, tracking systems may result in students dropping out of school or ending up in "dead-end" situations because they have not taken the courses required to go to college.

THE HIDDEN CURRICULUM According to conflict theorists, the *hidden curriculum* **is the transmission of cultural values and attitudes, such as conformity and obedience to authority, through implied demands found in the rules, routines, and regulations of schools** (Snyder, 1971).

Social Class and the Hidden Curriculum Although students from all social classes are subjected to the hidden curriculum, working-class and poverty-level students may be affected the most adversely (Cookson and Hodges Persell, 1985; Polakow, 1993; Ballantine, 2000). When teachers are from a higher class background than their students, they tend to use more structure in the classroom and to have lower expectations for students'

academic achievement (see Alexander, Entwisle, and Thompson, 1987). In a study of five elementary schools located in different communities, significant differences were found in the manner in which knowledge was transmitted to students even though the curriculum was organized similarly (Anyon, 1980). Schools for working-class students emphasize procedures and rote memorization without much decision making, choice, or explanation of why something is done a particular way. Schools for middle-class students stress the processes (such as figuring and decision making) involved in getting the right answer. Schools for affluent students focus on creative activities in which students express their own ideas and apply them to the subject under consideration. Schools for students from elite families work to develop students' analytical powers and critical thinking skills, applying abstract principles to problem solving.

Through the hidden curriculum, schools make working-class and poverty-level students aware that they will be expected to take orders from others, arrive at work punctually, follow bureaucratic rules, and experience high levels of boredom without complaining (Ballantine, 2000). Over time, these students may be disqualified from higher education and barred from obtaining the credentials necessary for well-paid occupations and professions (Bowles and Gintis, 1976). Educational credentials are extremely important in societies that emphasize *credentialism* **— a process of social selection in which class advantage and social status are linked to the possession of academic qualifications** (Collins,

---

Cultural capital Pierre Bourdieu's term for people's social assets, including values, beliefs, attitudes, and competencies in language and culture.

Tracking the assignment of students to specific courses and educational programs based on their test scores, previous grades, or both.

Hidden curriculum the transmission of cultural values and attitudes, such as conformity and obedience to authority, through implied demands found in the rules, routines, and regulations of schools.

Credentialism a process of social selection in which class advantage and social status are linked to the possession of academic qualifications.

1979; Marshall, 1998). Credentialism is closely related to *meritocracy*—previously defined as a social system in which status is assumed to be acquired through individual ability and effort (Young, 1994/1958). Persons who acquire the appropriate credentials for a job are assumed to have gained the position through what they know, not who they are or who they know. According to conflict theorists, the hidden curriculum determines in advance that the most valued credentials will primarily stay in the hands of the elites, so the United States is not actually as meritocratic as some might claim.

Gender Bias and the Hidden Curriculum According to conflict theorists, gender bias is embedded in both the formal and the hidden curricula of schools. Although most females in the United States have a greater opportunity for education than those living in developing nations, their educational opportunities are not equal to those of males in their social class (see AAUW, 1995; Orenstein, 1995). Through reading materials, classroom activities, and treatment by teachers and peers, female students learn that they are less important than male students. Over time, differential treatment undermines females' self-esteem and discourages them from taking certain courses, such as math and science, that are usually dominated by male teachers and students (Raffalli, 1994). Although special programs have been developed in some schools to encourage girls to pursue math and science as areas of study, girls and young women still take fewer courses in these areas or drop them because they are uninteresting.

Symbolic Interactionist Perspectives on Education

Unlike functionalist analysts, who focus on the functions and dysfunctions of education, and conflict theorists, who focus on the relationship between education and inequality, symbolic interactionists focus on classroom communication patterns and educational practices, such as labeling, that affect students' self-concept and aspirations.

LABELING AND THE SELF-FULFILLING PROPHECY Chapter 6 explains that *labeling* is the process whereby a person is identified by others as possessing a specific characteristic or exhibiting a certain pattern of behavior (such as being deviant). According to symbolic interactionists, the process of labeling is directly related to the power and status of those persons who do the labeling and those who are being labeled. In schools, teachers and administrators are empowered to label children in various ways, including grades, written comments on deportment (classroom behavior), and placement in classes. For example, based on standardized test scores or classroom performance, educators label some children as "special ed" or low achievers, whereas others are labeled as average or "gifted and talented." For some students, labeling amounts to a *self-fulfilling prophecy*—previously defined as an unsubstantiated belief or prediction resulting in behavior that makes the originally false belief come true (Merton, 1968). A classic form of labeling and the self-fulfilling prophecy occurs through the use of IQ (intelligence quotient) tests, which claim to measure a person's inherent intelligence, apart from any family or school influences on the individual. In many school systems, IQ tests are used as one criterion in determining student placement in classes and ability groups.

USING LABELING THEORY TO EXAMINE THE IQ DEBATE The relationship between IQ testing and labeling theory has been of special interest to sociologists. In the 1960s, two social scientists conducted an experiment in an elementary school during which they intentionally misinformed teachers about the intelligence test scores of students in their classes (Rosenthal and Jacobson, 1968). Despite the fact that the students were randomly selected for the study and had no measurable differences in intelligence, the researchers informed the teachers that some of the students had extremely high IQ test scores, whereas others had average to below-average scores. As the researchers observed, the teachers began to teach "exceptional" students in a different manner from other students. In turn, the "exceptional" students began to outperform their "average" peers and to excel in their classwork. This study called attention to the labeling effect of IQ scores.

However, experiments such as this also raise other important issues: What if a teacher (as a result of stereotypes based on the relationship between IQ and race) believes that some students of color are less capable of learning? Will that teacher (often without realizing it) treat such students as if they are incapable of learning? In their controversial book *The Bell Curve: Intelligence and Class Structure in American Life*, Richard J. Herrnstein and Charles Murray (1994) argue that intelligence is genetically inherited and that people cannot be "smarter" than they are born to be, regardless of their environment or education. According to Herrnstein and Murray, certain racial–ethnic groups differ in average IQ and are likely to differ in "intelligence genes" as well. For example, they point out that, on average, people living in Asia score higher on IQ tests than white Americans and that African Americans score 15 points lower on average than white Americans. Based on an all-white sample, the authors also concluded that low intelligence leads to social pathology, such as high rates of crime, dropping out of school, and winding up poor. In contrast, high intelligence typically leads to success, and family background plays only a secondary role.

Many scholars disagree with Herrnstein and Murray's research methods and conclusions. Two major flaws found in their approach were as follows: (1) the authors used biased statistics that underestimate the impact of hard-to-measure factors such as family background, and (2) they used scores from the Armed Forces Qualification Test, an exam that depends on the amount of schooling that people have completed. Thus, what the authors claim is immutable intelligence is actually acquired skills (Weinstein, 1997). Despite this refutation, the idea of inherited mental inferiority tends to take on a life of its own when people want to believe that such differences exist (Duster, 1995; Hauser, 1995; Taylor, 1995). According to researchers, many African American and Mexican American children are placed in special education classes on the basis of IQ scores when the students were not fluent in English and thus could not understand the directions given for the test. Moreover, when children are labeled as "special ed" students or as being *learning disabled*, these terms are social constructions that may lead to stigmatization and become a self-fulfilling prophecy (Carrier, 1986; Coles, 1987).

Labeling students based on IQ scores has been an issue for many decades. Immigrants from southern and eastern Europe—particularly from Italy, Poland, and Russia—who arrived in this country at the beginning of the twentieth century had lower IQ scores on average than did northern European immigrants who had arrived earlier from nations such as Great Britain. For many of the white ethnic students, IQ testing became a self-fulfilling prophecy: Teachers did not expect them to do as well as children from a northern European (WASP) family background and thus did not encourage them or give them an opportunity to overcome language barriers or other educational obstacles. Although many students persisted and achieved an education, the possibility that differences in IQ scores could be attributed to linguistic, cultural, and educational biases in the tests was largely ignored (Feagin and Feagin, 2003). Debates over the possible intellectual inferiority of white ethnic groups are unthinkable today, but arguments pertaining to African Americans and IQ continue to surface.

A self-fulfilling prophecy can also result from labeling students as gifted. Gifted students are considered to be those with above-average intellectual ability, academic aptitude, creative or productive thinking, or leadership skills (Ballantine, 2000). When some students are labeled as better than others, they may achieve at a higher level because of the label. Ironically, such labeling may also result in discrimination against these students: Boys and girls who are high in achievement may be the victims of *anti-intellectualism*—hostility or opposition toward persons assumed to have great mental ability or toward subject matter believed to necessitate significant intellectual ability or knowledge for its comprehension. Advocates of the teaching of creationism, textbook censorship, and organized prayer in public schools have been charged with anti-intellectualism. Removing evolution from the curriculum in order to teach creationism has been described as tantamount to

✓Checkpoints

Checkpoint 12.2

2. How do manifest and latent functions of education differ?
3. Why do conflict theorists believe that cultural capital is an important factor in class reproduction?
4. How is tracking related to social inequality?
5. What is the relationship among race, class, gender, and the hidden curriculum?
6. What is a key symbolic interactionist perspective on education?

"Rich" schools and "poor" schools are readily identifiable by their buildings and equipment. What are the long-term social consequences of unequal funding for schools?

"educational deprivation" for children (Scott, 1992). School prayer and censorship of textbooks continue to be topics of protracted debate in education and religion because they are issues with many macrolevel and microlevel implications.

Problems in Education

Education in kindergarten through high school is a microcosm of many of the issues and problems facing the United States. Today, there are almost 15,000 school districts in the United States in what is probably the most decentralized system of public education in any high-income, developed nation of the world. One of the biggest problems in public education today results from unequal funding.

Unequal Funding of Public Schools

Why does unequal funding in public education exist? Most educational funds come from state legislative appropriations and local property taxes. State and local governments contribute about 47 percent *each* toward educational expenses, and the federal government pays the remaining 6 percent, largely for special programs for students who are disadvantaged (e.g., the Head Start program) or have disabilities (Bagby, 1997). Although public school spending increased by 28 percent between

1991 and 1996, there was an 8.4-percent increase in enrollment during that same period, and special education programs received an increasing share of school budgets (*New York Times*, 1997). According to one economist, "The largest share of the new money has gone to special programs, not to regular education. We have shifted much of the . . . responsibility for children with disabilities from private charities and institutions to the public schools" (qtd. in *New York Times*, 1997: A15).

Per-capita spending on public and secondary education varies widely from state to state. In part, this is because the local property-tax base has been eroding in central cities as major industries have relocated or gone out of business. Many middle- and upper-income families have moved to suburban areas with their own property-tax base so their children can attend relatively new schools equipped with the latest textbooks and state-of-the-art computers—advantages that schools in central cities and poverty-ridden rural areas lack (see Kozol, 1991; Ballantine, 2000).

In recent years, some states have been held accountable by the courts for unequal funding that results in "rich" and "poor" school districts. However, proposals to improve educational funding have been relatively limited in scope. Recently, voucher systems—which would allow students and their families to spend a specified sum of government money to purchase education at the school of their choice—have gained many supporters who believe that this system might be an answer to some of the problems that plague public education today; however, opponents strongly disagree with this approach, believing that it would harm public education instead of helping it (see Box 12.2).

School Violence

Violence and fear of violence continue to be problems in schools throughout the United States. In the 1990s, violent acts resulted in numerous deaths in schools across the nation. Schools in communities such as Pearl, Mississippi, West Paducah, Kentucky, Jonesboro, Arkansas, Springfield, Oregon, and Littleton, Colorado, witnessed a series of killings in schools by students that shocked people across the world.

Throughout the nation, there has been a call for greater school security. As a result, more students are in an academic environment that is similar to a prison. However, many education analysts believe that technology and increased law enforcement should not be the primary tools for achieving school discipline and ending violence and crime on school campuses. Education scholar John Devine (1996) emphasizes that schools must be understood as part of a larger, tumultuous society: Violence in schools will not end until guns are eliminated from U.S. society. Given the fact that some schools do not motivate students to learn and are arranged much like a prison, is it any wonder that school dropout rates are high?

Dropping Out

Although there has been a decrease in the overall school dropout rate over the past two decades, about 10 percent of people between the ages of fourteen and twenty-four have left school before earning a high school diploma. However, ethnic and class differences are significant in dropout rates. For example, Latinos/as (Hispanics) have the highest dropout rate (34.4 percent), followed by African Americans (17.1 percent) and non-Hispanic whites (13.7 percent) (U.S. Census Bureau, 2001g). The dropout rate also varies by region. In New York City, for example, African American, Puerto Rican, and Italian American youths have the highest dropout rates (Pinderhughes, 1997).

Reflections

Reflections 12.2

What effect might a high dropout rate among low-income children in central-city and rural schools have on the nation as a whole?

Why are Latinos/as more likely to drop out of high school than are other racial and ethnic groups? First, the category of "Hispanic" or "Latino/a" incorporates a wide diversity of young people—including those who trace their origins to Mexico, Puerto Rico, Haiti, and countries in Central and South America—who may leave school for a variety of reasons. Second, some students may drop out of school partly because their teachers have labeled them as "troublemakers." Some students have been repeatedly expelled from school before they actually become "dropouts." Third, some critics disagree with the use of the term "dropout," believing that Latinos/as have been excluded from meaningful academic programs and discouraged from gaining the education necessary to move into the middle-income categories, and thus have been pushed out of the educational system rather than dropping out of it.

Students who drop out of school may be skeptical about the value of school even while they are still attending because they believe that school will not increase their job opportunities. Upon leaving school, many dropouts have high hopes of making some money and enjoying their newfound freedom. However, these feelings often turn to disappointment when they find that few jobs are available and that they do not meet the minimum educational requirements for any "good" jobs that exist (Pinderhughes, 1997).

Racial Segregation and Resegregation

In many areas of the United States, schools remain racially segregated or have become resegregated after earlier attempts at integration failed. In 1954 the U.S. Supreme Court ruled (in *Brown v. The Board of Education of Topeka, Kansas*) that "separate but equal" segregated schools are unconstitutional because they are inherently unequal. However, four

Box 12.2 Sociology and Social Policy
The Ongoing Debate Over School Vouchers

As discussed in Chapter 2 ("Culture"), one of the core American values is equality—at least equality of *opportunity,* an assumed equal opportunity to achieve success. However, with problems in public education such as are discussed in this chapter, does every child really have that equal opportunity? If it is at least possible that a child could get a better education at a private school than at an inadequately funded inner-city school, should the government provide that child's family with the resources to send the child to a private school in order to have the same opportunity as a child from a wealthier family? Should the government provide that funding even if the private school teaches specific religious beliefs as part of the academic curriculum? Questions such as these have been of concern to social policy makers, educators, parents, and others in the United States for decades, and it appears that the controversy is far from being resolved. To understand the issues involved in the school voucher controversy, let's briefly look at the intended purpose of vouchers and at the arguments for and against voucher programs.

The original idea, first proposed in 1955, was to provide a voucher (equal in value to the average amount spent per student by the local school district) to the family of any student who left the public schools to attend a private school (Henderson, 1997). The private school could exchange the voucher for that amount of money and apply it to the student's tuition at the private school; the school district would save an equivalent amount of money as a result of the student not enrolling in the public schools. Over the years, variations on the original idea have been proposed, such as allowing vouchers to be used for transfers between public schools in the same school district, transfers from one public school district to another, transfers only to schools with no religious connections, and vouchers for use only by students from low-income families.

Although the legislatures in at least twenty-six states have rejected school voucher programs (*New York Times,* 2002a), school districts in a number of cities have adopted various types of school choice programs. Some of those plans authorize only transfers between public schools in the same district; other plans authorize the use of vouchers for transfers to private schools. For example, the Cleveland, Ohio, school district allows students to attend participating public or private schools of their parents' choosing—even public schools in adjacent districts—and provides tuition aid to the parents according to financial need. The tuition aid goes to the public or

decades later, racial segregation remains a fact of life in education.

Efforts to bring about *desegregation*—the abolition of legally sanctioned racial–ethnic segregation— or *integration*—the implementation of specific action to change the racial–ethnic and/or class composition of the student body—have failed in many districts throughout the country. Some school districts have bused students across town to achieve racial integration. Others have changed school attendance boundaries or introduced magnet schools with specialized programs such as science or the fine arts to change the racial–ethnic composition of schools. But school segregation does not exist in isolation. Racially segregated housing patterns are associated with the high rate of school segregation experienced by African American, Latina/o, and other students of color.

Ethnic enclaves of recent immigrants also may result in a concentration of students in a particular school where they have no opportunity to interact with children from other income levels or family backgrounds.

Even in more-integrated schools, resegregation often occurs at the classroom level (Mickelson and Smith, 1995). Because of past racial discrimination and current socioeconomic inequalities, many children of color are placed in lower-level courses and special education classes. At the same time, non-Latina/o white and Asian American students are more likely to be enrolled in high-achievement courses and programs for the gifted and talented.

Despite the difficulties associated with U.S. public schools, many people have not lost hope that schools can be improved. Some individuals private school in which the student is enrolled. In 1999–2000, 96 percent of the students participating in this program were enrolled in religiously affiliated schools. The program was challenged in court on the basis that it violated the constitutional separation of church and state.

In *Zelman v. Simmons-Harris,* a divided U.S. Supreme Court upheld the Cleveland program against the constitutional challenge, with the majority's opinion holding that the plan was adopted for the purpose of providing educational assistance to poor children and was neutral toward religion. Would the same result have been reached if the plan was for *all* children instead of only those from low-income families?

According to some analysts, the Court's decision shifted the battle regarding voucher programs from a legal argument based on a constitutional question to a political debate involving social policy and politics (Nagourney, 2002). Advocates of the Cleveland program argue that the Court's decision was good because children from lower-income families can now have a wider range of educational choices, a range similar to that currently available to children from middle-income families. They assert that competition for students' vouchers will improve public education, forcing school administrators and teachers to perform at a higher level and thereby producing greater competencies in students. By contrast, opponents of school vouchers argue that the Court's

decision could be the beginning of the end of public schools, that public education may be unable to withstand the loss of funds and the transfer of gifted students to private schools.

What are the major social policy issues surrounding voucher systems? Many people still strongly believe that using public tax dollars (even though they are technically given to the students and their parents rather than the schools) for funding private, religiously based schools violates the separation of church and state. Despite the Court's decision, they will continue to make that argument, whether on constitutional grounds or on the basis of public policy. Ultimately, from a social policy standpoint, the future of school vouchers may depend on the willingness (or unwillingness) of political leaders to authorize vouchers and the willingness of voters to pay for them.

How the social issues are framed may be significant in shaping the future of vouchers. If vouchers are issued only to the families of low-income children, will they be thought of as a welfare program, and thus be stigmatized? If they are available to all students, would they in effect be a tax subsidy for private schools, becoming more costly than the public school systems that currently exist? What effect would vouchers have on children's educational opportunities and the already struggling public schools? What are your thoughts on this important social policy issue?

have begun their own initiatives to make a difference in education (see Box 12.3).

Class, Race, and Social Reproduction in Higher Education

Even for students who complete high school, access to colleges and universities is determined not only by prior academic record but also by the ability to pay. Although public institutions such as community colleges and state colleges and universities are funded primarily by tax dollars, the cost of attending such institutions has increased dramatically over the past decade. Even with the lower cost of junior and community colleges, the enrollment of low-income students has dropped since the 1980s as a result of declining scholarship funds and the necessity for many students to

work full time or part time to finance their own education. By contrast, students from affluent families are more likely to attend prestigious state

✓ Checkpoints

Checkpoint 12.3

7. What are some of the biggest problems facing public education today?
8. How have schools sought to cope with the problem of school violence?
9. Why are ethnic and class differences significant in dropout rates?
10. How does college enrollment differ by race and ethnicity?
11. What sociological factors contribute to these differences?

Box 12.3 You Can Make a Difference
Breaking the Rules to Change a School!

Why should that be unique? There's nothing [the principal] is doing that couldn't be done everywhere else. Why do we have to depend on a single charismatic individual?
—Robert Slavin, education researcher at Johns Hopkins University, describing the recent work of Craig Spilman, principal of Canton Middle School in Baltimore (qtd. in Larson, 1997: 92)

Like many schools in a working-class neighborhood overlooking an industrial area of a city, Canton Middle School, located on Baltimore's Inner Harbor, looks like a fortress. There are grates over the windows of the austere brick building; the entrance has steel doors. Of the 750 students at Canton, 55 percent are white ethnics who trace their ancestry to Poland, Italy, or Greece. African Americans make up 35 percent of the student body; of the remainder, 5 percent are Latinos/as, and 5 percent are Native Americans. Most of the students come from poverty-level family backgrounds; 86 percent qualify for the federal government's free lunch program. Is this school another urban jungle? Not according to one journalist who recently visited Canton Middle School:

> With classes in session, the halls are quiet. Bright. The floors gleam. More important: the school has become an academic showcase, with test scores advancing year to year—all this within a broader school system described by its own interim chief executive officer as "academically bankrupt." (Larson, 1997: 92)

Why is Canton different from other schools? Analysts attribute the success of this school to the difference that one person can make. Craig Spilman, the principal, took over at a time when absenteeism was at an all-time high and test scores were at an all-time low. Today, this situation has been reversed. To

bring about change, Spilman hired teachers who were not currently employed in the Baltimore school system. He also abandoned traditional classroom models and adopted an "Expeditionary Learning" protocol in which children work in teams conducting field research aimed at producing a final product such as an exhibition or a presentation. Thus far, presentations have included the application of mathematics, physics, and historical research techniques to interior design plans for an old Baltimore cannery that is being converted to office space. Although Spilman is a school principal, other analysts suggest that many people could accomplish similar results—if they believe change is possible and are willing to work to bring it about (Larson, 1997).

For example, firefighters at a fire station on Chicago's South Side have served as friends and mentors to the schoolchildren who live in a nearby housing project. When children began skipping school to hang out at the firehouse, the firefighters started rewarding the children for perfect school attendance or good grades with prizes and small presents (Fedarko, 1997). Each child adopts a firefighter, who helps with homework and imparts lessons on how to behave—for example, the children are not allowed to swear, and their faces and hands must be clean (Fedarko, 1997).

Today, many college students also tutor low-income children or serve as mentors or informal "big brothers" or "big sisters" for the children. Some do their volunteer work through sororities, fraternities, and campus service organizations; others link up with church groups, schools, or community organizations such as those that teach literacy to adults.

What can *you* do to enhance someone's educational opportunities?

---

universities or private colleges, where the annual cost may be more than $30,000 per year.

Thus, the ability to pay for a college education reproduces the class system. The disproportionately low number of people of color enrolled in colleges is reflected in the educational achievement of people, aged 25 and over, as shown in the "Census Profiles" feature.

At all levels of formal education in the United States, some schools are operated by the govern-

ment (primarily by state and local governments), and others are run by various types of nongovernmental entities. In the private sector, some of the schools either were founded by or are operated by religious organizations. By way of example, some of the best-known private universities in this country were initially run by religious groups that sought to integrate the principles of education with the teaching of the religious and moral beliefs of their group. Consequently, there is an overlap between education and religion in a variety of academic institutions at all levels. To gain a better understanding of how sociologists systematically examine religion in society, let's first look at religion from a historical perspective.

Religion in Historical Perspective

Religion is a system of beliefs, symbols, and rituals, based on some sacred or supernatural realm, that guides human behavior, gives meaning to life, and unites believers into a community (Durkheim, 1995/1912). For many people, religious beliefs provide the answers for seemingly unanswerable questions about the meaning of life and death.

Religion and the Meaning of Life

Religion seeks to answer important questions such as why we exist, why people suffer and die, and what happens when we die. Whereas science and medicine typically rely on existing scientific evidence to respond to these questions, religion seeks to explain suffering, death, and injustice in the realm of the sacred. According to Emile Durkheim, *sacred* refers to those aspects of life that are extraordinary or supernatural—in other words, those things that are set apart as "holy." People feel a sense of awe, reverence, deep respect, or fear for that which is considered sacred. Across cultures and in different eras, many things have been considered sacred, including invisible gods, spirits, specific animals or trees, altars, crosses, holy books, and special words or songs that only the initiated could speak or sing (Collins, 1982). Those things that people do not set apart as sacred are referred to as *profane*—the everyday, secular or "worldly" aspects of life (Collins, 1982). Thus, whereas sacred beliefs are rooted in the holy or supernatural, secular beliefs have their foundation in scientific knowledge or everyday explanations. In the debate between creationists and evolutionists, for example, advocates of creationism view it as a belief founded in sacred (Biblical) teachings whereas advocates of evolutionism assert that their beliefs are based on provable scientific facts.

In addition to beliefs, religion also comprises symbols and rituals. According to the anthropologist Clifford Geertz (1966), religion is a set of cultural symbols that establishes powerful and pervasive

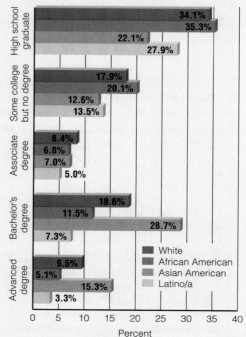

CENSUS PROFILES — Educational Achievement of Persons Aged 25 and Over

Census 2000 asked people to indicate the highest degree or level of schooling they had completed. Sixteen categories, ranging from "no schooling completed" to "doctorate degree," were set forth as responses; however, we are looking only at the categories of high school graduate and above:

What is the highest degree or level of school this person has COMPLETED? Mark [X] ONE box. *If currently enrolled, mark the previous grade or highest degree received.*

☐ HIGH SCHOOL GRADUATE — high school DIPLOMA or the equivalent *(for example: GED)*
☐ Some college credit, but less than 1 year
☐ 1 or more years of college, no degree
☐ Associate degree *(for example: AA, AS)*
☐ Bachelor's degree *(for example: BA, AB, BS)*
☐ Master's degree *(for example: MA, MS, MEng, MEd, MSW, MBA)*
☐ Professional degree *(for example: MD, DDS, DVM, LLB, JD)*
☐ Doctorate degree *(for example: PhD, EdD)*

As shown below, census data reflect that the highest levels of educational attainment are held by Asian Americans, followed by non-Hispanic white respondents. For these statistics to change significantly, greater educational opportunities and more-affordable higher education would need to be readily available to African Americans and Latinos/as, who historically have experienced racial discrimination and inadequately funded public schools with high dropout rates and low high school graduation rates.

Source: U.S. Census Bureau, 2001f.

AP/Wide World of Photos

Orthodox Jews at the Western Wall in Jerusalem—a wall that holds special significance for all Jews—express their faith in God and in the traditions of their ancestors.

moods and motivations to help people interpret the meaning of life and establish a direction for their behavior. People often act out their religious beliefs in the form of *rituals*—symbolic actions that represent religious meanings (McGuire, 2002). Rituals range from songs and prayers to offerings and sacrifices that worship or praise a supernatural being, an ideal, or a set of supernatural principles (Roberts, 1995). Rituals differ from everyday actions in that they have very strictly determined behavior. According to the sociologist Randall Collins

Reflections

Reflections 12.3

What is the relationship between education and religion in establishing and maintaining social values and goals in a society?

(1982: 34), "In rituals, it is the forms that count. Saying prayers, singing a hymn, performing a primitive sacrifice or a dance, marching in a procession, kneeling before an idol or making the sign of the cross—in these, the action must be done the right way."

CATEGORIES OF RELIGION   Although it is difficult to establish exactly when religious rituals first began, anthropologists have concluded that all known groups over the past 100,000 years have had some form of religion (Haviland, 1999). Religions have been classified into four main categories based on their dominant belief: simple supernaturalism, animism, theism, and transcendent idealism (McGee, 1975). In very simple preindustrial societies, religion often takes the form of *simple supernaturalism*—the belief that supernatural forces affect people's lives either positively or negatively. This type of religion does not acknowledge specific gods or supernatural spirits but focuses instead on impersonal forces that may exist in people or natural objects. By contrast, **animism is the belief that plants, animals, or other elements of the natural world are endowed with spirits or life forces having an impact on events in society.** Animism is identified with early hunting and gathering societies and with many Native American societies, in which everyday life was not separated from the elements of the natural world (Albanese, 1992).

The third category of religion is *theism*—a belief in a god or gods. Horticultural societies were among the first to practice *monotheism*—a belief in a single, supreme being or god who is responsible for significant events such as the creation of the world. Three of the major world religions—Christianity, Judaism, and Islam—are monotheistic. By contrast, Shinto and a number of indigenous religions of Africa are forms of *polytheism*—a belief in more than one god (see Table 12.1). The fourth category of religion, transcendent idealism, is *nontheistic* because it does not focus on worship of a god or gods. *Transcendent idealism* is a belief in sacred principles of thought and conduct. Principles such as truth, justice, affirmation of life, and tolerance for others are central tenets of transcendent idealists, who seek an elevated state of consciousness in which they can fulfill their true potential.

Table 12.1 Major World Religions

	Current Followers	Founder/Date	Beliefs
Christianity	1.7 billion	Jesus; 1st century C.E.	Jesus is the Son of God. Through good moral and religious behavior (and/or God's grace), people achieve eternal life with God.
Islam	1 billion	Muhammad; ca. 600 C.E.	Muhammad received the Qur'an (scriptures) from God. On Judgment Day, believers who have submitted to God's will, as revealed in the Qur'an, will go to an eternal Garden of Eden.
Hinduism	719 million	No specific founder; ca. 1500 B.C.E.	Brahma (creator), Vishnu (preserver), and Shiva (destroyer) are divine. Union with ultimate reality and escape from eternal reincarnation are achieved through yoga, adherence to scripture, and devotion.
Buddhism	309 million	Siddhartha Gautama; 500 to 600 B.C.E.	Through meditation and adherence to the Eightfold Path (correct thought and behavior), people can free themselves from desire and suffering, escape the cycle of eternal rebirth, and achieve nirvana (enlightenment).
Judaism	18 million	Abraham, Isaac, and Jacob; ca. 2000 B.C.E.	God's nature and will are revealed in the Torah (Hebrew scripture) and in His intervention in history. God has established a covenant with the people of Israel, who are called to a life of holiness, justice, mercy, and fidelity to God's law.
Confucianism	5.9 million	K'ung Fu-Tzu (Confucius); circa 500 B.C.E.	The sayings of Confucius (collected in the *Analects*) stress the role of virtue and order in the relationships among individuals, their families, and society.

Religion and Scientific Explanations

During the Industrial Revolution, scientific explanations began to compete with religious views of life. Rapid growth in scientific and technological knowledge gave rise to the idea that science would ultimately answer questions that previously had been in the realm of religion. Many scholars believed that increases in scientific knowledge would result in *secularization*—**the process by which religious beliefs, practices, and institutions lose their significance in sectors of society and culture** (Berger, 1967). Secularization involves a decline of religion in everyday life and a corresponding increase in organizations that are highly bureaucratized, fragmented, and impersonal (Chalfant, Beckley, and Palmer, 1994).

Religion   a system of beliefs, symbols, and rituals, based on some sacred or supernatural realm, that guides human behavior, gives meaning to life, and unites believers into a community.

Sacred   those aspects of life that are extraordinary or supernatural.

Profane   the everyday, secular, or "worldly" aspects of life.

Animism   the belief that plants, animals, or other elements of the natural world are endowed with spirits or life forces having an effect on events in society.

Secularization   the process by which religious beliefs, practices, and institutions lose their significance in sectors of society and culture.

✓**Checkpoints**

Checkpoint 12.4

12. What is religion?
13. How does "the sacred" differ from that which is defined as "profane"?
14. What is secularization?

In the United States, some people argue that science and technology have overshadowed religion, but others point to the resurgence of religious beliefs and an unprecedented development of alternative religions in recent years (Kosmin and Lachman, 1993; Roof, 1993; Singer and Lalich, 1995). For people who believe that secularization contributes to a decline in morality and traditional family values, public schools are seen as part of the problem, as described by a woman from Ohio:

> It bothers me what they are teaching kids in school. . . . They are changing history around. This country was founded on God. The people that came and founded this country were Godly people, and they have totally taken that out of history. They are trying to get rid of everything that ever says anything about God to please someone who is offended by it. That bothers me. They are teaching the kids sex education at too early an age. They know too much. If they have all that on their minds all the time, they are going to get into trouble. (qtd. in Roof, 1993: 98)

However, other parents strongly disagree with this mother's view on public schools and argue instead that religious organizations should be responsible for teaching religious beliefs.

Sociological Perspectives on Religion

According to the sociologist Meredith B. McGuire (2002), religion as a social institution is a powerful, deeply felt, and influential force in human society. Sociologists study the social institution of religion because of the importance religion holds for many people; they also want to know more about the influence of religion on society, and vice versa (McGuire, 2002).

The major sociological perspectives have different outlooks on the relationship between religion and society. Functionalists typically emphasize the ways in which religious beliefs and rituals can bind people together. Conflict explanations suggest that religion can be a source of false consciousness in society. Symbolic interactionists focus on the meanings that people give to religion in their everyday life.

Functionalist Perspectives on Religion

Emile Durkheim was one of the first sociologists to emphasize that religion is essential to the maintenance of society. He suggested that religion was a cultural universal found in all societies because it met basic human needs and served important societal functions.

For Durkheim, the central feature of all religions is the presence of sacred beliefs and rituals that bind people together in a collectivity. In his studies of the religion of the Australian aborigines, for example, Durkheim found that each clan had established its own sacred totem, which included kangaroos, trees, rivers, rock formations, and other animals or natural creations. To clan members, their totem was sacred; it symbolized some unique quality of their clan. People developed a feeling of unity by performing ritual dances around their totem, which caused them to abandon individual self-interest. Durkheim suggested that the correct performance of the ritual gives rise to religious conviction. Religious beliefs and rituals are *collective representations*—group-held meanings that express something important about the group itself (McGuire, 2002). Because of the intertwining of group consciousness and society, functionalists suggest that religion has three important functions in any society:

1. *Meaning and purpose.* Religion offers meaning for the human experience. Some events create a profound sense of loss on both an individual basis (such as injustice, suffering, and the death of a loved one) and a group basis (such as famine, earthquake, economic depression, or subjugation by an enemy). Inequality may cause people to wonder why their own situation is no better than it is. Most religions offer explanations for these concerns. Explanations may differ from one religion to another, yet each tells the individual or group

The shared experiences and beliefs associated with religion have helped many groups maintain a sense of social cohesion and a feeling of belonging in the face of prejudice and discrimination.

C. W. McKeen/Syracuse Newspapers/The Image Works

future generations) in a way that otherwise might not be possible (McGuire, 2002).

Religion also helps maintain social control in society by conferring supernatural legitimacy on the norms and laws of society. In some societies, social control occurs as a result of direct collusion between the dominant classes and the dominant religious organizations.

In the United States, the separation of church and state reduces religious legitimation of political power. Nevertheless, political leaders often use religion to justify their decisions, stating that they have prayed for guidance in deciding what to do (McGuire, 2002). This informal relationship between religion and the state has been referred to as *civil religion*—the set of beliefs, rituals, and symbols that makes sacred the values of the society and places the nation in the context of the ultimate system of meaning (Roberts, 1995: 381). Civil religion is not tied to any one denomination or religious group; it has an identity all its own. For example, many civil ceremonies in the United States have a marked religious quality. National values are celebrated on "high holy days" such as Memorial Day and the Fourth of July. Political inaugurations and courtroom trials both require people to place their hand on a Bible while swearing to do their duty or tell the truth, as the case may be. The United States flag is the primary sacred object of our civil religion, and the pledge of allegiance has included the phrase "one nation under God" for many years now. U.S. currency bears the inscription "In God We Trust."

Some critics have attempted to eliminate all vestiges of civil religion from public life. However, sociologist Robert Bellah (1967), who has studied civil religion extensively, argues that civil religion

that life is part of a larger system of order in the universe (McGuire, 2002). Some (but not all) religions even offer hope of an afterlife for persons who follow the religion's tenets of morality in this life. Such beliefs help make injustices in this life easier to endure.

2. *Social cohesion and a sense of belonging.* Religious teachings and practices, by emphasizing shared symbolism, help promote social cohesion. An example is the Christian ritual of communion, which not only commemorates a historical event but also allows followers to participate in the unity ("communion") of themselves with other believers (McGuire, 2002). All religions have some form of shared experiences that rekindle the group's consciousness of its own unity.

3. *Social control and support for the government.* All societies attempt to maintain social control through systems of rewards and punishments. Sacred symbols and beliefs establish powerful, pervasive, long-lasting motivations based on the concept of a general order of existence (Geertz, 1966). In other words, if individuals consider themselves to be part of a larger order that holds the ultimate meaning in life, they will feel bound to one another (and past and

Civil religion the set of beliefs, rituals, and symbols that makes sacred the values of the society and places the nation in the context of the ultimate system of meaning.

is not the same thing as Christianity; rather, it is limited to affirmations that members of any denomination can accept. As McGuire (1997: 193) explains,

> Civil religion is appropriate to actions in the official public sphere, and Christianity (and other religions) are granted full liberty in the sphere of personal piety and voluntary social action. This division of spheres of relevance is particularly important for countries such as the United States, where religious pluralism is both a valued feature of sociological life and a barrier to achieving a unified perspective for decision making.

However, Bellah's assertion does not resolve the problem for those who do not believe in the existence of God or for those who believe that *true* religion is trivialized by civil religion.

Conflict Perspectives on Religion

Many functionalists view religion, including civil religion, as serving positive functions in society, but some conflict theorists view religion negatively.

KARL MARX ON RELIGION   For Marx, *ideologies*—"systematic views of the way the world ought to be"—are embodied in religious doctrines and political values (Turner, Beeghley, and Powers, 1995: 135). These ideologies also serve to justify the status quo and retard social change. The capitalist class uses religious ideology as a tool of domination to mislead the workers about their true interests. For this reason, Marx wrote his now famous statement that religion is the "opiate of the masses." People become complacent because they have been taught to believe in an afterlife in which they will be rewarded for their suffering and misery in this life. Although these religious teachings soothe the masses' distress, any relief is illusory. Religion unites people under a "false consciousness" that they share common interests with members of the dominant class (Roberts, 1995).

MAX WEBER ON RELIGION   Whereas Marx believed that religion retarded social change, Weber argued just the opposite. For Weber, religion could be a catalyst to produce social change.

In *The Protestant Ethic and the Spirit of Capitalism* (1976/1904–1905), Weber asserted that the religious teachings of John Calvin were directly related to the rise of capitalism. Calvin emphasized the doctrine of *predestination*—the belief that, even before they are born, all people are divided into two groups, the saved and the damned, and only God knows who will go to heaven (the elect) and who will go to hell. Because people cannot know whether they will be saved, they tend to look for earthly signs that they are among the elect. According to the Protestant ethic, those who have faith, perform good works, and achieve economic success are more likely to be among the chosen of God. As a result, people work hard, save their money, and do not spend it on worldly frivolity; instead, they reinvest it in their land, equipment, and labor (Chalfant, Beckley, and Palmer, 1994).

The spirit of capitalism grew in the fertile soil of the Protestant ethic. Even as people worked ever harder to prove their religious piety, structural conditions in Europe led to the Industrial Revolution, free markets, and the commercialization of the economy—developments that worked hand in hand with Calvinist religious teachings. From this viewpoint, wealth was an unintended consequence of religious piety and hard work. With the contemporary secularizing influence of wealth, people often think of wealth and material possessions as the major (or only) reason to work. Although it is no longer referred to as the "Protestant ethic," many people still refer to the "work ethic" in somewhat the same manner that Weber did. For example, political and business leaders in the United States often claim that "the work ethic is dead."

Like Marx, Weber was acutely aware that religion could reinforce existing social arrangements, especially the stratification system. The wealthy can use religion to justify their power and privilege: It is a sign of God's approval of their hard work and morality (McGuire, 2002). As for the poor, if they work hard and live a moral life, they will be richly rewarded in another life.

From a conflict perspective, religion tends to promote conflict between groups and societies. According to conflict theorists, conflict may be *between* religious groups (for example, anti-Semitism), *within* a religious group (for example, when a splinter group leaves an existing denomination), or be-

According to Marx and Weber, religion serves to reinforce social stratification in a society. For example, according to Hindu belief, a person's social position in his or her current life is a result of behavior in a former life.

Mark O'Neill/Canada Wide

tween a religious group and the *larger society* (for example, the conflict over religion in the classroom). Conflict theorists assert that in attempting to provide meaning and purpose in life while at the same time promoting the status quo, religion is used by the dominant classes to impose their own control over society and its resources (McGuire, 2002). Many feminists object to the patriarchal nature of most religions; some advocate a break from traditional religions, whereas others seek to reform religious language, symbols, and rituals to eliminate the elements of patriarchy.

Symbolic Interactionist Perspectives on Religion

Thus far, we have been looking at religion primarily from a macrolevel perspective. Symbolic interactionists focus their attention on a microlevel analysis that examines the meanings that people give to religion in their everyday life.

RELIGION AS A REFERENCE GROUP For many people, religion serves as a reference group to help them define themselves. For example, religious symbols have meaning for large bodies of people. The Star of David holds special significance for Jews, just as the crescent moon and star do for Muslims and the cross does for Christians.

For individuals as well, a symbol may have a certain meaning beyond that shared by the group. For instance, a symbol given to a child may have special meaning when he or she grows up and faces war or other crises. It may not only remind the adult of a religious belief but also create a feeling of closeness with a relative who is now deceased. It has been said that the symbolism of religion is so very powerful because it "expresses the essential facts of our human existence" (Collins, 1982: 37).

HER RELIGION AND HIS RELIGION Early feminist Charlotte Perkins Gilman (1976/1923) believed that "men's religion" taught people to submit and obey rather than to think about and realistically confront situations. Consequently, she asserted, the monopolization of religious thoughts and doctrines by men contributed to intolerance and the subordination of women. Along with other feminist thinkers, Gilman concluded that "God the Mother" images could be useful as a means of encouraging people to be more cooperative and compassionate, rather than competitive and violent.

Not all people interpret religion in the same way. In virtually all religions, women have much less influence in establishing social definitions of appropriate gender roles both within the religious community and in the larger community. Therefore, women and men may belong to the same

Concept Table 12.A Sociological Perspectives on Education and Religion

	Education	**Religion**
Functionalist Perspective	One of the most important components of society: Schools teach not only content but also to put group needs ahead of the individual's.	Sacred beliefs and rituals bind people together and help maintain social control.
Conflict Perspective	Schools perpetuate class, racial–ethnic, and gender inequalities through what they teach to whom.	Religion may be used to justify the status quo (Marx) or to promote social change (Weber).
Symbolic Interactionist Perspective	Labeling and the self-fulfilling prophecy are an example of how students and teachers affect each other as they interpret their interactions.	Religion may serve as a reference group for many people, but because of race, class, and gender, people may experience it differently.

religious group, but their individual religion will not necessarily be a carbon copy of the group's entire system of beliefs. In fact, according to McGuire (2002), women's versions of a certain religion probably differ markedly from men's versions. For example, whereas an Orthodox Jewish man may focus on his public ritual roles and his discussion of sacred texts, women have few ritual duties and are more likely to focus on their responsibilities in the home. Consequently, the meaning of being Jewish may be different for women than for men.

Religious symbolism and language typically create a social definition of the roles of men and women. For example, religious symbolism may depict the higher deities as male and the lower deities as female. Sometimes, females are depicted as negative, or evil, spiritual forces. For instance, the Hindu goddess Kali represents men's eternal battle against the evils of materialism (Daly, 1973). Historically, language has defined women as being nonexistent in the world's major religions. Phrases such as "for all men" in Catholic and Episcopal services gradually have been changed to "for all"; however, some churches retain the traditional liturgy. Although there has been resistance, especially by women, to some of the terms, inclusive language is less common than older male terms for God (Briggs, 1987).

Many women resist the subordination they have experienced in organized religion. They have worked to change the existing rules that have ex-

✓ **Checkpoints**

Checkpoint 12.5

15. What are the three important functions that religion plays in any society?
16. Why is Karl Marx's view of religion a conflict perspective?
17. What was Max Weber's central belief about religion?
18. What central issues regarding religion are of particular interest to symbolic interactionists?

cluded them or placed them in a clearly subordinate position.

Concept Table 12.A summarizes the major sociological perspectives on education and religion.

Types of Religious Organization

Religious groups vary widely in their organizational structure. Although some groups are large and somewhat bureaucratically organized, others are small and have a relatively informal authority structure. Some require total commitment from their members; others expect members to have only a partial commitment. Sociologists have developed typologies or ideal types of religious organization to

Table 12.2 Characteristics of Churches and Sects

Characteristic	Church	Sect
Organization	Large, bureaucratic organization, led by a professional clergy	Small, faithful group, with high degree of lay participation
Membership	Open to all; members usually from upper and middle classes	Closely guarded membership, usually from lower classes
Type of Worship	Formal, orderly	Informal, spontaneous
Salvation	Granted by God, as administered by the church	Achieved by moral purity
Attitude Toward Other Institutions and Religions	Tolerant	Intolerant

enable them to study a wide variety of religious groups. The most common categorization sets forth four types: ecclesia, church, sect, and cult.

Ecclesia

Ecclesia **is a religious organization that is so integrated into the dominant culture that it claims as its membership all members of a society.** Membership in the ecclesia occurs as a result of being born into the society, rather than by any conscious decision on the part of individual members. The linkages between the social institutions of religion and government are often very strong in such societies. Although no true ecclesia exists in the contemporary world, the Anglican church (the official church of England), the Lutheran church in Sweden and Denmark, the Catholic church in Spain, and Islam in Iran and Pakistan come fairly close.

The Church–Sect Typology

To help explain the different types of religious organizations found in societies, Ernst Troeltsch (1960/1931) and his teacher, Max Weber (1963/1922), developed a typology that distinguishes between the characteristics of churches and sects (see Table 12.2). Unlike an ecclesia, a church is not con-

sidered to be a state religion; however, it may still have a powerful influence on political and economic arrangements in society. A *church* **is a large, bureaucratically organized religious organization that tends to seek accommodation with the larger society in order to maintain some degree of control over it.** Church membership is largely based on birth; children of church members are typically baptized as infants and become lifelong members of the church. Older children and adults may choose to join the church, but they are required to go through an extensive training program that culminates in a ceremony similar to the one that infants go through. Leadership is hierarchically arranged, and clergy generally have many years of formal education. Churches have very restrained services that appeal to the intellect rather than the emotions (Stark, 1992). Religious services are

Ecclesia a religious organization that is so integrated into the dominant culture that it claims as its membership all members of a society.

Church a large, bureaucratically organized religious organization that tends to seek accommodation with the larger society in order to maintain some degree of control over it.

Table 12.3 Major U.S. Denominations That Self-Identify as Christian

Religious Body	Members	Churches
Roman Catholic Church	62,018,000	19,584
Southern Baptist Convention	15,729,000	40,870
United Methodist Church	8,400,000	36,170
National Baptist Convention, U.S.A.[a]	8,200,000	33,000
Church of God in Christ[a]	5,500,000	15,300
Evangelical Lutheran Church in America	5,178,000	10,862
Church of Jesus Christ of Latter Day Saints[a]	4,923,000	10,811
Presbyterian Church (U.S.A.)	3,575,000	11,260
National Baptist Convention of America	3,500,000	2,500
Lutheran Church–Missouri Synod	2,594,000	6,218
Assemblies of God	2,526,000	11,937
African Methodist Episcopal Church	2,500,000	6,200
National Missionary Baptist Convention of America[a]	2,500,000	(N/A)
Progressive National Baptist Convention	2,500,000	2,000
Episcopal Church[a]	2,365,000	7,390
Greek Orthodox Church[a]	1,955,000	523
American Baptist Churches in U.S.A.	1,507,000	3,800
Churches of Christ	1,500,000	15,000
Pentecostal Assemblies of the World	1,500,000	1,750
United Church of Christ	1,421,000	6,017
African Methodist Episcopal Zion Church	1,252,000	3,098
Baptist Bible Fellowship, International[a]	1,200,000	4,500
Christian Churches and Churches of Christ	1,072,000	5,579
Jehovah's Witnesses	1,040,000	11,064
Orthodox Church in America	1,000,000	625

[a]Current data not available; prior data used may no longer be comparable.

Source: U.S. Census Bureau, 2000c.

highly ritualized; they are led by clergy who wear robes, enter and exit in a formal processional, administer sacraments, and read services from a prayer book or other standardized liturgical format.

Midway between the church and the sect is a *denomination*—a large, organized religion characterized by accommodation to society but frequently lacking in the ability or intention to dominate society (Niebuhr, 1929). Denominations have a trained ministry, and while involve-

ment by lay members is encouraged more than in the church, their participation is usually limited to particular activities, such as readings or prayers. Denominations tend to be more tolerant and less likely than churches to expel or excommunicate members. This form of organization is most likely to thrive in societies characterized by *religious pluralism*—a situation in which many religious groups exist because they have a special appeal to specific segments of the population. Perhaps because of its diversity, the United States has more denominations than any other country. Table 12.3 shows this diversity.

A *sect* **is a relatively small religious group that has broken away from another religious organization to renew what it views as the original version of the faith.** Unlike churches, sects offer members a more personal religion and an intimate relationship with a supreme being, depicted as taking an active interest in the individual's everyday life. Whereas churches use formalized prayers, often from a prayer book, sects have informal prayers composed at the time they are given. Whereas churches typically appeal to members of the upper classes, and denominations to members of the middle and upper-middle classes, sects seek to meet the needs of people who are low in the stratification system—that is, the masses (Stark, 1992).

According to the church–sect typology, as members of a sect become more successful economically and socially, their religious organization is also likely to focus more on this world and less on the next. If some members of the sect do not achieve financial success, they may feel left behind as other members and the ministers shift their priorities. Eventually, this process will weaken some organizations, and people will split off to create new, less worldly versions of the group that will be more committed to "keeping the faith." Those who defect to form a new religious organization may start another sect or form a cult (Stark and Bainbridge, 1981).

Cults

A *cult* **is a religious group with practices and teachings outside the dominant cultural and religious traditions of a society.** Although many people view cults negatively, some major religions (including Judaism, Islam, and Christianity) and some denominations (such as the Mormons) started as cults. Cult leadership is based on charismatic characteristics of the individual, including an unusual ability to form attachments with others (Stark, 1992). An example is the religious movement started by Reverend Sun Myung Moon, a Korean electrical engineer who believed that God had revealed to him that Judgment Day was rapidly approaching. Out of this movement, the Unification church, or "Moonies," grew and flourished, recruiting new members through their personal attachments to present members (Stark, 1992).

✓ Check
19. Wha organi
20. What is
21. What is th
22. What are c

Chapter 12

402

Table 12.4 U.S. Members
Bodies,

Trends in Relig
in the United Sta

As we have seen throughout this chapter, religion in the United States is very diverse. Pluralism and religious freedom are among the cultural values most widely espoused, and no state church or single denomination predominates. As shown in Table 12.4, Protestants constitute the largest religious body in the United States, followed by Roman Catholics, Jews, Eastern Churches, and others.

The rise of a new fundamentalism has occurred at the same time that a number of mainline denominations have been losing membership. Whereas "old" fundamentalism usually appealed to people from lower-income, rural, southern backgrounds, the "new" fundamentalism appears to have a much wider following among persons from all socioeconomic levels, geographical areas, and occupations. Some member of the political elite in Washington have vowed to bring religion "back" into schools and public life. "New-right" fundamentalists have been especially critical of *secular humanism*—a belief in the perfectibility of human beings through their own efforts rather than through a belief in God and a religious conversion. According to fundamentalists, "creeping"

Denomination a large organized religion characterized by accommodation to society but frequently lacking in ability or intention to dominate society.

Sect a relatively small religious group that has broken away from another religious organization to renew what it views as the original version of the faith.

Cult a religious group with practices and teachings outside the dominant cultural and religious traditions of a society.

Religous Body	Number of Members
Protestants	85,500,000
Roman Catholics	59,450,000
Muslims	6,000,000
Jews	5,602,000
Orthodox Christians	5,631,000
Buddhists	1,864,000
Hindus	795,000

Sources: Ash, 1997; J. Wright, 1997.

Should prayer be permitted in the classroom? On the school grounds? At school athletic events? Given the diversity of beliefs U.S. people hold, arguments and court cases over activities such as prayer around the school flagpole will no doubt continue in the future.

secular humanism has been most visible in the public schools, which, instead of offering children a fair and balanced picture, are teaching things that seem to children to prove that their parents' lifestyle and religion are inferior and perhaps irrational (Carter, 1994). The new-right fundamentalists claim that banning the teaching of Christian beliefs in the classroom while teaching things that are contrary to their faith is an infringement on their freedom of religion. Worse yet, they argue, it is the equivalent of establishing an unconstitutional state religion—a secular religion that does not recognize God (Jenkinson, 1979).

But how might students and teachers who come from very diverse religious and cultural backgrounds feel about religious instruction or organized prayer in public schools? Rick Nelson, a teacher in the Fairfax County, Virginia, public school system, explains his concern about the potential impact of group prayer on students in his classroom:

I think it really trivializes religion when you try to take such a serious topic with so many different viewpoints and cover it in the public schools. At my school we have teachers and students who are Hindu. They are really devout, but they are not monotheistic. . . . I am not opposed to individual prayer by students. I expect students to pray when I give them a test. They need to do that for my tests. . . . But when there is group prayer . . . who's going to lead the group? And if I had my Hindu students lead the prayer, I will tell you it will disrupt many of my students and their parents. It will disrupt the mission of my school, unfortunately . . . if my students are caused to participate in a group Hindu prayer. (CNN, 1994)

As we have seen in this chapter, the debate continues over what should be taught and what practices (such as Bible reading and prayer) should be permitted in public schools.

✓ Checkpoints

Checkpoint 12.7

23. What are some current conflicts between religion and public education?

Education and Religion in the Future

This chapter ends as it began, by noting that education and religion will remain important social institutions in the twenty-first century. Also remaining, however, are the controversies that we have discussed—controversies that your generation must attempt to resolve.

With regard to education, questions remain about what should be taught, not only in terms of preparing your children for their adult lives and the world of work but also in regard to the values to which you want your children exposed. Unlike many other countries, no central authority in the United States decides the curriculum to be taught to all students nationwide. Rather, each state enacts its own laws and regulations, with local school boards frequently making the final determination. Accordingly, no general standards exist as to what is to be *taught* to students or how it is to be taught, although many states have now adopted levels as to what (as a minimum) must be *learned* in order to graduate from high school.

Obviously, the debate over what should be taught is not limited to just religious and moral issues; rather, it includes the entire curriculum. If, as critics assert, academic achievement in the United States compares unfavorably with the level of achievement by students in many other countries, what can be done to change it? In 1994, Congress passed the Goals 2000: Educate America Act to challenge the nation's public school systems to meet certain education goals. Among those were the following:

- By the year 2000, American students will leave grades four, eight, and twelve having demonstrated competency in challenging subject matter, including English, mathematics, science, history, and geography; and every school in America will ensure that all students learn to use their minds well, so they may be prepared for responsible citizenship, further learning, and productive employment in our modern economy.
- By the year 2000, U.S. students will be first in the world in science and mathematics achievement.
- By the year 2000, every adult American will be literate and will possess the knowledge and

skills necessary to compete in a global economy and exercise the rights and responsibilities of citizenship.

Early in the twenty-first century, very little progress has been made on these important goals. The act left it up to the individual states to determine what they would do to meet the challenge in their own school systems. Likewise, the goals arguably are overly idealistic. For example, how does being first in the world in mathematics and science achievement help students in central-city schools and on reservations to gain a better education? Questions such as this are unanswered by Goals 2000.

According to Gary D. Fenstermacher (1994), the education needed for the future well-being of people in the United States is different from that of the past:

> For this education is not bent on assimilation, to the melting of different cultures and languages into some common American pot, or to merely readying today's children for tomorrow's workforce. In contrast, the education that must engage us today and in the future is how to form common space and common speech and common commitment while respecting and preserving our differences in heritage, race, language, culture, gender, sexual orientation, spiritual values, and political ideologies. It is a new challenge for America.

Analysts of religion in the United States have reached similar conclusions about the future of religion:

> Whatever the future holds, this generation's struggle over values and commitment is not over—if for no other reason than that the memories of the past live on as reminders of who they are and where they came from. . . . [People] know that religion, for all its institutional limitations, holds a vision of life's unity and meaningfulness, and for that reason will continue to have a place in their narrative. In a very basic sense, religion itself was never the problem, only social forms of religion that stifle the human spirit. The sacred lives on and is real to those who can access it. (Roof, 1993: 261)

Religious organizations will continue to be important in the lives of many people. However, the influence of religious beliefs and values will be felt even by those who claim no religious beliefs of their own. In many nations, the rise of *religious nationalism* has led to the blending of strongly held religious

Reflections

Reflections 12.4

Do you believe that the influence of religion on education will increase, decrease, or remain about the same over the next ten years? Why?

and political beliefs. Although the rise of religious nationalism is occurring throughout the world, it is especially strong in the Middle East, where Islamic nationalism has spread rapidly (Juergensmeyer, 1993). Many analysts believe that major conflicts around the globe are linked to religious nationalism as well as to other economic and political issues.

In the United States, the influence of religion will be evident in ongoing battles over school prayer, abortion, gay rights, and women's issues, among others. On some fronts, religion may unify people; on others, it may contribute to confrontations among individuals and groups. Legal scholar Stephen L. Carter (1994) suggests that tensions between religion and other social institutions are not necessarily negative. According to Carter, pervasive, totalitarian states find all conflict threatening and often quickly remove religious liberty when a statist dictator takes control. In the United States, however, one of our most cherished freedoms is religious liberty. Maintaining an appropriate balance between the social institutions of education and religion remains a challenge in the twenty-first century.

Chapter Review

How are the social institutions of education and religion similar?
Education and religion are powerful and influential forces in society. Both institutions impart values, beliefs, and knowledge considered essential to the social reproduction of individual personalities and entire cultures.

What is the primary function of education?
Education is the social institution responsible for the systematic transmission of knowledge, skills, and cultural values within a formally organized structure.

What is the functionalist perspective on education?
According to functionalists, education has both manifest functions (socialization, transmission of culture, so-

cial control, social placement, and change and innovation) and latent functions (keeping young people off the streets and out of the job market, matchmaking and producing social networks, and creating a generation gap).

What is the conflict perspective on education?
From a conflict perspective, education is used to perpetuate class, racial–ethnic, and gender inequalities through tracking, ability grouping, and a hidden curriculum that teaches subordinate groups conformity and obedience.

What is the symbolic interactionist perspective on education?
According to symbolic interactionists, education may be a self-fulfilling prophecy for some students, such that these students come to perform up—or down—to the expectations held for them by teachers.

What is religion?
Religion is a system of beliefs, symbols, and rituals, based on some sacred or supernatural realm, that guides human behavior, gives meaning to life, and unites believers into a community.

What is the functionalist perspective on religion?
According to functionalists, religion has three important functions in any society: (1) providing meaning and purpose to life, (2) promoting social cohesion and a sense of belonging, and (3) providing social control and support for the government.

What is the conflict perspective on religion?
From a conflict perspective, religion can have negative consequences in that the capitalist class uses religion as a tool of domination to mislead workers about their true interests. However, Max Weber believed that religion could be a catalyst for social change.

What is the symbolic interactionist perspective on religion?
Symbolic interactionists examine the meanings that people give to religion and the meanings that they attach to religious symbols in their everyday life.

What are the different types of religious organizations?
Religious organizations can be categorized as ecclesia, churches, denominations, sects, and cults. Some of the world's major religions started off as cults built around a charismatic leader and developed into sects, denominations, and churches. No true ecclesia, or government-sponsored religion, exists in the contemporary world.

Key Terms

animism 392
church 399
civil religion 395
credentialism 383
cult 401
cultural capital 382
denomination 400
ecclesia 399
education 378
hidden curriculum 383
profane 391
religion 391
sacred 391
sect 401
secularization 394
tracking 382

Questions for Critical Thinking

1. Why does so much controversy exist over what should be taught in U.S. public schools?
2. How are the values and attitudes you learned from your family reflected in your beliefs about education and religion?
3. How would you design a research project to study the effects of civil religion on everyday life? What kind of data would be most accessible?
4. If Durkheim, Marx, and Weber were engaged in a discussion about education and religion, on what topics might they agree? On what topics would they disagree?

 # Resources on the Internet

Chapter-Related Web Sites
The following web sites have been selected for their relevance to the topics in this chapter. These sites are among the more stable, but please note that web site addresses change frequently.

The Center for Education Reform
http://edreform.com
This site was designed to provide information to parents as consumers of education. Here you will find a great deal of resources about public education in America.

U.S. Department of Education
http://www.ed.gov/index.jsp
This official web site provides a wealth of information on public policy, along with grade-level resources.

National Education Association
http://www.nea.org
NEA has a long, proud history as the nation's leading organization committed to advancing the cause of public education. With its headquarters in Washington, D.C., NEA has 2.7 million members who work at every level of education, from preschool to university graduate programs. NEA has affiliates in every state, as well as in more than 13,000 local communities across the United States.

American Federation of Teachers
http://www.aft.org
From this teacher's union web site you can research teacher salaries in each state.

American Religion Data Archive
http://www.thearda.com/arda.asp?Show=Home
This site has a wealth of data on religious participation and behavior in American society. It provides access to the General Social Survey as well as the National Congregations Study.

Weberian Sociology of Religion
http://www.ne.jp/asahi/moriyuki/abukuma
For students who are interested in Max Weber's theories of religion, this web page provides online articles and selections from Weber's writings.

Hartford Institute for Religion Research: Sociology of Religion
http://hirr.hartsem.edu/sociology/sociology.html
This site is a resource for people interested in studying religion as a social phenomenon. Students will be interested in the online articles and the brief description of the field.

World Religions Index
http://wri.leaderu.com
This site offers detailed descriptions, articles, and data on the major religions in the world today.

Companion Web Site for This Book
Virtual Society: The Wadsworth Sociology Resource Center
Visit http://sociology.wadsworth.com and click on the page for Kendall, *Sociology in Our Times: The Essentials*, Fourth Edition, to access a wide range of enrichment material to aid in your study of sociology. Click on the Student Resources section of the web site. Next, select from the pull-down menu the chapter that you are presently studying. Among the useful options for self-study are chapter objectives, flashcards, practical study tips, and practice tests for each chapter.

MicroCase Online

From the Virtual Society home page, click on MicroCase Online to access book-specific MicroCase exercises that allow you to further explore sociological issues and principles presented in the text.

InfoTrac College Edition

Another unique option available to you at the Student Resources section of the companion web site is InfoTrac College Edition, an online library with access to hundreds of scholarly and popular periodicals. Below are suggested search terms for this chapter. Results from these and other searches are found at the site.

- Search keywords: *standardized tests*. More and more states are turning to standardized testing to assess proficiency as a measure of school account-ability and to determine whether a child should continue to the next grade. In the articles that you found, what are some of the social costs and benefits of standardized tests?
- Search keywords: *college costs*. Look for articles that address the increasing cost of obtaining a college education. How does this situation affect minorities and the economically disadvantaged?

Virtual Explorations CD-ROM

Go to the Virtual Explorations CD-ROM to begin the interactive exercise for this chapter. This Virtual Exploration will introduce you to some of the exciting resources for sociology on the World Wide Web. You will be guided through an exercise that employs related web sites on education and religion. Answer the questions, and e-mail your responses to your instructor.

Ken Condon/SIS

Politics and the Economy in Global Perspective

13

Politics, Power, and Authority
Power and Authority
Ideal Types of Authority

Political Systems in Global Perspective
Monarchy
Authoritarianism
Totalitarianism
Democracy

Perspectives on Power and Political Systems
Functionalist Perspectives: The Pluralist Model
Conflict Perspectives: Elite Models

The U.S. Political System
Political Parties and Elections
Political Participation and Voter Apathy
Governmental Bureaucracy

Economic Systems in Global Perspective
Preindustrial, Industrial, and Postindustrial Economies
Capitalism
Socialism
Mixed Economies

Work in the Contemporary United States
Professions
Other Occupations
Contingent Work
Unemployment
Labor Unions and Worker Activism
Employment Opportunities for Persons with a Disability

Politics and the Economy in the Future

"Police! Open the door! It's coming down."

Kicked hard, twice, the wooden door shattered at the lock. [The U.S. marshal and county deputy sheriff] lunged, guns drawn, into the bedroom.

A man, naked except for a sheet across his midriff, blinked in apparent confusion from the bed in the dawn light.

"Put your hands where I can see them," barked [one officer]. "Turn over, face down, FACE DOWN, mister!" Three officers swarmed over the man, cuffing his hands behind his back.

In seconds, it was over. The man, with a record of three convictions and 23 arrests on charges from drug possession to assault, was in custody for alleged violation of probation.

—Paul Valentine (1992: D1), a *Washington Post* reporter, describing a raid that he and a newspaper photographer recorded on a "media ride-along" with law enforcement officials

Ride-alongs occur when reporters ride with police to document the activities of law enforcement officials; however, they raise privacy issues that have been addressed by the U.S. Supreme Court. For example, what happens if the people confronted by the police are innocent but subjected to humiliation in their own homes due to the presence of media representatives, as well as law enforcement officials? Consider another raid participated in by Paul Valentine:

Charles and Geraldine Wilson were still in bed when they heard the pounding on their front door that April morning. Their 9-year-old granddaughter was . . . waiting for the school bus. They heard the child go to the door but couldn't tell what was happening.

When Charles Wilson reached the living room wearing only undershorts, he was confronted by three plain-

A. Ramey/PhotoEdit

Increasingly, drug raids and other police department activities are witnessed by journalists and photographers who might be considered quasi-participants. Does having representatives of the media present when the police conduct a search help to preserve democracy or might it be an invasion of peoples' right to privacy?

clothes officers with guns drawn. At their side were [Paul Valentine] and a woman who [was] a *Washington Post* photographer. After police forced Wilson to the floor, his wife appeared dressed only in a negligee.

The photographer sprang to action, taking pictures of the besieged couple, including one of Wilson with an officer's knee on his back and a gun to his head.

—Joan Biskupic (1999: A2), another *Washington Post* reporter, explaining the events that led up to the U.S. Supreme Court decision in *Wilson v. Layne,* which stated that the police violated the Fourth Amendment prohibition on unreasonable searches and seizures when they permitted the media to accompany them to a home to photograph the apprehension of suspects

I n a democratic society with a capitalist economy, what part should the media play in the daily lives of individuals? In the age of the cell phone, the Internet, and instantaneous global communications, do we as private citizens have a right to privacy in our own homes? Questions such as these arise in situations like the "media ride-alongs" described above, but they are also pressing issues that sometimes must be dealt with by officials in the legislative, executive, and judicial bodies of our nation. Likewise, members of the media must balance the private rights of individuals against the need to increase revenues by attracting more readers, viewers, and advertisers. In this chapter, we discuss the intertwining effect of politics, the economy, and the media as media members grapple with such issues as these. Before reading on, test your knowledge of the media by taking the quiz in Box 13.1.

Questions and Issues

Chapter Focus Question: What effect does the intertwining of politics, the economy, and the media have on the United States and other nations?

What are the major political systems around the world?

How does the center of power differ in the pluralist model and the elite model of the U.S. power structure?

What are the key characteristics of capitalism and socialism?

What will be the place of democracy and capitalism in the future?

Box 13.1 Sociology and Everyday Life

How Much Do You Know About the Media?

True	False		
T	F	1.	U.S. mass media industries earn about $220 billion a year.
T	F	2.	In the United States, no media are publicly owned.
T	F	3.	The top ten U.S. newspaper chains own 20 percent of the daily newspapers.
T	F	4.	The Federal Communications Commission (FCC) has regulations preventing vertical integration of media companies.
T	F	5.	Television's first live marathon broadcast coverage of a political scandal occurred during the Clinton administration in the 1990s.
T	F	6.	Thirty-minute nightly news programs on the major television networks (NBC, CBS, and ABC) typically contain twenty-six minutes of news and four minutes of commercials.
T	F	7.	Almost all movies in the United States are distributed by six large studios.
T	F	8.	Both major U.S. political parties have been accused of purchasing television commercials with questionable campaign contributions.
T	F	9.	Most journalists identify themselves as Republicans.
T	F	10.	Some analysts believe that "reality" TV shows such as *Cops* and *America's Most Wanted* blur the distinction between "news" and "entertainment."

Answers on page 410.

Politics, Power, and Authority

Politics **is the social institution through which power is acquired and exercised by some people and groups.** In contemporary societies, the government is the primary political system. *Government* **is the formal organization that has the legal and political authority to regulate the relationships among members of a society and between the society and those outside its borders.** Some social analysts refer to the government as the *state*—**the political entity that possesses a legitimate monopoly over the use of force within its territory to achieve its goals.**

Whereas political science focuses primarily on power and its distribution in different types of political systems, *political sociology* is the area of sociology that examines politics and the government. Political sociology primarily focuses on the *social circumstances* of politics and explores how the political arena and its actors are intertwined with so-

cial institutions such as the economy, religion, education, and the media. According to the political analyst Michael Parenti (1998: 7–8), many people underrate the significance of politics in daily life:

> Politics is something more than what politicians do when they run for office. Politics is concerned with the struggles that shape social relations within societies and affairs between nations. The taxes and

Politics   the social institution through which power is acquired and exercised by some people and groups.

Government   the formal organization that has the legal and political authority to regulate the relationships among members of a society and between the society and those outside its borders.

State   the political entity that possesses a legitimate monopoly over the use of force within its territory to achieve its goals.

Box 13.1

Answers to the Sociology Quiz on the Media

1. **True.** Mass media industries earn about $220 billion annually. Newspapers earn about $55 billion, television earns $49 billion, movies earn $35 billion, and book publishing, magazines, recordings, and radio account for the rest.

2. **False.** Although most media outlets are privately owned, Public Broadcasting Service (PBS) and National Public Radio (NPR) are funded by government support, grants from nonprofit foundations, and donations from viewers and listeners. However, corporate underwriters now play an increasing role in their funding.

3. **True.** The fact that the top ten newspaper chains own 20 percent (one-fifth) of the nation's daily newspapers is an example of *concentration of ownership*.

4. **False.** When the FCC began deregulating the broadcast media in the 1980s, a corresponding increase occurred in *vertical integration* (a company attempting to control several related aspects of a business at once). For example, AOL Time Warner owns Warner movie studios, is a major book and magazine publisher, and owns and operates many cable TV systems and channels such as Home Box Office (HBO) and CNN.

5. **False.** Previous marathon broadcast coverage occurred during the Johnson administration's involvement in the Vietnam War, the Nixon administration's involvement in the Watergate scandal, and the Reagan administration's involvement in Iran-Contra, a congressional investigation of how weapons were illegally supplied to the Nicaraguan Contras.

6. **False.** News programs such as *NBC Nightly News* typically have twenty-one minutes of news content, eight minutes of commercials, and one minute of "self-promotions" for other NBC programs and properties.

7. **True.** Six major studios—Columbia, Paramount, 20th Century-Fox, MCA/Universal, Time Warner, and Walt Disney—distribute not only the films they produce but also most of the films made by independent producers.

8. **True.** "Soft money" contributions, which are made outside the limits imposed by federal election laws, have allegedly been used by both parties for campaign-style ads, but leaders of the national political parties claimed that their own ads were about social issues, not candidates.

9. **False.** Forty-four percent of journalists identify themselves as Democrats, 16.3 percent as Republicans, and 34 percent as independents.

10. **True.** Since TV "reality shows" imitate news stories by reenacting events and interviewing crime victims, it is difficult for some viewers to determine whether they are watching "real" news.

Sources: Based on Bagby, 1998; Biagi, 1998; and *Brill's Content*, 1998.

prices we pay and the jobs available to us, the chances that we will live in peace or perish in war, the costs of education and the availability of scholarships, the safety of the airliner or highway we travel on, the quality of the food we eat and the air we breathe, the availability of affordable housing and medical care, the legal protections against racial and sexual discrimination—all the things that directly affect the quality of our lives are influenced in some measure by politics. . . . To say you are not interested in politics, then, is like saying you are not interested in your own well-being.

Using a power–conflict framework for his analysis, Parenti (1998) suggests that the media often distort—either intentionally or unintentionally—the information they provide to citizens. According to Parenti, the media have the power to influence public opinion in a way that favors management over labor, corporations over their critics, affluent whites over subordinate-group members in central cities, political officials over protestors, and free-market capitalism over public-sector development. Parenti's assertion raises an interesting issue about the distribution of power in the United States and other high-income nations: Do the media distort information to suit their own interests?

Power and Authority

Power is the ability of persons or groups to achieve their goals despite opposition from others (Weber, 1968/1922). Through the use of persuasion, authority, or force, some people are able to get others to acquiesce to their demands. Consequently, power is a *social relationship* that involves both leaders and followers. Power is also a dimension in the structure of social stratification. Persons in positions of power control valuable resources of society—including wealth, status, comfort, and safety—and are able to direct the actions of others while protecting and enhancing the privileged social position of their class (Domhoff, 1998). For example, the sociologist G. William Domhoff (1998) argues that the media tend to reflect "the biases of those with access to them—corporate leaders, government officials, and policy experts." However, although Domhoff believes the media can amplify the message of powerful people and marginalize the concerns of others, he does not think the media are as important as government officials and corporate leaders are in the U.S. power equation.

What about power on a global basis? Although the most basic form of power is physical violence or force, most political leaders do not want to base their power on force alone. Instead, they seek to legitimize their power by turning it into *authority*— power that people accept as legitimate rather than coercive.

Ideal Types of Authority

Who is most likely to accept authority as legitimate and adhere to it? People have a greater tendency to accept authority as legitimate if they are economically or politically dependent on those who hold power. They may also accept authority more readily if it reflects their own beliefs and values (Turner, Beeghley, and Powers, 1998). Weber's outline of three *ideal types* of authority—traditional, charismatic, and rational–legal—shows how different bases of legitimacy are tied to a society's economy.

TRADITIONAL AUTHORITY According to Weber, *traditional authority* is power that is legitimized on the basis of long-standing custom. In preindustrial societies, the authority of traditional leaders, such as kings, queens, pharaohs, emperors, and religious dignitaries, is usually grounded in religious beliefs and custom. For example, British kings and queens historically traced their authority from God. Members of subordinate classes obey a traditional leader's edicts out of economic and political dependency and sometimes personal loyalty. However, as societies industrialize, traditional authority is challenged by a more complex division of labor and by the wider diversity of people who now inhabit the area as a result of high immigration rates.

Gender, race, and class relations are closely intertwined with traditional authority. Political scientist Zillah R. Eisenstein (1994) suggests that *racialized patriarchy*—the continual interplay of race and gender—reinforces traditional structures of power in contemporary societies. According to Eisenstein (1994: 2), "Patriarchy differentiates women from men while privileging men. Racism simultaneously differentiates people of color from whites and privileges whiteness. These processes are distinct but intertwined."

CHARISMATIC AUTHORITY *Charismatic authority* is power legitimized on the basis of a leader's exceptional personal qualities or the demonstration of extraordinary insight and accomplishment that inspire loyalty and obedience from followers. Charismatic leaders may be politicians, soldiers, or entertainers, among others (Shils, 1965; Bendix, 1971).

Power the ability of persons or groups to achieve their goals despite opposition from others.

Authority power that people accept as legitimate rather than coercive.

Charismatic authority tends to be temporary and relatively unstable; it derives primarily from individual leaders (who may change their minds, leave, or die) and from an administrative structure usually limited to a small number of faithful followers. For this reason, charismatic authority often becomes routinized. The *routinization of charisma* occurs when charismatic authority is succeeded by a bureaucracy controlled by a rationally established authority or by a combination of traditional and bureaucratic authority (Turner, Beeghley, and Powers, 1998). According to Weber (1968/1922: 1148), "It is the fate of charisma to recede . . . after it has entered the permanent structures of social action."

RATIONAL–LEGAL AUTHORITY According to Weber, *rational–legal authority* is power legitimized by law or written rules and regulations. Rational–legal authority—also known as *bureaucratic authority*—is based on an organizational structure that includes a clearly defined division of labor, hierarchy of authority, formal rules, and impersonality. Power is legitimized by procedures; if leaders obtain their positions in a procedurally correct manner (such as by election or appointment), they have the right to act.

Rational–legal authority is held by elected or appointed government officials and by officers in a formal organization. However, authority is invested in the *office*, not in the *person* who holds the office. For example, although the U.S. Constitution grants rational–legal authority to the office of the presidency, a president who fails to uphold the public trust may be removed from office. In contemporary society, the media may play an important role in bringing to light allegations about presidents or other elected officials. Examples include the media blitzes surrounding the Watergate investigation of the 1970s that led to the resignation of President Richard M. Nixon and the late 1990s political firestorm over campaign fund-raising and the sex scandal involving President Bill Clinton.

In a rational–legal system, the governmental bureaucracy is the apparatus responsible for creating and enforcing rules in the public interest. Weber believed that rational–legal authority was the only means to attain "efficient, flexible, and competent regulation under a rule of law" (Turner, Beeghley, and Powers, 1995: 218). Weber's three types of authority are summarized in Concept Table 13.A.

✓ Checkpoints

Checkpoint 13.1

1. What is political sociology?
2. What are the ideal types of authority?
3. Who is most likely to hold traditional authority?
4. How does charismatic authority differ from rational–legal authority?

Political Systems in Global Perspective

Political systems as we know them today have evolved slowly. In the earliest societies, politics was not an entity separate from other aspects of life. Political institutions first emerged in agrarian societies as they acquired surpluses and developed greater social inequality. Elites took control of politics and used custom or traditional authority to justify their position. When cities developed circa 3500–3000 B.C.E., the *city-state*—a city whose power extended to adjacent areas—became the center of political power.

Nation-states as we know them began to develop in Europe between the twelfth and fifteenth centuries (see Tilly, 1975). A *nation-state* is a unit of political organization that has recognizable national boundaries and whose citizens possess specific legal rights and obligations. Nation-states emerge as countries develop specific geographic territories and acquire greater ability to defend their borders. Improvements in communication and transportation make it possible for people in a larger geographic area to share a common language and culture. As charismatic and traditional authority are superseded by rational–legal authority, legal standards come to prevail in all areas of life, and the nation-state claims a monopoly over the legitimate use of force (P. Kennedy, 1993).

Approximately 190 nation-states currently exist throughout the world; today, everyone is born, lives, and dies under the auspices of a nation-state (see Skocpol and Amenta, 1986). Four main types of political systems are found in nation-states: monarchy, authoritarianism, totalitarianism, and democracy.

Concept Table 13.A Weber's Three Types of Authority

Max Weber's three types of authority are shown here in global perspective. Sultan Ali Mirah of Ethiopia is an example of traditional authority sanctioned by custom. Charismatic authority is exemplified by Dr. Martin Luther King, Jr., of the United States, whose leadership was based on personal qualities. Australian Supreme Court justices represent rational–legal authority, which depends upon established rules and procedures.

	Description	**Examples**	
Traditional	Legitimized by longstanding custom Subject to erosion as traditions weaken	Patrimony (authority resides in traditional leader supported by larger social structures, as in old British monarchy) Patriarchy (rule by men occupying traditional positions of authority, as in the family)	 Betty Press/Woodfin Camp & Associates
Charismatic	Based on leader's personal qualities Temporary and unstable	Napoleon Adolf Hitler Martin Luther King, Jr. César Chávez Mother Teresa	 SuperStock
Rational–legal	Legitimized by rationally established rules and procedures Authority residing in the office, not the person	Modern British Parliament U.S. presidency, Congress, federal bureaucracy	 Cary Wolinsky/Stock Boston

Monarchy

Monarchy **is a political system in which power resides in one person or family and is passed from generation to generation through lines of inheritance.** Monarchies are most common in agrarian societies and are associated with traditional authority patterns. However, the relative power of monarchs has varied across nations, depending on religious, political, and economic conditions.

Routinization of charisma   the process by which charismatic authority is succeeded by a bureaucracy controlled by a rationally established authority or by a combination of traditional and bureaucratic authority.

Monarchy   a political system in which power resides in one person or family and is passed from generation to generation through lines of inheritance.

Through its many ups and downs, the British royal family has remained a symbol of that nation's monarchy. Monarchies typically pass power from generation to generation in one family. Shown here are the likely successors to the British throne in the twenty-first century.

Absolute monarchs claim a hereditary right to rule (based on membership in a noble family) or a divine right to rule (a God-given right to rule that legitimizes the exercise of power). In limited monarchies, rulers depend on powerful members of the nobility to retain their thrones. Unlike absolute monarchs, *limited monarchs* are not considered to be above the law. In *constitutional monarchies*, the royalty serve as symbolic rulers or heads of state while actual authority is held by elected officials in national parliaments. In present-day monarchies such as Great Britain, Sweden, Spain, and the Netherlands, members of royal families primarily perform ceremonial functions. In nations such as Great Britain, the media often focus large amounts of time and attention on the royal family, especially the personal lives of its members. The extensive coverage of the life and death of Diana, the Princess of Wales, is one of many examples.

Authoritarianism

Authoritarianism **is a political system controlled by rulers who deny popular participation in government.** A few authoritarian regimes have been absolute monarchies whose rulers claimed a hereditary

right to their position. Today, Saudi Arabia and Kuwait are examples of authoritarian absolute monarchies. In *dictatorships*, power is gained and held by a single individual. Pure dictatorships are rare; all rulers need the support of the military and the backing of business elites to maintain their position. *Military juntas* result when military officers seize power from the government, as has happened in recent decades in Argentina, Chile, and Haiti. Today, authoritarian regimes exist in Fidel Castro's Cuba and in the People's Republic of China.

Totalitarianism

Totalitarianism **is a political system in which the state seeks to regulate all aspects of people's public and private lives.** Totalitarianism relies on modern technology to monitor and control people; mass propaganda and electronic surveillance are widely used to influence people's thinking and control their actions. One example of a totalitarian regime was the National Socialist (Nazi) party in Germany during World War II; military leaders there sought to control all aspects of national life, not just government operations. Other examples include the former Soviet Union and contemporary Iraq under Saddam Hussein's regime.

To keep people from rebelling, totalitarian governments enforce conformity: People are denied the right to assemble for political purposes, access to information is strictly controlled, and secret police enforce compliance, creating an environment of constant fear and suspicion.

Many nations do not recognize totalitarian regimes as being the legitimate government for a

Reflections

Reflections 13.1

Why do authoritarian and totalitarian regimes often seek to limit media access or to influence how the regimes are depicted by the media in other nations?

particular country. Afghanistan in the year 2001 was an example. As the war on terrorism began in the aftermath of the September 11 terrorist attacks on the United States, many people developed a heightened awareness of the Taliban regime, which ruled most of Afghanistan and was engaged in fierce fighting to capture the rest of the country. The Taliban regime maintained absolute control over the Afghan people in most of that country. For example, it required that all Muslims take part in prayer five times each day and that men attend prayer at mosques, where women were forbidden (Marquis, 2001). Taliban leaders claimed that their actions were based on Muslim law and espoused a belief in never-ending *jihad*—a struggle against one's perceived enemies. Although the totalitarian nature of the Taliban regime was difficult for many people, it was particularly oppressive for women, who were viewed by this group as being "biologically, religiously and prophetically" inferior to men (McGeary, 2001: 41). Consequently, this regime had made the veil obligatory and banned women from public life. U.S. government officials believed that the Taliban regime was protecting Osama bin Laden, the man thought to have been the mastermind behind numerous terrorist attacks on U.S. citizens and facilities, both on the mainland and abroad. As a totalitarian regime, the Taliban leadership was recognized by only three other governments despite controlling most of Afghanistan.

Once the military action commenced in Afghanistan, most of what U.S. residents knew about the Taliban and about the war on terrorism was based on media accounts and "expert opinions" that were voiced on television. According to the political analyst Michael Parenti (1998), the media play a significant role in framing the information we receive about the political systems of other countries. *Framing* refers to how news is packaged, including the amount of exposure given to a story, its placement, the positive or negative tone of the story, the headlines and photographs, and the accompanying visual and auditory effects if the story is being broadcast. The war in Afghanistan, like other wars in the past, was typically framed as a fight to save freedom and democracy.

Democracy

Democracy **is a political system in which the people hold the ruling power either directly or through elected representatives.** The literal meaning of *democracy* is "rule by the people" (from the Greek words *demos*, meaning "the people," and *kratein*, meaning "to rule"). In an ideal-type democracy, people would actively and directly rule themselves. *Direct participatory democracy* requires that citizens be able to meet together regularly to debate and decide the issues of the day. However, if all 272 million people in the United States came together in one place for a meeting, they would occupy an area of more than seventy square miles, and a single round of five-minute speeches would require more than five thousand years (based on Schattschneider, 1969).

In countries such as the United States, Canada, Australia, and Great Britain, people have a voice in the government through *representative democracy*, whereby citizens elect representatives to serve as bridges between themselves and the government. The U.S. Constitution requires that each state have two senators and a minimum of one member in the House of Representatives. The current size of the House (435 seats) has not changed since the apportionment following the 1910 census. Therefore, based on Census 2000, those 435 seats have been reapportioned based on the increase or decrease in a state's population between 1990 and 2000 (see "Census Profiles: Political Representation and Shifts in the U.S. Population").

In a representative democracy, elected representatives are supposed to convey the concerns and interests of those they represent, and the government is expected to be responsive to the wishes of the people. Elected officials are held accountable to the people through elections. However, representative democracy is not always equally accessible to all people in a nation. Throughout U.S. history, members of subordinate racial–ethnic groups have been denied full participation in the democratic process. Gender and social class have also limited some people's democratic participation. For example, women

Authoritarianism a political system controlled by rulers who deny popular participation in government.

Totalitarianism a political system in which the state seeks to regulate all aspects of people's public and private lives.

Democracy a political system in which the people hold the ruling power either directly or through elected representatives.

CENSUS PROFILES — Political Representation and Shifts in the U.S. Population

The increase or decrease in the number of members of the House of Representatives elected by the voters in each state is a reflection of how the U.S. population has shifted over the last decade. Increases in population occurred in a number of states located in the Southeast, Southwest, and West, as contrasted with a relative decline in the population of other states, primarily in the Midwest and Northeast. What sociological factors do you think may have contributed to this shift in the U.S. population?

States with an Increase in House Seats

Arizona (2 more seats)
California (1 more seat)
Colorado (1 more seat)
Florida (2 more seats)
Georgia (2 more seats)
Nevada (1 more seat)
North Carolina (1 more seat)
Texas (2 more seats)

States with a Decrease in House Seats

Connecticut (1 fewer seat)
Illinois (1 fewer seat)
Indiana (1 fewer seat)
Michigan (1 fewer seat)
Mississippi (1 fewer seat)
New York (2 fewer seats)
Ohio (1 fewer seat)
Oklahoma (1 fewer seat)
Pennsylvania (2 fewer seats)
Wisconsin (1 fewer seat)

Source: U.S. Census Bureau, 2001a.

✓ Checkpoints

Checkpoint 13.2

5. What is a monarchy, and where do monarchies exist today?
6. How does authoritarianism compare with totalitarianism?
7. What is democracy?

---

would receive 40 seats in a hundred-seat legislative body, and a party receiving 20 percent of the votes would receive 20 seats.

Perspectives on Power and Political Systems

Is political power in the United States concentrated in the hands of the few or distributed among the many? Sociologists and political scientists have suggested many different answers to this question; however, two prevalent models of power have emerged: pluralist and elite.

Functionalist Perspectives: The Pluralist Model

The pluralist model is rooted in a functionalist perspective that assumes people share a consensus on central concerns, such as freedom and protection from harm, and that the government serves important functions no other institution can fulfill. According to Emile Durkheim (1933/1893), the purpose of government is to socialize people to be good citizens, to regulate the economy so that it operates effectively, and to provide necessary services for citizens. Contemporary functionalists state the four main functions as follows: (1) maintaining law and order, (2) planning and directing society, (3) meeting social needs, and (4) handling international relations, including warfare.

But what happens when people do not agree on specific issues or concerns? Functionalists suggest that divergent viewpoints lead to a system of political pluralism, in which the government functions as an arbiter between competing interests and viewpoints. According to the *pluralist model*, **power in political systems is widely dispersed**

---

have not always had the same rights as men. Full voting rights were not gained by women until ratification of the Nineteenth Amendment in 1920.

Even representative democracies are not all alike. As compared to the winner-takes-all elections in the United States, which are usually decided by who wins the most votes, the majority of European elections are based on a system of proportional representation, meaning that each party is represented in the national legislature according to the proportion of votes that party received. For example, a party that won 40 percent of the vote

throughout many competing interest groups (Dahl, 1961).

According to the pluralist model, the diverse needs of women and men, people of all religions and racial–ethnic backgrounds, and the wealthy, middle class, and poor are met by political leaders who engage in a process of bargaining, accommodation, and compromise. Competition among leadership groups in government, business, labor, education, law, medicine, and consumer organizations, among others, helps prevent abuse of power by any one group. Everyday people can influence public policy by voting in elections, participating in existing special interest groups, or forming new ones to gain access to the political system. In sum, power is widely dispersed, and leadership groups that wield influence on some decisions are not the same groups that may be influential in other decisions (Dye and Zeigler, 2000).

Paul Hosefros/NYT Pictures

Special interest groups help people advocate their interests and further their causes. Advocates may run for public office and gain a wider voice in the political process. An example is U.S. Senator Ben Nighthorse Campbell of Colorado (left), a Cheyenne chief, who is shown here as he recently participated in a ground-breaking ceremony for the National Museum of the American Indian in Washington, D.C.

SPECIAL INTEREST GROUPS *Special interest groups* are political coalitions made up of individuals or groups that share a specific interest they wish to protect or advance with the help of the political system (Greenberg, 1999). Examples of special interest groups include the AFL-CIO (representing the majority of labor unions) and public interest or citizens' groups such as the American Conservative Union and Zero Population Growth.

What purpose do special interest groups serve in the political process? According to some analysts, special interest groups help people advocate their own interests and further their causes. Broad categories of special interest groups include banking, business, education, energy, the environment, health, labor, persons with a disability, religious groups, retired persons, women, and those espousing a specific ideological viewpoint; obviously, many groups overlap in interests and membership. Special interest groups are also referred to as *pressure groups* (because they put pressure on political leaders) or *lobbies*. Lobbies are often referred to in terms of the organization they represent or the single issue on which they focus—for example, the "gun lobby" and the "dairy lobby." The people who are paid to influence legislation on behalf of specific clients are referred to as *lobbyists*.

Over the past four decades, special interest groups have become more involved in "single-issue politics," in which political candidates are often supported or rejected solely on the basis of their views on a specific issue—such as abortion, gun control, gay and lesbian rights, or the environment. Single-issue groups derive their strength from the intensity of their beliefs; leaders have little room to compromise on issues.

POLITICAL ACTION COMMITTEES Funding of lobbying efforts has become more complex in recent years. Reforms in campaign finance laws in the 1970s set limits on direct contributions to political candidates and led to the creation of

Pluralist model an analysis of political systems that views power as widely dispersed throughout many competing interest groups.

political action committees (PACs)—organizations of special interest groups that solicit contributions from donors and fund campaigns to help elect (or defeat) candidates based on their stances on specific issues.

As the cost of running for political office has skyrocketed, candidates have relied more on PACs for financial assistance. PACs contributed more than $245,000,000 to candidates for the U.S. House and Senate during 1999 and 2000, an increase of 19 percent over the preceding two-year period (Federal Election Commission, 2001). Advertising, staff, direct-mail operations, telephone banks, computers, consultants, travel expenses, office rentals, and other expenses incurred in political campaigns make PAC money vital to candidates.

Some PACs represent the "public interest" and ideological interest groups such as gay rights or the National Rifle Association. Other PACs represent the capitalistic interests of large corporations. Realistically, members of the least-privileged sectors of society are not represented by PACs. As one senator pointed out, "There aren't any Poor PACs or Food Stamp PACs or Nutrition PACs or Medicare PACs" (qtd. in Greenberg and Page, 1993: 240). Critics of pluralism argue that "Big Business" wields such disproportionate power in U.S. politics that it undermines the democratic process (see Lindblom, 1977; Domhoff, 1978).

As an outgrowth of record-setting campaign spending in the 1996 national election, campaign financing abuses were alleged by both Republicans and Democrats in Washington. At the center of the controversy was the issue of "soft money" contributions, which are made outside the limits imposed by federal election laws. Over $260 million in soft money raised by the two political parties and foreign contributions, particularly from Asian sources, came under intense legal and social scrutiny (Bagby, 1998). Investigations into campaign fund-raising and spending will no doubt continue into the future as the costs of running for public office continue to grow dramatically.

Conflict Perspectives: Elite Models

Although conflict theorists acknowledge that the government serves a number of important purposes in society, they assert that government exists for the benefit of wealthy or politically powerful elites who use the government to impose their will on the masses. According to the *elite model*, **power in political systems is concentrated in the hands of a small group of elites and the masses are relatively powerless.** The pluralist model and the elite model are compared in Figure 13.1.

Contemporary elite models are based on the assumption that decisions are made by the elites, who agree on the basic values and goals of society. However, the needs and concerns of the masses are not often given full consideration by those in the elite. According to this approach, power is highly concentrated at the top of a pyramid-shaped social hierarchy, and public policy reflects the values and preferences of the elite, not the preferences of the people (Dye and Ziegler, 2000).

C. WRIGHT MILLS AND THE POWER ELITE
Who makes up the U.S. power elite? According to the sociologist C. Wright Mills (1959a), the *power elite* comprises leaders at the top of business, the executive branch of the federal government, and the military. Of these three, Mills speculated that the "corporate rich" (the highest-paid officers of the biggest corporations) were the most powerful because of their unique ability to parlay the vast economic resources at their disposal into political power. At the middle level of the pyramid, Mills placed the legislative branch of government, special interest groups, and local opinion leaders. The bottom (and widest layer) of the pyramid is occupied by the unorganized masses, who are relatively powerless and are vulnerable to economic and political exploitation.

G. WILLIAM DOMHOFF AND THE RULING CLASS Sociologist G. William Domhoff (1998) asserts that in fact, this nation has a *ruling class*—the corporate rich, who make up less than 1 percent of the U.S. population. Domhoff uses the term *ruling class* to signify a relatively fixed group of privileged people who wield sufficient power to constrain political processes and serve underlying capitalist interests. Although the power elite controls the everyday operation of the political system, who *governs* is less important than who *rules*.

According to Domhoff (1998), the corporate rich influence the political process in three ways. First, they affect the candidate selection process by helping to finance campaigns and providing favors

Figure 13.1 Pluralist and Elite Models

PLURALIST MODEL

ELITE MODEL

- Decisions are made on behalf of the people by leaders who engage in bargaining, accommodation, and compromise.

- Competition among leadership groups makes abuse of power by any one group difficult.

- Power is widely dispersed, and people can influence public policy by voting.

- Public policy reflects a balance among competing interest groups.

- Decisions are made by a small group of elite people.

- Consensus exists among the elite on the basic values and goals of society.

- Power is highly concentrated at the top of a pyramid-shaped social hierachy.

- Public policy reflects the values and preferences of the elite.

to political candidates. Second, through participation in the special interest process, the corporate rich are able to obtain favors, tax breaks, and favorable regulatory rulings. Finally, the corporate rich gain access to the policy-making process through their appointments to governmental advisory committees, presidential commissions, and other governmental positions.

Power elite models call our attention to a central concern in contemporary U.S. society: the ability of democracy and its ideals to survive in the context of the increasingly concentrated power held by capitalist oligarchies such as the media giants we discuss in this chapter.

✓ Checkpoints

Checkpoint 13.3

8. Why is the pluralist model considered to be a functionalist perspective?
9. What are special interest groups, and what purpose do they serve in the political process?
10. Why are political action committees important in the U.S. political system?
11. How does the elite model of C. Wright Mills compare with that of G. William Domhoff?

Political action committees (PACs) organizations of special interest groups that solicit contributions from donors and fund campaigns to help elect (or defeat) candidates based on their stances on specific issues.

Elite model a view of society that sees power in political systems as being concentrated in the hands of a small group of elites whereas the masses are relatively powerless.

The cost of mounting a campaign for president of the United States has grown dramatically with each passing decade. Prior to celebrating at the presidential gala of 2001, President George W. Bush raised a record $44 million for Republicans at presidential dinners.

Manny Ceneta/AFP-CORBIS

The U.S. Political System

The U.S. political system is made up of formal elements, such as the legislative process and the duties of the president, and informal elements, such as the role of political parties in the election process. We now turn to an examination of these informal elements, including political parties, political socialization, and voter participation.

Political Parties and Elections

A *political party* is an organization whose purpose is to gain and hold legitimate control of government; it is usually composed of people with similar attitudes, interests, and socioeconomic status. A political party (1) develops and articulates policy positions, (2) educates voters about issues and simplifies the choices for them, and (3) recruits candidates who agree with those policies, helps those candidates win office, and holds the candidates responsible for implementing the party's policy positions. In carrying out these functions, a party may try to modify the demands of special interests, build a consensus that could win majority support, and provide simple and identifiable choices for the voters on election day. Political parties create a *platform*, a formal statement of the party's political positions on various social and economic issues.

Since the Civil War, the Democratic and Republican parties have dominated the U.S. political system. Although one party may control the presidency for several terms, at some point the voters elect the other party's nominee, and control shifts.

How well do the parties measure up to the ideal-type characteristics of a political party? Although both parties have been successful in getting their candidates elected at various times, they generally do not meet the ideal characteristics for a political party because they do not offer voters clear policy alternatives. Moreover, the two parties are oligarchies, dominated by active elites who hold views that are further from the center of the political spectrum than are those of a majority of members of their party. As a result, voters in primary elections (in which the nominees of political parties for most offices other than president and vice president are chosen) may select nominees whose views are closer to the center of the political spectrum and further away from the party's own platform. Likewise, party loyalties appear to be declining among voters, who may vote in one party's primary but then cast their ballot in general elections without total loyalty to that party, or cast a "split-ticket" ballot (voting for one party's candidate in one race and another party's candidate in another).

Political Participation and Voter Apathy

Why do some people vote and others not? How do people come to think of themselves as being conservative, moderate, or liberal? Key factors include individuals' political socialization and attitudes.

Political socialization **is the process by which people learn political attitudes, values, and behavior.** For young children, the family is the primary agent of political socialization, and children tend to learn and hold many of the same opinions held by their parents. By the time children reach school age, they typically identify with the political party (if any) of their parents (Burnham, 1983). As we grow older, other agents of socialization begin to affect our political beliefs, including our peers, teachers, and the media. Over time, these other agents may cause people's political attitudes and values to change.

In addition to the socialization process, people's socioeconomic status affects their political attitudes, values, and beliefs. For example, individuals who are very poor or are unable to find employment may believe that society has failed them and therefore tend to be indifferent toward the political system (Zipp, 1985; Pinderhughes, 1986). Believing that casting a ballot would make no difference to their circumstances, they do not vote.

Democracy in the United States has been defined as a government "of the people, by the people, and for the people." Accordingly, it would stand to reason that "the people" would actively participate in their government at any or all of four levels: (1) voting, (2) attending and taking part in political meetings, (3) actively participating in political campaigns, and (4) running for and/or holding political office. At most, about 10 percent of the voting-age population in this country participates at a level higher than simply voting, and over the past forty years, less than half of the voting-age population has voted in nonpresidential elections. Even in presidential elections, voter turnout is often relatively low. Slightly over 51 percent of the voting-age population (age eighteen and older) voted in the 2000 presidential election—an election so close that its outcome wasn't certain until more than a month after election day (see Box 13.2).

The United States has one of the lowest percentages of voter turnout of all Western nations. In many of the other Western nations, the average turnout is between 80 and 90 percent of all eligible voters. Why is it that so many eligible voters in this country stay away from the polls? During any election, millions of voting-age persons do not go to the polls due to illness, disability, lack of transportation, nonregistration, or absenteeism. However, these explanations do not account for why many other people do not vote. According to some conservative analysts, people may not vote because they are satisfied with the status quo or because they are apathetic and uninformed—they lack an understanding of both public issues and the basic processes of government.

By contrast, liberals argue that people stay away from the polls because they feel alienated from politics at all levels of government—federal, state, and local—due to political corruption and influence peddling by special interests and large corporations. Participation in politics is influenced by gender, age, race/ethnicity, and—especially—socioeconomic status (SES). The rate of participation increases as a person's SES increases. Table 13.1 on page 423 shows voting preferences by gender, race/ethnicity, age, sexual orientation, education, region, and income in the 2000 presidential election. One explanation for the higher rates of political participation at higher SES levels is that advanced levels of education may give people a better understanding of government processes, a belief that they have more at stake in the political process, and greater economic resources to contribute to the process. Some studies suggest that during their college years, many people develop assumptions about political participation that continue throughout their lives.

Governmental Bureaucracy

When most people think of political power, they overlook one of its major sources—the governmental bureaucracy. Negative feelings about

Reflections

Reflections 13.2

What role do the media play in political socialization in the United States? Do you believe that there are valid complaints about how that role is played?

Political party   an organization whose purpose is to gain and hold legitimate control of government.

Political socialization   the process by which people learn political attitudes, values, and behavior.

Box 13.2 Sociology and Social Policy

The Electoral College: Is School Over for This Body?

Sociologists who study political power are interested in the social and political conflicts that arise over the allocation of power in a society. Such a conflict occurred in the 2000 election for president of the United States, resulting in some social policy advocates calling for a change in how this important office is filled.

On Tuesday, November 7, 2000, more than 100 million U.S. citizens cast their votes in the presidential election. As is customary, the news media watched and reported: First, they reported that Texas governor George W. Bush had won the election, then that U.S. vice president Al Gore had overcome Bush's lead in various "key states," so the outcome was not certain.

In electing the U.S. president, why do "key states" matter? Why not just be bound by the total vote? If the result of the election were determined by who received the most votes, Gore—who received about a half-million more votes than did Bush—would have been elected president, yet Bush was sworn in as the next president. Why? The answer is that the framers of the Constitution wanted the president of the United States to be elected by the *states*, not by a plurality of the nation's voters. Accordingly, the U.S. Constitution provides that the president is elected by vote of "electors" selected by the various states and that the number of electors allocated to each state is equal to that state's number of U.S. senators and representatives. In 2000, Bush won more states with more electors than did Gore, although even that was not certain until December 12, 2000, when the U.S. Supreme Court ruled (in a seven-to-two decision in the case of *George W. Bush v. Albert Gore*) that a Florida Supreme Court order requiring a recount of certain inaccurately marked ballots to determine the voters' intent had "constitutional problems" and therefore had to be set aside. As a result, Bush received Florida's electoral votes. It was the first time since 1888 that the person receiving the highest number of votes nationwide was not the winner.

Numerous political analysts and social policy advocates began to call for a constitutional amendment that would eliminate the Electoral College and elect as president the candidate receiving the most votes. Obviously, many of these were people who had supported Gore, and their reaction was viewed by others as simply that of a poor loser. However, before either agreeing or disagreeing with them, it is important to look at the social policy reasons for having—or doing away with—the current system of electing the nation's president.

The delegates to the Constitutional Convention in 1787 realized that they needed to create a national government that was strong enough to provide national security, promote domestic tranquility, and regulate interstate commerce—factors that remain vitally important to us today. But the founders disagreed about how strong that central government should be. Delegates from the wealthiest and most populous states—Massachusetts, Pennsylvania, and Virginia—wanted a strong government and a legislature whose seats would be apportioned based on population. Delegates from the smaller states believed that such a government would be controlled by those populous states, so they wanted a legislature in which each state would have one seat. The Constitution represented a compromise between those two points of view: Each state would have equal representation in one house of the legislature, whereas seats in the other house would be apportioned on the basis of population (Greenberg, 1999). With presidential electors apportioned on the combined representation of each state in the legislature, neither the populous states nor the larger number of states with smaller populations would be able to elect a president without support from some of the others.

As a result of this provision, a presidential candidate has to have broad appeal to voters in many states in order to be elected. Critics of the current method of electing presidents argue that, with the rapid flow of information and people across state lines, state boundaries and interests are no longer relevant in electing the nation's chief executive. Supporters of the electoral system counter with the argument that state boundaries and interests remain important and that the current system is necessary in order to stop the election of a candidate who represents only the interests of some particular region of the nation.

What are the sociological implications of social policy questions such as this? How are the lives of everyday people affected by political processes such as the presidential election? What do you think?

Table 13.1 Voter Preferences in the 2000 Presidential Election, by Selected Characteristics

		Republican	Democrat
Gender	Men	52%	43%
	Women	43	54
Race/Ethnicity	Whites	53	42
	African Americans	8	90
	Latinos/as	32	64
	Asian Americans	38	57
Age	18–29	45	48
	30–44	48	48
	45–59	49	47
	60 and older	51	23
Sexual Orientation	Gay, lesbian, bisexual	26	67
Education	Did not graduate from high school	39	59
	High school graduate	49	48
	Some college	50	45
	College graduate	49	46
Region	Eastern United States	39	56
	Midwest	49	47
	Southern United States	52	45
	Western United States	45	48
Family Income	Under $15,000	37	57
	$15,000–$29,999	41	53
	$30,000–$49,999	46	49
	Over $50,000	51	46

Note: Due to the presence of other candidates on the ballot, columns do not total 100 percent.

Source: Lester, 2000.

bureaucracy are perhaps strongest when people are describing the "red tape" and "faceless bureaucrats" in government with whom they must deal. But who are these "faceless bureaucrats," and what do they do? Roughly 90 percent of the top-echelon positions in the federal government are held by white men (Weiner, 1994).

Because rising to the top of the bureaucracy may take as much as twenty years, some analysts argue that white women and all people of color simply have not held positions in the federal government long enough to reach the top positions. Others argue that the entrenched rules of this bureaucracy, one of the most powerful in the world,

weigh against the promotion of those individuals who are not white and male (Weiner, 1994).

The governmental bureaucracy has been able to perpetuate itself and expand because many of its employees have highly specialized knowledge and skills and cannot be replaced easily by "outsiders." In addition, as the United States has grown in size and complexity, public policy is made more by bureaucrats than by elected officials. For example, offices and agencies have been established to create rules, policies, and procedures for dealing with complex issues such as nuclear power, environmental protection, and drug safety; bureaucracies announce an estimated twenty rules or regulations

Figure 13.2 Example of the Iron Triangle of Power

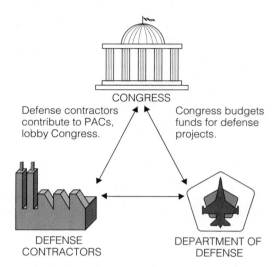

CONGRESS

Defense contractors contribute to PACs, lobby Congress.

Congress budgets funds for defense projects.

DEFENSE CONTRACTORS

DEPARTMENT OF DEFENSE

Defense Department awards contracts to contractors; personnel move back and forth between employment by Defense Department and by defense contractors.

for every one law passed by Congress (Dye and Ziegler, 2000).

The federal budget is the central ingredient in the bureaucracy. Preparing the annual federal budget is a major undertaking for the president and the Office of Management and Budget, one of the most important agencies in Washington. Getting the budget approved by Congress is an even more monumental task; however, even with the highly publicized wrangling over the budget by the president and Congress, the final congressional appropriations are usually within 2–3 percent of the budget originally proposed by the president (Dye and Ziegler, 2000).

In part, this is due to the *iron triangle of power*—a three-way arrangement in which a private interest group (usually a corporation), a congressional committee or subcommittee, and a bureaucratic agency make the final decision on a political issue that is to be decided by that agency. Figure 13.2 illustrates the alliance among the Defense Department (Pentagon), private military (or defense) contractors, and members of Congress. According to the sociologist Joe Feagin,

The Iron Triangle has a revolving door of money, influence, and jobs among these three sets of actors, involving trillions of dollars. Military contractors who receive contracts from the Defense Department serve on the advisory committees that recommend what weapons they believe are needed. Many people move around the triangle from job to job, serving in the military, then in the Defense Department, then in military industries. (Feagin and Feagin, 1994: 405)

The Iron Triangle is also referred to as the *military–industrial complex*—the mutual interdependence of the military establishment and private military contractors. According to the sociologist C. Wright Mills (1976), this alliance of economic, military, and political power amounts to a "permanent war economy" or "military economy." However, the economist John Kenneth Galbraith (1985) has argued that the threat of war is good for the economy because government money spent on military preparedness stimulates the private sector of the economy, creates jobs, and encourages consumer spending. For example, at the beginning of the war on terrorism, the Pentagon awarded what was expected to become the largest military contract in U.S. history to Lockheed Martin to build a new generation of supersonic stealth fighter jets. The contract was expected to be worth more than $200 billion, creating thousands of new jobs in an economy that was struggling at the time of the announcement (Dao, 2000). Overall, terrorist attacks on the United States resulted in large expenditures on national security and on domestic security. As a result of the sudden, unexpected attack on the United States on September 11, 2001, many political leaders believed that the ability of the government to quickly respond with troops and sophisticated weapons systems justified

✓ Checkpoints

Checkpoint 13.4

12. What is a political party?
13. What is political socialization?
14. Why is there concern about levels of political participation and voter apathy in the United States?
15. What is the military–industrial complex?

the expenditures that had been made on the military–industrial complex over the years. However, others believed that such expenditures were problematic. Issues regarding the military–industrial complex are intricately linked to problems in the economy, which we will examine now.

Economic Systems in Global Perspective

The *economy* **is the social institution that ensures the maintenance of society through the production, distribution, and consumption of goods and services.** *Goods* are tangible objects that are necessary (such as food, clothing, and shelter) or desired (such as VCRs and electric toothbrushes). *Services* are intangible activities for which people are willing to pay (such as dry cleaning, a movie, or medical care).

Preindustrial, Industrial, and Postindustrial Economies

In all societies, the specific method of producing goods is related to the technoeconomic base of the society, as discussed in Chapter 10. In each society, people develop an economic system, ranging from simple to very complex, for the sake of survival.

Preindustrial economies include hunting and gathering, horticultural and pastoral, and agrarian societies. Most workers engage in *primary sector production*—**the extraction of raw materials and natural resources from the environment.** These materials and resources are typically consumed or used without much processing. The production units in hunting and gathering societies are small; most goods are produced by family members. The division of labor is by age and by gender (Hodson and Sullivan, 2002). The potential for producing surplus goods increases as people learn to domesticate animals and grow their own food. In horticultural and pastoral societies, the economy becomes distinct from family life. The distribution process becomes more complex with the accumulation of a *surplus* such that some people can engage in activities other than food production. In agrarian societies, production is related primarily to producing food. However, workers have a greater variety of specialized tasks, such as warlord or priest; for ex-

Charles Orrico/SuperStock

The future of work has changed dramatically over the past century. Not all social scientists agree on whether robots will eventually replace many humans in the work force. What do you think work will be like in the future?

ample, warriors are necessary to protect the surplus goods from plunder by outsiders (Hodson and Sullivan, 2002).

Industrial economies result from sweeping changes to the system of production and distribution of goods and services during industrialization. Drawing on new forms of energy (such as steam, gasoline, and electricity) and machine technology, factories proliferate as the primary means of producing goods.

Military–industrial complex the mutual interdependence of the military establishment and private military contractors.

Economy the social institution that ensures the maintenance of society through the production, distribution, and consumption of goods and services.

Primary sector production the sector of the economy that extracts raw materials and natural resources from the environment.

Most workers engage in *secondary sector production*—**the processing of raw materials (from the primary sector) into finished goods.** For example, steel workers process metal ore; auto workers then convert the ore into automobiles, trucks, and buses. In industrial economies, work becomes specialized and repetitive, activities become bureaucratically organized, and workers primarily work with machines instead of with one another. With the emergence of mass production, larger surpluses are generated, typically benefiting some people and organizations but not others.

A *postindustrial economy* is based on *tertiary sector production*—**the provision of services rather than goods**—as a primary source of livelihood for workers and profit for owners and corporate shareholders. Tertiary sector production includes a wide range of activities, such as fast-food service, transportation, communication, education, real estate, advertising, sports, and entertainment.

Capitalism and socialism are the principal economic models in industrial and postindustrial countries. As we examine these two models, keep in mind that no society has a purely capitalist or socialist economy.

Capitalism

Capitalism **is an economic system characterized by private ownership of the means of production, from which personal profits can be derived through market competition and without government intervention.** Most of us think of ourselves as "owners" of private property because we own a car, a television, or other possessions. However, most of us are not capitalists; we *spend money* on the things we own rather than *make money* from them. Only a relatively few people own income-producing property from which a profit can be realized by producing and distributing goods and services; everyone else is a consumer. "Ideal" capitalism has four distinctive features: (1) private ownership of the means of production, (2) pursuit of personal profit, (3) competition, and (4) lack of government intervention.

PRIVATE OWNERSHIP OF THE MEANS OF PRODUCTION Capitalist economies are based on the right of individuals to own income-producing property, such as land, water, mines, and factories, and the right to "buy" people's labor. However,

under early monopoly capitalism (1890–1940), most ownership shifted from individuals to huge *corporations*—**organizations that have legal powers, such as the ability to enter into contracts and buy and sell property, separate from their individual owners.** In advanced monopoly capitalism (1940–present), ownership and control of major industrial and business sectors have become increasingly concentrated, and many corporations have become more global in scope. *Transnational corporations* **are large corporations that are headquartered in one country but sell and produce goods and services in many countries.** Twenty of the largest U.S. transnationals are listed in Table 13.2.

PURSUIT OF PERSONAL PROFIT A tenet of capitalism is the belief that people are free to maximize their individual gain through personal profit; in the process, the entire society will benefit from their activities (Smith, 1976/1776). Economic development is assumed to benefit both capitalists and workers, and the general public also benefits from public expenditures (such as for roads, schools, and parks) made possible through an increase in business tax revenues. However, critics assert that specific individuals and families (not the general public) have been the primary recipients of profits. In early monopoly capitalism, stockholders in companies such as American Tobacco, Westinghouse, and Sears, Roebuck derived massive profits from companies that held near-monopolies on specific goods and services (Hodson and Sullivan, 2002). In advanced (late) monopoly capitalism, profits have become even more concentrated.

COMPETITION In theory, competition acts as a balance to excessive profits. When producers vie with one another for customers, they must be able to offer innovative goods and services at competitive prices. However, from the time of early industrial capitalism, the trend has been toward less, rather than more, competition among companies. In early monopoly capitalism, competition was diminished by increasing concentration *within* a particular industry, a classic case being the virtual monopoly on oil held by John D. Rockefeller's Standard Oil Company (Tarbell, 1925/1904; Lundberg, 1988). Today, Microsoft Corporation has been the subject of federal investigations and lawsuits because it so dominates certain areas of the computer software

Table 13.2 The 20 Largest U.S. Transnational Corporations (ranked by foreign revenues)

Corporation	Product	Foreign Revenues ($ Billions)	Total Revenues ($ Billions)	Assets ($ Billions)
Exxon/Mobil	Oil refining	115.4	160.8	94.0
IBM	Office equipment	50.3	87.5	34.2
Ford	Motor vehicles	50.1	162.5	49.7
General Motors	Motor vehicles	46.4	176.5	33.2
General Electric	Electronics	35.3	111.6	41.0
Texaco	Oil refining	32.7	42.4	15.7
Citigroup	Banking	28.7	82.0	239.8
Hewlett-Packard	Office equipment	23.3	42.3	4.3
Wal-Mart Stores	Retail sales	22.7	165.0	70.3
Compaq Computer	Computer equipment	21.1	38.5	3.2
American International Group	Insurance	20.3	40.6	268.2
Chevron	Oil refining	20.0	45.1	45.6
Philip Morris	Tobacco products	19.6	61.7	23.6
Proctor & Gamble	Cleaning products	18.3	38.1	32.1
Motorola	Office equipment	17.7	30.9	37.3
Intel	Computer peripherals	16.6	29.3	11.7
E.I. du Pont	Chemicals	13.2	26.9	14.8
Xerox	Office products	12.6	23.1	3.6
Lucent Technologies	Communication technology	12.1	38.3	8.0
Coca-Cola	Beverages	12.1	19.8	14.8

Source: Forbes, 2000a.

industry that it has virtually no competitors in those areas. In other situations, several companies may dominate certain industries. An *oligopoly* **exists when several companies overwhelmingly control an entire industry.** An example is the music industry, in which a few giant companies are behind many of the labels and artists (see Table 13.3). More specifically, a *shared monopoly* **exists when four or fewer companies supply 50 percent or more of a particular market.** Examples include U.S. automobile manufacturers (referred to as the "Big Three") and cereal companies (three of which control 77 percent of the market).

In advanced monopoly capitalism, mergers also occur *across* industries: Corporations gain

Secondary sector production the sector of the economy that processes raw materials (from the primary sector) into finished goods.

Tertiary sector production the sector of the economy that is involved in the provision of services rather than goods.

Capitalism an economic system characterized by private ownership of the means of production, from which personal profits can be derived through market competition and without government intervention.

Corporations organizations that have legal powers, such as the ability to enter into contracts and buy and sell property, separate from their individual owners.

Transnational corporations large corporations that are headquartered in one country but sell and produce goods and services in many countries.

Table 13.3 The Music Industry's Big Five

Company	Country	Leading Artists
BMG Entertainment (RCA, Arista)	Germany	Kenny G Next Usher
Universal Music Group (MCA, Polygram)	Canada	K-Ci & Jojo George Strait Wynonna Judd
Sony Music Entertainment	Japan	Mariah Carey Michael Jackson Savage Garden
EMI Group	United Kingdom	Spice Girls Janet Smashing Pumpkins
Warner Music Group (Atlantic, Elektra)	United States	Brandy Hootie and the Blowfish Jewel

Sources: Biagi, 1994: 248, Biagi, 1998: 149.

near-monopoly control over all aspects of the production and distribution of a product by acquiring both the companies that supply the raw materials and the companies that are the outlets for the product. For example, an oil company may hold leases on the land where the oil is pumped out of the ground, own the plants that convert the oil into gasoline, and own the individual gasoline stations that sell the product to the public.

Corporations with control both within and across industries are often formed by a series of mergers and acquisitions across industries. These corporations are referred to as *conglomerates*—**combinations of businesses in different commercial areas, all of which are owned by one holding company.** Media ownership is a case in point; companies such as AOL Time Warner have extensive holdings in radio and television stations, cable television companies, book publishing firms, and film production and distribution companies, to name only a few. Today, most of our news originates from ten massive conglomerates that dominate the global media market. According to media scholars, three

factors triggered the rapid growth of the media conglomerates: (1) aggressive political and economic maneuvering by dominant media firms, (2) introduction of new technologies that increased the cost efficiency of global systems, and (3) policies of the U.S. government and organizations such as the World Bank, International Monetary Fund, and World Trade Organization (McChesney, 1998).

Although media titans such as Rupert Murdock, head of News Corporation—the conglomerate that owns over a hundred newspapers worldwide, major movie studios, publishing interests, the Fox TV network, and cable channels—seldom discuss the business dealings or political negotiations involved in mass media and telecommunications megamergers, journalists occasionally provide insights on the thinking of such individuals. For example, the journalist Ken Aulette describes a discussion he had with Rupert Murdock about the continual growth of Murdock's media empire:

Does [he] tire of the constant competition to get bigger, to win each war? When does Murdock say *enough?* I asked him this question at the end of a long night that started with a drink and a stroll over his six acre property in Beverly Hills. . . . After dinner in the dining room of his comfortable Spanish-style home, Murdock sipped a glass of California Chardonnay and treated the question as something so alien as to be incomprehensible. "You go on," he responded, opaquely. "There is a global village in some sense. You are competing everywhere. . . . And I just enjoy it." (Aulette, 1998: 287)

When questioned on another occasion regarding mergers, joint ventures, and cross-ownership of media firms, Murdock stated, "We [media and telecommunications firms] can join forces now, or we can kill each other and then join forces" (McChesney, 1998: 15). Murdock and other media CEOs acknowledge that two or three companies will eventually dominate the media and telecommunications industry; however, in their view this is not necessarily a bad thing. As Murdock commented, "Monopoly is a terrible thing until you have it" (McChesney, 1998: 54).

Competition is also reduced over the long run by *interlocking corporate directorates* — **members of the board of directors of one corporation who also sit on the board(s) of other corporations.** Although the Clayton Antitrust Act of 1914 made it illegal for a person to sit simultaneously on the boards of directors of two corporations that are in *direct* competition with each other, a person may serve simultaneously on the board of a financial institution (a bank, for example) and the board of a commercial corporation (a computer manufacturing company or a furniture store chain, for example) that borrows money from the bank. Directors of competing corporations may also serve together on the board of a third corporation that is not in direct competition with the other two. In 1996, former Defense Secretary Frank C. Carlucci sat on the corporate boards of fourteen Fortune 1,000 companies;

former Labor Secretary Ann D. McLaughlin sat on eleven; and former Health, Education, and Welfare Secretary Joseph A. Califano, Jr., sat on nine (Dobrzynski, 1996). An example of interlocking directorates is depicted in Figure 13.3. Interlocking directorates diminish competition by producing interdependence. Individuals who serve on multiple boards are often able to forge cooperative arrangements that benefit their corporations but not necessarily the general public. When several corporations are controlled by the same financial interests, they are more likely to cooperate with one another than to compete (Mintz and Schwartz, 1985).

LACK OF GOVERNMENT INTERVENTION
Ideally, capitalism works best without government intervention in the marketplace. The policy of *laissez-faire* (les-ay-FARE, which means "leave alone") was advocated by the economist Adam Smith in his 1776 treatise *An Inquiry into the Nature and Causes of the Wealth of Nations.* Smith argued that when people pursue their own selfish interests, they are guided "as if by an invisible hand" to promote the best interests of society (see Smith, 1976/1776). Today, terms such as *market economy* and *free enterprise* are often used, but the underlying assumption is the same: that free market competition, not the government, should regulate prices and wages. However, the "ideal" of unregulated markets benefiting all citizens has seldom been realized. Individuals and companies in pursuit of higher profits have run roughshod over weaker competitors, and small businesses have grown into large, monopolistic corporations. Accordingly, government regulations were implemented in an effort to curb the excesses of the marketplace brought about by laissez-faire policies.

Reflections

Reflections 13.3

Why is there widespread concern about the growth of media conglomerates in a democratic society? In your opinion, is this concern valid?

Oligopoly a condition existing when several companies overwhelmingly control an entire industry.

Shared monopoly a condition that exists when four or fewer companies supply 50 percent or more of a particular market.

Conglomerates combinations of businesses in different commercial areas, all of which are owned by one holding company.

Interlocking corporate directorates members of the board of directors of one corporation who also sit on the board(s) of other corporations.

Figure 13.3 The General Motors Board of Directors

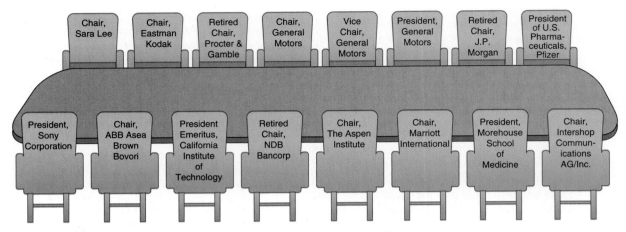

The 1999 General Motors Board of Directors shows the nature of interlocking directorates. On the chair representing each of the directors is the name of another entity each director is connected with, and his or her position with that entity.

Source: General Motors, 1999.

However, much of what is referred to as government intervention has been in the form of aid to business. Between 1850 and 1900, corporations received government assistance in the form of public subsidies and protection from competition by tariffs, patents, and trademarks. Government intervention in the 1990s included billions of dollars in subsidies to farmers, tax credits for corporations, and large subsidies or loan guarantees to automakers, aircraft companies, railroads, and others. Overall, most corporations have gained much more than they have lost as a result of government involvement in the economy.

Socialism

Socialism **is an economic system characterized by public ownership of the means of production, the pursuit of collective goals, and centralized decision making.** Like "pure" capitalism, "pure" socialism does not exist. Karl Marx described socialism as a temporary stage en route to an ideal communist society. Although the terms *socialism* and *communism* are associated with Marx and are often used interchangeably, they are not identical. Marx defined communism as an economic system characterized by common ownership of all economic resources (Marshall, 1994). In the *Communist Manifesto* and

Das Kapital, he predicted that the working class would become increasingly impoverished and alienated under capitalism. As a result, the workers would become aware of their own class interests, revolt against the capitalists, and overthrow the entire system (see Turner, Beeghley, and Powers, 1998). After the revolution, private property would be abolished and capital would be controlled by collectives of workers who would own the means of production. The government (previously used to further the interests of the capitalists) would no longer be necessary. People would contribute according to their abilities and receive according to their needs (Marx and Engels, 1967/1848; Marx, 1967/1867). "Ideal" socialism has three distinctive features: (1) public ownership of the means of production, (2) pursuit of collective goals, and (3) centralized decision making.

PUBLIC OWNERSHIP OF THE MEANS OF PRODUCTION In a truly socialist economy, the means of production are owned and controlled by a collectivity or the state, not by private individuals or corporations. For example, prior to the early 1990s, the state owned all the natural resources and almost all the capital in the Soviet Union. At least in theory, goods were produced to meet the needs of the people. Access to housing and medical care was considered to be a right.

Leaders of the Soviet Union and some Eastern European nations decided to abandon government ownership and control of the means of production because the system was unresponsive to the needs of the marketplace and offered no incentive for increased efficiency (Boyes and Melvin, 1994). Since the early 1990s, Russia and other states in the former Soviet Union have attempted to privatize ownership of production. China—previously the world's other major communist economy—announced in 1997 that it would privatize most state industries (Serrill, 1997). In *privatization*, resources are converted from state ownership to private ownership; the government takes an active role in developing, recognizing, and protecting private property rights (Boyes and Melvin, 1994). However, it appears that these measures may have failed in Russia: Shortages once again loom, and many people want to see a transformation to a more socialist form of economy.

PURSUIT OF COLLECTIVE GOALS Socialism is based on the pursuit of collective goals, rather than on personal profits. Equality in decision making replaces hierarchical relationships (such as between owners and workers or political leaders and citizens). Everyone shares in the goods and services of society, especially necessities such as food, clothing, shelter, and medical care, based on need, not on ability to pay. In reality, however, few societies pursue purely collective goals.

CENTRALIZED DECISION MAKING Another tenet of socialism is centralized decision making. In theory, economic decisions are based on the needs of society; the government is responsible for aiding the production and distribution of goods and services. Central planners set wages and prices to ensure that the production process works. When problems such as shortages and unemployment arise, they can be dealt with quickly and effectively by the central government (Boyes and Melvin, 1994).

Mixed Economies

As we have seen, no economy is truly capitalist or socialist; most economies are mixtures of both. A *mixed economy* combines elements of a market economy (capitalism) with elements of a com-

✓**Checkpoints**

Checkpoint 13.5

16. What is a preindustrial economy?
17. What are primary-sector, secondary-sector, and tertiary-sector production?
18. What are the four distinctive features of "ideal" capitalism?
19. What are transnational corporations?
20. How does an oligopoly differ from a shared monopoly?
21. What are conglomerates, and why are they of concern to some social analysts?
22. What is an interlocking corporate directorate?
23. What are the three distinctive features of "ideal" socialism?
24. How does socialism differ from capitalism?
25. What is meant by the terms *mixed economy* and *democratic socialism*?

mand economy (socialism). Sweden, Great Britain, and France have mixed economies, sometimes referred to as ***democratic socialism***—**an economic and political system that combines private ownership of some of the means of production, governmental distribution of some essential goods and services, and free elections.** For example, government ownership in Sweden is limited primarily to railroads, mineral resources, a public bank, and liquor and tobacco operations (Feagin and Feagin, 1997). Compared with capitalist economies, however, the government in a mixed economy plays a larger role in setting rules, policies, and objectives.

Socialism an economic system characterized by public ownership of the means of production, the pursuit of collective goals, and centralized decision making.

Mixed economy an economic system that combines elements of a market economy (capitalism) with elements of a command economy (socialism).

Democratic socialism an economic and political system that combines private ownership of some of the means of production, governmental distribution of some essential goods and services, and free elections.

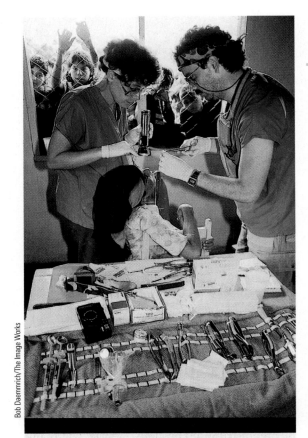

Bob Daemmrich/The Image Works

Professionals are expected to share their knowledge and display concern for others.

The government is also heavily involved in providing services such as medical care, child care, and transportation. In Sweden, for example, all residents have health insurance, housing subsidies, child allowances, paid parental leave, and day-care subsidies. National insurance pays medical bills associated with work-related injuries, and workplaces are specially adapted for persons with disabilities. College tuition is free, and public funds help subsidize cultural institutions such as theaters and orchestras ("General Facts on Sweden," 1988; Kelman, 1991). Recently, some analysts have suggested that the United States has assumed many of the characteristics of a *welfare state* (a state in which there is extensive government action to provide support and services to the citizens) as it has attempted to meet the basic needs of older persons, young children, unemployed people, and persons with a disability (Esping-Andersen, 1990).

Work in the Contemporary United States

The economy in the United States and other contemporary societies is partially based on the work (purposeful activity, labor, or toil) that people perform. However, work in high-income nations is highly differentiated and often fragmented because people have many kinds of occupations. Some occupations are referred to as professions.

Professions

Although sociologists do not always agree on exactly which occupations are professions, most of them agree that the term "professionals" includes most doctors, natural scientists, engineers, computer scientists, certified public accountants, economists, social scientists, psychotherapists, lawyers, policy experts of various sorts, professors, at least some journalists and editors, some clergy, and some artists and writers.

CHARACTERISTICS OF PROFESSIONS *Professions* are high-status, knowledge-based occupations that have five major characteristics (Freidson, 1970, 1986; Larson, 1977):

1. *Abstract, specialized knowledge.* Professionals have abstract, specialized knowledge of their field based on formal education and interaction with colleagues. Education provides the credentials, skills, and training that allow professionals to have job opportunities and assume positions of authority within organizations (Brint, 1994).

2. *Autonomy.* Professionals are autonomous in that they can rely on their own judgment in selecting the relevant knowledge or the appropriate technique for dealing with a problem. Consequently, they expect patients, clients, or students to respect that autonomy.

3. *Self-regulation.* In exchange for autonomy, professionals are theoretically self-regulating. All professions have licensing, accreditation, and regulatory associations that set professional standards and that require members to adhere to a code of ethics as a form of public accountability. Realistically, however, many professionals work within large-scale bureau-

Figure 13.4 SAT Scores by Parents' Income and Education, 2000

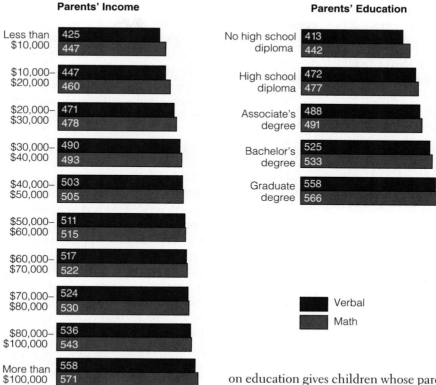

Source: College Board, 2000.

cracies that have rules, policies, and procedures to which everyone must adhere.

4. *Authority.* Because of their authority, professionals expect compliance with their directions and advice. Their authority is based on mastery of the body of specialized knowledge and on their profession's autonomy: Professionals do not expect the client to argue about the professional advice rendered.

5. *Altruism.* Ideally, professionals have concern for others, not just their own self-interest. The term *altruism* implies some degree of self-sacrifice whereby professionals go beyond their self-interest or personal comfort so that they can help a patient or client (Hodson and Sullivan, 2002).

SOCIAL REPRODUCTION OF PROFESSIONALS Although higher education is one of the primary qualifications for a profession, the emphasis

on education gives children whose parents are professionals a disproportionate advantage early in life (Brint, 1994). There is a direct linkage between parental education/income and children's scores on college admissions tests such as the SAT, as shown in Figure 13.4. In turn, test scores are directly related to students' ability to gain admission to colleges and universities, which serve as springboards to most professions.

Race and gender are also factors in access to the professions. Historically, people of color have been underrepresented. Today, as more persons from underrepresented groups have gained access to professions such as law and medicine, their children have also gained the educational and mentoring opportunities that are necessary for professional careers. Between 1980 and 1997, there was a 63-percent increase in the number of African Americans, aged 25 and above, with at least four years of college education. However, by 2000, only 16 percent of African

Professions high-status, knowledge-based occupations.

Americans and 11 percent of Latinos/as aged 25 and above had graduated from college, as compared with 28 percent of whites and 38 percent of Asian Americans (U.S. Census Bureau, 2001e).

DEPROFESSIONALIZATION   Certain professions are undergoing a process of *deprofessionalization* in which some of the characteristics of a profession are eliminated (Hodson and Sullivan, 2002). Occupations such as pharmacist have already been *deskilled*, as Nino Guidici explains:

> In the old days [people] took druggists as doctors. . . . All we do [now] is count pills. Count out twelve on the counter, put 'em in here, count out twelve more. . . . Doctors used to write out their own formulas and we made most of these things. Most of the work is now done in the laboratory. The real druggist is found in the manufacturing firms. They're the factory workers and they're the pharmacists. We just get the name of the drugs and the number and the directions. It's a lot easier. (qtd. in Terkel, 1990/1972)

However, colleges of pharmacy in many universities have fought against deprofessionalization by upgrading the degrees awarded to pharmacy graduates from the traditional B.S. in pharmacy to a Ph.D. This upgrading of degrees has also occurred over the past two decades in law schools, where the Bachelor of Laws (LL.B.) has been changed to the Juris Doctor (J.D.) degree.

Other Occupations

Occupations **are categories of jobs that involve similar activities at different work sites** (Reskin and Padavic, 1994). More than 500 different occupations and 21,000 occupational specialties, ranging from motion picture cartoonist to drop-hammer operator, are listed by the U.S. Department of Labor (1993). Historically, occupations have been classified as blue collar and white collar. Blue-collar workers were primarily factory and craft workers who did manual labor; white-collar workers were office workers and professionals. However, contemporary workers in the service sector do not easily fit into either of these categories; neither do the so-called pink-collar workers, primarily women, who are employed in occupations such as preschool teacher, dental assistant, secretary, and clerk (Hodson and Sullivan, 2002).

Sociologists establish broad occupational categories by distinguishing between employment in the primary labor market and in the secondary labor market. The *primary labor market* **consists of high-paying jobs with good benefits that have some degree of security and the possibility of future advancement.** By contrast, the *secondary labor market* **consists of low-paying jobs with few benefits and very little job security or possibility for future advancement** (Bonacich, 1972).

UPPER-TIER JOBS: MANAGERS AND SUPERVISORS   Managers are essential in contemporary bureaucracies, where work is highly specialized and authority structures are hierarchical. Workers at each level of the hierarchy take orders from their immediate superiors and perhaps give orders to a few subordinates. Upper-level managers are typically responsible for coordination of activities and control of workers. The *span of control*, or the number of workers a manager supervises, is affected by the organizational structure and by technology. Some analysts believe hierarchical organization is necessary to coordinate the activities of a large number of people; others suggest that it produces apathy and alienation among workers (Blauner, 1964).

LOWER-TIER AND MARGINAL JOBS   Positions in the lower tier of the service sector are part of the secondary labor market, characterized by low wages, little job security, few chances for advancement, higher unemployment rates, and very limited (if any) unemployment benefits. Typical lower-tier positions include janitor, waitress, messenger, sales clerk, typist, file clerk, migrant laborer, and textile worker. Large numbers of young people, people of color, recent immigrants, and white women are employed in this sector (Callaghan and Hartmann, 1991).

Marginal jobs **differ from the employment norms of the society in which they are located;** examples in the U.S. labor market include personal service and private household workers. Marginal jobs are frequently not covered by government work regulations—such as minimum standards of pay, working conditions, and safety standards—or do not provide sufficient hours of work each week to make a living (Hodson and Sullivan, 2002). Today, more than 11 million U.S. workers are employed in personal service industries such as eating and drinking places, hotels, laundries, beauty shops, and household service, primarily maid service (Hodson and Sullivan,

Spencer Grant/PhotoEdit

Occupational segregation by race and gender is clearly visible in personal service industries, such as restaurants and fast-food chains. Women and people of color are disproportionately represented in marginal jobs such as waitperson or fast-food server—jobs that typically do not meet societal norms for minimum pay, benefits, or security.

2002). Occupational segregation by race and gender is clearly visible in marginal jobs in personal service industries. Private household workers, including launderers, cooks, maids, housekeepers, gardeners, babysitters, and nannies, frequently must travel long distances to get to work, and many rely on public transportation, which can add hours to their workday. Moreover, household work is marginal: It lacks regularity, stability, and adequacy. The jobs are excluded from most labor legislation, employers often pay cash in order to avoid payroll taxes and Social Security, and the jobs typically provide no insurance or retirement benefits (Hodson and Sullivan, 2002).

Contingent Work

Contingent work **is part-time work, temporary work, or subcontracted work that offers advantages to employers but that can be detrimental to the welfare of workers.** Contingent work is found in every segment of the work force, including colleges and universities, where tenure-track positions are more scarce than in the past and a series of one-year, non-tenure-track appointments at the lecturer or instructor level has become a means of livelihood for many professionals (Mydans, 1995). The federal government is part of this trend, as is private enter-

prise. In the health care field, physicians, nurses, and other workers are increasingly employed through temporary agencies (Rosengarten, 1995).

Employers benefit by hiring workers on a part-time or temporary basis; they are able to cut costs, maximize profits, and have workers available only when they need them. As some companies have cut their work force, or downsized, they have replaced "staffing table" employees who had higher salaries and full benefit packages with part-time and hourly employees who receive lower wages and no benefits. Temporary workers are the fastest-growing segment of the contingent work force, and agencies that "place" them have increased dramatically in number in the last decade. The agencies provide workers on a contract basis to employers for an hourly fee; workers are paid a portion of this fee.

Subcontracted work is another form of contingent work that often cuts employers' costs at the expense of workers. Instead of employing a large work force, many companies have significantly reduced the size of their payrolls and benefit plans by *subcontracting*—**an agreement in which a corporation contracts with other (usually smaller) firms to provide specialized components, products, or services to the larger corporation.** Companies ranging from athletic shoe manufacturers to high-tech industries

Occupations categories of jobs that involve similar activities at different work sites.

Primary labor market the sector of the labor market that consists of high-paying jobs with good benefits that have some degree of security and the possibility of future advancement.

Secondary labor market the sector of the labor market that consists of low-paying jobs with few benefits and very little job security or possibility for future advancement.

Marginal jobs jobs that differ from the employment norms of the society in which they are located.

Contingent work part-time work, temporary work, or subcontracted work that offers advantages to employers but that can be detrimental to the welfare of workers.

Subcontracting an agreement in which a corporation contracts with other (usually smaller) firms to provide specialized components, products, or services to the larger corporation.

Rob Crandall/The Image Works

When the economy is booming, fewer people are unemployed. However, long lines at unemployment offices are a recurring sight throughout the United States in periods when workers are laid off from their jobs. As the sociological imagination reminds us, when one person is unemployed, it may be a personal problem, but when large numbers of people are laid off, the concern becomes a pressing social issue.

have found that it is more cost efficient (and profitable) to subcontract some of their work to small contractors. Hiring and paying workers becomes the responsibility of the subcontractor, not of the larger corporation. The employment practices of some subcontractors are unethical, if not illegal. For many contingent workers, however, the only alternative to such a job is unemployment.

Unemployment

There are three major types of unemployment—cyclical, seasonal, and structural. *Cyclical unemployment* occurs as a result of lower rates of production during recessions in the business cycle; a recession is a decline in an economy's total production that lasts at least six months. Although massive layoffs initially occur, some of the workers will eventually be rehired, largely depending on the length and severity of the recession. *Seasonal unemployment* results from shifts in the demand for workers based on conditions such as the weather (in agriculture, the construction industry, and tourism) or the season (holidays and summer vacations). Both of these

types of unemployment tend to be relatively temporary.

By contrast, structural unemployment may be permanent. *Structural unemployment* arises because the skills demanded by employers do not match the skills of the unemployed or because the unemployed do not live where the jobs are located (McEachern, 1994). This type of unemployment often occurs when a number of plants in the same industry are closed or when new technology makes certain jobs obsolete. Structural unemployment often results from capital flight—the investment of capital in foreign facilities, as previously discussed. Today, many workers fear losing their jobs, exhausting their unemployment benefits (if any), and still not being able to find another job.

The *unemployment rate* is the percentage of unemployed persons in the labor force actively seeking jobs. In 2000 the U.S. unemployment rate was 4.0 percent; however, with corporate scandals, the falling stock market, and global political uncertainty, among other economic concerns, the mid-year unemployment rate in 2002 was 5.2 percent for white Americans and 10.7 percent for African Americans (CNN, 2002). Like other types of "official" statistics, however, unemployment rates may be misleading. For example, Census Bureau unemployment statistics do not include all workers; they are based on a monthly sample survey of about sixty thousand households. In addition, individuals who become discouraged in their attempt to find work and are no longer actively seeking employment are not counted as unemployed.

Labor Unions and Worker Activism

In their individual and collective struggles to improve their work environment and gain some measure of control over their work-related activities, workers have used a number of methods to resist workplace alienation. Many have joined labor unions to gain strength through collective action.

LABOR UNIONS U.S. labor unions came into being in the mid-nineteenth century. Unions have been credited with gaining an eight-hour workday,

PhotoEdit

Seeking to improve economic and social opportunities for farm workers, the late César Chávez held rallies and engaged in other protest activities in an effort to better the workers' lives.

a five-day workweek, health and retirement benefits, sick leave and unemployment insurance, and workplace health and safety standards for many employees. Most of these gains have occurred through *collective bargaining*—negotiations between employers and labor union leaders on behalf of workers. In some cases, union leaders have called strikes to force employers to accept the union's position on wages and benefits. While on strike, workers may picket in front of the workplace to gain media attention, to fend off "scabs" (nonunion workers) who might take over their jobs, and in some cases to discourage customers from purchasing products made or sold by their employer. In recent years, strike activity has diminished significantly because many workers have feared losing their jobs. In 2000, only 39 strikes, or work stoppages, involving more than 1,000 workers were reported, as compared with significantly higher numbers in the 1960s and 1970s (U.S. Census Bureau, 2001g). The number of workers involved in the actions declined from a peak of more than 2.5 million in 1971 to 192,000 in 1995.

Although the overall *number* of union members in the United States has increased since the 1960s, primarily as a result of the growth of public employee unions (such as the American Federation of Teachers), the *proportion* of all employees who are union members has declined. Today, about 17 percent of all U.S. employees belong to unions or employee associations. Part of the decline in union participation has been attributed to the structural shift away from manufacturing and manual work and toward the service sector, where employees have been less able to unionize.

ABSENTEEISM AND SABOTAGE   Absenteeism is one means by which workers resist working conditions they consider to be oppressive. Other workers use sabotage to bring about informal work stoppages. The phrase "throwing a monkey wrench in the gears" originated with the practice of workers "losing" a tool in assembly-line machinery. Sabotage of this sort effectively brought the assembly line to a halt until the now-defective piece of machinery could be repaired. Although most workers do not sabotage machinery, a significant number do resist what they perceive to be oppression from supervisors and employers. In a study of Asian American women who work in low-status occupations such as hotel housekeepers, the sociologist Esther Ngan-Ling Chow (1994) found that many of the women felt they had little to lose if they chose to resist or rebel. Thus, resistance helps people in lower-tier service jobs survive at work (Dill, 1988; Chow, 1994).

Studies of worker resistance are very important to our understanding of how people deal with the social organization of work. Previously, workers (especially women) have been portrayed as passive "victims" of their work environment. However, this is not always the case for either women or men; many workers resist problematic situations and demand better wages and work conditions.

Employment Opportunities for Persons with a Disability

An estimated 49.7 million persons in the United States have one or more physical or mental disabilities that differentially affect their opportunities for employment. In 1990 the United States became the first nation to formally address the issue of equality for persons with a disability. When Congress passed the Americans with Disabilities Act (ADA), this law established "a clear and comprehensive prohibition of discrimination on the basis of disability."

Unemployment rate   the percentage of unemployed persons in the labor force actively seeking jobs.

Box 13.3 You Can Make a Difference

Creating Access to Information Technologies for Workers with Disabilities

People are shocked I can do this. They have never imagined a blind person could find a use for a computer, much less make a business out of it.

—Cathy Murtha, a partner in Magical Mist Creations, a company that designs and evaluates sites on the World Wide Web (qtd. in Sreenivasan, 1996: C7)

If you are a person with a visual impairment or a person who plans to supervise workers—some of whom may be visually impaired—you can make a difference in the workplace by learning about new technologies that make jobs more accessible for workers with disabilities. For example, new computer hardware and software programs and the Internet are having a profound effect on workers who are visually impaired. According to Larry Scadden, technological consultant for persons with disabilities at the National Science Foundation, "The Internet has changed forever the lives of blind people, mainly because it provides independent access to information" (qtd. in Sreenivasan, 1996: C7). The goal of making the Internet and the World Wide Web accessible to visually impaired persons requires several things:

- Text must be in such a form that it can be read aloud by a speech synthesizer with text-to-speech software or be output in a Braille format.
- The user must be able to navigate the screen with a keyboard, not a mouse.
- Good descriptions must be given of photos and graphics that the user will not be able to see.

To find out more about creating greater access to information for persons with visual impairment and other disabilities, contact the following agencies:

- American Council of the Blind, 1155 15th St., NW, Suite 720, Washington, DC 20005. (202) 393-3666 or (800) 424-8666. Online:
http://www.acb.org
- The National Foundation of the Blind. Online:
http://www.nfb.org

Combined with previous disability rights laws (such as those that provide for the elimination of architectural barriers from new, federally funded buildings and for the maximum integration of schoolchildren with disabilities), the ADA is a legal mandate for the full equality of people with disabilities.

Despite this law, about two-thirds of working-age persons with a disability are unemployed today (U.S. Census Bureau, 2002). Most persons with a disability believe they could work if they were offered the opportunity. However, even when persons with a disability are able to find jobs, they earn less than persons without a disability (Yelin, 1992). On the average, workers with a disability make 85 percent (for men) and 70 percent (for women) of what their co-workers without disabilities earn, and the gap is growing (U.S. Census Bureau, 2002). Studies have shown a thirty-year overall decline in the economic condition of persons with disabilities (Yelin, 1992; Burkhauser, Haveman, and Wolfe, 1993). The problem has been particularly severe for African Americans and Latinos/as with disabilities (Kirkpatrick, 1994). Among Latinos/as with dis-

abilities, only 10 percent are employed full time; those who work full time earn 73 percent of what white (non-Latino) persons with disabilities earn.

What does it cost to "mainstream" persons with disabilities? Disability expenditures have been estimated to be $169.4 billion per year for the working-age population, taking into account money spent by both the government and the private sector (Albrecht, 1992). However, between 1978 and 1992, Sears, Roebuck found that the average cost of an accommodation for a worker with a disability was only $126, with some accommodations involving little or no cost (such as flexible schedules, back-support belts, rest periods, and changes in employee work stations). Other accommodations cost about $1,000, with the largest single cost being $14,500 each for special Braille computer displays for visually impaired employees (Noble, 1995). As employment opportunities change in the global economy, it is important that employers use common sense and a creative approach to thinking about how persons with disabilities can fit into the marketplace (see Box 13.3).

✓**Checkpoints**

Checkpoint 13.6

26. What are the key characteristics of professions?
27. What factors contribute to the social reproduction of professionals?
28. How does the primary labor market differ from the secondary labor market?
29. How do marginal jobs reflect occupational segregation by race and gender?
30. What is contingent work?
31. What are the major types of unemployment?
32. What have been the major issues of labor unions and worker activists over the past hundred years?

Politics and the Economy in the Future

Will the government of the United States and other nation-states become obsolete in the future? According to analysts, there has been some erosion of the powers of governments in high-income nations such as the United States, along with a corresponding increase in transnational politics and the economy. Trade agreements are an example: Consider the North American Free Trade Agreement (NAFTA), which unites the United States, Canada, Mexico, and portions of South America in an economic alliance; the European Union (EU), which binds many European nations together in a political and economic partnership; and the General Agreement on Tariffs and Trade (GATT), which brings together the 124 nations in the World Trade Organization.

Do such agreements signal the end of individual governments? Despite trade alliances, many analysts believe that nation-states remain of great significance. According to scholar Paul Kennedy (1993: 134), individual nations will remain the primary locus of identity for most people. Regardless of who their employer is and what they do for a living, individuals pay taxes to the state, are subject to its laws, serve (if need be) in its armed forces, and can travel abroad only by having its passport. Therefore, as new challenges—such as terrorism, war, and economic instability—increase, most people in democracies will turn to their own governments and demand solutions.

What is the future of democracy? Today, millions of people around the globe remain committed to what they believe are the "ideals of democracy." For example, recent U.S. immigrants who have escaped ethnic conflicts and poverty in their own nations often seek a life of dignity in a society they believe to be governed by democratic principles. Similarly, millions of people in Eastern Europe, the former Soviet Union, Latin America, and China have put their lives on the line for democracy, with varying degrees of success. Ironically, people who believe most in the ideals of democracy and freedom may be those who have experienced the least of it, either in other nations or in our own. In the United States, for example, many working-class and poor people, white women, people of color, and members of religious minorities who have experienced prejudice and discrimination still believe that a reenvisioning of democratic ideals is possible (Eisenstein, 1994).

Politics and the economy are so intertwined in the United States and on a global basis that many social scientists speak of the two as a single entity, the *political economy*. As we have seen, both political power and economic power are concentrated in the hands of a few in the "new" global economy. Most futurists predict that transnational corporations will become even more significant in the global economy of the future. As they continue to compete for world market shares, these corporations will become even less aligned with the values of any one nation. However, those who advocate increased globalization typically focus on its potential impact on high-income nations, not the effect it may have on the 80 percent of the world's population that resides in lower-income nations. Persons in lower-income nations may become increasingly baffled and resentful when they are bombarded with media images of affluence and consumption that are not attainable by them.

The chasm between rich and poor nations probably will widen in the future as industrialized nations purchase fewer raw materials from industrializing countries and more products and services from one another. This change will take place because raw materials of all sorts are no longer as important to manufacturers in industrialized nations. For example, oil may become less important as a power source with the development of solar power and controlled nuclear fusion.

Mike Blank/Stone-Getty Images

The fate of high-income nations is increasingly linked to the globalization of markets. Dramatic changes in stock markets in nations such as Japan and the United States create reverberations that are felt in financial markets around the world.

What effect will global political and economic changes have on the average worker? In recent years, the average worker in the United States and other high-income nations has benefited more than have workers in industrializing nations from global competition and economic growth. For example, the average citizen of Switzerland has an income several hundred times that of a resident of Ethiopia. More than a billion of the world's people live in abject poverty; for many, this means attempting to survive on less than $370 a year (P. Kennedy, 1993).

In the twenty-first century, terrorism and war have influenced how many people think about politics and the economy within and across nations. Until 2001, most international acts of terrorism that targeted U.S. interests, for example, took place out-side of this country. The events of September 2001 were a frightening and shocking experience for people in the United States, who realized that the terrorists—whose objectives were not clearly known and whose linkage to any one government or political regime was vague—had wreaked immense destruction within our nation's borders, leaving about 3,000 dead and the entire nation fearful for its long-term safety.

Around the world, acts of terrorism extract a massive toll by producing rampant fear, widespread loss of human life, and extensive destruction of property. For the first time, people in the United States have been confronted with threats of terrorism and problems associated with war that have been facts of life in some other countries for many years. In times such as this, many people rely on government officials to calm their fears; others seek to learn more about current events through media coverage of national and international issues. Ironically, some who are highly critical of the intertwining relationships we have discussed regarding the media, the government, and the economy find that they look to television coverage, Internet reports, and other information technologies to learn what is going on minute by minute in their nation and around the globe. Often they do not question the factual content or the potential media bias that may be included in such reports because the media provide the best—and sometimes the only—source of information about our rapidly changing nation and world.

Chapter Review

What are the three types of authority?
Max Weber identified three types of authority: charismatic, traditional, and rational–legal. Charismatic authority is power based on a leader's personal qualities. Traditional authority is based on long-standing custom. Rational–legal authority is based on law or written rules and regulations, as found in contemporary bureaucracies.

What are the main types of political systems?
The main types of political systems are monarchies, authoritarian systems, totalitarian systems, and democratic systems.

Reflections

Reflections 13.4

Will extensive coverage of events by global media conglomerates substantially affect political and economic conditions in nations around the world?

How do pluralist and power elite perspectives view power in the United States?
According to the pluralist (functionalist) model, power is widely dispersed throughout many competing interest groups. People influence policy by voting, joining special interest groups and political action campaigns, and forming new groups. According to the elite (conflict) model, power is concentrated in a small group of elites, while the masses are relatively powerless.

Who are the power elite, and why are they important?
According to C. Wright Mills, the power elite is composed of influential business leaders, key government leaders, and the military. The elites possess greater resources than the masses, and public policy reflects their preferences.

What is the primary function of the economy?
The economy is the social institution that ensures the maintenance of society through the production, distribution, and consumption of goods and services.

How do the major contemporary economic systems differ?
Capitalism, socialism, and mixed economies are the main systems in industrialized countries. Capitalism is characterized by ownership of the means of production, pursuit of personal profit, competition, and limited government intervention. Socialism is characterized by public ownership of the means of production, the pursuit of collective goals, and centralized decision making. In mixed economies, elements of a capitalist, market economy are combined with elements of a command, socialist economy. These mixed economies are often referred to as democratic socialism.

What are the characteristics of professions?
Professions are high-status, knowledge-based occupations characterized by abstract, specialized knowledge, autonomy, authority over clients and subordinate occupational groups, and a degree of altruism.

What is contingent work?
Contingent work is part-time work, temporary work, or subcontracted work that offers advantages to employers but may be detrimental to workers. Through the use of contingent workers, employers are able to cut costs and maximize profits, but workers have little or no job security.

Key Terms

authoritarianism 414
authority 411
capitalism 426
conglomerates 428
contingent work 435
corporations 426
democracy 415
democratic socialism 431
economy 425
elite model 418
government 409
interlocking corporate directorates 429
marginal jobs 434
military–industrial complex 424
mixed economy 431
monarchy 413
occupations 434
oligopoly 427
pluralist model 416
political action committees 418
political party 420
political socialization 421
politics 409
power 411
primary labor market 434
primary sector production 425
professions 432
routinization of charisma 412
secondary labor market 434
secondary sector production 426
shared monopoly 427
socialism 430
state 409
subcontracting 435
tertiary sector production 426
totalitarianism 414
transnational corporations 426
unemployment rate 436

Questions for Critical Thinking

1. Who is ultimately responsible for decisions and policies that are made in a democracy such as the United States, the people or their elected representatives?
2. How would you design a research project that studied the relationship between campaign contributions to elected representatives and their subsequent voting records? What would be your hypothesis? What kinds of data would you need to gather? How would you gather accurate data?

 ## Resources on the Internet

Chapter-Related Web Sites
The following web sites have been selected for their relevance to the topics in this chapter. These sites are

among the more stable, but please note that web site addresses change frequently.

Political Science
http://www.trinity.edu/~mkearl/polisci.html
This page contains links to a wealth of information. The field is described in a way that is easy to understand. This is a great place to start with a study of the political system.

The 2002 Economic Census Is Underway
http://www.census.gov/epcd/www/econ2002.html
An economic census of American business is collected every five years. Take a look at the plans under way to collect the latest data.

1997 Economic Census Data
http://www.census.gov/epcd/www/econ97.html
From here, you can access the data collected during the latest economic survey of American business.

Explore the political parties in U.S. society by searching these web pages for information, albeit biased, about the U.S. political system:

Democratic National Committee
http://www.democrats.org

Republican National Committee
http://www.rnc.org

American Independent Party
http://www.aipca.org

Reform Party
http://www.reformparty.org

Green Party of the United States
http://www.greenpartyus.org

VoteNader.org
http://www.votenader.com/issues/politicalreform.html
The web site of Ralph Nader's 2000 presidential campaign.

The United Nations
http://www.un.org
This is the UN's web site and also a gateway to hundreds of other information sources about global politics and economic conditions.

Companion Web Site for This Book

Virtual Society: The Wadsworth Sociology Resource Center
Visit http://sociology.wadsworth.com and click on the page for Kendall, *Sociology in Our Times: The Essentials*, Fourth Edition, to access a wide range of enrichment material to aid in your study of sociology. Click on the Student Resources section of the web site. Next, select from the pull-down menu the chapter that you are presently studying. Among the useful options for self-study are chapter objectives, flashcards, practical study tips, and practice tests for each chapter.

MicroCase Online
From the Virtual Society home page, click on MicroCase Online to access book-specific MicroCase exercises that allow you to further explore sociological issues and principles presented in the text.

InfoTrac College Edition
Another unique option available to you at the Student Resources section of the companion web site is InfoTrac College Edition, an online library with access to hundreds of scholarly and popular periodicals. Below are suggested search terms for this chapter. Results from these and other searches are found at the site.

- Search keywords: *campaign contributions*. Find articles that discuss campaign finance reform. What are the problems associated with campaign contributions?
- Search keywords: *political participation*. Find articles that specifically address voter apathy. Is voter apathy a problem only in the United States, or does it exist in other countries? What solutions have been proposed to tackle this problem?

Explorations CD-ROM
Go to the Virtual Explorations CD-ROM to begin the interactive exercise for this chapter. This Virtual Exploration will introduce you to some of the exciting resources for sociology on the World Wide Web. You will be guided through an exercise that employs related web sites on politics and the economy in global perspective. Answer the questions, and e-mail your responses to your instructor.

Jeff Fowler/SIS

Health, Health Care, and Disability

14

Health in Global Perspective

Health in the United States
Social Epidemiology
Lifestyle Factors

Health Care in the United States
The Rise of Scientific Medicine and Professionalism
Medicine Today
Paying for Medical Care in the United States
Paying for Medical Care in Other Nations
Social Implications of Advanced Medical Technology
Holistic Medicine and Alternative Medicine

Sociological Perspectives on Health and Medicine
A Functionalist Perspective: The Sick Role
A Conflict Perspective: Inequalities in Health
 and Health Care
A Symbolic Interactionist Perspective: The Social
 Construction of Illness
A Postmodernist Perspective: The Clinical Gaze

Disability
Sociological Perspectives on Disability
Social Inequalities Based on Disability

Health Care in the Future

My name is Heather Neumann and I'm an 18-year-old freshman at the University of Texas at Austin. When I was nine, I was diagnosed with Juvenile Rheumatoid Arthritis and given all sorts of different medications. I learned that if I didn't go into a remission, my life wouldn't be too swell. Well, I was pretty upset about it all! I played piano, loved all sports, was good at art and was being told that I might not be able to do these things when I got older. That's when I decided to take charge of my life. I took things one day at a time and pushed myself to give 110% at everything.

I continued to take piano until I went to high school, even though it was not always possible to practice when my hands got really bad. And I played every sport offered at my school, however tough it was on my body. Every time someone said, "You can't do that, you're sick" I strived not only to do it, but to do it better than anyone else. . . . It was frustrating for

me when my mind wanted to do more than my body would let it. Sometimes my hands would swell and I couldn't wear my class ring. . . . Sometimes I had to ask my boyfriend to carry my books because my shoulder hurt or because my knees decided not to work half-way down the hall. I was often weak. . . .

I am still keeping myself busy in college. I still have [juvenile rheumatoid arthritis] and I still take two million pills. There are still days when I can't get out of bed, but I accomplish so much in the meantime. I am playing Varsity Field Hockey, I am the service chair for my sorority, a member of student government, an editor of the literary arts journal and an honors student. On days when it feels like I can't cope, I remind myself that there is nothing I cannot do if I put my mind to it.

—Heather Neumann describing her experience with chronic illness (*Band-Aides & Blackboards*, 1999)

Like many other young people, Heather Neumann has learned how to live with chronic medical problems. She has chosen to share her story, "Forever Heather," with other young people, particularly those with chronic illnesses or disabilities, on *Band-Aides and Blackboards,* a web site maintained by Joan Fleitas, a nursing professor and pediatric clinician. According to Fleitas (1999), the goal of this web site is to communicate with children and young adults so that they can understand medical problems and increase their acceptance of them. The site is also designed to sensitize others to the harmful effects of everyday practices, such as teasing a child living with a chronic illness or attaching a stigma to illness or disability (Fleitas, 1999). This web site is only one of the many ways in which health professionals and laypersons are transforming the Internet into a source of information about illness and disability and helping those who seek moral and psychological support when coping with such conditions.

In this chapter, we will explore the dynamics of health, health care, and disability from a sociological perspective, as well as look at issues through the eyes of those who have experienced medical problems. Before reading on, test your knowledge about health, illness, and health care by taking the quiz in Box 14.1.

Questions and Issues

Chapter Focus Question: Why are health, health care, and disability significant concerns not only for individuals but also for entire societies?

What is the relationship between the social environment and health and illness?

What are the major issues in U.S. health care?

How do functionalist, conflict, and symbolic interactionist approaches differ in their analysis of health and health care?

What are some of the consequences of disability?

What does the concept of health mean to you? At one time, health was considered to be simply the absence of disease. However, the World Health Organization (1946: 3) defines *health* as **a state of complete physical, mental, and social well-being.** According to this definition, health involves not only the absence of disease, but also a positive sense of wellness. In other words, health is a multidimensional phenomenon: It includes physical, social, and psychological factors.

What is illness? Illness refers to an interference with health; like health, illness is socially defined

Box 14.1 Sociology and Everyday Life

How Much Do You Know About Health, Illness, and Health Care?

True	False	
T	F	1. Some social scientists view sickness as a special form of deviant behavior.
T	F	2. The field of epidemiology focuses primarily on how individuals acquire disease and bodily injury.
T	F	3. The primary reason that African Americans have shorter life expectancies than whites is the high rate of violence in central cities and the rural South.
T	F	4. Native Americans have shown dramatic improvement in their overall health level since the 1950s.
T	F	5. Health care in most high-income, developed nations is organized on a fee-for-service basis as it is in the United States.
T	F	6. The medical–industrial complex has operated in the United States with virtually no regulation, and allegations of health care fraud have largely been overlooked by federal and state governments.
T	F	7. Media coverage of chronic depression and other mental conditions focuses almost exclusively on these problems as "women's illnesses."
T	F	8. It is extremely costly for employers to "mainstream" persons with disabilities in the workplace.

Answers on page 446.

and may change over time and between cultures. For example, in the United States and Canada, obesity is viewed as unhealthy, whereas in other times and places, obesity signaled that a person was prosperous and healthy. A disease, by comparison, is an objective reality: It is a particular destructive process in the body, with specific causes and characteristic symptoms. There are specific medical criteria for identifying a disease.

What happens when a person is perceived to have an illness or disease? Healing involves both personal and institutional responses to perceived illness and disease. One aspect of institutional healing is health care and the health care delivery system in a society. *Health care is any activity intended to improve health.* When people experience illness, they often seek medical attention in hopes of having their health restored. A vital part of health care is *medicine—an institutionalized system for the scientific diagnosis, treatment, and prevention of illness.*

Health in Global Perspective

Studying health and health care issues around the world offers insights on illness and how political and economic forces shape health care in nations. Disparities in health are glaringly apparent between high-income and low-income nations when we examine factors such as the prevalence of life-threatening diseases, rates of life expectancy and

Health a state of complete physical, mental, and social well-being.

Health care any activity intended to improve health.

Medicine an institutionalized system for the scientific diagnosis, treatment, and prevention of illness.

Box 14.1

Answers to the Sociology Quiz on Health, Illness, and Health Care

1. **True.** Some social scientists view sickness as a special form of deviant behavior. However, it is not equivalent to other forms of deviance such as crime or violent behavior. Unlike many who are defined as criminal, the sick are provided with therapeutic care so that their health will be restored and they can fulfill their roles in society (Weiss and Lonnquist, 2003).

2. **False.** The primary focus of the epidemiologist is on the health problems of social aggregates or large groups of people, not on individuals as such (Cockerham, 1998).

3. **False.** The lower life expectancy of African Americans as a category is due to a higher prevalence of life-threatening illnesses, such as cancer, heart disease, hypertension, and AIDS. However, it should be noted that African American males do have the highest death rates from homicide of any racial–ethnic category in the United States (Cockerham, 1998).

4. **True.** Native Americans (including American Indians and native Alaskans) as a category have had significant improvement in health in recent decades. Some analysts attribute this change to better nutrition and health care services. However, other analysts point out that Native Americans continue to have high rates of mortality from diabetes, alcohol-related illnesses, and suicide (Cockerham, 1998).

5. **False.** The United States is one of only two high-income, developed nations that do not have some form of universal health coverage. In the United States, health care has traditionally been purchased by the patient. In most other high-income nations, health care is provided or purchased by the government (Cockerham, 1998).

6. **False.** In the mid-to-late 1990s, government investigations focused on rising health care payments and allegations of fraud in the health care delivery system. Thus far, billing frauds have been found in Medicare and Medicaid payments to physicians, hospitals, nursing homes, home health agencies, medical labs, and medical equipment manufacturers (Findlay, 1997).

7. **False.** Until recently, chronic depression and other mental conditions were most often depicted as "female" problems. However, in the late 1990s, male celebrities such as Mike Wallace, the news correspondent who is co-editor of *60 Minutes,* have made the general public more aware of male depression. Wallace produced a documentary on chronic depression, and author Terrence Real, a psychotherapist, wrote a book about depression in men: *I Don't Want to Talk About It: Overcoming the Secret Legacy of Male Depression* (Brody, 1997). However, it should be noted that women average higher levels of depression than men do, perhaps because women and men have unequal adult statuses (see Mirowsky, 1996).

8. **False.** Although disability expenditures nationwide may be costly, individual employers often find that they can accommodate the workplace needs of a worker with a disability for costs ranging from zero to several thousand dollars, thus opening up new opportunities for people previously excluded from certain types of jobs and careers (Noble, 1995).

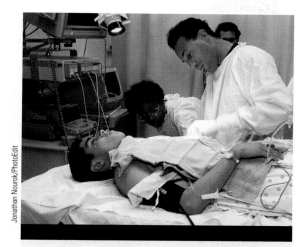

Life expectancy, life chances, and access to health care are much better for some people than for others. How are race, class, and gender related to these issues in your community?

Jonathan Nourok/PhotoEdit

The *infant mortality rate* is the number of deaths of infants under 1 year of age per 1,000 live births in a given year. The infant mortality rate in some low-income nations is staggering: 149 infants under 1 year of age die per 1,000 live births in Sierra Leone, 196 die in Angola, and 122 die in Malawi (U.S. Census Bureau, 2001h). In fact, almost 14 percent of all children born in low-income nations die before they reach their first birthday. The World Health Organization (1999) reports that about five million babies born in low-income countries during 1998 died during the *first month* of life. A child born in Latin America or Asia can expect to live between 7 and 13 fewer years, on average, than one born in North America or Western Europe (Epidemiological Network for Latin America and the Caribbean, 2000).

There are many reasons for these differences in life expectancy and infant mortality. Many people in low-income countries have insufficient or contaminated food; lack access to pure, safe water; and do not have adequate sewage and refuse disposal. Added to these hazards is lack of information about how to maintain good health. Many of these nations also lack qualified physicians and health care facilities with up-to-date equipment and medical procedures.

Nevertheless, tremendous progress has been made in saving the lives of children and adults over the past 15 years. Life expectancy at birth has risen to more than 70 years in 84 countries, up from only 55 countries in 1990. Life expectancy in low-income nations increased on average from 53 to 62 years, and mortality of children under 5 years of age dropped from 149 to 85 per 1,000 live births. Although this increase has been attributed to a number of factors, an especially important advance has been the development of a safe water supply. The percentage of the world's population with access to safe water nearly doubled between 1990 and 2000 (United Nations Development Programme, 2001).

infant mortality, and access to health services. In regard to global health, for example, the number of people infected with HIV/AIDS more than doubled between 1990 and 2000 (from fewer than 15 million to more than 34 million). AIDS has cut life expectancy by five years in Nigeria, eighteen years in Kenya, and thirty-three years in Zimbabwe (U.S. Census Bureau, 2001h). *Life expectancy* **refers to an estimate of the average lifetime of people born in a specific year.** AIDS results in higher mortality rates in childhood and young adulthood, stages in the life course when mortality is otherwise low. However, AIDS is not the only disease reducing life expectancy in some nations. Most deaths in low- and middle-income nations are linked to infectious and parasitic diseases that are now rare in high-income, industrialized nations. Among these diseases are tuberculosis, polio, measles, diphtheria, meningitis, hepatitis, malaria, and leprosy (United Nations Development Programme, 2001). Although it is estimated that only 13 percent of U.S. citizens and 9 percent of Canadians will die prior to age sixty, health experts estimate that more than 1.5 billion people around the world will die prior to age sixty. This is particularly true in low-income nations such as Zambia, where 80 percent of the people are not expected to see their sixtieth birthday.

Life expectancy   an estimate of the average lifetime of people born in a specific year.

Infant mortality rate   the number of deaths of infants under 1 year of age per 1,000 live births in a given year.

Will improvements in health around the world continue to occur? Organizations such as the United Nations argue that both public-sector and private-sector initiatives will be required to improve global health conditions. For example, a United Nations report states that in the era of globalization and dominance by transnational corporations, "money talks louder than need" when "cosmetic drugs and slow-ripening tomatoes come higher on the list [of priorities] than a vaccine against malaria or drought-resistant crops for marginal lands" (United Nations Development Programme, 1999: 68).

Recently, pressing questions have arisen about the availability of new technologies and lifesaving drugs around the world. An example is the problem of providing access to drugs in countries with high rates of HIV/AIDS. Many people cannot afford to pay for drugs, such as the three-drug combination therapy that prolongs the life of many AIDS patients. Transnational pharmaceutical companies fear that if they provide their name-brand drugs at a lower price in low-income countries, that might undercut their major sales base in high-income countries if those drugs become available as generic products (which are less costly and can be made by more than one manufacturer) or are re-imported into the high-income countries at a reduced price. The companies claim that they need the money generated from sales of their name-brand drugs in order to fund research on other products that will reduce suffering and sometimes prolong human life. Pharmaceutical companies that hold the patents on various drugs see their products as something that needs to be protected by law, whereas people in human relief agencies around the world are concerned about the fact that one-third of the world's population does not have access to essential medicines and that, even worse, this figure rises to one-half in the poorest parts of Africa and Asia (United Nations Development Programme, 2001). If we are to see a significant improvement in life expectancy and health among people in all of the nations of the world, improvements are needed in the availability to them of new medical technologies and lifesaving drugs.

How about improvements in health and health care within one nation? Is there a positive relationship between the amount of money that a society spends on health care and the overall physical, mental, and social well-being of its people? Not necessarily. If there were such a relationship,

✓Checkpoints

Checkpoint 14.1

1. Why are sociologists interested in studying health and medicine?
2. What is life expectancy, and why is it a key indicator of health in a nation?
3. What is the infant mortality rate?

people in the United States would be among the healthiest and most physically fit people in the world. We spend more than one trillion dollars—the equivalent of $3,800 per person—on health care each year, and the amount has almost doubled in the past 10 years (U.S. Census Bureau, 2000c). But by comparing health care expenditures in Sweden and the United States with infant mortality rates in these two countries, we can see that large expenditures for health care do not always produce better health care for individuals. Sweden spends an average of $1,348 per person on health care and has an infant mortality rate of 4.5; by contrast, the United States has an infant mortality rate of 6.7 (U.S. Census Bureau, 2000c). Similarly, a child born in Sweden in 1999 had a life expectancy of 79.2 years, whereas a child born in the United States that same year had a life expectancy of 76.7 years (U.S. Census Bureau, 2000c).

Health in the United States

Even if we limit our discussion (for the moment) to people in the United States, why are some of us healthier than others? Is it biology—our genes—that accounts for this difference? Does the environment within which we live have an effect? How about our own individual lifestyle?

Social Epidemiology

The field of social epidemiology attempts to answer questions such as these. *Social epidemiology* is the study of the causes and distribution of health, disease, and impairment throughout a population (Weiss and Lonnquist, 2003). Typically, the target of the investigation is disease agents, the environment, and the human host. *Disease agents* include biological agents such as insects, bacteria, and viruses that carry or cause disease; nutrient agents such as fats and carbohydrates; chemical agents such as gases and pollutants in the air; and physical agents such as temperature, humidity, and radiation. The *environment* includes the physical (geography and climate), biological (presence or absence of known disease agents), and social (socioeconomic status, occupation, and location of home) environments. The human *host* takes into account demographic factors (age, sex, and race/ethnicity), physical condition, habits and customs, and lifestyle (Weiss and Lonnquist, 2003). Let's look briefly at some of these factors.

AGE Rates of illness and death are highest among the old and the young. Mortality rates drop shortly after birth and begin to rise significantly during middle age. After age 65, rates of chronic diseases and mortality increase rapidly. **Chronic diseases are illnesses that are long term or lifelong and that develop gradually or are present from birth;** in contrast, **acute diseases are illnesses that strike suddenly and cause dramatic incapacitation and sometimes death** (Weitz, 2001).

Two of the most common sources of chronic disease and premature death are tobacco use, which increases mortality among both smokers and people who breathe the tobacco smoke of others, and alcohol abuse, both of which are discussed later in this chapter. The fact that rates of chronic diseases increase rapidly after age 65 has obvious implications not only for people reaching that age (and their families) but also for society. The Census Bureau projects that about 20 percent of the U.S. population will be at least age 65 by the year 2050 and that the population of persons aged 85 and over will have tripled from about 4 million (1.5 percent) in 1995 to about 12 million (5 percent). The cost of caring for many of these people—especially those who must be institutionalized—will increase in at least direct proportion to their numbers.

SEX Prior to the twentieth century, women had lower life expectancies than men because of high mortality rates during pregnancy and childbirth. Preventive measures have greatly reduced this cause of female mortality, and women now live longer than men. For babies born in the United States in 2000, for example, life expectancy at birth was 74.2 years for males and 79.9 years for females. Fe-

males have a slight biological advantage over males in this regard from the beginning of life, as can be seen in the fact that they have lower morality rates both in the prenatal stage and in the first month of life (Weiss and Lonnquist, 2003). However, the sociologist Ingrid Waldron (1994) notes that gender roles and gender socialization also contribute to the difference in life expectancy. Men are more likely to work in dangerous occupations such as commercial fishing, mining, construction, and public safety/firefighting. As a result of gender roles, males may be more likely than females to engage in risky behavior such as drinking alcohol, smoking cigarettes (there is more social pressure on women not to smoke), using drugs, driving dangerously, and engaging in fights. Finally, women are more likely to use the health care system, with the result that health problems are identified and treated earlier (while there is a better chance of a successful outcome), whereas many men are more reluctant to consult doctors.

Because women on average live longer than men, it is easy to jump to the conclusion that they are healthier than men. However, although men at all ages have higher rates of fatal diseases, women have higher rates of chronic illness (Waldron, 1994).

RACE/ETHNICITY AND SOCIAL CLASS Is race/ethnicity or social class a greater determinant of health and mortality? Although some health problems (for example, sickle cell anemia and hypertension in African Americans) may have a genetic basis, class is generally more of a factor than race or ethnicity (Navarro, 1991). In a study of the mortality experience over a 28-year period of a random sample of African American and white males 35 years of age or older when the study began, race had no significant effect on mortality, whereas socioeconomic status appeared to be a major factor (Keil et al., 1992). As discussed in prior chapters,

---

Social epidemiology the study of the causes and distribution of health, disease, and impairment throughout a population.

Chronic diseases illnesses that are long term or lifelong and that develop gradually or are present from birth.

Acute diseases illnesses that strike suddenly and cause dramatic incapacitation and sometimes death.

people of color are more likely to have incomes below the poverty line, and the poorest people typically receive less preventive care and less optimal management of chronic diseases than do other people (Leary, 1996). People living in central cities, where there are high levels of poverty and crime, or in remote rural areas generally have greater difficulty in getting health care because most doctors prefer to locate their practice in a "safe" area, particularly one with a patient base that will produce a high income. Although rural Americans make up 20 percent of the U.S. population, only 9 percent of the nation's physicians practice in rural areas, and fewer specialists such as cardiologists are available in these areas (Ricketts, 1999).

Another factor is occupation. People with lower incomes are more likely to be employed in jobs that expose them to danger and illness—working in the construction industry or around heavy equipment in a factory, for example, or holding a job as a convenience store clerk or other position that exposes a person to the risk of armed robbery. Finally, people of color and poor people are more likely to live in areas that contain environmental hazards.

However, although Latinas/os are more likely than non-Latino/a whites to live below the poverty line, they have lower death rates from heart disease, cancer, accidents, and suicide, and an overall lower death rate. One explanation may be dietary factors and the strong family life and support networks found in many Latina/o families (Weiss and Lonnquist, 2003). Obviously, more research is needed on this point, for the answer might be beneficial to all people.

Lifestyle Factors

As noted previously, social epidemiologists also examine lifestyle choices as a factor in health, disease, and impairment. We will examine three life-

style factors as they relate to health: drugs, sexually transmitted diseases, and diet and exercise.

DRUG USE AND ABUSE   What is a drug? There are many different definitions, but for our purposes, a **drug is any substance—other than food and water—that, when taken into the body, alters its functioning in some way.** Drugs are used for either therapeutic or recreational purposes. *Therapeutic* use occurs when a person takes a drug for a specific purpose such as reducing a fever or controlling a cough. In contrast, *recreational* drug use occurs when a person takes a drug for no purpose other than achieving a pleasurable feeling or psychological state. Alcohol and tobacco are examples of drugs that are primarily used for recreational purposes; their use by people over a fixed age (which varies from time to time and place to place) is lawful. Other drugs—such as some anti-anxiety drugs and tranquilizers—may be used legally only if prescribed by a physician for therapeutic use but are frequently used illegally for recreational purposes.

Alcohol   The use of alcohol is considered an accepted part of the dominant culture in the United States. Adults in this country consume an average of 3 gallons of wine, 32 gallons of beer, and 2 gallons of liquor a year. In fact, adults consume more beer on average than milk (24 gallons) or coffee (26 gallons) (U.S. Census Bureau, 2001g). However, these statistics overlook the fact that among people who drink, 10 percent account for roughly half the total alcohol consumption in this country (Levinthal, 1996).

Although the negative short-term effects of alcohol are usually overcome, chronic heavy drinking or alcoholism can cause permanent damage to the brain or other parts of the body (Fishbein and Pease, 1996). For alcoholics, the long-term negative health effects include *nutritional deficiencies* resulting from poor eating habits (chronic heavy drinking contributes to high caloric consumption but low nutritional intake); *cardiovascular problems* such as inflammation and enlargement of the heart muscle, high blood pressure, and stroke; and eventually *alcoholic cirrhosis*—a progressive development of scar tissue that chokes off blood vessels in the liver and destroys liver cells by interfering with their use of oxygen (Levinthal, 1996). Alco-

Reflections

Reflections 14.1

Are lifestyle factors purely choices and responsibilities of each individual, or do societal factors such as advertising contribute to people's lifestyle decisions?

Despite a variety of warnings from the U.S. Surgeon General about the potentially harmful effects of smoking, many people continue to light up cigarettes. Even those who do not smoke may be affected by environmental tobacco smoke.

holic cirrhosis is the ninth most frequent cause of death in the United States. The social consequences of heavy drinking are not always limited to the person doing the drinking. For example, abuse of alcohol and other drugs by a pregnant woman can damage the unborn fetus.

Nicotine (Tobacco) The nicotine in tobacco is a toxic, dependency-producing psychoactive drug that is more addictive than heroin. It is classified as a stimulant because it stimulates central nervous system receptors and activates them to release adrenaline, which raises blood pressure, speeds up the heartbeat, and gives the user a temporary sense of alertness. Although the overall proportion of smokers in the general population has declined somewhat since the 1964 Surgeon General warning that smoking is linked to cancer and other serious diseases, tobacco is still responsible for about one in every five deaths in this country (Akers, 1992). Even people who never light up a cigarette are harmed by *environmental tobacco smoke*—the smoke in the air inhaled by nonsmokers as a result of other people's tobacco smoking (Levinthal, 1996). Researchers have found that environmental smoke is especially hazardous for nonsmokers who carpool or work with heavy smokers.

Illegal Drugs Marijuana is the most extensively used illegal drug in the United States. About one-third of all people over age twelve have tried marijuana at least once. Although most marijuana users are between the ages of eighteen and twenty-five, use by teenagers has more than doubled during the past decade. High doses of marijuana smoked during pregnancy can disrupt the development of a fetus and result in congenital abnormalities and neurological disturbances (Fishbein and Pease, 1996). Furthermore, some studies have found an increased risk of cancer and other lung problems associated with marijuana because its smokers are believed to inhale more deeply than tobacco users.

Another widely used illegal drug is cocaine: About 23 million people over the age of 12 in the United States report that they have used cocaine at least once, and about 1 million acknowledge having used it during the past month (Substance Abuse and Mental Health Services Administration, 2000). People who use cocaine over extended periods of time have higher rates of infection, heart problems, internal bleeding, hypertension, stroke, and other neurological and cardiovascular disorders than do nonusers. Intravenous cocaine users who share contaminated needles are also at risk for contracting AIDS.

Each of the drugs discussed in these few paragraphs represents a lifestyle choice that affects health. Whereas age, race/ethnicity, sex, and—at least to some degree—social class are ascribed characteristics, taking drugs is a voluntary action on a person's part.

SEXUALLY TRANSMITTED DISEASES The circumstances under which a person engages in sexual activity is another lifestyle choice with health implications. Although most people find sexual activity enjoyable, it can result in transmission of certain *sexually transmitted diseases* (STDs), including AIDS, gonorrhea, syphilis, and genital herpes. Prior to 1960, the incidence of STDs in this country had been reduced sharply by barrier-type contraceptives (e.g., condoms) and the use of

Drug any substance—other than food and water—that, when taken into the body, alters its functioning in some way.

Michael Newman/PhotoEdit

Drug abuse is a problem throughout the life course. Here, teenagers receive counseling from a drug treatment specialist and share their ideas and concerns with others in their peer group.

body, it stays there for the rest of a person's life, regardless of treatment. However, the earlier that treatment is received, the more likely it is that the severity of the symptoms will be reduced. About 40 percent of persons infected with genital herpes have only a first attack of symptoms of the disease; the remaining 60 percent may have attacks four or five times a year for several years.

AIDS AIDS (acquired immunodeficiency syndrome), which is caused by HIV (human immunodeficiency virus), is among the most significant health problems that this nation—and the world—faces today. Although AIDS almost inevitably ends in death, no one actually dies *of* AIDS. Rather, AIDS reduces the body's ability to fight diseases, making a person vulnerable to many diseases—such as pneumonia—that result in death.

AIDS was first identified in 1981, and the total number of AIDS-related deaths in the United States through 1985 was only 12,493; however, the numbers rose rapidly and precipitously after that, and the number of *new* reported cases of AIDS in this country in 1993 was 103,533. Fortunately, awareness of the syndrome and how it can be acquired began to produce some results: The number of new AIDS cases per year has declined since the mid-1990s to a low of 41,113 in 2000. Likewise, advances in treatment have reduced the number of deaths from AIDS-related diseases in the United States. The 16,169 deaths in 2000 were less than half the number of such deaths in 1995 (National Centers for Disease Control, 2002; U.S. Census Bureau, 2000c).

Worldwide, however, the number of people with HIV or AIDS is increasing at an alarming rate. The United Nations estimates that in 2001, a total of 40 million people had HIV/AIDS; 15 percent of all new cases worldwide are children, and children account for about 580,000 AIDS-related deaths each year (Joint United Nations Programme on HIV/AIDS, 2001). Two-thirds of the people with HIV/AIDS live in sub-Saharan Africa; 20 percent of people with HIV/AIDS live in south and southeast Asia.

penicillin as a cure. However, in the 1960s and 1970s the number of cases of STDs increased rapidly with the introduction of the birth control pill, which led to women having more sexual partners and couples being less likely to use barrier contraceptives.

Gonorrhea and Syphilis Until the 1960s, gonorrhea (today the second-most-common STD) and syphilis (which can be acquired not only by sexual intercourse but also by kissing or coming into intimate bodily contact with an infected person) were the principal STDs in this country. Today, however, they constitute less than 15 percent of all cases of STDs reported in U.S. clinics. Untreated gonorrhea may spread from the sexual organs to other parts of the body, among other things negatively affecting fertility; it can also spread to the brain or heart and cause death. Untreated syphilis can, over time, cause cardiovascular problems, brain damage, and even death. Penicillin can cure most cases of either gonorrhea or syphilis as long as the disease has not spread.

Genital Herpes This sexually transmitted disease produces a painful rash on the genitals. Genital herpes cannot be cured: Once the virus enters the

✓Checkpoints

Checkpoint 14.2

4. What do social epidemiologists study?
5. What is the relationship among age, sex, race/ethnicity, and health in the United States?
6. What is the key difference between chronic disease and acute disease?
7. How do lifestyle factors contribute to health, disease, and impairment?

HIV is transmitted through unprotected (or inadequately protected) sexual intercourse with an infected partner (either male or female), by sharing a contaminated hypodermic needle with someone who is infected, by exposure to blood or blood products (usually from a transfusion), and by an infected woman who passes the virus on to her child during pregnancy, childbirth, or breast feeding.

STAYING HEALTHY: DIET AND EXERCISE
Lifestyle choices also include positive actions such as a healthy diet and good exercise. Over the past several decades, many people in the United States have begun to improve their dietary habits. More individuals now eat larger amounts of vegetables, fruits, and cereals, and substitute unsaturated fats and oils for saturated fats. These changes have contributed to a marked decrease in the incidence of heart disease and of some types of cancer. However, recent studies have raised concern that a significant percentage of children and adults in this country are overweight or obese to an extent that may decrease their life expectancy.

Exercise is another factor. Regular exercise (at least three times a week) keeps the heart, lungs, muscles, and bones in good health and slows the aging process.

Health Care in the United States

Understanding health care as it exists in the United States today requires a brief examination of its history. During the nineteenth century, people became doctors in this country either through apprenticeships, purchasing a mail-order diploma, completing high school and attending a series of lectures, or obtaining bachelor's and M.D. degrees and studying abroad for a number of years. At that time, medical schools were largely proprietary institutions, and their officials were often more interested in acquiring students than in enforcing standards. The state licensing boards established to improve medical training and stop the proliferation of "irregular" practitioners failed to slow the growth of medical schools, and their number increased from 90 in 1880 to 160 in 1906. Medical school graduates were largely poor and frustrated because of the overabundance of doctors and quasi-medical practitioners, so doctors became highly competitive and anxious to limit the number of new practitioners. The obvious way to accomplish this was to reduce the number of medical schools and set up licensing laws to eliminate unqualified or irregular practitioners (Kendall, 1980).

The Rise of Scientific Medicine and Professionalism

Although medicine had been previously viewed more as an art than as a science, several significant discoveries during the nineteenth century in areas such as bacteriology and anesthesiology began to give medicine increasing credibility as a science (Nuland, 1997). At the same time that these discoveries were occurring, the ideology of science was being advocated in all areas of life, and people came to believe that almost any task could be done better if the appropriate scientific methods were used. To make medicine in the United States more scientific (and more profitable), the Carnegie Foundation (at the request of the American Medical Association and the forerunner of the Association of American Medical Colleges) commissioned an official study of medical education. The "Flexner report" that resulted from this study has been described as the catalyst of modern medical education but has also been criticized for its lack of objectivity.

THE FLEXNER REPORT To conduct his study, Abraham Flexner met with the leading faculty at the Johns Hopkins University School of Medicine to develop a model of how medical education should take place; he next visited each of the 155 medical schools then in existence, comparing

Flash! Light/Stock Boston

Although women today make up about half of the first-year class in U.S. medical schools, they are underrepresented in tenured faculty positions and in specializations such as surgery.

them with the model. Included in the model was the belief that a medical school should be a full-time, research-oriented, laboratory facility that devoted all of its energies to teaching and research, not to the practice of medicine (Kendall, 1980). It should employ "laboratory men" to train students in the "science" of medicine, and the students should then apply the principles they had learned in the sciences to the illnesses of patients (Brown, 1979). Only a few of the schools Flexner visited were deemed to be equipped to teach scientific medicine; nonetheless, his model became the standard for the profession (Duffy, 1976).

As a result of the Flexner report (1910), all but two of the African American medical schools then in existence were closed, and only one of the medical schools for women survived. As a result, white women and people of color were largely excluded from medical education for the first half of the twentieth century. Until the civil rights movement and the women's movement of the 1960s and 1970s, virtually all physicians were white, male, and upper- or upper-middle class.

THE PROFESSIONALIZATION OF MEDICINE

Despite its adverse effect on people of color and women who might desire a career in medicine, the Flexner report did help professionalize medicine. When we compare post-Flexner medicine with the characteristics of professions (see Chapter 13), we find that it meets those characteristics:

1. *Abstract, specialized knowledge.* Physicians undergo a rigorous education that results in a theoretical understanding of health, illness, and medicine. This education provides them with the credentials, skills, and training associated with being a professional.
2. *Autonomy.* Physicians are autonomous and (except as discussed subsequently in this chapter) rely on their own judgment in selecting the appropriate technique for dealing with a problem. They expect patients to respect that autonomy.
3. *Self-regulation.* Theoretically, physicians are self-regulating. They have licensing, accreditation, and regulatory boards and associations that set professional standards and require members to adhere to a code of ethics as a form of public accountability.
4. *Authority.* Because of their authority, physicians expect compliance with their directions and advice. They do not expect clients to argue about the advice rendered (or the price to be charged).
5. *Altruism.* Physicians perform a valuable service for society rather than acting solely in their own self-interest. Many physicians go beyond their self-interest or personal comfort so that they can help a patient.

However, with professionalization, licensed medical doctors gained control over the entire medical establishment, a situation that has continued until the present and—despite current efforts at cost control by insurance companies and others—may continue into the future.

Medicine Today

Throughout its history in the United States, medical care has been on a *fee-for-service* basis: Patients are billed individually for each service they receive, including treatment by doctors, laboratory work, hospital visits, prescriptions, and other health-related expenses. Fee-for-service is an expensive way to deliver health care because there are few restrictions on the fees charged by doctors, hospitals, and other medical providers.

There are both good sides and bad sides to the fee-for-service approach. The good side is that in the "true spirit" of capitalism, coupled with the hard work and scholarship of many people, this approach has resulted in remarkable advances in medicine. Among recent medical innovations are the following:

- *Bloodless surgery*—operations that previously would have required blood transfusions are in some instances now performed without transfusions, using combinations of iron supplements and vitamins, large doses of a blood-building drug (synthetic erythropoietin) that stimulates the bone marrow to produce red blood cells, and intravenous fluids to increase circulation (Langone, 1997).
- *Robotdoc*—a computer-controlled robot used in hip replacement surgery to bore precision holes in a patient's bone, a task that surgeons must otherwise do manually (and with much less precision) with a mallet and chisel—is an example of the future use of robots in medical operations (Olmos, 1997).

The bad side of fee-for-service medicine is its inequality of distribution. In effect, the United States has a two-tier system of medical care. Those who can afford it are able to get top-notch medical treatment. And where they receive it may not be much like the hospitals that most of us have visited:

> Every afternoon, between three and five, high above New York's Fifth Avenue, the usual quiet of Eleven West is broken by the soft rustle of white linen cloths and the clink of silver and china as high tea is served room by room. . . . Down the hall, a concierge waits to take your dinner order, provide a video from a list of over 950 titles, arrange for a manicure or massage, or send up that magazine or best-seller that you wanted to read. No, this is not a hitherto unknown outpost of the Four Seasons or the Ritz, but a 19-room wing of the Mt. Sinai Medical Center, one of the nation's leading hospitals. . . .
>
> Here at Mt. Sinai and a few other top hospitals . . . sheets are 250-count cotton, the bathrooms are marble and stocked with toiletries, and there is ample room to accommodate a nice seating group of leather wing chairs and a brocade sofa. And here no call button is pressed in vain. Hospital personnel not only

Reflections

Reflections 14.2

How does the hospital wing at Mt. Sinai Medical Center that is described here compare with the facilities available to typical patients at most hospitals? How are income and wealth related to quality of life—and even length of life—issues?

> come when summoned but are eagerly waiting to cater to your every need. . . .
>
> On these select floors, multi-tiered food service carts and their clattering, plastic trays are gone. "Room service" is in full force. Meals are presented, often course by course, by bow-tied, black-jacketed waiters from rolling, linen-covered tables. (Winik, 1997)

However, the charges for rooms on floors such as those described above may run from $250 to $1,000 per night more than the cost of the standard private room (Winik, 1997). Obviously, this sort of medical care is not within the budget of most of us. However, the cost of health care per person in the United States rose from $141 in 1960 to $4,200 (almost thirty times as much) in 1999 (U.S. Census Bureau, 2001g) and is still increasing. Figure 14.1 reflects this cost increase. Keeping in mind the issues that have been raised throughout this text regarding income disparity in the United States, the questions to be considered at this point are "Who pays for medical care, and how?" and "What about the people who simply cannot afford adequate medical care?"

Paying for Medical Care in the United States

The United States and the Union of South Africa are the only developed nations without some form of universal health coverage for all citizens. Before we examine the health care systems of several other nations, however, let's look more closely at our own system.

PRIVATE HEALTH INSURANCE Part of the reason that the cost of fee-for-service health care in the United States escalated rapidly beginning

Figure 14.1 Increase in Cost of Health Care, 1970–2000

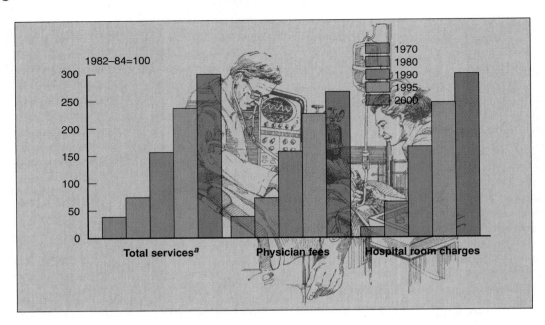

a Includes other medical care services.

Source: Computed by author based on U.S. Health Care Financing Administration, 2000.

in the 1970s (see Figure 14.1) was the expansion of medical insurance programs at that time. Third-party providers (public and private insurers) began picking up large portions of doctor and hospital bills for insured patients. With third-party fee-for-service payment, patients pay premiums into a fund that in turn pays doctors and hospitals for each treatment the patient receives. According to the medical sociologist Paul Starr (1982), third-party fee-for-service is the main reason for medical inflation because it gives doctors and hospitals an incentive to increase medical services. In other words, the more services they provide, the more fees they charge and the more money they make. Medical sociologists Gregory L. Weiss and Lynne E. Lonnquist (2003) note that the principle of supply and demand (health care providers, in competition with one another, should be motivated to offer the best possible service at the lowest price in order to attract customers) works ineffectively in health care. Patients have no incentive to limit their visits to doctors or hospitals because they have already paid their premiums and feel entitled to medical care, regardless of the cost. Likewise, many patients simply depend on the advice of their physicians to determine what treatment

to have; they do not independently decide on what medical care they need.

PUBLIC HEALTH INSURANCE   The United States has two nationwide public health insurance programs, Medicare and Medicaid (see Box 14.2). Medicare is a program for persons age 65 or older who are covered by Social Security or who are eligible and "buy into" the program by paying a monthly premium (Atchley, 2000). Medicare pays part of the health care costs of these people. Medicaid, a jointly funded federal–state–local program, was established to make health care more available to the poor. However, both the Medicaid program and the Medicare program are in financial difficulty. Current projections call for Medicaid spending to double and Medicare spending to triple in the next few years (Weiss and Lonnquist, 2003).

HEALTH MAINTENANCE ORGANIZATIONS (HMOS)   Created in an effort to provide workers with health coverage by keeping costs down, *health maintenance organizations (HMOs) provide, for a set monthly fee, total care with an emphasis on prevention to avoid costly treatment later.* The

Box 14.2 Sociology and Social Policy

Medicare and Medicaid: The Pros and Cons of Government-Funded Medical Care

Should people receive health care even if they cannot afford to pay for it? This question has been a concern in the United States, particularly for older adults who live on a fixed annual income. Consequently, in 1965 Congress enacted the Medicare program. This federal program for people age sixty-five and over (who are covered by Social Security or railroad retirement insurance or who have been permanently and totally disabled for two years or more) is primarily funded through Social Security taxes paid by current workers. We refer to Medicare as an entitlement program because people who receive benefits under the plan must have paid something to be covered. At the time of this writing, Medicare Part A provides coverage for some inpatient hospital expenses (for up to about ninety days) and limited coverage for skilled nursing care at home. By contrast, Part B (which pays some outpatient expenses) requires that the person who is covered must pay an additional premium amounting to about $50 per month.

How successful has Medicare been as a social policy? Medicare has created greater access to health care services for many older people. However, it also has serious limitations, including the fact that it primarily provides for acute, short-term care and is limited to partial payment for ninety days of hospital care. With the aging population, it is problematic that this plan covers only a restricted amount of skilled nursing care and home health services. Patients must pay a deductible and co-payments (fees paid by people each time they see a health care provider), and the plan pays only 80 percent of allowable charges, not the actual amount charged by health care providers. Furthermore, Medicare does not include many of the things that some older people need the most, such as dental care, hearing aids, eyeglasses, long-term care in nursing homes, or custodial or nonmedical service, including adult day care or homemaker services. Medicare also does not cover the cost of prescription drugs in most instances, although legislation that is currently pending may cover at least part of the cost for some people (those with incomes below a specific figure) in the future. There are such serious gaps in Medicare coverage that many older people who can afford to do so buy medigap policies, which are supplementary private insurance policies that fill the gaps in the Medicare coverage.

As we have seen, one form of government-funded health care for older people—Medicare—is considered to be an entitlement program. The other one that we will examine—Medicaid—is more often thought of as a welfare program because it primarily serves low-income people. Unlike Medicare recipients, who are seen as "worthy" of their health care benefits, some Medicaid recipients have been stigmatized for participation in this "welfare program." Medicaid is the joint federal–state means-tested welfare program that provides health care insurance for low-income people who meet specific eligibility requirements, such as being aged, blind, disabled, or pregnant, or elderly (Weitz, 2001). Although people over age sixty-five constitute about 13 percent of Medicaid recipients, they account for more than 40 percent of the program's total expenditures (Hooyman and Kiyak, 1999). Today, many physicians refuse to take Medicaid patients because the administrative paperwork is burdensome and reimbursements are low—typically less than one-half of what private insurance companies pay for the same services.

When we analyze Medicare and Medicaid as social policies, we find that both programs provide broad coverage for older people. However, both programs have serious flaws, since they are extremely expensive and are highly vulnerable to costly fraud by health care providers and some recipients. Finally, across lines of race/ethnicity and class, some studies have shown that these medical assistance programs typically provide higher-quality health care to older people who are white and more affluent than they do for people of color, very low-income individuals, people with disabilities, and those who live in rural areas (Pear, 1994).

How will the aging of the U.S. population affect government-funded health care programs in the future? The Congressional Budget Office estimates that the number of people receiving Medicare benefits in 2030 will be 95 percent higher than today but that the number of workers paying into Medicare will only be about 15 percent higher (Crippen, 2001). Are there better solutions than Medicare and Medicaid for providing better health care for older people at less cost? What do you think?

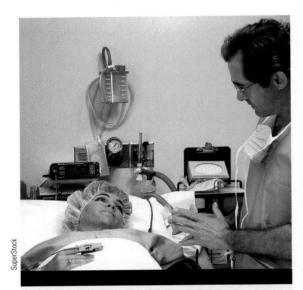

With the introduction of managed care and significant growth in the number of health maintenance organizations in recent years, many U.S. physicians fear that the professional autonomy and authority historically guaranteed to members of their profession are rapidly evaporating.

doctors do not work on a fee-for-service basis, and patients are encouraged to get regular checkups and to practice good health practices (e.g., exercise and eat right). As long as patients use only the doctors and hospitals that are affiliated with their HMO, they pay no fees, or only small co-payments, beyond their insurance premiums (Anders, 1996). Seeing HMOs as a potential source of high profits because of their emphasis on preventing serious illness, many for-profit corporations moved into the HMO business in the 1980s (Anders, 1996). However, research shows that preventive care is good for the individual's health but does not necessarily lower total costs.

Recent concerns about physicians being used as gatekeepers who might prevent some patients from obtaining referrals to specialists or from getting needed treatment have resulted in changes in the policies of some HMOs, which now allow patients to visit health care providers outside an HMO's network or to receive other previously unauthorized services by paying a higher co-payment. However, critics charge that those HMOs whose primary-care physicians are paid on a capitation basis—meaning that they receive only a fixed amount per patient that they see, regardless of how long they spend with that patient—in effect encourage doctors to undertreat patients (Weitz, 2001).

MANAGED CARE Another approach to controlling health care costs in the United States is known as *managed care*—any system of cost containment that closely monitors and controls health care providers' decisions about medical procedures, diagnostic tests, and other services that should be provided to patients (Weitz, 2001). In most managed-care programs, patients choose a primary-care physician from a list of participating doctors. When patients need medical services, they must first contact the primary-care physician; if a specialist is needed for treatment, the primary-care physician refers the patient to a specialist who participates in the program. Doctors must get approval before they perform certain procedures or admit a patient to a hospital; if they fail to obtain such advance approval, the insurance company has the right to refuse to pay for the treatment or hospital stay.

THE UNINSURED AND THE UNDERINSURED Despite public and private insurance programs, about one-third of all U.S. citizens are without health insurance or had difficulty getting or paying for medical care at some time in the last year. As shown on Map 14.1, the number of people not covered by health insurance varies from state to state. Of the people not covered by health insurance, 8.5 million are children (Mills, 2001). An estimated 38.7 million people in the United States had no health insurance in 2000—approximately 14 percent of the nation's population (Mills, 2001). The working poor constitute a substantial portion of this category, and it is estimated that 18 million of the uninsured hold full-time jobs. They make too little to afford health insurance but too much to qualify for Medicaid, and their employers do not provide health insurance coverage. What happens when they need medical treatment? "They do without," explains Ray Hanley, medical services director for the Arkansas Department of Human Services (qtd. in Kilborn, 1997: A10).

Paying for Medical Care in Other Nations

Other industrialized and industrializing countries do not leave their citizens in the situation in which some people in the United States find

Map 14.1 Persons Not Covered by Health Insurance, by State

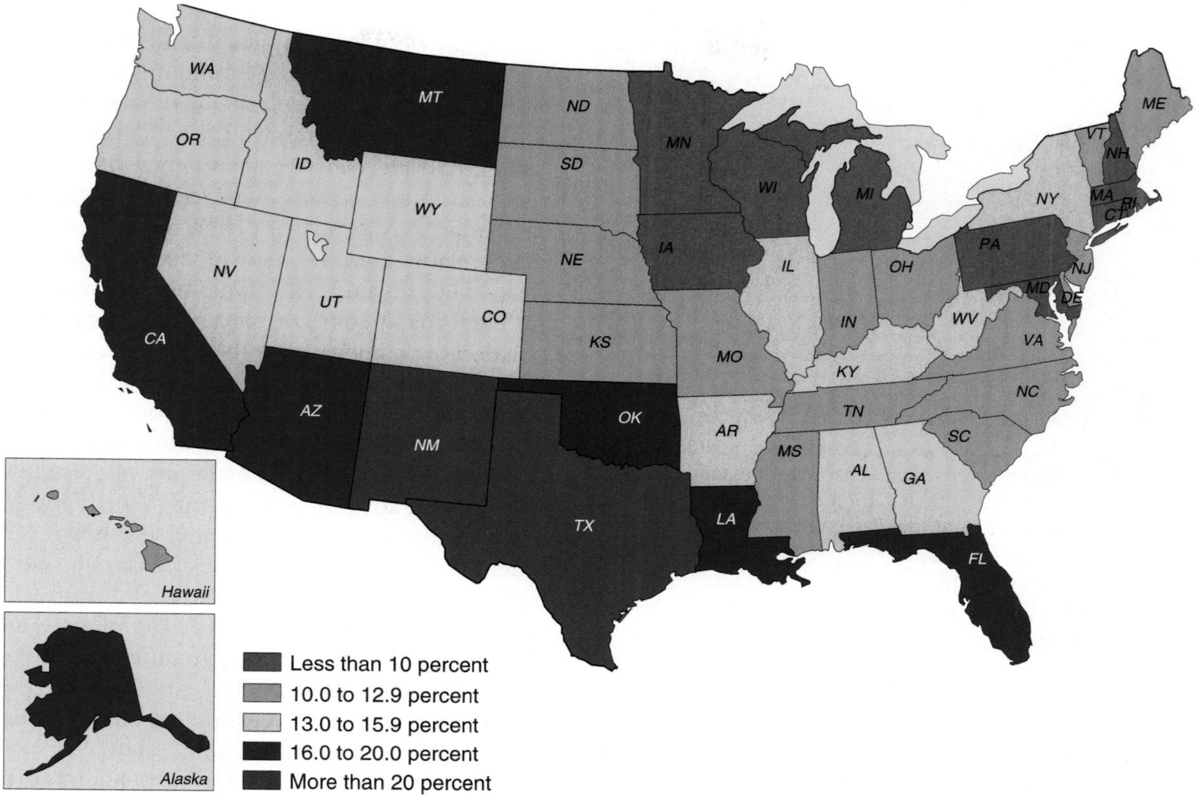
WA
OR
CA
NV
ID
MT
WY
UT
CO
AZ
NM
ND
SD
NE
KS
OK
TX
MN
IA
MO
AR
LA
WI
IL
MI
IN
OH
KY
TN
MS
AL
GA
FL
SC
NC
VA
WV
PA
NY
VT
NH
ME
MA
RI
CT
NJ
DE
MD
Hawaii
Alaska
Less than 10 percent
10.0 to 12.9 percent
13.0 to 15.9 percent
16.0 to 20.0 percent
More than 20 percent

Source: Mills, 2001.

themselves. Let's examine how other nations pay for health care.

CANADA Prior to the 1960s, Canada's health care system was similar to that of the United States today. However, in 1962 the government of the province of Saskatchewan implemented a health insurance plan despite opposition from doctors, who went on strike to protest the program. The strike was not successful, as the vast majority of citizens supported the government, which maintained health services by importing doctors from Great Britain. The Saskatchewan program proved itself viable in the years following the strike, and by 1972 all Canadian provinces and territories had coverage for medical and hospital services (Kendall, Lothian Murray, and Linden, 2000). As a result, Canada has a ***universal health care system*—a health care system in which all citizens receive medical services paid for by tax revenues.** In Canada, these revenues are supplemented by insurance premiums paid by all taxpaying citizens.

One major advantage of the Canadian system over that in the United States is a significant

Health maintenance organizations (HMOs) companies that provide, for a set monthly fee, total care with an emphasis on prevention to avoid costly treatment later.

Managed care any system of cost containment that closely monitors and controls health care providers' decisions about medical procedures, diagnostic tests, and other services that should be provided to patients.

Universal health care system a health care system in which all citizens receive medical services paid for by tax revenues.

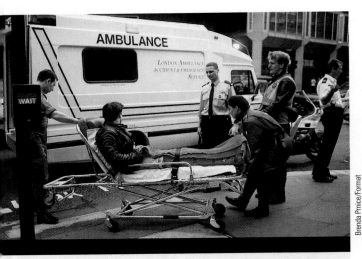

Brenda Prnice/Format

Unlike the United States, Great Britain provides health care services at no charge to the nation's entire population. What are the greatest strengths and weaknesses of socialized medicine?

reduction in administrative costs. Whereas more than 20 percent of the U.S. health care dollar represents administrative costs, in Canada the corresponding figure is 10 percent (Weiss and Lonnquist, 2003). However, the system is not without its critics, who claim that it is costly and often wasteful. For example, Canadians are allowed unlimited trips to the doctor, and doctors can increase their income by ordering extensive tests and repeat visits, just as in the United States (Kendall, Lothian Murray, and Linden, 2000).

The Canadian health care system does not constitute what is referred to as *socialized medicine*—a **health care system in which the government owns the medical care facilities and employs the physicians.** Rather, Canada has maintained the private nature of the medical profession. Although the government pays most health care costs, the physicians are not government employees and have much greater autonomy than physicians in the health care system in Great Britain.

GREAT BRITAIN   In 1946, Great Britain passed the National Health Service Act, which provided for all health care services to be available at no charge to the entire population. Although physicians work out of offices or clinics—as in the

United States or Canada—the government sets health care policies, raises funds and controls the medical care budget, owns health care facilities, and directly employs physicians and other health care personnel (Weiss and Lonnquist, 2003). Unlike the Canadian model, the health care system in Great Britain *does* constitute socialized medicine. Physicians receive capitation payments from the government: a fixed annual fee for each patient in their practice regardless of how many times they see the patient or how many procedures they perform. They also receive supplemental payments for each low-income or elderly patient in their practice, to compensate for the extra time such patients may require; bonus payments if they meet targets for providing preventive services, such as immunizations against disease; and financial incentives if they practice in medically underserved areas (Weitz, 2001). Physicians may accept private patients, but such patients rarely constitute more than a small fraction of a physician's practice; hospitals reserve a small number of beds for private patients (Weiss and Lonnquist, 2003). Why would anyone want to be a private patient who pays for her or his own care or hospital bed? The answer is primarily found in the desire to avoid the long waits ("queues") that the general population encounters and the fact that private patients can enter the hospital for surgery at times convenient to the consumer rather than wait upon the convenience of the system (Gill, 1994).

CHINA   After a lengthy civil war, in 1949 the Communist Party won control of mainland China but found itself in charge of a vast nation with a population of one billion people, most of whom lived in poverty and misery. Malnutrition was prevalent, life expectancies were short, and infant and maternal mortality rates were high. In the cities, only the elite could afford medical care; in the rural areas where most of the population resided, Western-style health care barely existed (Weitz, 2001). With a lack of both financial resources and trained health care personnel, China needed to adopt innovative strategies in order to improve the health of its populace. One such policy was developing a large number of *physician extenders* and sending them out into the cities and rural areas to educate the public regarding health and health care and to treat illness and disease. Referred to as *street doctors* in urban areas and *barefoot doctors* in the countryside, these individuals had little formal training and worked under the supervision of trained physicians (Weitz, 2001).

Over the past three decades, Chinese medical training has become more rigorous and supervision has increased. All doctors receive training in both Western and traditional Chinese medicine. Doctors who work in hospitals receive a salary; all other doctors now work on a fee-for-service basis. In urban areas, the cost of health care is paid by employers; however, 78 percent of the population lives in rural areas, where most work on family farms and are expected to pay for their own health care. The cost of health care generally remains low, but the cost of hospital care has risen; accordingly, many Chinese—if they can afford it—purchase health care insurance to cover the cost of hospitalization (Weitz, 2001). As a low-income country, China spends only 2 percent to 3 percent of its gross domestic product on health care, but the health of its citizens is only slightly below that of most industrialized nations.

Regardless of which system of delivering medical care that a nation may have, the health care providers and the general population of the nation are having to face new issues that arise as a result of new technology.

Social Implications of Advanced Medical Technology

Advances in medical technology are occurring at a speed that is almost unbelievable; however, sociologists and other social scientists have identified specific social implications of some of the new technologies (see Weiss and Lonnquist, 2003):

1. *The new technologies create options for people and for society, but options that alter human relationships.* An example is the ability of medical personnel to sustain a life that in earlier times would have ended as the result of disease or an accident. Although this can be beneficial, technologically advanced equipment that can sustain life after consciousness is lost and there is no likelihood that the person will recover can create a difficult decision for the family of that person if he or she has not left a *living will*—a document stating the

person's wishes regarding the medical circumstances under which his or her life should be terminated. Federal law requires all hospitals and other medical facilities to honor the terms of a living will.

2. *The new technologies increase the cost of medical care.* For example, the computerized axial tomography (CT or CAT) scanner—which combines a computer with X-rays that are passed through the body at different angles—produces clear images of the interior of the body that are invaluable in investigating disease. However, the cost of such a scanner is around $1 million. Magnetic resonance imaging (MRI) equipment that allows pictures to be taken of internal organs ranges in cost from $1 million to $2.5 million. Can the United States afford such equipment in every hospital for every patient? The money available for health care is not unlimited, and when it is spent on high-tech equipment and treatment, it is being reallocated from other health care programs that might be of greater assistance to more people.

3. *The new technologies raise provocative questions about the very nature of life.* In Chapter 11, we briefly discuss in vitro fertilization—a form of assisted reproductive technology. But during 1997, Dr. Ian Williams and his associates in Scotland took in vitro fertilization a step further: They cloned a lamb (that they named Dolly) from the DNA of an adult sheep. Although for many years scientists had been able to clone animals from embryonic cells, what the Scottish team accomplished had far greater ramifications: If scientists can duplicate mammals from adult DNA, is it possible to clone perfect (whatever that may be) human beings instead of taking a chance on a child that is born to a couple? If it is possible, would it be ethical? For example, if—as discussed earlier in this text—most parents prefer a boy over a girl if they are going to have only one child, would the world suddenly

Socialized medicine   a health care system in which the government owns the medical care facilities and employs the physicians.

Michael Newman/PhotoEdit

Use of herbal therapies is a form of alternative medicine that is increasing in popularity in the United States. Here, a Korean American doctor measures out herbs for one of his patients.

have substantially more boys than girls? Would everyone start to look alike, eliminating diversity? If a child were born other than through cloning and had some sort of biological defect, would the child have the right to sue his or her parents for negligence?

However, at the same time that high-tech medicine is becoming a major part of overall health care, many people are turning to holistic medicine and alternative healing practices.

Holistic Medicine and Alternative Medicine

When examining the subject of medicine, it is easy to think only in terms of conventional (or mainstream) medical treatment. By contrast, *holistic medicine* is an approach to health care that focuses on prevention of illness and disease and is aimed at treating the whole person—body and mind—rather than just the part or parts in which symptoms occur. Under this approach, it is important that people not look solely to medicine and doctors for their health, but rather that people engage in health-promoting behavior. Likewise, med-

ical professionals must not only treat illness and disease but also work with the patient to promote a healthy lifestyle and self-image.

Many practitioners of *alternative medicine*—healing practices inconsistent with dominant medical practice—take a holistic approach, and today many people are turning to alternative medicine either in addition to or in lieu of traditional medicine. Jill Neimark tells how she feels about the two types of medicine:

> I'd like to make a confession: I believe. I believe in the brilliance of Western medicine, in the technological wonders of organ transplants, brain scans, and laparoscopic surgery—and I also believe that shamanic healers from indigenous cultures have discovered treatments for illnesses that traditional medicine can't touch, that the mind alone can trigger and then reverse illness, that dietary changes, along with vitamin and herbal supplements, can powerfully impact the course of disease. I believe in merging the best of both worlds—the orthodox and the alternative.
>
> I'm not alone in my belief. This is a season of unprecedented possibility in medicine. And yet in spite of the tremendous changes, there is still some resistance from mainstream medicine, as well as the government and the pharmaceutical industry. One can almost feel the tectonic plates of conventional and holistic medicine crashing up against each other and realigning the landscape of health care. (Neimark, 1997)

Are medical doctors as opposed to alternative medicine as Neimark indicates? In understanding the medical establishment's reaction to alternative medicine, it is important to keep in mind the philosophy of scientific medicine—that medicine is a

Reflections

Reflections 14.3

Why is there organized resistance by some physicians to holistic and alternative medical approaches? Could there be both "good" and "bad" reasons for such resistance?

✓ Checkpoints

Checkpoint 14.3

8. What major factors contributed to the rise of scientific medicine and professionalism?
9. What is meant by fee-for-service health care?
10. How does private health insurance differ from public health insurance?
11. How has managed care been used to control health care costs in the United States?
12. How does health care in Canada, Great Britain, and China differ from that in the United States?

science, not an art. Thus, to the extent to which alternative medicine is "nonscientific," it must be quackery and therefore something that is undoubtedly worthless and possibly harmful. Undoubtedly, self-interest is also involved in mainstream medicine's reaction to alternative medicine: If the public can be persuaded that scientific medicine is the only legitimate healing practice, fewer health care dollars will be spent on a form of medical treatment that is (at least to some extent) in competition with the medical establishment (Weiss and Lonnquist, 2003). But if all forms of alternative medicine (including chiropractic, massage, and spiritual) are taken into account, people spend more money on unconventional therapies than they do for all hospitalizations (Weiss and Lonnquist, 2003).

Sociological Perspectives on Health and Medicine

Functionalist, conflict, symbolic interactionist, and postmodernist perspectives focus on different aspects of health and medicine; each provides us with significant insights on the problems associated with these pressing social concerns.

A Functionalist Perspective: The Sick Role

According to the functionalist approach, if society is to function as a stable system, it is important for people to be healthy and to contribute to their society. Consequently, sickness is viewed as a form of deviant behavior that must be controlled by society. This view was initially set forth by the sociologist Talcott Parsons (1951) in his concept of the *sick role*—**the set of patterned expectations that defines the norms and values appropriate for individuals who are sick and for those who interact with them.** According to Parsons, the sick role has four primary characteristics:

1. People who are sick are not responsible for their condition. It is assumed that being sick is not a deliberate and knowing choice of the sick person.
2. People who assume the sick role are temporarily exempt from their normal roles and obligations. For example, people with illnesses are typically not expected to go to school or work.
3. People who are sick must want to get well. The sick role is considered to be a temporary role that people must relinquish as soon as their condition improves sufficiently. Those who do not return to their regular activities in a timely fashion may be labeled as hypochondriacs or malingerers.
4. People who are sick must seek competent help from a medical professional to hasten their recovery.

As these characteristics show, Parsons believed that illness is dysfunctional for both individuals and the larger society. Those who assume the sick role are unable to fulfill their necessary social roles, such as being parents or employees. Similarly, people who are ill lose days from their productive roles in society, thus weakening the ability of groups and organizations to fulfill their functions. During the winter months, for instance, the media often carry stories about how much illnesses such as flu are "costing" U.S. employers because many workers are at home ill. The sick role affects the bottom

Holistic medicine an approach to health care that focuses on prevention of illness and disease and is aimed at treating the whole person—body and mind—rather than just the part or parts in which symptoms occur.

Sick role the set of patterned expectations that defines the norms and values appropriate for individuals who are sick and for those who interact with them.

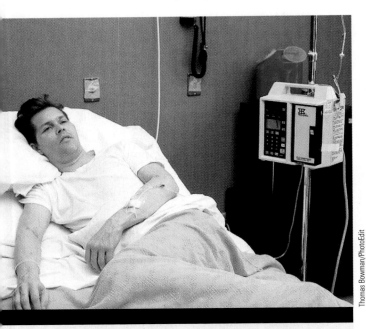

Thomas Bowman/PhotoEdit

According to the functionalist perspective, the sick role exempts the patient from routine activities for a period of time but assumes that the individual will seek appropriate medical attention and get well as soon as possible.

line in business because the employee who is not at work contributes to a loss in productivity.

According to Parsons, it is important for the society to maintain social control over people who enter the sick role. Physicians are empowered to determine who may enter this role and when patients are ready to exit it. Because physicians spend many years in training and have specialized knowledge about illness and its treatment, they are certified by the society to be "gatekeepers" of the sick role. When patients seek the advice of a physician, they enter into the patient–physician relationship, which does not contain equal power for both parties. The patient is expected to follow the "doctor's orders" by adhering to a treatment regime, recovering from the malady, and returning to a normal routine as soon as possible.

What are the major strengths and weaknesses of Parsons's model and, more generally, of the functionalist view of health and illness? Parsons's analysis of the sick role was pathbreaking when it was introduced. Some social analysts believe that Parsons made a major contribution to our knowledge of how society explains illness-related behav-

ior and how physicians have attained their gatekeeper status. In contrast, other analysts believe that the sick-role model does not take into account racial–ethnic, class, and gender variations in the ways that people view illness and interpret this role. For example, this model does not take into account the fact that many individuals in the working class may choose not to accept the sick role unless they are seriously ill—because they cannot afford to miss time from work and lose a portion of their earnings. Moreover, people without health insurance may not have the option of assuming the sick role.

A Conflict Perspective: Inequalities in Health and Health Care

Unlike the functionalist approach, conflict theory emphasizes the political, economic, and social forces that affect health and the health care delivery system. Among the issues of concern to conflict theorists are the ability of all people to obtain health care; how race, class, and gender inequalities affect health and health care; power relationships between doctors and other health care workers; the dominance of the medical model of health care; and the role of profit in the health care system.

Who is responsible for problems in the U.S. health care system? According to many conflict theorists, problems in U.S. health care delivery are rooted in the capitalist economy, which views medicine as a commodity that is produced and sold by the medical–industrial complex. The *medical–industrial complex* **encompasses both local physicians and hospitals as well as global health-related industries such as insurance companies and pharmaceutical and medical supply companies** (Relman, 1992).

The United States is one of the few industrialized nations that relies almost exclusively on the medical–industrial complex for health care delivery and does not have universal health coverage, which provides some level of access to medical treatment for all people. Consequently, access to high-quality medical care is linked to people's ability to pay and to their position within the class structure. Those who are affluent or have good medical insurance may receive high-quality, state-of-the-art care in the medical–industrial complex because of its elaborate technologies and treatments. However, people below the poverty level and those just above it have greater difficulty gaining access to medical care. Referred to as the *medically indigent,* these individuals do not earn enough to afford private medical care but earn just enough money to keep them from qualifying for Medicaid (Weiss and Lonnquist, 2003). In the profit-oriented capitalist economy, these individuals are said to "fall between the cracks" in the health care system.

Who benefits from the existing structure of medicine? According to conflict theorists, physicians—who hold a legal monopoly over medicine—benefit from the existing structure because they can charge inflated fees. Similarly, clinics, pharmacies, laboratories, hospitals, supply manufacturers, insurance companies, and many other corporations derive excessive profits from the existing system of payment in medicine. In recent years, large drug companies and profit-making hospital corporations have come to occupy a larger and larger part of health care delivery. As a result, medical costs have risen rapidly, and the federal government and many insurance companies have placed pressure for cost containment on other players in the medical–industrial complex (Tilly and Tilly, 1998).

Theorists using a radical conflict framework for their analysis believe that the only way to reduce inequalities in the U.S. health care structure is to eliminate capitalism or curb the medical–industrial complex. From this approach, capitalism is implicated in both the rates of illness and how health care is delivered. An example of how rates of illness are related to capitalism is the predominance of workplace hazards and environmental pollution that are dangerous but are often not corrected because corrective measures would be too costly, reducing corporate profits. Capitalistic forms of health care delivery are found in recent medical entrepreneurship on the Internet, whereby companies promote products and medical devices, some of which may not have approval from regulatory agencies such as the Food and Drug Administration.

Conflict theorists increase our awareness of inequalities of race, class, and gender as these statuses influence people's access to health care. They also inform us about the problems associated with health care becoming "big business." However, some analysts believe that the conflict approach is unduly pessimistic about the gains that have been made in health status and longevity—gains that are at least partially due to large investments in research and treatment by the medical–industrial complex.

A Symbolic Interactionist Perspective: The Social Construction of Illness

Symbolic interactionists attempt to understand the specific meanings and causes that we attribute to particular events. In studying health, symbolic interactionists focus on the meanings that social actors give their illness or disease and how these affect people's self-concept and relationships with others. According to symbolic interactionists, we socially construct "health" and "illness" and how both should be treated. For example, some people explain disease by blaming it on those who are ill. If we attribute cancer to the acts of a person, we can assume that we will be immune to that disease if we do not engage in the same behavior. Nonsmokers who learn that a lung cancer victim had a two-pack-a-day habit feel comforted that they are unlikely to suffer the same fate. Similarly, victims of AIDS are often blamed for promiscuous sexual conduct or intravenous drug use, regardless of how they contracted HIV. In this case, the social definition of the illness leads to the stigmatization of individuals who suffer from the disease.

Although biological characteristics provide objective criteria for determining medical conditions such as heart disease, tuberculosis, or cancer, there is also a subjective component to how illness is defined. This subjective component is very important when we look at conditions such as childhood hyperactivity, mental illness, alcoholism, drug abuse, cigarette smoking, and overeating, all of which have been medicalized. The term *medicalization* refers to the process whereby nonmedical problems become defined and treated

Medical–industrial complex local physicians, local hospitals, and global health-related industries such as insurance companies and pharmaceutical and medical supply companies that deliver health care today.

Medicalization the process whereby nonmedical problems become defined and treated as illnesses or disorders.

For most people, occasional gambling is simply a leisure pursuit. However, when gambling is medicalized, it becomes defined as a disease for which there may be treatment.

Jodi Buren/Woodfin Camp & Associates

as illnesses or disorders. Medicalization may occur on three levels: (1) the conceptual level (e.g., the use of medical terminology to define the problem), (2) the institutional level (e.g., physicians are supervisors of treatment and gatekeepers to applying for benefits), and (3) the interactional level (e.g., when physicians treat patients' conditions as medical problems) (Conrad and Schneider, 1992). For example, the sociologists Deborah Findlay and Leslie Miller (1994: 277) explain how gambling has been medicalized:

> Habitual gambling . . . has been regarded by a minority as a sin, and by most as a leisure pursuit—perhaps wasteful but a pastime nevertheless. Lately, however, we have seen gambling described as a psychological illness—"compulsive gambling." It is in the process of being medicalized. The consequences of this shift in discourse (that is, in the way of thinking and talking) about gambling are considerable for doctors, who now have in gamblers a new market for their services or "treatment"; perhaps for gambling halls, which may find themselves subject to new regulations, insofar as they are deemed to contribute to the "disease"; and not least, for gamblers themselves, who are no longer treated as sinners or wastrels, but as patients, with claims on our sympathy, and to our medical insurance plans as well.

Sociologists often refer to this form of medicalization as the *medicalization of deviance* because it

gives physicians and other medical professionals greater authority to determine what should be considered "normal" and "acceptable" behavior and to establish the appropriate mechanisms for controlling "deviant behaviors."

According to symbolic interactionists, medicalization is a two-way process: Just as conditions can be medicalized, so can they be demedicalized. ***Demedicalization*** **refers to the process whereby a problem ceases to be defined as an illness or a disorder.** Examples include the removal of certain behaviors (such as homosexuality) from the list of mental disorders compiled by the American Psychiatric Association and the deinstitutionalization of mental health patients. The process of demedicalization also continues in women's health as advocates seek to redefine childbirth and menopause as natural processes rather than as illnesses (Conrad, 1996).

In addition to how health and illness are defined, symbolic interactionists examine how doctors and patients interact in health care settings. Some physicians may hesitate to communicate certain kinds of medical information to patients, such as why they are prescribing certain medications or what side-effects or drug interactions may occur (Kendall, 1998).

Symbolic interactionist perspectives on health and health care provide us with new insights on the social construction of illness and how health and illness cannot be strictly determined by medical criteria. Symbolic interactionists also make us aware of the importance of communication between physicians and patients, including factors that may reduce effective medical treatment for some individuals. However, these approaches have been criticized for suggesting that few objective medical criteria exist for many illnesses and for overemphasizing microlevel issues without giving adequate recognition to macrolevel issues such as the effects on health care of managed care, health maintenance organizations, and for-profit hospital chains.

A Postmodernist Perspective: The Clinical Gaze

In *The Birth of the Clinic* (1994/1963), postmodern theorist Michel Foucault questioned existing assumptions about medical knowledge and the power that doctors have gained over other medical personnel and everyday people. Foucault asserted that truth in medicine—like all other areas of life—is a social construction, in this instance one that doctors have created. Foucault believed that doctors gain power through the *clinical* (or "observing") *gaze*, which they use to gather information. Doctors develop the clinical gaze through their observation of patients; as the doctors begin to diagnose and treat medical conditions, they also start to speak "wisely" about everything. As a result, other people start to believe that doctors can "penetrate illusion and see . . . the hidden truth" (Shawver, 1998).

According to Foucault, the prestige of the medical establishment was further enhanced when it became possible to categorize all illnesses within a definitive network of disease classification under which physicians can claim that they know why patients are sick. Moreover, the invention of new tests made it necessary for physicians to gaze upon the naked body, to listen to the human heart with an instrument, and to run tests on the patient's body fluids. Patients who objected were criticized by the doctors for their "false modesty" and "excessive restraint" (Foucault, 1994/1963: 163). As the new rules allowed for the patient to be touched and prodded, the myth of the doctor's diagnostic wisdom was further enhanced, and "medical gestures, words, gazes took on a philosophical density that had formerly belonged only to mathe-

matical thought" (Foucault, 1994/1963: 199). For Foucault, the formation of clinical medicine was merely one of the more-visible ways in which the fundamental structures of human experience have changed throughout history.

Foucault's work provides new insights on medical dominance, but it has been criticized for its lack of attention to alternative viewpoints. Among these is the possibility that medical breakthroughs and new technologies actually help physicians become wiser and more scientific in their endeavors. Another criticism is that Foucault's approach is based upon the false assumption that people are passive individuals who simply comply with doctors' orders—he does not take into account that people (either consciously or unconsciously) may resist the myth of the "wise doctor" and not follow "doctors' orders" (Lupton, 1997).

Concept Table 14.A summarizes the major sociological perspectives on health and medicine.

Disability

What is a disability? There are many different definitions. In business and government, disability is often defined in terms of work—for instance, "an inability to engage in gainful employment." Medical professionals tend to define it in terms of organically based impairments—the problem being entirely within the body (Albrecht, 1992). However, not all disabilities are visible to others or necessarily

Demedicalization the process whereby a problem ceases to be defined as an illness or a disorder.

✓ Checkpoints

Checkpoint 14.4

13. What are the four primary characteristics of the sick role?
14. How might the medical–industrial complex contribute to inequalities in health care?
15. What is the social construction of illness?
16. What are medicalization and demedicalization?
17. What is meant by the clinical gaze?

Concept Table 14.A Sociological Perspectives on Health and Medicine

A functionalist perspective: the sick role	People who are sick are temporarily exempt from normal obligations, but must want to get well and seek competent help.
A conflict perspective: inequalities in health and health care	Problems in health care are rooted in the capitalist system, exemplified by the medical–industrial complex.
A symbolic interactionist perspective: the social construction of illness	People socially construct both "health" and "illness," and how both should be treated.
A postmodernist perspective: the clinical gaze	Doctors gain power through observing patients to gather information, thus appearing to speak "wisely."

limit people physically. *Disability* **refers to a reduced ability to perform tasks one would normally do at a given stage of life and that may result in stigmatization or discrimination against the person with disabilities.** In other words, the notion of disability is based not only on physical conditions but also on social attitudes and the social and physical environments in which people live. In an elevator, for example, the buttons may be beyond the reach of persons using a wheelchair. In this context, disability derives from the fact that certain things have been made inaccessible to some people (Weitz, 2001). According to disability rights advocates, disability must be thought of in terms of how society causes or contributes to the problem—not in terms of what is "wrong" with the person with a disability.

An estimated 49.7 million persons in the United States have one or more physical or mental disabilities. This number continues to increase for several reasons. First, with advances in medical technology, many people who once would have died from an accident or illness now survive, although with an impairment. Second, as more people live longer, they are more likely to experience diseases (such as arthritis) that may have disabling consequences (Albrecht, 1992). Third, persons born with serious disabilities are more likely to survive infancy because of medical technology. However, less than 15 percent of persons with a disability today were born with it; accidents, disease, and war account for most disabilities in this country.

Although anyone can become disabled, some people are more likely to be or to become disabled than others. African Americans have higher rates of disability than whites, especially more serious disabilities; persons with lower incomes also have higher rates of disability (Weitz, 2001). However, "disability knows no socioeconomic boundaries. You can become disabled from your mother's poor nutrition or from falling off your polo pony," says Patrisha Wright, a spokesperson for the Disability Rights Education and Defense Fund (qtd. in Shapiro, 1993: 10).

For persons with chronic illness and disability, life expectancy may take on a different meaning. Knowing that they will likely not live out the full life expectancy for persons in their age cohort, they may come to "treasure each moment," as did the late James Keller, a well-respected former college baseball coach:

In December 1992, I found out I have Lou Gehrig's disease—amyotrophic lateral sclerosis, or ALS. I learned that this disease destroys every muscle in the body, that there's no known cure or treatment and that the average life expectancy for people with ALS is two to five years after diagnosis.

Those are hard facts to accept. Even today, nearly two years after my diagnosis, I see myself as 42-year-old career athlete who has always been blessed with excellent health. Though not an hour goes by in which I don't see or hear in my mind that phrase "two to five years," I still can't quite believe it. Maybe my resistance to those words is exactly what gives me the strength to live with them and the will to make the best of every day in every way. (Keller, 1994)

As Keller's comments illustrate, disease and disability are intricately linked.

Michael Newman/PhotoEdit

One of the most compelling concerns for many people with a disability is access to buildings and other public accommodations. For persons using a wheelchair, steps such as these are a formidable obstacle in everyday life.

Reflections

Reflections 14.4

Throughout this text, the personal narratives of many individuals are used as examples to help illustrate the issue being discussed. What can we learn about health, health care, and disability by examining the lived experiences of patients?

under high levels of stress that may result in neuroses and other mental health problems (Albrecht, 1992). As shown in Table 14.1, nearly one out of four people in the United States (23 percent) has a "chronic health condition which, given the physical, attitudinal, and financial barriers built into the social system, makes it difficult to perform one or more activities generally considered appropriate for persons of their age" (Weitz, 2001).

Can a person in a wheelchair have equal access to education, employment, and housing? If public transportation is not accessible to those in wheelchairs, the answer is certainly no. As disability rights activist Mark Johnson put it, "Black people fought for the right to ride in the front of the bus. We're fighting for the right to get on the bus" (qtd. in Shapiro, 1993: 128).

Many disability rights advocates argue that persons with a disability have been kept out of the mainstream of society. They have been denied equal opportunities in education by being consigned to special education classes or special schools. For example, people who grow up deaf are often viewed as disabled; however, many members of the deaf community instead view themselves as a "linguistic minority" that is part of a unique culture (Lane, 1992; Cohen, 1994). They believe that they have been restricted from entry into schools and the work force not due to their own limitations, but by societal barriers.

Living with disabilities is a long-term process. For infants born with certain types of congenital

Environment, lifestyle, and working conditions may all contribute to either temporary or chronic disability. For example, air pollution in automobile-clogged cities leads to a higher incidence of chronic respiratory disease and lung damage, which may result in severe disability for some people. Eating certain types of food and smoking cigarettes increase the risk for coronary and cardiovascular diseases (Albrecht, 1992). In contemporary industrial societies, workers in the second tier of the labor market (primarily recent immigrants, white women, and people of color) are at the greatest risk for certain health hazards and disabilities. Employees in data processing and service-oriented jobs may also be affected by work-related disabilities. The extensive use of computers has been shown to harm some workers' vision; to produce joint problems such as arthritis, low-back pain, and carpal tunnel syndrome; and to place employees

Disability   a reduced ability to perform tasks one would normally do at a given stage of life and that may result in stigmatization or discrimination against the person with disabilities.

Table 14.1 Percentage of U.S. Population with Disabilities, 1997

Characteristic	Percentage[a]
With a disability	23.0
Severe	14.8
Not severe	8.3
Has difficulty or is unable to:	
See words and letters	3.7
Hear normal conversation	3.8
Have speech understood	1.1
Lift or carry ten pounds	7.3
Use stairs	9.5
Walk	9.4
Has difficulty or needs assistance with:	
Getting around inside the house	1.8
Getting in/out of bed or a chair	3.0
Taking a bath or shower	2.4
Dressing	1.7
Eating	0.7
Getting to or using the toilet	1.1
Has difficulty or needs assistance with:	
Going outside the home alone	4.1
Managing money and bills	2.2
Preparing meals	2.3
Doing light housework	3.1
Using the telephone	1.4

[a]Percentage of persons aged 15 and older.

Source: Stern, 2001.

(present at birth) problems, their disability first acquires social significance for their parents and caregivers. In a study of children with disabilities in Israel, the sociologist Meira Weiss (1994) challenged the assumption that parents automatically bond with infants, especially those born with visible disabilities. She found that an infant's appearance may determine how parents will view the child. Parents are more likely to be bothered by external, openly visible disabilities than by internal or disguised ones; some parents are more willing to consent to or even demand the death of an "appearance-impaired" child (Weiss, 1994). According to Weiss, children born with internal (concealed) disabilities are at least initially more acceptable to parents because they do not violate the parents' perceived body images of their children. Weiss's study provides insight into the social significance that people attach to congenital disabilities.

Among persons who acquire disabilities through disease or accidents later in life, the social significance of their disability can be seen in how they initially respond to their symptoms and diagnosis, how they view the immediate situation and their future, and how the illness and disability affect their lives. When confronted with a disability, most people adopt one of two strategies—avoidance or vigilance. Those who use the avoidance strategy deny their condition in order to maintain hopeful images of the future and elude depression; for example, some individuals refuse to participate in rehabilitation following a traumatic injury because they want to pretend that the disability does not exist. By contrast, those using the vigilance strategy actively seek knowledge and treatment so that they can respond appropriately to the changes in their bodies (Weitz, 2001).

Sociological Perspectives on Disability

How do sociologists view disability? Those using the functionalist framework often apply Parsons's sick-role model, which is referred to as the *medical model* of disability. According to the medical model, people with disabilities become, in effect, chronic patients under the supervision of doctors and other medical personnel, subject to a doctor's orders or a program's rules, and not to their own judgment (Shapiro, 1993). From this perspective, disability is deviance.

The deviance framework is also apparent in some symbolic interactionist perspectives. According to symbolic interactionists, people with a disability experience *role ambiguity* because many people equate disability with deviance (Murphy et al., 1988). By labeling individuals with a disability as "deviant," other people can avoid them or treat them as outsiders. Society marginalizes people with a disability because they have lost old roles and statuses and are labeled as "disabled" persons. According to the sociologist Eliot Freidson (1965), how the people are labeled results from three factors: (1) their degree of responsibility for their impairment, (2) the apparent seriousness of their condition, and (3) the perceived legitimacy of the condition. Freidson concluded that the definitions of and expectations for people with a disability are socially constructed factors.

Finally, from a conflict perspective, persons with a disability are members of a subordinate group in conflict with persons in positions of power in the government, in the health care industry, and in the rehabilitation business, all of whom are trying to control their destinies (Albrecht, 1992). Those in positions of power have created policies and artificial barriers that keep people with disabilities in a subservient position (Asch, 1986; Hahn, 1987). Moreover, in a capitalist economy, disabilities are big business. When people with disabilities are defined as a social problem and public funds are spent to purchase goods and services for them, rehabilitation becomes a commodity that can be bought and sold by the medical–industrial complex (Albrecht, 1992). From this perspective, persons with a disability are objectified. They have an economic value as consumers of goods and services that will allegedly make them "better" people. Many persons with a disability endure the same struggle for resources faced by people of color, women, and older persons. Individuals who hold more than one of these ascribed statuses, combined with experiencing disability, are doubly or triply oppressed by capitalism.

Social Inequalities Based on Disability

People with visible disabilities are often the objects of prejudice and discrimination, which interfere with their everyday life. For example, Marylou Breslin, executive director of the Disability Rights Education and Defense Fund, was wearing a businesswoman's suit, sitting at the airport in her battery-powered wheelchair, and drinking a cup of coffee while waiting for a plane. A woman walked by and plunked a coin in the coffee cup that Breslin held in her hand, splashing coffee on Breslin's blouse (Shapiro, 1993). Why did the woman drop the coin in Breslin's cup? The answer to this question is found in stereotypes built on lack of knowledge about or exaggeration of the characteristics of people with a disability. Some stereotypes project the image that persons with disabilities are deformed individuals who may also be

horrible deviants. For example, slasher movies such as the *Nightmare on Elm Street* series show a villain who was turned into a hateful, sadistic killer because of disfigurement resulting from a fire. Lighter fare such as the *Batman* movies depict villains as individuals with disabilities: the Joker, disfigured by a fall into a vat of acid, and the Penguin, born with flippers instead of arms (Shapiro, 1993). Other stereotypes show persons with disabilities as pathetic individuals to be pitied. Fund-raising activities by many charitable organizations—such as "poster child" campaigns showing a photograph of a friendly-looking child with a visible disability—sometimes contribute to this perception. Even apparently positive stereotypes become harmful to people with a disability. An example is what some disabled persons refer to as "supercrips"—people with severe disabilities who seem to excel despite the impairment and who receive widespread press coverage in the process. Disability rights advocates note that such stereotypes do not reflect the day-to-day reality of most persons with disabilities, who must struggle constantly with smaller challenges (Shapiro, 1993).

Today, many working-age persons with a disability in the United States are unemployed (see "Census Profiles: Disability and Employment Status"). Most of them believe that they could and would work if offered the opportunity. However, even when persons with a severe disability are able to find jobs, they typically earn less than persons without a disability (Yelin, 1992). On average, workers with a severe disability make 50 percent (for men) and 64 percent (for women) of what their co-workers without disabilities earn, and the gap is growing (U.S. Census Bureau, 2000a). The problem has been particularly severe for African Americans and Latinos/as with disabilities. Among Latinos/as with a severe disability, only 26 percent are employed; those who work earn 80 percent of what white (non-Latino) persons with a severe disability earn (U.S. Census Bureau, 2000a).

Employment, poverty, and disability are related. On the one hand, people may become economically disadvantaged as a result of chronic illness or disability. On the other hand, poor people are less likely to be educated and more likely to be malnourished and have inadequate access to health care—all of which contribute to the risk of chronic illness, physical and mental disability, and

CENSUS PROFILES
Disability and Employement Status

One of the questions on Census 2000 asked respondents about long-lasting conditions such as a physical, mental, or emotional condition that substantially limited important basic activities. It also asked if they were working at a job or business. The answers allowed the Census Bureau to determine how many people with a disability and how many people without a disability were employed at the time the census was conducted. As you can see from the figure below, for the population aged 21 to 64 years (the period during which people are most likely to be employed), less than 50 percent of persons with a disability are employed, compared with the almost 80 percent of persons without a disability who are employed. Is employment among persons with a disability related primarily to their disability status, or do other factors—such as prejudice or lack of willingness to make the necessary accommodations that would allow such persons to hold a job—play a significant part in the high rate of unemployment of persons with a disability?

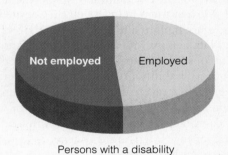

Persons with a disability

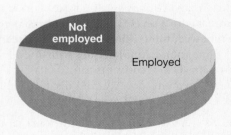

Persons without a disability

Source: U.S. Census Bureau, 2001f.

the inability to participate in the labor force. In addition, the type of employment available to people with limited resources increases their chances of becoming disabled. They may work in hazardous places such as mines, factory assembly lines, and chemical plants, or in the construction industry, where the chance of becoming seriously disabled is much higher (Albrecht, 1992).

Generalizations about the relationship between disability and income are difficult to make for at least three reasons. First, most research on disability is organized around specific conditions or impairments, making it problematic to reach conclusions about how much income a person loses because of a disability (DeJong, Batavia, and Griss, 1989). Second, generalizations about entire categories of people (such as whites, African Americans, or Latinos/as) tend to be inaccurate because of differences *within* each group. For example, the sociologist Ronald Angel (1984) found significant differences in the effect of disabilities on Latinos depending on whether they were Mexican American, Puerto Rican, or Cuban American. Angel concluded that although both Mexican American and Puerto Rican men had lower rates of full-time employment and lower hourly wages relative to (non-Latino) whites, Puerto Ricans were worse off than Mexican Americans in terms of earnings and number of work hours. Third, there is the problem of determining which factor occurred first. For example, some of the problems for Puerto Ricans with disabilities are tied to their overall position of economic disadvantage. Although for both males and females, disability is associated with lower earnings and higher rates of unemployment, a disability has a stronger negative effect on women's labor force participation than it does on men's.

✓ Checkpoints

Checkpoint 14.5

18. Why is disability a sociological issue?
19. How many people in the United States have one or more disabilities?
20. What are the unique problems faced by persons with chronic illness and disability?
21. What are the key sociological perspectives on disability?

Compared to men, women with disabilities are overrepresented as clerical and service workers and underrepresented as managers and administrators. Women with disabilities are also much less likely to be covered by pension and health plans than are men (Russo and Jansen, 1988).

Health Care in the Future

A central question regarding the future of health care is the role that advanced technologies will play in medical diagnosis and treatment. Technology is a major stimulus for social change, and the health care systems in high-income nations such as the United States reflect the rapid rate of technological innovation that has occurred in the last few decades. In the future, advanced health care technologies will no doubt provide even more accurate and quicker diagnosis, improved treatment techniques, and increased life expectancy. However, technology alone cannot solve many of the problems confronting us in health and health care delivery. In fact, some aspects of technological innovation may be dysfunctional for individuals and society. As we have seen, some technological "advances" raise new ethical concerns, such as the moral and legal issues surrounding the cloning of human life. Some "advances" also may fail: A new prescription drug may be found to cause side-effects that are more serious than the illness that it was supposed to remedy.

Whether advanced technology succeeds or fails in some areas, it will probably continue to increase the cost of health care in the future. As a result, the gap between the rich and the poor in the United States will contribute to inequalities of access to vital medical services. On a global basis, new technologies may lower the death rate in some low-income countries, but it will primarily be the wealthy in those nations who will have access to the level of health care that many people in higher-income countries take for granted.

The organization of U.S. health care will continue to change in the future. Despite recent legal challenges, many managed-care companies continue their cost-containment strategies and intervene in medical decisions formerly reserved for physicians and other health care professionals. Consequently, managed care and new technologies will continue to change the traditional doctor–patient relationship, even in how people communicate with one another.

Finally, to a degree, health care in the future will be up to each of us. What measures will we take to safeguard ourselves against illness and disorders? How can we help others who are the victims of acute and chronic diseases or disabilities? Although we cannot change global or national health problems, there are some small (but not insignificant) things we can do to make a difference (see Box 14.3).

Chapter Review

What is health, and why are sociologists interested in studying health and medicine?
According to the World Health Organization, health is a state of complete physical, mental, and social well-being. In other words, health is not only a biological issue but also a social issue. For this reason, sociologists are interested in studying health and medicine. As a social institution, medicine is one of the most important components of quality of life.

Do nations that spend the most on health care have the healthiest citizens?
There is not always a direct relationship between health care expenditures and people's health, particularly in terms of factors such as infant mortality and life expectancy. For example, on average the United States spends more per person on health care than Sweden, but that country has a lower infant mortality rate and greater life expectancy than the United States.

How do acute and chronic diseases differ, and which is most closely linked to spiraling health care costs?
Acute diseases are illnesses that strike suddenly and cause dramatic incapacitation and sometimes death. Chronic diseases, such as arthritis, diabetes, and heart disease, are long-term or lifelong illnesses that develop gradually or are present at birth. Treatment of chronic diseases is typically more costly because of the duration of these problems.

What are some lifestyle factors that contribute to illness?
Social epidemiologists identify a number of lifestyle factors that contribute to illness, including abuse of alcohol and other drugs, use of nicotine (tobacco), and abuse of illegal drugs such as marijuana and cocaine.

Box 14.3 You Can Make a Difference
Joining the Fight Against Illness and Disease!

Each day trained volunteers nurture female patients in hospitals and long-term care facilities by administering complimentary makeup and moisturizer. Among the benefits to patients receiving this caring social interaction and gentle touch are higher self-esteem and shorter recovery times. (Fiffer and Fiffer, 1994: 120)

Beverly Barnes did not intend to start the program now known as Patient Pride when she went to the hospital to visit a friend who was recuperating from surgery. However, by the time she left the hospital, having helped her friend put on makeup and freshen her appearance, Barnes realized how much seemingly small acts of kindness can mean to people who are ill or recuperating from surgery. Eventually, Patient Pride, Inc., became a nonprofit corporation that relies on volunteers to visit patients, and new chapters servicing hospitals and long-term health care facilities have opened throughout the

country. Recently, the program has sought to include male patients as well as women.

Patient Pride is only one of many examples of how everyday people can make a difference in the lives of people experiencing illness or disability. Here are some Internet sources that you can check out to learn more about volunteering and participating in the fight against diseases such as cancer and cardiovascular disease:

- The American Cancer Society's home page (which has links to various volunteer activities):

http://www.cancer.org

- The American Heart Association's home page:

http://www.amhrt.org

You can also learn more about the health problems discussed in this chapter. To learn more about addictive behavior, go to this site:

http://www.cts.com:80/~habtsmrt

Other lifestyle choices contribute to contracting sexually transmitted diseases, including HIV/AIDS.

How is health care paid for in the United States?
Throughout most of the past hundred years, medical care in the United States has been paid for on a fee-for-service basis. The term *fee for service* means that patients are billed individually for each service they receive. This approach to paying for medical services is expensive because few restrictions are placed on the fees that doctors, hospitals, and other medical providers can charge patients. Recently, there have been efforts at cost containment, and HMOs and managed care have produced both positive and negative results in the contemporary practice of medicine. Health maintenance organizations (HMOs) provide, for a set monthly fee, total care with an emphasis on prevention to avoid costly treatment later. Managed care refers to any system of cost containment that closely monitors and controls health care providers' decisions about medical procedures, diagnostic tests, and other services that should be provided to patients.

What is holistic medicine, and how does it differ from traditional health care?
Holistic medicine is an approach to health care that focuses on prevention of illness and disease and is aimed at treating the whole person—body and mind—rather than just the part or parts in which symptoms occur. Under this approach, it is important that people not look solely to medicine and doctors for their health, but rather that people engage in health-promoting behavior.

What is the functionalist perspective on health and illness?
According to the functionalist approach, if society is to function as a stable system, it is important for people to be healthy and to contribute to their society. Consequently, sickness is viewed as a form of deviant behavior that must be controlled by society. Sociologist Talcott Parsons (1951) described the sick role—the set of patterned expectations that defines the norms and values appropriate for individuals who are sick and for those who interact with them. Although individuals are given permission to not perform their usual activities for a period of time, they are expected to seek medical attention and get well as soon as possible so that they can go about their normal routine.

What is the conflict perspective on health and illness?
Conflict theory tends to emphasize the political, economic, and social forces that affect health and the health

care delivery system. Among these issues are the ability of all people to obtain health care; how race, class, and gender inequalities affect health and health care; power relations between doctors and other health care workers; the dominance of the medical model of health care; and the role of profit in the health care system.

What is the symbolic interactionist perspective on health and illness?

In studying health, symbolic interactionists focus on the fact that the meaning that social actors give their illness or disease will affect their self-concept and their relationships with others. According to symbolic interactionists, we socially construct "health" and "illness" and how both should be treated. Symbolic interactionists also examine medicalization—the process whereby nonmedical problems become defined and treated as illnesses or disorders.

What is the postmodernist perspective on health and illness?

Postmodern theorists such as Michel Foucault argue that doctors and the medical establishment have gained control over illness and patients at least partly because of the physicians' clinical gaze, which replaces all other systems of knowledge. The myth of the wise doctor was also supported by the development of disease classification systems and new tests.

What is a disability, and how prevalent are disabilities in the United States?

Disability is a physical or health condition that stigmatizes or causes discrimination. An estimated 49.7 million persons in the United States have one or more physical or mental disabilities. This number continues to increase for several reasons. First, with advances in medical technology, many people who once would have died from an accident or illness now survive, although with an impairment. Second, as more people live longer, they are more likely to experience diseases (such as arthritis) that may have disabling consequences. Third, persons born with serious disabilities are more likely to survive infancy because of medical technology. However, less than 15 percent of persons with a disability today were born with it; accidents, disease, and war account for most disabilities in this country.

Key Terms

acute diseases 449
chronic diseases 449
demedicalization 466
disability 468

drug 450
health 444
health care 445
health maintenance organization (HMO) 456
holistic medicine 462
infant mortality rate 447
life expectancy 447
managed care 458
medical–industrial complex 464
medicalization 465
medicine 445
sick role 463
social epidemiology 448
socialized medicine 460
universal health care system 459

Questions for Critical Thinking

1. Why is it important to explain the social, as well as the biological, aspects of health and illness in societies?
2. In what ways are race, class, and gender intertwined with physical and mental disorders?
3. How would functionalists, conflict theorists, and symbolic interactionists suggest that health care delivery might be improved in the United States?
4. Based on this chapter, how do you think illness and disability will be handled in the United States in the near future? Are there things that we can learn from other nations regarding the delivery of health care? Why or why not?

 ## Resources on the Internet

Chapter-Related Web Sites

The following web sites have been selected for their relevance to the topics in this chapter. These sites are among the more stable, but please note that web site addresses change frequently.

United States Census Bureau: Disability

http://www.census.gov/hhes/www/disability.html
Census figures have been collected on persons with a disability. From this page you can access information about income and demographic data related to disability.

United States Census Bureau Facts & Features: 12th Anniversary of the Americans with Disabilities Act

http://www.census.gov/Press-Release/www/2002/cb02ff11.html
This site offers brief details about the status of persons with a disability in the United States.

HealthcareSource.com
http://www.healthcaresource.com
This site provides information about jobs and careers in the health care profession.

Health Care and Social Assistance
http://www.census.gov/prod/ec97/97s62-us.pdf
This report is from the 1997 Economic Census. It contains detailed reports and a careful analysis of health and economic conditions in America.

The Americans with Disabilities Act:
ADA Home Page
http://www.usdoj.gov/crt/ada/adahom1.htm
This web page disseminates information to persons with a disability. From this site you can learn about the connections among disability, health, work, and full participation in U.S. society.

U.S. Department of Health and Human Services
http://www.hhs.gov
This official web site offers the latest health-related headlines, links to government agencies, and opportunities for working on health-related projects.

Parsons's Theory of the Sick Role
http://www.ahs.cqu.edu.au/psysoc/units/53270/pdf/m4c11.pdf
This document has a chapter that details Talcott Parsons's theory of the sick role.

Companion Web Site for This Book
Virtual Society: The Wadsworth Sociology Resource Center
Visit **http://sociology.wadsworth.com** and click on the page for Kendall, *Sociology in Our Times: The Essentials*, Fourth Edition, to access a wide range of enrichment material to aid in your study of sociology.

Click on the Student Resources section of the web site. Next, select from the pull-down menu the chapter that you are presently studying. Among the useful options for self-study are chapter objectives, flashcards, practical study tips, and practice tests for each chapter.

MicroCase Online
From the Virtual Society home page, click on MicroCase Online to access book-specific MicroCase exercises that allow you to further explore sociological issues and principles presented in the text.

InfoTrac College Edition
Another unique option available to you at the Student Resources section of the companion web site is InfoTrac College Edition, an online library with access to hundreds of scholarly and popular periodicals. Below are suggested search terms for this chapter. Results from these and other searches are found at the site.

- Search keywords: *HMO (health maintenance organization)*. Look for articles that question the effectiveness of HMOs. What types of criticisms are discussed?
- Search keywords: *uninsured health*. Find recent statistics on the number of people without health insurance in the United States. Is the number of uninsured Americans increasing or decreasing?

Virtual Explorations CD-ROM
Go to the Virtual Explorations CD-ROM to begin the interactive exercise for this chapter. This Virtual Exploration will introduce you to some of the exciting resources for sociology on the World Wide Web. You will be guided through an exercise that employs related web sites on health, health care, and disability. Answer the questions, and e-mail your responses to your instructor.

Sandra Dionisi/SIS

Population and Urbanization

Demography: The Study of Population
Fertility
Mortality
Migration
Population Composition

Population Growth in Global Context
The Malthusian Perspective
The Marxist Perspective
The Neo-Malthusian Perspective
Demographic Transition Theory
Other Perspectives on Population Change

A Brief Glimpse at International Migration Theories

Urbanization in Global Perspective
Emergence and Evolution of the City
Preindustrial Cities
Industrial Cities
Postindustrial Cities

Perspectives on Urbanization and the Growth of Cities
Functionalist Perspectives: Ecological Models
Conflict Perspectives: Political Economy Models
Symbolic Interactionist Perspectives: The Experience of City Life

Problems in Global Cities

Urban Problems in the United States
Divided Interests: Cities, Suburbs, and Beyond
The Continuing Fiscal Crises of the Cities

Population and Urbanization in the Future

In another half hour of walking, I'll arrive at the highway where I'll catch a bus to take me to Oaxaca City. From there another bus will carry me to Mexico City, then yet another will take me to Nuevo Laredo, on the border. My plan is to go to the United States as a *mojado,* or wetback.

It didn't take a lot of thinking for me to decide to make this trip. It was a matter of following the tradition of the village. . . . For several decades, Macuiltiangus—that's the name of my village—has been an emigrant village, and our people have spread out like the roots of a tree under the earth, looking for sustenance. My people have had to emigrate to survive. First, they went to Oaxaca City, then to Mexico City, and for the past thirty years up to the present, the compass has always pointed to the United States.

I, too, joined the emigrant stream. For a year I worked in Mexico City as a night watchman in a parking garage. I earned the minimum wage and could barely pay living expenses. A lot of the time I had to resort to severe diets and other limitations, just to pay the rent on the apartment where I lived, so that one day I wouldn't come home and find that the owner had put my belongings outside.

After that year, I quit as a night watchman and came back to work at my father's side in the little carpentry shop that supplies the village with simple items of furniture. During the years when I worked at carpentry, I noticed that going to the United States was a routine of village people. People went so often that it was like they were visiting a nearby city. I'd see them leave and come home as changed people. The trips erased for a while the lines that the sun, the wind and the dust put in a peasant's skin. People came home with good haircuts, good clothes, and most of all, they brought dollars in their pockets. . . . Today when somebody says "I'm going to *Los,*" everybody understands that he's referring to Los Angeles, California, the most common destination of us villagers.

But I'm not going to Los Angeles, at least not now. This time, I want to try my luck in the state of Texas, specifically, in Houston, where a friend of mine has been living for several years. He's lent me money for the trip.

—Ramón "Tianguis" Pérez, explaining why he decided to become an undocumented immigrant worker in the United States (Pérez, 1991: 12–14)

Pérez is one of the many migrants to the United States who come from Mexico. It is estimated that 20 percent of legal migrants and slightly over 50 percent of undocumented migrants to the United States come from—or at least through—Mexico (U.S. Immigration and Naturalization Service, 2000).

When many people think about population growth, they primarily focus on the number of births minus the number of deaths in a specific region or in the world. However, international population flows—such as Pérez describes in talking about his decision to migrate to the United States—have significantly changed the size and composition of cities and rural areas in many nations, particularly those that are identified as high-income, industrialized countries. In recent years, new technologies such as the Internet and cellular telephones have provided those who wish to migrate with an expanded network of friends and contacts that makes it possible for them to plan for either temporary or permanent immigration to countries such as the United States. Sociologists are among the scholars who study migration and its effect on human population and on the urban areas of the world.

In this chapter, we explore the dynamics of migration as it affects population growth and urban change in societies. Before reading on, test your knowledge about issues regarding migration by taking the quiz in Box 15.1.

Questions and Issues

Chapter Focus Question: What effect does migration have on cities and on shifts in the global population?

How are people affected by population changes?

How do ecological/functionalist models and political economy/conflict models differ in their explanations of urban growth?

What is meant by the experience of urban life, and how do sociologists seek to explain this experience?

What are the best-case and worst-case scenarios regarding population and urban growth in the twenty-first century, and how might some of the worst-case scenarios be averted?

Box 15.1 Sociology and Everyday Life
How Much Do You Know About Migration?

True	False		
T	F	1.	Adults in their thirties and forties are more likely to migrate from one region or country to another than are younger individuals.
T	F	2.	Migration adds more people to a population than just the initial number of individuals who move into an area.
T	F	3.	The United States is one of only a few nations that have "guest worker" programs or work-related visas that allow individuals to legally enter the country for a period of time to work and then return to their country of origin.
T	F	4.	The number of people living outside their native country is lower today than it has been in the past.
T	F	5.	In the past 200 years, the majority of migrants to the United States were from Europe.
T	F	6.	With regard to intentional migration, most people move to escape political or religious persecution.
T	F	7.	Census 2000, conducted by the U.S. government, made little effort to count those people residing in this country without official documentation.
T	F	8.	Trends in migration within the United States have shown a decidedly westward movement over the past fifty years.

Answers on page 480.

Demography: The Study of Population

Although population growth has slowed in the United States, the world's population of 6.2 billion in 2000 is increasing by more than 76 million people per year as a result of the larger number of births than deaths worldwide (see Figure 15.1). Between 2000 and 2030, almost all of the world's 1.4-percent annual population growth will occur in low-income countries in Africa, Asia, and Latin America (Population Reference Bureau, 2001).

What causes the population to grow rapidly in some nations? This question is of interest to scholars who specialize in the study of *demography*—a **subfield of sociology that examines population size, composition, and distribution.** Many sociological studies use demographic analysis as a component of the research design because all aspects of social life are affected by demography. For example, an important relationship exists between population size and the availability of food, water, energy, and housing. Population size, composition, and distribution are also connected to issues such as poverty, racial and ethnic diversity, shifts in the age structure of society, and concerns about environmental degradation (Weeks, 2002).

Increases or decreases in population can have a powerful impact on the social, economic, and political structures of societies. As used by demographers, a *population* is a group of people who live in a specified geographic area. Changes in populations occur as a result of three processes: fertility (births), mortality (deaths), and migration.

> **Demography** a subfield of sociology that examines population size, composition, and distribution.

Box 15.1

Answers to the Sociology Quiz on Migration

1. **False.** Young adults are more likely to be migrants than are older individuals. The peak ages for migration are between twenty and twenty-nine years of age.

2. **True.** Over time, migration to a specific area contributes more than just the initial number of immigrants who move into the area. Children and grandchildren born to the immigrant population add several times the original number to the population base.

3. **False.** Many nations, including England, France, Germany, and the Netherlands, as well as countries in the Middle East and in Africa, have guest worker programs of varying kinds.

4. **False.** Intentional migration reached an all-time high in terms of absolute numbers at the beginning of the twenty-first century. It is estimated that about 145 million people lived outside their country of origin in the mid-1990s and that this number increases by anywhere from two to four million people annually.

5. **True.** Although most contemporary U.S. immigrants are from Latin America and Asia—with only 10 percent from Europe—the majority of immigrants over the past 200 years came from Europe. In the early decades of the twentieth century, more than 90 percent of immigrants to the United States were from Europe.

6. **False.** Most people move for economic reasons; however, political persecution and religious persecution are also factors in why some people leave certain countries.

7. **False.** Many organized efforts were made by officials conducting the census enumeration to locate people residing in this country, whether or not those people possessed valid visas or citizenship papers.

8. **True.** Migration within the United States has shown a decidedly westward movement over the past few decades as the share of the U.S. population living in the Northeast fell from 26 percent in 1950 to 19 percent in 2000.

Sources: Based on Population Research Bureau, 2001; Schaechter, 2001; and Weeks, 2002.

Fertility

Fertility **is the actual level of childbearing for an individual or a population.** The level of fertility in a society is based on biological and social factors, the primary biological factor being the number of women of childbearing age (usually between ages 15 and 45). Other biological factors affecting fertility include the general health and level of nutrition of women of childbearing age. Social factors influencing the level of fertility include the roles available to women in a society and prevalent viewpoints regarding what constitutes the "ideal" family size.

Based on biological capability alone, most women could produce twenty or more children during their childbearing years. *Fecundity* is the potential number of children who could be born if every woman reproduced at her maximum biological capacity. Fertility rates are not as high as fecundity rates because people's biological capabilities are limited by social factors such as practicing voluntary abstinence and refraining from sexual intercourse until an older age, as well as by contraception, voluntary sterilization, abortion, and infanticide. Additional social factors affecting fertility include significant changes in the number of available partners for sex

Figure 15.1 Growth in the World's Population

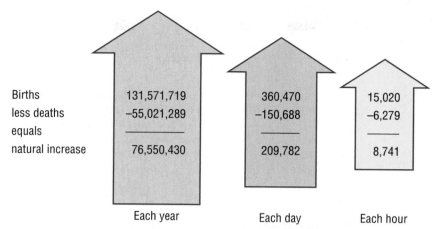

	Each year	Each day	Each hour
Births	131,571,719	360,470	15,020
less deaths	−55,021,289	−150,688	−6,279
equals			
natural increase	76,550,430	209,782	8,741

Every single day, the world's population increases by more than 200,000 people.

Source: Weeks, 2002.

and/or marriage (as a result of war, for example), increases in the number of women of childbearing age in the work force, and high rates of unemployment. In some countries, governmental policies also affect the fertility rate. For example, China's two-decades-old policy of allowing only one child per family in order to limit population growth will result in that country's population starting to decline in 2042, according to United Nations projections (Beech, 2001). Over time, China will be faced with a declining fertility rate combined with a rapidly aging population, many of whom will be dependent on younger people for their care and economic contributions.

The most basic measure of fertility is the *crude birth rate*—**the number of live births per 1,000 people in a population in a given year.** In 2000 the crude birth rate in the United States was 14.2 per 1,000, as compared with an all-time high rate of 27 per 1,000 in 1947 (following World War II). This measure is referred to as a "crude" birth rate because it is based on the entire population and is not "refined" to incorporate significant variables affecting fertility, such as age, marital status, religion, and race/ethnicity.

In most areas of the world, women are having fewer children. Women who have six or seven children tend to live in agricultural regions of the world, where children's labor is essential to the

family's economic survival and child mortality rates are very high. For example, Uganda has a crude birth rate of 48.0 per 1,000, as compared with 14.2 per 1,000 in the United States (U.S. Census Bureau, 2000c). However, in Uganda and some other African nations, families need to have many children in order to ensure that one or two will live to adulthood due to high rates of poverty, malnutrition, and disease.

Mortality

The primary cause of world population growth in recent years has been a decline in *mortality*—**the incidence of death in a population.** The simplest measure of mortality is the *crude death rate*—**the number of deaths per 1,000 people in a population in a given year.** In 2000 the U.S. crude death

Fertility the actual level of childbearing for an individual or a population.

Crude birth rate the number of live births per 1,000 people in a population in a given year.

Mortality the incidence of death in a population.

Crude death rate the number of deaths per 1,000 people in a population in a given year.

Betty Press/Woodfin Camp & Associates

Women tend to have more children in agricultural regions of the world, such as Kenya, where children's labor is essential to the family's economic survival and child mortality rates are very high.

care for infants. Differential levels of access to these services are reflected in the divergent infant mortality rates for African Americans and whites. In 1998, for example, the U.S. mortality rate for white infants was 6.0 per 1,000 live births, as compared with 14.1 per 1,000 live births for African American infants (U.S. Census Bureau, 2000c).

As discussed in Chapter 14, *life expectancy* is an estimate of the average lifetime in years of people born in a specific year. For persons born in the United States in 2000, for example, life expectancy at birth was 76.7 years, as compared with 80.2 years in Japan and 50 years or less in the African nations of Congo, Ethiopia, Kenya, and Uganda. Life expectancy varies by sex; for instance, females born in the United States in 2000 could expect to live about 79.4 years as compared with 73.9 years for males. Life expectancy also varies by race; for example, African American men have a life expectancy at birth of about 67.8 years, compared to 74.6 years for white males (U.S. Census Bureau, 2000c).

Migration

Migration **is the movement of people from one geographic area to another for the purpose of changing residency.** Migration affects the size and distribution of the population in a given area. *Distribution* refers to the physical location of people throughout a geographic area. In the United States, people are not evenly distributed throughout the country; many of us live in densely populated areas. *Density* is the number of people living in a specific geographic area. In urbanized areas, density may be measured by the number of people who live per room, per block, or per square mile.

Migration may be either international (movement between two nations) or internal (movement within national boundaries). Internal migration has occurred throughout U.S. history and has significantly changed the distribution of the population over time. In the late nineteenth and early twentieth centuries, a major population shift occurred as thousands of people moved from rural to urban areas. In 1800, about 5 percent of the population resided in urban areas; by 2000, almost 80 percent did so (U.S. Census Bureau, 2000c).

Migration involves two types of movement: immigration and emigration. *Immigration* is the movement of people into a geographic area to take

rate was 8.8 per 1,000 (U.S. Census Bureau, 2000c). In high-income, developed nations, mortality rates have declined dramatically as diseases such as malaria, polio, cholera, tetanus, typhoid, and measles have been virtually eliminated by vaccinations and improved sanitation and personal hygiene (Weeks, 2002). Just as smallpox appeared to be eradicated, however, HIV/AIDS rapidly rose to surpass the 30-percent fatality rate of smallpox. (The ten leading causes of death are shown in Table 15.1.) In low-income, less-developed nations, infectious diseases remain the leading cause of death; in some areas, mortality rates are increasing rapidly as a result of HIV/AIDS. Children under age fifteen constitute a growing number of those who are infected with HIV.

But many children do not survive long enough to contract communicable diseases. On a global basis, large numbers of newborn infants do not live to see their first birthday. The measure of these deaths is referred to as the *infant mortality rate*, which is defined in Chapter 14 as the number of deaths of infants under 1 year of age per 1,000 live births in a given year. The infant mortality rate is an important reflection of a society's level of preventive (prenatal) medical care, maternal nutrition, childbirth procedures, and neonatal

Table 15.1 The Ten Leading Causes of Death in the United States, 1900 and 1997

Cause of Death—1900	Rank	Cause of Death—1997
Influenza/pneumonia	1	Heart disease
Tuberculosis	2	Cancer
Stomach/intestinal disease	3	Stroke
Heart disease	4	Chronic lung disease
Cerebral hemorrhage	5	Accidents
Kidney disease	6	Pneumonia and influenza
Accidents	7	Diabetes
Cancer	8	HIV
Diseases in early infancy	9	Suicide
Diptheria	10	Homicide

Sources: Hostetler, 1994; Cockerham, 1995; and U.S. Census Bureau, 2000c.

up residency. Each year, about one million people enter the United States, primarily from Latin America and Asia; however, immigration rates are not an accurate reflection of the actual number of immigrants who enter a country. The U.S. Immigration and Naturalization Service records only legal immigration based on entry visas and change-of-immigration-status forms. Similarly, few records are maintained regarding *emigration*—the movement of people out of a geographic area to take up residency elsewhere. To determine the net migration in a geographic area, the number of people leaving that area to take up permanent or semi-permanent residence elsewhere (emigrants) is subtracted from the number of people entering that area to take up residence there (immigrants), unless more people are moving out of the area than into it, in which case the mathematical process is reversed.

People migrate either voluntarily or involuntarily. *Pull* factors at the international level, such as a democratic government, religious freedom, employment opportunities, or a more temperate climate, may draw voluntary immigrants into a nation. Within nations, people from large cities may be pulled to rural areas by lower crime rates, more space, and a lower cost of living. People such as Ramón Pérez, whose decision to migrate to the United States is described at the beginning of this chapter, are drawn by pull factors such as greater economic opportunities

Reflections

Reflections 15.1

How has migration affected cities with which you are familiar? Has recent immigration changed the composition of the population in those cities? How?

at their destination and are pushed by factors such as low wages and few employment opportunities in their previous place of residence. *Push* factors at the international level, such as political unrest, violence, war, famine, plagues, and natural disasters, may encourage people to leave one area and relocate elsewhere. Push factors in regional U.S. migration include unemployment, harsh weather conditions, a high cost of living, inadequate school systems, and high crime rates.

Involuntary, or forced, migration usually occurs as a result of political oppression, such as when Jews fled Nazi Germany in the 1930s or

Migration the movement of people from one geographic area to another for the purpose of changing residency.

Jon Burbank/The Image Works

Political unrest, violence, and war are "push" factors that encourage people, such as these former residents of Southeast Asia, to leave their country of origin. By contrast, job opportunities in the United States—including backbreaking manual labor—are a major "pull" factor for people in developing nations such as Mexico.

Paul Conklin/PhotoEdit

when Afghans left their country to escape oppression there in the early 2000s. Slavery is the most striking example of involuntary migration; for example, the 10–20 million Africans forcibly transported to the Western Hemisphere prior to 1800 did not come by choice (see Chapter 9).

Population Composition

Changes in fertility, mortality, and migration affect the *population composition*—**the biological and social characteristics of a population, including age, sex, race, marital status, education, occupation, income, and size of household.**

One measure of population composition is the *sex ratio*—**the number of males for every hundred females in a given population.** A sex ratio of 100 indicates an equal number of males and females in the population. If the number is greater than 100, there are more males than females; if it is less than 100, there are more females than males. In the United States, the estimated sex ratio for 2000 was 96.3, which means there were about 96 males per 100 females. Although approximately 124 males are conceived for every 100 females, male fetuses miscarry at a higher rate. From birth to age 14, the sex ratio is 105; in the age 35–44 category, however, the ratio shifts to 98.9, and from this point on, women outnumber men.

By age 65, the sex ratio is about 82.3—that is, there are 82 men for every 100 women. As "Census Profiles: Sex Ratios of the U.S. Population Compared by Race and Ethnicity" demonstrates, the ratio of males to females varies among racial and ethnic categories in addition to varying by age.

For demographers, sex and age are significant population characteristics; they are key indicators of fertility and mortality rates. The age distribution of a population has a direct bearing on the demand for schooling, health, employment, housing, and pensions. The current distribution of a population can be depicted in a *population pyramid*—**a graphic representation of the distribution of a population by sex and age.** Population pyramids are a series of bar graphs divided into five-year age cohorts; the left side of the pyramid shows the

✓ Checkpoints

Checkpoint 15.1

1. What is demography?
2. What is the most basic measure of fertility?
3. What is the simplest measure of mortality?
4. What are the major push and pull factors of migration?
5. What is the sex ratio?

CENSUS PROFILES

Sex Ratios of the U.S. Population Compared by Race and Ethnicity

The U.S. Census Bureau asks respondents to indicate the sex of everyone living in their household. Census 2000 found that the U.S. population was made of 143.4 million females (50.9 percent of the population) and 138.1 million males (49.1 percent of the population). Using this data, the Census Bureau computed the sex ratio (which it refers to as the male–female ratio) for various classifications of people by multiplying the number of males times 100, divided by the number of females.

When the sex ratios for various racial or ethnic categories of people are compared, some pronounced differences are evident, as shown below. As you can see, the categories with the highest male–female ratios (more males than females) are Latino/a and Native Hawaiian or other Pacific Islander. To determine reasons for differences in the sex ratio, it would be important to take into account a number of factors, including the age of the people in the various categories. For example, up to age 24, the sex ratio of the U.S. population is about 105, reflecting the fact that more boys than girls are born every year and that boys continue to outnumber girls until the age 35 to 44 category, when the ratio slips to 98.9. More than 63 percent of the U.S. Latino/a population is younger than 35 years of age, compared to slightly less than 47 percent of the U.S. white population. However, more than 57 percent of the African American population in the United States is younger than age 35; if age were the only factor involved in these differences, the African American sex ratio would be between that for Latinos/as and whites. Since the African American male–female ratio is lower than that for the other categories shown, age is obviously not the only factor involved in these differences.

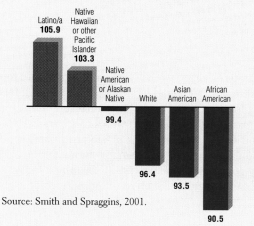

Source: Smith and Spraggins, 2001.

number or percentage of males in each age bracket; the right side provides the same information for females. The age/sex distribution in the United States and other high-income nations does not have the appearance of a classic pyramid, but rather is more rectangular or barrel-shaped. By contrast, low-income nations, such as Mexico and Iran, which have high fertility and mortality rates, do fit the classic population pyramid. Figure 15.2 compares the demographic composition of the United States, France, Mexico, and Iran.

Population Growth in Global Context

What are the consequences of global population growth? Scholars do not agree on the answer to this question. Some biologists have warned that Earth is a finite ecosystem that cannot support the 10 billion people predicted by 2050; however, some economists have emphasized that free-market capitalism is capable of developing innovative ways to solve such problems. The debate is not a new one; for several centuries, strong opinions have been voiced about the effects of population growth on human welfare.

The Malthusian Perspective

English clergyman and economist Thomas Robert Malthus (1766–1834) was one of the first scholars to systematically study the effects of population. Displeased with societal changes brought about by the Industrial Revolution in England, Malthus (1965/1798: 7) anonymously published *An Essay on the Principle of Population, As It Affects the Future Improvement of Society*, in which he argued that "the power of population is infinitely greater

Population composition the biological and social characteristics of a population, including age, sex, race, marital status, education, occupation, income, and size of household.

Sex ratio a term used by demographers to denote the number of males for every hundred females in a given population.

Population pyramid a graphic representation of the distribution of a population by sex and age.

Figure 15.2 Population Pyramids for Mexico, Iran, the United States, and France

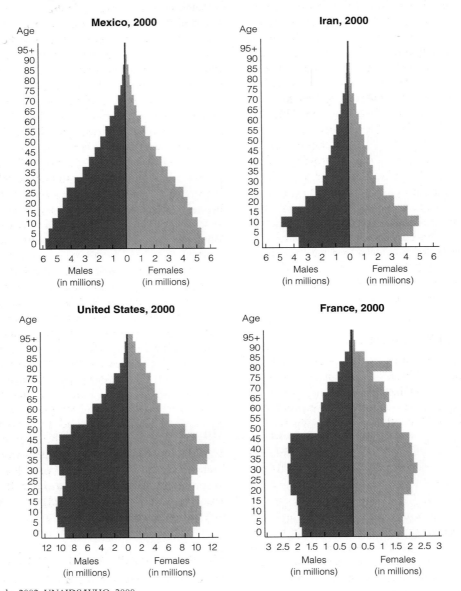

Source: Weeks, 2002; UNAIDS/WHO, 2000.

than the power of the earth to produce subsistence [food] for man."

According to Malthus, the population, if left unchecked, would exceed the available food supply. He argued that the population would increase in a geometric (exponential) progression (2, 4, 8, 16 . . .) while the food supply would increase only by an arithmetic progression (1, 2, 3, 4 . . .). In other words,

a *doubling effect* occurs: Two parents can have four children, sixteen grandchildren, and so on, but food production increases by only one acre at a time. Thus, population growth inevitably surpasses the food supply, and the lack of food ultimately ends population growth and perhaps eliminates the existing population (Weeks, 2002). Even in a best-case scenario, overpopulation results in poverty.

However, Malthus suggested that this disaster might be averted by either positive or preventive checks on population. *Positive checks* are mortality risks such as famine, disease, and war; *preventive checks* are limits to fertility. For Malthus, the only acceptable preventive check was *moral restraint*; people should practice sexual abstinence before marriage and postpone marriage as long as possible in order to have only a few children. Although Malthus later found data disproving his model and wrote essays modifying his earlier statements, his original text is more widely cited (Keyfitz, 1994).

Malthus has had a lasting impact on the field of population studies. Most demographers refer to his dire predictions when they examine the relationship between fertility and subsistence needs (Davis, 1955). Overpopulation is still a daunting problem that capitalism and technological advances thus far have not solved, especially in middle- and low-income nations with rapidly growing populations and very limited resources.

The Marxist Perspective

Among those who attacked the ideas of Malthus were Karl Marx and Frederick Engels. According to Marx and Engels, the food supply is not threatened by overpopulation; technologically, it is possible to produce the food and other goods needed to meet the demands of a growing population. Marx and Engels viewed poverty as a consequence of exploitation of workers by the owners of the means of production. They argued, for example, that England had poverty because the capitalists skimmed off some of the workers' wages as profits. The labor of the working classes was used by capitalists to earn profits, which, in turn, were used to purchase machinery that could replace the workers, rather than supply food for all.

From this perspective, overpopulation occurs because capitalists desire to have a surplus of workers (an industrial reserve army) so as to suppress wages and force workers concerned about losing their livelihoods to be more productive. Marx believed that overpopulation would contribute to the eventual destruction of capitalism: Unemployment would make the workers dissatisfied, resulting in a class consciousness based on their shared oppression and the eventual overthrow of the system.

According to some contemporary economists, the greatest crisis today facing low-income nations

is capital shortage, not fo... technological advances, agricult... reached the level at which it can ... needs of the world if food is distributed ... Capital shortage refers to the lack of ade... money or property to maintain a business; it is ... problem because the physical capital of the past no longer meets the needs of modern economic development. In the past, self-contained rural economies survived on local labor, using local materials to produce the capital needed for other laborers. For example, in a typical village a carpenter made the loom needed by the weaver to make cloth. Today, in the global economy, the one-to-one exchange between the carpenter and the weaver is lost. With an antiquated, locally made loom, the weaver cannot compete against electronically controlled, mass-produced looms. Therefore, the village must purchase capital from the outside, using its own meager financial resources. In the process, the complementary relationship between labor and capital is lost; modern technology brings with it steep costs and results in village noncompetitiveness and underemployment (see Keyfitz, 1994).

Marx and Engels made a significant contribution to the study of demography by suggesting that poverty, not overpopulation, is the most important issue with regard to food supply in a capitalist economy. Although Marx and Engels offer an interesting counterpoint to Malthus, some scholars argue that the Marxist perspective is self-limiting because it attributes the population problem solely to capitalism. In actuality, nations with socialist economies also have demographic trends similar to those in capitalist societies.

The Neo-Malthusian Perspective

More recently, *neo-Malthusians* (or "new Malthusians") have reemphasized the dangers of overpopulation. To neo-Malthusians, Earth is "a dying planet" with too many people and too little food, compounded by environmental degradation. Overpopulation and rapid population growth result in global environmental problems ranging from global warming and rain-forest destruction to famine and vulnerability to epidemics (Ehrlich, Ehrlich, and Daily, 1995). Unless significant changes are made, including improving the status of women, reducing racism and religious prejudice, reforming the agriculture

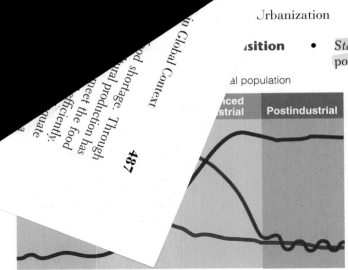

system, and shrinking the growing gap between rich and poor, the consequences will be dire (Ehrlich, Ehrlich, and Daily, 1995).

Early neo-Malthusians published birth control handbooks, and widespread acceptance of birth control eventually reduced the connection between people's sexual conduct and fertility (Weeks, 2002). Later neo-Malthusians have encouraged people to be part of the solution to the problem of overpopulation by having only one or two children in order to bring about *zero population growth*—**the point at which no population increase occurs from year to year** because the number of births plus immigrants is equal to the number of deaths plus emigrants (Weeks, 2002).

Demographic Transition Theory

Some scholars who disagree with the neo-Malthusian viewpoint suggest that the theory of demographic transition offers a more accurate picture of future population growth. *Demographic transition is the process by which some societies have moved from high birth and death rates to relatively low birth and death rates as a result of technological development.* Demographic transition is linked to four stages of economic development (see Figure 15.3):

- *Stage 1: Preindustrial societies.* Little population growth occurs because high birth rates are offset by high death rates. Food shortages, poor sanitation, and lack of adequate medical care contribute to high rates of infant and child mortality.

- *Stage 2: Early industrialization.* Significant population growth occurs because birth rates are relatively high whereas death rates decline. Improvements in health, sanitation, and nutrition produce a substantial decline in infant mortality rates. Overpopulation is likely to occur because more people are alive than the society has the ability to support.

- *Stage 3: Advanced industrialization and urbanization.* Very little population growth occurs because both birth rates and death rates are low. The birth rate declines as couples control their fertility through contraceptives and become less likely to adhere to religious directives against their use. Children are not viewed as an economic asset; they consume income rather than produce it. Societies in this stage attain zero population growth, but the actual number of births per year may still rise due to an increased number of women of childbearing age.

- *Stage 4: Postindustrialization.* Birth rates continue to decline as more women gain full-time employment and the cost of raising children continues to increase. The population grows very slowly, if at all, because the decrease in birth rates is coupled with a stable death rate.

Debate continues as to whether this evolutionary model accurately explains the stages of population growth in all societies. Advocates note that demographic transition theory highlights the relationship between technological development and population growth, thus making Malthus's predictions obsolete. Scholars also point out that demographic transitions occur at a faster rate in now-low-income nations than they previously did in the nations that are already developed. For example, nations in the process of development have higher birth rates and death rates than the now-developed societies did when they were going through the transition. The death rates declined in the now-developed nations as a result of internal economic development—not, as is the case today, through improved methods of disease control (Weeks, 2002). Critics suggest that this theory best explains development in Western societies.

Other Perspectives on Population Change

In recent decades, other scholars have continued to develop theories about how and why changes in population growth patterns occur. Some have studied the relationship between economic development and a decline in fertility; others have focused on the process of *secularization*—the decline in the significance of the sacred in daily life—and how a change from believing that otherworldly powers are responsible for one's life to a sense of responsibility for one's own well-being is linked to a decline in fertility. Based on this premise, some analysts argue that the processes of industrialization and economic development are typically accompanied by secularization, but the relationship between these factors is complex when it comes to changes in fertility.

Shifting from the macrolevel to the microlevel, education and social psychological factors also play into the decisions that individuals make about how many children to have. Family planning information is more readily available to people with more years of formal education and may cause them to engage in decision making in accord with *rational choice theory*, which is based on the assumption that people make decisions based on a calculated cost–benefit analysis ("What do I gain and lose from a specific action?"). In low-income countries or other settings in which children are identified as an economic resource for their parents throughout life, fertility rates are higher than in higher-income countries. However, as modernization and urbanization occur in such societies, the positive economic effects of having more children may be offset by the cost of caring for those children and the lowered economic advantage gained from having children in an industrialized nation.

As demographers have reformulated the demographic transition theory, they have highlighted additional factors that are likely to be causes of fertility decline, and they have suggested that demographic transition is not just one process, but rather is a set of intertwined transitions. One is the epidemiological transition—the shift from deaths at younger ages due to acute, communicable diseases. Another is the fertility transition—the shift from natural fertility to controlled fertility, resulting in a decrease in the fertility rate. Other transitions include the migration transition, the urban

✓Checkpoints

Checkpoint 15.2

6. What is the Malthusian perspective on population growth?
7. What is the Marxist perspective on population?
8. Why do some neo-Malthusians emphasize zero population growth?
9. What is demographic transition?
10. How have concepts such as secularization and rational choice theory been used to explain population growth patterns?

transition, the age transition, and the family and household transition, which occur as a result of lower fertility, longer life, an older age structure, and predominantly urban residence.

As the demographer John R. Weeks (2002) points out, we can best understand demographic events and behavior by studying the context of global change to determine how factors such as political change, economic development, and perhaps the process of "westernization" may influence population growth and patterns of migration.

A Brief Glimpse at International Migration Theories

Why do people relocate from one nation to another? Several major theories have been developed in an attempt to explain international migration. The *neoclassical economic approach* assumes that migration patterns occur based on geographic differences in the supply of and demand for labor. The United States and other high-income countries that have had growing economies and a limited supply of workers for certain types of jobs have paid higher wages than are available in areas with

Zero population growth   the point at which no population increase occurs from year to year.

Demographic transition   the process by which some societies have moved from high birth rates and death rates to relatively low birth rates and death rates as a result of technological development.

a less-developed economy and a large labor force. As a result, people move to gain higher wages and sometimes better living conditions. They also may take jobs in other countries so that they can send money to their families in their country of origin. It is estimated, for example, that Mexican workers in the United States send about half of what they earn to their families across the border, an amount that may total nearly $7 billion per year in good economic times (Ferriss, 2001).

Unlike the neoclassical explanation of migration, which focuses on individual decision making, the *new households economics of migration approach* emphasizes the part that entire families or households play in the migration process. From this approach, the previous example of Mexican workers' temporary migration to the United States would be examined not only from the perspective of the individual worker but also on what the entire family gains from the process of having one or more migrant family members work in another country. By having a diversity of family income (originating from more than one source), the family is cushioned from the economic woes of the nation that most of the family members think of as "home."

Two conflict perspectives on migration add to our knowledge of why people migrate. Split-labor-market theory (as previously discussed in Chapter 9) suggests that immigrants from low-income countries are often recruited for secondary-labor-market positions: dead-end jobs with low wages, unstable employment, and sometimes hazardous working conditions. By contrast, migrants from higher-income countries may migrate for primary-sector employment—jobs in which well-educated workers are paid high wages and receive benefits such as health insurance and a retirement plan. The global migration of some high-tech workers is an example of this process, whereas the migration of farm workers and construction helpers is an example of secondary-labor-market migration.

Finally, world systems theory (discussed later in this chapter) views migration as linked to the problems caused by capitalist development around the world (Massey et al., 1993). As the natural resources, land, and work force in low-income countries with little or no industrialization have come under the influence of international markets, there has been a corresponding flow of migrants from those nations to the highly industrialized, high-income countries, especially those with which the

✓**Checkpoints**

Checkpoint 15.3

11. How does the new households economics of migration approach contribute to our understanding of migration processes?
12. What is the network theory of migration?
13. How does the institutional theory of migration explain streams of immigration?

poorer nations have had the most economic, political, or military contact.

After flows of migration commence, the pattern may continue because potential migrants have personal ties with relatives and friends who now live in the country of destination and can serve as a source of stability when the potential migrants relocate to the new country. Known as *network theory*, this approach suggests that once migration has commenced, it takes on a life of its own and that the migration pattern which ensues may be different from the original *push* or *pull* factors that produced the earlier migration. Another approach, *institutional theory*, suggests that migration may be fostered by groups—such as humanitarian aid organizations relocating refugees or smugglers bringing people into a country illegally—and that the actions of these groups may produce a larger stream of migrants than would otherwise be the case.

As you can see from these diverse approaches to explaining contemporary patterns of migration, the reasons that people migrate are numerous and complex, involving processes occurring at the individual, family, and societal levels.

Urbanization in Global Perspective

Urban sociology is a subfield of sociology that examines social relationships and political and economic structures in the city. According to urban sociologists, a *city* is a relatively dense and permanent settlement of people who secure their livelihood primarily through nonagricultural activities. Although cities have existed for thousands of years, only about 3 percent of the world's population

An increasing proportion of the world's population lives in cities. How does this street scene in Hong Kong compare with major U.S. cities?

Paul Conklin/PhotoEdit

lived in cities 200 years ago, as compared with almost 50 percent today. Current estimates suggest that two out of every three people around the world will live in urban areas by 2050 (United Nations, 2000). Thus, the process of urbanization continues on a global basis.

Emergence and Evolution of the City

Cities are a relatively recent innovation when compared with the length of human existence. The earliest humans are believed to have emerged anywhere from 40,000 to 1 million years ago, and permanent human settlements are believed to have begun first about 8000 B.C.E. However, some scholars date the development of the first city between 3500 and 3100 B.C.E., depending largely on whether a formal writing system is considered as a requisite for city life (Flanagan, 1999; Sjoberg, 1965; Weeks, 2002).

According to the sociologist Gideon Sjoberg (1965), three preconditions must be present in order for a city to develop:

1. A *favorable physical environment*, including climate and soil favorable to the development of plant and animal life and an adequate water supply to sustain both.
2. An *advanced technology* (for that era) that could produce a social surplus in both agricultural and nonagricultural goods.
3. A *well-developed social organization*, including a power structure, in order to provide social stability to the economic system.

Based on these prerequisites, Sjoberg places the first cities in the Middle Eastern region of Mesopotamia or in areas immediately adjacent to it at about 3500 B.C.E. However, not all scholars concur; some place the earliest city in Jericho (located in present-day Jordan) at about 8000 B.C.E. with a population of about 600 people (see Kenyon, 1957).

The earliest cities were not large by today's standards. The population of the larger Mesopotamian centers was between five and ten thousand (Sjoberg, 1965). The population of ancient Babylon (probably founded around 2200 B.C.E.) may have grown as large as 50,000 people; Athens may have held 80,000 people (Weeks, 2002). Four to five thousand years ago, cities with at least 50,000 people existed in the Middle East (in what today is Iraq and Egypt) and Asia (in what today is Pakistan and China), as well as in Europe. About 3,500 years ago, cities began to reach this size in Central and South America.

Preindustrial Cities

The largest preindustrial city was Rome; by 100 C.E., it may have had a population of 650,000 (Chandler and Fox, 1974). With the fall of the Roman Empire in 476 C.E., the nature of European cities changed. Seeking protection and survival, those persons who lived in urban settings typically did so in walled cities containing no more than 25,000 people. For the next 600 years, the

Tony Freeman/PhotoEdit

During the industrial era, people not only moved from the countryside into cities, but some people also moved from the cities to the suburbs after transportation became available to make getting from home to work and back again an easier process.

urban population continued to live in walled enclaves as competing warlords battled for power and territory during the "dark ages." Slowly, as trade increased, cities began to tear down their walls.

Preindustrial cities were limited in size by a number of factors. For one thing, crowded housing conditions and a lack of adequate sewage facilities increased the hazards from plagues and fires, and death rates were high. For another, food supplies were limited. In order to generate food for each city resident, at least fifty farmers had to work in the fields (Davis, 1949), and animal power was the only means of bringing food to the city. Once foodstuffs arrived in the city, there was no effective way to preserve them. Finally, migration to the city was difficult. Many people were in serf, slave, and caste systems whereby they were bound to the land. Those able to escape such restrictions still faced several weeks of travel to reach the city, thus making it physically and financially impossible for many people to become city dwellers.

In spite of these problems, many preindustrial cities had a sense of *community*—a set of social relationships operating within given spatial boundaries or locations that provides people with a sense of identity and a feeling of belonging. The cities were full of people from all walks of life, both rich and poor, and they felt a high degree of social integration. You will recall that Ferdinand Tönnies (1940/1887) described such a community as *Gemeinschaft*—a society in which social relationships are based on personal bonds of friendship and kinship and on intergenerational stability, such that people have a commitment to the entire group and feel a sense of togetherness. By contrast, industrial cities were characterized by Tönnies as *Gesellschaft*—societies exhibiting impersonal and specialized relationships, with little long-term commitment to the group or consensus on values (see Chapter 4). In *Gesellschaft* societies, even neighbors are "strangers" who perceive that they have little in common with one another.

Industrial Cities

The Industrial Revolution changed the nature of the city. Factories sprang up rapidly as production shifted from the primary, agricultural sector to the secondary, manufacturing sector. With the advent of factories came many new employment opportunities not available to people in rural areas. Emergent technology, including new forms of transportation and agricultural production, made it easier for people to leave the countryside and move to the city. Between 1700 and 1900, the population of many European cities mushroomed. For example, the population of London increased from 550,000 to almost 6.5 million. Although the Industrial Revolution did not start in the United States until the mid-nineteenth century, the effect was similar. Between 1870 and 1910, for example, the population of New York City grew by 500 percent. In fact, New York City became the first U.S. *metropolis*—one or more central cities and their surrounding suburbs that dominate the economic and cultural life of a region. Nations, such as Japan and Russia, that became industrialized after England and the United States experienced a delayed pattern of urbanization, but this process moved quickly once it commenced in those countries.

Postindustrial Cities

Since the 1950s, postindustrial cities have emerged in nations such as the United States as their economies have gradually shifted from secondary (manufacturing) production to tertiary (service and information-processing) production. Postindustrial cities increasingly rely on an economic structure that is based on scientific knowledge rather than industrial production, and as a result, a class of professionals and technicians grows in size and influence. Postindustrial cities are dominated by "light" industry, such as software manufacturing; information-processing services, such as airline and hotel reservation services; educational complexes; medical centers; convention and entertainment centers; and retail trade centers and shopping malls. Most families do not live close to a central business district. Technological advances in communication and transportation make it possible for middle- and upper-income individuals and families to have more work options and to live greater distances from the workplace; however, these options are not often available to people of color and those at the lower end of the class structure.

Although fast-food restaurants create entry-level job opportunities for recent immigrants to postindustrial cities and for other people at the bottom of the class structure, these positions typically do not become stepping stones toward upward social mobility. A recent study by the sociologist Jennifer Parker Talwar (2002) examined how immigrant fast-food workers in New York City's ethnic neighborhoods adjusted to living in a postindustrial city. After spending four years working in fast-food restaurants there, Talwar concluded that many immigrants create bonds with co-workers from diverse backgrounds at the restaurant but that the available jobs do not aid their hopes for upward mobility.

On a global basis, cities such as New York, London, and Tokyo appear to fit the model of the postindustrial city (see Sassen, 2001). These cities have experienced a rapid growth in knowledge-based industries such as financial services. London, Tokyo,

and New York have—at least until recently—experienced an increase in the number of highly paid professional jobs, and more workers have been in high-income categories. Many people have benefited for a number of years from these high incomes and have created a lifestyle that is based on materialism and gentrification of urban spaces. Meanwhile, those persons outside the growing professional categories have seen their own quality of life further deteriorate and their job opportunities become increasingly restricted to secondary labor markets in their respective "global" cities.

Perspectives on Urbanization and the Growth of Cities

Urban sociology follows in the tradition of early European sociological perspectives that compared social life with biological organisms or ecological processes. For example, Auguste Comte pointed out that cities are the "real organs" that make a society function. Emile Durkheim applied natural ecology to his analysis of *mechanical solidarity*, characterized by a simple division of labor and shared religious beliefs such as are found in small, agrarian societies, and *organic solidarity*, characterized by interdependence based on the elaborate division of labor found in large, urban societies (see Chapter 4). These early analyses became the foundation for ecological models/functionalist perspectives in urban sociology.

Functionalist Perspectives: Ecological Models

Functionalists examine the interrelations among the parts that make up the whole; therefore, in studying the growth of cities, they emphasize the life cycle of urban growth. Like the social philosophers and sociologists before him, the University of Chicago sociologist Robert Park (1915) based his analysis of the city on *human ecology*—the study of the relationship between people and their physical environment. According to Park (1936), economic competition produces certain regularities in land-use patterns and population distributions. Applying Park's idea to the study of urban land-use patterns, the sociologist Ernest W. Burgess (1925) developed the concentric zone

✓ Checkpoints

Checkpoint 15.4

14. What major factors contributed to the emergence of the city?
15. How do preindustrial, industrial, and postindustrial cities differ?
16. What are some examples of postindustrial cities?

Bill Varie/CORBIS

Despite an increase in telecommuting and more diverse employment opportunities in the high-tech economy, our highways have grown increasingly congested. Can we implement measures to reduce the problems of urban congestion and environmental pollution, or will these problems grow worse with each passing year?

model, an ideal construct that attempted to explain why some cities expand radially from a central business core.

THE CONCENTRIC ZONE MODEL Burgess's *concentric zone model* is a description of the process of urban growth that views the city as a series of circular areas or zones, each characterized by a different type of land use, that developed from a central core (see Figure 15.4a). *Zone 1* is the central business district and cultural center. In *Zone 2*, houses formerly occupied by wealthy families are divided into rooms and rented to recent immigrants and poor persons; this zone also contains light manufacturing and marginal businesses (such as

secondhand stores, pawnshops, and taverns). *Zone 3* contains working-class residences and shops and ethnic enclaves. *Zone 4* comprises homes for affluent families, single-family residences of white-collar workers, and shopping centers. *Zone 5* is a ring of small cities and towns populated by persons who commute to the central city to work and by wealthy people living on estates.

Two important ecological processes are involved in the concentric zone theory: invasion and succession. *Invasion* **is the process by which a new category of people or type of land use arrives in an area previously occupied by another group or type of land use** (McKenzie, 1925). For example, Burgess noted that recent immigrants and low-income individuals "invaded" Zone 2, formerly occupied by wealthy families. *Succession* **is the process by which a new category of people or type of land use gradually predominates in an area formerly dominated by another group or activity** (McKenzie, 1925). In Zone 2, for example, when some of the single-family residences were sold and subsequently divided into multiple housing units, the remaining single-family owners moved out because the "old" neighborhood had changed. As a result of their move, the process of invasion was complete and succession had occurred.

Invasion and succession theoretically operate in an outward movement: Those who are unable to "move out" of the inner rings are those without upward social mobility, so the central zone ends up being primarily occupied by the poorest residents—except when gentrification occurs. *Gentrification* **is the process by which members of the middle and upper-middle classes, especially whites, move into the central-city area and renovate existing properties.** Centrally located, naturally attractive areas are the most likely candidates for gentrification. To urban ecologists, gentrification is the solution to revitalizing the central city. To conflict theorists, however, gentrification creates additional hardships for the poor by depleting the amount of affordable housing available and by "pushing" them out of the area (Flanagan, 1999).

The concentric zone model demonstrates how economic and political forces play an important part in the location of groups and activities, and it shows how a large urban area can have internal differentiation (Gottdiener, 1985). However, the model is most applicable to older cities that experienced high levels of immigration early in the

Figure 15.4 Three Models of the City

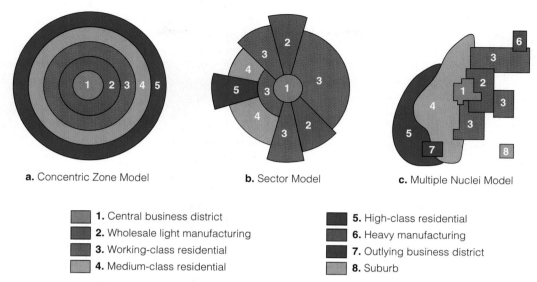

a. Concentric Zone Model   **b.** Sector Model   **c.** Multiple Nuclei Model

1. Central business district
2. Wholesale light manufacturing
3. Working-class residential
4. Medium-class residential
5. High-class residential
6. Heavy manufacturing
7. Outlying business district
8. Suburb

Source: Adapted from Harris and Ullman, 1945.

twentieth century and to a few midwestern cities such as St. Louis (Queen and Carpenter, 1953). No city, including Chicago (on which the model is based), entirely conforms to this model.

THE SECTOR MODEL In an attempt to examine a wider range of settings, urban ecologist Homer Hoyt (1939) studied the configuration of 142 cities. Hoyt's *sector model* emphasizes the significance of terrain and the importance of transportation routes in the layout of cities. According to Hoyt, residences of a particular type and value tend to grow outward from the center of the city in wedge-shaped sectors, with the more-expensive residential neighborhoods located along the higher ground near lakes and rivers or along certain streets that stretch in one direction or another from the downtown area (see Figure 15.4b). By contrast, industrial areas tend to be located along river valleys and railroad lines. Middle-class residential zones exist on either side of the wealthier neighborhoods. Finally, lower-class residential areas occupy the remaining space, bordering the central business area and the industrial areas. Hoyt (1939) concluded that the sector model applied to cities such as Seattle, Minneapolis, San Francisco, Charleston (South Carolina), and Richmond (Virginia).

THE MULTIPLE NUCLEI MODEL According to the *multiple nuclei model* developed by urban ecologists Chauncey Harris and Edward Ullman (1945), cities do not have one center from which all growth radiates, but rather have numerous centers of development based on specific urban needs or activities (see Figure 15.4c). As cities began to grow rapidly, they annexed formerly outlying and independent townships that had been communities in their own right. In addition to the central business district, other nuclei developed around entities such as an educational institution, a medical complex, or a government center. Residential neighborhoods may exist close to or far away from these nuclei.

Invasion the process by which a new category of people or type of land use arrives in an area previously occupied by another group or land use.

Succession the process by which a new category of people or type of land use gradually predominates in an area formerly dominated by another group or activity.

Gentrification the process by which members of the middle and upper-middle classes, especially whites, move into a central-city area and renovate existing properties.

A wealthy residential enclave may be located near a high-priced shopping center, for instance, whereas less-expensive housing must locate closer to industrial and transitional areas of town. This model may be applicable to cities such as Boston. However, critics suggest that it does not provide insights about the uniformity of land-use patterns among cities and relies on an after-the-fact explanation of why certain entities are located where they are (Flanagan, 1999).

CONTEMPORARY URBAN ECOLOGY   Urban ecologist Amos Hawley (1950) revitalized the ecological tradition by linking it more closely with functionalism. According to Hawley, urban areas are complex and expanding social systems in which growth patterns are based on advances in transportation and communication. For example, commuter railways and automobiles led to the decentralization of city life and the movement of industry from the central city to the suburbs (Hawley, 1981).

Other urban ecologists have continued to refine the methodology used to study the urban environment. *Social area analysis* examines urban populations in terms of economic status, family status, and ethnic classification (Shevky and Bell, 1966). For example, middle- and upper-middle-class parents with school-aged children tend to cluster together in "social areas" with a "good" school district; young single professionals may prefer to cluster in the central city for entertainment and nightlife.

The influence of human ecology on the field of urban sociology is still very strong today (see Frisbie and Kasarda, 1988). Contemporary research on European and North American urban patterns is often based on the assumption that spatial arrangements in cities conform to a common, most efficient design (Flanagan, 1999). However, some critics have noted that ecological models do

According to conflict theorists, exploitation by the capitalist class increasingly impoverishes low-income families. However, many people seek dignity and autonomy despite the obstacles that they face.

not take into account the influence of powerful political and economic elites on the development process in urban areas (Feagin and Parker, 1990).

Conflict Perspectives: Political Economy Models

Conflict theorists argue that cities do not grow or decline by chance. Rather, they are the product of specific decisions made by members of the capitalist class and political elites. These far-reaching decisions regarding land use and urban development benefit the members of some groups at the expense of others (see Castells, 1977/1972). Karl Marx suggested that cities are the arenas in which the intertwined processes of class conflict and capital accumulation take place; class consciousness and worker revolt are more likely to develop when workers are concentrated in urban areas (Flanagan, 1999).

According to the sociologists Joe R. Feagin and Robert Parker (1990), three major themes prevail in political economy models of urban growth. First, both economic *and* political factors affect patterns of urban growth and decline. Economic factors include capitalistic investments in production, workers, workplaces, land, and buildings. Political factors include governmental protection of

Reflections

Reflections 15.2

Karl Marx asserted that class consciousness and worker revolt are more likely to develop when workers are concentrated in urban areas. Does large-scale immigration of new workers make this prediction more or less likely to occur?

the right to own and dispose of privately held property as owners see fit and the role of government officials in promoting the interests of business elites and large corporations.

Second, urban space has both an exchange value and a use value. *Exchange value* refers to the profits that industrialists, developers, bankers, and others make from buying, selling, and developing land and buildings. By contrast, *use value* is the utility of space, land, and buildings for everyday life, family life, and neighborhood life. In other words, land has purposes other than simply for generating profits—for example, for homes, open spaces, and recreational areas. Today, class conflict exists over the use of urban space, as is evident in battles over rental costs, safety, and development of large-scale projects (see Tabb and Sawers, 1984).

Third, both structure and agency are important in understanding how urban development takes place. *Structure* refers to institutions such as state bureaucracies and capital investment circuits that are involved in the urban development process. *Agency* refers to human actors, including developers, business elites, and activists protesting development, who are involved in decisions about land use.

CAPITALISM AND URBAN GROWTH IN THE UNITED STATES According to political economy models, urban growth is influenced by capital investment decisions, power and resource inequality, class and class conflict, and government subsidy programs. Members of the capitalist class choose corporate locations, decide on sites for shopping centers and factories, and spread the population that can afford to purchase homes into sprawling suburbs located exactly where the capitalists think they should be located (Feagin and Parker, 1990).

Today, a few hundred financial institutions and developers finance and construct most major and many smaller urban development projects around the country, including skyscrapers, shopping malls, and suburban housing projects. These decision makers set limits on the individual choices of the ordinary citizen with regard to real estate, just as they do with regard to other choices (Feagin and Parker, 1990). They can make housing more affordable or totally unaffordable for many people. Ultimately, their motivation rests not in benefiting the community, but rather in making a profit; the cities that they produce reflect this mindset.

One of the major results of these urban development practices is *uneven development*—the tendency of some neighborhoods, cities, or regions to grow and prosper whereas others stagnate and decline (Perry and Watkins, 1977). Conflict theorists argue that uneven development reflects inequalities of wealth and power in society. The problem not only affects areas in a state of decline but also produces external costs, even in "boom" areas, that are paid by the entire community. Among these costs are increased pollution, increased traffic congestion, and rising rates of crime and violence. According to the sociologist Mark Gottdiener (1985: 214), these costs are "intrinsic to the very core of capitalism, and those who profit the most from development are not called upon to remedy its side effects."

GENDER REGIMES IN CITIES Feminist perspectives have only recently been incorporated in urban studies (Garber and Turner, 1995). From this perspective, urbanization reflects the workings not only of the political economy but also of patriarchy. According to the sociologist Lynn M. Appleton (1995), different kinds of cities have different *gender regimes*—prevailing ideologies of how women and men should think, feel, and act; how access to social positions and control of resources should be managed; and how relationships between men and women should be conducted. The higher density and greater diversity found in central cities such as New York City serve as a challenge to the private patriarchy found in the home and workplace in lower-density, homogeneous areas such as suburbs and rural areas. *Private patriarchy* is based on a strongly gendered division of labor in the home, gender-segregated paid employment, and women's dependence on men's income. At the same time, cities may foster *public patriarchy* in the form of women's increasing dependence on paid work and the state for income and their decreasing emotional interdependence with men. At this point, gender often intersects with class and race as a form of oppression because lower-income women of color often reside in central cities. Public patriarchy may be perpetuated by cities through policies that limit women's access to paid work and public transportation. However, such cities may also be a forum for challenging patriarchy; all residents who differ in marital status, paternity, sexual orientation, class, and/or race/ethnicity tend to live close to one another and may hold a common belief that both

Festive occasions such as this New York City street fair provide opportunities for urban villagers to mingle with others, enjoying entertainment and social interaction.

public and private patriarchy should be eliminated (Appleton, 1995).

Symbolic Interactionist Perspectives: The Experience of City Life

Symbolic interactionists examine the *experience* of urban life. How does city life affect the people who live in a city? Some analysts answer this question positively; others are cynical about the effects of urban living on the individual.

SIMMEL'S VIEW OF CITY LIFE According to the German sociologist Georg Simmel (1950/1902–1917), urban life is highly stimulating, and it shapes people's thoughts and actions. Urban residents are influenced by the quick pace of the city and the pervasiveness of economic relations in everyday life. Due to the intensity of urban life, people have no choice but to become somewhat insensitive to events and individuals around them. Many urban residents avoid emotional involvement with one another and try to ignore events taking place around them. Urbanites feel wary toward other people because most interactions in the city are economic rather than social. Simmel suggests that attributes such as punctuality and exactness are rewarded but that friendliness and warmth in interpersonal relations are viewed as personal weaknesses. Some people act reserved to cloak their deeper feelings of distrust or dislike toward others. However, Simmel did not view city life as completely negative; he also pointed out that urban living could have a liberating effect on people because they had opportunities for individualism and autonomy (Flanagan, 1999).

URBANISM AS A WAY OF LIFE Based on Simmel's observations on social relations in the city, the early Chicago School sociologist Louis Wirth (1938) suggested that urbanism is a "way of life." *Urbanism* refers to the distinctive social and psychological patterns of life typically found in the city. According to Wirth, the size, density, and heterogeneity of urban populations typically result in an elaborate division of labor and in spatial segregation of people by race/ethnicity, social class, religion, and/or lifestyle. In the city, primary-group ties are largely replaced by secondary relationships; social interaction is fragmented, impersonal, and often superficial. Even though people gain some degree of freedom and privacy by living in the city, they pay a price for their autonomy, losing the group support and reassurance that come from primary-group ties.

From Wirth's perspective, people who live in urban areas are alienated, powerless, and lonely. A sense of community is obliterated and replaced by the "mass society"—a large-scale, highly institutionalized society in which individuality is supplanted by mass messages, faceless bureaucrats, and corporate interests.

GANS'S URBAN VILLAGERS In contrast to Wirth's gloomy assessment of urban life, the sociologist Herbert Gans (1982/1962) suggested that not everyone experiences the city in the same way.

Based on research in the west end of Boston in the late 1950s, Gans concluded that many residents develop strong loyalties and a sense of community in central-city areas that outsiders may view negatively. According to Gans, there are five major categories of adaptation among urban dwellers. *Cosmopolites* are students, artists, writers, musicians, entertainers, and professionals who choose to live in the city because they want to be close to its cultural facilities. *Unmarried people and childless couples* live in the city because they want to be close to work and entertainment. *Ethnic villagers* live in ethnically segregated neighborhoods; some are recent immigrants who feel most comfortable within their own group. The *deprived* are poor individuals with dim future prospects; they have very limited education and few, if any, other resources. The *trapped* are urban dwellers who can find no escape from the city; this group includes persons left behind by the process of invasion and succession, downwardly mobile individuals who have lost their former position in society, older persons who have nowhere else to go, and individuals addicted to alcohol or other drugs. Gans concluded that the city is a pleasure and a challenge for some urban dwellers and an urban nightmare for others.

GENDER AND CITY LIFE   In their everyday lives, do women and men experience city life differently? According to the scholar Elizabeth Wilson (1991), some men view the city as *sexual space* in which women, based on their sexual desirability and accessibility, are categorized as prostitutes, lesbians, temptresses, or virtuous women in need of protection. Wilson suggests that more-affluent, dominant-group women are more likely to be viewed as virtuous women in need of protection by their own men or police officers. Cities offer a paradox for women: On the one hand, cities offer more freedom than is found in comparatively isolated rural, suburban, and domestic settings; on the other, women may be in greater physical danger in the city. For Wilson, the answer to women's vulnerability in the city is not found in offering protection for them, but rather in changing people's perceptions that they can treat women as sexual objects because of the impersonality of city life (Wilson, 1991).

CITIES AND PERSONS WITH A DISABILITY
Chapter 14 describes how disability rights advocates believe that structural barriers create a "disabling" environment for many people, particularly in large urban settings. Many cities have made their streets and sidewalks more user-friendly for persons in wheelchairs and individuals with visual disability by constructing concrete ramps with slide-proof surfaces at intersections or installing traffic lights with sounds designating when to "Walk." However, both urban and rural areas have a long way to go before many persons with disabilities will have access to the things they need to become productive members of the community: educational and employment opportunities. Because some persons with disabilities cannot navigate the streets and sidewalks of their communities or face obstacles getting into buildings that marginally, at best, meet the accessibility standards of the Americans with Disabilities Act, many persons with a disability are unemployed.

Political scientist Harlan Hahn (1997: 177–178) traces the problem of lack of access to the beginnings of industrialism:

> The rise of industrialism produced extensive changes in the lives of disabled as well as nondisabled people. As factories replaced private dwellings as the primary sites of production, routines and architectural configurations were standardized to suit nondisabled workers. Both the design of worksites and of the products that were manufactured gave virtually no attention to the needs of people with disabilities. As a result, patterns of aversion and avoidance toward disabled persons were embedded in the construction of commodities, landscapes, and buildings that would remain for centuries. . . .
>
> The social and economic changes fostered by industrialization may have been exacerbated by the accompanying process of urbanization. As workers increasingly moved from farms and rural villages to live near the institutions of mass production, the character of community life appeared to shift perceptibly. Deviant or atypical personal characteristics that may have gradually become familiar in a small community seemed bizarre or disturbing in an urban milieu.

As Hahn's statement suggests, historical patterns in the dynamics of industrial capitalism contributed to discrimination against persons with disabilities, and this legacy remains evident in contemporary

Concept Table 15.A Perspectives on Urbanism and the Growth of Cities

Functionalist Perspectives: Ecological Models	Concentric zone model	Due to invasion, succession, and gentrification, cities are a series of circular zones, each characterized by a particular land use.
	Sector model	Cities consist of wedge-shaped sectors, based on terrain and transportation routes, with the most-expensive areas occupying the best terrain.
	Multiple nuclei model	Cities have more than one center of development, based on specific needs and activities.
Conflict Perspectives: Political Economy Models	Capitalism and urban growth	Members of the capitalist class choose locations for skyscrapers and housing projects, limiting individual choices by others.
	Gender regimes in cities	Different cities have different prevailing ideologies regarding access to social positions and resources for men and women.
	Global patterns of growth	Capital investment decisions by core nations result in uneven growth in peripheral and semiperipheral nations.
Symbolic Interactionist Perspectives: The Experience of City Life	Simmel's view of city life	Due to the intensity of city life, people become somewhat insensitive to individuals and events around them.
	Urbanism as a way of life	Size, density, and heterogeneity of urban population result in elaborate division of labor and space.
	Gans's urban villagers	Five categories of adaptation occur among urban dwellers, ranging from cosmopolites to trapped city dwellers.
	Gender and city life	Cities offer women a paradox: more freedom than in more isolated areas, yet greater potential danger.

cities. Structural barriers are further intensified when other people do not respond favorably toward persons with disabilities. Scholar and disability rights advocate Sally French (1999: 25–26), who is visually disabled, describes her own experience:

> I have lived in the same house for 16 years and yet I cannot recognize my neighbors. I know nothing about them at all; which children belong to whom, who has come and gone, who is old or young, ill or well, black or white. . . . On moving to my present house I informed several neighbors that, because of my inability to recognize them, I would doubtless pass them by in the street without greet-

ing them. One neighbor, who had previously seen me striding confidently down the road, refused to believe me, but the others said they understood and would talk to me if our paths crossed. For the first couple of weeks it worked and I was surprised how often we met, but after that their greetings rapidly decreased and then ceased altogether. Why this happened I am not sure, but I suspect that my lack of recognition strained the interaction and limited the social reward they received from the encounter.

Concept Table 15.A examines the multiple perspectives on urban growth and urban living.

Figure 15.5 The World's Ten Largest Metropolises

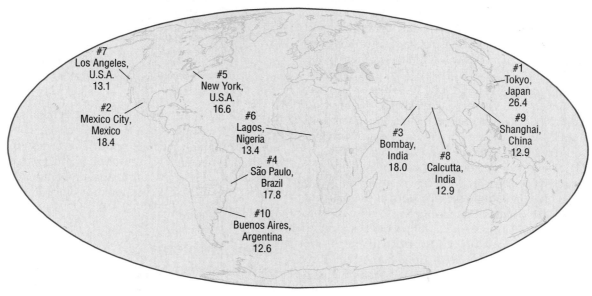

Note: 2000 population in millions.

Source: Population Reference Bureau, 2001.

✓ Checkpoints

Checkpoint 15.5

17. Why are functionalist perspectives of urban growth based on ecological models?
18. What is the political economy model of urban growth?
19. What is Georg Simmel's view of city life?
20. What is urbanism?
21. What are the various different types of urban dwellers as identified by Herbert Gans?

Problems in Global Cities

As we have seen, although people have lived in cities for thousands of years, the time is rapidly approaching when more people worldwide will live in or near a city than live in a rural area. In the middle-income and low-income regions of the world, Latin America is becoming the most urbanized: Four megacities—Mexico City (18 million), Buenos Aires (12 million), Lima (7 million), and Santiago (5 million)—already contain more than half of the region's population and continue to grow rapidly. By 2010, Rio de Janeiro and São Paulo are expected to have a combined population of about 40 million people living in a 350-mile-long megalopolis (Petersen, 1994). By 2015, New York City will be the only U.S. city among the world's ten most populous (Figure 15.5 shows current populations).

Natural increases in population (higher birth rates than death rates) account for two-thirds of new urban growth, and rural-to-urban migration accounts for the rest. Some people move from rural areas to urban areas because they have been displaced from their land. Others move because they are looking for a better life. No matter what the reason, migration has caused rapid growth in cities in sub-Saharan Africa, India, Algeria, and Egypt. At the same time that the population is growing rapidly, the amount of farmland available for growing crops to feed people is decreasing. In Egypt, for example, land that was previously used for growing crops is now used for petroleum refineries, food-processing plants, and other factories (Kaplan, 1996).

Rapid global population growth in Latin America and other regions is producing a wide variety of urban problems, including overcrowding, environmental pollution, and the disappearance of farmland. In fact, many cities in middle- and low-income nations are quickly reaching the point at which food, housing, and basic public services are available to only a limited segment of the population (Crossette, 1996). With urban populations growing at a rate of 170,000 people a day, cities such as Cairo, Lagos, Dhaka, Beijing, and São Paulo are likely to soon have acute water shortages; Mexico City is already experiencing a chronic water shortage (*New York Times*, 1996).

As global urbanization has increased over the past three decades, differences in urban areas based on economic development at the national level have become apparent. Some cities in what Immanuel Wallerstein's (1984) world systems theory describes as core nations (see Chapter 7) are referred to as *global cities*—interconnected urban areas that are centers of political, economic, and cultural activity. New York, Tokyo, and London are generally considered the largest global cities. These global cities are the sites of new and innovative product development and marketing, and they are often the "command posts" for the world economy (Sassen, 2001). But economic prosperity is not shared equally by all of the people in the core-nation global cities. Sometimes the living conditions of workers in low-wage service sector jobs or in assembly production jobs more closely resemble the living conditions of workers in semiperipheral nations than they resemble the conditions of middle-class workers in their own country.

Most African countries and many countries in South America and the Caribbean are *peripheral* nations, previously defined as nations that depend on core nations for capital, have little or no industrialization (other than what may be brought in by core nations), and have uneven patterns of urbanization. According to Wallerstein (1984), the wealthy in peripheral nations support the exploitation of poor workers by core-nation capitalists in return for maintaining their own wealth and position. Poverty is thus perpetuated, and the problems worsen because of the unprecedented population growth in these countries.

In regard to the semiperipheral nations, only two cities are considered to be global cities: São Paulo, Brazil, the center of the Brazilian economy, and Singapore, the economic center of a multi-country region in Southeast Asia (Friedmann,

✓ Checkpoints

Checkpoint 15.6

22. What are some of the major problems in cities around the world?

1995). Like peripheral nations, semiperipheral nations—such as India, Iran, and Mexico—are confronted with unprecedented population growth. In addition, a steady flow of rural migrants to large cities is creating enormous urban problems (see Box 15.2). What is the outlook for cities in the United States?

Urban Problems in the United States

Even the most optimistic of observers tends to agree that cities in the United States have problems brought on by years of neglect and deterioration. As we have seen in previous chapters, poverty, crime, racism, sexism, homelessness, inadequate public school systems, alcoholism and other drug abuse, gangs and guns, and other social problems are most visible and acute in urban settings. Issues of urban growth and development are intertwined with many of these problems.

Divided Interests: Cities, Suburbs, and Beyond

Since World War II, a dramatic population shift has occurred in this country as thousands of families have moved from cities to suburbs. Even though some people lived in suburban areas prior to the twentieth century, it took the involvement of the federal government and large-scale development to spur the dramatic shift that began in the 1950s (Palen, 1995). According to urban historian Kenneth T. Jackson (1985), postwar suburban growth was fueled by aggressive land developers, inexpensive real estate and construction methods, better transportation, abundant energy, government subsidies, and racial stress in the cities. However, the sociologist J. John Palen (1995) suggests that the Baby Boom following World War II and the liberalization of lending policies by federal

Box 15.2 Sociology in Global Perspective
Urban Migration and the "Garbage Problem"

Why do people around the world move from one location to another within their own country? Like people in the United States, individuals and families around the globe are more likely to move from rural to urban areas than vice versa. In some low-income countries, people move to cities primarily because they have been displaced from their land. However, others move in hopes of finding new opportunities and a better quality of life. No matter what the reason, rural-to-urban migration has produced rapid growth in many cities, including ones located in Latin America, sub-Saharan Africa, India, and Egypt. As the population continues to grow rapidly in these regions, the amount of farmland available for growing crops to feed people decreases.

Although rapid global population growth and strong patterns of rural-to-urban migration have produced a wide variety of urban problems—including overcrowding, environmental pollution, and the disappearance of farmland—one pressing problem in many cities around the world is the collection of household garbage, something that people in high-income countries simply take for granted. We put our garbage in the apartment dumpster or place our garbage can at curbside outside our residence, and the garbage is picked up on the appointed day, never to be seen by us again. Obviously, we are not without garbage problems even in the United States and other high-income countries, as our landfills overflow and our city "dumps" become a source of environmental pollution and shame. However, our problems are not as serious as those in urban areas in low-income countries, where some large cities have inadequate household-garbage-collection facilities. Consider the fact that less than 10 percent of the population in some cities benefits from the regular collection of household wastes. Estimates show that the proportion of garbage *not* collected by official means in Accra (Ghana) and Kampala (Uganda) is 90 percent and is 65 percent in Dar es Salaam (Tanzania) and 50 percent in Bogotá (Columbia) (see Mid-

dleton, 1999). Of course, these figures do not mean that garbage is not collected or "recycled" at all.

In the United States, we tend to think of recycling in terms of formal programs that encourage people to separate papers, cans, and bottles from other forms of trash so that these recyclables can be processed and used again. In low-income countries, however, the process is different. In Dar es Salaam, for example, informal scavenging at city dumps is a source of employment for many people. The scavenged resources are used as inputs to small-scale manufacturing industries that produce low-cost buckets, charcoal stoves, and lamps, which are then sold to residents of the city's squatter settlements for a lower cost than if the products had been made from imported raw materials (Middleton, 1999). This informal pattern of recycling and reuse takes place because of individual initiative and necessity rather than urban policies directed toward recycling. However, Dar es Salaam residents are not the only ones who deal with some aspects of the "garbage problem" in this manner. Mexico City and Cairo also have large squatter communities whose residents support themselves by living and working at official or unofficial rubbish dump sites.

Overall, the picture of the global garbage problem is not rosy: Uncollected garbage is a major problem in that it can be a serious fire hazard and a health hazard, attracting pests and becoming a breeding ground for certain diseases. In Kampala, carnivorous Marabou storks live on the garbage and the rodents that are attracted to it. Today, demographers and other social analysts are concerned—just as analysts have been concerned since the days of Thomas Malthus—about rapid population growth and the patterns of migration to the largest cities of the world, many of which have already far exceeded their capacity to provide a safe urban environment for existing residents.

agencies such as the Veterans Administration (VA) and the Federal Housing Authority (FHA) were significant factors in mass suburbanization.

Regardless of its causes, mass suburbanization has created a territorial division of interests between cities and suburban areas (Flanagan, 1999). Al-

though many suburbanites rely on urban centers for their employment, entertainment, and other services, they pay their property taxes to suburban governments and school districts. Some affluent suburbs have state-of-the-art school districts, police and fire departments, libraries, and infrastructures

AP/Wide World Photos

Monika Graff/The Image Works

Affluent gated communities and enclaves of million-dollar homes stand in sharp contrast to low-income housing when we see them on the urban landscape. What sociological theories help us to describe the disparity of lifestyles and life chances shown in these two settings?

(such as roads, sewers, and water treatment plants). By contrast, central-city services and school districts languish for lack of funds. Affluent families living in "gentrified" properties typically send their children to elite private schools whereas the children of poor families living in racially segregated public housing projects attend underfunded (and often substandard) public schools.

RACE, CLASS, AND SUBURBS The intertwining impact of race and class is visible in the division between central cities and suburbs. About 41 percent of central-city residents are persons of color, although they constitute a substantially smaller portion of the nation's population; just 27 percent of all African Americans live in suburbs. For most African American suburbanites, class is more important than race in determining one's neighbors. According to Vincent Lane, chairman of the Chicago Housing Authority, "Suburbanization isn't about race now; it's about class. Nobody wants to be around poor people, because of all the problems that go along with poor people: poor schools, unsafe streets, gangs" (qtd. in De Witt, 1994: A12).

Nationally, most suburbs are predominantly white, and many upper-middle- and upper-class suburbs remain virtually white. For example, only 5 percent of the population in northern Fulton County (adjoining Atlanta, Georgia) is African American. Likewise, in Plano (adjoining Dallas, Texas), nearly nine out of ten students in the public schools are white, whereas the majority of students in the Dallas Independent School District are African American, Latina/o, or Asian American. In the suburbs, people of color (especially African Americans) often become resegregated (see Feagin and Sikes, 1994). An example is Chicago, which remains one of the most-segregated metropolitan areas in the country in spite of its fair-housing ordinance. African Americans who have fled the high crime of Chicago's South Side primarily reside in nearby suburbs such as Country Club Hill and Chicago Heights whereas suburban Asian Americans are most likely to live in Skokie and Naperville and suburban Latinos/as to reside in Maywood, Hillside, and Bellwood (De Witt, 1994). Similarly, suburban Latinas/os are highly concentrated in eight metropolitan areas in California, Texas, and Florida; by far the largest such racial–ethnic concentration is found in the Los Angeles–Long Beach metropolitan area, with over 1.7 million Latinas/os. Like other groups, affluent Latinas/os live in affluent suburbs whereas poorer Latinas/os remain segregated in less desirable central-city areas (Palen, 1995).

Some analysts argue that the location of one's residence is a matter of personal choice. However, other analysts suggest that residential segregation reflects discriminatory practices by landlords, homeowners, and white realtors and their agents, who engage in *steering* people of color to different neighborhoods than those shown to their white counterparts. Lending practices of banks (including the *redlining* of certain properties so that acquiring a loan is virtually impossible) and the behavior of neighbors further intensify these problems (see Feagin and Sikes, 1994). In a study of suburban property taxes, the sociologist Andrew A. Beveridge found that African American homeowners are taxed more than whites on comparable homes in 58 percent of the suburban regions and 30 percent of the cities (cited in Schemo, 1994). Some analysts suggest that African Americans are more likely to move to suburbs with declining tax bases because they have limited finances, because they are steered there by real estate agents, or because white flight occurs as African American homeowners move in, leaving a heavier tax burden for the newcomers and those who remain behind. Longer-term residents may not see their property reassessed or their taxes go up for some period of time; in some cases, reassessment does not occur until the house is sold (Schemo, 1994).

BEYOND THE SUBURBS: EDGE CITIES New urban fringes (referred to as *edge cities*) have been springing up beyond central cities and suburbs in recent years (Garreau, 1991). The Massachusetts Turnpike corridor west of Boston and the Perimeter area north and east of Atlanta are examples. Edge cities initially develop as residential areas; then retail establishments and office parks move into the area, creating the unincorporated edge city. Commuters from the edge city are able to travel around (rather than in and out of) the metropolitan region's center and can avoid its rush-hour traffic quagmires. Edge cities may not have a governing body or correspond to municipal boundaries; however, they drain taxes from central cities and older suburbs. Many businesses and industries have moved physical plants and tax dollars to these areas; land is cheaper, and utility rates and property taxes are lower.

Lower taxes are a contributing factor to another recent development in the United States—the growth of Sunbelt cities in the southern and western states. In the 1970s, millions of people moved from the north-central and northeastern states to this area. Four reasons are generally given for this population shift: (1) more jobs and higher wages; (2) lower taxes; (3) pork-barrel programs that funneled federal money into projects in the Sunbelt, creating jobs and encouraging industry; and (4) easier transition to new industry, since most industry in the northern states was based on heavy manufacturing rather than high technology.

The Continuing Fiscal Crises of the Cities

The largest cities in the United States have faced periodic fiscal crises for many years. In the twenty-first century, these crises have been intensified by terrorist attacks that have at least temporarily reduced major sources of revenue in many urban areas. However, earlier fiscal crises can be traced to the 1980s, when federal aid to cities was cut drastically. Many cities were faced with cutting services or raising taxes (or both) at a time when the tax base was already shrinking because of suburban flight.

The fiscal crisis of the cities became quite visible in the decaying infrastructure. Old water mains threatened to burst, roads became plagued by potholes, and sections of highways and bridges crumbled. In the economic boom of the 1990s, many cities regained their economic footing as growth took place in many sectors of the marketplace and massive development and gentrification occurred. However, unequal development resulted in even greater disparity between the "good" areas of cities and those areas that were considered "less desirable." Among other things, the increase in high-priced commercial and residential property in central cities produced a corresponding increase in homelessness and the significant concentration of poor people in other areas of the cities (Sassen, 2001).

Reflections

Reflections 15.3

Why do some analysts view immigration as a possible solution for contemporary urban problems whereas other analysts argue that immigration is a major factor in urban problems?

✓Checkpoints

Checkpoint 15.7

23. How has suburbanization affected large urban areas in the United States?
24. What part do race and class play in the process of suburbanization?
25. Why do cities continue to experience fiscal crises?

Shortly after the twenty-first century began, however, the nation's economy headed into troubled waters, a situation that was worsened by the terrorist attacks of September 11, 2001. Consider, for example, that tourism in New York City prior to those attacks was a $25-billion-per-year industry employing more than 280,000 people. The revenue received from tourism was the equivalent of $3,100 for every person living in New York City although it was obviously not distributed that way (Herbert, 2001). Other cities, including Las Vegas, Nevada; San Francisco, California; and Orlando, Florida—all of which are top tourist destinations—experienced significant slumps in revenue as people chose to remain at home in the aftermath of the attacks. Loss of tax revenues (such as from hotel and restaurant taxes), loss of jobs, and the need for heightened security to prevent future terrorist attacks have taken a significant toll on many cities that had been able to make some progress during the earlier periods of economic growth. Consider the effect on New York City alone, where 80,000 people became unemployed as a result of the 2001 attacks there and another 75,000 people (although still employed) had their work schedules cut back or could no longer rely on tips, which often accounted for half of their income (Greenhouse, 2001). The economic effect of the terrorist attacks on cities around the world remains to be assessed in the years to come, but the immediate picture appears to be somewhat bleak.

Population and Urbanization in the Future

In the future, rapid global population growth is inevitable. Although death rates have declined in many low-income nations, there has not been a corresponding decrease in birth rates. Between 1985 and 2025, 93 percent of all global population growth will have occurred in Africa, Asia, and Latin America; 83 percent of the world's population will live in those regions by 2025 (Petersen, 1994). Perhaps even more amazing is the fact that in the five-year span between 1995 and 2000, 21 percent of the entire world's population increase occurred in two countries: China and India. Figure 15.6a shows the net annual additions to the populations of six countries during that five-year period. Figure 15.6b shows the growth of the world's population from 1927 to 1999 and the expected growth to eight billion by the year 2025.

In the future, low-income countries will have an increasing number of poor people. While the world's population will *double*, the urban population will *triple* as people migrate from rural to urban areas in search of food, water, and jobs. Of all low-income regions, Latin America is becoming the most urbanized: Four megacities—Mexico City (18 million), Buenos Aires (12 million), Lima (7 million), and Santiago (5 million)—already contain more than half of this region's population and continue to grow rapidly. By 2010, Rio de Janeiro and São Paulo are expected to have a combined population of about 40 million people living in a 350-mile-long *megalopolis*—a continuous concentration of two or more cities and their suburbs that have grown until they form an interconnected urban area (Petersen, 1994).

One of the many effects of urbanization is greater exposure of people to the media. Increasing numbers of poor people in less-developed nations will see images from the developed world that are beamed globally by news networks such as CNN. As futurist John L. Petersen (1994: 119) notes, "For the first time in history, the poor are beginning to understand how relatively poor they are compared to the rich nations. They see, in detail, how the rest of the world lives and feel their increasing disenfranchisement."

Infants born today will be teenagers in the year 2017. By then, in a worst-case scenario, central cities and nearby suburbs in the United States will have experienced bankruptcy exacerbated by sporadic race- and class-oriented violence. The infrastructure will be beyond the possibility of repair. Families and businesses with the ability to do so will have long since moved to "new cities," where

Figure 15.6 Growth of the World's Population

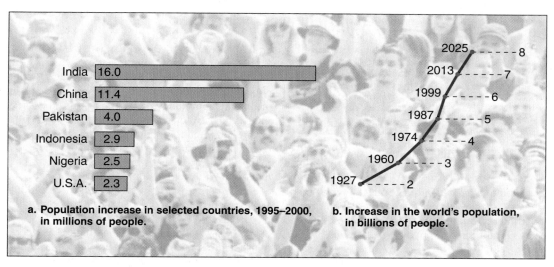

a. Population increase in selected countries, 1995–2000, in millions of people.

India	16.0
China	11.4
Pakistan	4.0
Indonesia	2.9
Nigeria	2.5
U.S.A.	2.3

b. Increase in the world's population, in billions of people.

Year	Billions
1927	2
1960	3
1974	4
1987	5
1999	6
2013	7
2025	8

Source: United Nations Population Division, 1999.

they will inevitably diminish the quality of life that they originally sought there. Areas that we currently think of as being relatively free from such problems will be characterized by depletion of natural resources and greater air and water pollution (see Ehrlich, Ehrlich, and Daily, 1995). If social and environmental problems become too great in one nation, members of the capitalist class may simply move to another country.

By contrast, in a best-case scenario, the problems brought about by rapid population growth in low-income nations will be remedied by new technologies that make goods readily available to people. International trade agreements such as NAFTA (North American Free Trade Agreement) and GATT (General Agreement on Trade and Tariffs) are removing trade barriers and making it possible for all nations to fully engage in global trade. People in low-income nations will benefit by gaining jobs and opportunities to purchase goods at lower prices. Of course, the opposite also may occur: People may be exploited as inexpensive labor, and their country's natural resources may be depleted as transnational corporations buy up raw materials without contributing to the long-term economic stability of the nation.

In the United States, a best-case scenario for the future might include improvements in how tax revenues are collected and spent. Some analysts

have suggested that regional governments must be developed. Regionalization of government would create wider service areas than existing city and county (or parish or borough) governments. Regional governments would be responsible for water, wastewater (sewage), transportation, schools, parks, hospitals, and other public services over a wider area. Revenues would be shared among central cities, affluent suburbs, and edge cities based on the assumption that everyone will benefit if the quality of life is improved throughout the region. With regard to pollution in urban areas, some futurists predict that environmental activism will increase dramatically as people see irreversible changes in the atmosphere and experience firsthand the effects of environmental hazards and pollution on their health and well-being. Environmental activism is discussed in Chapter 16 ("Collective Behavior, Social Movements, and Social Change").

At the macrolevel, we may be able to do little about population and urbanization; however, at the microlevel, we may be able to exercise some degree of control over our communities and our lives (see Box 15.3). In both cases, futurists suggest that as we approach the future, we must "leave the old ways and invent new ones" (Petersen, 1994: 340). What aspects of our "old ways" do you think we should discard? Can you help invent new ones?

Box 15.3 You Can Make a Difference
Working Toward Better Communities

We've definitely touched a nerve. People believe that sprawl and traffic are out of control, and the vast majority want more open space, reliable public transit and neighborhood reinvestment. All the evidence shows that Americans support smarter growth, and our elected officials better start paying attention.
—Don Chen, director of the Smart Growth America coalition (qtd. in Smart Growth America, 2001)

According to a recent study by Smart Growth America, a new nationwide coalition of more than sixty public interest groups, many people in the United States support the concept of "smart growth" to reduce traffic congestion, preserve their communities, and protect the environment and open space. Advocates of smart growth suggest that many citizens are ahead of their public officials in wanting to find a solution to problems such as urban sprawl, and believe that it is possible for everyday people to have a voice in how cities give priority to improving services—such as schools, roads, affordable housing, and public transportation—in existing communities, rather than encouraging new housing, commercial development, and highways in the countryside.

The Smart Growth Network began in 1996, when the U.S. Environmental Protection Agency joined with several nonprofit and government organizations to address the issue of community concerns about boosting the economy, protecting the environment, and enhancing community vitality (Smart Growth Network, 2001). Among the "principles of smart growth" are the importance of providing quality housing for people of all income levels, creating walkable neighborhoods, and encouraging community collaboration on how and where a city should grow. (To see all of the principles, visit the Smart Growth web site [http://www.smartgrowth.org].)

How can you make a difference regarding issues of urban planning and change? In most cities, urban planning is community based in that elected bodies such as the city council or planning commission invite "citizen participation" in their dialogues regarding community development. In addition to official channels, many people continue to use grassroots efforts to encourage the development of adequate housing, protection of open space and the environment, and good public-health and community services.

As the urban population in the United States continues to grow, each of us has a unique opportunity to learn about the issues that will affect the quality of life and profoundly influence what kind of lives we and our neighbors throughout the entire community will be able to live. There is no single "right" answer or set of answers on how to achieve smart growth in any particular city; rather, the citizens of each city must find their own best solutions, taking into account the existing structure of their community, the extent of likely population growth, and the resources that are available to the city's government. But how your city will handle future growth and change is something that should be important to you. Today, most large cities—and some smaller cities—have a web site with links to other sites that discuss these issues regarding that particular city; such information can also be found in local newspapers and on city-sponsored television stations. You can make a difference in your community by learning about these issues and what solutions are being proposed, and by making sure that your opinions regarding these issues get taken into account.

Chapter Review

What are the processes that produce population changes?
Populations change as the result of fertility (births), mortality (deaths), and migration.

What is the Malthusian perspective?
Over two hundred years ago, Thomas Malthus warned that overpopulation would result in poverty, starvation, and other major problems that would limit the size of the population. According to Malthus, the population would increase geometrically while the food supply would increase only arithmetically, resulting in a critical food shortage and poverty.

What are the views of Karl Marx and the neo-Malthusians on overpopulation?
According to Karl Marx, poverty is the result of capitalist greed, not overpopulation. More recently, neo-Malthusians have reemphasized the dangers of overpopulation and encouraged zero population growth—

the point at which no population increase occurs from year to year.

What are the stages in demographic transition theory?

Demographic transition theory links population growth to four stages of economic development: (1) the preindustrial stage, with high birth rates and death rates; (2) early industrialization, with relatively high birth rates and a decline in death rates; (3) advanced industrialization and urbanization, with low birth rates and death rates; and (4) postindustrialization, with additional decreases in the birth rate coupled with a stable death rate.

How do preindustrial cities differ from industrial and postindustrial cities?

Because of their limited size, preindustrial cities tend to provide a sense of community and a feeling of belonging. The Industrial Revolution changed the size and nature of the city; people began to live close to factories and to one another, resulting in overcrowding and poor sanitation. In postindustrial cities, some people live and work in suburbs or outlying edge cities.

What are the three functionalist models of urban growth?

Functionalists view urban growth in terms of ecological models. The concentric zone model sees the city as a series of circular areas, each characterized by a different type of land use. The sector model describes urban growth in terms of terrain and transportation routes. The multiple nuclei model views cities as having numerous centers of development from which growth radiates.

What is the political economy model/conflict perspective on urban growth?

According to political economy models/conflict perspectives, urban growth is influenced by capital investment decisions, power and resource inequality, class and class conflict, and government subsidy programs. At the global level, capitalism also influences the development of cities in core, peripheral, and semiperipheral nations.

How do symbolic interactionists view urban life?

Symbolic interactionist perspectives focus on how people experience urban life. Some analysts view the urban experience positively; others believe that urban dwellers become insensitive to events and to people around them.

How did the U.S. population change during the second half of the twentieth century?

During the 1950s, a dramatic population shift occurred in the United States as people moved from cities to suburbs; almost 80 percent of the U.S. population lives in urban areas today. Edge cities have also developed beyond central cities and suburbs, first with residential areas and then with retail establishments and office parks.

Key Terms

crude birth rate 481
crude death rate 481
demographic transition 488
demography 479
fertility 480
gentrification 494
invasion 494
migration 482
mortality 481
population composition 484
population pyramid 484
sex ratio 484
succession 494
zero population growth 488

Questions for Critical Thinking

1. What impact might a high rate of immigration have on culture and personal identity in the United States?
2. If you were designing a study of growth patterns for the city where you live (or one you know well), which theoretical model(s) would provide the most useful framework for your analysis?
3. What do you think everyday life in U.S. cities, suburbs, and rural areas will be like in 2020? Where would you prefer to live? What does your answer reflect about the future of U.S. cities?

 ## Resources on the Internet

Chapter-Related Web Sites

The following web sites have been selected for their relevance to the topics in this chapter. These sites are among the more stable, but please note that web site addresses change frequently.

United States Census Bureau: Census 2000 Gateway

http://www.census.gov/main/www/cen2000.html
This web page is the starting point for studying the census and the population data that have been collected and released thus far.

United States Census Bureau: POPClocks

http://www.census.gov/main/www/popclock.html
Go to this site to see the latest update on the growing world population.

The State of World Population 2001

http://www.unfpa.org/swp/swpmain.htm

This UN-funded site provides links to information related to gender, health, and economic development, all within the context of a growing world population.

National Urban League

http://www.nul.org

The mission of the National Urban League is to enable African Americans to secure economic self-reliance, parity and power, and civil rights.

Population Reference Bureau

http://www.prb.org

This site organizes a rich variety of data related to global population.

Patterns of World Urbanization

http://www.prb.org/Content/NavigationMenu/
PRB/Educators/Human_Population/Urbanization2/
Patterns_of_World_Urbanization1.htm

It's a long address, but this site from the Population Reference Bureau contains a global map showing urban and economic development along with careful analysis.

World Population Foundation

http://www2.tribute.nl/wpf/uk/main.html

WPF (the World Population Foundation) is a Dutch nonprofit organization working for the improvement of reproductive and sexual health in developing countries.

Children in Urban America

http://134.48.55.172:8000/cuap/index.html

This site is an online archive developed by Marquette University and funded by the National Endowment for the Humanities that shows the many ways children experienced city life during the last century and a half. Designed for teachers, students, historians, and general users, the site features hundreds of documents and images about children in Milwaukee County, Wisconsin, drawn from newspapers, government and other official records, oral histories and memoirs, and many other sources.

Companion Web Site for This Book

Virtual Society: The Wadsworth Sociology Resource Center

Visit http://sociology.wadsworth.com and click on the page for Kendall, *Sociology in Our Times: The Essentials,* Fourth Edition, to access a wide range of enrichment material to aid in your study of sociology. Click on the Student Resources section of the web site. Next, select from the pull-down menu the chapter that you are presently studying. Among the useful options for self-study are chapter objectives, flashcards, practical study tips, and practice tests for each chapter.

MicroCase Online

From the Virtual Society home page, click on MicroCase Online to access book-specific MicroCase exercises that allow you to further explore sociological issues and principles presented in the text.

InfoTrac College Edition

Another unique option available to you at the Student Resources section of the companion web site is InfoTrac College Edition, an online library with access to hundreds of scholarly and popular periodicals. Below are suggested search terms for this chapter. Results from these and other searches are found at the site.

- Search keywords: *population growth.* Find articles that discuss the negative social consequences of population growth. What efforts are being made to address these problems?
- Search keywords: *urban sprawl.* What problems are associated with urban sprawl? What solutions are currently being proposed to address this issue?

Virtual Explorations CD-ROM

Go to the Virtual Explorations CD-ROM to begin the interactive exercise for this chapter. This Virtual Exploration will introduce you to some of the exciting resources for sociology on the World Wide Web. You will be guided through an exercise that employs related web sites on population and urbanization. Answer the questions, and e-mail your responses to your instructor.

Stephanie Carter/SIS

Collective Behavior, Social Movements, and Social Change

16

Collective Behavior
Conditions for Collective Behavior
Dynamics of Collective Behavior
Distinctions Regarding Collective Behavior
Types of Crowd Behavior
Explanations of Crowd Behavior
Mass Behavior

Social Movements
Types of Social Movements
Stages in Social Movements

Social Movement Theories
Relative Deprivation Theory
Value-Added Theory
Resource Mobilization Theory
Social Constructionist Theory: Frame Analysis
New Social Movement Theory

Social Change in the Future
The Physical Environment and Change
Population and Change
Technology and Change
Social Institutions and Change
A Few Final Thoughts

I was raised along the [Hudson] river. I was in the first generation that was taught the river was unsafe—not because of tides that might pull you down but because of water quality. As a young adult, I found a legacy I had been kept from inheriting. The lives of my family had swirled around the river; my grandfather was a fisherman; that's where families gathered. I discovered that connection. But then there was a larger connection. It seemed that every community on the river had lost touch with it and with the notion that the river was their home. The greatest single tragedy on the Hudson is that hundreds of years of history are disappearing. It's like burning down a museum or trashing a library. The loss is devastating and profound.

—John Cronin, a former commercial fisherman who was appointed Hudson Riverkeeper, describing how he became

an activist in an effort to save a river he loves from environmental degradation (qtd. in Rosenblatt, 1999: 76)

To me [the movement to save the Hudson River] is a struggle of good and evil—between short-term greed and ignorance and a long-term vision of building communities that are dignified and enriching and that meet the obligations to future generations. There are two visions of America. One is that this is just a place where you make a pile for yourself and keep moving. And the other is that you put down roots and build communities that are examples to the rest of humanity.

—Robert F. Kennedy, Jr., chief prosecuting attorney for the Hudson Riverkeepers and senior attorney for the Natural Resources Defense Council, explaining why he believes it is important to crack down on those who pollute the environment (qtd. in Rosenblatt, 1999: 76)

In their best-selling book *The Riverkeepers* (1999), John Cronin and Robert F. Kennedy, Jr., describe how they became environmental activists seeking social change. Sociologists define *social change* **as the alteration, modification, or transformation of public policy, culture, or social institutions over time;** such change is usually brought about by collective behavior and social movements. Although the early endeavors of Cronin and Kennedy to protect the Hudson River constituted collective behavior, their actions led to an environmental movement that now includes twenty Riverkeepers on waterways throughout the United States.

In this chapter, we will examine collective behavior, social movements, and social change from a sociological perspective. We will use environmental activism as an example of how people may use social movements as a form of mass mobilization and social transformation (Buechler, 2000). Before reading on,

test your knowledge about collective behavior and environmental issues by taking the quiz in Box 16.1.

Questions and Issues

Chapter Focus Question: Can collective behavior and social movements make people aware of important social issues such as environmental pollution?

What causes people to engage in collective behavior?

What are some common forms of collective behavior?

How can different types of social movements be distinguished from one another?

What draws people into social movements?

What factors contribute to social change?

Collective Behavior

Collective behavior **is voluntary, often spontaneous activity that is engaged in by a large number of people and typically violates dominant-group norms and values.** Unlike the *organizational behavior* found in corporations and voluntary associations (such as labor unions and environmental organizations), collective behavior lacks an official division of labor, hierarchy of authority, and established rules and procedures. Unlike *institutional behav-*

ior (in education, religion, or politics, for example), it lacks institutionalized norms to govern behavior. Collective behavior can take various forms, including crowds, mobs, riots, panics, fads, fashions, and public opinion.

According to the sociologist Steven M. Buechler (2000), early sociologists studied collective behavior because they lived in a world that was responding to the processes of modernization, including urbanization, industrialization, and proletarianization of workers. Contemporary forms of

Box 16.1 Sociology and Everyday Life

How Much Do You Know About Collective Behavior and Environmental Issues?

True	False		
T	F	1.	The environmental movement in the United States started in the 1960s.
T	F	2.	People who hold strong attitudes regarding the environment are very likely to be involved in social movements to protect the environment.
T	F	3.	Groups may engage in civil disobedience or use symbolic gestures to call attention to their issue.
T	F	4.	Most sociologists believe that people act somewhat irrationally when they are in large crowds.
T	F	5.	People are most likely to believe rumors when no other information is readily available on a topic.
T	F	6.	Influencing public opinion is a very important activity for many social movements.
T	F	7.	Social movements are more likely to flourish in democratic societies.
T	F	8.	Most social movements in the United States seek to improve society by changing some specific aspect of the social structure.
T	F	9.	Sociologists have found that people in a community respond very similarly to natural disasters and to disasters caused by technological failures.
T	F	10.	People have the capacity to change the environment for better or for worse.

Answers on page 514.

collective behavior, particularly social protests, are variations on the themes that originated during the transition from feudalism to capitalism and the rise of modernity in Europe (Buechler, 2000). Today, some forms of collective behavior and social movements are directed toward public issues such as air pollution, water pollution, and the exploitation of workers in global sweatshops by transnational corporations (see Shaw, 1999). For example, in 1995, Riverkeeper, the group of environmental activists who protect the Hudson River, mounted local campaigns against a member of Congress, including phone calls, a letter-writing campaign, and a picket line maintained by environmentalists, fishermen, and citizens, to call attention to her expected vote on bills that would limit the Clean Water Act and slash the Environmental Protection Agency's funding. Over time, Riverkeeper claims to have "turned" the congresswoman into an environmentalist through various forms of collective behavior (Cronin and Kennedy, 1999).

Conditions for Collective Behavior

Collective behavior occurs as a result of some common influence or stimulus that produces a response from a collectivity. A *collectivity* is a number of people who act together and may mutually transcend,

Social change   the alteration, modification, or transformation of public policy, culture, or social institutions over time.

Collective behavior   voluntary, often spontaneous activity that is engaged in by a large number of people and typically violates dominant-group norms and values.

Box 16.1

Answers to the Sociology Quiz on Collective Behavior and Environmental Issues

1. **False.** The environmental movement in the United States is the result of more than 100 years of collective action. The first environmental organization, the American Forestry Association (now American Forests), originated in 1875.

2. **False.** Since the 1980s, public opinion polls have shown that the majority of people in the United States have favorable attitudes regarding protection of the environment and banning nuclear weapons; however, far fewer individuals are actually involved in collective action to further these causes.

3. **True.** Environmental groups have held sit-ins, marches, boycotts, and strikes, which sometimes take the form of civil disobedience. Others have hanged political leaders in effigy or held officials hostage. Still others have dressed as grizzly bears to block traffic in Yellowstone National Park or created a symbolic "crack" (made of plastic) on the Glen Canyon Dam on the Colorado River to denounce development in the area.

4. **False.** Although some early social psychological theories were based on the assumption of "crowd psychology" or "mob behavior," most sociologists believe that individuals act quite rationally when they are a part of a crowd.

5. **True.** Rumors are most likely to emerge and circulate when people have very little information on a topic that is important to them. For example, rumors abound in times of technological disasters, when people are fearful and often willing to believe a worst-case scenario.

6. **True.** Many social movements, including grassroots environmental activism, attempt to influence public opinion so that local decision makers will feel obliged to correct a specific problem through changes in public policy.

7. **True.** Having a democratic process available is important for dissenters. Grassroots movements have used the democratic process to bring about change even when elites have sought to discourage such activism.

8. **True.** Most social movements are reform movements that focus on improving society by changing some specific aspect of the social structure. Examples are environmental movements and the disability rights movement.

9. **False.** Most sociological studies have found that people respond differently to natural disasters, which usually occur very suddenly, and to technological disasters, which may occur gradually. One of the major differences is the communal bonding that tends to occur following natural disasters, as compared with the extreme social conflict that may follow technological disasters.

10. **True.** One of the goals of most environmental movements is to stress the importance of "thinking globally and acting locally" to protect the environment.

Sources: Based on Adams, 1991; Gamson, 1990; Hynes, 1990; Worster, 1985; and Young, 1990.

bypass, or subvert established institutional patterns and structures. Three major factors contribute to the likelihood that collective behavior will occur: (1) structural factors that increase the chances of people responding in a particular way, (2) timing, and (3) a breakdown in social control mechanisms and a corresponding feeling of normlessness (McPhail, 1991; Turner and Killian, 1993).

A common stimulus is an important factor in collective behavior. For example, the publication of *Silent Spring* (1962) by former Fish and Wildlife Service biologist Rachel Carson is credited with triggering collective behavior directed at demanding a clean environment and questioning how much power large corporations should have. Carson described the dangers of pesticides such as DDT, which was then being promoted by the chemical industry as the miracle that could give the United States the unchallenged position as food supplier to the world (Cronin and Kennedy, 1999). Carson's activism has been described in this way:

> Carson was not a wild-eyed reformer intent on bringing the industrial age to a grinding halt. She wasn't even opposed to pesticides per se. She was a careful scientist and brilliant writer whose painstaking research on pesticides proved that the "miraculous" bursts of agricultural productivity had long-term costs undisclosed in the chemical industry's exaggerated puffery. Americans were losing things—their health, many birds and fishes, and the purity of their waterways—that they should value more than modest savings at the grocery store. (Cronin and Kennedy, 1999: 151)

Timing is another significant factor in bringing about collective behavior. For example, in the 1960s smog had started staining the skies in this country; in Europe, birds and fish were dying from environmental pollution; and oil spills from tankers were provoking public outrage worldwide (Cronin and Kennedy, 1999). People in this country were ready to acknowledge that problems existed. By writing *Silent Spring*, Carson made people aware of the hazards of chemicals in their foods and the destruction of wildlife. However, that is not all she produced: As a consequence of her careful research and writing, she also produced anger in people at a time when they were beginning to wonder if they were being deceived by the very industries that they had entrusted with their lives and their resources. Once aroused to action, many people began demanding an honest, compre-

hensive accounting of where pollution was occurring and how it might be endangering public health and environmental resources. Public outcries also led to investigations in courts and legislatures throughout the United States as people began to demand legal recognition of the right to a clean environment (Cronin and Kennedy, 1999).

A breakdown in social control mechanisms has been a powerful force in triggering collective behavior regarding environmental protection and degradation. During the 1970s, people in the "Love Canal" area of Niagara Falls, New York, became aware that their neighborhood and their children's school had been built over a canal where tons of poisonous waste had been dumped by a chemical company between 1930 and 1950. After the company closed the site, covered it with soil, and sold it (for $1) to the city of Niagara Falls, homes and a school were built on the sixteen-acre site. Over the next two decades, an oily black substance began oozing into the homes in the area and killing the trees and grass on the lots; schoolchildren reported mysterious illnesses and feelings of malaise. Tests indicated that the dump site contained more than two hundred different chemicals, many of which could cause cancer or other serious health problems. Upon learning this information, Lois Gibbs, a mother of one of the schoolchildren, began a grassroots campaign to force government officials to relocate community members injured by seepages from the chemical dump. The collective behavior of neighborhood volunteers was not only successful in eventually bringing about social change but also inspired others to engage in collective behavior regarding environmental problems in their communities. After the Love Canal occurrence, Lois Gibbs founded Citizens' Clearinghouse for Hazardous Waste, which has assisted more than 1,700 community groups in fighting pollution in their neighborhoods (see Cable and Cable, 1995; Cronin and Kennedy, 1999). Similarly, the Hudson Valley Riverkeeper originated in a grassroots movement that was fueled by "people power and never lost touch with its power base" (Cronin and Kennedy, 1999: 277).

Dynamics of Collective Behavior

To better understand the dynamics of collective behavior, let's briefly examine several questions. First, how do people come to transcend, bypass, or subvert established institutional patterns and structures? Some environmental activists have found that they cannot get their point across unless they go outside established institutional patterns and organizations. For example, Lois Gibbs and other Love Canal residents initially tried to work within established means through the school administration and state health officials to clean up the problem. However, they quickly learned that their problems were not being solved through "official" channels. As the problem appeared to grow worse, organizational responses became more defensive and obscure. Accordingly, some residents began acting outside of established norms by holding protests and strikes (Gibbs, 1982). Some situations are more conducive to collective behavior than others. When people can communicate quickly and easily with one another, spontaneous behavior is more likely (Turner and Killian, 1993). When people are gathered together in one general location (whether lining the streets or assembled in a massive stadium), they are more likely to respond to a common stimulus.

Second, how do people's actions compare with their attitudes? People's attitudes (as expressed in public opinion surveys, for instance) are not always reflected in their political and social behavior. Issues pertaining to the environment are no exception. For example, people may indicate in survey research that they believe the quality of the environment is very important, but the same people may not turn out on election day to support propositions that protect the environment or candidates who promise to focus on environmental issues. Likewise, individuals who indicate on a questionnaire that they are concerned about increases in ground-level ozone—the primary component of urban smog—often drive single-occupant, oversized vehicles that government studies have shown to be "gas guzzlers" that contribute to lowered air quality in urban areas. As a result, smog levels increase, contributing to human respiratory problems and dramatically reduced agricultural crop yields (Voynick, 1999).

Third, why do people act collectively rather than singly? As the sociologists Ralph H. Turner and Lewis M. Killian (1993: 12) note, people believe that there is strength in numbers: "the rhythmic stamping of feet by hundreds of concert-goers in unison is different from isolated, individual cries of 'bravo.'" Likewise, people may act as a collectivity when they believe it is the only way to fight those with greater

AP/Wide World Photos

What kind of crowd is this? Based on the crowd's purpose and the form of interaction that takes place, sociologists would identify these 1,000 shoppers, who are trying to beat the holiday rush, as a casual crowd. How do casual crowds differ from other types of crowds?

power and resources. Collective behavior is not just the sum total of a large number of individuals acting at the same time; rather, it reflects people's joint response to some common influence or stimulus.

Distinctions Regarding Collective Behavior

People engaging in collective behavior may be divided into crowds and masses. A *crowd* is a relatively large number of people who are in one another's immediate vicinity (Lofland, 1993). By contrast, a *mass* is a number of people who share an interest in a specific idea or issue but who are not in one another's immediate vicinity (Lofland, 1993). To further distinguish between crowds and masses, think of the difference between a riot and a

AP/Wide World Photos

Protest movements take place around the world when people believe that an injustice has occurred or that great inequalities exist in the distribution of societal resources. These Zapatista rebels in Mexico are part of a movement seeking political inclusion and greater access to economic opportunities.

rumor: People who participate in a riot must be in the same general location; those who spread a rumor may be thousands of miles apart, communicating by telephone or the Internet.

Collective behavior may also be distinguished by the dominant emotion expressed. According to the sociologist John Lofland (1993: 72), the *dominant emotion* refers to the "publicly expressed feeling perceived by participants and observers as the most prominent in an episode of collective behavior." Lofland suggests that fear, hostility, and joy are three fundamental emotions found in collective behavior; however, grief, disgust, surprise, or shame may also predominate in some forms of collective behavior.

Types of Crowd Behavior

When we think of a crowd, many of us think of *aggregates,* previously defined as a collection of people who happen to be in the same place at the same time but who share little else in common. However, the presence of a relatively large number of people in the same location does not necessarily produce collec-

tive behavior. Sociologist Herbert Blumer (1946) developed a typology in which crowds are divided into four categories: casual, conventional, expressive, and acting. Other scholars have added a fifth category, protest crowds.

CASUAL AND CONVENTIONAL CROWDS *Casual crowds* are relatively large gatherings of people who happen to be in the same place at the same time; if they interact at all, it is only briefly. People in a shopping mall or a subway car are examples of casual crowds. Other than sharing a momentary interest, such as a clown's performance or a small child's fall, a casual crowd has nothing in common. The casual crowd plays no active part in the event—such as the child's fall—which likely would have occurred whether or not the crowd was present; the crowd simply observes.

Conventional crowds are made up of people who come together for a scheduled event and thus share a common focus. Examples include religious services, graduation ceremonies, concerts, and college lectures. Each of these events has preestablished schedules and norms. Because these events occur regularly, interaction among participants is much more likely; in turn, the events would not occur without the crowd, which is essential to the event.

EXPRESSIVE AND ACTING CROWDS *Expressive crowds* provide opportunities for the expression of some strong emotion (such as joy, excitement, or grief). People release their pent-up emotions in conjunction with other persons experiencing similar emotions. Examples include worshippers at religious revival services; mourners lining the streets when a celebrity, public official, or religious leader has died; and revelers assembled at Mardi Gras or on New Year's Eve at Times Square in New York.

Crowd a relatively large number of people who are in one another's immediate vicinity.

Mass a number of people who share an interest in a specific idea or issue but who are not in one another's immediate vicinity.

Box 16.2 Changing Times: Media and Technology
How the Internet Has Changed Environmental Activism

We had to keep the media's interest. That was the only way we got anything done. . . . They kept Love Canal in the public consciousness. They educated the public about toxic chemical wastes. One day, we decided we'd take a child's coffin and an adult coffin to the state capital and give it to [New York] Governor Carey. It was a way of keeping us in the news. It would demonstrate our plight. (Gibbs, 1982: 96–97)

Prior to the development of the Internet and the World Wide Web, environmental activists such as Lois Gibbs worked continually for media coverage of environmental problems such as toxic waste in their community. The activists believed that the top priorities for the environmental movement included educating people about the problem and pressuring elected officials to take swift action to alleviate the situation (A. Levine, 1982).

Today, environmental groups have e-mail addresses and maintain web sites that provide a vast array of information about environmental concerns. Web sites include Lois Gibbs's information about herself and the Center for Health, Environment, and Justice (http://www.chej.org) and the Institute for Global Communications' environmental alerts, activist resources, and information on military toxins and nuclear waste (http://www.igc.org). More-specialized environmental activist groups online include "Right to Know," an organization that focuses on environmental justice issues (http://www.rtk.net/). Activist groups often ask their members to send e-mail messages to the president, members of Congress, and state or local officials. Sometimes they inform people of demonstrations or protest marches that will take place across the country after corporate or governmental officials have made decisions that may have an adverse effect on the environment. Even suggestions for how to design effective environmental action alerts are provided on the Internet (http://dlis.gseis.ucla.edu/people/pagre/alerts.html).

At the beginning of the twenty-first century, environmental activists continue to use both print and electronic media to put forth their message; however, the Internet and the World Wide Web offer unsurpassed ease in communicating environmental concerns around the world. For example, two years before the 2000 Olympic Games in Sydney, Australia, the Greenpeace web site (http://www.greenpeace.org) reported that the organization had removed thirty drums of toxic sediment allegedly produced by the chemical giant Orica (formerly ICA Australia) from Homebush Bay. This area was not only the gateway to the 2000 Olympic Games, but also a future site for residential redevelopment. By publishing this information on its web site, the Greenpeace organization sought to inform others about the problem and pressure the corporation to "immediately clean up its toxic mess."

Is it possible to use the Internet to mobilize people for action on pressing social issues such as environmental pollution? How can we determine which organization's claims are most valid? What do you think?

Acting crowds are collectivities so intensely focused on a specific purpose or object that they may erupt into violent or destructive behavior. Mobs, riots, and panics are examples of acting crowds, but casual and conventional crowds may become acting crowds under some circumstances. A *mob* **is a highly emotional crowd whose members engage in, or are ready to engage in, violence against a specific target—a person, a category of people, or physical property.** Mob behavior in this country has included lynchings, fire bombings, effigy hangings, and hate crimes. Mob violence tends to dissipate relatively quickly once a target has been injured, killed, or destroyed. Sometimes, actions such as an effigy hanging are used symbolically by groups that are not violent otherwise. For example, Lois Gibbs and other Love Canal residents called attention to their problems with the chemical dump site by staging a protest in which they "burned in effigy" the governor and the health commissioner to emphasize their displeasure with the lack of response from these public officials (A. Levine, 1982). Activities such as this have been used by environmental activists and others who seek social change in hopes of keeping media attention focused on the problem they want to highlight. Recently, e-mail and web sites on the Internet have changed national grassroots campaigns, including those associated with environmental activism (see Box 16.2).

Compared with mob actions, riots may be of somewhat longer duration. A *riot* **is violent crowd behavior that is fueled by deep-seated emotions but not directed at one specific target.** Riots are often triggered by fear, anger, and hostility; however, not all riots are caused by deep-seated hostility and hatred—people may be expressing joy and exuberance when rioting occurs. Examples include celebrations after sports victories such as those that occurred in Montreal, Canada, following a Stanley Cup win and in Vancouver following a playoff victory (Kendall, Lothian Murray, and Linden, 2000).

A *panic* **is a form of crowd behavior that occurs when a large number of people react to a real or perceived threat with strong emotions and self-destructive behavior.** The most common type of panic occurs when people seek to escape from a perceived danger, fearing that few (if any) of them will be able to get away from that danger. Panics can also arise in response to events that people believe are beyond their control—such as a major disruption in the economy. Although panics are relatively rare, they receive massive media coverage because they provoke strong feelings of fear in readers and viewers, and the number of casualties may be large.

PROTEST CROWDS Sociologists Clark McPhail and Ronald T. Wohlstein (1983) added protest crowds to the four types of crowds identified by Blumer. *Protest crowds* engage in activities intended to achieve specific political goals. Examples include sit-ins, marches, boycotts, blockades, and strikes. In 1997 an International Day of Protest was staged against Nike's labor practices in countries such as Vietnam (Shaw, 1999). Some protests take the form of *civil disobedience*—**nonviolent action that seeks to change a policy or law by refusing to comply with it.** Acts of civil disobedience may become violent, as in a confrontation between protesters and police officers; in this case, a protest crowd becomes an *acting crowd.* In the 1960s, African American students and sympathetic whites used sit-ins to call attention to racial injustice and demand social change (see Morris, 1981; McAdam, 1982). Some of these protests can escalate into violent confrontations even when violence was not the intent of the organizers.

As you may recall, collective action often puts individuals into a situation where they engage in

Reflections

Reflections 16.1

Why is collective behavior, rather than individual protest, typically a more effective method for bringing about social change to deal with problems such as environmental degradation?

some activity as a group that they might not do on their own. Does this mean that people's actions are produced by some type of "herd mentality"? Some analysts have answered this question affirmatively; however, sociologists typically do not agree with that assessment.

Explanations of Crowd Behavior

What causes people to act collectively? How do they determine what types of action to take? One of the earliest theorists to provide an answer to these questions was Gustave Le Bon, a French scholar who focused on crowd psychology in his contagion theory.

CONTAGION THEORY *Contagion theory* focuses on the social–psychological aspects of collective behavior; it attempts to explain how moods, attitudes, and behavior are communicated rapidly and why they are accepted by others (Turner and Killian, 1993). Le Bon (1841–1931) argued that

Mob a highly emotional crowd whose members engage in, or are ready to engage in, violence against a specific target—a person, a category of people, or physical property.

Riot violent crowd behavior that is fueled by deep-seated emotions but is not directed at one specific target.

Panic a form of crowd behavior that occurs when a large number of people react to a real or perceived threat with strong emotions and self-destructive behavior.

Civil disobedience nonviolent action that seeks to change a policy or law by refusing to comply with it.

Convergence theory is based on the assumption that crowd behavior involves shared emotions, goals, and beliefs, such as the importance of protecting the environment. An example is the Earth Day events that brought together these children carrying this banner to foster environmental causes

David Young-Wolff/PhotoEdit

people are more likely to engage in antisocial behavior in a crowd because they are anonymous and feel invulnerable. Le Bon (1960/1895) suggested that a crowd takes on a life of its own that is larger than the beliefs or actions of any one person. Because of its anonymity, the crowd transforms individuals from rational beings into a single organism with a collective mind. In essence, Le Bon asserted that emotions such as fear and hate are contagious in crowds because people experience a decline in personal responsibility; they will do things as a collectivity that they would never do when acting alone.

Le Bon's theory is still used by many people to explain crowd behavior. However, critics argue that the "collective mind" has not been documented by systematic studies.

SOCIAL UNREST AND CIRCULAR REACTION
Sociologist Robert E. Park was the first U.S. sociologist to investigate crowd behavior. Park believed that Le Bon's analysis of collective behavior lacked several important elements. Intrigued that people could break away from the powerful hold of culture and their established routines to develop a new social order, Park added the concepts of social unrest and circular reaction to contagion theory. According to Park, social unrest is transmitted by a process of *circular reaction*—the interactive communication between persons such that the dis-

content of one person is communicated to another, who, in turn, reflects the discontent back to the first person (Park and Burgess, 1921).

CONVERGENCE THEORY *Convergence theory* focuses on the shared emotions, goals, and beliefs that many people may bring to crowd behavior. Because of their individual characteristics, many people have a predisposition to participate in certain types of activities (Turner and Killian, 1993). From this perspective, people with similar attributes find a collectivity of like-minded persons with whom they can express their underlying personal tendencies. Although people may reveal their "true selves" in crowds, their behavior is not irrational; it is highly predictable to those who share similar emotions or beliefs.

Convergence theory has been applied to a wide array of conduct, from lynch mobs to environmental movements. In social psychologist Hadley Cantril's (1941) study of one lynching, he found that the participants shared certain common attributes: They were poor and working-class whites who felt that their status was threatened by the presence of successful African Americans. Consequently, the characteristics of these individuals made them susceptible to joining a lynch mob even if they did not know the target of the lynching.

Convergence theory adds to our understanding of certain types of collective behavior by pointing out how individuals may have certain attributes—such as racial hatred or fear of environmental problems that directly threaten them—that initially bring them together. However, this theory does not explain how the attitudes and characteristics of individuals who take some collective action differ from those who do not.

EMERGENT NORM THEORY Unlike contagion and convergence theories, *emergent norm theory* emphasizes the importance of social norms in shaping crowd behavior. Drawing on the symbolic interactionist perspective, the sociologists Ralph Turner and Lewis Killian (1993: 12) asserted that crowds develop their own definition of a situation and establish norms for behavior that fit the occasion:

> Some shared redefinition of right and wrong in a situation supplies the justification and coordinates the action in collective behavior. People do what they would not otherwise have done when they panic collectively, when they riot, when they engage in civil disobedience, or when they launch terrorist campaigns, because they find social support for the view that what they are doing is the right thing to do in the situation.

According to Turner and Killian (1993: 13), emergent norms occur when people define a new situation as highly unusual or see a long-standing situation in a new light.

Sociologists using the emergent norm approach seek to determine how individuals in a given collectivity develop an understanding of what is going on, how they construe these activities, and what type of norms are involved. For example, in a study of audience participation, the sociologist Steven E. Clayman (1993) found that members of an audience listening to a speech applaud promptly and independently but wait to coordinate their booing with other people; they do not wish to "boo" alone.

Some emergent norms are permissive—that is, they give people a shared conviction that they may disregard ordinary rules such as waiting in line, taking turns, or treating a speaker courteously. Collective activity such as mass looting may be defined (by participants) as taking what rightfully belongs to them and punishing those who have been exploitative. For example, following the Los Angeles riots of 1992, some analysts argued

that Korean Americans were targets of rioters because they were viewed by Latinos/as and African Americans as "callous and greedy invaders" who became wealthy at the expense of members of other racial–ethnic groups (Cho, 1993). Thus, rioters who used this rationalization could view looting and burning as a means of "paying back" Korean Americans or of gaining property (such as TV sets and microwave ovens) from those who had already taken from them. Once a crowd reaches some agreement on the norms, the collectivity is supposed to adhere to them. If crowd members develop a norm that condones looting or vandalizing property, they will proceed to cheer for those who conform and ridicule those who are unwilling to abide by the collectivity's new norms.

Emergent norm theory points out that crowds are not irrational. Rather, new norms are developed in a rational way to fit the immediate situation. However, critics note that proponents of this perspective fail to specify exactly what constitutes a norm, how new ones emerge, and how they are so quickly disseminated and accepted by a wide variety of participants. One variation of this theory suggests that no single dominant norm is accepted by everyone in a crowd; instead, norms are specific to the various categories of actors rather than to the collectivity as a whole (Snow, Zurcher, and Peters, 1981). For example, in a study of football victory celebrations, the sociologists David A. Snow, Louis A. Zurcher, and Robert Peters (1981) found that each week, behavioral patterns were changed in the postgame revelry, with some being modified, some added, and some deleted.

Mass Behavior

Not all collective behavior takes place in face-to-face collectivities. *Mass behavior* is collective behavior that takes place when people (who often are geographically separated from one another) respond to the same event in much the same way. For people to respond in the same way, they

Mass behavior collective behavior that takes place when people (who often are geographically separated from one another) respond to the same event in much the same way.

typically have common sources of information that provoke their collective behavior. The most frequent types of mass behavior are rumors, gossip, mass hysteria, public opinion, fashions, and fads. Under some circumstances, social movements constitute a form of mass behavior. However, we will examine social movements separately because they differ in some important ways from other types of dispersed collectivities.

RUMORS AND GOSSIP *Rumors* **are unsubstantiated reports on an issue or subject** (Rosnow and Fine, 1976). Whereas a rumor may spread through an assembled collectivity, rumors may also be transmitted among people who are dispersed geographically. Although rumors may initially contain a kernel of truth, they may be modified as they spread to serve the interests of those repeating them. Rumors thrive when tensions are high and when little authentic information is available on an issue of great concern.

People are willing to give rumors credence when no opposing information is available. For example, as residents of Love Canal waited for information from health department officials about their exposure to toxic chemicals and from the government about possible relocation at state expense to another area, new waves of rumors spread through the community daily. By the time a meeting was called by health department officials to provide homeowners with the results of air-sample tests for hazardous chemicals (such as chloroform and benzene) performed on their homes, already fearful residents were ready to believe the worst, as Lois Gibbs (1982: 25. Reprinted with permission) describes:

> Next to the names [of residents] were some numbers. But the numbers had no meaning. People stood there looking at the numbers, knowing nothing of what they meant but suspecting the worst.
>
> One woman, divorced and with three sick children, looked at the piece of paper with numbers and started crying hysterically: "No wonder my children are sick. Am I going to die? What's going to happen to my children?" No one could answer. . . .
>
> The night was very warm and humid, and the air was stagnant. On a night like that, the smell of Love Canal is hard to describe. It's all around you. It's as though it were about to envelop you and smother you. By now, we were outside, standing in the parking lot. The woman's panic caught on, starting a chain reaction. Soon, many people there were hysterical.

Once a rumor begins to circulate, it seldom stops unless compelling information comes to the forefront that either proves the rumor false or makes it obsolete.

In industrialized societies with sophisticated technology, rumors come from a wide variety of sources and may be difficult to trace. Print media (newspapers and magazines) and electronic media (radio and television), fax machines, cellular networks, satellite systems, and the Internet aid the rapid movement of rumors around the globe. In addition, modern communications technology makes anonymity much easier. In a split second, messages (both factual and fictitious) can be disseminated to thousands of people through e-mail, computerized bulletin boards, and Internet newsgroups.

Whereas rumors deal with an issue or a subject, *gossip* **refers to rumors about the personal lives of individuals.** Charles Horton Cooley (1963/1909) viewed gossip as something that spread among a small group of individuals who personally knew the person who was the object of the rumor. Today, this is frequently not the case; many people enjoy gossiping about people whom they have never met. Tabloid newspapers and magazines such as the *National Inquirer* and *People*, and television "news" programs that purport to provide "inside" information on the lives of celebrities, are sources of contemporary gossip, much of which has not been checked for authenticity.

MASS HYSTERIA AND PANIC *Mass hysteria* is a form of dispersed collective behavior that occurs when a large number of people react with strong emotions and self-destructive behavior to a real or perceived threat. Does mass hysteria actually occur? Although the term has been widely used, many sociologists believe that this behavior is best described as a panic with a dispersed audience.

An example of mass hysteria or a panic with a widely dispersed audience was actor Orson Welles's 1938 Halloween eve radio dramatization of H. G. Wells's science fiction classic *The War of the Worlds*. A CBS radio dance music program was interrupted suddenly by a news bulletin informing the audience that Martians had landed in New Jersey and were in the process of conquering Earth. Some listeners became extremely frightened even though an announcer had indicated before, during, and after the performance that the broadcast was a fictitious dramatization. According to some reports, as many as one million of the estimated ten million listeners believed that this astonishing event had occurred. Thousands were reported to have hidden in their storm cellars or to have gotten in their cars so that they could flee from the Martians (see Brown, 1954). In actuality, the program probably did not generate mass hysteria, but rather a panic among gullible listeners. Others switched stations to determine if the same "news" was being broadcast elsewhere. When they discovered that it was not, they merely laughed at the joke being played on listeners by CBS. In 1988, on the fiftieth anniversary of the broadcast, a Portuguese radio station rebroadcast the program; once again, a panic ensued.

The "Harry Potter phenomenon," as it is referred to by marketers and retailers, became a popular fad when the first Harry Potter film was released. The fad included merchandise such as action figures, games, puzzles, and trading cards. Is advertising a factor in determining what will become a fad and how long the fad will last?

FADS AND FASHIONS As you will recall from Chapter 2, a *fad* is a temporary but widely copied activity enthusiastically followed by large numbers of people. Fads can be embraced by widely dispersed collectivities; news networks such as CNN may bring the latest fad to the attention of audiences around the world. The United States has witnessed a number of fads. One especially remembered by faculty who have been on college campuses for several decades was the 1970s fad of "streaking"—students taking off their clothes and running naked in public. Regardless of how it may sound, this activity was not purely spontaneous. Streakers had to calculate and plan their activity so that an audience (often including members of the media) would be present. Streaking had no meaning if it was not widely publicized; for this reason, some students chose graduation ceremonies and other highly visible occasions for their streaking escapades. Other fads, such as exercise regimes and health practices, tend to be taken more seriously.

Unlike fads, fashions tend to be longer lasting. In Chapter 2, *fashion* is defined as a currently valued style of behavior, thinking, or appearance. Fashion also applies to art, music, drama, literature, architecture, interior design, and automobiles, among other things. However, most sociological research on fashion has focused on clothing, especially women's apparel (Davis, 1992).

In preindustrial societies, clothing styles remained relatively unchanged. With the advent of industrialization, items of apparel became readily available at low prices because of mass production. Fashion became more important as people embraced the "modern" way of life and as advertising encouraged "conspicuous consumption."

Georg Simmel, Thorstein Veblen, and Pierre Bourdieu have all viewed fashion as a means of status differentiation among members of different social classes. Simmel (1957/1904) suggested a classic "trickle-down" theory (although he did not use those exact words) to describe the process by which members of the lower classes emulate the fashions of the upper class. As the fashions descend through the status hierarchy, they are watered down and "vulgarized" so that they are no longer recognizable to members of the upper class, who then regard them as unfashionable and in bad taste (Davis, 1992). Veblen (1967/1899) asserted that fashion serves

Rumors unsubstantiated reports on an issue or subject.

Gossip rumors about the personal lives of individuals.

mainly to institutionalize conspicuous consumption among the wealthy. Almost eighty years later, Bourdieu (1984) similarly (but more subtly) suggested that "matters of taste," including fashion sensibility, constitute a large share of the "cultural capital" possessed by members of the dominant class.

Herbert Blumer (1969) disagreed with the trickle-down approach, arguing that "collective selection" best explains fashion. Blumer suggested that people in the middle and lower classes follow fashion because it is *fashion*, not because they desire to emulate members of the elite class. Blumer thus shifts the focus on fashion to collective mood, tastes, and choices: "Tastes are themselves a product of experience. . . . They are formed in the context of social interaction, responding to the definitions and affirmation given by others. People thrown into areas of common interaction and having similar runs of experience develop common tastes" (qtd. in Davis, 1992: 116). Perhaps one of the best refutations of the trickle-down approach is the way in which fashion today often originates among people in the lower social classes and is mimicked by the elites. In the mid-1990s, the so-called grunge look was a prime example of this.

PUBLIC OPINION   *Public opinion* **consists of the attitudes and beliefs communicated by ordinary citizens to decision makers** (Greenberg, 1999). It is measured through polls and surveys, which use research methods such as interviews and questionnaires, as described in Chapter 1. Many people are not interested in all aspects of public policy but are concerned about issues they believe are relevant to themselves. Even on a single topic, public opinion will vary widely based on race/ethnicity, religion, region, social class, education level, gender, age, and so on.

Scholars who examine public opinion are interested in the extent to which the public's attitudes are communicated to decision makers and the effect (if any) that public opinion has on policy making (Turner and Killian, 1993). Some political scientists argue that public opinion has a substantial effect on decisions at all levels of government (see Greenberg, 1999); others strongly disagree.

Today, people attempt to influence elites and vice versa. Consequently, a two-way process occurs with the dissemination of *propaganda*—**information provided by individuals or groups that have a vested interest in furthering their own cause or damaging an opposing one.** Although many of us think of propaganda in negative terms, the information provided can be correct and can have a positive effect on decision making.

In recent decades, grassroots environmental activists (including the Love Canal residents and the Riverkeepers) have attempted to influence public opinion. In a study of public opinion on environmental issues, the sociologist Riley E. Dunlap (1992) found that public awareness of the seriousness of environmental problems and public support for environmental protection increased dramatically between the late 1960s and the 1990s. However, it is less clear that public opinion translates into action by either decision makers in government and industry or by individuals (such as a willingness to adopt a more ecologically sound lifestyle).

Initially, most grassroots environmental activists attempt to influence public opinion so that local decision makers will feel the necessity of correcting a specific problem through changes in public policy. Although activists usually do not start out

Reflections

Reflections 16.2

How can propaganda be used in an effort to sway public opinion both by environmental groups and by those opposed to a particular environmental proposal?

✓Checkpoints

Checkpoint 16.1

1. What conditions are most conducive for collective behavior?
2. What are the five types of crowd behavior?
3. What is civil disobedience, and what part does it play in collective behavior?
4. How do each of these theories explain crowd behavior: (a) contagion theory, (b) social unrest and circular reaction, (c) convergence theory, and (d) emergent norm theory?
5. What are fashions and fads?
6. What is public opinion, and how is it communicated to decision makers?

seeking broader social change, they often move in that direction when they become aware of how widespread the problem is in the larger society or on a global basis. One of two types of social movements often develops at this point—one focuses on NIMBY ("not in my backyard"), whereas the other focuses on NIABY ("not in anyone's backyard") (Freudenberg and Steinsapir, 1992).

Social Movements

Although collective behavior is short-lived and relatively unorganized, social movements are longer lasting, are more organized, and have specific goals or purposes. A *social movement* is an **organized group that acts consciously to promote or resist change through collective action** (Goldberg, 1991). Because social movements have not become institutionalized and are outside the political mainstream, they offer "outsiders" an opportunity to have their voices heard.

Social movements are more likely to develop in industrialized societies than in preindustrial societies, where acceptance of traditional beliefs and practices makes such movements unlikely. Diversity and a lack of consensus (hallmarks of industrialized nations) contribute to demands for social change, and people who participate in social movements typically lack power and other resources to bring about change without engaging in collective action. Social movements are most likely to spring up when people come to see their personal troubles as public issues that cannot be solved without a collective response.

Social movements make democracy more available to excluded groups (see Greenberg, 1999). Historically, people in the United States have worked at the grassroots level to bring about changes even when elites sought to discourage activism (Adams, 1991). For example, the civil rights movement brought into its ranks African Americans in the South who had never been allowed to participate in politics (see Killian, 1984). The women's suffrage movement gave voice to women who had been denied the right to vote (Rosenthal et al., 1985). Similarly, a grassroots environmental movement gave the working-class residents of Love Canal a way to "fight city hall" and Hooker Chemicals (A. Levine, 1982), as Lois Gibbs (1982: 38–40. Reprinted with permission.) explains:

People were pretty upset. They were talking and stirring each other up. I was afraid there would be violence. We had a meeting at my house to try to put everything together [and] decided to form a homeowners' association. We got out the word as best we could and told everyone to come to the Frontier Fire Hall on 102d Street. . . . The firehouse was packed with people, and more were outside. . . .

I was elected president. . . . I took over the meeting but I was scared to death. It was only the second time in my life I had been in front of a microphone or a crowd. . . . We set four goals right at the beginning—(1) get all the residents within the Love Canal area who wanted to be evacuated, evacuated and relocated, especially during the construction and repair of the canal; (2) do something about propping up property values; (3) get the canal fixed properly; and (4) have air sampling and soil and water testing done throughout the whole area, so we could tell how far the contamination had spread. . . .

Most social movements rely on volunteers like Lois Gibbs to carry out the work. Traditionally, women have been strongly represented in both the membership and the leadership of many grassroots movements (A. Levine, 1982; Freudenberg and Steinsapir, 1992).

The Love Canal activists set the stage for other movements that have grappled with the kind of issues that the sociologist Kai Erikson (1994) refers to as a "new species of trouble." Erikson describes the "new species" as environmental problems that "contaminate rather than merely damage . . . they pollute, befoul, taint, rather than just create wreckage . . . they penetrate human tissue indirectly rather than just wound the surfaces by assaults of a more straightforward kind. . . . And the evidence is growing that they scare human beings in new and special ways, that they elicit an uncanny fear in us" (Erikson, 1991: 15). The chaos that Erikson (1994: 141) describes is the result of technological disasters—"meaning everything that can go wrong when systems fail, humans

Public opinion the attitudes and beliefs communicated by ordinary citizens to decision makers.

Propaganda information provided by individuals or groups that have a vested interest in furthering their own cause or damaging an opposing one.

Social movement an organized group that acts consciously to promote or resist change through collective action.

Martin Luther King, Jr., a leader of the civil rights movement in the 1950s and 1960s, advocated nonviolent protests that sometimes took the form of civil disobedience. Here, he marches alongside his wife, Coretta Scott King, who took over many of Dr. King's activities after he was assassinated.

Flip Schulke/Black Star

err, designs prove faulty, engines misfire, and so on." A recent example of such a disaster occurred in Japan, where more than 300,000 residents living within six miles of the nuclear plant at Tokaimura were told to stay indoors in the aftermath of three workers' mishandling of stainless-steel pails full of uranium, causing the worst nuclear accident in Japan's history (Larimer, 1999). Although no lives were immediately lost, workers in the plant soaked up potentially lethal doses of radiation, some of which also leaked from the plant into the community. The 52 nuclear power plants in Japan have been plagued by other accidents and radiation leaks, causing concern for many people around the globe (Larimer, 1999).

Social movements provide people who otherwise would not have the resources to enter the game of politics a chance to do so. We are most familiar with those movements that develop around public policy issues considered newsworthy by the media, ranging from abortion and women's rights to gun control and environmental justice. However, a number of other types of social movements exist as well.

Types of Social Movements

Social movements are difficult to classify; however, sociologists distinguish among movements on the basis of their *goals* and the *amount of change* they seek to produce (Aberle, 1966; Blumer, 1974).

Some movements seek to change people whereas others seek to change society.

REFORM MOVEMENTS Grassroots environmental movements are an example of *reform movements*, which seek to improve society by changing some specific aspect of the social structure. Members of reform movements usually work within the existing system to attempt to change existing public policy so that it more adequately reflects their own value system. Examples of reform movements (in addition to the environmental movement) include labor movements, animal rights movements, antinuclear movements, Mothers Against Drunk Driving, and the disability rights movement.

Sociologist Lory Britt (1993) suggested that some movements arise specifically to alter social responses to and definitions of stigmatized attributes. From this perspective, social movements may bring about changes in societal attitudes and practices while at the same time causing changes in participants' social emotions. For example, the civil rights and gay rights movements helped replace shame with pride (Britt, 1993). Such a sense of pride may extend beyond current members of a reform movement. The late César Chávez, organizer of a Mexican American farm workers' movement that developed into the United Farm Workers Union, noted that the "consciousness and pride raised by our union is alive and thriving inside millions of young Hispanics who will never work on a farm!" (qtd. in Ayala, 1993: E4).

REVOLUTIONARY MOVEMENTS Movements seeking to bring about a total change in society are referred to as *revolutionary movements*. These movements usually do not attempt to work within the existing system; rather, they aim to remake the system by replacing existing institutions with new ones. Revolutionary movements range from utopian groups seeking to establish an ideal society to radical terrorists who use fear tactics to intimidate those with whom they disagree ideologically (see Alexander and Gill, 1984; Berger, 1988; Vetter and Perlstein, 1991).

Movements based on terrorism often use tactics such as bombings, kidnappings, hostage taking, hijackings, and assassinations (Vetter and Perlstein, 1991). Over the past thirty years, a number of movements in the United States have engaged in terrorist activities or supported a policy of violence. However, the terrorist attacks in New York City and Washington, D.C., on September 11, 2001, and the events that followed those attacks proved to all of us that terrorism within this country can originate from the activities of revolutionary terrorists from outside the country as well.

RELIGIOUS MOVEMENTS Social movements that seek to produce radical change in individuals are typically based on spiritual or supernatural belief systems. Also referred to as *expressive movements*, *religious movements* are concerned with renovating or renewing people through "inner change." Fundamentalist religious groups seeking to convert nonbelievers to their belief system are an example of this type of movement. Some religious movements are *millenarian*—that is, they forecast that "the end is near" and assert that an immediate change in behavior is imperative. Relatively new religious movements in industrialized Western societies have included Hare Krishnas, the Unification Church, Scientology, and the Divine Light Mission, all of which tend to appeal to the psychological and social needs of young people seeking meaning in life that mainstream religions have not provided for them.

ALTERNATIVE MOVEMENTS Movements that seek limited change in some aspect of people's behavior are referred to as *alternative movements*.

For example, early in the twentieth century the Women's Christian Temperance Union attempted to get people to abstain from drinking alcoholic beverages. Some analysts place "therapeutic social movements" such as Alcoholics Anonymous in this category; however, others do not, due to their belief that people must change their lives completely in order to overcome alcohol abuse (see Blumberg, 1977). More recently, a variety of "New Age" movements have directed people's behavior by emphasizing spiritual consciousness combined with a belief in reincarnation and astrology. Such practices as vegetarianism, meditation, and holistic medicine are often included in the self-improvement category. Beginning in the 1990s, some alternative movements have included the practice of yoga (usually without its traditional background in the Hindu religion) as a means by which the self can be liberated and union can be achieved with the supreme spirit or universal soul.

RESISTANCE MOVEMENTS Also referred to as *regressive movements*, *resistance movements* seek to prevent change or to undo change that has already occurred. Virtually all of the proactive social movements previously discussed face resistance from one or more reactive movements that hold opposing viewpoints and want to foster public policies that reflect their own beliefs. Examples of resistance movements are groups organized since the 1950s to oppose school integration, civil rights and affirmative action legislation, and domestic partnership initiatives. However, perhaps the most widely known resistance movement includes many who label themselves "pro-life" advocates—such as Operation Rescue, which seeks to close abortion clinics and make abortion illegal under all circumstances (Gray, 1993; Van Biema, 1993). Protests by some radical anti-abortion groups have grown violent, resulting in the death of several doctors and clinic workers, and creating fear among health professionals and patients seeking abortions (Belkin, 1994).

Stages in Social Movements

Do all social movements go through similar stages? Not necessarily, but there appear to be identifiable stages in virtually all movements that succeed beyond their initial phase of development.

As the issue of access to abortion services illustrates, clashes are inevitable between members of proactive social movements and their counterparts in resistance movements.

In the *preliminary* (or *incipiency*) *stage*, widespread unrest is present as people begin to become aware of a problem. At this stage, leaders emerge to agitate others into taking action. In the *coalescence stage*, people begin to organize and to publicize the problem. At this stage, some movements become formally organized at local and regional levels. In the *institutionalization* (or *bureaucratization*) *stage*, an organizational structure develops, and a paid staff (rather than volunteers) begins to lead the group. When the movement reaches this stage, the initial zeal and idealism of members may diminish as administrators take over management of the organization. Early grassroots supporters may become disillusioned and drop out; they may also start another movement to address some as-yet-unsolved aspect of the original problem. For example, some national environmental organizations—such as the Sierra Club, the National Audubon Society, and the National Parks and Conservation Association—that started as grassroots conservation movements are currently viewed by many people as being unresponsive to local environmental problems (Cable and Cable, 1995). As a result, new movements have arisen.

Social Movement Theories

What conditions are most likely to produce social movements? Why are people drawn to these movements? Sociologists have developed several theories to answer these questions.

Relative Deprivation Theory

According to relative deprivation theory, people who are satisfied with their present condition are less likely to seek social change. Social movements arise as a response to people's perception that they have been deprived of their "fair share" (Rose, 1982). Thus, people who suffer relative deprivation are more likely to feel that change is necessary and to join a social movement in order to bring about that change. *Relative deprivation* refers to the discontent that people may feel when they compare their achievements with those of similarly situated persons and find that they have less than they think they deserve (Orum and Orum, 1968). Karl Marx captured the idea of relative

✓ Checkpoints

Checkpoint 16.2

7. Why are social movements more likely to emerge in industrialized societies?
8. What are the major types of social movements?
9. Why is terrorism considered to be a revolutionary movement?
10. What are examples of major alternative and resistance movements?

deprivation in this description: "A house may be large or small; as long as the surrounding houses are small it satisfies all social demands for a dwelling. But let a palace arise beside the little house, and it shrinks from a little house to a hut" (qtd. in Ladd, 1966: 24). Movements based on relative deprivation are most likely to occur when an upswing in the standard of living is followed by a period of decline, such that people have *unfulfilled rising expectations*—newly raised hopes of a better lifestyle that are not fulfilled as rapidly as the people expected or are not realized at all.

Although most of us can relate to relative deprivation theory, it does not fully account for why people experience social discontent but fail to join a social movement. Even though discontent and feelings of deprivation may be necessary to produce certain types of social movements, they are not sufficient to bring movements into existence. In fact, the sociologist Anthony Orum (1974) found the best predictor of participation in a social movement to be prior organizational membership and involvement in other political activities.

Value-Added Theory

The *value-added theory* developed by sociologist Neil Smelser (1963) is based on the assumption that certain conditions are necessary for the development of a social movement. Smelser called his theory the "value-added" approach based on the concept (borrowed from the field of economics) that each step in the production process adds something to the finished product. For example, in the process of converting iron ore into automobiles, each stage "adds value" to the final product (Smelser, 1963). Similarly, Smelser asserted, six conditions are necessary and sufficient to produce social movements when they combine or interact in a particular situation:

1. *Structural conduciveness*. People must become aware of a significant problem and have the opportunity to engage in collective action. According to Smelser, movements are more likely to occur when a person, class, or agency can be singled out as the source of the problem; when channels for expressing grievances either are not available or fail; and when the aggrieved have a chance to communicate among themselves.

2. *Structural strain*. When a society or community is unable to meet people's expectations that something should be done about a problem, strain occurs in the system. The ensuing tension and conflict contribute to the development of a social movement based on people's belief that the problem would not exist if authorities had done what they were supposed to do.

3. *Spread of a generalized belief*. For a movement to develop, there must be a clear statement of the problem and a shared view of its cause, effects, and possible solution.

4. *Precipitating factors*. To reinforce the existing generalized belief, an inciting incident or dramatic event must occur. With regard to technological disasters, some (including Love Canal) gradually emerge from a long-standing environmental threat whereas others (including the Three Mile Island nuclear power plant) involve a suddenly imposed problem.

5. *Mobilization for action*. At this stage, leaders emerge to organize others and give them a sense of direction.

6. *Social control factors*. If there is a high level of social control on the part of law enforcement officials, political leaders, and others, it becomes more difficult to develop a social movement or engage in certain types of collective action.

Value-added theory takes into account the complexity of social movements and makes it possible to test Smelser's assertions regarding the necessary and sufficient conditions that produce such movements. However, critics note that the approach is rooted in the functionalist tradition and views structural strains as disruptive to society. Smelser's theory has been described as a mere variant of convergence theory, which, you will recall, is based on the assumption that people with similar predispositions will be activated by a common event or object (Quarantelli and Hundley, 1993).

Resource Mobilization Theory

Smelser's value-added theory tends to underemphasize the importance of resources in social movements. By contrast, *resource mobilization theory* focuses on the ability of members of a social movement to acquire resources and mobilize people in order to advance their cause (Oberschall, 1973; McCarthy and Zald, 1977). Resources

How is the issue of immigration framed in these photos? Research based on frame analysis often investigates how social issues are framed and what names they are given.

include money, people's time and skills, access to the media, and material goods such as property and equipment. Assistance from outsiders is essential for social movements. For example, reform movements are more likely to succeed when they gain the support of political and economic elites (Oberschall, 1973).

Resource mobilization theory is based on the assumption that participants in social movements are rational people. According to the sociologist Charles Tilly (1973, 1978), movements are formed and dissolved, mobilized and deactivated, based on rational decisions about the goals of the group, available resources, and the cost of mobilization and collective action. Resource mobilization theory also assumes that participants must have some degree of economic and political resources to make the movement a success. In other words, widespread discontent alone cannot produce a social movement; adequate resources and motivated people are essential to any concerted social action (Aminzade, 1973; Gamson, 1990). Based on an analysis of fifty-three U.S. social protest groups ranging from labor unions to peace movements between 1800 and 1945, the sociologist William

Gamson (1990) concluded that the organization and tactics of a movement strongly influence its chances of success. However, critics note that this theory fails to account for social changes brought about by groups with limited resources.

Scholars continue to modify resource mobilization theory and to develop new approaches for investigating the diversity of movements at the beginning of the twenty-first century (see Buechler, 2000). For example, emerging perspectives based on resource mobilization theory emphasize the ideology and legitimacy of movements as well as material resources (see Zald and McCarthy, 1987; McAdam, McCarthy, and Zald, 1988).

Social Constructionist Theory: Frame Analysis

Recent theories based on a symbolic interactionist perspective focus on the importance of the symbolic presentation of a problem to both participants and the general public (see Snow et al., 1986; Capek, 1993). *Social constructionist theory* is based on the assumption that social movements are an interactive, symbolically defined, and negotiated process that involves participants, opponents, and bystanders (Buechler, 2000).

Research based on this perspective often investigates how problems are framed and what names they are given. This approach reflects the influence of the sociologist Erving Goffman's *Frame Analysis* (1974), in which he suggests that our interpretation of the particulars of events and activities is dependent on the framework from which we perceive them. According to Goffman (1974: 10), the purpose of frame analysis is "to try to isolate some of the basic frameworks of understanding available in our society for making sense out of events and to analyze the special vulnerabilities to which these frames of reference are subject." In other words, various "realities" may be simultaneously occurring among participants engaged in the same set of activities. Sociologist Steven M. Buechler (2000: 41) explains the relationship between frame analysis and social movement theory:

> Framing means focusing attention on some bounded phenomenon by imparting meaning and significance to elements within the frame and setting them apart from what is outside the frame. In the context of social movements, framing refers to the interactive, collective ways that movement actors assign meanings to their activities in the conduct of social movement activism. The concept of framing is designed for discussing the social construction of grievances as a fluid and variable process of social interaction—and hence a much more important explanatory tool than resource mobilization theory has maintained.

Sociologists have identified at least three ways in which grievances are framed. First, *diagnostic framing* identifies a problem and attributes blame or causality to some group or entity so that the social movement has a target for its actions. Second, *prognostic framing* pinpoints possible solutions or remedies, based on the target previously identified. Third, *motivational framing* provides a vocabulary of motives that compel people to take action (Benford, 1993; Snow and Benford, 1988). When successful framing occurs, the individual's vague dissatisfactions are turned into well-defined grievances, and people are compelled to join the movement in an effort to reduce or eliminate those grievances (Buechler, 2000).

Beyond motivational framing, additional frame alignment processes are necessary in order to supply a continuing sense of urgency to the movement. *Frame alignment* is the linking together of interpretive orientations of individuals and social movement organizations so that there is congruence between individuals' interests, beliefs, and values and the movement's ideologies, goals, and activities (Snow et al., 1986). Four distinct frame alignment processes occur in social movements: (1) *frame bridging* is the process by which movement organizations reach individuals who already share the same world view as the organization, (2) *frame amplification* occurs when movements appeal to deeply held values and beliefs in the general population and link those to movement issues so that people's preexisting value commitments serve as a "hook" that can be used to recruit them, (3) *frame extension* occurs when movements enlarge the boundaries of an initial frame to incorporate other issues that appear to be of importance to potential participants, and (4) *frame transformation* refers to the process whereby the creation and maintenance of new values, beliefs, and meanings induce movement participation by redefining activities and events in such a manner that people believe they must become involved in collective action (Buechler, 2000). Some or all of these frame alignment processes are used by social movements as they seek to define grievances and recruit participants.

Sociologist William Gamson (1995) believes that social movements borrow, modify, or create frames as they seek to advance their goals. For example, social activism regarding nuclear disasters involves differential framing over time. When the Fermi reactor had a serious partial meltdown in 1966, there was no sustained social opposition. However, when Pennsylvania's Three Mile Island nuclear power plant experienced a similar problem in 1979, the situation was defined as a technological disaster. At that time, social movements opposing nuclear power plants grew in number, and each intensified its efforts to reduce or eliminate such plants. This example shows that people react toward something on the basis of the meaning that thing has for them, and such meanings typically emerge from the political culture, including the social activism, of that specific era (Buechler, 2000).

Frame analysis provides new insights on how social movements emerge and grow when people are faced with problems such as technological disasters, about which greater ambiguity typically

Andrew Lichtenstein/The Image Works

Referred to as "Cancer Alley," this area of Baton Rouge, Louisiana, is home to a predominantly African American population and also to many refineries that heavily pollute the region. Sociologists suggest that environmental racism is a significant problem in the United States and other nations. What do you think?

exists, and when people are attempting to "name" the problems associated with things such as nuclear or chemical contamination. However, frame analysis has been criticized for its "ideational biases" (McAdam, 1996). According to the sociologist Doug McAdam (1996), frame analyses of social movements have looked almost exclusively at ideas and their formal expression, whereas little attention has been paid to other significant factors such as movement tactics, mobilizing structures, and changing political opportunities that influence the signifying work of movements. In this context, "political opportunity" means government structure, public policy, and political conditions that set the boundaries for change and political action. These boundaries are crucial variables in explaining why various social movements have different outcomes (Meyer and Staggenborg, 1996; Gotham, 1999).

New Social Movement Theory

New social movement theory looks at a diverse array of collective actions and the manner in which those actions are based on politics, ideology, and culture. It also incorporates sources of identity, including, race, class, gender, and sexuality, as sources of collective action and social movements. Examples of "new social movements" include ecofeminism and environmental justice movements. Ecofeminism emerged in the late 1970s and early 1980s out of the feminist, peace, and ecology movements. Prompted by the near-meltdown at the Three Mile Island nuclear power plant, ecofeminists established World Women in Defense of the Environment. *Ecofeminism* is based on the belief that patriarchy is a root cause of environmental problems. According to ecofeminists, patriarchy not only results in the domination of women by men but also contributes to a belief that nature is to be possessed and dominated, rather than treated as a partner (see Ortner, 1974; Merchant, 1983, 1992; Mies and Shiva, 1993).

Another "new social movement" focuses on environmental justice and the intersection of race and class in the environmental struggle. Sociologist Stella M. Capek (1993) investigated a contaminated landfill in the Carver Terrace neighborhood of Texarkana, Texas, and found that residents were able to mobilize for change and win a federal buyout and relocation by symbolically linking their issue to a larger *environmental justice* framework. Since the 1980s, the emerging environmental justice movement has focused on the issue of **environmental racism—the belief that a disproportionate number of hazardous facilities (including industries such as waste disposal/treatment and chemical plants) are placed in low-income areas populated primarily by people of color** (Bullard and Wright, 1992). These areas have been left out of most of the environmental cleanup that has taken place in the last two decades (Schneider, 1993). Capek concludes that linking Carver Terrace with environmental justice led to it being designated as a cleanup site. She also views this as an important turning point in new social movements: "Carver Terrace is significant not only as a federal buyout and relocation of a minority community, but also as a marker of the emergence of environmental racism as a major new component of environmental social movements in the United States" (Capek, 1993: 21).

Sociologist Steven M. Buechler (2000) has argued that theories pertaining to twenty-first-century social movements should be oriented toward the

Concept Table 16.A Social Movement Theories

	Key Components
Relative Deprivation	People who are discontent when they compare their achievements with those of others consider themselves relatively deprived and join social movements in order to get what they view as their "fair share," especially when there is an upswing in the economy followed by a decline.
Value-Added	Certain conditions are necessary for a social movement to develop: (1) structural conduciveness, such that people are aware of a problem and have the opportunity to engage in collective action; (2) structural strain, such that society or the community cannot meet people's expectations for taking care of the problem; (3) growth and spread of a generalized belief as to causes and effects of and possible solutions to the problem; (4) precipitating factors, or events that reinforce the beliefs; (5) mobilization of participants for action; and (6) social control factors, such that society comes to allow the movement to take action.
Resource Mobilization	A variety of resources (money, members, access to media, and material goods such as equipment) are necessary for a social movement; people participate only when they feel the movement has access to these resources.
Social Construction Theory: Frame Analysis	Based on the assumption that social movements are an interactive, symbolically defined, and negotiated process involving participants, opponents, and bystanders, frame analysis is used to determine how people assign meaning to activities and processes in social movements.
New Social Movement	The focus is on sources of social movements, including politics, ideology, and culture. Race, class, gender, sexuality, and other sources of identity are also factors in movements such as ecofeminism and environmental justice.

structural, macrolevel contexts in which movements arise. These theories should incorporate both political and cultural dimensions of social activism:

> Social movements are historical products of the age of modernity. They arose as part of a sweeping social, political, and intellectual change that led a significant number of people to view society as a social construction that was susceptible to social reconstruction through concerted collective effort. Thus, from their inception, social movements have had a dual focus. Reflecting the political, they have always involved some form of challenge to prevailing forms of authority. Reflecting the cultural, they have always operated as symbolic laboratories in which re-

flexive actors pose questions of meaning, purpose, identity, and change. (Buechler, 2000: 211)

Concept Table 16.A summarizes the main theories of social movements.

As we have seen, social movements may be an important source of social change. Throughout this

Environmental racism   the belief that a disproportionate number of hazardous facilities (including industries such as waste disposal/treatment and chemical plants) are placed in low-income areas populated primarily by people of color.

✓**Checkpoints**

Checkpoint 16.3

11. What is relative deprivation theory?
12. What conditions are necessary to produce social movements?
13. What is the primary focus of resource mobilization theory?
14. What is the social constructionist approach to examining social movements?
15. What are examples of newer social movements?

text, we have examined a variety of social problems that have been the focus of one or more social movements during the past hundred years. In the process of bringing about change, most movements initially develop innovative ways to get their ideas across to decision makers and the public. Some have been successful in achieving their goals; others have not. As the historian Robert A. Goldberg (1991) has suggested, gains made by social movements may be fragile, acceptance brief, and benefits minimal and easily lost. For this reason, many groups focus on preserving their gains while simultaneously fighting for those they believe they still desire.

Social Change in the Future

In this chapter, we have focused on collective behavior and social movements as potential forces for social change in contemporary societies. A number of other factors also contribute to social change, including the physical environment, population trends, technological development, and social institutions.

The Physical Environment and Change

Changes in the physical environment often produce changes in the lives of people; in turn, people can make dramatic changes in the physical environment, over which we have only limited control. Throughout history, natural disasters have taken their toll on individuals and societies. Major natural disasters—including hurricanes, floods, and tornados—can devastate an entire population. Even comparatively "small" natural disasters change the lives of many people. As the sociologist Kai Erikson (1976, 1994)

has suggested, the trauma that people experience from disasters may outweigh the actual loss of physical property—memories of such events can haunt people for many years.

Some natural disasters are exacerbated by human decisions. For example, floods are viewed as natural disasters, but excessive development may contribute to a flood's severity. As office buildings, shopping malls, industrial plants, residential areas, and highways are developed, less land remains as groundcover to absorb rainfall. When heavier-than-usual rains occur, flooding becomes inevitable; some regions of the United States have remained under water for days or even weeks in recent years. Clearly, humans cannot control the rain, but human decisions can worsen the consequences.

Since flooding is one of the many problems that face the world today, it may seem strange that one of the major concerns in the twenty-first century is the availability of water. Many experts warn that water is a finite resource that is necessary for both human survival and the production of goods. However, water is being wasted and polluted, and the supply of *potable* (drinkable) water is limited.

People contribute to changes in the Earth's physical condition. Through soil erosion and other degradation of grazing land, often at the hands of people, an estimated 24 billion tons of topsoil is lost annually. As people clear forests to create farmland and pastures and to acquire lumber and firewood, the Earth's tree cover continues to diminish. As millions of people drive motor vehicles, the amount of carbon dioxide in the environment continues to rise each year, possibly resulting in global warming.

Just as people contribute to changes in the physical environment, human activities must also be adapted to changes in the environment. For example, we are being warned to stay out of the sunlight because of increases in ultraviolet rays, a cause of skin cancer, as a result of the accelerating depletion of the ozone layer. If the ozone warnings are accurate, the change in the physical environment will dramatically affect those who work or spend their leisure time outside.

Population and Change

Changes in population size, distribution, and composition affect the culture and social structure of a society and change the relationships among nations. As discussed in Chapter 15, the countries experiencing the most rapid increases in population have a less-developed infrastructure to deal with those changes. How will nations of the world deal with population growth as the global population continues to move toward seven billion? Only time will provide a response to this question.

In the United States, a shift in population distribution from central cities to suburban and exurban areas has produced other dramatic changes. Central cities have experienced a shrinking tax base as middle- and upper-middle-income residents and businesses have moved to suburban and outlying areas. As a result, schools and public services have declined in many areas, leaving those people with the greatest needs with the fewest public resources and essential services. The changing composition of the U.S. population has resulted in children from more diverse cultural backgrounds entering school, producing a demand for new programs and changes in curricula. An increase in the birth rate has created a need for more child care; an increase in the older population has created a need for services such as medical care, placing greater demands on programs such as Social Security.

As we have seen in previous chapters, population growth and the movement of people to urban areas have brought profound changes to many regions and intensified existing social problems. Among other factors, growth in the global population is one of the most significant driving forces behind environmental concerns such as the availability and use of natural resources.

Technology and Change

Technology is an important force for change; in some ways, technological development has made our lives much easier. Advances in communication and transportation have made instantaneous worldwide communication possible but have also brought old belief systems and the status quo into question as never before. Today, we are increasingly moving information instead of people—and doing it almost instantly. Advances in science and medicine have made significant changes in people's lives in high-income countries. The light bulb, the automobile, the airplane, the assembly line, and recent high-tech developments have contributed to dramatic changes in people's lives. Individuals in high-income nations have benefited from the use of the technology; those in low-income nations

Jeff Greenberg/PhotoEdit

Pollution of lakes, rivers, and other bodies of water has an adverse effect on food supplies, air quality, and the entire environment. What influence does a "business as usual" approach have on environmental quality in your city?

may have paid a disproportionate share of the cost of some of these inventions and discoveries.

Scientific advances will continue to affect our lives, from the foods we eat to our reproductive capabilities. Genetically engineered plants have been developed and marketed in recent years, and biochemists are creating potatoes, rice, and cassava with the same protein value as meat (Petersen, 1994). Advances in medicine have made it possible for those formerly unable to have children to procreate; women well beyond menopause are now able to become pregnant with the assistance of medical technology. Advances in medicine have also increased the human lifespan, especially for white and middle- or upper-class individuals in high-income nations; medical advances have also contributed to the declining death rate in low-income nations, where birth rates have not yet been curbed.

Just as technology has brought about improvements in the quality and length of life for many, it has also created the potential for new disasters, ranging from global warfare to localized technological disasters at toxic waste sites. As the sociologist William Ogburn (1966) suggested, when a change in the material culture occurs in society, a period of *cultural lag* follows in which the nonmaterial (ideological) culture has not caught up with material development. The rate of technological advance at the level of material culture today is mind-boggling. Many of us can never hope to understand technological advances in the areas of artificial intelligence, holography, virtual reality, biotechnology, cold fusion, and robotics.

One of the ironies of twenty-first-century high technology is the increased vulnerability that results from the increasing complexity of such systems. As futurist John L. Petersen (1994: 70) notes, "The more complex a system becomes, the more likely the chance of system failure. There are unknown secondary effects and particularly vulnerable nodes." He also asserts that most of the world's population will not participate in the technological revolution that is occurring in high-income nations (Petersen, 1994).

Technological disasters may result in the deaths of tens of thousands, especially if we think of modern warfare as a technological disaster. Nuclear power, which can provide energy for millions, can also be the source of a nuclear war that could devastate the planet. As a government study on even limited nuclear war concluded,

> Natural resources would be destroyed; surviving equipment would be designed to use materials and skills that might no longer exist; and indeed some regions might be almost uninhabitable. Furthermore, pre-war patterns of behavior would surely change, though in unpredictable ways. (U.S. Congress, 1979, qtd. in Howard, 1990: 320)

Even when lives are not lost in technological disasters, families are uprooted and communities cease to exist as people are relocated. In many cases, the problem is not solved; people are moved either temporarily or permanently away from the site.

Social Institutions and Change

Many changes occurred in the family, religion, education, the economy, and the political system during the twentieth century and early in the twenty-first century. As we saw in Chapter 11, the size and composition of families in the United States changed with the dramatic increase in the number of single-person and single-parent households. Changes in families produced changes in the socialization of children, many of whom spend large amounts of time in front of a television set or in child-care facilities outside their own homes. Although some political and religious leaders advocate a return to "traditional" family life, numerous scholars argue that such families never worked quite as well as some might wish to believe.

Public education changed dramatically in the United States during the twentieth century. This country was one of the first to provide "universal" education for students regardless of their ability to pay. As a result, at least until recently, the United States has had one of the most highly educated populations in the world. Today, the United States still has one of the best public education systems in the world for the top 15 percent of the students, but it badly fails the bottom 25 percent. As the nature of the economy changes, schools almost inevitably will have to change, if for no other reason than demands from leaders in business and industry for an educated work force that allows U.S. companies to compete in a global economic environment.

Political systems experienced tremendous change and upheaval in some parts of the world during the twentieth century. The United States participated in two world wars and numerous other "conflicts," the largest and most divisive being in Vietnam and the most recent being the war on terrorism. The cost of these wars and conflicts, and of increased security measures as a result of terrorism, is staggering.

A new concept of world security is emerging, requiring the cooperation of high-income countries in halting the proliferation of weapons of mass destruction and combating terrorism. That concept also requires the cooperation of *all* nations in reducing the plight of the poorest people in the low-income countries of the world.

Although we have examined changes in the physical environment, population, technology, and social institutions separately, they all operate together in a complex relationship, sometimes producing large, unanticipated consequences. As we move further into the twenty-first century, we need new ways of conceptualizing social life at both the macrolevel

Reflections

Reflections 16.3

How might social science research be used to produce positive changes in our physical and social environments?

Box 16.3 You Can Make a Difference
Recycling for Tomorrow

Numerous social movements have called our attention to the fact that our use of resources, and hence the production of waste, is a major source of pollution in industrialized nations.

Environmental advocates have suggested that recycling is an important first step that any of us can take to improve our communities. Here are a few questions about the basic facts of recycling (the answers appear below):

1. How many times can a glass bottle be recycled?
   a. 1
   b. 5
   c. 10
   d. an unlimited number of times
2. How much paper (paper products) is currently recycled in the United States?
   a. about 25%
   b. about 50%
   c. about 75%
   d. almost 100%
3. Which of the following is not one of the three Rs of recycling?
   a. reclaim
   b. reduce
   c. recycle
   d. reuse
4. Which type of soft-drink container is the most recycled?
   a. aluminum cans
   b. boxes
   c. glass
   d. plastic bottles
5. What is the most recycled product on Earth (by percentage)?
   a. appliances
   b. automobiles
   c. boats
   d. food

How did you score on this brief quiz? If each of us not only learns about recycling but also implements plans in our household, we can become part of the solution rather than part of the environmental problem.

Environmental sociologists have determined that in many cases, recycling helps diminish waste disposal problems by reducing the amount of waste hauled off to landfills. Paper products (magazines, newspapers, and packaging), glass and plastic bottles, aluminum and steel cans, and appliances (refrigerators, home computers, and televisions) are all recyclable.

Advocates believe that recycling is worth the time and effort it takes because the process reduces the amount of waste and saves energy and natural resources. From a sociological perspective, recycling may also create a sense of community among people as they work toward a cleaner city and, in some cases, use recycling programs as a way to raise funds for schools, youth groups, churches, and other community-based organizations. To learn more about starting a recycling program, visit the following web site:

http://environment.about.com/library/weekly/blrcy4.htm

Answers: 1—d, 2—b, 3—a, 4—a, 5—b

and the microlevel. The sociological imagination helps us think about how personal troubles—regardless of our race, class, gender, age, sexual orientation, or physical abilities and disabilities—are intertwined with the public issues of our society and the global community of which we are a part. As one analyst noted regarding Lois Gibbs and Love Canal,

If Love Canal has taught Lois Gibbs—and the rest of us—anything, it is that ordinary people become very smart very quickly when their lives are threatened. They become adept at detecting absurdity, even when it is concealed in bureaucratese and scientific jargon. Lois Gibbs learned that one cannot always rely on government to act in the best interests of ordinary citizens—at least, not without considerable prodding. She determined that she would prod them until her objectives were attained. She led one of the most successful, single-purpose grass roots efforts of our time. (M. Levine, 1982: xv)

Taking care of the environment is an example of something that government and each of us as individuals can do to help (see Box 16.3). And it is vitally important that we all do everything that we can in order to protect the environment.

A Few Final Thoughts

In this text, we have covered a substantial amount of material, examined different perspectives on a wide variety of social issues, and have suggested different methods by which to deal with them. The purpose of this text is not to encourage you to take any particular point of view; rather, it is to allow you to understand different viewpoints and ways in which they may be helpful to you and to society in dealing with the issues of the twenty-first century. Possessing that understanding, we can hope that the future will be something we can all look forward to—producing a better way of life, not only in this country but worldwide as well.

Chapter Review

What is the relationship between social change and collective behavior?

Social change—the alteration, modification, or transformation of public policy, culture, or social institutions over time—is usually brought about by collective behavior, which is defined as relatively spontaneous, unstructured activity that typically violates established social norms.

When is collective behavior likely to occur?

Collective behavior occurs when some common influence or stimulus produces a response from a relatively large number of people.

What is a crowd?

A crowd is a relatively large number of people in one another's immediate presence. Sociologist Herbert Blumer divided crowds into four categories: (1) casual crowds, (2) conventional crowds, (3) expressive crowds, and (4) acting crowds (including mobs, riots, and panics). A fifth type of crowd is a protest crowd.

What causes crowd behavior?

Social scientists have developed several theories to explain crowd behavior. Contagion theory asserts that a crowd takes on a life of its own as people are transformed from rational beings into part of an organism that acts on its own. A variation on this is social unrest and circular reaction—people express their discontent to others, who communicate back similar feelings, resulting in a conscious effort to engage in the crowd's behavior. Convergence theory asserts that people with similar attributes find other like-minded persons with whom they can release underlying personal tendencies. Emergent norm theory asserts that as a crowd develops, it comes up with its own norms that replace more conventional norms of behavior.

What are the primary forms of mass behavior?

Mass behavior is collective behavior that occurs when people respond to the same event in the same way even if they are not in geographic proximity to one another. Rumors, gossip, mass hysteria, fads and fashions, and public opinion are forms of mass behavior.

What are the major types of social movements, and what are their goals?

A social movement is an organized group that acts consciously to promote or resist change through collective action. Reform, revolutionary, religious, and alternative movements are the major types identified by sociologists. Reform movements seek to improve society by changing some specific aspect of the social structure. Revolutionary movements seek to bring about a total change in society—sometimes by the use of terrorism. Religious movements seek to produce radical change in individuals based on spiritual or supernatural belief systems. Alternative movements seek limited change of some aspect of people's behavior. Resistance movements seek to prevent change or to undo change that has already occurred.

How do social movements develop?

Social movements typically go through three stages: (1) a preliminary stage (unrest results from a perceived problem), (2) coalescence (people begin to organize), and (3) institutionalization (an organization is developed, and paid staff replaces volunteers in leadership positions).

How do relative deprivation theory, value-added theory, and resource mobilization theory explain social movements?

Relative deprivation theory asserts that if people are discontented when they compare their accomplishments with those of others similarly situated, they are more likely to join a social movement than are people who are relatively content with their status. According to value-added theory, six conditions are required for a social movement: (1) a perceived problem, (2) a perception that the authorities are not resolving the problem, (3) a spread of the belief to an adequate number of people, (4) a precipitating incident, (5) mobilization of other people by leaders, and (6) a lack of social control. By contrast, resource mobilization theory asserts that successful social movements can occur only when they gain the support of political and economic elites, who provide access to the resources necessary to maintain the movement.

What is the primary focus of research based on frame analysis and new social movement theory? Research based on frame analysis often highlights the social construction of grievances through the process of social interaction. Various types of framing occur as problems are identified, remedies are sought, and people feel compelled to take action. Like frame analysis, new social movement theory has been used in research that looks at technological disasters and cases of environmental racism.

Key Terms

civil disobedience 519
collective behavior 512
crowd 516
environmental racism 532
gossip 522
mass 516
mass behavior 521
mob 518
panic 519
propaganda 524
public opinion 524
riot 519
rumors 522
social change 512
social movement 525

Questions for Critical Thinking

1. What types of collective behavior in the United States do you believe are influenced by inequalities based on race/ethnicity, class, gender, age, or disabilities? Why?
2. Which of these four explanations—contagion theory, social unrest and circular reaction, convergence theory, and emergent norm theory—do you believe best explains crowd behavior? Why?
3. In the text, the Love Canal environmental movement is analyzed in terms of the value-added theory. How would you analyze that movement under (a) the relative deprivation theory and (b) the resource mobilization theory?

4. Using the sociological imagination that you have gained in this course, what are some positive steps that you believe might be taken in the United States to make our society a better place for everyone? What types of collective behavior and/or social movements might be required in order to take those steps?

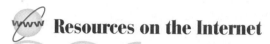

Resources on the Internet

Chapter-Related Web Sites
The following web sites have been selected for their relevance to the topics in this chapter. These sites are among the more stable, but please note that web site addresses change frequently.

Women and Social Movements in the United States, 1775–2000
http://womhist.binghamton.edu
This page contains historical information related to U.S. social movements and women.

Civil Rights Movement Veterans
http://www.crmvet.org
This page chronicles the U.S. civil rights movement from the perspective of people who were there and experienced it firsthand.

The Peace Movement in Historical Perspective: Conversation with Nobel Laureate Linus Pauling
http://globetrotter.berkeley.edu/conversations/Pauling
This online interview is part of the Institute of International Studies' "Conversations with History" series, which uses Internet technology to share with the public the University of California–Berkeley's distinction as a global forum for ideas. Dr. Linus Pauling received the Nobel Prize in Chemistry in 1954 and the Nobel Peace Prize in 1963.

Collective Behavior: Crowds/Mobs/Riots
http://www.geocities.com/collectivebehavior2
Here is a concise explanation of collective behavior along with links to several examples of historical events.

Collective Behavior and Social Movements
http://www2.pfeiffer.edu/~lridener/DSS/collbeh.html
This page contains an online reference list to articles and other web sites.

Collective Behavior and the Social Psychologies of Social Institutions
http://www.trinity.edu/~mkearl/socpsy-8.html
The most extensive site yet on collective behavior. Dozens and dozens of links are included in this online course.

Introduction to Future Studies
http://www.csudh.edu/global_options/
IntroFS.HTML
This page presents the study of the future as an academic discipline; it includes a solid introduction and links to very useful sites.

Companion Web Site for This Book

Virtual Society: The Wadsworth Sociology Resource Center
Visit http://sociology.wadsworth.com and click on the page for Kendall, *Sociology in Our Times: The Essentials*, Fourth Edition, to access a wide range of enrichment material to aid in your study of sociology. Click on the Student Resources section of the web site. Next, select from the pull-down menu the chapter that you are presently studying. Among the useful options for self-study are chapter objectives, flashcards, practical study tips, and practice tests for each chapter.

MicroCase Online
From the Virtual Society home page, click on MicroCase Online to access book-specific MicroCase exercises that allow you to further explore sociological issues and principles presented in the text.

InfoTrac College Edition
Another unique option available to you at the Student Resources section of the companion web site is InfoTrac College Edition, an online library with access to hundreds of scholarly and popular periodicals. Below are suggested search terms for this chapter. Results from these and other searches are found at the site.

- Search keywords: *Love Canal.* Look for articles that discuss the Love Canal disaster and its impact on current environmental problems. How did the Love Canal situation result in social change?
- Search keyword: *terrorism.* Conduct a limit search with the word *nuclear.* Look for articles that address how the United States can prepare for nuclear terrorism, as well as articles that assess its probability.

Virtual Explorations CD-ROM
Go to the Virtual Explorations CD-ROM to begin the interactive exercise for this chapter. This Virtual Exploration will introduce you to some of the exciting resources for sociology on the World Wide Web. You will be guided through an exercise that employs related web sites on collective behavior, social movements, and social change. Answer the questions, and e-mail your responses to your instructor.

Glossary

absolute poverty a level of economic deprivation that exists when people do not have the means to secure the most basic necessities of life.

achieved status a social position that a person assumes voluntarily as a result of personal choice, merit, or direct effort.

acute diseases illnesses that strike suddenly and cause dramatic incapacitation and sometimes death.

ageism prejudice and discrimination against people on the basis of age, particularly against older people.

agents of socialization the persons, groups, or institutions that teach us what we need to know in order to participate in society.

aggregate a collection of people who happen to be in the same place at the same time but share little else in common.

alienation a feeling of powerlessness and estrangement from other people and from oneself.

animism the belief that plants, animals, or other elements of the natural world are endowed with spirits or life forces having an effect on events in society.

anomie Emile Durkheim's designation for a condition in which social control becomes ineffective as a result of the loss of shared values and of a sense of purpose in society.

anticipatory socialization the process by which knowledge and skills are learned for future roles.

ascribed status a social position conferred at birth or received involuntarily later in life based on attributes over which the individual has little or no control, such as race/ethnicity, age, and gender.

assimilation a process by which members of subordinate racial and ethnic groups become absorbed into the dominant culture.

authoritarian leaders people who make all major group decisions and assign tasks to members.

authoritarian personality a personality type characterized by excessive conformity, submissiveness to authority, intolerance, insecurity, a high level of superstition, and rigid, stereotypic thinking.

authoritarianism a political system controlled by rulers who deny popular participation in government.

authority power that people accept as legitimate rather than coercive.

bilateral descent a system of tracing descent through both the mother's and father's sides of the family.

body consciousness a term that describes how a person perceives and feels about his or her body.

bureaucracy an organizational model characterized by a hierarchy of authority, a clear division of labor, explicit rules and procedures, and impersonality in personnel matters.

bureaucratic personality a psychological construct that describes those workers who are more concerned with following correct procedures than they are with getting the job done correctly.

capitalism an economic system characterized by private ownership of the means of production, from which personal profits can be derived through market competition and without government intervention.

capitalist class (or **bourgeoisie**) Karl Marx's term for the class that consists of those who own and control the means of production.

caste system a system of social inequality in which people's status is permanently determined at birth based on their parents' ascribed characteristics.

category a number of people who may never have met one another but share a similar characteristic, such as education level, age, race, or gender.

chronic diseases illnesses that are long term or lifelong and that develop gradually or are present from birth.

church a large, bureaucratically organized religious organization that tends to seek accommodation with the larger society in order to maintain some degree of control over it.

civil disobedience nonviolent action that seeks to change a policy or law by refusing to comply with it.

civil religion the set of beliefs, rituals, and symbols that makes sacred the values of the society and places the nation in the context of the ultimate system of meaning.

class conflict Karl Marx's term for the struggle between the capitalist class and the working class.

class system a type of stratification based on the ownership and control of resources and on the type of work that people do.

cohabitation a situation in which two people live together, and think of themselves as a couple, without being legally married.

collective behavior voluntary, often spontaneous activity that is engaged in by a large number of people and typically violates dominant-group norms and values.

comparable worth (or **pay equity**) the belief that wages ought to reflect the worth of a job, not the gender or race of the worker.

conflict perspectives the sociological approach that views groups in society as engaged in a continuous power struggle for control of scarce resources.

541

conformity the process of maintaining or changing behavior to comply with the norms established by a society, subculture, or other group.

conglomerate a combination of businesses in different commercial areas, all of which are owned by one holding company.

content analysis the systematic examination of cultural artifacts or various forms of communication to extract thematic data and draw conclusions about social life.

contingent work part-time work, temporary work, or subcontracted work that offers advantages to employers but that can be detrimental to the welfare of workers.

conventional (street) crime all violent crime, certain property crimes, and certain morals crimes.

core nations according to world systems theory, dominant capitalist centers characterized by high levels of industrialization and urbanization.

corporate crime illegal acts committed by corporate employees on behalf of the corporation and with its support.

corporations organizations that have legal powers, such as the ability to enter into contracts and buy and sell property, separate from their individual owners.

correlation a relationship that exists when two variables are associated more frequently than could be expected by chance.

counterculture a group that strongly rejects dominant societal values and norms and seeks alternative lifestyles.

credentialism a process of social selection in which class advantage and social status are linked to the possession of academic qualifications.

crime behavior that violates criminal law and is punishable with fines, jail terms, and other sanctions.

criminology the systematic study of crime and the criminal justice system, including the police, courts, and prisons.

crowd a relatively large number of people who are in one another's immediate vicinity.

crude birth rate the number of live births per 1,000 people in a population in a given year.

crude death rate the number of deaths per 1,000 people in a population in a given year.

cult a religious group with practices and teachings outside the dominant cultural and religious traditions of a society.

cultural capital Pierre Bourdieu's term for people's social assets, including values, beliefs, attitudes, and competencies in language and culture.

cultural imperialism the extensive infusion of one nation's culture into other nations.

cultural lag William Ogburn's term for a gap between the technical development of a society (material culture) and its moral and legal institutions (nonmaterial culture).

cultural relativism the belief that the behaviors and customs of any culture must be viewed and analyzed by the culture's own standards.

cultural universals customs and practices that occur across all societies.

culture the knowledge, language, values, customs, and material objects that are passed from person to person and from one generation to the next in a human group or society.

culture shock the disorientation that people feel when they encounter cultures radically different from their own and believe they cannot depend on their own taken-for-granted assumptions about life.

demedicalization the process whereby a problem ceases to be defined as an illness or a disorder.

democracy a political system in which the people hold the ruling power either directly or through elected representatives.

democratic leaders leaders who encourage group discussion and decision making through consensus building.

democratic socialism an economic and political system that combines private ownership of some of the means of production, governmental distribution of some essential goods and services, and free elections.

demographic transition the process by which some societies have moved from high birth rates and death rates to relatively low birth rates and death rates as a result of technological development.

demography a subfield of sociology that examines population size, composition, and distribution.

denomination a large, organized religion characterized by accommodation to society but frequently lacking in ability or intention to dominate society.

dependency theory the belief that global poverty can at least partially be attributed to the fact that the low-income countries have been exploited by the high-income countries.

deviance any behavior, belief, or condition that violates significant social norms in the society or group in which it occurs.

differential association theory the proposition that individuals have a greater tendency to deviate from societal norms when they frequently associate with persons who are more favorable toward deviance than conformity.

disability a reduced ability to perform tasks one would normally do at a given stage of life and that may result in stigmatization or discrimination against the person with disabilities.

discrimination actions or practices of dominant-group members (or their representatives) that have a harmful effect on members of a subordinate group.

domestic partnerships household partnerships in which an unmarried couple lives together in a committed, sexually intimate relationship and is granted the same rights and benefits as those accorded to married heterosexual couples.

dramaturgical analysis the study of social interaction that compares everyday life to a theatrical presentation.

drug any substance—other than food and water—that, when taken into the body, alters its functioning in some way.

dual-earner marriages marriages in which both spouses are in the labor force.

dyad a group composed of two members.

ecclesia a religious organization that is so integrated into the dominant culture that it claims as its membership all members of a society.

economy the social institution that ensures the maintenance of society through the production, distribution, and consumption of goods and services.

education the social institution responsible for the systematic transmission of knowledge, skills, and cultural values within a formally organized structure.

egalitarian family a family structure in which both partners share power and authority equally.

ego according to Sigmund Freud, the rational, reality-oriented component of personality that imposes restrictions on the innate pleasure-seeking drives of the id.

elite model a view of society that sees power in political systems as being concentrated in the hands of a small group of elites whereas the masses are relatively powerless.

endogamy the practice of marrying within one's own social group or category.

environmental racism the belief that a disproportionate number of hazardous facilities (including industries such as waste disposal/treatment and chemical plants) are placed in low-income areas populated primarily by people of color.

ethnic group a collection of people distinguished, by others or by themselves, primarily on the basis of cultural or nationality characteristics.

ethnic pluralism the coexistence of a variety of distinct racial and ethnic groups within one society.

ethnocentrism the assumption that one's own culture and way of life are superior to all others.

ethnography a detailed study of the life and activities of a group of people by researchers who may live with that group over a period of years.

ethnomethodology the study of the commonsense knowledge that people use to understand the situations in which they find themselves.

exogamy the practice of marrying outside one's own social group or category.

experiment a research method involving a carefully designed situation in which the researcher studies the impact of certain variables on subjects' attitudes or behavior.

expressive leadership an approach to leadership that provides emotional support for members.

extended family a family unit composed of relatives in addition to parents and children who live in the same household.

families relationships in which people live together with commitment, form an economic unit and care for any young, and consider their identity to be significantly attached to the group.

family of orientation the family into which a person is born and in which early socialization usually takes place.

family of procreation the family that a person forms by having or adopting children.

feminism the belief that all people—both women and men—are equal and that they should be valued equally and have equal rights.

feminization of poverty the trend in which women are disproportionately represented among individuals living in poverty.

fertility the actual level of childbearing for an individual or a population.

folkways informal norms or everyday customs that may be violated without serious consequences within a particular culture.

formal organization a highly structured group formed for the purpose of completing certain tasks or achieving specific goals.

functionalist perspectives the sociological approach that views society as a stable, orderly system.

Gemeinschaft (guh-MINE-shoft) a traditional society in which social relationships are based on personal bonds of friendship and kinship and on intergenerational stability.

gender the culturally and socially constructed differences between females and males found in the meanings, beliefs, and practices associated with "femininity" and "masculinity."

gender bias behavior that shows favoritism toward one gender over the other.

gender identity a person's perception of the self as female or male.

gender role the attitudes, behavior, and activities that are socially defined as appropriate for each sex and are learned through the socialization process.

gender socialization the aspect of socialization that contains specific messages and practices concerning the nature of being female or male in a specific group or society.

generalized other George Herbert Mead's term for the child's awareness of the demands and expectations of the society as a whole or of the child's subculture.

genocide the deliberate, systematic killing of an entire people or nation.

gentrification the process by which members of the middle and upper-middle classes, especially whites, move into a central-city area and renovate existing properties.

Gesellschaft (guh-ZELL-shoft) a large, urban society in which social bonds are based on impersonal and specialized relationships, with little long-term commitment to the group or consensus on values.

goal displacement a process that occurs in organizations when the rules become an end in themselves rather than a means to an end, and organizational survival becomes more important than achievement of goals.

gossip rumors about the personal lives of individuals.

government the formal organization that has the legal and political authority to regulate the relationships among members of a society and between the society and those outside its borders.

groupthink the process by which members of a cohesive group arrive at a decision that many individual members privately believe is unwise.

health a state of complete physical, mental, and social well-being.

health care any activity intended to improve health.

health maintenance organizations (HMOs) companies that provide, for a set monthly fee, total care with an emphasis on prevention to avoid costly treatment later.

hermaphrodite a person in whom sexual differentiation is ambiguous or incomplete.

hidden curriculum the transmission of cultural values and attitudes, such as conformity and obedience to authority, through implied demands found in the rules, routines, and regulations of schools.

high-income countries (sometimes referred to as **industrial countries**) nations with highly industrialized economies; technologically advanced industrial, administrative, and service occupations; and relatively high levels of national and personal income.

holistic medicine an approach to health care that focuses on prevention of illness and disease and is aimed at treating the whole person—body and mind—rather than just the part or parts in which symptoms occur.

homogamy the pattern of individuals marrying those who have similar characteristics, such as race/ethnicity, religious background, age, education, or social class.

homophobia extreme prejudice directed at gays, lesbians, bisexuals, and others who are perceived as not being heterosexual.

id Sigmund Freud's term for the component of personality that includes all of the individual's basic biological drives and needs that demand immediate gratification.

ideal type an abstract model that describes the recurring characteristics of some phenomenon (such as bureaucracy).

illegitimate opportunity structures circumstances that provide an opportunity for people to acquire through illegitimate activities what they cannot achieve through legitimate channels.

impression management (presentation of self) Erving Goffman's term for people's efforts to present themselves to others in ways that are most favorable to their own interests or image.

income the economic gain derived from wages, salaries, income transfers (governmental aid), and ownership of property.

individual discrimination behavior consisting of one-on-one acts by members of the dominant group that harm members of the subordinate group or their property.

industrial societies societies based on technology that mechanizes production.

industrialization the process by which societies are transformed from dependence on agriculture and handmade products to an emphasis on manufacturing and related industries.

infant mortality rate the number of deaths of infants under 1 year of age per 1,000 live births in a given year.

informal structure those aspects of participants' day-to-day activities and interactions that ignore, bypass, or do not correspond with the official rules and procedures of the bureaucracy.

ingroup a group to which a person belongs and with which the person feels a sense of identity.

institutional discrimination the day-to-day practices of organizations and institutions that have a harmful impact on members of subordinate groups.

instrumental leadership goal- or task-oriented leadership.

intergenerational mobility the social movement (upward or downward) experienced by family members from one generation to the next.

interlocking corporate directorates members of the board of directors of one corporation who also sit on the board(s) of other corporations.

internal colonialism according to conflict theorists, a practice that occurs when members of a racial or ethnic group are conquered or colonized and forcibly placed under the economic and political control of the dominant group.

interview a research method using a data collection encounter in which an interviewer asks the respondent questions and records the answers.

intragenerational mobility the social movement (upward or downward) of individuals within their own lifetime.

invasion the process by which a new category of people or type of land use arrives in an area previously occupied by another group or land use.

iron law of oligarchy according to Robert Michels, the tendency of bureaucracies to be ruled by a few people.

job deskilling a reduction in the proficiency needed to perform a specific job that leads to a corresponding reduction in the wages for that job.

juvenile delinquency a violation of law or the commission of a status offense by young people.

kinship a social network of people based on common ancestry, marriage, or adoption.

labeling theory the proposition that deviants are those people who have been successfully labeled as such by others.

laissez-faire leaders leaders who are only minimally involved in decision making and who encourage group members to make their own decisions.

language a set of symbols that expresses ideas and enables people to think and communicate with one another.

latent functions unintended functions that are hidden and remain unacknowledged by participants.

laws formal, standardized norms that have been enacted by legislatures and are enforced by formal sanctions.

life chances Max Weber's term for the extent to which individuals have access to important societal resources such as food, clothing, shelter, education, and health care.

life expectancy an estimate of the average lifetime of people born in a specific year.

looking-glass self Charles Horton Cooley's term for the way in which a person's sense of self is derived from the perceptions of others.

low-income countries (sometimes referred to as **underdeveloped countries**) nations with little industrialization and low levels of national and personal income.

macrolevel analysis an approach that examines whole societies, large-scale social structures, and social systems.

majority (dominant) group a group that is advantaged and has superior resources and rights in a society.

managed care any system of cost containment that closely monitors and controls health care providers' decisions about medical procedures, diagnostic tests, and other services that should be provided to patients.

manifest functions functions that are intended and/or overtly recognized by the participants in a social unit.

marginal jobs jobs that differ from the employment norms of the society in which they are located.

marriage a legally recognized and/or socially approved arrangement between two or more individuals that carries certain rights and obligations and usually involves sexual activity.

mass a number of people who share an interest in a specific idea or issue but who are not in one another's immediate vicinity.

mass behavior collective behavior that takes place when people (who often are geographically separated from one another) respond to the same event in much the same way.

master status the most important status that a person occupies.

material culture a component of culture that consists of the physical or tangible creations (such as clothing, shelter, and art) that members of a society make, use, and share.

matriarchal family a family structure in which authority is held by the eldest female (usually the mother).

matriarchy a hierarchical system of social organization in which cultural, political, and economic structures are controlled by women.

matrilineal descent a system of tracing descent through the mother's side of the family.

matrilocal residence the custom of a married couple living in the same household (or community) as the wife's parents.

mechanical solidarity Emile Durkheim's term for the social cohesion in preindustrial societies, in which there is minimal division of labor and people feel united by shared values and common social bonds.

medical–industrial complex local physicians, local hospitals, and global health-related industries such as insurance companies and pharmaceutical and medical supply companies that deliver health care today.

medicalization the process whereby nonmedical problems become defined and treated as illnesses or disorders.

medicine an institutionalized system for the scientific diagnosis, treatment, and prevention of illness.

meritocracy a hierarchy in which all positions are rewarded based on people's ability and credentials.

microlevel analysis sociological theory and research that focus on small groups rather than on large-scale social structures.

middle-income countries (sometimes referred to as **developing countries**) nations with industrializing economies and moderate levels of national and personal income.

migration the movement of people from one geographic area to another for the purpose of changing residency.

military–industrial complex the mutual interdependence of the military establishment and private military contractors.

minority (subordinate) group a group whose members, because of physical or cultural characteristics, are disadvantaged and subjected to unequal treatment by the dominant group and who regard themselves as objects of collective discrimination.

mixed economy an economic system that combines elements of a market economy (capitalism) with elements of a command economy (socialism).

mob a highly emotional crowd whose members engage in, or are ready to engage in, violence against a specific target—a person, a category of people, or physical property.

modernization theory a perspective that links global inequality to different levels of economic development and suggests that low-income economies can move to middle- and high-income economies by achieving self-sustained economic growth.

monarchy a political system in which power resides in one person or family and is passed from generation to generation through lines of inheritance.

monogamy a marriage between two partners, usually a woman and a man.

mores strongly held norms with moral and ethical connotations that may not be violated without serious consequences in a particular culture.

mortality the incidence of death in a population.

neolocal residence the custom of a married couple living in their own residence apart from both the husband's and the wife's parents.

nonmaterial culture a component of culture that consists of the abstract or intangible human creations of society (such as attitudes, beliefs, and values) that influence people's behavior.

nonverbal communication the transfer of information between persons without the use of speech.

norms established rules of behavior or standards of conduct.

nuclear family a family composed of one or two parents and their dependent children, all of whom live apart from other relatives.

occupational (white-collar) crime illegal activities committed by people in the course of their employment or financial affairs.

occupations categories of jobs that involve similar activities at different work sites.

oligopoly a condition existing when several companies overwhelmingly control an entire industry.

organic solidarity Emile Durkheim's term for the social cohesion found in industrial societies, in which people perform very specialized tasks and feel united by their mutual dependence.

organized crime a business operation that supplies illegal goods and services for profit.

outgroup a group to which a person does not belong and toward which the person may feel a sense of competitiveness or hostility.

panic a form of crowd behavior that occurs when a large number of people react to a real or perceived threat with strong emotions and self-destructive behavior.

participant observation a research method in which researchers collect data while being part of the activities of the group being studied.

patriarchal family a family structure in which authority is held by the eldest male (usually the father).

patriarchy a hierarchical system of social organization in which cultural, political, and economic structures are controlled by men.

patrilineal descent a system of tracing descent through the father's side of the family.

patrilocal residence the custom of a married couple living in the same household (or community) as the husband's family.

pay gap a term used to describe the disparity between women's and men's earnings.

peer group a group of people who are linked by common interests, equal social position, and (usually) similar age.

peripheral nations according to world systems theory, nations that are dependent on core nations for capital, have little or no industrialization (other than what may be brought in by core nations), and have uneven patterns of urbanization.

personal space the immediate area surrounding a person that the person claims as private.

pink-collar occupations relatively low-paying, nonmanual, semiskilled positions primarily held by women, such as day-care workers, checkout clerks, cashiers, and waitpersons.

pluralist model an analysis of political systems that views power as widely dispersed throughout many competing interest groups.

political action committees (PACs) organizations of special interest groups that solicit contributions from donors and fund campaigns to help elect (or defeat) candidates based on their stances on specific issues.

political crime illegal or unethical acts involving the usurpation of power by government officials, or illegal/unethical acts perpetrated against the government by outsiders seeking to make a political statement, undermine the government, or overthrow it.

political party an organization whose purpose is to gain and hold legitimate control of government.

political socialization the process by which people learn political attitudes, values, and behavior.

politics the social institution through which power is acquired and exercised by some people and groups.

polyandry the concurrent marriage of one woman with two or more men.

polygamy the concurrent marriage of a person of one sex with two or more members of the opposite sex.

polygyny the concurrent marriage of one man with two or more women.

popular culture the component of culture that consists of activities, products, and services that are assumed to appeal primarily to members of the middle and working classes.

population composition the biological and social characteristics of a population, including age, sex, race, marital status, education, occupation, income, and size of household.

population pyramid a graphic representation of the distribution of a population by sex and age.

positivism a term describing Auguste Comte's belief that the world can best be understood through scientific inquiry.

postindustrial societies societies in which technology supports a service- and information-based economy.

postmodern perspectives the sociological approach that attempts to explain social life in modern societies that are characterized by postindustrialization, consumerism, and global communications.

power according to Max Weber, the ability of people or groups to achieve their goals despite opposition from others.

prejudice a negative attitude based on faulty generalizations about members of selected racial, ethnic, and other groups.

prestige the respect or regard with which a person or status position is regarded by others.

primary deviance the initial act of rule-breaking.

primary group Charles Horton Cooley's term for a small, less specialized group in which members engage in face-to-face, emotion-based interactions over an extended period of time.

primary labor market the sector of the labor market that consists of high-paying jobs with good benefits that have some degree of security and the possibility of future advancement.

primary sector production the sector of the economy that extracts raw materials and natural resources from the environment.

primary sex characteristics the genitalia used in the reproductive process.

profane the everyday, secular, or "worldly" aspects of life.

professions high-status, knowledge-based occupations.

propaganda information provided by individuals or groups that have a vested interest in furthering their own cause or damaging an opposing one.

public opinion the attitudes and beliefs communicated by ordinary citizens to decision makers.

punishment any action designed to deprive a person of things of value (including liberty) because of some offense the person is thought to have committed.

qualitative research sociological research methods that use interpretive description (words) rather than statistics (numbers) to analyze underlying meanings and patterns of social relationships.

quantitative research sociological research methods that are based on the goal of scientific objectivity and that focus on data that can be measured numerically.

race a category of people who have been singled out as inferior or superior, often on the basis of physical characteristics such as skin color, hair texture, and eye shape.

racial socialization the aspect of socialization that contains specific messages and practices concerning the nature of one's racial or ethnic status.

racism a set of attitudes, beliefs, and practices that is used to justify the superior treatment of one racial or ethnic group and the inferior treatment of another racial or ethnic group.

rationality the process by which traditional methods of social organization, characterized by informality and spontaneity, are gradually replaced by efficiently administered formal rules and procedures.

reference group a group that strongly influences a person's behavior and social attitudes, regardless of whether that individual is an actual member.

relative poverty a condition that exists when people may be able to afford basic necessities but are still unable to maintain an average standard of living.

reliability in sociological research, the extent to which a study or research instrument yields consistent results when applied to different individuals at one time or to the same individuals over time.

religion a system of beliefs, symbols, and rituals, based on some sacred or supernatural realm, that guides human behavior, gives meaning to life, and unites believers into a community.

resocialization the process of learning a new and different set of attitudes, values, and behaviors from those in one's background and previous experience.

riot violent crowd behavior that is fueled by deep-seated emotions but is not directed at one specific target.

role a set of behavioral expectations associated with a given status.

role conflict a situation in which incompatible role demands are placed on a person by two or more statuses held at the same time.

role exit a situation in which people disengage from social roles that have been central to their self-identity.

role expectation a group's or society's definition of the way that a specific role *ought* to be played.

role performance how a person *actually* plays a role.

role strain a condition that occurs when incompatible demands are built into a single status that a person occupies.

role-taking the process by which a person mentally assumes the role of another person in order to understand the world from that person's point of view.

routinization of charisma the process by which charismatic authority is succeeded by a bureaucracy controlled by a rationally established authority or by a combination of traditional and bureaucratic authority.

rumor an unsubstantiated report on an issue or subject.

sacred those aspects of life that are extraordinary or supernatural.

sanctions rewards for appropriate behavior or penalties for inappropriate behavior.

Sapir–Whorf hypothesis the proposition that language shapes the view of reality of its speakers.

scapegoat a person or group that is incapable of offering resistance to the hostility or aggression of others.

second shift Arlie Hochschild's term for the domestic work that employed women perform at home after they complete their workday on the job.

secondary analysis a research method in which researchers use existing material and analyze data that were originally collected by others.

secondary deviance the process that occurs when a person who has been labeled a deviant accepts that new identity and continues the deviant behavior.

secondary group a larger, more specialized group in which members engage in more impersonal, goal-oriented relationships for a limited period of time.

secondary labor market the sector of the labor market that consists of low-paying jobs with few benefits and very little job security or possibility for future advancement.

secondary sector production the sector of the economy that processes raw materials (from the primary sector) into finished goods.

secondary sex characteristics the physical traits (other than reproductive organs) that identify an individual's sex.

sect a relatively small religious group that has broken away from another religious organization to renew what it views as the original version of the faith.

secularization the process by which religious beliefs, practices, and institutions lose their significance in sectors of society and culture.

segregation the spatial and social separation of categories of people by race, ethnicity, class, gender, and/or religion.

self-concept the totality of our beliefs and feelings about ourselves.

self-fulfilling prophecy the situation in which a false belief or prediction produces behavior that makes the originally false belief come true.

semiperipheral nations according to world systems theory, nations that are more developed than peripheral nations but less developed than core nations.

sex the biological and anatomical differences between females and males.

sex ratio a term used by demographers to denote the number of males for every hundred females in a given population.

sexism the subordination of one sex, usually female, based on the assumed superiority of the other sex.

sexual orientation a person's preference for emotional–sexual relationships with members of the opposite sex (heterosexuality), the same sex (homosexuality), or both (bisexuality).

shared monopoly a condition that exists when four or fewer companies supply 50 percent or more of a particular market.

sick role the set of patterned expectations that defines the norms and values appropriate for individuals who are sick and for those who interact with them.

significant others those persons whose care, affection, and approval are especially desired and who are most important in the development of the self.

slavery an extreme form of stratification in which some people are owned by others.

small group a collectivity small enough for all members to be acquainted with one another and to interact simultaneously.

social bond theory the proposition that the probability of deviant behavior increases when a person's ties to society are weakened or broken.

social change the alteration, modification, or transformation of public policy, culture, or social institutions over time.

social construction of reality the process by which our perception of reality is largely shaped by the subjective meaning that we give to an experience.

social control systematic practices developed by social groups to encourage conformity to norms, rules, and laws and to discourage deviance.

social Darwinism Herbert Spencer's belief that those species of animals, including human beings, best adapted to their environment survive and prosper, whereas those poorly adapted die out.

social devaluation a situation in which a person or group is considered to have less social value than other individuals or groups.

social epidemiology the study of the causes and distribution of health, disease, and impairment throughout a population.

social facts Emile Durkheim's term for patterned ways of acting, thinking, and feeling that exist *outside* any one individual but that exert social control over each person.

social group a group that consists of two or more people who interact frequently and share a common identity and a feeling of interdependence.

social institution a set of organized beliefs and rules that establishes how a society will attempt to meet its basic social needs.

social interaction the process by which people act toward or respond to other people; the foundation for all relationships and groups in society.

social mobility the movement of individuals or groups from one level in a stratification system to another.

social movement an organized group that acts consciously to promote or resist change through collective action.

social stratification the hierarchical arrangement of large social groups based on their control over basic resources.

social structure the complex framework of societal institutions and the social practices that make up a society and that organize and establish limits on people's behavior.

socialism an economic system characterized by public ownership of the means of production, the pursuit of collective goals, and centralized decision making.

socialization the lifelong process of social interaction through which individuals acquire a self-identity and the physical, mental, and social skills needed for survival in society.

socialized medicine a health care system in which the government owns the medical care facilities and employs the physicians.

society a large social grouping that shares the same geographical territory and is subject to the same political authority and dominant cultural expectations.

sociobiology the systematic study of how biology affects social behavior.

socioeconomic status (SES) a combined measure that attempts to classify individuals, families, or households in terms of factors such as income, occupation, and education to determine class location.

sociological imagination C. Wright Mills's term for the ability to see the relationship between individual experiences and the larger society.

sociology the systematic study of human society and social interaction.

sociology of family the subdiscipline of sociology that attempts to describe and explain patterns of family life and variations in family structure.

split labor market a term used to describe the division of the economy into two areas of employment, a primary sector or upper tier, composed of higher-paid (usually dominant-group) workers in more-secure jobs, and a secondary sector or lower tier, composed of lower-paid (often subordinate-group) workers in jobs with little security and hazardous working conditions.

state the political entity that possesses a legitimate monopoly over the use of force within its territory to achieve its goals.

status a socially defined position in a group or society characterized by certain expectations, rights, and duties.

status symbol a material sign that informs others of a person's specific status.

stereotypes overgeneralizations about the appearance, behavior, or other characteristics of members of particular groups.

strain theory the proposition that people feel strain when they are exposed to cultural goals that they are unable to obtain because they do not have access to culturally approved means of achieving those goals.

subcontracting an agreement in which a corporation contracts with other (usually smaller) firms to provide specialized components, products, or services to the larger corporation.

subculture a group of people who share a distinctive set of cultural beliefs and behaviors that differs in some significant way from that of the larger society.

succession the process by which a new category of people or type of land use gradually predominates in an area formerly dominated by another group or activity.

superego Sigmund Freud's term for the conscience, consisting of the moral and ethical aspects of personality.

survey a poll in which the researcher gathers facts or attempts to determine the relationships among facts.

symbol anything that meaningfully represents something else.

symbolic interactionist perspectives the sociological approach that views society as the sum of the interactions of individuals and groups.

taboos mores so strong that their violation is considered to be extremely offensive and even unmentionable.

technology the knowledge, techniques, and tools that allow people to transform resources into a usable form and the knowledge and skills required to use what is developed.

tertiary deviance deviance that occurs when a person who has been labeled a deviant seeks to normalize the behavior by relabeling it as nondeviant.

tertiary sector production the sector of the economy that is involved in the provision of services rather than goods.

theory a set of logically interrelated statements that attempts to describe, explain, and (occasionally) predict social events.

total institution Erving Goffman's term for a place where people are isolated from the rest of society for a set period of time and come under the control of the officials who run the institution.

totalitarianism a political system in which the state seeks to regulate all aspects of people's public and private lives.

tracking the assignment of students to specific courses and educational programs based on their test scores, previous grades, or both.

transnational corporations large corporations that are headquartered in one country but sell and produce goods and services in many countries.

transsexual a person who believes that he or she was born with the body of the wrong sex.

transvestite a male who lives as a woman or a female who lives as a man but does not alter the genitalia.

triad a group composed of three members.

unemployment rate the percentage of unemployed persons in the labor force actively seeking jobs.

universal health care system a health care system in which all citizens receive medical services paid for by tax revenues.

urbanization the process by which an increasing proportion of a population lives in cities rather than in rural areas.

validity in sociological research, the extent to which a study or research instrument accurately measures what it is supposed to measure.

values collective ideas about what is right or wrong, good or bad, and desirable or undesirable in a particular culture.

variable in sociological research, any concept with measurable traits or characteristics that can change or vary from one person, time, situation, or society to another.

wealth the value of all of a person's or family's economic assets, including income, personal property, and income-producing property.

working class those who must sell their labor to the owners in order to earn enough money to survive.

zero population growth the point at which no population increase occurs from year to year.

References

AAUW (American Association of University Women). 1995. *How Schools Shortchange Girls/The AAUW Report: A Study of Major Findings on Girls and Education.* New York: Marlowe.

Aberle, D. F., A. K. Cohen, A. K. Davis, M. J. Leng, Jr., and F. N. Sutton. 1950. "The Functional Prerequisites of Society." *Ethics,* 60 (January): 100–111.

Aberle, David F. 1966. *The Peyote Religion Among the Navaho.* Chicago: Aldine.

Achenbaum, W. Andrew. 1978. *Old Age in the New Land: The American Experience Since 1870.* Baltimore: Johns Hopkins University Press.

Adams, Tom. 1991. *Grass Roots: How Ordinary People Are Changing America.* New York: Citadel.

Adler, Jerry. 1999. "The Truth About High School." *Newsweek* (May 10): 56–58.

Adler, Patricia A., and Peter Adler. 1997. *Constructions of Deviance: Social Power, Context, and Interaction* (2nd ed.). Belmont, CA: Wadsworth.

——. 1998. *Peer Power: Preadolescent Culture and Identity.* New Brunswick, NJ: Rutgers University Press.

Adorno, Theodor W., Else Frenkel-Brunswick, Daniel J. Levinson, and R. Nevitt Sanford. 1950. *The Authoritarian Personality.* New York: Harper & Row.

Agger, Ben. 1993. *Gender, Culture, and Power: Toward a Feminist Postmodern Critical Theory.* Westport, CT: Praeger.

Agnew, Robert. 1985. "Social Control Theory and Delinquency: A Longitudinal Test." *Criminology,* 23: 47–61.

Agonafir, Rebecca. 2002. "Workplace Surveillance of Internet and E-Mail Usage." Retrieved July 6, 2002. Online: http://firstclass.wellesley.edu/~ragonafi/cs100.rp1.html

Aguirre, Adalberto, and David V. Baker. 1994. "Racial Prejudice and the Death Penalty: A Research Note." *Social Justice,* 20 (1/2): 150–155.

Aiello, John R., and S. E. Jones. 1971. "Field Study of Proxemic Behavior of Young School Children in Three Subcultural Groups." *Journal of Personality and Social Psychology,* 19: 351–356.

Akers, Ronald. 1992. *Drugs, Alcohol, and Society: Social Structure, Process, and Policy.* Belmont, CA: Wadsworth.

Akers, Ronald L. 1990. "Rational Choice, Deterrence and Social Learning Theory in Criminology: The Path Not Taken." *Journal of Criminal Law and Criminology,* 81: 653–676.

Albanese, Catherine L. 1992. *America, Religions and Religion.* Belmont, CA: Wadsworth.

Albas, Daniel, and Cheryl Albas. 1988. "Aces and Bombers: The Post-Exam Impression Management Strategies of Students." *Symbolic Interaction,* 11 (Fall): 289–302.

Albrecht, Gary L., 1992. *The Disability Business: Rehabilitation in America.* Newbury Park, CA: Sage.

Alexander, Karl L., Doris Entwisle, and Maxine Thompson. 1987. "School Performance, Status Relations, and the Structure of Sentiment: Bringing the Teacher Back In." *American Sociological Review,* 52: 665–682.

Alexander, Peter, and Roger Gill (Eds.). 1984. *Utopias.* London: Duckworth.

Allen, Michael Patrick. 1987. *The Founding Fortunes: A New Anatomy of the Super-Rich Families in America.* New York: Dutton.

Allen, Paula Gunn. 1997. "Angry Women Are Building: Issues and Struggles Facing American Indian Women Today." In Estelle Disch (Ed.), *Reconstructing Gender: A Multicultural Anthology.* Mountain View, CA: Mayfield, pp. 41–45.

Allport, Gordon. 1958. *The Nature of Prejudice* (abridged ed.). New York: Doubleday/Anchor.

Altemeyer, Bob. 1981. *Right-Wing Authoritarianism.* Winnipeg: University of Manitoba Press.

——. 1988. *Enemies of Freedom: Understanding Right-Wing Authoritarianism.* San Francisco: Jossey-Bass.

Alwin, Duane, Philip Converse, and Steven Martin. 1985. "Living Arrangements and Social Integration." *Journal of Marriage and the Family,* 47: 319–334.

American Academy of Child and Adolescent Psychiatry. 1997. "Children and Watching TV." Retrieved June 22, 1999. Online: http://www.aacap.org/factsfam/tv.htm

American Anthropological Association. 2001. "What Is Anthropology?" Retrieved July 21, 2001. Online: http://www.aaanet.org/anthbroc.htm

American Management Association. 2001. Workplace Monitoring and Surveillance Survey, 2001. Retrieved July 8, 2002. Online: http://www.amanet.org/research/pdfs/emsfu_short.pdf

American Sociological Association. 1997. *Code of Ethics.* Washington, DC: American Sociological Association (orig. pub. 1971).

Aminzade, Ronald. 1973. "Revolution and Collective Political Violence: The Case of the Working Class of Marseille, France, 1830–1871." Working Paper #86, Center for Research on Social Organization. Ann Arbor: University of Michigan, October 1973.

Amiri, Rina. 2001. "Muslim Women as Symbols-and Pawns." *New York Times* (Nov. 27): A21.

Amott, Teresa, and Julie Matthaei. 1996. *Race, Gender, and Work: A Multicultural Economic History of Women in the United States* (rev. ed.). Boston: South End.

Anders, George. 1996. *Health Against Wealth: HMOs and the Breakdown of Medical Trust.* Boston: Houghton Mifflin.

Andersen, Margaret L., and Patricia Hill Collins (Eds.). 1998. *Race, Class, and Gender: An Anthology* (3rd ed.). Belmont, CA: Wadsworth.

Anderson, David C. 1993. "Ellen Baxter." *New York Times Magazine* (Dec. 19): 36–39.

Anderson, Elijah. 1990. *Streetwise: Race, Class, and Change in an Urban Community.* Chicago: University of Chicago Press.

——. 1999. *Code of the Street: Decency, Violence, and the Moral Life of the Inner City.* New York: Norton.

Angel, Ronald. 1984. "The Costs of Disability for Hispanic Males." *Social Science Quarterly,* 65: 426–443.

Angier, Natalie. 1993. "'Stopit!' She Said. 'Nomore!'" *New York Times Book Review* (Apr. 25): 12.

Anyon, Jean. 1980. "Social Class and the Hidden Curriculum of Work." *Journal of Education,* 162: 67–92.

APA Online 2000. "Psychiatric Effects of Violence." Public Information: APA Fact Sheet Series. Washington, DC: American Psychological Association. Retrieved Apr. 5, 2000. Online: http://www.psych.org/psych/htdocs/public_info/media_violence.html

Apple, Michael W. 1980. "Analyzing Determinations: Understanding and Evaluating the Production of Social Outcomes in Schools." *Curriculum Inquiry,* 10: 55–76.

Appleton, Lynn M. 1995. "The Gender Regimes in American Cities." In Judith A. Garber and Robyne S. Turner (Eds.), *Gender in Urban Research.* Thousand Oaks, CA: Sage, pp. 44–59.

Argyris, Chris. 1960. *Understanding Organizational Behavior.* Homewood, IL: Dorsey.

——. 1962. *Interpersonal Competence and Organizational Effectiveness.* Homewood, IL: Dorsey.

Arnold, Regina A. 1990. "Processes of Victimization and Criminalization of Black Women." *Social Justice,* 17 (3): 153–166.

Asch, Adrienne. 1986. "Will Populism Empower Disabled People?" In Harry G. Boyle and Frank Reissman (Eds.), *The New Populism: The Power of Empowerment.* Philadelphia: Temple University Press, pp. 213–228.

Asch, Solomon E. 1955. "Opinions and Social Pressure." *Scientific American,* 193 (5): 31–35.

———. 1956. "Studies of Independence and Conformity: A Minority of One Against a Unanimous Majority." *Psychological Monographs*, 70 (9) (Whole No. 416).

Ash, Russell. 1997. *The Top 10 of Everything, 1998*. New York: DK Publishing.

Ashe, Arthur, and Arnold Rampersad. 1994. *Days of Grace: A Memoir*. New York: Ballantine.

Ashe, Arthur R., Jr. 1988. *A Hard Road to Glory: A History of the African-American Athlete*. New York: Warner.

Associated Press. 1997. "Survey Shows Makeup of Gangs." *Austin American-Statesman* (July 6): B2.

———. 1999. "Consumer Groups Accuse Credit Card Industry of Hooking College Students." Retrieved June 1, 1999. Online: http://abcnews.go.com/politics/AP19990609_16.html

Atchley, Robert C. 2000. *Social Forces and Aging: An Introduction to Social Gerontology* (9th ed.). Belmont, CA: Wadsworth.

Aulette, Judy Root. 1994. *Changing Families*. Belmont, CA: Wadsworth.

Aulette, Ken. 1998. *The Highwaymen: Warriors of the Information Superhighway*. San Diego: Harvest/Harcourt.

Austin American-Statesman. 2001. "High-Tech Manufacturing in Mexico." (Sept. 24): D1.

Axtell, Roger E. 1991. *Gestures: The Do's and Taboos of Body Language Around the World*. New York: Wiley.

Ayala, Elaine. 1993. "Unfinished Work: Cesar Chavez Is Dead But His Struggle Has Been Reborn, Followers Say." *Austin American-Statesman* (Sept. 6): E1, E4. Quoted from a speech by Cesar Chavez to the Commonwealth Club of San Francisco, Nov. 9, 1984.

Babbie, Earl. 2001. *The Practice of Social Research* (9th ed.). Belmont, CA: Wadsworth.

Baber, Kristine M., and Katherine R. Allen. 1992. *Women and Families: Feminist Reconstructions*. New York: Guilford.

Bachman, Ronet. 1992. *Death and Violence on the Reservation: Homicide, Family Violence, and Suicide in American Indian Populations*. New York: Auburn.

Bachu, Amara, and Martin O'Connell. 2001. *Fertility of American Women: June 2000*. Current Population Reports, P20–543RV. Washington, DC: U.S. Census Bureau.

Bagby, Meredith (Ed.). 1997. *Annual Report of the United States of America 1997*. New York: McGraw-Hill.

Bagby, Meredith. 1998. *Annual Report of the United States of America: 1998 Edition*. New York: McGraw-Hill.

Baker, Robert. 1993. "'Pricks' and 'Chicks': A Plea for 'Persons.'" In Anne Minas (Ed.), *Gender Basics: Feminist Perspectives on Women and Men*. Belmont, CA: Wadsworth, pp. 66–68.

Bales, Kevin. 1999. *Disposable People: New Slavery in the Global Economy*. Berkeley: University of California Press.

Ballantine, Jeanne H. 2000. *The Sociology of Education: A Systematic Analysis* (5th ed.). Englewood Cliffs, NJ: Prentice Hall.

Band-Aides & Blackboards. 1999. "Forever Heather." Retrieved Oct. 2, 1999. Online: http://funrsc.fairfield.edu/~jfleitas//heather.html

Bane, Mary Jo. 1986. "Household Composition and Poverty: Which Comes First?" In Sheldon H. Danziger and Daniel H. Weinberg (Eds.), *Fighting Poverty: What Works and What Doesn't*. Cambridge, MA: Harvard University Press.

Bane, Mary Jo, and David T. Ellwood. 1994. *Welfare Realities: From Rhetoric to Reform*. Cambridge, MA: Harvard University Press.

Barakat, Matthew. 2000. "Survey: Women's Salaries Beat Men's in Some Fields." *Austin American-Statesman* (July 4): D1, D3.

Bardwell, Jill R., Samuel W. Cochran, and Sharon Walker. 1986. "Relationship of Parental Education, Race, and Gender to Sex Role Stereotyping in Five-Year-Old Kindergarteners." *Sex Roles*, 15: 275–281.

Barlett, Donald L., and James B. Steele. 1994. *America: Who Really Pays the Taxes?* New York: Touchstone.

Barlow, Hugh D. 1987. *Introduction to Criminology* (4th ed.). Boston: Little, Brown.

Barnard, Chester. 1938. *The Functions of the Executive*. Cambridge, MA: Harvard University Press.

Barnett, Harold. 1979. "Wealth, Crime and Capital Accumulation." *Contemporary Crises*, 3: 171–186.

Baron, Dennis. 1986. *Grammar and Gender*. New Haven, CT: Yale University Press.

Barovick, Harriet. 2001. "Hope in the Heartland." *Time* (special edition, July): G1–G3.

Barrymore, Drew, with Todd Gold. 1994. *Little Girl Lost*. New York: Pocket. In Jay David (Ed.), *The Family Secret: An Anthology*. New York: Morrow, pp. 171–199.

Basow, Susan A. 1992. *Gender Stereotypes and Roles* (3rd ed.). Pacific Grove, CA: Brooks/Cole.

Baudrillard, Jean. 1983. *Simulations*. New York: Semiotext.

———. 1998. *The Consumer Society: Myths and Structures*. London: Sage (orig. pub. 1970).

Baxter, J. 1970. "Interpersonal Spacing in Natural Settings." *Sociology*, 36 (3): 444–456.

BBC News. 1999. "World: Asia-Pacific-Japan on Suicide Alert." *BBC Online Network* (July 2). Retrieved Aug. 25, 2001. Online: http://news.bbc.co.uk/hi/english/world/asia-pacific/newsid_383000/383823.stm

Becker, Howard S. 1963. *Outsiders: Studies in the Sociology of Deviance*. New York: Free Press.

Beech, Hannah. 2001. "China's Lifestyle Choice." *Time* (Aug. 6): 32.

Beeghley, Leonard. 1996. *The Structure of Social Stratification in the United States* (2nd ed.). Boston: Allyn & Bacon.

Belkin, Lisa. 1994. "Kill for Life?" *New York Times Magazine* (Oct. 30): 47–51, 62–64, 76, 80.

Bell, Inge Powell. 1989. "The Double Standard: Age." In Jo Freeman (Ed.), *Women: A Feminist Perspective* (4th ed.). Mountain View, CA: Mayfield, pp. 236–244.

Bellah, Robert N. 1967. "Civil Religion." *Daedalus*, 96: 1–21.

Belsky, Janet. 1990. *The Psychology of Aging: Theory, Research, and Interventions* (2nd ed.). Pacific Grove, CA: Brooks/Cole.

Bendix, Reinhard. 1971. "Charismatic Leadership." In Reinhard Bendix and Guenther Roth (Eds.), *Scholarship and Partisanship: Essays on Max Weber*. Berkeley: University of California Press, pp. 170–187.

Benford, Robert D. 1993. "'You Could Be the Hundredth Monkey': Collective Action Frames and Vocabularies of Motive Within the Nuclear Disarmament Movement." *Sociological Quarterly*, 34: 195–216.

Benjamin, Lois. 1991. *The Black Elite: Facing the Color Line in the Twilight of the Twentieth Century*. Chicago: Nelson-Hall.

Bennahum, David S. 1999. "For Kosovars, an On-Line Phone Directory of a People in Exile." *New York Times* (July 15): D7.

Benokraitis, Nijole V. 1997. *Marriages and Families: Changes, Choices, and Constraints* (2nd ed.). Englewood Cliffs, NJ: Prentice Hall.

———. 1999. *Marriages and Families: Changes, Choices, and Constraints* (3rd ed.). Upper Saddle River, NJ: Prentice Hall.

Benokraitis, Nijole V., and Joe R. Feagin. 1995. *Modern Sexism: Blatant, Subtle, and Covert Discrimination* (2nd ed.). Englewood Cliffs, NJ: Prentice Hall.

Benson, Susan Porter. 1983. "The Customers Ain't God: The Work Culture of Department Store Saleswomen, 1890–1940." In Michael H. Frisch and Daniel J. Walkowitz (Eds.), *Working Class America:*

Essays on Labor, Community, and American Society. Urbana: University of Illinois Press, pp. 185–211.

Berenson, Alex. 2002. "From Dream Team at Tyco to a Refrain of Dennis Who?" *New York Times* (June 6): C1–C5.

Berger, Bennett M. 1988. "Utopia and Its Environment." *Society* (January/February): 37–41.

Berger, Peter. 1963. *Invitation to Sociology: A Humanistic Perspective*. New York: Anchor.

———. 1967. *The Sacred Canopy: Elements of a Sociological Theory of Religion*. New York: Doubleday.

Berger, Peter, and Hansfried Kellner. 1964. "Marriage and the Construction of Reality." *Diogenes*, 46: 1–32.

Berger, Peter, and Thomas Luckmann. 1967. *The Social Construction of Reality: A Treatise in the Sociology of Knowledge*. Garden City, NY: Anchor.

Bergmann, Barbara R. 1986. *The Economic Emergence of Women*. New York: Basic.

Berliner, David C., and Bruce J. Biddle. 1995. *The Manufactured Crisis: Myths, Fraud, and the Attack on America's Public Schools*. Reading, MA: Addison-Wesley.

Bernard, Jessie. 1982. *The Future of Marriage*. New Haven, CT: Yale University Press (orig. pub. 1973).

Biagi, Shirley. 1994. *Media/Impact: An Introduction to Mass Media* (2nd ed.). Belmont, CA: Wadsworth.

———. 1998. *Media/Impact: An Introduction to Mass Media* (3rd ed.). Belmont, CA: Wadsworth.

Biblarz, Arturo, R. Michael Brown, Dolores Noonan Biblarz, Mary Pilgram, and Brent F. Baldree. 1991. "Media Influence on Attitudes Toward Suicide." *Suicide and Life-Threatening Behavior*, 21 (4): 374–385.

Biskupic, Joan. 1999. "High Court to Review Reporter Ride-Alongs." *Washington Post* (Mar. 21): A2. Retrieved July 29, 2000. Online: http://www.washingtonpost.com

Blanchard, Kendall. 1980. "Sport and Ritual in Choctaw Society: Structure and Perspective." In Helen Schwartzman (Ed.), *Play and Culture*. Champaign, IL: Leisure, pp. 83–91.

Blau, Judith R. 1984. *Architects and Firms: A Sociological Perspective on Architectural Practice*. Cambridge, MA: MIT Press.

Blau, Peter M., and Marshall W. Meyer. 1987. *Bureaucracy in Modern Society* (3rd ed.). New York: Random House.

Blauner, Robert. 1964. *Alienation and Freedom*. Chicago: University of Chicago Press.

———. 1972. *Racial Oppression in America*. New York: Harper & Row.

Bluestone, Barry, and Bennett Harrison. 1982. *The Deindustrialization of America*. New York: Basic.

Blumberg, Leonard. 1977. "The Ideology of a Therapeutic Social Movement: Alcoholics Anonymous." *Journal of Studies on Alcohol*, 38: 2122–2143.

Blumer, Herbert G. 1946. "Collective Behavior." In Alfred McClung Lee (Ed.), *A New Outline of the Principles of Sociology*. New York: Barnes & Noble, pp. 167–219.

———. 1969. *Symbolic Interactionism: Perspective and Method*. Englewood Cliffs, NJ: Prentice Hall.

———. 1974. "Social Movements." In R. Serge Denisoff (Ed.), *The Sociology of Dissent*. New York: Harcourt, pp. 74–90.

Bohls, Kirk. 1999. "Wheels of Change Turn Too Slowly for Black Coaches." *Austin American-Statesman* (Aug. 27): C1, C9.

Bologh, Roslyn Wallach. 1992. "The Promise and Failure of Ethnomethodology from a Feminist Perspective: Comment on Rogers." *Gender & Society*, 6 (2): 199–206.

Bonacich, Edna. 1972. "A Theory of Ethnic Antagonism: The Split Labor Market." *American Sociological Review*, 37: 547–549.

———. 1976. "Advanced Capitalism and Black–White Relations in the United States: A Split Labor Market Interpretation." *American Sociological Review*, 41: 34–51.

Bordewich, Fergus M. 1996. *Killing the White Man's Indian*. New York: Anchor/Doubleday.

Bordo, Susan. 1993. *Unbearable Weight: Feminism, Western Culture, and the Body*. Berkeley: University of California Press.

Bourdieu, Pierre. 1984. *Distinction: A Social Critique of the Judgement of Taste*. Trans. Richard Nice. Cambridge, MA: Harvard University Press.

Bourdieu, Pierre, and Jean-Claude Passeron. 1990. *Reproduction in Education, Society and Culture*. Newbury Park, CA: Sage.

Bowers, William J. 1984. *Legal Homicide: Death as Punishment in America, 1864–1982*. Boston: Northeastern University Press.

Bowles, Samuel. 1977. "Unequal Education and the Reproduction of the Social Division of Labor." In Jerome Karabel and A. H. Halsey (Eds.), *Power and Ideology in Education*. New York: Oxford University Press, pp. 137–153.

Bowles, Samuel, and Herbert Gintis. 1976. *Schooling in Capitalist America: Education and the Contradictions of Economic Life*. New York: Basic.

Boyes, William, and Michael Melvin. 1994. *Economics* (2nd ed.). Boston: Houghton Mifflin.

Bozett, Frederick. 1988. "Gay Fatherhood." In Phyllis Bronstein and Carolyn Pape Cowan (Eds.), *Fatherhood Today: Men's Changing Role in the Family*. New York: Wiley, pp. 60–71.

Bramlett, Matthew D., and William D. Mosher. 2001. "First Marriage Dissolution, Divorce, and Remarriage: United States." DHHS publication no. 2001–1250 01–0384 (5/01). Hyattsville, MD: Department of Health and Human Services.

Braun, Denny. 1991. *The Rich Get Richer: The Rise of Income Inequality in the United States and the World*. Chicago: Nelson-Hall.

Braverman, Harry. 1974. *Labor and Monopoly Capital*. New York: Monthly Review Press.

Breault, K. D. 1986. "Suicide in America: A Test of Durkheim's Theory of Religious and Family Integration, 1933–1980." *American Journal of Sociology*, 92 (3): 628–656.

Bremner, Brian. 2000. "A Japanese Way of Death." *Business Week* (Aug. 22). Retrieved Aug. 25, 2001. Online: http://www.businessweek.com/bwdaily/dnflash/aug2000/nf20000822_176.htm

Briggs, Sheila. 1987. "Women and Religion." In Beth B. Hess and Myra Marx Ferree (Eds.), *Analyzing Gender: A Handbook of Social Science Research*. Newbury Park, CA: Sage, pp. 408–441.

Brill's Content. 1998. "Tale of the Videotapes." (September): 89.

Brint, Steven. 1994. *In an Age of Experts: The Changing Role of Professionals in Politics and Public Life*. Princeton, NJ: Princeton University Press.

Brinton, Mary E. 1989. "Gender Stratification in Contemporary Urban Japan." *American Sociological Review*, 54 (August): 549–564.

Britt, Lory. 1993. "From Shame to Pride: Social Movements and Individual Affect." Paper presented at the 88th annual meeting of the American Sociological Association, Miami, August.

Brody, Jane E. 1997. "Despite the Despair of Depression, Few Men Seek Treatment." *New York Times* (Dec. 30): F7.

Brooke, James. 1993a. "Attack on Brazilian Indians Is Worst Since 1910." *New York Times* (Aug. 21): Y3.

———. 1993b. "Slavery on Rise in Brazil, As Debt Chains Workers." *New York Times* (May 23): 3.

Brooks-Gunn, Jeanne. 1986. "The Relationship of Maternal Beliefs About Sex Typing to Maternal and Young Children's Behavior." *Sex Roles*, 14: 21–35.

Brown, E. Richard. 1979. *Rockefeller Medicine Men*. Berkeley: University of California Press.

Brown, Robert W. 1954. "Mass Phenomena." In Gardner Lindzey (Ed.), *Handbook of Social Psychology* (vol. 2). Reading, MA: Addison-Wesley, pp. 833–873.

Brustad, Robert J. 1996. "Attraction to Physical Activity in Urban Schoolchildren: Parental Socialization and Gender Influence." *Research Quarterly for Exercise and Sport*, 67: 316–324.

Buckingham, Simon. 1999. "Shift Down to Gear Up." Retrieved June 16, 1999. Online: http://www.unorg.com/a46.htm

Buechler, Steven M. 2000. *Social Movements in Advanced Capitalism: The Political Economy and Cultural Construction of Social Activism*. New York: Oxford University Press.

Buia, Carole. 2001. "The Best of Both Worlds." *Time* (Special Issue: "Music Goes Global," Fall 2001): 10–13.

Bullard, Robert B., and Beverly H. Wright. 1992. "The Quest for Environmental Equity: Mobilizing the African-American Community for Social Change." In Riley E. Dunlap and Angela G. Mertig (Eds.), *American Environmentalism: The U.S. Environmental Movement, 1970–1990*. New York: Taylor & Francis, pp. 39–49.

Bumpass, Larry, James E. Sweet, and Andrew J. Cherlin. 1991. "The Role of Cohabitation in Declining Rates of Marriage." *Journal of Marriage and the Family*, 53: 913–927.

Burciaga, Jose Antonio. 1993. *Drink Cultura*. Santa Barbara, CA: Capra.

Burgess, Ernest W. 1925. "The Growth of the City." In Robert E. Park and Ernest W. Burgess (Eds.), *The City*. Chicago: University of Chicago Press, pp. 47–62.

Burkhauser, Richard V., Robert H. Haveman, and Barbara L. Wolfe. 1993. "How People with Disabilities Fare When Public Policies Change." *Journal of Policy Analysis and Management*, 12: 251–269.

Burnham, Walter Dean. 1983. *Democracy in the Making: American Government and Politics*. Englewood Cliffs, NJ: Prentice Hall.

Burns, John F. 1994. "India Fights Abortion of Female Fetuses." *New York Times* (Aug. 27): 5.

Burns, Tom. 1992. *Erving Goffman*. New York: Routledge.

Burros, Marian. 1994. "Despite Awareness of Risks, More in U.S. Are Getting Fat." *New York Times* (July 17): 1, 8.

———. 1996. "Eating Well: A Law to Encourage Sharing in a Land of Plenty." *New York Times* (Dec. 11): B6.

Burt, Martha R. 1992. *Over the Edge: The Growth of Homelessness in the 1980s*. New York: Russell Sage Foundation.

Busch, Ruth C. 1990. *Family Systems: Comparative Study of the Family*. New York: Lang.

Butler, John S. 1991. *Entrepreneurship and Self-Help Among Black Americans*. New York: SUNY Press.

Butler, Robert N. 1975. *Why Survive? Being Old in America*. New York: Harper & Row.

Butterfield, Fox. 1998. "Prison Population Growing Although Crime Rate Drops." *New York Times* (Aug. 9): 14.

———. 1999. "Indians Are Crime Victims at Rate Above U.S. Average." *New York Times* (Feb. 15): A12.

Buvinić, Mayra. 1997. "Women in Poverty: A New Global Underclass." *Foreign Policy* (Fall): 38–53.

Byrne, John A. 1993. "The Horizontal Corporation: It's About Managing Across, Not Up and Down." *Business Week* (Dec. 20): 76–81.

Cable, Sherry, and Charles Cable. 1995. *Environmental Problems, Grassroots Solutions: The Politics of Grassroots Environmental Conflict*. New York: St. Martin's.

Cahill, Spencer E. 1986. "Language Practices and Self Definition: The Case of Gender Identity Acquisition." *Sociological Quarterly*, 27 (September): 295–312.

Callaghan, Polly, and Heidi Hartmann. 1991. *Contingent Work*. Washington, DC: Economic Policy Institute.

Campbell, Anne. 1984. *The Girls in the Gang* (2nd ed.). Cambridge, MA: Basil Blackwell.

Cancian, Francesca M. 1990. "The Feminization of Love." In C. Carlson (Ed.), *Perspectives on the Family: History, Class, and Feminism*. Belmont, CA: Wadsworth, pp. 171–185.

———. 1992. "Feminist Science: Methodologies That Challenge Inequality." *Gender & Society*, 6 (4): 623–642.

Canetto, Silvia Sara. 1992. "She Died for Love and He for Glory: Gender Myths of Suicidal Behavior." *OMEGA*, 26 (1): 1–17.

Canter, R. J., and S. S. Ageton. 1984. "The Epidemiology of Adolescent Sex-Role Attitudes." *Sex Roles*, 11: 657–676.

Cantor, Muriel G. 1980. *Prime-Time Television: Content and Control*. Newbury Park, CA: Sage.

———. 1987. "Popular Culture and the Portrayal of Women: Content and Control." In Beth B. Hess and Myra Marx Ferree (Eds.), *Analyzing Gender: A Handbook of Social Science Research*. Newbury Park, CA: Sage, pp. 190–214.

Cantril, Hadley. 1941. *The Psychology of Social Movements*. New York: Wiley.

Capek, Stella M. 1993. "The 'Environmental Justice' Frame: A Conceptual Discussion and Application." *Social Problems*, 40 (1): 5–23.

Cargan, Leonard, and Matthew Melko. 1982. *Singles: Myths and Realities*. Newbury Park, CA: Sage.

Carmichael, Stokely, and Charles V. Hamilton. 1967. *Black Power: The Politics of Liberation in American Education*. New York: Random House.

Carrier, James G. 1986. *Social Class and the Construction of Inequality in American Education*. New York: Greenwood.

Carson, Rachel. 1962. *Silent Spring*. Boston: Houghton Mifflin.

Carter, David L. 1986. "Hispanic Police Officers' Perception of Discrimination." *Police Studies*, 9 (4): 204–210.

Carter, Stephen L. 1994. *The Culture of Disbelief: How American Law and Politics Trivializes Religious Devotion*. New York: Anchor/Doubleday.

Cashmore, E. Ellis. 1996. *Dictionary of Race and Ethnic Relations* (4th ed.). London: Routledge.

Castells, Manuel. 1977. *The Urban Question*. London: Edward Arnold (orig. pub. 1972 as *La Question Urbaine*, Paris).

———. 1998. *End of Millennium*. Malden, MA: Blackwell.

Cavender, Gray. 1995. "Alternative Theory: Labeling and Critical Perspectives." In Joseph F. Sheley (Ed.), *Criminology: A Contemporary Handbook* (2nd ed.). Belmont, CA: Wadsworth, pp. 349–371.

Celis, William, III. 1994. "Nations Envied for Schools Share Americans' Worries." *New York Times* (July 13): B4.

Chafetz, Janet Saltzman. 1984. *Sex and Advantage: A Comparative, Macro-Structural Theory of Sex Stratification*. Totowa, NJ: Rowman & Allanheld.

———. 1989. "Marital Intimacy and Conflict: The Irony of Spousal Equality." In Jo Freeman (Ed.), *Women: A Feminist Perspective* (4th ed.). Mountain View, CA: Mayfield, pp. 149–156.

Chagnon, Napoleon A. 1992. *Yanomamo: The Last Days of Eden*. New York: Harcourt (rev. from 4th ed., *Yanomamo: The Fierce People*,

published by Holt, Rinehart & Winston).

Chalfant, H. Paul, Robert E. Beckley, and C. Eddie Palmer. 1994. *Religion in Contemporary Society* (3rd ed.). Itasca, IL: Peacock.

Chambliss, William J. 1973. "The Saints and the Roughnecks." *Society*, 11: 24–31.

Chandler, Tertius, and Gerald Fox. 1974. *3000 Years of Urban History*. New York: Academic Press.

Chen, Hsiang-shui. 1992. *Chinatown No More: Taiwan Immigrants in Contemporary New York*. Ithaca, NY: Cornell University Press.

Cherlin, Andrew J. 1992. *Marriage, Divorce, Remarriage*. Cambridge, MA: Harvard University Press.

Chesney-Lind, Meda. 1986. "Women and Crime: The Female Offender." *Signs*, 12: 78–96.

———. 1989. "'Girls' Crime and Woman's Place: Toward a Feminist Model of Female Delinquency." *Crime and Delinquency*, 35 (1): 5–29.

———. 1997. *The Female Offender*. Thousand Oaks, CA: Sage.

Children's Defense Fund. 1999. *The State of America's Children: Yearbook 1999*. Boston: Beacon.

———. 2001. *The State of America's Children: Yearbook 2001*. Boston: Beacon.

Cho, Sumi K. 1993. "Korean Americans vs. African Americans: Conflict and Construction." In Robert Gooding-Williams (Ed.), *Reading Rodney King, Reading Urban Uprising*. New York: Routledge, pp. 196–211.

Chow, Esther Ngan-Ling. 1994. "Asian American Women at Work." In Maxine Baca Zinn and Bonnie Thornton Dill (Eds.), *Women of Color in U.S. Society*. Philadelphia: Temple University Press, pp. 203–227.

Christians, Clifford G. G., Kim B. Rotzoll, and Mark Fackler. 1987. *Media Ethics*. New York: Longman.

Chronicle of Higher Education. 2001. "Almanac Issue: 2001–2002 (Aug. 30): 7–36.

Churchill, Ward. 1994. *Indians Are Us? Culture and Genocide in Native North America*. Monroe, ME: Common Courage.

Clayman, Steven E. 1993. "Booing: The Anatomy of a Disaffiliative Response." *American Sociological Review*, 58 (1): 110–131.

Clinard, Marshall B., and Peter C. Yeager. 1980. *Corporate Crime*. New York: Free Press.

Clines, Francis X. 1993. "An Unfettered Milken Has Lessons to Teach." *New York Times* (Oct. 16): 1, 9.

———. 1996. "A Chef's Training Program That Feeds Hope as Well as

Hunger." *New York Times* (Dec. 11): B1, B6.

Cloward, Richard A., and Lloyd E. Ohlin. 1960. *Delinquency and Opportunity: A Theory of Delinquent Gangs*. New York: Free Press.

Clymer, Adam. 1993. "A Daughter of Slavery Makes the Senate Listen." *New York Times* (July 23): A10.

CNN. 1994. "Both Sides: School Prayer." (Nov. 26).

———. 2002. "Urban League: Blacks 'Not in the End Zone Yet.'" Retrieved July 24, 2002. Online: http://www.cnn.com/2002/ALLPOLITICS/07/21/urban.league.report/index.html

Coakley, Jay J. 1998. *Sport in Society: Issues and Controversies* (6th ed.). New York: McGraw-Hill.

Cockerham, William C. 1995. *Medical Sociology* (6th ed.). Englewood Cliffs, NJ: Prentice Hall.

———. 1998. *Medical Sociology* (7th ed.). Upper Saddle River, NJ: Prentice Hall.

Cohen, Adam. 1999. "A Curse of Cliques." *Time* (May 3): 44–45.

Cohen, Leah Hager. 1994. *Train Go Sorry: Inside a Deaf World*. Boston: Houghton Mifflin.

Colatosti, Camille. 1992. "Making 65 Cents on the Dollar." *Labor Notes*, Detroit.

Coleman, Richard P., and Lee Rainwater. 1978. *Social Standing in America: New Dimensions of Class*. New York: Basic.

Coles, Gerald. 1987. *The Learning Mystique: A Critical Look at "Learning Disabilities."* New York: Pantheon.

Coles, Robert. 1979. *Work Mobility and Participation: A Comparative Study of American and Japanese Industry*. Berkeley: University of California Press.

College Board. 2000. "SAT Program Information-2000 Profile of College-Bound Seniors." Retrieved Sept. 2, 2000. Online: http://www.collegeboard.org/sat/cbsenior/yr2000/nat/natbk400.html#income

Collins, Patricia Hill. 1990. *Black Feminist Thought: Knowledge, Consciousness, and the Politics of Empowerment*. London: HarperCollins Academic.

———. 1991. "The Meaning of Motherhood in Black Culture." In Robert Staples (Ed.), *The Black Family: Essays and Studies*. Belmont, CA: Wadsworth, pp. 169–178. Orig. pub. in *SAGE: A Scholarly Journal on Black Women*, 4 (Fall 1987): 3–10.

———. 1998. *Fighting Words: Black Women and the Search for Justice*. Minneapolis: University of Minnesota Press.

Collins, Randall. 1971. "A Conflict Theory of Sexual Stratification." *Social Problems*, 19 (1): 3–21.

———. 1979. *The Credential Society: An Historical Sociology of Education*. New York: Academic Press.

———. 1982. *Sociological Insight: An Introduction to Non-Obvious Sociology*. New York: Oxford University Press.

Collins, Sharon M. 1989. "The Marginalization of Black Executives." *Social Problems*, 36: 317–331.

Colt, George Howe. 1991. *The Enigma of Suicide*. New York: Summit.

Coltrane, Scott. 1989. "Household Labor and the Routine Production of Gender." *Social Problems*, 36: 473–490.

Comer, James P. 1988. *Maggie's American Dream: The Life and Times of a Black Family*. New York: New American Library/Penguin.

Comfort, Alex. 1976. "Age Prejudice in America." *Social Policy*, 7 (3): 3–8.

Condry, Sandra McConnell, John C. Condry, Jr., and Lee Wolfram Pogatshnik. 1983. "Sex Differences: A Study of the Ear of the Beholder." *Sex Roles*, 9: 697–704.

Conrad, Peter. 1996. "Medicalization and Social Control." In Phil Brown (Ed.), *Perspectives in Medical Sociology* (2nd ed.). Prospect Heights, IL: Waveland, pp. 137–162.

Conrad, Peter, and Joseph W. Schneider. 1992. *Deviance and Medicalization: From Badness to Sickness*. Philadelphia: Temple University Press.

Cook, P. 1990. "Robbery in the United States: An Analysis of Recent Trends and Patterns." In N. Weiner, M. Zahn, and R. Sagi (Eds.), *Violence: Patterns, Causes, Public Policy*. New York: Harcourt, pp. 85–97.

Cook, Sherburn F. 1973. "The Significance of Disease in the Extinction of the New England Indians." *Human Biology*, 45: 485–508.

Cookson, Peter W., Jr., and Caroline Hodges Persell. 1985. *Preparing for Power: America's Elite Boarding Schools*. New York: Basic.

Cooley, Charles Horton. 1963. *Social Organization: A Study of the Larger Mind*. New York: Schocken (orig. pub. 1909).

———. 1998. "The Social Self—the Meaning of 'I.'" In Hans-Joachim Schubert (Ed.), *On Self and Social Organization-Charles Horton Cooley*. Chicago: University of Chicago Press, pp. 155–175. Reprinted from Charles Horton Cooley, *Human Nature and the Social Order*. New York: Schocken, 1902.

Coontz, Stephanie. 1992. *The Way We Never Were: American Families and the Nostalgia Trap*. New York: Basic.

———. 1997. *The Way We Really Are: Coming to Terms with America's Changing Families*. New York: Basic.

Corsaro, William A. 1985. *Friendship and Peer Culture in the Early Years.* Norwood, NJ: Ablex.

———. 1992. "Interpretive Reproduction in Children's Peer Cultures." *Social Psychology Quarterly,* 55 (2): 160–177.

———. 1997. *Sociology of Childhood.* Thousand Oaks, CA: Pine Forge.

Cortese, Anthony J. 1999. *Provocateur: Images of Women and Minorities in Advertising.* Latham, MD: Rowman & Littlefield.

Cose, Ellis. 1993. *The Rage of a Privileged Class.* New York: HarperCollins.

Coser, Lewis A. 1956. *The Functions of Social Conflict.* Glencoe, IL: Free Press.

Coverman, Shelley. 1989. "Women's Work Is Never Done. The Division of Domestic Labor." In Jo Freeman (Ed.), *Women: A Feminist Perspective.* Mountain View, CA: Mayfield, pp. 356–368.

Cox, Oliver C. 1948. *Caste, Class, and Race.* Garden City, NY: Doubleday.

Craig, Steve. 1992. "Considering Men and the Media." In Steve Craig (Ed.), *Men, Masculinity, and the Media.* Newbury Park, CA: Sage, pp. 1–7.

Cressey, Donald. 1969. *Theft of the Nation.* New York: Harper & Row.

Crippen, Dan L. 2001. "Prescription Drugs and Medicare Financing." Retrieved July 21, 2002. Online: http://www.cbo.gov/showdoc.cfm?index=2760&sequence=0&from=7

Cronin, John, and Robert F. Kennedy, Jr. 1999. *The Riverkeepers: Two Activists Fight to Reclaim Our Environment as a Basic Human Right.* New York: Touchstone.

Crossette, Barbara. 1996. "Hope, and Pragmatism, for U.S. Cities Conference." *New York Times* (June 3): A3.

———. 1997. "The 21st Century Belongs to. . . ." *New York Times* (Oct. 19): WK3.

Crow Dog, Mary, and Richard Erdoes. 1991. *Lakota Woman.* New York: HarperPerennial.

Currie, Elliott. 1998. *Crime and Punishment in America.* New York: Metropolitan.

Curtiss, Susan. 1977. *Genie: A Psycholinguistic Study of a Modern Day "Wild Child."* New York: Academic Press.

Cyrus, Virginia. 1993. *Experiencing Race, Class, and Gender in the United States.* Mountain View, CA: Mayfield.

Dahl, Robert A. 1961. *Who Governs?* New Haven, CT: Yale University Press.

Dahrendorf, Ralf. 1959. *Class and Class Conflict in an Industrial Society.* Stanford, CA: Stanford University Press.

Daly, Kathleen, and Meda Chesney-Lind. 1988. "Feminism and Criminology." *Justice Quarterly,* 5: 497–533.

Daly, Mary. 1973. *Beyond God the Father.* Boston: Beacon.

———. 1978. *Gyn/ecology: The Meta-Ethics of Radical Feminism.* Boston: Beacon.

Daniels, Arlene Kaplan. 1988. *Invisible Careers: Women Civic Leaders from the Volunteer World.* Chicago: University of Chicago Press.

Daniels, Cora. 2002. "Most Powerful Black Executives." Retrieved July 13, 2002. Online: http://www.fortune.com/indext.jhtml?channel=print_article.jhtml&doc_id=208625

Daniels, Roger. 1993. *Prisoners Without Trial: Japanese-Americans in World War II.* New York: Hill & Wang.

Danziger, Sheldon, and Peter Gottschalk. 1995. *American Unequal.* Cambridge, MA: Harvard University Press.

Dao, James. 2000. "Lockheed Wins $200 Billion Deal for Fighter Jet." *New York Times* (Oct. 27): A1, A9.

Darley, John M., and Thomas R. Shultz. 1990. "Moral Rules: Their Content and Acquisition." *Annual Review of Psychology,* 41: 525–556.

Dart, Bob. 1999. "Kids Get More Screen Time than Book Time." *Austin American-Statesman* (June 28): A1, A5.

Davis, F. James. 1991. *Who Is Black?* University Park: Pennsylvania State University Press.

Davis, Fred. 1992. *Fashion, Culture, and Identity.* Chicago: University of Chicago Press.

Davis, Kingsley. 1940. "Extreme Social Isolation of a Child." *American Journal of Sociology,* 45 (4): 554–565.

———. 1949. *Human Society.* New York: Macmillan.

———. 1955. "Malthus and the Theory of Population." In Paul Lazarsfeld and M. Rosenberg (Eds.), *The Language of Social Research.* New York: Free Press, pp. 540–553.

Davis, Kingsley, and Wilbert Moore. 1945. "Some Principles of Stratification." *American Sociological Review,* 7 (April): 242–249.

Dean, L. M., F. N. Willis, and J. N. la Rocco. 1976. "Invasion of Personal Space as a Function of Age, Sex and Race." *Psychological Reports,* 38 (3) (pt. 1): 959–965.

Death Penalty Institute of Oklahoma. 2001. "Death Row Population per Capita by State as of April 1, 2001." Retrieved Oct. 31, 2001. Online: http://www.dpio.org/death_row/States_per_Capita.html

Deaux, Kay, and Mary E. Kite. 1987. "Thinking About Gender." In Beth B. Hess and Myra Marx Ferree (Eds.), *Analyzing Gender: A Handbook of Social Science Research.* Newbury Park, CA: Sage, pp. 92–117.

Deegan, Mary Jo. 1988. *Jane Addams and the Men of the Chicago School, 1892–1918.* New Brunswick, NJ: Transaction.

Degher, Douglas, and Gerald Hughes. 1991. "The Identity Change Process: A Field Study of Obesity." *Deviant Behavior,* 12: 385–402.

DeJong, Gerben, Andrew I. Batavia, and Robert Griss. 1989. "America's Neglected Health Minority: Working-Age Persons with Disabilities." *Milbank Quarterly,* 67 (suppl. 2): 311–351.

Delgado, Richard. 1995. "Introduction." In Richard Delgado (Ed.), *Critical Race Theory: The Cutting Edge.* Philadelphia: Temple University Press, pp. xiii–xvi.

Derber, Charles. 1983. *The Pursuit of Attention: Power and Individualism in Everyday Life.* New York: Oxford University Press.

Devine, John. 1996. *Maximum Security: The Culture of Violence in Inner-City Schools.* Chicago: University of Chicago Press.

De Witt, Karen. 1994. "Wave of Suburban Growth Is Being Fed by Minorities." *New York Times* (Aug. 15): A1, A12.

Dikotter, Frank. 1996. "Culture, 'Race' and Nation: The Formation of Identity in Twentieth Century China." *Journal of International Affairs* (Winter): 590–605.

Dill, Bonnie Thornton. 1988. "'Making Your Job Good Yourself': Domestic Service and the Construction of Personal Dignity." In Ann Bookman and Sandra Morgen (Eds.), *Women and the Politics of Empowerment.* Philadelphia: Temple University Press, pp. 33–52.

Dobrzynski, Judith H. 1996. "When Directors Play Musical Chairs." *New York Times* (Nov. 17): F1, F8, F9.

Dollard, John. 1957. *Caste and Class in a Southern Town.* Garden City, NY: Doubleday (orig. pub. 1937).

Dollard, John, Neal E. Miller, Leonard W. Doob, O. H. Mowrer, and Robert R. Sears. 1939. *Frustration and Aggression.* New Haven, CT: Yale University Press.

Domhoff, G. William. 1974. *The Bohemian Grove and Other Retreats.* New York: Harper and Row.

———. 1978. *The Powers That Be: Processes of Ruling Class Domination in America.* New York: Random House.

———. 1983. *Who Rules America Now? A View for the '80s.* Englewood Cliffs, NJ: Prentice Hall.

———. 1990. *The Power Elite and the State: How Policy Is Made in America.* New York: Aldine De Gruyter.

———. 1998. *Who Rules America? Power and Politics in the Year 2000.* Mountain View, CA: Mayfield.

Drake, Jeffrey. 2002. "Communication Keys for Success." Retrieved July 5, 2002. Online: http://www.achievemax.com/newsletter/00issue/communication-keys.htm

Du Bois, W. E. B. 1967. *The Philadelphia Negro: A Social Study.* New York: Schocken (orig. pub. 1899).

Dubowitz, Howard, Maureen Black, Raymond H. Starr, Jr., and Susan Zuravin. 1993. "A Conceptual Definition of Child Neglect." *Criminal Justice and Behavior,* 20 (1): 8–26.

Duff, Robert W., and Lawrence K. Hong. 1984. "Self-Images of Women Bodybuilders." *Sociology of Sport Journal,* 2: 374–380.

Duffy, John. 1976. *The Healers.* New York: McGraw-Hill.

Dugger, Celia W. 1997. "In Historical Surge to Be Americans, 5,000 Take Oath." *New York Times* (July 5): 1, 27.

Dunlap, Riley E. 1992. "Trends in Public Opinion Toward Environmental Issues: 1965–1990." In Riley E. Dunlap and Angela G. Mertig (Eds.), *American Environmentalism: The U.S. Environmental Movement, 1970–1990.* New York: Taylor & Francis, pp. 89–113.

Dupre, Roslyn, and Paul Gains. 1997. "Fundamental Differences." *Women's Sports & Fitness* (October): 63–68.

Durkheim, Emile. 1933. *The Division of Labor in Society.* Trans. George Simpson. New York: Free Press (orig. pub. 1893).

———. 1956. *Education and Sociology.* Trans. Sherwood D. Fox. Glencoe, IL: Free Press.

———. 1964a. *The Rules of Sociological Method.* Trans. Sarah A. Solovay and John H. Mueller. New York: Free Press (orig. pub. 1895).

———. 1964b. *Suicide.* Trans. John A. Sparkling and George Simpson. New York: Free Press (orig. pub. 1897).

———. 1995. *The Elementary Forms of Religious Life.* Trans. Karen E. Fields. New York: Free Press (orig. pub. 1912).

Durning, Alan. 1993. "Life on the Brink." In William Dan Perdue (Ed.), *Systemic Crisis: Problems in Society, Politics, and World Order.* Fort Worth: Harcourt, pp. 274–282.

Duster, Troy. 1995. "Symposium: The Bell Curve." *Contemporary Sociology: A Journal of Reviews,* 24 (2): 158–161.

Dworkin, Rosalind J. 1982. "A Woman's Report: Numbers Are Not Enough." In Anthony Dworkin and Rosalind Dworkin (Eds.), *The Minority Report* (2nd ed.). New York: Holt, Rinehart & Winston, pp. 375–400.

Dye, Thomas R., and Harmon Zeigler. 2000. *The Irony of Democracy: An Uncommon Introduction to American Politics* (millennial ed.). Belmont, CA: Wadsworth.

Early, Kevin E. 1992. *Religion and Suicide in the African-American Community*. Westport, CT: Greenwood.

Ebaugh, Helen Rose Fuchs. 1988. *Becoming an EX: The Process of Role Exit*. Chicago: University of Chicago Press.

Eccles, Jacquelynne S., Janis E. Jacobs, and Rena D. Harold. 1990. "Gender Role Stereotypes, Expectancy Effects, and Parents' Socialization of Gender Difference." *Journal of Social Issues*, 46: 183–201.

Eder, Donna. 1995. *School Talk: Gender and Adolescent Culture* (with Catherine Colleen Evans and Stephen Parker). New Brunswick, NJ: Rutgers University Press.

Ehrenreich, Barbara. 1989. *Fear of Falling: The Inner Life of the Middle Class*. New York: HarperPerennial.

———. 2001. *Nickel and Dimed: On (Not) Getting by in America*. New York: Metropolitan.

Ehrlich, Paul R., Anne H. Ehrlich, and Gretchen C. Daily. 1995. *The Stork and the Plow: The Equity Answer to the Human Dilemma*. New Haven, CT: Yale University Press.

Eighner, Lars. 1993. *Travels with Lizbeth*. New York: St. Martin's.

Eisenstein, Zillah R. 1994. *The Color of Gender: Reimaging Democracy*. Berkeley: University of California Press.

Eisler, Benita. 1983. *Class Act: America's Last Dirty Secret*. New York: Franklin Watts.

Eitzen, D. Stanley, and George H. Sage. 1997. *The Sociology of North American Sport* (6th ed.). Dubuque, IA: Brown.

Elkin, Frederick, and Gerald Handel. 1989. *The Child and Society: The Process of Socialization* (5th ed.). New York: Random House.

Elkind, David. 1995. "School and Family in the Postmodern World." *Phi Delta Kappan* (September): 8–21.

Elliott, Jennifer A. 1994. *An Introduction to Sustainable Development: The Developing World*. London and New York: Routledge.

Emling, Shelley. 1997a. "Haiti Held in Grip of Another Drought." *Austin American-Statesman* (Sept. 19): A17, A18.

———. 1997b. "In Haiti, It's Resort vs. Reality." *Austin American-Statesman* (Sept. 27): A17, A19.

Engels, Friedrich. 1970. *The Origins of the Family, Private Property, and the State*. New York: International (orig. pub. 1884).

Engerman, Stanley L. 1995. "The Extent of Slavery and Freedom Throughout the World as a Whole and in Major Subareas." In Julian L. Simon (Ed.), *The State of Humanity*. Cambridge, MA: Blackwell, pp. 171–177.

England, Paula, and Melissa S. Herbert. 1993. "The Pay of Men in 'Female' Occupations: Is Comparable Worth Only for Women?" In Christine L. Williams (Ed.), *Doing "Women's Work": Men in Nontraditional Occupations*. Newbury Park, CA: Sage, pp. 28–48.

Epidemiological Network for Latin America and the Caribbean. 2000. "HIV and AIDS in the Americas: An Epidemic with Many Faces." Retrieved Nov. 23, 2001. Online: http://www.census.gov/ipc/www/hivaidinamerica.pdf

Epstein, Cynthia Fuchs. 1988. *Deceptive Distinctions: Sex, Gender, and the Social Order*. New Haven, CT: Yale University Press.

Erikson, Kai T. 1962. "Notes on the Sociology of Deviance." *Social Problems*, 9: 307–314.

———. 1964. "Notes on the Sociology of Deviance." In Howard S. Becker (Ed.), *The Other Side: Perspectives on Deviance*. New York: Free Press, pp. 9–21.

———. 1976. *Everything in Its Path: Destruction of Community in the Buffalo Creek Flood*. New York: Simon & Schuster.

———. 1991. "A New Species of Trouble." In Stephen Robert Couch and J. Stephen Kroll-Smith (Eds.), *Communities at Risk: Collective Responses to Technological Hazards*. New York: Land, pp. 11–29.

———. 1994. *A New Species of Trouble: Explorations in Disaster, Trauma, and Community*. New York: Norton.

Esping-Andersen, Gosta. 1990. *The Three Worlds of Welfare Capitalism*. Cambridge, MA: Polity.

Espiritu, Yen Le. 1995. *Filipino American Lives*. Philadelphia: Temple University Press.

Essed, Philomena. 1991. *Understanding Everyday Racism*. Newbury Park, CA: Sage.

Esterberg, Kristin G. 1997. *Lesbian and Bisexual Identities: Constructing Communities, Constructing Self*. Philadelphia: Temple University Press.

Etkind, Marc. 1997. *. . . Or Not to Be: A Collection of Suicide Notes*. New York: Riverhead.

Etzioni, Amitai. 1975. *A Comparative Analysis of Complex Organizations: On Power, Involvement, and Their Correlates* (rev. ed.). New York: Free Press.

———. 1994. *The Spirit of Community: The Reinvention of American Society*. New York: Touchstone.

Evans, Glen, and Norman L. Farberow. 1988. *The Encyclopedia of Suicide*. New York: Facts on File.

Evans, Peter B., and John D. Stephens. 1988. "Development and the World Economy." In Neil J. Smelser (Ed.), *Handbook of Sociology*. Newbury Park, CA: Sage, pp. 739–773.

Ezell, William Bruce, Jr. 1993. "Letters to the Editor: The Symbolism of the Confederate Battle Flag." *Chronicle of Higher Education* (Nov. 3): B2.

Fabes, Richard A., and Carol L. Martin. 1991. "Gender and Age Stereotypes of Emotionality." *Personality and Social Psychology Bulletin*, 17: 532–540.

Fagot, Beverly I. 1984. "Teacher and Peer Reactions to Boys' and Girls' Play Styles." *Sex Roles*, 11: 691–702.

Falk, Patricia. 1989. "Lesbian Mothers: Psychological Assumptions in Family Law." *American Psychologist*, 44: 941–947.

Fallon, Patricia, Melanie A. Katzman, and Susan C. Wooley. 1994. *Feminist Perspectives on Eating Disorders*. New York: Guilford.

Faludi, Susan. 1999. *Stiffed: The Betrayal of the American Man*. New York: Morrow.

Farb, Peter. 1973. *Word Play: What Happens When People Talk*. New York: Knopf.

Farley, John E. 2000. *Majority–Minority Relations* (4th ed.). Englewood Cliffs, NJ: Prentice Hall.

"The Favored Infants." 1976. *Human Behavior* (June): 49–50.

Feagin, Joe R. 1991. "The Continuing Significance of Race: Antiblack Discrimination in Public Places." *American Sociological Review*, 56 (February): 101–116.

Feagin, Joe R., and Clairece Booher Feagin. 1994. *Social Problems: A Critical Power–Conflict Perspective* (4th ed.). Englewood Cliffs, NJ: Prentice Hall.

———. 1997. *Social Problems: A Critical Power–Conflict Perspective* (5th ed.). Englewood Cliffs, NJ: Prentice Hall.

———. 2003. *Racial and Ethnic Relations* (7th ed.). Upper Saddle River, NJ: Prentice Hall.

Feagin, Joe R., Anthony M. Orum, and Gideon Sjoberg (Eds.). 1991. *A Case for the Case Study*. Chapel Hill: University of North Carolina Press.

Feagin, Joe R., and Robert Parker. 1990. *Building American Cities: The Urban Real Estate Game* (2nd ed.). Englewood Cliffs, NJ: Prentice Hall.

Feagin, Joe R., and Melvin P. Sikes. 1994. *Living with Racism: The Black Middle-Class Experience*. Boston: Beacon.

Feagin, Joe R., and Hernán Vera. 1995. *White Racism: The Basics*. New York: Routledge.

Fedarko, Kevin. 1997. "How a Few Firemen Created a Safe Haven for Some Chicago Kids." *Time* (Nov. 17): 72–73.

Federal Bureau of Investigation (FBI). 2001. *Crime in the United States: 2000*. Washington, DC: U.S. Government Printing Office.

Federal Election Commission, 2001. "FEC Reports on Congressional Financial Activity for 2000." Retrieved Oct. 24, 2001. Online: http://www.fec.gov/press/051501congfinact/051501congfinact.html

Federal Trade Commission. 2000. "Marketing Violent Entertainment to Children: A Review of Self-Regulation and Industry Practices in the Motion Picture, Music Recording & Electronic Game Industries." Retrieved Sept. 9, 2001. Online: http://www.ftc.gov/opa/2000/09/youthviol.htm

Fenstermacher, Gary D. 1994. "The Absence of Democratic and Educational Ideals from Contemporary Educational Reform Initiatives." The Elam Lecture, presented to the Educational Press Association of America, Chicago, June 10.

Ferriss, Susan. 2001. "Cold Spell: U.S. Recession Chills Mexico's Economic Hot Spot." *Austin American-Statesman* (Nov. 25): E1, E4 (based on data from the Bank of Mexico).

Fields, Jason, and Lynne M. Casper. 2001. *America's Family and Living Arrangements: March 2000*. Current Population Reports, P20–537. Washington, DC: U.S. Census Bureau.

Fiffer, Steve, and Sharon Sloan Fiffer. 1994. *50 Ways to Help Your Community*. New York: Mainstream/Doubleday.

Findlay, Deborah A., and Leslie J. Miller. 1994. "Through Medical Eyes: The Medicalization of Women's Bodies and Women's Lives." In B. Singh Bolaria and Harley D. Dickinson (Eds.), *Health, Illness, and Health Care in Canada* (2nd ed.). Toronto: Harcourt, pp. 276–306.

Findlay, Steven. 1997. "Health Care Industry Feeling a Little Discomfort." *USA Today* (Aug. 6): 4B.

Fine, Michelle, and Lois Weis. 1998. *The Unknown City: The Lives of Poor and Working-Class Young People*. Boston: Beacon.

Finley, M. I. 1980. *Ancient Slavery and Modern Ideology.* New York: Viking.

Firestone, Shulamith. 1970. *The Dialectic of Sex.* New York: Morrow.

Fishbein, Diana H., and Susan E. Pease. 1996. *The Dynamics of Drug Abuse.* Boston: Allyn & Bacon.

Fisher-Thompson, Donna. 1990. "Adult Sex-Typing of Children's Toys." *Sex Roles,* 23: 291–303.

Fjellman, Stephen M. 1992. *Vinyl Leaves: Walt Disney World & America.* Boulder, CO: Westview.

Flanagan, William G. 1995. *Urban Sociology: Images and Structure* (2nd ed.). Needham Heights, MA: Allyn & Bacon.

——. 1999. *Urban Sociology: Images and Structure* (3rd ed.). Boston: Allyn & Bacon.

Fleitas, Joan. 1999. *Band-Aides & Blackboards.* Retrieved Oct. 2, 1999. Online: http://funrsc.fairfield.edu/~jfleitas.html

Flexner, Abraham. 1910. *Medical Education in the United States and Canada.* New York: Carnegie Foundation.

Florida, Richard, and Martin Kenney. 1991. "Transplanted Organizations: The Transfer of Japanese Industrial Organization to the U.S." *American Sociological Review,* 56 (3): 381–398.

Forbes. 1996. "The 100 Largest U.S. Multinationals." (July 15): 288–290.

——. 2000a. "The 100 Largest U.S. Multinationals." Retrieved Sept. 4, 2000. Online: http://www.forbes.com/tool/toolbox/int500

——. 2000b. "200 Global Billionaires." (July 3): 162–263.

——. 2001. "The Forbes 400." (Oct. 8): 127–298.

Ford, Clyde W. 1994. *We Can All Get Along: 50 Steps You Can Take to Help End Racism.* New York: Dell.

Fortune. 2001. "Most Powerful Women in Business." Retrieved July 20, 2002. Online: http://www.fortune.com/lists/women/index.html?_requestid=13939

Foucault, Michel. 1979. *Discipline and Punish: The Birth of the Prison.* New York: Vintage.

——. 1994. *The Birth of the Clinic: An Archeology of Medical Perception.* New York: Vintage (orig. pub. 1963).

Fox, Mary Frank. 1989. "Women and Higher Education: Gender Differences in the Status of Students and Scholars." In Jo Freeman (Ed.), *Women: A Feminist Perspective.* Mountain View, CA: Mayfield, pp. 217–235.

Fox News. 1999. "Divorce Rates for Children of Broken Families on Decline." Retrieved Sept. 18, 1999. Online: http://www.foxnews.com/js_index.sml?content=/news/national/0812/d_ap_0812_11.sml

Frankenberg, Ruth. 1993. *White Women, Race Matters: The Social Construction of Whiteness.* Minneapolis: University of Minnesota Press.

Franklin, John Hope. 1980. *From Slavery to Freedom: A History of Negro Americans.* New York: Vintage.

Freidson, Eliot. 1965. "Disability as Social Deviance." In Marvin B. Sussman (Ed.), *Sociology and Rehabilitation.* Washington, DC: American Sociology Association, pp. 71–99.

——. 1970. *Profession of Medicine.* New York: Dodd, Mead.

——. 1986. *Professional Powers.* Chicago: University of Chicago Press.

French, Sally. 1999. "The Wind Gets in My Way." In Mairian Corker and Sally French (Eds.), *Disability Discourse.* Buckingham, England: Open University Press, pp. 21–27.

Freud, Sigmund. 1924. *A General Introduction to Psychoanalysis* (2nd ed.). New York: Boni & Liveright.

Freudenberg, Nicholas, and Carl Steinsapir. 1992. "Not in Our Backyards: The Grassroots Environmental Movement." In Riley E. Dunlap and Angela G. Mertig (Eds.), *American Environmentalism: The U.S. Environmental Movement, 1970–1990.* New York: Taylor & Francis, pp. 27–37.

Friedan, Betty. 1993. *The Fountain of Age.* New York: Simon & Schuster.

Friedmann, John. 1995. "The World City Hypothesis." In Paul L. Knox and Peter J. Taylor (Eds.), *World Cities in a World-System.* Cambridge, England: Cambridge University Press, pp. 317–331.

Friedrichs, David O. 1996. *Trusted Criminals: White Collar Crime in Contemporary Society.* Belmont, CA: Wadsworth.

Frisbie, W. Parker, and John D. Kasarda. 1988. "Spatial Processes." In Neil Smelser (Ed.), *The Handbook of Sociology.* Newbury Park, CA: Sage, pp. 629–666.

Furstenberg, Frank F., Jr., and Andrew J. Cherlin. 1991. *Divided Families: What Happens to Children When Parents Part.* Cambridge, MA: Harvard University Press.

Gabriel, Trip. 1995. "A Generation's Heritage: After the Boom, a Boomlet." *New York Times* (Feb. 12): 1, 15.

——. 1996. "High-Tech Pregnancies Test Hope's Limits." *New York Times* (Jan. 7): 1, 10–11.

Gailey, Christine Ward. 1987. "Evolutionary Perspectives on Gender Hierarchy." In Beth B. Hess and Myra Marx Ferree (Eds.), *Analyzing Gender: A Handbook of Social Science Research.* Newbury Park, CA: Sage, pp. 32–67.

Gaines, Donna. 1991. *Teenage Wasteland: Suburbia's Dead-End Kids.* New York: HarperPerennial.

Galbraith, John Kenneth. 1985. *The New Industrial State* (4th ed.). Boston: Houghton Mifflin.

Gambino, Richard. 1975. *Blood of My Blood.* New York: Doubleday/Anchor.

Gamson, William. 1990. *The Strategy of Social Protest* (2nd ed.). Belmont, CA: Wadsworth.

——. 1995. "Constructing Social Protest." In Hank Johnston and Bert Klandermans (Eds.), *Social Movements and Culture.* Minneapolis: University of Minnesota Press, pp. 85–106.

Gandara, Ricardo. 1995. "*Dichos de la Vida:* Homespun Proverbs Link Hispanic Culture's Past with the Present." *Austin American-Statesman* (Jan. 21): E1, E10.

Gans, Herbert. 1974. *Popular Culture and High Culture: An Analysis and Evaluation of Tastes.* New York: Basic.

——. 1982. *The Urban Villagers: Group and Class in the Life of Italian Americans* (updated and expanded ed.; orig. pub. 1962). New York: Free Press.

Garbarino, James. 1989. "The Incidence and Prevalence of Child Maltreatment." In L. Ohlin and M. Tonry (Eds.), *Family Violence.* Chicago: University of Chicago Press, pp. 219–261.

Garber, Judith A., and Robyne S. Turner. 1995. "Introduction." In Judith A. Garber and Robyne S. Turner (Eds.), *Gender in Urban Research.* Thousand Oaks, CA: Sage, pp. x–xxvi.

Garcia Coll, Cynthia T. 1990. "Developmental Outcomes of Minority Infants: A Process-Oriented Look into Our Beginnings." *Child Development,* 61: 270–289.

Gardner, Carol Brooks. 1989. "Analyzing Gender in Public Places: Rethinking Goffman's Vision of Everyday Life." *American Sociologist,* 20 (Spring): 42–56.

Garfinkel, Harold. 1967. *Studies in Ethnomethodology.* Englewood Cliffs, NJ: Prentice Hall.

Garfinkel, Irwin, and Sara S. McLanahan. 1986. *Single Mothers and Their Children: A New American Dilemma.* Washington, DC: Urban Institute Press.

Gargan, Edward A. 1996. "An Indonesian Asset Is Also a Liability." *New York Times* (Mar. 16): 17, 18.

Garreau, Joel. 1991. *Edge City: Life on the New Frontier.* New York: Doubleday.

——. 1993. "GAK Attack." *Austin American-Statesman* (Jan. 9): D1, D6.

Gaylin, Willard. 1992. *The Male Ego.* New York: Viking/Penguin.

Gecas, Viktor. 1982. "The Self-Concept." In Ralph H. Turner and James F. Short, Jr. (Eds.), *Annual Review of Sociology, 1982.* Palo Alto, CA: Annual Reviews, pp. 1–33.

Geertz, Clifford. 1966. "Religion as a Cultural System." In Michael Banton (Ed.), *Anthropological Approaches to the Study of Religion.* London: Tavistock, pp. 1–46.

Gelles, Richard J., and Murray A. Straus. 1988. *Intimate Violence: The Definitive Study of the Causes and Consequences of Abuse in the American Family.* New York: Simon & Schuster.

"General Facts on Sweden." 1988. *Fact Sheets on Sweden.* Stockholm: Swedish Institute.

General Motors. 1999. *General Motors Annual Report.* Retrieved Sept. 2, 2000. Online: http://www.generalmotors.com/company/investors/ar1999/bod/bod_frameset.htm

George, Susan. 1993. "A Fate Worse Than Debt." In William Dan Perdue (Ed.), *Systemic Crisis: Problems in Society, Politics, and World Order.* Fort Worth: Harcourt, pp. 85–96.

Gerbner, George, Larry Gross, Michael Morton, and Nancy Signorielli. 1987. "Charting the Mainstream: Television's Contributions to Political Orientations." In Donald Lazere (Ed.), *American Media and Mass Culture: Left Perspectives.* Berkeley: University of California Press, pp. 441–464.

Gereffi, Gary. 1994. "The International Economy and Economic Development." In Neil J. Smelser and Richard Swedberg (Eds.), *The Handbook of Economic Sociology.* Princeton, NJ: Princeton University Press, pp. 206–233.

Gergen, Kenneth J. 1991. *The Saturated Self: Dilemmas of Identity in Contemporary Life.* New York: Basic.

Gerschenkron, Alexander. 1962. *Economic Backwardness in Historical Perspective.* Cambridge, MA: Harvard University Press.

Gerson, Kathleen. 1993. *No Man's Land: Men's Changing Commitment to Family and Work.* New York: Basic.

Gerstel, Naomi, and Harriet Engel Gross. 1995. "Gender and Families in the United States: The Reality of Economic Dependence." In Jo Freeman (Ed.), *Women: A Feminist Perspective* (5th ed.). Mountain View, CA: Mayfield, pp. 92–127.

Gibbs, Lois Marie, as told to Murray Levine. 1982. *Love Canal: My Story.* Albany: SUNY Press.

Gibbs, Nancy. 1994. "Home Sweet School." *Time* (Oct. 31): 62–63.

———. 1999. "The Littleton Massacre." *Time* (May 3): 25–36.

Giffin, Mary, and C. Felsenthal. 1983. *A Cry for Help.* Garden City, NY: Doubleday.

Gilbert, Dennis L. 2003. *The American Class Structure in an Age of Growing Inequality* (6th ed.). Belmont, CA: Wadsworth.

Gilder, George F. 1986. *Men and Marriage.* New York: Pelican.

Gill, Derek. 1994. "A National Health Service: Principles and Practice." In Peter Conrad and Rochelle Kern (Eds.), *The Sociology of Health and Illness* (4th ed.). New York: St. Martin's, pp. 480–494.

Gilligan, Carol. 1982. *In a Different Voice: Psychological Theory and Women's Development.* Cambridge, MA: Harvard University Press.

Gilman, Charlotte Perkins. 1976. *His Religion and Hers.* Westport, CT: Hyperion (orig. pub. 1923).

Gilmore, David D. 1990. *Manhood in the Making: Cultural Concepts of Masculinity.* New Haven, CT: Yale University Press.

Glastris, Paul. 1990. "The New Way to Get Rich." *U.S. News & World Report* (May 7): 26–36.

Gleick, Elizabeth, Susan Reed, and Susan Schindehette. 1994. "The Baby Trap." *People* (Oct. 24): 39–56.

Goffman, Erving. 1956. "The Nature of Deference and Demeanor." *American Anthropologist,* 58: 473–502.

———. 1959. *The Presentation of Self in Everyday Life.* Garden City, NY: Doubleday.

———. 1961a. *Asylums: Essays on the Social Situation of Mental Patients and Other Inmates.* Chicago: Aldine.

———. 1961b. *Encounters: Two Studies in the Sociology of Interaction.* London: Routledge and Kegan Paul.

———. 1963a. *Behavior in Public Places: Notes on the Social Structure of Gatherings.* New York: Free Press.

———. 1963b. *Stigma: Notes on the Management of Spoiled Identity.* Englewood Cliffs, NJ: Prentice Hall.

———. 1967. *Interaction Ritual: Essays on Face to Face Behavior.* Garden City, NY: Anchor.

———. 1974. *Frame Analysis: An Essay on the Organization of Experience.* Boston: Northeastern University Press.

Gold, Rachel Benson, and Cory L. Richards. 1994. "Securing American Women's Reproductive Health." In Cynthia Costello and Anne J. Stone (Eds.), *The American Woman 1994–95.* New York: Norton, pp. 197–222.

Goldberg, Robert A. 1991. *Grassroots Resistance: Social Movements in Twentieth Century America.* Belmont, CA: Wadsworth.

Golden, Carla. 1987. "Diversity and Variability in Women's Sexual Identities." In the Boston Lesbian Psychologies Collective (Eds.), *Lesbian Psychologies.* Urbana: University of Illinois Press, pp. 18–34.

Golden, Stephanie. 1992. *The Women Outside: Meanings and Myths of Homelessness.* Berkeley: University of California Press.

Goode, Erich. 1996. "The Stigma of Obesity." In Erich Goode (Ed.), *Social Deviance.* Boston: Allyn & Bacon, pp. 332–340.

Goode, William J. 1960. "A Theory of Role Strain." *American Sociological Review,* 25: 483–496.

———. 1982. "Why Men Resist." In Barrie Thorne with Marilyn Yalom (Eds.), *Rethinking the Family: Some Feminist Questions.* New York: Longman, pp. 131–150.

Goodman, Mary Ellen. 1964. *Race Awareness in Young Children* (rev. ed.). New York: Collier.

Goodman, Peter S. 1996. "The High Cost of Sneakers." *Austin American-Statesman* (July 7): F1, F6.

Gordon, Milton. 1964. *Assimilation in American Life: The Role of Race, Religion, and National Origins.* New York: Oxford University Press.

Gordon, Philip L. 2001. "Federal Judge's Victory Just the First Shot in the Battle Over Workplace Monitoring." Retrieved July 8, 2002. Online: http://www.privacyfoundation.org/workplace/law/law_show.asp?id=75&action=0

Gotham, Kevin Fox. 1999. "Political Opportunity, Community Identity, and the Emergence of a Local Anti-Expressway Movement." *Social Problems,* 46: 332–354.

Gottdiener, Mark. 1985. *The Social Production of Urban Space.* Austin: University of Texas Press.

———. 1997. *The Theming of America.* Boulder, CO: Westview.

Gottlieb, Lori. 2000. *Stick Figure: A Diary of My Former Self.* New York: Simon & Schuster.

Gouldner, Alvin W. 1970. *The Coming Crisis of Western Sociology.* New York: Basic.

Gray, Paul. 1993. "Camp for Crusaders." *Time* (Apr. 19): 40.

Green, Donald E. 1977. *The Politics of Indian Removal: Creek Government and Society in Crisis.* Lincoln: University of Nebraska Press.

Greenberg, Edward S. 1999. *The Struggle for Democracy* (3rd ed.). New York: Addison-Wesley.

Greenberg, Edward S., and Benjamin I. Page. 1993. *The Struggle for Democracy.* New York: HarperCollins.

Greenhouse, Steven. 2001. "Also Hurt by Sept. 11: A Legion Still Working

But Making Much Less." *New York Times* (Nov. 29): A28.

Hacker, Helen Mayer. 1951. "Women as a Minority Group." *Social Forces,* 30 (October): 60–69.

———. 1974. "Women as a Minority Group: Twenty Years Later." In Florence Denmark (Ed.), *Who Discriminates Against Women?* Beverly Hills, CA: Sage, pp. 124–134.

Hadden, Richard W. 1997. *Sociological Theory: An Introduction to the Classical Tradition.* Peterborough, Ontario: Broadview.

Hahn, Harlan. 1987. "Civil Rights for Disabled Americans: The Foundation of a Political Agenda." In Alan Gartner and Tom Joe (Eds.), *Images of the Disabled, Disabling Images.* New York: Praeger, pp. 181–203.

———. 1997. "Advertising the Acceptably Employable Image." In Lennard J. Davis (Ed.), *The Disability Studies Reader.* New York: Routledge, pp. 172–186.

Haines, Valerie A. 1997. "Spencer and His Critics." In Charles Camic (Ed.), *Reclaiming the Sociological Classics: The State of the Scholarship.* Malden, MA: Blackwell, pp. 81–111.

Halberstadt, Amy G., and Martha B. Saitta. 1987. "Gender, Nonverbal Behavior, and Perceived Dominance: A Test of the Theory." *Journal of Personality and Social Psychology,* 53: 257–272.

Hale-Benson, Janice E. 1986. *Black Children: Their Roots, Culture and Learning Styles* (rev. ed.). Provo, UT: Brigham Young University Press.

Hall, Edward. 1966. *The Hidden Dimension.* New York: Anchor/Doubleday.

Hall, Roberta M., with Bernice R. Sandler. 1982. *The Classroom Climate: A Chilly One for Women?* Washington, DC: Association of American Colleges, Project on the Status and Education of Women.

———. 1984. *Out of the Classroom: A Chilly Campus Climate for Women.* Washington, DC: Association of American Colleges, Project on the Status and Education of Women.

Halle, David. 1993. *Inside Culture: Art and Class in the American Home.* Chicago: University of Chicago Press.

Haraway, Donna. 1994. "A Cyborg Manifesto: Science, Technology, and Socialist-Feminism in the Late Twentieth Century." In Anne C. Herrmann and Abigail J. Stewart (Eds.), *Theorizing Feminism: Parallel Trends in the Humanities and Social Sciences.* Boulder, CO: Westview, pp. 424–457.

Harding, Sandra. 1986. *The Science Question in Feminism.* Ithaca, NY: Cornell University Press.

Harlow, Harry F., and Margaret Kuenne Harlow. 1962. "Social Deprivation in Monkeys." *Scientific American,* 207 (5): 137–146.

———. 1977. "Effects of Various Mother–Infant Relationships on Rhesus Monkey Behaviors." In Brian M. Foss (Ed.), *Determinants of Infant Behavior* (vol. 4). London: Methuen, pp. 15–36.

Harrington, Michael. 1985. *The New American Poverty.* New York: Viking/Penguin.

Harris, Anthony R. 1991. "Race, Class, and Crime." In Joseph F. Sheley (Ed.), *Criminology: A Contemporary Handbook.* Belmont, CA: Wadsworth, pp. 94–119.

Harris, Chauncey D., and Edward L. Ullman. 1945. "The Nature of Cities." *Annals of the Academy of Political and Social Sciences* (November): 7–17.

Harris, Diana K. 1990. *Sociology of Aging* (2nd ed.). New York: Harper & Row.

Harris, Marvin. 1974. *Cows, Pigs, Wars, and Witches.* New York: Random House.

———. 1985. *Good to Eat: Riddles of Food and Culture.* New York: Simon & Schuster.

Harrison, Algea O., Melvin N. Wilson, Charles J. Pine, Samuel Q. Chan, and Raymond Buriel. 1990. "Family Ecologies of Ethnic Minority Children." *Child Development,* 61 (2): 347–362.

Harrison, Janine. 2001. "Welfare Reports Document Increasing Homelessness in Australia." *World Socialist Web Site.* Retrieved Sept. 11, 2001. Online: http://wsws.orgarticles/2001/jun2001/home-j07_prn.shtml

Hartmann, Heidi. 1976. "Capitalism, Patriarchy, and Job Segregation by Sex." *Signs: Journal of Women in Culture and Society,* 1 (Spring): 137–169.

———. 1981. "The Unhappy Marriage of Marxism and Feminism." In Lydia Sargent (Ed.), *Women and Revolution.* Boston: South End.

Harvard Law School Seminar. 1996. "Cultural Imperialism." Retrieved June 23, 1999. Online: http://roscoe.law.harvard.edu/courses/techseminar96/course/sessions/culturalimperialism/selene2.html

Harvey, David. 1989. *The Condition of Postmodernity.* Cambridge, MA: Blackwell.

Hauchler, Ingomar, and Paul M. Kennedy (Eds.). 1994. *Global Trends: The World Almanac of Development and Peace.* New York: Continuum.

Hauser, Robert M. 1995. "Symposium: The Bell Curve." *Contemporary Sociology: A Journal of Reviews,* 24 (2): 149–153.

Hauser, Robert M., and David L. Featherman. 1976. "Equality of Schooling: Trends and Prospects." *Sociology of Education*, 49: 99–120.

Haviland, William A. 1993. *Cultural Anthropology* (7th ed.). Orlando, FL: Harcourt.

———. 1999. *Cultural Anthropology* (9th ed.). Orlando, FL: Harcourt.

Hawley, Amos. 1950. *Human Ecology*. New York: Ronald.

———. 1981. *Urban Society* (2nd ed.). New York: Wiley.

Healey, Joseph F. 1995. *Race, Ethnicity, Gender, and Class: The Sociology of Group Conflict and Change*. Thousand Oaks, CA: Pine Forge.

Henderson, Rick. 1997. "Schools of Thought." *Reasonline* (January). Retrieved July 24, 2002. Online: http://reason.com/9701/fe.rick.shtml

Henley, Nancy. 1977. *Body Politics: Power, Sex, and Nonverbal Communication*. Englewood Cliffs, NJ: Prentice Hall.

Henneberger, Melinda. 1993. "Gang Membership Grows in Middle-Class Suburbs." *New York Times* (July 24): 1, 12.

Henry, William A., III. 1990. "Beyond the Melting Pot." *Time* (Apr. 9): 28–35.

Herbert, Bob. 2001. "The Tourism Crisis." *New York Times* (Nov. 29): A31.

Heritage, John. 1984. *Garfinkel and Ethnomethodology*. Cambridge, MA: Polity.

Herrnstein, Richard J., and Charles Murray. 1994. *The Bell Curve: Intelligence and Class Structure in American Life*. New York: Free Press.

Heshka, Stanley, and Yona Nelson. 1972. "Interpersonal Speaking Distances as a Function of Age, Sex, and Relationship." *Sociometry*, 35 (4): 491–498.

Hesse-Biber, Sharlene. 1996. *Am I Thin Enough Yet? The Cult of Thinness and the Commercialization of Identity*. New York: Oxford University Press.

Hibbard, David R., and Duane Buhrmester. 1998. "The Role of Peers in the Socialization of Gender-Related Social Interaction Styles." *Sex Roles*, 39: 185–203.

Higginbotham, Elizabeth. 1994. "Black Professional Women: Job Ceilings and Employment Sectors." In Maxine Baca Zinn and Bonnie Thornton Dill (Eds.), *Women of Color in U.S. Society*. Philadelphia: Temple University Press, pp. 113–131.

Hight, Bruce. 1994. "A Level Playing Field: Critics Say Minorities Need a Shot Off the Field." *Austin American-Statesman* (June 8): A1, A11.

Higley, Stephen Richard. 1997. "Privilege, Power, and Place: The Geography of the American Upper Class."

In Diana Kendall (Ed.), *Race, Class, and Gender in a Diverse Society: A Text-Reader*. Boston: Allyn & Bacon, pp. 70–82.

Hills, Stuart L. 1971. *Crime, Power, and Morality*. Scranton, PA: Chandler.

Hilsenrath, Jon. 1996. "In China, a Taste of Buy-Me TV." *New York Times* (Nov. 17): F1, F11.

Hindelang, Michael J., Travis Hirschi, and Joseph Weis. 1981. *Measuring Delinquency*. Beverly Hills, CA: Sage.

Hirschi, Travis. 1969. *Causes of Delinquency*. Berkeley: University of California Press.

Hochschild, Arlie Russell. 1973. "A Review of Sex Role Research." *American Journal of Sociology*, 78 (January): 1011–1029.

———. 1983. *The Managed Heart: Commercialization of Human Feeling*. Berkeley: University of California Press.

———. 1997. *The Time Bind: When Work Becomes Home and Home Becomes Work*. New York: Metropolitan.

Hochschild, Arlie Russell, with Ann Machung. 1989. *The Second Shift: Working Parents and the Revolution at Home*. New York: Viking/ Penguin.

Hochschild, Jennifer L. 1995. *Facing Up to the American Dream: Race, Class, and the Soul of the Nation*. Princeton, NJ: Princeton University Press.

Hodson, Randy, and Robert E. Parker. 1988. "Work in High Technology Settings: A Review of the Empirical Literature." *Research in the Sociology of Work*, 4: 1–29.

Hodson, Randy, and Teresa A. Sullivan. 2002. *The Social Organization of Work* (3rd ed.). Belmont, CA: Wadsworth.

Hoecker-Drysdale, Susan. 1992. *Harriet Martineau: First Woman Sociologist*. Oxford, England: Berg.

Hoffnung, Michele. 1995. "Motherhood: Contemporary Conflict for Women." In Jo Freeman (Ed.), *Women: A Feminist Perspective* (5th ed.). Mountain View, CA: Mayfield, pp. 162–181.

Holland, Dorothy C., and Margaret A. Eisenhart. 1981. *Women's Peer Groups and Choice of Career*. Final report for the National Institute of Education. ERIC ED 199 328. Washington, DC.

———. 1990. *Educated in Romance: Women, Achievement, and College Culture*. Chicago: University of Chicago Press.

Hoose, Phillip M. 1989. *Necessities: Racial Barriers in American Sports*. New York: Random House.

Hoover, Eric. 2001. "The Lure of Easy Credit Leaves More Students Struggling with Debt." *Chronicle of Higher Education* (June 15): A35–A36.

Hooyman, Nancy R. R., and H. Asuman Kiyak. 1999. *Social Gerontology: A Multidisciplinary Perspective* (5th ed.). Boston: Allyn & Bacon.

Horan, Patrick M. 1978. "Is Status Attainment Research Atheoretical?" *American Sociological Review*, 43: 534–541.

Hostetler, A. J. 1994. "U.S. Death Rate Falls to Lowest Level Ever Despite Rise in AIDS." *Austin American-Statesman* (Dec. 16): A4.

Howard, Michael E. 1990. "On Fighting a Nuclear War." In Francesca M. Cancian and James William Gibson (Eds.), *Making War, Making Peace: The Social Foundations of Violent Conflict*. Belmont, CA: Wadsworth, pp. 314–322.

Hoyt, Homer. 1939. *The Structure and Growth of Residential Neighborhoods in American Cities*. Washington, DC: Federal Housing Administration.

Hsiung, Ping-Chun. 1996. *Living Rooms as Factories: Class, Gender, and the Satellite Factory System in Taiwan*. Philadelphia: Temple University Press.

Hughes, Everett C. 1945. "Dilemmas and Contradictions of Status." *American Journal of Sociology*, 50: 353–359.

Hull, Gloria T., Patricia Bell-Scott, and Barbara Smith. 1982. *All the Women Are White, All the Blacks Are Men, But Some of Us Are Brave*. Old Westbury, NY: Feminist.

Hurst, Charles E. 1998. *Social Inequality: Forms, Causes, and Consequences* (3rd ed.). Boston: Allyn & Bacon.

Hurtada, Aida. 1996. *The Color of Privilege: Three Blasphemies on Race and Feminism*. Ann Arbor: University of Michigan Press.

Huston, Aletha C. 1985. "The Development of Sex Typing: Themes from Recent Research." *Developmental Review*, 5: 2–17.

Huyssen, Andreas. 1984. *After the Great Divide*. Bloomington: Indiana University Press.

Hwang, S. S., R. Saenz, and B. E. Aguirre. 1995. "The SES-Selectivity of Interracially Married Asians." *International Migration Review*, 29: 469–491.

Hynes, H. Patricia. 1990. *Earth Right: Every Citizen's Guide*. Rocklin, CA: Prima.

Ignico, Arlene A., and Barbara J. Mead. 1990. "Children's Perceptions of the Gender-Appropriateness of Physical Activities." *Perceptual and Motor Skills*, 71: 1275–1281.

Inciardi, James A., Ruth Horowitz, and Anne E. Pottieger. 1993. *Street Kids, Street Drugs, Street Crime: An Examination of Drug Use and Serious Delinquency in Miami*. Belmont, CA: Wadsworth.

International Monetary Fund. 1992. *World Economic Outlook*. Washington, DC: International Development Fund.

Iovine, Julie V. 1997. "An Industry Monitors Child Labor." *New York Times* (Oct. 16): B1, B9.

Irons, Edward D., and Gilbert W. Moore. 1985. *Black Managers: The Case of the Banking Industry*. New York: Praeger.

Irvine, Jacqueline Jordan. 1986. "Teacher–Student Interactions: Effects of Student Race, Sex, and Grade Level." *Journal of Educational Psychology*, 78 (1): 14–21.

Ishikawa, Kaoru. 1984. "Quality Control in Japan." In Naoto Sasaki and David Hutchins (Eds.), *The Japanese Approach to Product Quality: Its Applicability to the West*. Oxford: Permagon, pp. 1–5.

Jackson, Kenneth T. 1985. *Crabgrass Frontier: The Suburbanization of the United States*. New York: Oxford University Press.

Jacobs, Gloria. 1994. "Where Do We Go from Here? An Interview with Ann Jones." *Ms.* (September/ October): 56–63.

Jacobs, Jerry A. 1993. "Men in Female-Dominated Fields." In Christine L. Williams (Ed.), *Doing "Women's Work": Men in Nontraditional Occupations*. Newbury Park, CA: Sage, pp. 49–63.

Jameson, Fredric. 1984. "Postmodernism, or, The Cultural Logic of Late Capitalism." *New Left Review*, 146: 59–92.

Janis, Irving. 1972. *Victims of Groupthink*. Boston: Houghton Mifflin.

———. 1989. *Crucial Decisions: Leadership in Policymaking and Crisis Management*. New York: Free Press.

Jankowski, Martin Sanchez. 1991. *Islands in the Street: Gangs and American Urban Society*. Berkeley: University of California Press.

Jaramillo, P. T., and Jesse T. Zapata. 1987. "Roles and Alliances Within Mexican-American and Anglo Families." *Journal of Marriage and the Family*, 49 (November): 727–735.

Jary, David, and Julia Jary. 1991. *The Harper Collins Dictionary of Sociology*. New York: HarperPerennial.

Jayson, Sharon. 2000. "U.T. Law School Doubles Its Ranks of New Black Students." *Austin American-Statesman* (July 8): B1, B6.

Jehl, Douglas. 1999. "The Internet's 'Open Sesame' Is Answered Warily." *New York Times* (Mar. 18): A4.

Jenkinson, Edward B. 1979. *Censors in the Classroom: The Mind Benders.* Carbondale: Southern Illinois University Press.

Jensen, Robert. 1995. "Men's Lives and Feminist Theory." *Race, Gender & Class*, 2 (2): 111–125.

Jewell, K. Sue. 1993. *From Mammy to Miss America and Beyond: Cultural Images and the Shaping of US Social Policy.* New York: Routledge.

Johnson, Allan. 1995. *The Blackwell Dictionary of Sociology.* Malden, MA: Blackwell.

Johnson, Claudia. 1994. *Stifled Laughter: One Woman's Story About Fighting Censorship.* Golden, CO: Fulcrum.

Johnson, Earvin "Magic," with William Novak. 1992. *My Life.* New York: Fawcett Crest.

Johnson, Justin M. 1994. "'Flotsam on the Sea of Humanity': A View from the Bench on Class, Race, and Gender." *Social Justice*, 20 (1/2): 140–149.

Johnston, David Cay. 1999. "Gap Between Rich and Poor Found Substantially Wider." *New York Times* (Sept. 5): A1.

———. 2002. "As Salary Grows, So Does a Gender Gap." *New York Times* (May 12): BU8.

Joint United Nations Programme on HIV/AIDS. 2000. "Report on the Global HIV/AIDS Epidemic-June 2000." Retrieved Nov. 22, 2001. Online: http://www.unaids.org/epidemic_update/report/Epi_report.htm

———. 2001. "Global Summary of the HIV/AIDS Epidemic, December 2001." Retrieved July 21, 2002. Online: http://www.unaids.org/worldaidsday/2001/EPIgraphics2001/EPIgraphic1_en.gif

Jones, Charisse. 1995. "Family Struggles on Brink of Comfort." *New York Times* (Feb. 18): 1, 9.

Juergensmeyer, Mark. 1993. *The New Cold War? Religious Nationalism Confronts the Secular State.* Berkeley: University of California Press.

Kalmijn, Matthijs. 1998. "Intermarriage and Homogamy: Causes, Patterns, Trends." *Annual Review of Sociology*, 24: 395–422.

Kanter, Rosabeth Moss. 1983. *The Change Masters: Innovation and Entrepreneurship in the American Corporation.* New York: Simon & Schuster.

———. 1985. "All That Is Entrepreneural Is Not Gold." *Wall Street Journal* (July 22): 18.

———. 1993. *Men and Women of the Corporation.* New York: Basic (orig. pub. 1977).

Kaplan, Robert D. 1996. "Cities of Despair." *New York Times* (June 6): A19.

Karmen, Andrew W. 1995. "Crime Victims." In Joseph F. Sheley (Ed.), *Criminology: A Contemporary Handbook* (2nd ed.). Belmont, CA: Wadsworth, pp. 45–164.

Kaspar, Anne S. 1986. "Consciousness Re-evaluated: Interpretive Theory and Feminist Scholarship." *Sociological Inquiry*, 56 (1): 30–49.

Katz, Michael B. 1989. *The Undeserving Poor: From the War on Poverty to the War on Welfare.* New York: Pantheon.

Katzer, Jeffrey, Kenneth H. Cook, and Wayne W. Crouch. 1991. *Evaluating Information: A Guide for Users of Social Science Research.* New York: McGraw-Hill.

Kaufman, Gayle. 1999. "The Portrayal of Men's Family Roles in Television Commercials." *Sex Roles*, 313: 439–451.

Keil, Julian E., Susan E. Sutherland, Rebecca G. Knapp, and Herman A. Tyroler. 1992. "Does Equal Socioeconomic Status in Black and White Men Mean Equal Risk of Mortality?" *American Journal of Public Health*, 82: 1133–1136.

Keister, Lisa A. 2000. *Wealth in America: Trends in Wealth Inequality.* Cambridge, U.K.: Cambridge University Press.

Keller, James. 1994. "'I Treasure Each Moment.'" *Parade* (Sept. 4): 4–5.

Kelly, John, and David Stark. 2002. "Crisis, Recovery, Innovation: Learning from 9/11." *Working Papers: Center on Organizational Innovation.* New York: Center on Organizational Innovation, Columbia University. Retrieved: June 26, 2002. Online: http://www.coi.columbia.edu/pdf/kelly_stark_cri.pdf

Kelman, Steven. 1991. "Sweden Sour? Downsizing the 'Third Way.'" *New Republic* (July 29): 19–23.

Kemp, Alice Abel. 1994. *Women's Work: Degraded and Devalued.* Englewood Cliffs, NJ: Prentice Hall.

Kendall, Diana. 1980. Square Pegs in Round Holes: Non-Traditional Students in Medical Schools. Unpublished doctoral dissertation, Department of Sociology, the University of Texas at Austin.

———. 1998. *Social Problems in a Diverse Society.* Needham Heights, MA: Allyn & Bacon.

Kendall, Diana, and Joe R. Feagin. 1983. "Blatant and Subtle Patterns of Discrimination: Minority Women in Medical Schools." *Journal of Intergroup Relations*, 9 (Summer): 21–27.

Kendall, Diana, Jane Lothian Murray, and Rick Linden. 2000. *Sociology in Our Times* (2nd Canadian ed.). Scarborough, Ontario: Nelson Thomson Learning.

Kennedy, Paul. 1993. *Preparing for the Twenty-First Century.* New York: Random House.

Kennedy, Randy. 1993. "To Homelessness and Back: A Man's Journey to Respect." *New York Times* (Nov. 28): A15.

Kenyon, Kathleen. 1957. *Digging Up Jericho.* London: Benn.

Kephart, William M., and William W. Zellner. 1994. *Extraordinary Groups: An Examination of Unconventional Life-Styles* (2nd ed.). New York: St. Martin's.

Kerbo, Harold R. 2000. *Social Stratification and Inequality: Class Conflict in Historical, Comparative, and Global Perspective* (4th ed.). New York:McGraw-Hill.

Kessler-Harris, Alice. 1990. *A Woman's Wage: Historical Meanings and Social Consequences.* Lexington: University Press of Kentucky.

Keyfitz, Nathan. 1994. "The Scientific Debate: Is Population Growth a Problem? An Interview with Nathan Keyfitz." *Harvard International Review* (Fall): 10–11, 74.

Kidron, Michael, and Ronald Segal. 1995. *The State of the World Atlas.* New York: Penguin.

Kilborn, Peter T. 1997. "Illness Is Turning into Financial Catastrophe for More of the Uninsured." *New York Times* (Aug. 1): A10.

Kilbourne, Jean. 1999. *Deadly Persuasion: The Addictive Power of Advertising.* New York: Simon & Schuster.

Killian, Lewis. 1984. "Organization, Rationality, and Spontaneity in the Civil Rights Movement." *American Sociological Review*, 49: 770–783.

Kimmel, Michael S., and Michael A. Messner. 1998. *Men's Lives* (4th ed.). Boston: Allyn & Bacon.

King, Gary, Robert O. Keohane, and Sidney Verba. 1994. *Designing Social Inquiry: Scientific Inference in Qualitative Research.* Princeton, NJ: Princeton University Press.

King, Leslie, and Madonna Harrington Meyer. 1997. "The Politics of Reproductive Benefits: U.S. Insurance Coverage of Contraceptive and Infertility Treatments." *Gender and Society*, 11 (1): 8–30.

Kirkpatrick, P. 1994. "Triple Jeopardy: Disability, Race and Poverty in America." *Poverty and Race*, 3: 1–8.

Kitsuse, John I. 1980. "Coming Out All Over: Deviance and the Politics of Social Problems." *Social Problems*, 28: 1–13.

Klein, Alan M. 1993. *Little Big Men: Bodybuilding Subculture and Gender Construction.* Albany: SUNY Press.

Klockars, Carl B. 1979. "The Contemporary Crises of Marxist Criminology." *Criminology*, 16: 477–515.

Kluckhohn, Clyde. 1961. "The Study of Values." In Donald N. Barrett (Ed.), *Values in America.* South Bend, IN: University of Notre Dame Press, pp. 17–46.

Knox, Paul L., and Peter J. Taylor (Eds.). 1995. *World Cities in a World-System.* Cambridge, England: Cambridge University Press.

Knudsen, Dean D. 1992. *Child Maltreatment: Emerging Perspectives.* Dix Hills, NY: General Hall.

Kohlberg, Lawrence. 1969. "Stage and Sequence: The Cognitive–Developmental Approach to Socialization." In David A. Goslin (Ed.), *Handbook of Socialization Theory and Research.* Chicago: Rand McNally, pp. 347–480.

———. 1981. *The Philosophy of Moral Development: Moral Stages and the Idea of Justice*, vol. 1: *Essays on Moral Development.* San Francisco: Harper & Row.

Kohn, Melvin L. 1977. *Class and Conformity: A Study in Values* (2nd ed.). Homewood, IL: Dorsey.

Kohn, Melvin L., Atsushi Naoi, Carrie Schoenbach, Carmi Schooler, and Kazimierz M. Slomczynski. 1990. "Position in the Class Structure and Psychological Functioning in the United States, Japan, and Poland." *American Journal of Sociology*, 95: 964–1008.

Kolata, Gina. 1993. "Fear of Fatness: Living Large in a Slimfast World." *Austin American-Statesman* (Jan. 3): C1, C6.

Korsmeyer, Carolyn. 1981. "The Hidden Joke: Generic Uses of Masculine Terminology." In Mary Vetterling-Braggin (Ed.), *Sexist Language: A Modern Philosophical Analysis.* Totowa, NJ: Littlefield, Adams, pp. 116–131.

Korten, David C. 1996. *When Corporations Rule the World.* West Hartford, CT: Kumarian.

Kosmin, Barry A., and Seymour P. Lachman. 1993. *One Nation Under God: Religion in Contemporary American Society.* New York: Crown.

Kozol, Jonathan. 1988. *Rachael and Her Children: Homeless Families in America.* New York: Fawcett Columbine.

———. 1991. *Savage Inequalities: Children in America's Schools.* New York: Crown.

Kraft, Scott. 1994. "Agog at Euro Disney-land." *Los Angeles Times* (Jan. 18): H1.

Kroloff, Charles A. 1993. *54 Ways You Can Help the Homeless*. Southport, CT: Hugh Lauter Levin Associates; and West Orange, NJ: Behrman.

Kuisel, Richard. 1993. *Seducing the French: The Dilemma of Americanization*. Berkeley: University of California Press.

Kurtz, Lester. 1995. *Gods in the Global Village: The World's Religions in Sociological Perspective*. Thousand Oaks, CA: Sage.

Kurz, Demie. 1995. *For Richer, for Poorer: Mothers Confront Divorce*. New York: Routledge.

Lacayo, Richard. 2001. "About Face: An Inside Look at How Women Fared Under Taliban Oppression and What the Future Holds for Them Now." *Time* (Dec. 3): 36–49.

Ladd, E. C., Jr. 1966. *Negro Political Leadership in the South*. Ithaca, NY: Cornell University Press.

Lamanna, Marianne, and Agnes Riedmann. 2003. *Marriages and Families: Making Choices in a Diverse Society* (8th ed.). Belmont, CA: Wadsworth.

Lamar, Joe. 2000. "Suicides in Japan Reach a Record High." *British Medical Journal* (Sept. 2). Retrieved Aug. 25, 2001. Online: http://www.findarticles.com/cf_dls/m0999/7260_321/66676910/p1/article.jhtml

Lane, Harlan. 1992. *The Mask of Benevolence: Disabling the Deaf Community*. New York: Vintage.

Langone, John. 1997. "Bloodless Surgery." *Time* (Fall special issue): 74–76.

Lapham, Lewis H. 1988. *Money and Class in America: Notes and Observations on Our Civil Religion*. New York: Weidenfeld & Nicolson.

Lapierre, Dominique. 1997. "'Jewel of Calcutta' Aided Poorest of Poor." *Austin American-Statesman* (Sept. 6): A10.

Lappé, Frances Moore, and Paul Martin Du Bois. 1994. *The Quickening of America: Rebuilding Our Nation, Remaking Our Lives*. San Francisco: Jossey-Bass.

Lapsley, Daniel K., 1990. "Continuity and Discontinuity in Adolescent Social Cognitive Development." In Raymond Montemayor, Gerald R. Adams, and Thomas P. Gullota (Eds.), *From Childhood to Adolescence: A Transitional Period?* (Advances in Adolescent Development, vol. 2). Newbury Park, CA: Sage.

Larimer, Tim. 1999. "The Japan Syndrome." *Time* (Oct. 11): 50–51.

Larson, Erik. 1997. "It's Not the Money, It's the Principal." *Time* (Oct. 27): 92–93.

Larson, Magali Sarfatti. 1977. *The Rise of Professionalism: A Sociological Analysis*. Berkeley: University of California Press.

Lasch, Christopher. 1977. *Haven in a Heartless World*. New York: Basic.

Lash, Scott, and John Urry. 1994. *Economies of Signs and Space*. London: Sage.

"Latino Legends in Sports." 1999. "Sports News." Retrieved Aug. 15, 1999. Online: http://www.latinosportslegends.com/news.htm

Latouche, Serge. 1992. "Standard of Living." In Wolfgang Sachs (Ed.), *The Development Dictionary*. Atlantic Highlands, NJ: Zed, pp. 250–263.

Laumann, Edward O., John H. Gagnon, Robert T. Michael, and Stuart Michaels. 1994. *The Social Organization of Sexuality*. Chicago: University of Chicago Press.

Le Bon, Gustave. 1960. *The Crowd: A Study of the Popular Mind*. New York: Viking (orig. pub. 1895).

Leary, Warren E. 1996. "Even When Covered by Insurance, Black and Poor People Receive Less Health Care." *New York Times* (Sept. 12): A10.

Lee, Felicia R. 1993. "Where Guns and Lives Are Cheap." *New York Times* (Mar. 21): 21.

Lee, Sharon M. 1993. "Racial Classifications in the U.S. Census: 1890–1990." *Ethnic and Racial Studies*, 16 (1): 75–94.

Leenaars, Antoon A. 1988. *Suicide Notes: Predictive Clues and Patterns*. New York: Human Sciences Press.

Leenaars, Antoon A. (Ed.). 1991. *Life Span Perspectives of Suicide: Time-Lines in the Suicide Process*. New York: Plenum.

Lefrançois, Guy R. 1999. *The Lifespan* (6th ed.). Belmont, CA: Wadsworth.

Lehmann, Jennifer M. 1994. *Durkheim and Women*. Lincoln: University of Nebraska Press.

Leidner, Robin. 1993. *Fast Food, Fast Talk: Service Work and the Routinization of Everyday Life*. Berkeley: University of California Press.

Lemert, Charles. 1997. *Postmodernism Is Not What You Think*. Malden, MA: Blackwell.

Lemert, Edwin M. 1951. *Social Pathology*. New York: McGraw-Hill.

Lengermann, Patricia Madoo, and Jill Niebrugge-Brantley. 1998. *The Women Founders: Sociology and Social Theory, 1830–1930*. New York: McGraw-Hill.

Lengermann, Patricia Madoo, and Ruth A. Wallace. 1985. *Gender in America: Social Control and Social Change*. Englewood Cliffs, NJ: Prentice Hall.

Lenzer, Gertrud (Ed.). 1998. *The Essential Writings: Auguste Comte and Positivism*. New Brunswick, NJ: Transaction.

Leonard, Andrew. 1999. "We've Got Mail—Always." *Newsweek* (Sept. 20): 58–61.

Leonard, Wilbert M., and Jonathan E. Reyman. 1988. "The Odds of Attaining Professional Athlete Status: Refining the Computations." *Sociology of Sport Journal*: 162–169.

Lerner, Gerda. 1986. *The Creation of Patriarchy*. New York: Oxford University Press.

Lester, David. 1992. *Why People Kill Themselves: A 1990s Summary of Research Findings of Suicidal Behavior* (3rd ed.). Springfield, IL: Thomas.

Lester, Will. 2000. "Voters Say What Drove the Choice for a Leader." *Austin American-Statesman* (Nov. 8): A11.

Lev, Michael A. 1998. "Suicide Imbedded in Japan's Culture." *Chicago Tribune* (Feb. 27). Retrieved Aug. 25, 2001. Online: http://seattletimes.nwsource.com/news/nation-world/html98/altjpan_022798.html

Lever, Janet. 1978. "Sex Differences in the Complexity of Children's Play and Games." *American Sociological Review*, 43: 471–483.

Levine, Adeline Gordon. 1982. *Love Canal: Science, Politics, and People*. Lexington, MA: Lexington.

Levine, Murray. 1982. "Introduction." In Lois Marie Gibbs, *Love Canal: My Story*. Albany: SUNY Press.

Levine, Peter. 1992. *Ellis Island to Ebbets Field: Sport and the American Jewish Experience*. New York: Oxford University Press.

Levinthal, Charles F. 1996. *Drugs, Behavior, and Modern Society*. Boston: Allyn & Bacon.

Levy, Janice C., and Eva Y. Deykin. 1989. "Suicidality, Depression, and Substance Abuse in Adolescence." *American Journal of Psychiatry*, 146 (11): 1462–1468.

Lewin, Tamar. 1997. "Wage Difference Between Women and Men Widens." *New York Times* (Sept. 15): A1.

Lewis, Anthony. 1997. "Violins in the Wings." *New York Times* (Nov. 28): A23.

Lewis, Paul. 1996. "World Bank Moves to Cut Poorest Nations' Debts." *New York Times* (Mar. 16): 17, 18.

Liebow, Elliot. 1993. *Tell Them Who I Am: The Lives of Homeless Women*. New York: Free Press.

Lightblau, Eric. 1999. "U.S. Crime Statistics Drop Again, Hitting 25-Year Low." *Austin American-Statesman* (July 19): A3.

Lindblom, Charles. 1977. *Politics and Markets*. New York: Basic.

Linton, Ralph. 1936. *The Study of Man*. New York: Appleton-Century-Crofts.

Lippa, Richard A. 1994. *Introduction to Social Psychology*. Pacific Grove, CA: Brooks/Cole.

Lips, Hilary M. 1989. "Gender-Role Socialization: Lessons in Femininity." In Jo Freeman (Ed.), *Women: A Feminist Perspective* (4th ed.). Mountain View, CA: Mayfield, pp. 197–216.

——. 1993. *Sex and Gender: An Introduction* (2nd ed.). Mountain View, CA: Mayfield.

——. 1997. *Sex and Gender: An Introduction* (3rd ed.). Mountain View, CA: Mayfield.

Loeb, Paul Rogat. 1994. *Generation at the Crossroads: Apathy and Action on the American Campus*. New Brunswick, NJ: Rutgers University Press.

Lofland, John. 1993. "Collective Behavior: The Elementary Forms." In Russell L. Curtis, Jr., and Benigno E. Aguirre (Eds.), *Collective Behavior and Social Movements*. Boston: Allyn & Bacon, pp. 70–75.

Lombardo, William K., Gary A. Cretser, Barbara Lombardo, and Sharon L. Mathis. 1983. "For Cryin' Out Loud—There Is a Sex Difference." *Sex Roles*, 9: 987–995.

London, Kathryn A. 1991. "Advance Data Number 194: Cohabitation, Marriage, Marital Dissolution, and Remarriage: United States 1988." U.S. Department of Health and Human Services: Vital and Health Statistics of the National Center, January 4.

Lorber, Judith. 1994. *Paradoxes of Gender*. New Haven, CT: Yale University Press.

Lott, Bernice. 1994. *Women's Lives: Themes and Variations in Gender Learning* (2nd ed.). Pacific Grove, CA: Brooks/Cole.

Lummis, C. Douglas. 1992. "Equality." In Wolfgang Sachs (Ed.), *The Development Dictionary*. Atlantic Highlands, NJ: Zed, pp. 38–52.

Lund, Kristina. 1990. "A Feminist Perspective on Divorce Therapy for Women." *Journal of Divorce*, 13 (3): 57–67.

Lundberg, Ferdinand. 1988. *The Rich and the Super-Rich: A Study in the Power of Money Today*. Secaucus, NJ: Lyle Stuart.

Lupton, Deborah. 1997. "Foucault and the Medicalisation Critique." In Alan Petersen and Robin Bunton (Eds.), *Foucault: Health and Medicine*. London: Routledge, pp. 94–110.

Lurie, Alison. 1981. *The Language of Clothes*. New York: Random House.

Maccoby, Eleanor E., and Carol Nagy Jacklin. 1987. "Gender Segregation in Childhood." *Advances in Child Development and Behavior*, 20: 239–287.

MacDonald, Kevin, and Ross D. Parke. 1986. "Parental–Child Physical Play: The Effects of Sex and Age of Children and Parents." *Sex Roles*, 15: 367–378.

Mack, Raymond W., and Calvin P. Bradford. 1979. *Transforming America: Patterns of Social Change* (2nd ed.). New York: Random House.

MacSween, Morag. 1993. *Anorexic Bodies: A Feminist and Sociological Perspective on Anorexia Nervosa*. New York: Routledge.

Maggio, Rosalie. 1988. *The Non-Sexist Word Finder: A Dictionary of Gender-Free Usage*. Boston: Beacon.

Malinowski, Bronislaw. 1922. *Argonauts of the Western Pacific*. New York: Dutton.

Malthus, Thomas R. 1965. *An Essay on Population*. New York: Augustus Kelley (orig. pub. 1798).

Mangione, Jerre, and Ben Morreale. 1992. *La Storia: Five Centuries of the Italian American Experience*. New York: HarperPerennial.

Mann, Coramae Richey. 1993. *Unequal Justice: A Question of Color*. Bloomington: Indiana University Press.

Mann, Patricia S. 1994. *Micro-Politics: Agency in a Postfeminist Era*. Minneapolis: University of Minnesota Press.

Mansfield, Alan, and Barbara McGinn. 1993. "Pumping Irony: The Muscular and the Feminine." In Sue Scott and David Morgan (Eds.), *Body Matters: Essays on the Sociology of the Body*. London: Falmer, pp. 49–58.

Marger, Martin N. 1994. *Race and Ethnic Relations: American and Global Perspectives*. Belmont, CA: Wadsworth.

———. 2000. *Race and Ethnic Relations: American and Global Perspectives* (5th ed.). Belmont, CA: Wadsworth.

Marquart, James W., Sheldon Ekland-Olson, and Jonathan R. Sorensen. 1994. *The Rope, the Chair, and the Needle*. Austin: University of Texas Press.

Marquis, Christopher. 2001. "An American Report Finds the Taliban's Violation of Religious Rights 'Particularly Severe.'" *New York Times* (Oct. 27): B3.

Marshall, Gordon. 1998. *A Dictionary of Sociology* (2nd ed.). New York: Oxford University Press.

Marshall, Susan E. 1994. "True Women or New Women? Status

Maintenance and Antisuffrage Mobilization in the Gilded Age." Paper presented at the American Sociological Association 89th annual meeting, Los Angeles.

Martin, Carol L. 1989. "Children's Use of Gender-Related Information in Making Social Judgments." *Developmental Psychology*, 25: 80–88.

Martin, Susan Ehrlich, and Nancy C. Jurik. 1996. *Doing Justice, Doing Gender*. Thousand Oaks, CA: Sage.

Martineau, Harriet. 1962. *Society in America* (edited, abridged). Garden City, NY: Doubleday (orig. pub. 1837).

Marx, Karl. 1967. *Capital: A Critique of Political Economy*. Ed. Friedrich Engels. New York: International (orig. pub. 1867).

Marx, Karl, and Friedrich Engels. 1967. *The Communist Manifesto*. New York: Pantheon (orig. pub. 1848).

———. 1970. *The German Ideology*, Part 1. Ed. C. J. Arthur. New York: International (orig. pub. 1845–1846).

Massey, Douglas J., G. Hugo Arango, A. Kowasuci, A. Pellegrino, and J. E. Taylor. 1993. "Theories of International Migration: A Review and Appraisal." *Population and Development Review*, 19: 431–466.

Massey, James L., and Marvin D. Krohn. 1986. "A Longitudinal Examination of an Integrated Social Process Model of Deviant Behavior." *Social Forces*, 65: 106–134.

Maynard, R. A. 1996. *Kids Having Kids: A Robin Hood Foundation Special Report on the Costs of Adolescent Childbearing*. New York: Robin Hood Foundation.

McAdam, Doug. 1982. *Political Process and the Development of Black Insurgency*. Chicago: University of Chicago Press.

———. 1996. "Conceptual Origins, Current Problems, Future Directions." In Doug McAdam, John D. McCarthy, and Meyer N. Zald (Eds.), *Comparative Perspectives on Social Movements*. New York: Cambridge University Press, pp. 23–40.

McAdam, Doug, John D. McCarthy, and Mayer N. Zald. 1988. "Social Movements." In Neil J. Smelser (Ed.), *Handbook of Sociology*. Newbury Park, CA: Sage, pp. 695–737.

McCall, George J., and Jerry L. Simmons. 1978. *Identities and Interactions: An Explanation of Human Associations in Everyday Life*. New York: Free Press.

McCall, Nathan. 1994. *Makes Me Wanna Holler: A Young Black Man in America*. New York: Random House.

McCann, Lisa M. 1997. "Patrilocal Co-Residential Units (PCUs) in Al-Barba: Dual Household Structure in

a Provincial Town in Jordan." *Journal of Comparative Family Studies* (Summer): 113–136.

McCarthy, John D., and Mayer N. Zald. 1977. "Resource Mobilization and Social Movements: A Partial Theory." *American Journal of Sociology*, 82: 1212–1241.

McCarthy, Terry. 2001. "Stirrings of a Woman's Movement." *Time* (Dec. 3): 46.

McChesney, Robert W. 1998. "The Political Economy of Global Communication." In Robert W. McChesney, Ellen Meiksins Wood, and John Bellamy Foster (Eds.), *Capitalism and the Information Age: The Political Economy of the Global Communication Revolution*. New York: Monthly Review Press, pp. 1–26.

McDonnell, Janet A. 1991. *The Dispossession of the American Indian, 1887–1934*. Bloomington: Indiana University Press.

McEachern, William A. 1994. *Economics: A Contemporary Introduction*. Cincinnati: South-Western.

McGeary, Johanna. 2001. "The Taliban Troubles." *Time* (Oct. 1): 36–43.

McGee, Reece. 1975. *Points of Departure*. Hinsdale, IL: Dryden.

McGuire, Meredith B. 1997. *Religion: The Social Context* (4th ed.). Belmont, CA: Wadsworth.

———. 2002. *Religion: The Social Context* (5th ed.). Belmont, CA: Wadsworth.

McIntyre, L. D., and E. Pernell. 1985. "The Impact of Race on Teacher Recommendations for Special Education Placement." *Journal of Multicultural Counseling and Development*, 13: 112–120.

McKenzie, Roderick D. 1925. "The Ecological Approach to the Study of the Human Community." In Robert Park, Ernest Burgess, and Roderick D. McKenzie, *The City*. Chicago: University of Chicago Press.

McLanahan, Sara, and Karen Booth. 1991. "Mother-Only Families." In Alan Booth (Ed.), *Contemporary Families: Looking Forward, Looking Backward*. Minneapolis: National Council on Family Relations, pp. 405–428.

McPhail, Clark. 1991. *The Myth of the Maddening Crowd*. New York: Aldine de Gruyter.

McPhail, Clark, and Ronald T. Wohlstein. 1983. "Individual and Collective Behavior Within Gatherings, Demonstrations, and Riots." In Ralph H. Turner and James F. Short, Jr. (Eds.), *Annual Review of Sociology*, vol. 9. Palo Alto, CA: Annual Reviews, pp. 579–600.

McPherson, Barry D., James E. Curtis, and John W. Loy. 1989. *The Social*

Significance of Sport: An Introduction to the Sociology of Sport. Champaign, IL: Human Kinetics.

McPherson, J. Miller, and Lynn Smith-Lovin. 1982. "Women and Weak Ties: Differences by Sex in the Size of Voluntary Organizations." *American Journal of Sociology*, 87 (January): 883–904.

———. 1986. "Sex Segregation in Voluntary Associations." *American Sociological Review*, 51 (February): 61–79.

Mead, George Herbert. 1934. *Mind, Self, and Society*. Chicago: University of Chicago Press.

Medved, Michael. 1992. *Hollywood vs. America: Popular Culture and the War on Traditional Values*. New York: HarperPerennial.

Mental Medicine. 1994. "Wealth, Health, and Status." *Mental Medicine Update*, 3 (2): 7.

Merchant, Carolyn. 1983. *The Death of Nature: Women, Ecology, and the Scientific Revolution*. San Francisco: Harper & Row.

———. 1992. *Radical Ecology: The Search for a Livable World*. New York: Routledge.

Merton, Robert King. 1938. "Social Structure and Anomie." *American Sociological Review*, 3 (6): 672–682.

———. 1949. "Discrimination and the American Creed." In Robert M. MacIver (Ed.), *Discrimination and National Welfare*. New York: Harper & Row, pp. 99–126.

———. 1968. *Social Theory and Social Structure* (enlarged ed.). New York: Free Press.

Messner, Michael A., Margaret Carlisle Duncan, and Kerry Jensen. 1993. "Separating the Men from the Girls: The Gendered Language of Televised Sports." *Gender & Society*, 7 (1): 121–137.

Meyer, David S., and Suzanne Staggenborg. 1996. "Movements, Countermovements, and the Structure of Political Opportunity." *American Journal of Sociology*, 101: 1628–1660.

Miall, Charlene. 1986. "The Stigma of Involuntary Childlessness." *Social Problems*, 33 (4): 268–282.

Michael, Robert T., John H. Gagnon, Edward O. Laumann, and Gina Kolata. 1994. *Sex in America*. Boston: Little, Brown.

Michels, Robert. 1949. *Political Parties*. Glencoe, IL: Free Press (orig. pub. 1911).

Mickelson, Roslyn Arlin, and Stephen Samuel Smith. 1995. "Education and the Struggle Against Race, Class, and Gender Inequality." In Margaret L. Andersen and Patricia Hill Collins (Eds.), *Race, Class, and Gender* (2nd ed.). Belmont, CA: Wadsworth, pp. 289–304.

Middleton, Nick. 1999. *The Global Casino: An Introduction to Environmental Issues* (2nd ed.). London: Arnold.

Mies, Maria, and Vandana Shiva. 1993. *Ecofeminism*. Highlands, NJ: Zed.

Miethe, Terance, and Charles Moore. 1987. "Racial Differences in Criminal Processing: The Consequences of Model Selection on Conclusions About Differential Treatment." *Sociological Quarterly*, 27: 217–237.

Milgram, Stanley. 1963. "Behavioral Study of Obedience." *Journal of Abnormal and Social Psychology*, 67: 371–378.

———. 1974. *Obedience to Authority*. New York: Harper & Row.

Miller, Casey, and Kate Swift. 1991. *Words and Women: New Language in New Times* (updated ed.). New York: HarperCollins.

Miller, Dan E. 1986. "Milgram Redux: Obedience and Disobedience in Authority Relations." In Norman K. Denzin (Ed.), *Studies in Symbolic Interaction*. Greenwich, CT: JAI, pp. 77–106.

Miller, L. Scott. 1995. *An American Imperative: Accelerating Minority Educational Advancement*. New Haven, CT: Yale University Press.

Mills, C. Wright. 1956. *White Collar*. New York: Oxford University Press.

———. 1959a. *The Power Elite*. Fair Lawn, NJ: Oxford University Press.

———. 1959b. *The Sociological Imagination*. London: Oxford University Press.

Mills, Robert. 2001. "Health Insurance Coverage: 2000." U.S. Census Bureau. Retrieved Nov. 22, 2001. Online: http://www.census.gov/hhes/hlthins/hlthin00/hlt00asc.html

Min, Pyong Gap. 1988. "The Korean American Family." In Charles H. Mindel, Robert W. Habenstein, and Roosevelt Wright, Jr. (Eds.), *Ethnic Families in America: Patterns and Variations* (3rd ed.). New York: Elsevier, pp. 199–229.

Mindel, Charles H., Robert W. Habenstein, and Roosevelt Wright, Jr. (Eds.). 1988. *Ethnic Families in America: Patterns and Variations* (3rd ed.). New York: Elsevier.

Mintz, Beth, and Michael Schwartz. 1985. *The Power Structure of American Business*. Chicago: University of Chicago Press.

Mirowsky, John. 1996. "Age and the Gender Gap in Depression." *Journal of Health and Social Behavior*, 37 (December): 362–380.

Money, John, and Anke A. Ehrhardt. 1972. *Man and Woman, Boy and Girl*. Baltimore: Johns Hopkins University Press.

Montaner, Carlos Alberto. 1992. "Talk English—You Are in the United States." In James Crawford (Ed.), *Language Loyalties: A Source Book on the Official English Controversy*. Chicago: University of Chicago Press, pp. 163–165.

Moody, Harry R. 1998. *Aging: Concepts and Controversy* (2nd ed.). Thousand Oaks, CA: Pine Forge.

Moore, K. A., A. K. Driscoll, and L. D. Lindberg. 1998. *A Statistical Portrait of Adolescent Sex, Contraception, and Childbearing*. Washington, DC: National Campaign to Prevent Teen Pregnancy.

Moore, Patricia, with C. P. Conn. 1985. *Disguised*. Waco, TX: Word.

Moore, Wilbert E. 1968. "Occupational Socialization." In David A. Goslin (Ed.), *Handbook on Socialization Theory and Research*. Chicago: Rand McNally, pp. 861–883.

Morris, Aldon. 1981. "Black Southern Student Sit-In Movement: An Analysis of Internal Organization." *American Sociological Review*, 46: 744–767.

Morselli, Henry. 1975. *Suicide: An Essay on Comparative Moral Statistics*. New York: Arno (orig. pub. 1881).

"Mouse Makes the Man." 1995. *New York Times* (Nov. 19): E2.

Mujica, Mauro E. 1993. "Why a Hispanic Heads an Organization Called U.S. English." *The New Yorker* (Dec. 27): 101.

Murdock, George P. 1945. "The Common Denominator of Cultures." In Ralph Linton (Ed.), *The Science of Man in the World Crisis*. New York: Columbia University Press, pp. 123–142.

Murphy, Robert E., Jessica Scheer, Yolanda Murphy, and Richard Mack. 1988. "Physical Disability and Social Liminality: A Study in the Rituals of Adversity." *Social Science and Medicine*, 26: 235–242.

Mydans, Seth. 1993. "A New Tide of Immigration Brings Hostility to the Surface, Poll Finds." *New York Times* (June 27): A1.

———. 1995. "Part-Time College Teaching Rises, as Do Worries." *New York Times* (Jan. 4): B6.

———. 1997. "Its Mood Dark as the Haze, Southeast Asia Aches." *New York Times* (Oct. 26): 3.

Myrdal, Gunnar. 1970. *The Challenge of World Poverty: A World Anti-Poverty Program in Outline*. New York: Pantheon/Random House.

Nacos, Brigitte. 2002. "Terrorism, the Mass Media, and the Events of 9-11." *Phi Kappa Phi Forum* (Spring): 13–19.

Naffine, Ngaire. 1987. *Female Crime: The Construction of Women in Criminology*. Boston: Allen & Unwin.

Nagourney, Adam. 2002. "The Battleground Shifts." *New York Times* (June 28): A1, A17.

National Campaign to Prevent Teen Pregnancy. 1997. *Whatever Happened to Childhood? The Problem of Teen Pregnancy in the United States*. Washington, DC: National Campaign to Prevent Teen Pregnancy.

National Center for Health Statistics. 1997. "Report of Final Mortality Statistics, 1995." Retrieved July 26, 1998. Online: http://www.cdc.gov/nchswww/release/97facts/97sheets/95morrel.htm

———. 1999. "Report of Final Mortality Statistics, 1998." Retrieved Aug. 26, 2000. Online: http://www.cdc.gov/nchs/data/nvs48_11.pdf

National Center for Injury Prevention and Control. 1999. "10 Leading Causes of Death, United States, 1997." Retrieved Aug. 26, 2000. Online: http://webapp.cdc.gov/sasweb/ncipc/leadcaus.html

National Centers for Disease Control. 2001. "43 Percent of First Marriages Break Up Within 15 Years." Retrieved July 14, 2002. Online: http://www.cdc.gov/nchs/releases/01news/firstmarr.htm

———. 2002. "Update: AIDS—United States, 2000." Retrieved July 21, 2002. Online: http://www.cdc.gov/mmwr/preview/mmwrhtml/mm5127a2.htm#top

National Council on Crime and Delinquency. 1969. *The Infiltration into Legitimate Business by Organized Crime*. Washington, DC: National Council on Crime and Delinquency.

National Institute of Mental Health. 2002. "Suicide Facts." Online: http://www.nimh.nih.gov/research/suifact.htm. Retrieved: May 24, 2002.

Navarrette, Ruben, Jr. 1997. "A Darker Shade of Crimson." In Diana Kendall (Ed.), *Race, Class, and Gender in a Diverse Society*. Boston: Allyn & Bacon, 1997: 274–279. Reprinted from Ruben Navarrette, Jr., *A Darker Shade of Crimson*. New York: Bantam, 1993.

Navarro, Vidente. 1991. "Race or Class or Race and Class: Growing Mortality Differentials in the United States." *International Journal of Health Services*, 21: 229–235.

Neimark, Jill. 1997. "On the Front Lines of Alternative Medicine." *Psychology Today* (January/February): 51–68.

Nelson, Margaret K., and Joan Smith. 1999. *Working Hard and Making Do: Surviving in Small Town America*. Berkeley: University of California Press.

Nelson, Mariah Burton. 1994. *The Stronger Women Get, the More Men Love Football: Sexism and the American Culture of Sports*. New York: Harcourt.

Neuborne, Ellen. 1995. "Imagine My Surprise." In Barbara Findlen (Ed.), *Listen Up: Voices from the Next Feminist Generation*. Seattle: Seal, pp. 29–35.

New York Times. 1996. "The Megacity Summit." (Apr. 8): A14.

———. 1997. "Shift in Schools' Spending." (Dec. 12): A15.

———. 2002a. "The Landscape on Vouchers." (June 28): A17.

———. 2002b. "Text: Senate Judiciary Committee Hearing, June 6, 2002." Retrieved June 9, 2002. Online: http://www.nytimes.com/2002/06/06.../06TEXT-INQ2.html

New York Times Magazine. 1998. "Child-Care Caste System." (Apr. 5): 41.

Newburger, Eric C. 2001. "Home Computers and Internet Use in the United States, August 2000." *Current Population Reports*, P23–207. U.S. Census Bureau. Retrieved Sept. 8, 2001. Online: http://www.census.gov/population/www/socdemo/computer.html

Newitz, Annalee. 1993. "I Was a Credit Card Virgin." *Bad Subjects: Political Education for Everyday Life* (e-magazine). Retrieved May 30, 1999. Online: http://english-www.hss.cmu.edu/bs/08/Newitz-Rubio-Caffrey.html

Newman, Katherine S. 1988. *Falling from Grace: The Experience of Downward Mobility in the American Middle Class*. New York: Free Press.

———. 1993. *Declining Fortunes: The Withering of the American Dream*. New York: Basic.

———. 1999. *No Shame in My Game: The Working Poor in the Inner City*. New York: Knopf and the Russell Sage Foundation.

Niebuhr, H. Richard. 1929. *The Social Sources of Denominationalism*. New York: Meridian.

Nielsen, Joyce McCarl. 1990. *Sex and Gender in Society: Perspectives on Stratification* (2nd ed.). Prospects Heights, IL: Waveland.

Nisbet, Robert. 1979. "Conservatism." In Tom Bottomore and Robert Nisbet (Eds.), *A History of Sociological Analysis*. London: Heinemann, pp. 81–117.

Noble, Barbara Presley. 1995. "A Level Playing Field, for Just $121." *New York Times* (Mar. 5): F21.

Noel, Donald L. 1972. *The Origins of American Slavery and Racism*. Columbus, OH: Merrill.

Nolan, Patrick, and Gerhard E. Lenski. 1999. *Human Societies: An Introduction to Macrosociology* (8th ed.). New York: McGraw-Hill.

Norman, Michael. 1993. "One Cop, Eight Square Blocks." *New York Times Magazine* (Dec. 12): 62–90, 96.

NOW (National Organization of Women). 2002. "Stop the Abuse of Women and Girls in Afghanistan!" Retrieved July 14, 2002. Online: http://www.nowfoundation.org/global/taliban.html

Nuland, Sherwin B. 1997. "Heroes of Medicine." *Time* (Fall special edition): 6–10.

Oakes, Jeannie. 1985. *Keeping Track: How High Schools Structure Inequality.* New Haven, CT: Yale University Press.

Oberschall, Anthony. 1973. *Social Conflict and Social Movements.* Englewood Cliffs, NJ: Prentice Hall.

Oboler, Suzanne. 1995. *Ethnic Labels, Latino Lives: Identity and the Politics of (Re)presentation in the United States.* Minneapolis: University of Minnesota Press.

O'Connell, Helen. 1994. *Women and the Family.* Prepared for the UN-NGO Group on Women and Development. Atlantic Highlands, NJ: Zed.

Odendahl, Teresa. 1990. *Charity Begins at Home: Generosity and Self-Interest Among the Philanthropic Elite.* New York: Basic.

Ogburn, William F. 1966. *Social Change with Respect to Culture and Original Nature.* New York: Dell (orig. pub. 1922).

Olmos, David R. 1997. "Dr. Robot in the OR." *Austin American-Statesman* (July 21): D1, D8.

Omi, Michael, and Howard Winant. 1994. *Racial Formation in the United States: From the 1960s to the 1990s.* New York: Routledge.

Orenstein, Peggy, in association with the American Association of University Women. 1995. *SchoolGirls: Young Women, Self-Esteem, and the Confidence Gap.* New York: Anchor/Doubleday.

Ortner, Sherry B. 1974. "Is Female to Male as Nature Is to Culture?" In Michelle Rosaldo and Louise Lamphere (Eds.), *Women, Culture, and Society.* Stanford, CA: Stanford University Press.

Ortner, Sherry B., and Harriet Whitehead (Eds.). 1981. *Sexual Meanings: The Cultural Construction of Gender and Sexuality.* Cambridge, MA: Cambridge University Press.

Orum, Anthony M. 1974. "On Participation in Political Protest Movements." *Journal of Applied Behavioral Science,* 10: 181–207.

Orum, Anthony M., and Amy W. Orum. 1968. "The Class and Status Bases of Negro Student Protest." *Social Science Quarterly,* 49 (December): 521–533.

Ostrower, Francie. 1997. *Why the Wealthy Give: The Culture of Elite Philanthropy.* Princeton, NJ: Princeton University Press.

Ouchi, William. 1981. *Theory Z: How American Business Can Meet the Japanese Challenge.* Reading, MA: Addison-Wesley.

Oxendine, Joseph B. 1995. *American Indian Sports Heritage* (rev. ed.). Lincoln: University of Nebraska Press.

Padilla, Felix M. 1993. *The Gang as an American Enterprise.* New Brunswick, NJ: Rutgers University Press.

Page, Charles H. 1946. "Bureaucracy's Other Face." *Social Forces,* 25 (October): 89–94.

Palen, J. John. 1995. *The Suburbs.* New York: McGraw-Hill.

Parenti, Michael. 1994. *Land of Idols: Political Mythology in America.* New York: St. Martin's.

———. 1998. *America Besieged.* San Francisco: City Lights.

Park, Robert E. 1915. "The City: Suggestions for the Investigation of Human Behavior in the City." *American Journal of Sociology,* 20: 577–612.

———. 1928. "Human Migration and the Marginal Man." *American Journal of Sociology,* 33.

———. 1936. "Human Ecology." *American Journal of Sociology,* 42: 1–15.

Park, Robert E., and Ernest W. Burgess. 1921. *Human Ecology.* Chicago: University of Chicago Press.

Parker, Robert Nash. 1995. "Violent Crime." In Joseph F. Sheley (Ed.), *Criminology: A Contemporary Handbook* (2nd ed.). Belmont, CA: Wadsworth, pp. 169–185.

Parrish, Dee Anna. 1990. *Abused: A Guide to Recovery for Adult Survivors of Emotional/Physical Child Abuse.* Barrytown, NY: Station Hill.

Parsons, Talcott. 1951. *The Social System.* Glencoe, IL: Free Press.

———. 1955. "The American Family: Its Relations to Personality and to the Social Structure." In Talcott Parsons and Robert F. Bales (Eds.), *Family, Socialization and Interaction Process.* Glencoe, IL: Free Press, pp. 3–33.

Patros, Philip G., and Tonia K. Shamoo. 1989. *Depression and Suicide in Children and Adolescents: Prevention, Intervention, and Postvention.* Boston: Allyn & Bacon.

PBS. 1992a. "The Glory and the Power, Part I."

———. 1992b. "Sex, Power, and the Workplace."

Pear, Robert. 1994. "Health Advisers See Peril in Plan to Cut Medicare." *New York Times* (Aug. 31): A1, A10.

Pearce, Diana. 1978. "The Feminization of Poverty: Women, Work, and Welfare." *Urban and Social Change Review,* 11 (1/2): 28–36.

Pearson, Judy C. 1985. *Gender and Communication.* Dubuque, IA: Brown.

Pedrick-Cornell, Claire, and Richard J. Gelles. 1982. "Elderly Abuse: The Status of Current Knowledge." *Family Relations,* 31: 457–465.

People. 1999. "Ready, Set, Go Buy Something." (June 21): 106–114.

Pérez, Ramón ("Tianguis"). 1991. *Diary of an Undocumented Immigrant.* Houston: Arte Publico.

Perrow, Charles. 1986. *Complex Organizations: A Critical Essay* (3rd ed.). New York: Random House.

Perry, David C., and Alfred J. Watkins (Eds.). 1977. *The Rise of the Sunbelt Cities.* Beverly Hills, CA: Sage.

Perry-Jenkins, Maureen, and Ann C. Crouter. 1990. "Men's Provider Role Attitudes: Implications for Household Work and Marital Satisfaction." *Journal of Family Issues,* 11: 136–156.

Peters, John F. 1985. "Adolescents as Socialization Agents to Parents." *Adolescence,* 20 (Winter): 921–933.

Petersen, John L. 1994. *The Road to 2015: Profiles of the Future.* Corte Madera, CA: Waite Group.

Peterson, Robert. 1992. *Only the Ball Was White: A History of Legendary Black Players and All-Black Professional Teams.* New York: Oxford University Press (orig. pub. 1970).

Peyser, Marc. 2002. "The Insiders." *Newsweek* (July 1): 38–43.

Phillips, John C. 1993. *Sociology of Sport.* Boston: Allyn & Bacon.

Piaget, Jean. 1954. *The Construction of Reality in the Child.* Trans. Margaret Cook. New York: Basic.

Pietilä, Hilkka, and Jeanne Vickers. 1994. *Making Women Matter: The Role of the United Nations.* Atlantic Highlands, NJ: Zed.

Pillard, Richard C., and James D. Weinrich. 1986. "Evidence of Familial Nature of Male Homosexuality." *Archives of General Psychiatry,* 43 (8): 800–812.

Pillemer, Karl A. 1985. "The Dangers of Dependency: New Findings on Domestic Violence Against the Elderly." *Social Problems,* 33 (December): 146–158.

Pillemer, Karl A., and David Finkelhor. 1988. "The Prevalence of Elder Abuse: A Random Sample Survey." *Gerontologist,* 28 (1): 51–57.

Pinderhughes, Dianne M. 1986. "Political Choices: A Realignment in Partisanship Among Black Voters?" In James D. Williams (Ed.), *The State of Black America 1986.* New York: National Urban League, pp. 85–113.

Pinderhughes, Howard. 1997. *Race in the Hood: Conflict and Violence Among Urban Youth.* Minneapolis: University of Minnesota Press.

Pines, Maya. 1981. "The Civilizing of Genie." *Psychology Today,* 15 (September): 28–29, 31–32, 34.

Polakow, Valerie. 1993. *Lives on the Edge: Single Mothers and Their Children in the Other America.* Chicago: University of Chicago Press.

Polanyi, Karl. 1944. *The Great Transformation: The Political and Economic Origins of Our Time.* New York: Beacon.

Popenoe, David, and Barbara Dafoe Whitehead. 1999. "The State of Our Unions: The Social Health of Marriage in America" (June). Retrieved Sept. 21, 1999. Online: http://marriage.rutgers.edu/State.html

Population Reference Bureau. 2001. "Human Population: Fundamentals of Growth Patterns of World Urbanization." Retrieved Nov. 22, 2001. Online: http://www.prb.org/Content/NavigationMenu/PRB/E.../Patterns_of_World_Urbanization.html

Porter, Judith D. R. 1971. *Black Child, White Child.* Cambridge, MA: Harvard University Press.

Portes, Alejandro, and Rubén G. Rumbaut. 1996. *Immigrant America: A Portrait* (2nd ed.). Berkeley: University of California Press.

Powell, Brian, and Douglas B. Downey. 1997. "Living in Single-Parent Households: An Investigation of the Same-Sex Hypothesis." *American Sociological Review,* 62 (August): 521–539.

Presthus, Robert. 1978. *The Organizational Society.* New York: St. Martin's.

Pryor, John, and Kathleen McKinney (Eds.). 1991. "Sexual Harassment." *Basic and Applied Social Psychology,* 17 (4). Marketed as a book; Hillsdale, NJ: Erlbaum.

Puffer, J. Adams. 1912. *The Boy and His Gang.* Boston: Houghton Mifflin.

Quarantelli, E. L., and James R. Hundley, Jr. 1993. "A Test of Some Propositions About Crowd Formation and Behavior." In Russell L. Curtis, Jr., and Benigno E. Aguirre (Eds.), *Collective Behavior and Social Movements.* Boston: Allyn & Bacon, pp. 183–193.

Queen, Stuart A., and David B. Carpenter. 1953. *The American City.* New York: McGraw-Hill.

Quinney, Richard. 1974. *Critique of the Legal Order.* Boston: Little, Brown.

———. 1979. *Class, State, and Crime.* New York: McKay.

———. 1980. *Class, State, and Crime* (2nd ed.). New York: Longman.

Qvortrup, Jens. 1990. *Childhood as a Social Phenomenon.* Vienna:

European Centre for Social Welfare Policy and Research.

Radcliffe-Brown, A. R. 1952. *Structure and Function in Primitive Society*. New York: Free Press.

Raffalli, Mary. 1994. "Why So Few Women Physicists?" *New York Times Supplement* (January): Sect. 4A, 26–28.

Ramirez, Marc. 1999. "A Portrait of a Local Muslim Family." *Seattle Times* (Jan. 24). Retrieved Aug. 16, 1999. Online: http://archives.seattletimes.com/cgi-bin/texis.mummy/web/vortex/display?storyID=36d4d218

Reckless, Walter C. 1967. *The Crime Problem*. New York: Meredith.

Reich, Robert. 1993. "Why the Rich Are Getting Richer and the Poor Poorer." In Paul J. Baker, Louis E. Anderson, and Dean S. Dorn (Eds.), *Social Problems: A Critical Thinking Approach* (2nd ed.). Belmont, CA: Wadsworth, pp. 145–149. Adapted from *The New Republic*, May 1, 1989.

Reid, Sue Titus. 1987. *Criminal Justice*. St. Paul, MN: West.

Reiman, Jeffrey. 1998. *The Rich Get Richer and the Poor Get Prison: Ideology, Class, and Criminal Justice* (5th ed.). Boston: Allyn & Bacon.

Reinharz, Shulamit. 1992. *Feminist Methods in Social Research*. New York: Oxford University Press.

Reinisch, June. 1990. *The Kinsey Institute New Report on Sex: What You Must Know to Be Sexually Literate*. New York: St. Martin's.

Reissman, Catherine. 1991. *Divorce Talk: Women and Men Make Sense of Personal Relationships*. New Brunswick, NJ: Rutgers University Press.

Relman, Arnold S. 1992. "Self-Referral—What's at Stake?" *New England Journal of Medicine*, 327 (Nov. 19): 1522–1524.

Reskin, Barbara F., and Heidi Hartmann. 1986. *Women's Work, Men's Work: Sex Segregation on the Job*. Washington, DC: National Academy Press.

Reskin, Barbara F., and Irene Padavic. 1994. *Women and Men at Work*. Thousand Oaks, CA: Pine Forge.

Richardson, Laurel. 1993. "Inequalities of Power, Property, and Prestige." In Virginia Cyrus (Ed.), *Experiencing Race, Class, and Gender in the United States*. Mountain View, CA: Mayfield, pp. 229–236.

Richardson, Lynda. 1994. "Minority Students Languish in Special Education System." *New York Times* (Apr. 6): A1, B8.

Ricketts, Thomas C. III. 1999. "Preface." In Thomas C. Ricketts III (Ed.), *Rural Health in the United States*. New York: Oxford University Press, pp. vii–viii.

Riggs, Robert O., Patricia H. Murrell, and JoAnne C. Cutting. 1993. *Sexual Harassment in Higher Education: From Conflict to Community*. ASHE-ERIC Higher Education Reports, 93–2. Washington, DC: George Washington University.

Rigler, David. 1993. "Letters: A Psychologist Portrayed in a Book About an Abused Child Speaks Out for the First Time in 22 Years." *New York Times Book Review* (June 13): 35.

Risman, Barbara J. 1987. "Intimate Relationships from a Microstructural Perspective: Men Who Mother." *Gender & Society*, 1: 6–32.

Ritzer, George. 1995. *Expressing America: A Critique of the Global Credit Card Society*. Thousand Oaks, CA: Pine Forge.

———. 1997. *Postmodern Society Theory*. New York: McGraw-Hill.

———. 1999. *Enchanting a Disenchanted World: Revolutionizing the Means of Consumption*. Thousand Oaks, CA: Pine Forge.

———. 2000a. *Modern Sociological Theory* (5th ed.). New York: McGraw-Hill.

———. 2000b. *Sociological Theory* (5th ed.). New York: McGraw-Hill.

Rizzo, Thomas A., and William A. Corsaro. 1995. "Social Support Processes in Early Childhood Friendships: A Comparative Study of Ecological Congruences in Enacted Support." *American Journal of Community Psychology*, 23: 389–418.

Roberts, Keith A. 1995. *Religion in Sociological Perspective*. Belmont, CA: Wadsworth.

Robinson, Brian E. 1988. *Teenage Fathers*. Lexington, MA: Lexington.

Robson, Ruthann. 1992. *Lesbian (Out)law: Survival Under the Rule of Law*. New York: Firebrand.

Rockwell, John. 1994. "The New Colossus: American Culture as Power Export." *New York Times* (Jan. 30): Sect. 2, pp. 1, 30.

Rodriguez, Clara E. 1989. *Puerto Ricans: Born in the U.S.A.* New York: Unwin Hyman.

Roethlisberger, Fritz J., and William J. Dickson. 1939. *Management and the Worker*. Cambridge, MA: Harvard University Press.

Rogers, Harrell R. 1986. *Poor Women, Poor Families: The Economic Plight of America's Female-Headed Households*. Armonk, NY: Sharpe.

Rollins, Judith. 1985. *Between Women: Domestics and Their Employers*. Philadelphia: Temple University Press.

Romero, Mary. 1997. "Introduction." In Mary Romero, Pierrette Hondagneu-Sotelo, and Vilma Ortiz (Eds.), *Challenging Fronteras: Structuring Latina and Latino Lives in the U.S.* New York: Routledge, pp. 3–5.

Roof, Wade Clark. 1993. *A Generation of Seekers: The Spiritual Journeys of the Baby Boom Generation*. San Francisco: HarperSanFrancisco.

Ropers, Richard H. 1991. *Persistent Poverty: The American Dream Turned Nightmare*. New York: Plenum.

Rose, Jerry D. 1982. *Outbreaks*. New York: Free Press.

Rosenblatt, Roger. 1999. "Let Rivers Run Deep." *Time* (Aug. 2): 74–77.

Rosengarten, Ellen M. 1995. Communication to author.

Rosenthal, Naomi, Meryl Fingrutd, Michele Ethier, Roberta Karant, and David McDonald. 1985. "Social Movements and Network Analysis: A Case Study of Nineteenth-Century Women's Reform in New York State." *American Journal of Sociology*, 90: 1022–1054.

Rosenthal, Robert, and Lenore Jacobson. 1968. *Pygmalion in the Classroom: Teacher Expectation and Student's Intellectual Development*. New York: Holt, Rinehart, and Winston.

Rosnow, Ralph L., and Gary Alan Fine. 1976. *Rumor and Gossip: The Social Psychology of Hearsay*. New York: Elsevier.

Ross, Dorothy. 1991. *The Origins of American Social Science*. Cambridge, England: Cambridge University Press.

Rossi, Alice S. 1992. "Transition to Parenthood." In Arlene Skolnick and Jerome Skolnick (Eds.), *Family in Transition*. New York: HarperCollins, pp. 453–463.

Rossi, Peter H. 1989. *Down and Out in America: The Origins of Homelessness*. Chicago: University of Chicago Press.

Rossides, Daniel W. 1986. *The American Class System: An Introduction to Social Stratification*. Boston: Houghton Mifflin.

Rostow, Walt W. 1971. *The Stages of Economic Growth: A Non-Communist Manifesto* (2nd ed.). Cambridge: Cambridge University Press (orig. pub. 1960).

———. 1978. *The World Economy: History and Prospect*. Austin: University of Texas Press.

Roth, Guenther. 1988. "Marianne Weber and Her Circle." In Marianne Weber, *Max Weber*. New Brunswick, NJ: Transaction, p. xv.

Rothchild, John. 1995. "Wealth: Static Wages, Except for the Rich." *Time* (Jan. 30): 60–61.

Rothman, Robert A. 1993. *Inequality and Stratification: Class, Color, and Gender* (2nd ed.). Englewood Cliffs, NJ: Prentice Hall.

Rousseau, Ann Marie. 1981. *Shopping Bag Ladies: Homeless Women Speak About Their Lives*. New York: Pilgrim.

Rubin, Lillian B. 1986. "A Feminist Response to Lasch." *Tikkun*, 1 (2): 89–91.

———. 1994. *Families on the Fault Line*. New York: HarperCollins.

Russell, Joel. 1997. "Early Lessons." *Hispanic Business* (June): 96–102.

Russo, Nancy Felipe, and Mary A. Jansen. 1988. "Women, Work, and Disability: Opportunities and Challenges." In Michelle Fine and Adrienne Asch (Eds.), *Women with Disabilities: Essays in Psychology, Culture, and Politics*. Philadelphia: Temple University Press.

Rutstein, Nathan. 1993. *Healing in America*. Springfield, MA: Whitcomb.

Rymer, Russ. 1993. *Genie: An Abused Child's Flight from Silence*. New York: HarperCollins.

Sadker, David, and Myra Sadker. 1985. "Is the OK Classroom OK?" *Phi Delta Kappan*, 55: 358–367.

———. 1986. "Sexism in the Classroom: From Grade School to Graduate School." *Phi Delta Kappan*, 68: 512–515.

Sadker, Myra, and David Sadker. 1994. *Failing at Fairness: How America's Schools Cheat Girls*. New York: Scribner.

Safilios-Rothschild, Constantina. 1969. "Family Sociology or Wives' Family Sociology? A Cross-Cultural Examination of Decision-Making." *Journal of Marriage and the Family*, 31 (2): 290–301.

Samovar, Larry A., and Richard E. Porter. 1991. *Communication Between Cultures*. Belmont, CA: Wadsworth.

Samuelson, Paul A., and William D. Nordhaus. 1989. *Economics* (13th ed.). New York: McGraw-Hill.

Sanger, David E. 1994. "Cutting Itself Down to Size: Japan's Inferiority Complex." *New York Times* (Feb. 6): E5.

Sapir, Edward. 1961. *Culture, Language and Personality*. Berkeley: University of California Press.

Sargent, Margaret. 1987. *Sociology for Australians* (2nd ed.). Melbourne, Australia: Longman Cheshire.

Sassen, Saskia. 1991. *The Global City: New York, London, Tokyo*. Princeton, NJ: Princeton University Press.

———. 1995. "On Concentration and Centrality in the Global City." In Paul L. Knox and Peter J. Taylor (Eds.), *World Cities in a World System*. Cambridge, England: Cambridge University Press.

———. 2001. *The Global City: New York, London, Tokyo* (2nd ed.). Princeton, NJ: Princeton University Press.

Schachter, Jason. 2001. "Geographical Mobility: March 1999 to March 2000." U.S. Census Bureau. Current Population Reports P20–538. Washington, DC: U.S. Government Printing Office.

Schaefer, Richard T. 1995. *Race and Ethnicity in the United States.* New York: HarperCollins.

———. 2000. *Racial and Ethnic Groups* (8th ed.). New York: Harper-Collins.

Schattschneider, Elmer Eric. 1969. *Two Hundred Americans in Search of a Government.* New York: Holt, Rinehart & Winston.

Schemo, Diana Jean. 1994. "Suburban Taxes Are Higher for Blacks, Analysis Shows." *New York Times* (Aug. 17): A1, A16.

———. 1996. "Indians in Brazil, Estranged from Their Land, Suffer an Epidemic of Suicide." *New York Times* (Aug. 25): 7.

Schmidt, Peter. 2002. "Next Stop, Supreme Court? Appeals Court Upholds Affirmative Action at University of Michigan Law School." *Chronicle of Higher Education* (May 24): A24.

Schneider, Keith. 1993. "The Regulatory Thickets of Environmental Racism." *New York Times* (Dec. 19): E5.

Schor, Juliet B. 1999. *The Overspent American: Upscaling, Downshifting, and the New Consumer.* New York: HarperPerennial.

Schubert, Hans-Joachim (Ed.). 1998. "Introduction." In *On Self and Social Organization-Charles Horton Cooley.* Chicago: University of Chicago Press, pp. 1–31.

Schur, Edwin M. 1965. *Crimes Without Victims: Deviant Behavior and Public Policy.* Englewood Cliffs, NJ: Prentice Hall.

———. 1983. *Labeling Women Deviant: Gender, Stigma, and Social Control.* Philadelphia: Temple University Press.

Schutz, Alfred. 1967. *The Phenomenology of the Social World.* Evanston, IL: Northwestern University Press (orig. pub. 1932).

Schwartz, John. 2002. "Too Much Information, Not Enough Knowledge." *New York Times* (June 9): WK5.

Schwarz, John E., and Thomas J. Volgy. 1992. *The Forgotten Americans.* New York: Norton.

Scott, Alan. 1990. *Ideology and the New Social Movements.* Boston: Unwin & Hyman.

Scott, Eugenie C. 1992. "The Evolution of Creationism." In Art Must, Jr. (Ed.), *Why We Still Need Public Schools: Church/State Relations, and Visions of Democracy.* Buffalo, NY: Prometheus.

Scott, Joan W. 1986. "Gender: A Useful Category of Historical Analysis." *American Historical Review,* 91 (December): 1053–1075.

Seccombe, Karen. 1991. "Assessing the Costs and Benefits of Children: Gender Comparisons Among Child-free Husbands and Wives." *Journal of Marriage and the Family,* 53 (1): 191–202.

Seegmiller, B. R., B. Suter, and N. Duviant. 1980. *Personal, Socioeconomic, and Sibling Influences on Sex-Role Differentiation.* Urbana: ERIC Clearinghouse of Elementary and Early Childhood Education, ED 176 895, College of Education, University of Illinois.

Seid, Roberta P. 1994. "Too 'Close to the Bone': The Historical Context for Women's Obsession with Slenderness." In Patricia Fallon, Melanie A. Katzman, and Susan C. Wooley (Eds.), *Feminist Perspectives on Eating Disorders.* New York: Guilford, pp. 3–16.

Sengoku, Tamotsu. 1985. *Willing Workers: The Work Ethic in Japan, England, and the United States.* Westport, CT: Quorum.

Serbin, Lisa A., Phyllis Zelkowitz, Anna-Beth Doyle, Dolores Gold, and Bill Wheaton. 1990. "The Socialization of Sex-Differentiated Skills and Academic Performance: A Mediational Model." *Sex Roles,* 23: 613–628.

Serrill, Michael S. 1992. "Struggling to Be Themselves." *Time* (Nov. 9): 52–54.

———. 1997. "Socialism Dies Again." *Time* (Sept. 22): 44.

Shapiro, Joseph P. 1993. *No Pity: People with Disabilities Forging a New Civil Rights Movement.* New York: Times/Random House.

Shaw, Randy. 1999. *Reclaiming America: Nike, Clean Air, and the New National Activism.* Berkeley: University of California Press.

Shawver, Lois. 1998. "Notes on Reading Foucault's *The Birth of the Clinic.*" Retrieved Oct. 2, 1999. Online: http://www.california.com/~rathbone/foucbc.htm

Sheen, Fulton J. 1995. *From the Angel's Blackboard: The Best of Fulton J. Sheen.* Ligouri, MO: Triumph.

Sheff, David. 1995. "If It's Tuesday, It Must Be Dad's House." *New York Times Magazine* (Mar. 26): 64–65.

Sheff, Nick. 1999. "My Long-Distance Life." *Newsweek* (Feb. 15): 16.

Sheley, Joseph F. 1991. *Criminology: A Contemporary Handbook.* Belmont, CA: Wadsworth.

Shenon, Philip. 1994. "China's Mania for Baby Boys Creates Surplus of Bachelors." *New York Times* (Aug. 16): A1, A4.

———. 1997. "Army Puts Blame on Officers." *Austin American-Statesman* (Sept. 12): A1, A6.

Sherman, Suzanne (Ed.). 1992. "Frances Fuchs and Gayle Remick." In *Lesbian and Gay Marriage: Private Commitments, Public Ceremonies.* Philadelphia: Temple University Press, pp. 189–201.

Shevky, Eshref, and Wendell Bell. 1966. *Social Area Analysis: Theory, Illustrative Application and Computational Procedures.* Westport, CT: Greenwood.

Shils, Edward A. 1965. "Charisma, Order, and Status." *American Sociological Review,* 30: 199–213.

Shorris, Earl. 1992. *Latinos: A Biography of the People.* New York: Norton.

Shum, Tedd. 1997. "Olympic Gymnast Chow Makes Impact on All Americans." Retrieved Aug. 15, 1999. Online: http://www.dailybruin.ucla.edu/DB/issues/97/05.30/view.shum.html

Sidel, Ruth. 1986. *Women and Children Last: The Plight of Poor Women in Affluent America.* New York: Viking.

Siegel, Larry J. 1998. *Criminology: Theories, Patterns, and Typologies* (6th ed.). Belmont, CA: West/Wadsworth.

Silversten, Scott. 1998. "Texas A&M LB Nguyen Overcomes Adversity in Several Areas." SLAM! Sports: College Football Notes. Retrieved Aug. 15, 1999. Online: http://www.canoe.ca/statsFBC/BC-FBC-LGNS-CNNSIALBTS-R.html

Simmel, Georg. 1950. *The Sociology of Georg Simmel.* Trans. Kurt Wolff. Glencoe, IL: Free Press (orig. written in 1902–1917).

———. 1957. "Fashion." *American Journal of Sociology,* 62 (May 1957): 541–558. Orig. pub. 1904.

———. 1990. *The Philosophy of Money.* Ed. David Frisby. New York: Routledge (orig. pub. 1907).

Simon, David R. 1996. *Elite Deviance* (5th ed.). Boston: Allyn & Bacon.

Simons, Marlise. 1993. "Prosecutor Fighting Girl-Mutilation." *New York Times* (Nov. 23): A4.

Simpson, George Eaton, and Milton Yinger. 1972. *Racial and Cultural Minorities: An Analysis of Prejudice and Discrimination* (4th ed.). New York: Harper & Row.

Simpson, Sally S. 1989. "Feminist Theory, Crime, and Justice." *Criminology,* 27: 605–632.

Singer, Margaret Thaler, with Janja Lalich. 1995. *Cults in Our Midst.* San Francisco: Jossey-Bass.

Sjoberg, Gideon. 1965. *The Preindustrial City: Past and Present.* New York: Free Press.

Skocpol, Theda, and Edwin Amenta. 1986. "States and Social Policies." In Ralph H. Turner and James F. Short, Jr. (Eds.), *Annual Review of Sociology,* 12: 131–157.

Skolnick, Jerome H. 1975. *Justice Without Trial* (2nd ed.). New York: Wiley.

Smart Growth America. 2001. "Americans Want Growth and Green; Demand Solutions to Traffic, Haphazard Development." Retrieved Nov. 30, 2001. Online: http://www.smartgrowthamerica.com/release.htm

Smart Growth Network. 2001. "About Smart Growth." Retrieved Nov. 30, 2001. Online: http://www.smart-growth.org/about/default.asp

Smelser, Neil J. 1963. *Theory of Collective Behavior.* New York: Free Press.

———. 1988. "Social Structure." In Neil J. Smelser (Ed.), *Handbook of Sociology.* Newbury Park, CA: Sage, pp. 103–129.

Smith, Adam. 1976. *An Inquiry into the Nature and Causes of the Wealth of Nations.* Ed. Roy H. Campbell and Andrew S. Skinner. Oxford, England: Clarendon (orig. pub. 1776).

Smith, Allen C., III, and Sheryl Kleinman. 1989. "Managing Emotions in Medical School: Students' Contacts with the Living and the Dead." *Social Science Quarterly,* 52 (1): 56–69.

Smith, Denise I., and Renee T. Spraggins. 2001. "Gender: 2000." U.S. Census Bureau. Retrieved Nov. 28, 2001. Online: http://www.census.gov/prod/2001pubs/c2kbr01-9.pdf

Smith, Dorothy E. 1999. *Writing the Social: Critique, Theory, and Investigations.* Toronto: University of Toronto Press.

Smith, Douglas, Christy Visher, and Laura Davidson. 1984. "Equity and Discretionary Justice: The Influence of Race on Police Arrest Decisions." *Journal of Criminal Law and Criminology,* 75: 234–249.

Smith, Wes. 2001. *Hope Meadows: Real-Life Stories of Healing and Caring from an Inspiring Community.* New York: Berkley.

Smyke, Patricia. 1991. *Women and Health.* Atlantic Highlands, NJ: Zed.

Snow, David A., and Leon Anderson. 1993. *Down on Their Luck: A Case Study of Homeless Street People.* Berkeley: University of California Press.

Snow, David A., and Robert Benford. 1988. "Ideology, Frame Resonance, and Participant Mobilization." In Bert Klandermans, Hanspeter Kriesi, and Sidney Tarrow (Eds.), *International Social Movement Research,* Vol. 1, *From Structure to Action.* Greenwich, CT: JAI, pp. 133–155.

Snow, David A., E. Burke Rochford, Jr., Steven K. Worden, and Robert D. Benford. 1986. "Frame Alignment Processes, Micromobilization, and Movement Participation." *American Sociological Review*, 51: 464–481.

Snow, David A., Louis A. Zurcher, and Robert Peters. 1981. "Victory Celebrations as Theater: A Dramaturgical Approach to Crowd Behavior." *Symbolic Interaction*, 4 (1): 21–41.

Snyder, Benson R. 1971. *The Hidden Curriculum*. New York: Knopf.

Sommers, Ira, and Deborah R. Baskin. 1993. "The Situational Context of Violent Female Offending." *Journal of Research in Crime and Delinquency*, 30 (2): 136–162.

South, Scott J., Charles M. Bonjean, Judy Corder, and William T. Markham. 1982. "Sex and Power in the Federal Bureaucracy." *Work and Occupations*, 9 (2): 233–254.

Sowell, Thomas. 1981. *Ethnic America*. New York: Basic.

Spence, Jan. 1997. "Homeless in Russia: A Visit with Valery Sokolov." Share International. Retrieved Sept. 15, 2001. Online: http://www.shareintl.org/archives/homelessness/hl-jsRussia.htm

Spindel, Carol. 2000. *Dancing at Halftime: Sports and the Controversy Over American Indian Mascots*. New York: New York University Press.

Sreenivasan, Sreenath. 1996. "Blind Users Add Access on the Web." *New York Times* (Dec. 2): C7.

Stannard, David E. 1992. *American Holocaust: Columbus and the Conquest of the New World*. New York: Oxford University Press.

Stark, Rodney. 1992. *Sociology* (4th ed.). Belmont, CA: Wadsworth.

Stark, Rodney, and William Sims Bainbridge. 1981. "American-Born Sects: Initial Findings." *Journal for the Scientific Study of Religion*, 20: 130–149.

Starr, Paul. 1982. *The Social Transformation of Medicine: The Rise of a Sovereign Profession and the Making of a Vast Industry*. New York: Basic.

Steffensmeier, Darrell, and Emilie Allan. 1995. "Criminal Behavior: Gender and Age." In Joseph F. Sheley (Ed.), *Criminology: A Contemporary Handbook*. Belmont, CA: Wadsworth, pp. 83–113.

Steffensmeier, Darrell, and Cathy Streifel. 1991. "Age, Gender, and Crime Across Three Historical Periods: 1935, 1960, and 1985." *Social Forces*, 69: 869–894.

Stein, Peter J. 1976. *Single*. Englewood Cliffs, NJ: Prentice Hall.

Stein, Peter J. (Ed.). 1981. *Single Life: Unmarried Adults in Social Context*. New York: St. Martin's.

Steinmetz, Suzanne K. 1987. "Elderly Victims of Domestic Violence." In Carl D. Chambers, John H. Lindquist, O. Z. White, and Michael T. Harter (Eds.), *The Elderly: Victims and Deviants*. Athens: Ohio University Press, pp. 126–141.

Stern, Sharon. 2001. "Americans with Disabilities." U.S. Census Bureau. Retrieved Nov. 23, 2001. Online: http://www.census.gov/hhes/www.disability.html

Stevenson, Mary Huff. 1988. "Some Economic Approaches to the Persistence of Wage Differences Between Men and Women." In Ann H. Stromberg and Shirley Harkess (Eds.), *Women Working: Theories and Facts in Perspective* (2nd ed.). Mountain View, CA: Mayfield, pp. 87–100.

Stevenson, Richard W. 1997a. "The Chief Banker for the Nations at the Bottom of the Heap." *New York Times* (Sept. 14): BU1, BU12.

———. 1997b. "World Bank Report Sees Era of Emerging Economies." *New York Times* (Sept. 10): C7.

Stewart, Abigail J. 1994. "Toward a Feminist Strategy for Studying Women's Lives." In Carol E. Franz and Abigail J. Stewart (Eds.), *Women Creating Lives: Identities, Resilience, and Resistance*. Boulder, CO: Westview, pp. 11–35.

Stier, Deborah S., and Judith A. Hall. 1984. "Gender Differences in Touch: An Empirical and Theoretical Review." *Journal of Personality and Social Psychology*, 47 (2): 440–459.

Substance Abuse and Mental Health Services Administration. 2000. *National Household Survey on Drug Abuse, 2000*. Retrieved Nov. 20, 2000. Online: http://www.samhsa.gov/oas/NHSDA/2kNHSDA/chapter2.htm

Sullivan, Teresa A., Elizabeth Warren, and Jay Lawrence Westbrook. 2000. *The Fragile Middle Class: Americans in Debt*. New Haven, CT: Yale University Press.

Sumner, William G. 1959. *Folkways*. New York: Dover (orig. pub. 1906).

Sutherland, Edwin H. 1939. *Principles of Criminology*. Philadelphia: Lippincott.

———. 1949. *White Collar Crime*. New York: Dryden.

Swidler, Ann. 1986. "Culture in Action: Symbols and Strategies." *American Sociological Review*, 51 (April): 273–286.

Symonds, William C. 1990. "Is Sex Discrimination Still Par for the Course?" *Business Week* (Dec. 24): 56.

Tabb, William K., and Larry Sawers. 1984. *Marxism and the Metropolis:*

New Perspectives in Urban Political Economy (2nd ed.). New York: Oxford University Press.

Takagi, Paul. 1979. "Death by Police Intervention." In U.S. Department of Justice, A *Community Concern: Police Use of Deadly Force*. Washington, DC: U.S. Government Printing Office.

Takaki, Ronald. 1989. *Strangers from a Different Shore: A History of Asian Americans*. New York: Penguin.

———. 1993. *A Different Mirror: A History of Multicultural America*. Boston: Little, Brown.

Talwar, Jennifer Parker. 2002. *Fast Food, Fast Track: Immigrants, Big Business, and the American Dream*. Boulder, CO: Westview.

Tannen, Deborah. 1993. "Commencement Address, State University of New York at Binghamton." Reprinted in *Chronicle of Higher Education* (June 9): B5.

Tanner, Robert. 1994. "Angry Words Fly as Flag Debated." *Austin American-Statesman* (Apr. 22): A6.

Tanzer, Andrew. 1996. "The Pacific Century." *Forbes* (July 15): 108–113.

Tarbell, Ida M. 1925. *The History of Standard Oil Company*. New York: Macmillan (orig. pub. 1904).

Tavris, Carol. 1993. *The Mismeasure of Woman*. New York: Touchstone.

Taylor, Carl S. 1993. *Girls, Gangs, Women, and Drugs*. East Lansing: Michigan State University Press.

Taylor, Chris. 1999. "We're Goths and Not Monsters." *Time* (May 3): 45.

Taylor, Howard F. 1995. "Symposium: The Bell Curve." *Contemporary Sociology: A Journal of Reviews*, 24 (2): 153–157.

Taylor, Steve. 1982. *Durkheim and the Study of Suicide*. New York: St. Martin's.

Terkel, Studs. 1990. *Working: People Talk About What They Do All Day and How They Feel About What They Do*. New York: Ballantine (orig. pub. 1972).

Thomas, R. David. 1992. *Dave's Way*. New York: Berkley.

Thompson, Becky W. 1994. *A Hunger So Wide and So Deep: American Women Speak Out on Eating Problems*. Minneapolis: University of Minnesota Press.

Thornberry, Terence P., Marvin D. Krohn, Alan J. Lizotte, and Deborah Chard-Wierschem. 1993. "The Role of Juvenile Gangs in Facilitating Delinquent Behavior." *Journal of Research in Crime and Delinquency*, 30 (1): 55–87.

Thorne, Barrie. 1993. *Gender Play: Girls and Boys in School*. New Brunswick, NJ: Rutgers University Press.

Thorne, Barrie, Cheris Kramarae, and Nancy Henley. 1983. *Language, Gender, and Society*. Rowley, MA: Newbury.

Thornton, Michael C., Linda M. Chatters, Robert Joseph Taylor, and Walter R. Allen. 1990. "Sociodemographic and Environmental Correlates of Racial Socialization by Black Parents." *Child Development*, 61: 401–409.

Thornton, Russell. 1984. "Cherokee Population Losses During the Trail of Tears: A New Perspective and a New Estimate." *Ethnohistory*, 31: 289–300.

Tierney, John. 1993. "Fernando, 16, Finds a Sanctuary in Crime." *New York Times* (Apr. 13): A1, A11.

Tilly, Charles. 1973. "Collective Action and Conflict in Large-Scale Social Change: Research Plans, 1974–78." Center for Research on Social Organization. Ann Arbor: University of Michigan, October.

———. 1975. *The Formation of National States in Western Europe*. Princeton, NJ: Princeton University Press.

———. 1978. *From Mobilization to Revolution*. Reading, MA: Addison-Wesley.

Tilly, Chris, and Charles Tilly. 1998. *Work Under Capitalism*. Boulder, CO: Westview.

Time. 1997. "An Encouraging Report Card." (June 23): 67.

———. 1999. "Numbers." (May 3): 20–52.

Tiryakian, Edward A. 1978. "Emile Durkheim." In Tom Bottomore and Robert Nisbet (Eds.), *A History of Sociological Analysis*. New York: Basic, pp. 187–236.

Tong, Rosemarie. 1989. *Feminist Thought: A Comprehensive Introduction*. Boulder, CO: Westview.

Tönnies, Ferdinand. 1940. *Fundamental Concepts of Sociology (Gemeinschaft und Gesellschaft)*. Trans. Charles P. Loomis. New York: American Book Company (orig. pub. 1887).

———. 1963. *Community and Society (Gemeinschaft and Gesellschaft)*. New York: Harper & Row (orig. pub. 1887).

Tracy, C. 1980. "Race, Crime and Social Policy: The Chinese in Oregon, 1871–1885." *Crime and Social Justice*, 14: 11–25.

Troeltsch, Ernst. 1960. *The Social Teachings of the Christian Churches*, vols. 1 and 2. Trans. O. Wyon. New York: Harper & Row (orig. pub. 1931).

Tumin, Melvin. 1953. "Some Principles of Stratification: A Critical Analysis." *American Sociological Review*, 18 (August): 387–393.

Turk, Austin. 1969. *Criminality and Legal Order.* Chicago: Rand McNally.

——. 1977. "Class, Conflict and Criminology." *Sociological Focus,* 10: 209–220.

Turner, Jonathan, Leonard Beeghley, and Charles H. Powers. 1995. *The Emergence of Sociological Theory* (3rd ed.). Belmont, CA: Wadsworth.

——. 1998. *The Emergence of Sociological Theory* (4th ed.). Belmont, CA: Wadsworth.

Turner, Ralph H., and Lewis M. Killian. 1993. "The Field of Collective Behavior." In Russell L. Curtis, Jr., and Benigno E. Aguirre (Eds.), *Collective Behavior and Social Movements.* Boston: Allyn & Bacon, pp. 5–20.

Twitchell, James B. 1999. *Lead Us into Temptation: The Triumph of American Materialism.* New York: Columbia University Press.

UNAIDS/WHO. 2000. "Report on the Global HIV/AIDS Epidemic—June 2000." Retrieved Aug. 5, 2000. Online: http://www.unaids.org/epidemic_update/report/glo_estim.pdf

UNICEF (United Nations Children's Fund). 1991. *Zur Situation der Kinder in der Welt.* Cologne. Quoted in Ingomar Hauchler and Paul M. Kennedy (Eds.). 1994. *Global Trends: The World Almanac of Development and Peace.* New York: Continuum.

United Nations. 1997. "Global Change and Sustainable Development: Critical Trends." United Nations Department for Policy Coordination and Sustainable Development. Posted online (Jan. 20).

——. 2000. *Long-Range World Population Projections.* New York: United Nations Population Division.

United Nations Development Programme. 1996. *Human Development Report: 1995.* New York: Oxford University Press.

——. 1997. *Human Development Report: 1996.* New York: Oxford University Press.

——. 1999. *Human Development Report: 1999.* New York: Oxford University Press.

——. 2001. *Human Development Report: 2001.* New York: Oxford University Press.

United Nations DPCSD. 1997. "Report of Commission on Sustainable Development, April 1997." New York: United Nations Department for Policy Coordination and Sustainable Development. Posted online.

United Nations Population Division. 1999. Retrieved Oct. 17, 1999.

Online: gopher:/gopher.undp.org/00/ungophers/popin/wdtrends

U.S. Bureau of Labor Statistics. 2000. "Labor Force Statistics from the Current Population Survey." Retrieved: Sept. 29, 2001. Online: http://www.bls.gov/news.release/work.t04.htm

U.S. Census Bureau. 1998a. Current Population Reports, P60–200, *Money Income in the United States: 1997.* Washington, DC: U.S. Government Printing Office.

——. 1998b. *Statistical Abstract of the United States: 1998.* Washington, DC: U.S. Government Printing Office.

——. 2000a. "Employment, Earnings, and Disability." Retrieved Aug. 6, 2000. Online: http://www.census.gov/hhes/www/disable/emperndis.pdf

——. 2000b. "Historical Income Tables-Households." Retrieved Sept. 25, 2001. Online: http://www.census.gov/hhes/income/histinc/h08.html

——. 2000c. *Statistical Abstract of the United States: 2000.* Washington, DC: U.S. Government Printing Office.

——. 2001a. "Apportionment Population and Number of Representatives, by State: Census 2000." Retrieved Oct. 27, 2001. Online: http://www.census.gov/population/cen2000/tab01.pdf

——. 2001b. "Census 2000 Supplementary Survey for United States." Retrieved Sept. 18, 2001. Online: http://www.census.gov/c2ss/www/Products/Profiles/2000/Tabular.htm

——. 2001c. "Health Insurance Coverage: 2000." Retrieved Sept. 29, 2001. Online: http://www.census.gov/hhes/hlthins/hlthin00/hi00tb.html

——. 2001d. "Overview of Race and Hispanic Origin: Census 2000 Brief." Washington, DC: U.S. Department of Commerce.

——. 2001e. "Poverty in the United States: 2000." Retrieved Sept. 27, 2001. Online: http://www.census.gov/hhes/poverty/poverty00.html

——. 2001f. "Profile of General Demographic Characteristics: 2000." Retrieved Sept. 26, 2001. Online: http://www.census.gov/Press_Release/www/2001/tables/dp_us_2000.pdf

——. 2001g. *Statistical Abstract of the United States: 2001.* Washington, DC: U.S. Government Printing Office.

——. 2001h. "Trends and Patterns of HIV Infection in Selected Developing Countries." Retrieved Nov. 23,

2001. Online: http://www.census.gov/ipc/www/hivcf.html

——. 2002. "Disability Data from March Current Population Survey." Retrieved July 24, 2002. Online: http://www.census.gov/hhes/www/disable/disabcps.html

U.S. Conference of Mayors. 2000. "A Status Report on Hunger and Homelessness in America's Cities, 2000." Retrieved Sept. 11, 2001. Online: http://usmayors.org/uscm/news/press_releases/documents/hunger_release.htm

U.S. Congress. 1990. *Indian Adolescent Mental Health.* OTA-H-446. Washington, DC: U.S. Government Printing Office.

U.S. Congress, Office of Technology Assessment. 1979. "The Effects of Nuclear War." Report quoted in Michael E. Howard. 1990. "On Fighting a Nuclear War." In Francesca M. Cancian and James William Gibson (Eds.), *Making War, Making Peace: The Social Foundations of Violent Conflict.* Belmont, CA: Wadsworth, pp. 314–322.

U.S. Department of Agriculture. 1996. *Rural Homelessness: Focusing on the Needs of the Rural Homeless.* Washington, DC: U.S. Department of Agriculture, Rural Housing Service, Rural Economic and Community Development.

U.S. Department of Justice. 2000a. "Capital Punishment Statistics." Retrieved Oct. 31, 2001. Online: http://www.ojp.usdoj.gov/bjs/cp.htm

——. 2000b. "Intimate Partner Violence." Retrieved Nov. 3, 2001. Online: http://www.ojp.usdoj.gov/bjs/pub/press/ipv.htm

U.S. Department of Labor, Employment and Training Administration. 1993. *Dictionary of Occupational Titles.* Washington, DC: U.S. Government Printing Office.

U.S. Health Care Financing Administration. 2000. "National Health Expenditure Amounts, and Average Annual Percent Change, by Type of Expenditure: Selected Calendar Years 1970–2008." Retrieved Sept. 3, 2000. Online: http://www.hcfa.gov/stats/NHE-Proj/proj1998/tables/table2.htm

U.S. Immigration and Naturalization Service. 2000. *1998 Statistical Yearbook of the Immigration and Naturalization Service.* Washington, DC: U.S. Department of Justice, Immigration and Naturalization Service.

Valentine, Paul W. 1992. "Early Morning Raids by Marshals Exploit the Element of Surprise." *Washington Post* (May 1): D1. Retrieved

July 29, 2000. Online: http://www.washingtonpost.com

Van Biema, David. 1993. "But Will It End the Abortion Debate?" *Time* (June 14): 52–54.

Vanneman, Reeve, and Lynn Weber Cannon. 1987. *The American Perception of Class.* Philadelphia: Temple University Press.

Vaughan, Diane. 1985. "Uncoupling: The Social Construction of Divorce." In James M. Henslin (Ed.), *Marriage and Family in a Changing Society* (2nd ed.). New York: Free Press, pp. 429–439.

Vaughan, Ted R., Gideon Sjoberg, and Larry T. Reynolds (Eds.). 1993. *A Critique of Contemporary American Sociology.* Dix Hills, NY: General Hall.

Veblen, Thorstein. 1967. *The Theory of the Leisure Class.* New York: Viking (orig. pub. 1899).

Vetter, Harold J., and Gary R. Perlstein. 1991. *Perspectives on Terrorism.* Pacific Grove, CA: Brooks/Cole.

Vissing, Yvonne. 1996. *Out of Sight, Out of Mind: Homeless Children and Families in Small Town America.* Lexington: University Press of Kentucky.

Vito, Gennaro F., and Ronald M. Holmes. 1994. *Criminology: Theory, Research and Policy.* Belmont, CA: Wadsworth.

Voynick, Steve. 1999. "Living with Ozone." *The World & I* (July): 192–199.

Wagner, Elvin, and Allen E. Stearn. 1945. *The Effects of Smallpox on the Destiny of the American Indian.* Boston: Bruce Humphries.

Waldman, Amy. 2001. "Behind the Burka: Women Subtly Fought Taliban." *New York Times* (Nov. 19): A1, B4.

Waldron, Ingrid. 1994. "What Do We Know About Causes of Sex Differences in Mortality? A Review of the Literature." In Peter Conrad and Rachelle Kern (Eds.), *The Sociology of Health and Illness: Critical Perspectives* (4th ed.). New York: St. Martin's.

Wallace, Harvey. 1996. *Family Violence: Legal, Medical, and Social Perspectives.* Boston: Allyn & Bacon.

Wallace, Walter L. 1971. *The Logic of Science in Sociology.* New York: Aldine de Gruyter.

Wallerstein, Immanuel. 1979. *The Capitalist World-Economy.* Cambridge, England: Cambridge University Press.

——. 1984. *The Politics of the World Economy.* Cambridge, England: Cambridge University Press.

——. 1991. *Unthinking Social Science: The Limits of Nineteenth-*

Century Paradigms. Cambridge, England: Polity.

Warner, W. Lloyd, and Paul S. Lunt. 1941. *The Social Life of a Modern Community.* New Haven, CT: Yale University Press.

Warr, Mark. 1985. "Fear of Rape Among Urban Women." *Social Problems*, 32: 238–250.

———. 1993. "Age, Peers, and Delinquency." *Criminology*, 31 (1): 17–40.

———. 1995. "America's Perceptions of Crime and Punishment." In Joseph F. Sheley (Ed.), *Criminology: A Contemporary Handbook* (2nd ed.). Belmont, CA: Wadsworth, pp. 15–31.

Waters, Malcolm. 1995. *Globalization.* London and New York: Routledge.

Watson, Paul. 1997. "Richer, Poorer." *Toronto Star* (Aug. 10): F1, F5.

Watson, Tracey. 1987. "Women Athletes and Athletic Women: The Dilemmas and Contradictions of Managing Incongruent Identities." *Sociological Inquiry*, 57 (Fall): 431–446.

Waxman, Laura, and Sharon Hinderliter. 1996. *A Status Report on Hunger and Homelessness in America's Cities: 1996.* Washington, DC: U.S. Conference of Mayors.

Weber, Max. 1963. *The Sociology of Religion.* Trans. E. Fischoff. Boston: Beacon (orig. pub. 1922).

———. 1968. *Economy and Society: An Outline of Interpretive Sociology.* Trans. G. Roth and G. Wittich. New York: Bedminster (orig. pub. 1922).

———. 1976. *The Protestant Ethic and the Spirit of Capitalism.* Trans. Talcott Parsons. Introduction by Anthony Giddens. New York: Scribner (orig. pub. 1904–1905).

Weeks, John R. 2002. *Population: An Introduction to Concepts and Issues* (8th ed.). Belmont, CA: Wadsworth.

Weigel, Russell H., and P. W. Howes. 1985. "Conceptions of Racial Prejudice: Symbolic Racism Revisited." *Journal of Social Issues*, 41: 124–132.

Weiner, Tim. 1994. "The Men in the Gray Federal Bureaucracy." *New York Times* (Apr. 10): E4.

Weinstein, Michael M. 1997. "'The Bell Curve,' Revisited by Scholars." *New York Times* (Oct. 11): A20.

Weisner, Thomas S., Helen Garnier, and James Loucky. 1994. "Domestic Tasks, Gender Egalitarian Values and Children's Gender Typing in Conventional and Nonconventional Families." *Sex Roles* (January): 23–55.

Weiss, Gregory L., and Lynne E. Lonnquist. 2003. *The Sociology of Health, Healing, and Illness* (4th ed.). Upper Saddle River, NJ: Prentice Hall.

Weiss, Meira. 1994. *Conditional Love: Attitudes Toward Handicapped*

Children. Westport, CT: Bergin & Garvey.

Weitz, Rose. 2001. *The Sociology of Health, Illness, and Health Care: A Critical Approach* (2nd ed.). Belmont, CA: Wadsworth.

Wellhousen, Karyn, and Zenong Yin. 1997. "Peter Pan Isn't a Girls' Part: An Investigation of Gender Bias in a Kindergarten Classroom." *Women and Language*, 20: 35–40.

Weston, Kath. 1991. *Families We Choose: Lesbians, Gays, Kinship.* New York: Columbia University Press.

Westrum, Ron. 1991. *Technologies and Society: The Shaping of People and Things.* Belmont, CA: Wadsworth.

White, Jack E. 1997. "I'm Just Who I Am." *Time* (May 5): 32–36.

White, Ralph, and Ronald Lippitt. 1953. "Leader Behavior and Member Reaction in Three 'Social Climates.'" In Dorwin Cartwright and Alvin Zander (Eds.), *Group Dynamics.* Evanston, IL: Row, Peterson, pp. 586–611.

White, Richard W. 1992. *Rude Awakening: What the Homeless Crisis Tells Us.* San Francisco: ICS.

Whorf, Benjamin Lee. 1956. *Language, Thought and Reality.* Ed. John B. Carroll. Cambridge, MA: MIT Press.

Whyte, William H., Jr. 1957. *The Organization Man.* Garden City, NY: Anchor.

Williams, Christine L. 1989. *Gender Differences at Work.* Berkeley: University of California Press.

Williams, Christine L. (Ed.). 1993. *Doing "Women's Work": Men in Nontraditional Occupations.* Newbury Park, CA: Sage.

Williams, Robin M., Jr. 1970. *American Society: A Sociological Interpretation* (3rd ed.). New York: Knopf.

Wilson, David (Ed.). 1997. "Globalization and the Changing U.S. City." *Annals of the American Academy of Political and Social Sciences*, 551 (May special issue).

Wilson, Edward O. 1975. *Sociobiology: A New Synthesis.* Cambridge, MA: Harvard University Press.

Wilson, Elizabeth. 1991. *The Sphinx in the City: Urban Life, the Control of Disorder, and Women.* Berkeley: University of California Press.

Wilson, William Julius. 1978. *The Declining Significance of Race: Blacks and Changing American Institutions.* Chicago: University of Chicago Press.

———. 1996. *When Work Disappears: The World of the New Urban Poor.* New York: Knopf.

Winik, Lyric Wallwork. 1997. "Oh Nurse, More Beluga Please." *Forbes*

FYI: The Good Life (Winter): 157–166.

Winn, Maria. 1985. *The Plug-in Drug: Television, Children, and the Family.* New York: Viking.

Wirth, Louis. 1938. "Urbanism as a Way of Life." *American Journal of Sociology*, 40: 1–24.

———. 1945. "The Problem of Minority Groups." In Ralph Linton (Ed.), *The Science of Man in the World Crisis.* New York: Columbia University Press, p. 38.

Wiseman, Jacqueline. 1970. *Stations of the Lost: The Treatment of Skid Row Alcoholics.* Chicago: University of Chicago Press.

Witt, Susan D. 1997. "Parental Influence on Children's Socialization to Gender Roles." *Adolescence*, 32: 253–260.

Wollstonecraft, Mary. 1974. *A Vindication of the Rights of Woman.* New York: Garland (orig. pub. 1797).

Wonders, Nancy. 1996. "Determinate Sentencing: A Feminist and Postmodern Story." *Justice Quarterly*, 13: 610–648.

Wood, Julia T. 1994. *Gendered Lives: Communication, Gender, and Culture.* Belmont, CA: Wadsworth.

———. 1999. *Gendered Lives: Communication, Gender, and Culture* (3rd ed.). Belmont, CA: Wadsworth.

Woodward, Danny. 2002. "Williams Sisters Spark a Net Gain." *Dallas Morning News* (July 10): B1, B6.

Wooley, Susan C. 1994. "Sexual Abuse and Eating Disorders: The Concealed Debate." In Patricia Fallon, Melanie A. Katzman, and Susan C. Wooley (Eds.), *Feminist Perspectives on Eating Disorders.* New York: Guilford, pp. 171–211.

World Bank. 1996. *World Development Report 1996: From Plan to Market.* New York: Oxford University Press for the World Bank.

———. 2001. *World Development Report 2002: Building Institutions for Markets.* Retrieved Sept. 17, 2001. Online: http: econ.worldbank.org/ wdr/subpage.php?pr=2391

World Health Organization. 1946. *Constitution of the World Health Organization.* New York: World Health Organization Interim Commission.

———. 1999. *The World Health Report 1999.* Retrieved Oct. 2, 1999. Online: http://www.who.int/whr/ 1999/en/pdf/burden.pdf

Worster, Donald. 1985. *Natures Economy: A History of Ecological Ideas.* New York: Cambridge University Press.

Wouters, Cas. 1989. "The Sociology of Emotions and Flight Attendants: Hochschild's Managed Heart." *Theory, Culture & Society*, 6: 95–123.

Wright, Erik Olin. 1978. "Race, Class, and Income Inequality." *American Journal of Sociology*, 83 (6): 1397.

———. 1979. *Class Structure and Income Determination.* New York: Academic Press.

———. 1985. *Class.* London: Verso.

———. 1997. *Class Counts: Comparative Studies in Class Analysis.* Cambridge, England: Cambridge University Press.

Wright, Erik Olin, Karen Shire, Shu-Ling Hwang, Maureen Dolan, and Janeen Baxter. 1992. "The Non-Effects of Class on the Gender Division of Labor in the Home: A Comparative Study of Sweden and the U.S." *Gender & Society*, 6 (2): 252–282.

Wright, John W. (Ed.). 1997. *The New York Times 1998 Almanac.* New York: Penguin Reference.

Yablonsky, Lewis. 1997. *Gangsters: Fifty Years of Madness, Drugs, and Death on the Streets of America.* New York: New York University Press.

Yelin, Edward H. 1992. *Disability and the Displaced Worker.* New Brunswick, NJ: Rutgers University Press.

Yinger, J. Milton. 1960. "Contraculture and Subculture." *American Sociological Review*, 25 (October): 625–635.

———. 1982. *Countercultures: The Promise and Peril of a World Turned Upside Down.* New York: Free Press.

Young, John. 1990. *Sustaining the Earth: The Story of the Environmental Movement—Its Past Efforts and Future Challenges.* Cambridge, MA: Harvard University Press.

Young, Michael Dunlap. 1994. *The Rise of the Meritocracy.* New Brunswick, NJ: Transaction (orig. pub. 1958).

Zald, Mayer N., and John D. McCarthy (Eds.). 1987. *Social Movements in an Organizational Society.* New Brunswick, NJ: Transaction.

Zaldivar, R. A. 1996. "1 in 5 Go Without Insurance for a Time, Report Says." *Austin American-Statesman* (June 24): A5.

Zavella, Patricia. 1987. *Women's Work and Chicano Families: Cannery Workers of the Santa Clara Valley.* Ithaca, NY: Cornell University Press.

Zeitlin, Irving M. 1997. *Ideology and Development of Sociological Theory* (6th ed.). Upper Saddle River, NJ: Prentice Hall.

Zelizer, Viviana. 1985. *Pricing the Priceless Child: The Changing Social Value of Children.* New Haven, CT: Yale University Press.

Zellner, William M. 1978. Vehicular Suicide: In Search of Incidence. Unpublished M.A. thesis, Western Illinois University, Macomb. Quoted in Richard T. Schaefer and Robert P. Lamm. 1992. *Sociology* (4th ed.). New York: McGraw-Hill, pp. 54–55.

Zimmerman, Don H. 1992. "They Were All Doing Gender, But They Weren't All Passing: Comment on Rogers." *Gender & Society*, 6 (2): 192–198.

Zipp, John F. 1985. "Perceived Representativeness and Voting: An Assessment of the Impact of 'Choices' vs. 'Echoes.'" *American Political Science Review*, 60 (3): 738–759.

Zuboff, Shoshana. 1988. *In the Age of the Smart Machine: The Future of Work and Power.* New York: Basic.

Zuravin, Susan J. 1991. "Research Definitions of Child Physical Abuse and Neglect: Current Problems." In Raymond H. Starr, Jr., and David A. Wolfe (Eds.), *The Effects of Child Abuse and Neglect.* London: Guilford, pp. 100–128.

Zurcher, Louis A. 1983. *Social Roles: Conformity, Conflict, and Creativity.* Beverly Hills, CA: Sage.

Credits

Name Index

Subject Index